3rd International Symposium on Engineering and Earth Sciences (ISEES 2020)

IOP Conference Series: Materials Science and Engineering
Volume 905

Grozny, Russia
28 - 29 February 2020

ISBN: 978-1-7138-4753-3
ISSN: 1757-8981

Printed from e-media with permission by:

Curran Associates, Inc.
57 Morehouse Lane
Red Hook, NY 12571

proceedings.com

Some format issues inherent in the e-media version may also appear in this print version.

Institute of Physics
Dirac House, Temple Back
Bristol BS1 6BE UK

Phone: 44 1 17 929 7481
Fax: 44 1 17 920 0979

techtracking@iop.org

Additional copies of this publication are available from:

Curran Associates, Inc.
57 Morehouse Lane
Red Hook, NY 12571 USA
Phone: 845-758-0400
Fax: 845-758-2633
Email: curran@proceedings.com
Web: www.proceedings.com

TABLE OF CONTENTS

IOP Conf. Series: Materials Science and Engineering **905** (2020) 011001 doi:10.1088/1757-899X/905/1/011001

PREFACE

International Symposium on Engineering and Earth Sciences: Applied and Fundamental Research dedicated to the 75th anniversary of professor Abdul-Hamid Mahmoudovich Bisliyev (ISEES 2020)

These conference proceedings include peer-reviewed original research papers of the 2020 International Symposium on Engineering and Earth Sciences (ISEES) that was held on 28-29 February 2020, in Grozny, Chechen Republic, Russia.

The event was dedicated to the 75th anniversary of the renowned Chechen scientist and enlightener, professor Abdul-Khamid Bisliev. Doctor of science Abdul-Khamid Bisliev is known for his outsanding expertise in magnetic fields physics. As a research fellow graduated from Moscow State University he had continued his scientific career in Grozny State University. As a globally recognized scientist and university manager he established a state-of-art physics lab in Grozny shortly after. Professor Bisliev tragically died in Grozny in course of a civil war, while protecting his colleague assaulted by armed villians.

The goals of the ISEES 2020 are

- to establish a pan-Caucasian scientific forum to attract collaborators from abroad and from the Russian mainland,
- to provide the local scientists with the opportunity to present their research results at the international level,
- to develop academic communities in the Chechen Republic,
- to make the Chechen Republic globally recognized as the science and academic center.

The ISEES 2020 proceedings include up-to-date, comprehensive, and valuable research papers in such fields as civil and construction engineering, materials science, engineering geology, meteorology and natural resource exploration. Several included contributions are discussing general aspects of engineering as science and profession, as well as study engineering practices.

All the accepted papers were submitted to rigorous single-blind peer-review by 2 expert referees. The papers were selected based on their originality, significance and clarity to meet the purposes of the conference. We sincerely hope that this volume will deliver to the readers not only a broad overview of the latest research results on Earth, environmental and materials sciences, but also provide a unique opportunity to learn more about the North Caucasus region and its academic community.

In addition, we would like to reflect on and take the opportunity to thank the participants and sponsors of the great annual congress in Grozny, who supported the participants and helped in holding the symposium throughout June 12-13. First and foremost, we want to express our deep gratitude to the international attendees for their dedication to the event, and the reviewers for their efforts and invaluable help in preparation of this volume. We would like to show appreciation of all the authors, participants and editors of the ISEES 2019 for their critical contribution to the work, organization, and holding of the symposium and to the preparation of the papers.

With the best regards,
The organizers

IOP Conf. Series: Materials Science and Engineering **905** (2020) 011002 doi:10.1088/1757-899X/905/1/011002

Peer review declaration

All papers published in this volume of *IOP Conference Series: Materials Science and Engineering* have been peer reviewed through processes administered by the Editors. Reviews were conducted by expert referees to the professional and scientific standards expected of a proceedings journal published by IOP Publishing.

- **Type of peer review:** Single-blind peer review with the mandatory spell and style check
- **Conference submission management system:** submissions were received and handled via e-mail
- **Number of submissions received:** 117
- **Number of submissions sent for review:** 111
- **Number of submissions accepted:** 78
- **Acceptance Rate (Number of Submissions Accepted / Number of Submissions Received X 100):** 70%
- **Average number of reviews per paper:** 2
- **Total number of reviewers involved:** 18
- **Any additional info on review process:** the core of the review team is comprised from educators and researchers from universities and research institutions in the Chechen Republic
- **Contact person for queries:** Ali R. Salgiriev, secretary of the meeting, 010585@list.ru

On the analysis of the stress state of a pipeline's bend by its height position

F K Abdrazakov, E N Mirkina, O V Mikheeva, S S Orlova and T A Pankova

Saratov State Agrarian University named after N.I. Vavilova, 1, Teatralnaya sq., Saratov, 410012, Russia

E-mail: abdrazakov.fk@mail.ru, docentmirkina@rambler.ru, omuk@inbox.ru, orlovass77@mail.ru, vtanja@mail.ru

Abstract. When underground laying, the pipeline bends in accordance with the terrain under the influence of its own weight and an overlying layer of soil. The article provides calculations of the radius of curvature of the pipe and its comparison with the allowable one. Thus, it can be seen from the above calculations that the obtained radii are less than the minimum permissible radius of curvature. Calculations showed that the radii in the horizontal and vertical planes are acceptable.

1. Introduction

When underground piping, the pipeline bends in accordance with the terrain under the influence of its own weight and an overlying layer of soil. The nature of the bending depends on the degree of fit to the bottom of the trench, the bending stiffness of pipe and the mechanical properties of soil. During the operation, the pre-bent state of pipeline changes under the influence of weight and pressure, products of temperature difference, subsidence of the base. The analysis of the stress state of pipeline's bend during the installation and operation according to its height position is of interest. There are works (1), (2) in which similar studies are carried out considering the features of the terrain.

2. Methods and materials

In the methodology for calculating the stress in a bent pipeline, which is considered below, it is determined by the formula:

$$G_i = \frac{E_k \cdot D}{2} \cdot y'' \quad (1)$$

where D is the outer diameter of pipe, E_k is the modulus of elasticity; y'' is the approximate value of the curvature of the axis of pipe at a point.

The second derivative in an arbitrary cross section of the pipeline y'' will be determined by the array $\{X_i, Y_i\}$ (Fig. 1) of measurements of the altitude of the points of the upper generatrix of the pipeline. At the same time, we approximate the known curve of the bend of the axis $Yf(x)$ by polynomials.

With simultaneously determined distances between space elements,

$$\breve{Y}_m(x) = a_0 + a_1 x + a_2 x^2 + ... + a_n x^m \quad (2)$$

the deviation measure of which from the table function $Y_i f(X_i)$ is written according to the least squares method:

$$S = \sum_{i=0}^{n} \left[\breve{Y}_m(x_i) - f(x_i) \right]^2 \quad (3)$$

Here, n is the number of measurements taken of the altitude position of the upper generatrix.

The approximation by polynomial (2) is not carried out "rigidly" by the measured points, since in the latter case the curve will describe well only the axes of pipeline, and the second arbitrary ones can very significantly, even up to the sign, differ from the true values.

The degree of polynomial (2) is assumed to be uncertain in advance and selected from the condition of least mean square deviation of the two subsequent polynomials between themselves. As we know (3), (4) is a condition corresponded to the smallest deviation of the selected polynomial from the unknown approximate function $Yf(x)$.

With such a choice of the polynomial degree, the condition $m \prec n$ holds, the bending curve does not pass strictly through the measured points.

The latter circumstance seems natural, given that the measured values are not absolutely accurate and were obtained with some error.

We formulate the criterion for determining the degree of polynomial (2), considering the desired polynomials and function $Yf(x)$ as elements of the normalized Hilbert function space $L_2[x_0,x_n]$ with the norm

$$II\breve{Y}_m(x)II = \sqrt{f_{x=o}^{x_n} \left[\breve{Y}_m(x) \right]^2 \cdot dx} \quad (4)$$

calculate the sequence of differences of polynomials

$$\delta Y_m(x) = \breve{Y}_m + 1(x) - \breve{Y}_m(x) \quad (5)$$

at $[1] = 2{,}3$. until at some $m = m^{\times}$ the first sharp local or "blurred" minimum of difference norms is found (Fig. 2)

$$II\delta\breve{Y}_m(x)II = \sqrt{f_{x0}^{xn} \left[\breve{Y}_m + 1(x) - Y_m(x) \right]^2 \cdot dx} \quad (6)$$

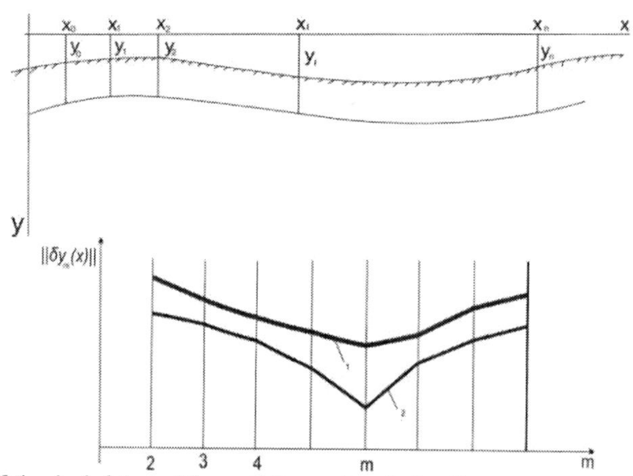

Figure 1. Measurement of the height position of the axis of pipelines

The appearance of a minimum is confirmed in Ref. 4 on a large amount of experimental material. The norms of differences (6) correspond to the mean-square deviation of two polynoids from each other and are simultaneously determined distances between elements of space $L_2[x_0,x_n]$.

The dependence of the norms of the differences on the degree of the polynomial with a "blurred" (1) and a sharp local (2) minimum ...

In connection with the described selection of the optimal degree of polynomial (2), we can suggest that its second derivative will be close to the actual curvature of the axis of pipeline. The degree of this proximity is checked by comparing the experimentally measured bending stresses on models and natural pipelines with theoretical values calculated by the formula (7).

The main parameters of pipeline that determine technical solutions for the technology of construction and operation are: pipe diameter, working pressure, pipe wall thickness, mass 1 kg of pipe length, characterizing the moment of inertia of the section, buoyancy of the pipe, necessary loads per 1 m of length, pipe deflections that cause stress state.

In direct proportion to the stiffness of pipes are the ability of pipeline to elastically bend and occupy a certain position with respect to the profile of trench's bottom.

3. Results

A pipeline of any diameter and purpose has curved sections in horizontal and vertical planes, which makes it necessary to bypass various obstacles, terrain and transitions through natural and artificial obstacles. Turns are carried out by elastic bending (free) or by cooking curved pipes.

The radius of elastic bending is calculated from the strength conditions of the metal pipe by the formulas:

$$R_{hor} \geq \frac{Ed_o}{4\sigma_b} \tag{7}$$

– in the horizontal plane

$$R_{con} \geq \frac{3}{5}\frac{Ed_o}{\sigma_b} \tag{8}$$

– in a vertical plane on a convex relief

$$\sigma_b = 0{,}15\frac{Pd_i}{\delta} - a_t E\Delta t \pm \psi_t \frac{C}{K_r} R_2^r \tag{9}$$

– bending stress.

where d_o – pipe outer diameter; P – pipe working pressure; δ – pipe wall thickness; a_t – coefficient of linear expansion of metal pipes; Δ_t – design temperature difference; ψ_t – coefficient of compressive (tensile) stresses; C – pipeline category coefficient; K_r – reliability factor; R_2^r – standard resistance of metal pipes.

$$R_{hor} = \frac{2{,}1\cdot 10^6 \cdot 72}{4\cdot 900} = 42000cm = 420m \tag{10}$$

$$R_{con} = \frac{3}{5}\cdot\frac{2{,}1\cdot 10^6 \cdot 72}{900} = 10080cm = 100{,}8m \tag{11}$$

$$R_{exp} = \frac{1}{2}\cdot\frac{2{,}1\cdot 10^6 \cdot 72}{900} = 84000cm = 840m \tag{12}$$

Table 1. To the calculation of the standard resistance of the metal [5]

E, kgf/ cm^2	d_o, cm	δ, cm	σb, kgf/cm^2	d_i, cm	a_t, 1/^{0}C	Δt, ^{0}C	Ψ_t	C	K_r	R_2^r, kgf/cm^2
$2{,}1\cdot 10^6$	72	1	900	70	$12\cdot 10^{-6}$	50	0,4	1	105	3300

Given the flexibility of the soil, the minimum allowable radius of curvature of the elastically curved section at P = 30 kgf/cm^2 is:

$$R_{\min} \geq \frac{Ed_o}{2[R_2^h - 0{,}15\frac{Pd_i}{\delta} \pm (1-\beta)a_t\Delta tE - \beta\frac{pd_o}{4\delta}]} = 490m \tag{13}$$

where β is the coefficient considering the decrease in stresses.

By changing the length of the curved section during its transverse movement.

$$\beta = \frac{1+5m^2}{1+\frac{15}{16}\frac{m^2}{\eta^2}} = 0{,}06 \tag{14}$$

where m is the slope coefficient of trench; η is the ratio of the deflection of the pipe to the length of the curve. [5]

4. Conclusion

Thus, it can be seen from the above calculations that the obtained radii are less than the minimum permissible radius of curvature. Calculations showed that the radii in the horizontal and vertical planes are acceptable.

References

[1] Kolosova N M, Mikheeva O V and Shmagina E Yu 2016 To the analysis of the pressure state of the bending of the pipeline in the high position *Techn. Regulat. in transp.construct.* **16**

[2] Dusaeva A S and Mikheeva O V 2016 Quality control of operation of the linear part of the pipeline *Trends in the development of construction, heat and gas supply and energy supply* Mater. of the int. sci. and pract. Conf. ed F K Abdrazakova pp 96–8

[3] Mikheeva O V and Dusaeva A S 2016 To the question of quality control of operation of the linear part of the pipeline *Agricult. J.* **10** 69–71

[4] Mikheeva O V, Kolosova N M and Shmagina E Yu 2016 The interaction of the underground pipeline under the influence of longitudinal movement *Techn. Regulat. in transp. construct.* **1(16)**

[5] Mikheeva O V, Kolosova N M and Illichyova I A 2014 Prediction of the structural reliability of the pipeline *Techn. Regulati. in transp. construct.* **4(8)** 58–62

[6] Ovchinnikov I I and Ovchinnikov I G 2012 Mechanics of structures with damage: nonlinear models and methods for determining the durability of structures operating in aggressive environments *Bull. of the Central Reg. Branch of RAASN* 120–7

[7] Ovchinnikov I G 1995 *Thin-walled structures under conditions of corrosion wear", Calculation and optimization* (Dnepropetrovsk: Publ. House of the Dagestan State Univer.) 192 p

[8] Frank-Kamenetsky D A 2008 *Diffusion and heat transfer in chemical kinetics* (Moscow: Intellect)

[9] *Official website of the group of companies ZAO Orengroup* Retrieved from: http://www.orenburgpm.ru/

[10] *Official website of company Ariel Plastkomplekt* Retrieved from: https: //www.arielplast.ru/files/pdf/Arielplast_PE_fittings_publication.pdf

[11] Orlova S S 2016 Evaluation of the operational reliability of pipelines transporting waste water *Agrar. Sci. J.* **3** 59–62

[12] Orlova S S, Pankova T A and Kochetkov A V 2016 A differential study of the kinetics of corrosion processes in pipelines transporting waste water *Hydrotechn. construct.* **4** 30–6

IOP Conf. Series: Materials Science and Engineering **905** (2020) 012002 doi:10.1088/1757-899X/905/1/012002

Materials used in the construction of roads

L S Aleroeva, I Z Magomadov, S Kh Mamasurov, R A Mamatsuev, Kh Kh Aptaev, N D Aisungurov, A M Istamulov and A L Bekishev

Department of Transport Systems, Grozny State Oil Technical University named after acad. M.D. Millionshchikov, Grozny, Russia

E-mail: liza.bonk.85@mail.ru, magomadov.isa@bk.ru, adlan.mamasurov@mail.ru, mamatsuyev@bk.ru, halid.aptaev@mail.ru, aysungurov91@mail.ru, istamylov48@mail.ru

Abstract. There is a huge variety of materials and technologies used in the implementation of construction and restoration work. They are also used in the construction of roads. It is impossible to consider all their varieties and give them a description. Therefore, we consider only the most effective in our opinion. This article describes and reviews some types of modern building materials and technologies used in the restoration of roads. Some varieties of these materials are listed. Their brief description is given. Their advantages and disadvantages are considered. The introduction contains the significance of materials in construction work. The optimal and most economical materials to use are listed in the summaries and conclusion. Certain types of materials and technologies are described in the article.

1. Introduction

Humidity is a natural destroyer of any roads. It destroys the foundation of the road structure and, when frozen in microcracks, breaks the coating. Therefore, when designing, the level of groundwater is always considered. It is possible to build a solid highway in marshy areas. However, special advanced technologies and appropriate building materials are needed here. Under normal climatic conditions, there are the following restrictions: at high humidity it is not recommended to use materials such as dusty sand, fine sandy loam, and dusty loam in the construction of roads [1]. Such types of soil are used exclusively in dry places, since when wet, they quite reduce the bearing property of the roadway and lead to its deformation.

2. Methods and materials

2.1. Road cake

According to the construction technology, the road is a kind of cake from several layers, each of which has its own task.

A layer of soil. The lowest layer is the soil layer. It is the basis on which the whole structure is based, the road service life and its condition depend on its reliability and stability.

Soils are divided into two large groups:

- incoherent or loose, which include – sand, gravel, rocky;
- cohesive, which are clay soils.

For the construction of the road embankment, it is necessary to use soils that do not change under the influence of natural factors, while maintaining stability during freezing and thawing. The following types of soil are included in the list of suitable for road construction:

- large sandy loam, draining sand, gravel, coarse, rocky soils;

- light sandy loam, silty, non-draining sands;
- clay.

Do not use too wet and excessively saline clay, silt, peat, and the upper soil layer with many plant roots to organize roads.

Sand pillow. This layer of sand is laid on the ground. It has several functions. For example, drainage and depreciation. Sand cushion softens the impact of vehicles on the roadway and increases its service life.

To increase the strength of roads, under the sand layer, you can lay a geogrid on it, which will not allow sand and gravel to mix, increasing the service life of roads.

Layers of macadam. The main supporting layer in the road cake consists of crushed stone. Its thickness and structure of the material depend on the type of road and the load on it. As a rule, during the construction of highways, three layers of crushed stone are laid: lower, middle, and upper.

The prepared road base is poured with a mixture of bitumen. The final stage is the laying of the final coating, thickness and characteristics of which depend on the load on the road. The better the coating, the longer the service life of the road.

Basically, asphalt is used for topcoat; cement concrete is chosen for roads with intensive operation and high load [1].

2.2. Stone materials

The need for road construction materials is very high. So, for the construction of 1 km. 1,600 tons of asphalt mix, 170 tons of black gravel (covered with bitumen), as well as 2,400 m^3 of white or ordinary gravel and 2,500 m^3 of sand are consumed in category II roads with asphalt concrete pavement. For the manufacture of asphalt concrete and black gravel, in addition, 190 tons of bitumen and 320 tons of mineral powder will be required [8].

Up to 50–60 % of the estimated cost of the constructed road is spent on the purchase and transportation of building materials to the place of work [8].

Natural stone materials include rocks, which are used in road construction in the form of loose and crushed masses. The following types of loose rocks: *sand-gravel raw material, pit gravel material, regular gravel* – sand-gravel material with sand and gravel grains. There is *medium regular gravel* and *large*; *gravel varietal (fractionated)* obtained by crushing ordinary gravel through a sieve with holes; crushed stone obtained by mechanical crushing from rock fragments. Crushed stone after crushing is sorted into fractions. Depending on the size, the crushed stone differs in the following fractions: *very large, large, medium, small, wedge, stone fines, sowing*. Crushed stone of these fractions is used in the following structural layers of the roadway: *very large* – when arranging the lower layers of the base; *large* – in the upper layers of the base; *middle* – in the upper layers of the base and lower layers of the coating; *a small wedge, stone fines and seeding* – in the upper layers of coatings, as well as for the installation of road wear mats. Crushed stone is prepared from different rocks – *hard and soft* [1]. *Hard rocks* include boundaries, diorites, basalts, gabbros, andesites, porphyries, and soft rocks include dolomites, limestones and sandstones. *Mineral binders*, depending on the hardening conditions, are divided into two groups: hydraulic and air. *Hydraulic binders* include *Portland cement*, hydraulic lime; to air – air lime, gypsum [2]. *Portland cement* is an astringent that hardens in water and air. It is obtained by fine grinding a marl of a certain composition and mixture, calcined before sintering, and a mixture consisting of limestone and clay components, which ensure the formation of calcium silicates in the resulting clinker. The fineness of grinding is very large. *Construction air lime* is an astringent obtained by roasting not before sintering carbonate rocks (calcium and dolomitic limestones, chalk, dolomites), followed by quenching with water or grinding to powder. *Hydraulic lime* is a hydraulic binder obtained by moderate firing, which does not lead to sintering of marly limestones containing from 8 to 20% clay impurities. *Organic binders* used in road construction include oil and shale bitumen, tar. It should be noted that in the Chechen Republic, fine-grained asphalt is used for the construction of pavement. In addition, apply sand, gravel, gravel, bitumen mixture.

2.3. New materials in road construction

Materials to strengthen the base of the pavement. Strengthening the bottom layer of the roadway and the lateral sides of the soil can be done effectively using geotextiles and polymer geosynthetics. The range of these materials is very wide: woven and non-woven geotextiles, geogrids, fiberglass, plastic grilles, meshes, geomembranes, drainage geocomposite (geodrainage) and other geosynthetic materials. Road construction in Russia using geotextile materials has been actively developing recently. A few years ago, almost no one knew about the existence of the above-listed geosynthetics. Nowadays, these materials are used both by individuals for laying tracks on a summer cottage, and by large road companies for the construction of highways. Geosynthetics – a type of building materials designed to improve the physical, mechanical and hydraulic properties of soils. The main purpose of the use of geosynthetic materials is to ensure the reliable functioning of the road, its individual elements in difficult natural and climatic conditions of construction and operation, as well as in the presence of technical and economic advantages in relation to traditional technologies. The main starting polymers for most geosynthetics are polyester, polyethylene, polypropylene, polyamide, polyaramide.

The main functions of geosynthetics:

- stiffening is a reinforcement of road structures due to the redistribution of stress forces arising in the soil mass and in pavement when exposed to loads from vehicles and their own weight;
- separation is an obstacle to the interpenetration of particles of contacting materials in the technological layers of the roadway;
- protection is a prevention or inhibition of erosion of soil particles and other particles along the slope surface;
- filtration is a prevention of penetration of soil particles into drains and their removal, reverse filtration;
- drainage is an acceleration of water drainage from layers of pavement and soil massifs [3].

The most popular material in this category is geogrids, which are flat polymer web materials with a mesh structure. The mesh is produced from high-strength bundles of threads by fastening in knots with the help of a stitching thread, weaving, gluing, fusion. In this case, cells, whose dimensions are larger than the edges that form the grid, are formed. To improve properties and increase stability, they are treated with special compounds.

Currently, geogrids are divided into different types:

- according to the quality of the material (polyethylene, polyamide, polyester);
- by scope of application (geogrids for asphalt concrete and soil);
- according to the method of cell orientation (mono-oriented grids, dual-orientation grids).

Worldwide, non-woven geotextiles are made from polypropylene. Polyester non-woven geotextile is offered as one of the varieties of the base brand based on polypropylene.

Modification of bitumen by emulsifiers. Since the main organic binder used in the production of asphalt concrete is petroleum bitumen, the most common way to improve its properties is to modify various active substances. Bitumen emulsion is a dark brown liquid, which is obtained by dispersing bitumen in water with the addition of an emulsifier. Bitumen emulsions are more often used in Russia as a binder, film-forming material in the construction and repair of roads. Bitumen emulsions are classified as *direct emulsions* in which the mass fraction of bitumen (30–70 %) in the form of tiny droplets is distributed in a continuous dispersed medium – water. In *reverse type emulsions*, water is dispersed in astringent bitumen or tar, the mass fraction of which is 70–80 %. Emulsifiers are substances that give stability to emulsions. In other words, stabilizers. As emulsifiers can be *surface-active substances* (*SAS*), soluble in both phases of the emulsions or in one of them, or *solid highly dispersed mineral powders* (clays, oxides, carbonates and sulfates, cement, soot). Solid emulsifiers are used mainly in the manufacture of bituminous pastes and rarely for road emulsions. Emulsions on solid emulsifiers in their composition contain 50–60 % bitumen (tar), 30–45 % water and 6–12 % solid emulsifier. Water-soluble emulsifiers, i.e. surfactants, which include *anionic* and *cationic surfactants*, are most often used for the manufacture of road emulsions. When using *anionic substances*, *anionic* and *alkaline emulsions* are obtained, when using *cationic* substances, *cationic* and *acidic* ones are obtained. Bitumen emulsion has its own advantages over bitumen. These include profitability, manufacturability, environmental friendliness. The emulsion requires less cash investment, allows you to save bitumen by 30–40 %, electricity and half the time. Bitumen emulsion can remain in liquid form and, unlike bitumen, is fire and explosion proof. The emulsion can be used on wet mineral materials, which allows to extend the construction season. But bitumen emulsions are not suitable for the construction of highways with high load; their qualities are most suitable for local repair work [4–6].

Modification of bitumen with polymers. Significantly improve the quality of bitumen when modified with polymers. *Polymer-bituminous binders* (PBB) mean binders obtained by combining bitumen and polymers. Recommendations for the use of modified bitumen instead of conventional bitumen are explained by their improved properties. Modified bitumens have increased elasticity, this important quality allows to withstand large loads, resist cracking and slows down the process of wear of asphalt concrete pavement. Also, polymer bitumen has a wide range of operating temperatures: the difference between the softening temperature and the brittle temperature reaches 100 °C (ordinary bitumen up to 60 °C).

Currently, due to the variety of polymer compounds offered by petrochemical industries, there is a wide selection of bitumen used for modification. They can be divided into three groups:

- thermoplastics or plastomers;
- elastomers;
- thermoelastic materials.

From the category of thermoplastics, polyethylene and atactic (stereo-random) polypropylene (APP) are often used [7, 9]. Thermoplastics consist of linear or low-branched polymers that soften when heated. A prerequisite for obtaining high-quality material is the compatibility of bitumen with APP, which is determined by the ratio of the components of bitumen. Elastomers consist of long polymer chains with wide branches. They are flexible over a wide temperature range: from low to 200 °C. When elastomers are added to bitumen, its viscosity increases, and elasticity improves. However, these systems are also unstable during storage. Constant mixing is required to prevent phase separation between bitumen and artificial material. Bitumen modified with elastomers can be called *bitumen with an elastic filler*. As elastomers, it is customary to use natural or regenerated rubber and polybutadiene. Thermoelastic polymers soften at temperatures above ordinary workers and are well deformed in this state. The most famous representative of the group of thermoelastic materials is *styrene-butadiene-styrene* (SBS) or *artificial rubber*. It is a block copolymer consisting of styrene and polybutadiene blocks. SBS gives bitumen excellent flexibility at low temperatures. Heat resistance is slightly lower than the APP, but when using a high-quality SBS modifier, it can reach 100 °C. SBS polymers have good adhesion and excellent resistance to cyclic alternating loads. Bitumens modified with SBS polymers correspond to climatic features of the main part of the territory of our country. The addition of this material to bitumen is from 3 to 6 of the total mass of the finished binder. The required amount of added material depends on the dispersed state of the introduced substance: if SBS is introduced into bitumen in a finely dispersed form, then the flow rate decreases, if in a finely dispersed form, then more modifier is required. Both in Russia and in world practice, it is precisely *the SBS class polymers* that are the main starting polymer to produce PBBs. The use of other polymers for these purposes is much lower. For example, in France, in addition to SBS, *the polymer EVA (ethyl vinyl acetate)* is used, but its use does not significantly increase the brittle temperature, which is very important for the climatic conditions of Russia. It is worth noting that in the Russian Federation bitumen of the same brand, but released at refineries in different places, may differ from each other. Thus, bitumen with the addition of high-quality polymers in practical studies show an obvious advantage compared to unenriched materials. Particularly large operational effect is shown by asphalt concrete on PBB in coatings of heavily loaded roads such as autobahns. In Russia, the use of PBB is spreading very slowly, not universally and amounts to several percent of the total consumption of road binders [10].

Modification of bitumen with rubber crumb. As a bitumen modifier, crumb rubber or a rubber asphalt modifier can be used. Small particles of 0.5–1.5 mm in size are added to bitumen with this system from 7 to 12 % by volume. At the outer working layer of the roadbed with interspersed rubber, physical and mechanical indicators are significantly improved: increased resistance to cracking and elasticity, high level of noise absorption, coefficient of frost resistance, which has a positive effect on the material, whose service life is increased by 2–3 times. Depending on the production method, a mixture of rubber with bitumen exhibits various properties. This mixture may have the following properties: a high degree of viscosity compared with ordinary bitumen, low heat sensitivity, a high degree of adhesion [10]. A mixture of fine rubber dust and small granules in the outer layer of the coating provides high stability during night frosts on the street. The rubber layer remains elastic and destroys the ice layer lying on it as soon as a vehicle passes over it, providing contact between the tire and the coating. The number of accidents on the streets equipped with such coatings during night frosts is much lower. If the pavement is made of coarse rubber granules or gravel, the risk of slipping on water is reduced, because, due to porosity, precipitation is easier and faster to absorb. A mixture of rubber and bitumen is also an ideal material for healing cracks on roads, while repaired sections of the road may be available for immediate movement. The volume of consumption of rubber crumb is 15–20 tons per 1 km of the roadbed. The first attempts to change the properties of bitumen by adding rubber granules were made in the late 1960s – early 1970s. In the late 1970s in Europe and especially in Belgium there was a special interest in the process of combining bitumen and rubber. There, this technology was first introduced in the early 1980s. And in 1979 there was applied porous asphalt from a mixture of rubber with bitumen as a street coating. This experiment clearly showed all the advantages expected from the use of new technology. Further technologies in this area have been developed. Very fine rubber dust was mixed with bitumen using experimental equipment. This method has opened the possibility of batching already in the production of bitumen, and not directly in the road-building machine. The difference between this classic wet method is lower temperatures, smaller particle size, and shorter reaction times. In 1994, a cement coating was developed for the streets with the addition of rubber particles of worn tires. This mixture consists of 5–15 % homogeneous and easily modified rubber, the presence of which does not interfere with the final kneading. Later, a mixture of rubber and bitumen (porous asphalt) began to be widely used in other European countries. Now this technology is making its way in Russia [10].

3. Results
The road must fulfill its purpose and be durable. This requires materials, both structural and finishing. In the technical and economic evaluation of planning and design solutions, design options must be compared. The most effective material, in our opinion, is geogrids, since they are the most stable and have high strength.

4. Conclusion
Geogrids are also an effective type of material for use in the Chechen Republic in terms of climatic conditions. Moreover, in this region the most elevated climate humidity, there are phenomena of frequent rainfall. Due to this fact, there occurs a sudden formation of cracks in the roadway and its deformation. And the iron mesh contained inside the canvas will just contribute to enhancing the quality of the road.

References
[1] *Taxi-pesok* Retrieved from: https://taxi-pesok.ru/stati/materialy-dlya-stroitelstva-dorog
[2] *Stroy-technics* Retrieved from: http://stroy-technics.ru/article/materialy-primenyaemye-dlya-stroitelstva-avtomobilnykh-dorog
[3] *Vse dlya stroiki* Retrieved from: https://vsedlyastroiki.ru/ru/ publikatsii/novyie-materialyi-v-dorojnom-stroitelstve/

[4] Anping P 2012 A study on A Study on Residents' Risk Perception in Abrupt Geological Hazard *J. of Risk Analysis and Crisis Response* **2(1)** 44–55

[5] Mushtakov R E, Silnov D S, Tarakanov O.V and Bukharov V A 2018 Investigation of modern attacks using proxy honeypot *ElConRus* no 3 Proc. of the 2018 IEEE Conf. of Russ. Young Res. in Electrical and Electronic Engineer. pp 86–9

[6] Mushtakov R E and Silnov D S 2017 New approach to detect suspicious activity using http-proxy honeypots *ElConRus* no 1 Proc. of the 2017 IEEE Russ. Sect. Young Res. in Electrical and Electronic Engineer. Conf. pp 187–9

[7] Win S S M and Aung T N 2017 Target oriented tweets monitoring system during natural disasters Int. J. of Networked and Distrib. Comput. **5(3)** 133–42

[8] Gokhman V A and Romadanov G A 2006 *General course of automobile roads* Textbook for high schools (Moscow: Higher School) 207 p

[9] Khalin V G, Shirokov S S and Maksimkin M V 2015 Innovation as a factor in the development of regional industrial complexes *Transp. busin. of Russ. 127(6)* 90–2

[10] Epstein A D 2007 Classification of innovations in housing and communal services Features of innovation in utilities and housing *Econ. sci.* **27(2)** 87–91

IOP Conf. Series: Materials Science and Engineering **905** (2020) 012003 doi:10.1088/1757-899X/905/1/012003

Adsorption of aromatic nitro and amino compounds with modified bentonite

O V Atamanova, E I Tikhomirova, A A Podoxenov, N V Vedeneeva and M A Grishin

Yuri Gagarin State Technical University of Saratov, Saratov, Russia

E-mail: O_V_Atamanjva@mail.ru

Abstract. Wastewater of many types of industries contribute to the entry into open water bodies of toxic organic compounds, the concentrations of which in some cases exceed the Maximum Permissible Concentration for Water in Commercial Fishery Reservoirs. Typical representatives of toxic organic pollutants in the aquatic environment are aromatic nitro and amino compounds. Currently, a significant number of wastewater treatment methods have been developed. One of the most promising methods of water treatment are adsorption methods. Of considerable interest is the use of natural clay minerals as sorption materials for wastewater treatment. The object of the study is model solutions of nitro and amino compounds, the adsorption of which was studied on different versions of modified bentonites. In this work, the adsorption capacity of modified bentonite with five different acid activation variants upon extraction of *n*-dinitrobenzene and *o*-phenylenediamine from solutions was studied. The adsorption parameters of the studied variants of bentonite, such as the adsorption value G, mg/g; interfacial distribution coefficient Kd, mg/dm^3; degree of adsorption recovery S, %. The constructed adsorption isotherms made it possible to identify the most effective variants of modified bentonite with additional activation by acid to extract aromatic nitro and amino compounds from solutions. The best adsorption ability upon extraction of p-dinitrobenzene from a solution was shown by glycerin-modified bentonite, subjected to heat treatment at 550 °C, the average fraction of granules, after activation with 9 % HCl solution and 30 % $CaCl_2$ solution. The best adsorption ability upon extraction of o-phenylenediamine from a solution was shown by glycerin-modified bentonite, subjected to heat treatment at 550 °C, the average fraction of granules, after activation with 18 % HCl solution and 38 % $CaCl_2$ solution.

1. Introduction

In conditions of constant pollution of the aquatic environment with various chemicals, high-quality wastewater treatment before it is discharged into natural water bodies or urban sewers is of particular importance.

Wastewater from chemical, light, pharmaceutical, paint and varnish and other industries usually contain a range of organic pollutants, including aromatic nitro and amino compounds.

Monitoring of natural water bodies of the Saratov region in places of waste water discharge from enterprises showed excess of maximum permissible concentration. aromatic substances in water by 5–28 %.

Aromatic nitro and amino compounds are among the common and highly toxic pollutants of the aquatic environment. Their widespread use is associated with active industrial use [1]. Nitro and amine compounds are most characteristic of the wastewater of enterprises producing dyes, pharmaceuticals, varnishes and paints, textile, synthetic polymers, plastics, rubbers, pesticides, cosmetic and rubber products [2, 3]. In addition, aromatic amino compounds can be formed during the destruction of soil organic matter, as well as some antiseptics and pesticides [1].

The main representatives of aromatic amino compounds include aniline, toluidines, phenylenediamines, xylidines, toluene diamines, etc. and their various derivatives. Aromatic nitro compounds are, first of all, nitrobenzenes, nitrotoluenes, dinitro- and trinitrobenzenes. Aromatic nitro and amino compounds are toxic. So, the ingress of amino derivatives of benzene into sources used for the preparation of drinking water can lead to negative consequences for the population and the environment. Their reactivity with respect to chlorine is very high, and at the stage of water disinfection with chlorine, the formation of various chlorine-substituted amines can occur – for example, aniline at a minimum dose of chlorine forms significant amounts of *n*- and *o*-chloroalanines and dichloranilines [1]. In this case, the toxicity of amines, as a rule, increases with the appearance of chlorine atoms in the molecule [1, 4].

The most characteristic manifestation of the toxic effect of aromatic amino compounds is selective damage to red blood cells; it is also possible to damage the central nervous system, liver and kidneys, although in general they do not possess selective hepato- and nephrotropy [3].

Aromatic nitro compounds are also characteristic of wastewater from the production of explosives and various nitro products [2, 4]. For fish, they are all very toxic; symptoms of poisoning are most often manifested in the form of apathy, stunning and anesthesia. A number of dinitro compounds affect the metabolism, exerting a uncoupling effect on the processes of conjugated oxidative phosphorylation. Aromatic nitro compounds, as well as amino compounds can cause dermatitis, eczema, allergies [4].

Among others, the adsorption method is included in the list of physico-chemical methods for wastewater treatment and water treatment.

The advantage of the adsorption method over others is that it has high efficiency at a low concentration of impurities in water, i.e. when the use of other methods becomes economically impractical or even impossible. The essence of the method is the concentration of impurities contained in water on the surface of a solid sorbent. This method can be used as an independent one for deep water purification in circulating and drainless water supply systems. The effectiveness of the use of sorption materials depends on their nature, specific surface area, the ratio of micro- and macropores and other factors that determine their different adsorption capacity [5–9].

The efficiency of adsorption purification reaches 80–95 % and depends on the chemical nature of the adsorbent, the size of the adsorption surface and its availability, on the chemical structure of the substance and its state in an aqueous solution [6].

Recently, bentonite adsorbents have gained particular popularity among sorption materials [9, 10].

Sorption materials based on natural clay minerals have significant advantages over synthetic aluminosilicates. These advantages include significant resistance to high temperatures, intense radioactive radiation, and the effects of oxidizing environments [11, 12]. Natural clay materials are widely distributed in the Russian Federation and abroad. Therefore, the prospect of their use as adsorbents for the purification of water from aromatic nitro and amino compounds is promising, which is the focus of this article.

2. Methods and materials

Sorption materials, which are bentonite modified with glycerin with various activation options, were accepted for research. The initial bentonite was brought from the Sarigyukh deposit. The modification of bentonite was carried out on the basis of the LISCON Research and Production Enterprise (Saratov).

For experimental studies, bentonite granules of the middle fraction (size 0.5–1.0 mm) were selected. The preference of the middle fraction of bentonite granules is justified by the results of earlier experiments [9].

The composition of the studied sorption materials included:

- No. 1 – bentonite, modified with glycerin, subjected to heat treatment at 550 °C, the average fraction of granules, without activation,

- No. 2 – glycerin modified bentonite, subjected to heat treatment at 550 °C, the average fraction of granules, activation with 9 % HCl solution,
- No. 3 – glycerin-modified bentonite, subjected to heat treatment at 550 °C, the average granule fraction, activation with an 18 % HCl solution,
- No. 4 – glycerin-modified bentonite, subjected to heat treatment at 550 °C, the average granule fraction, activation with 9 % HCl solution and 30 % CaCl2 solution,
- No. 5 – glycerin-modified bentonite, subjected to heat treatment at 550 °C, the average granule fraction, activation with 18 % HCl solution and 38 % $CaCl_2$ solution.

The laboratory experiment was carried out with 3 and 4-fold repetition in accordance with standard methods. To assess the static sorption, solutions of *n*-dinitrobenzene of the following concentrations were prepared: 5, 10, 20, 30, 40, 50, 55 and 60 mg/dm^3. Solutions of o-phenylenediamine of the following concentrations were also prepared: 100, 300, 500, 700, 900, 1500, 2000, 2500 and 3000 mg/dm^3.

The activation method of the sorption material was as follows. Before the start of the experiment, a weighed portion of 60 g of the initial bentonite granules was placed in a 400 ml previously washed and thoroughly dried measuring cup, after which the bentonite granules were soaked in water for 24 hours. Then, the sorption material was transferred to a Petri dish and dried at room temperature to obtain an air-dry sample, from which a sample weighing 10 g was taken for studies (Fig. 1).

Figure 1. Air-dry sample of sorption material

For acid activation after the operations, either 10 ml of a 9 % HCl solution (hereinafter activation option No. 1) or 10 ml of an 18 % HCl solution (hereinafter activation option No. 2) was poured onto a sample, which depended on the method chosen. After that, bentonite granules in an acid solution were intensively mixed for an hour and left for 6 hours in a glass with an acid solution. After keeping in acid, the solution was removed, and the granules were washed until the acidity of the water obtained after washing was completely neutralized and left in air to obtain an air-dry sample.

The granules were also activated by preliminary acid activation and final salt using a CaCl2 solution. Such a change in the activation procedure was chosen on the basis of the assumption that the best sorption of organic substances occurs when ions with a large core radius are in the interlayer space of aluminosilicate. The stage of acid activation is similar to that described above, but after acid removal, without receiving an air-dry sample, the obtained bentonite was kept for one hour with periodic stirring in 40 ml of either 30 % $CaCl_2$ solution (hereinafter activation option No. 3) or 38 % $CaCl_2$ solution (hereinafter, activation option No. 4). The activation procedure with a CaCl2 solution was repeated three times for the largest possible replacement of cations in the interlayer space. After washing the granules from the remainder of the $CaCl_2$ solution, the presence of Ca_2 + ions in the washed water was checked by titration of the washed water with Trilon B solution according to the procedure in accordance with [13], since the presence of calcium ions would not allow determination of the substance due to the reaction with NaOH.

To study the adsorption capacity of modified bentonites, *n*-dinitrobenzene and *o*-phenylenediamine were chosen as typical representatives of aromatic nitro and amino compounds. Thus, studies of the adsorption capacity of bentonite were carried out on model solutions of these representatives of benzene derivatives.

The methods for determining the concentration of p-dinitrobenzene and *o*-phenylenediamine are adopted in accordance with the recommendations of I.M. Korenman [14].

3. Results

The sorption ability of modified bentonites with respect to p-dinitrobenzene and *o*-phenylenediamine was evaluated under static conditions [15, 16]. According to the results of a laboratory experiment, the following were established: adsorption value G, mg/g; interfacial distribution coefficient Kd, mg/dm3; degree of adsorption recovery S, % on the studied variants of modified bentonite.

The calculated sorption parameters of p-dinitrobenzene on the studied sorption materials are presented in Table 1.

The general adsorption isotherms of *n*-dinitrobenzene on modified bentonite are shown in Fig. 2 and fig. 3. Here, the residual concentration of p-dinitrobenzene in the solution is denoted by Ci, and the equilibrium concentration is denoted by Cpi, mg/dm^3.

Table 1. Sorption parameters of p-dinitrobenzene on the studied variants of bentonite

Substance	No. of sorption material	Γ, mg/g	K_d, mg/dm^3	S, %
n-dinitrobenzene	1	0.60±0.05	0.01±0.00	25.12±1.80
	2	0.72±0.06	0.02±0.00	30.00±2.28
	3	0.91±0.07	0.02±0.00	38.4±3.10
	4	1.50±0.12	0.07±0.00	62.50±6.00
	5	1.00±0.09	0.03±0.00	41.67±3.63

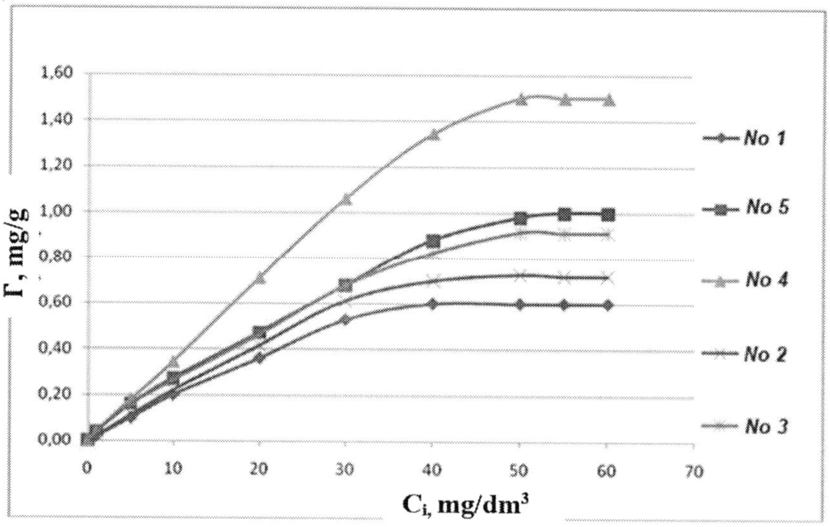

Figure 2. Sorption isotherms Γ = f (Ci) p-dinitrobenzene on bentonite of various activation options.

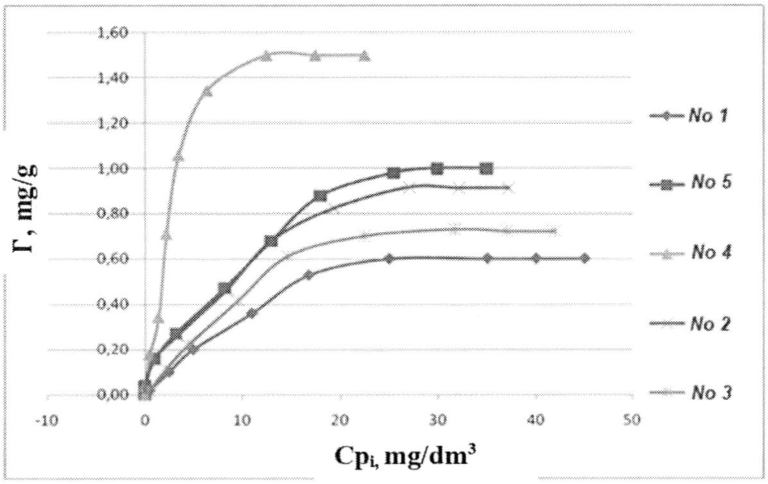

Figure 3. Sorption isotherms G = f (Cpi) of p-dinitrobenzene on bentonite of various activation options

The results of the study and the obtained sorption characteristics of the studied variants of bentonite with respect to o-phenylenediamine are presented in table. 2.

Table 2. Parameters of sorption of *o*-phenylenediamine on the studied bentonite variants

Substance	No. of sorption material	Γ, mg/g	K_d, mg/dm³	S, %
	1	38.43±2.84	0.01±0.00	32.03±2.63
	2	46.12±3.36	0.01±0.00	38.43±3.07
o-phenylenediamine	3	48.79±3.61	0.02±0.00	40.66±3.17
	4	50.84±3.86	0.02±0.00	42.36±3.47
	5	52.52±4.04	0.02±0.00	43.76±3.60

The general adsorption isotherms of o-phenylenediamine on modified bentonite are shown in Fig. 4 and 5.

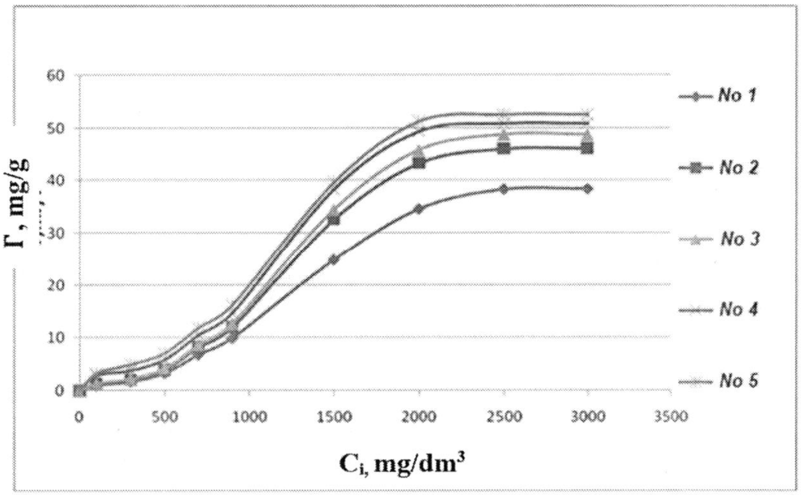

Figure 4. Sorption isotherms G = f (Ci) o-phenylenediamine on bentonite of various activation options

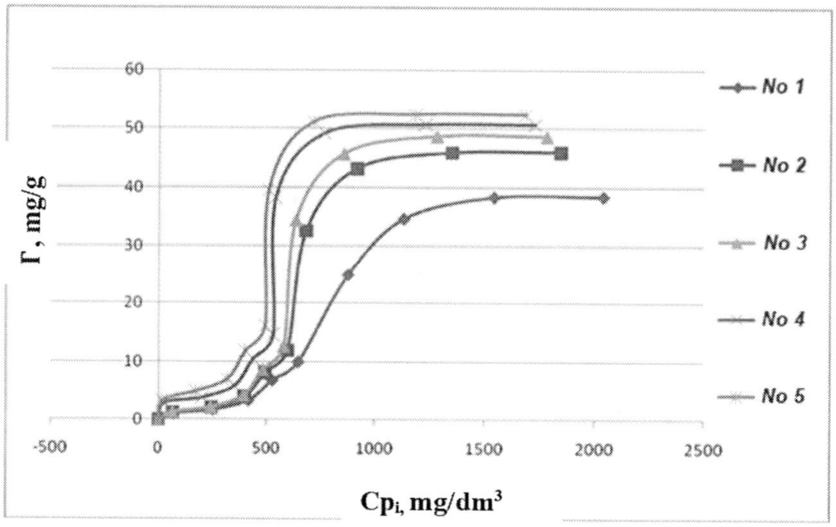

Figure 5. Sorption isotherms G = f (Cpi) of o-phenylenediamine on bentonite of various activation options.

An analysis of the obtained adsorption parameters of p-dinitrobenzene and *o*-phenylenediamine on different versions of modified bentonite under static conditions, as well as the constructed adsorption isotherms made it possible to perform a comparative assessment of the sorption ability of the studied adsorbents.

4. Conclusion

A laboratory experiment and the constructed adsorption isotherms Γ = f (Ci) showed (see Fig. 2) that the best adsorption ability with respect to p-dinitrobenzene from all the adsorbents studied is shown by glycerin-modified bentonite, which is subjected to heat treatment at 550 °C, middle fraction of granules, activation with 9 % HCl and 30 % $CaCl_2$. It exceeds the closest value of the sorption limit value to modified bentonite with an activation option of 18 % HCl solution and 38 % $CaCl_2$ solution by 1.5 times, which in absolute terms is 0.5 mg/g.

Modified bentonite, which has not undergone activation, exhibits the worst adsorption capacity upon extraction of p-dinitrobenzene from a solution under static conditions. In terms of the limiting value of sorption, it is inferior to the nearest bentonite with the activation option of 9 % HCl by 17 %, and to the variant with the best adsorption efficiency, it is 60 %.

Additional activation of modified bentonite with $CaCl_2$ causes a significant increase in the limiting value of sorption for the variant with activation of 9 % HCl and insignificant for the variant with activation of 18 % HCl. So for the first of these options, the increase of the limiting value of sorption is 52 %, which is 0.78 mg/dm^3 in absolute terms, for the second option, these indicators are 9 % and 0.09 mg/dm^3, respectively.

The values of the limiting value of sorption of p-dinitrobenzene for all variants of modified bentonite under static conditions are reached only at sufficiently large values of the concentration of the model solution in the concentration range from 40 to 55 mg/dm^3.

Activation of modified bentonite with an 18 % acid solution shows the best result of the adsorption ability of bentonite with respect to *p*-dinitrobenzene compared with activation of bentonite with a 9 % solution. However, upon subsequent activation of the sorption material with a $CaCl_2$ solution, this cannot be said, because a significant superiority of the bentonite variant with activation of 9 % HCl solution and 30 % $CaCl_2$ solution is observed in comparison with other variants. This fact can be explained on the basis of published data [17] by the fact that upon activation in bentonite many macro- and micropores are formed, macropores appear due to leaching of metal clay from the crystal lattice, which leads to the formation of a certain amount of silica gel, which causes an increase sorption of non-polar substances. When using high acid concentrations, a disorder of the aluminosilicate layers can be observed, therefore, subsequent activation with a $CaCl_2$ solution has a greater effect after activation with a 9 % HCl solution than after activation with an 18 % HCl solution, since in the second case, due to large disturbances in the structure of the material, to obtain a smaller volume of Ca-bentonite than in the case of activation with a weaker acid solution [9, 17].

The best adsorption effect of *o*-phenylenediamine was shown by the variant of activation of modified bentonite with 18 % HCl solution and subsequent activation of 38 % CaC12 solution. Its advantage over the second most effective version of the activation of bentonite modified with glycerin and subjected to heat treatment at 550 ° C, followed by activation with 9 % HCl solution and 30 % CaC12 solution, is 3.2 % or 1.64 mg/dm^3 in absolute values.

The worst results are achieved when using modified bentonite without activation (adsorbent No. 1), however, it is still quite effective and the value of the maximum sorption value for it is 38.43 mg/dm^3, which is 27.83 % or 14.09 mg/dm^3 lower better rate.

Analysis of graphical dependencies in Fig. 4 shows that at the point with an abscissa C = 100 mg/dm^3, the value of the maximum sorption of *o*-phenylenediamine for activation options No. 3 and No. 4 is higher than for the closest in effectiveness version of activation No. 2 by 1.5 and 1,9 mg/g, respectively, as in Fig. 5 is expressed as a significant difference between the equilibrium concentrations of o-phenylenediamine for different variants of modified bentonite. So for options No. 3 and No. 4, the value of the equilibrium concentration after sorption of a solution of a concentration of 100 mg/dm^3 is 28.1 mg/dm^3 and 18.23 mg/dm^3, respectively, and for activation option No. 2 it is 67 mg/dm^3, therefore, the efficiency sorption at this stage is much higher than when full saturation is achieved.

In absolute values, the sorption of the secondary aromatic amine (*o*-phenylenediamine) turned out to be many times higher than the sorption of the secondary nitro compound (p-dinitrobenzene): G = 52.52 mg/dm^3 against G = 1.5 mg/dm^3 for better results after activation. It can be assumed that for amines there are fewer obstacles to penetration of bentonite into the interlayer space than for nitro compounds, and this is expressed in a lesser effect of steric and energy effects of adsorption centers,

The values of the interfacial distribution coefficient and the degree of adsorption extraction in the case of complete saturation of the sorption material for both substances are approximately the same and make up slightly more than 40 % for better results, while the activation option by 9 % HCl solution and 30 % $CaCl_2$ solution with the degree of adsorption extraction of p- dinitrobenzene 62.5 %.

Both studied substances saturated the adsorbent completely only at very high concentrations of the substance solution (for *n*-dinitrobenzene – 55–60 mg/dm^3; for *o*-phenylenediamine – 2500–3000 mg/dm^3).

All the considered options for the activation of modified bentonite had a lesser effect on the ability of the adsorbent to adsorb *o*-phenylenediamine than on the ability to adsorb *n*-dinitrobenzene. Such a difference can be explained by differences in the necessary conditions for establishing a strong bond of the substance with the sorbent. So, unlike nitro compounds, the interaction of amines with modified aluminosilicates is realized by the ion exchange mechanism, which results in the release of counterions to the aqueous phase of the solution and the binding of amines in the structure of aluminosilicate, as well as due to the molecular sieve mechanism, which assumes the key role of the steric factor in establishing the interaction "adsorbent – adsorbate" [9].

Acknowledgment
The results were obtained in the framework of implementing Federal grant No. 5.3922.2017/PCh of the Russian Ministry of Education.

References

[1] Tran V V, Park D and Lee Y-C September 2018 Hydrogel applications for adsorption of contaminants in water and wastewater treatment *Environ. Sci. and Pollut. Res.* **25(25)** 24569–99

[2] Grushko Ya M 1982 *Harmful Organic Compounds in Industrial Wastewater* Handbook 2rd ed (Leningrad: Chemistry) 216 p

[3] Khizbullin F F and Muslimova I M 2001 The problem of the accumulation of polychlorinated dibenzo-p dioxins and dibenzofurans in sludge collectors of organochlorine synthesis plants *Ecol. Chem.* **4** 269–74

[4] Harmful substances in industry. Handbook for chemists, engineers and doctors In 3 volumes, vol I *Organic substances* ed by N V Lazareva and E N Levina vol I *Organic matter* (Leningrad: Chemistry) 197, 592 p

[5] Khokhryakova E N and Resnick Y E 2007 *Water Treatment: Reference* (Moscow: Aqua-Term Ed.) 240 p

[6] Bonilla-Petriciolet A, Mendoza-Castillo D-I and Reynel-Ávila H-E 2017 *Adsorption Processes for Water Treatment and Purification* (Springer) 256 p

[7] Lowell S and Shields J E 1991 Chemisorption. Powder Surface Area and Porosity pp. 210–6

[8] Yang H and Zheng Q October 2003 Structure stability of organic montmorillonite used for preparing polyethylene/montmorillonite nanocomposite *J. of Mater. Sci. Letters* **22(20)** 1431–3

[9] Kosarev A V, Atamanova O V, Tikhomirova, E I and Istrashkina M V 2018 Kinetics of adsorption of 2-methylalinine by modified bentonite at sewage treatment *Water and Ecol.* **3** 24–31

[10] Tikhomirova E I, Plotnikova O A, Atamanova O V, Koshelev A V and Podolsky A L 2019 The use of multicomponent adsorption filters in water purification systems and luminescent control of ecotoxicant content *Theoret. and Appl. Ecol.* **1** 73–81

[11] Bährle-Rapp M 2007 *Montmorillonite* (Springer Lexikon Kosmetik und Körperpflege) 360 p

[12] Cloos P, Bafot C and Herbillon A 1981 Interlayer formation of humin in smectites *Nature* **289** 391–3

[13] Standard 31954-2012 *Drinking water. Methods for determining stiffness* 2013 (Moscow: Standartinform) 7 p

[14] Korenman I M 2012 *Photometric analysis. Methods for the determination of organic compounds* (Moscow: The book on demand) 342 p

[15] Sabio E, Zamora F, Gañán J, González J F June 2006 Characterisation under static and dynamic conditions of commercial activated carbons for their use in wastewater plants *Appl. Surface Sci.* **252(17)** 6058–63

[16] Zohral M, Rose J and Borschneck D 2014 Urban Wastewater Treatment by Adsorption of Organic Matters on Modified Bentonite by (Iron-Aluminum) *J. of Encapsulat. and Adsorpt. Sci.* **4** 71–9

[17] Li H, Teppen B J and Boyd S A 2004 Thermodynamics of Nitroaromatic Compound Adsorption from Water by Smectite Clay *Environ. Sci. & Technol.* **38(20)** 5433–42

IOP Conf. Series: Materials Science and Engineering **905** (2020) 012004 doi:10.1088/1757-899X/905/1/012004

On the relationship between parameters of melt spreading over the surface of substrates of variable composition and state diagrams

A M Bagov[1], A I Khasanov[2,3], M-A V Zubhadzhiev[2] and L Kh Umarova[2]

[1] Kabardino-Balkarian State University, 173, Chernyshevsky St., Nalchik, 360004, Russia

[2] Chechen State University, 17a, Dudayev Boulevard, Grozny, 364060, Russia

[3] Integrated Research Institute n.a. Kh. I. Ibragimov, Russian Academy of Sciences of Grozny, 21A, Staropromyslovsky avenue, Grozny, 364051, Russia

E-mail: vegros@rambler.ru, aslan2001@rambler.ru, zmagamed-ali@mail.ru, lipa-um@yandex.ru

Abstract. On of metered-dose drops of various concentrations on a substrate of variable composition. The basis for the substrates and droplets is the contact melting (CM) method related to state diagrams which are used to study the concentration dependence of surface characteristics (spot shape and size, wetting and spreading, surface tension, etc.) under the same conditions which improves accuracy of the characteristics. It was found that the microstructure of the transition zone of the substrate-drop depends on the location of the drop on the substrate, i.e. on the concentration of the substrate.

1. Introduction

The aim of the study is to analyze approaches to the formation and application of metered droplets (bubbles) on solid substrates of variable composition; to develop a less time-consuming approach to obtain drops of metal liquids (alloys) of the same size, under the same conditions, in the right amount from pure metals and alloys of various concentrations; to show the relationship between the state diagram and surface characteristics, the configuration and the diameter of the spreading spot.

This article continues the studies described in [1] which showed that the configuration and the size of the spreading spot depend on the concentration of the substrate of a variable composition.

The first solution to the problem was suggested by M. Simon. In 1831, he developed an accurate method of applying drops of the same size. The method was substantiated in the works by M. Cantor and E. Schrödinger [2, 3]. For the practical implementation of the maximum pressure method, various instruments and devices have been developed (M. Simon, S. Sanden, F. M. Jaeger, P. A. Rebinder, Pugachevich, etc.) [4–7].

Further development of the method for studying surface characteristics was continued in [8–10 et al.]. The methods can be used to study surface characteristics at high temperatures.

Using the ideas suggested in [11], we developed a method for producing metered drops from pure metals and alloys as well as substrates of variable composition. To apply this method on interlayers produced by contact melting, a special holder was used.

The main node of the holder is a set of plates of equal thicknesses and with equal holes, which ensure the separation of the liquid diffusion zone into separate parts (plates, tablets). Each plate has a hole equal to the diameter of the sample, which allows you to adjust the size of future drops. The hole is filled with metal or a composition of various concentrations along the cylindrical sample obtained by contact melting, from which it is necessary to obtain droplets. The cell is placed in an appropriate medium with a temperature sufficient to melt metal (alloy) in the hole. After that, a "layer" of the liquid metal is produced, leading to the formation of disks (tablets) (Figure 1).

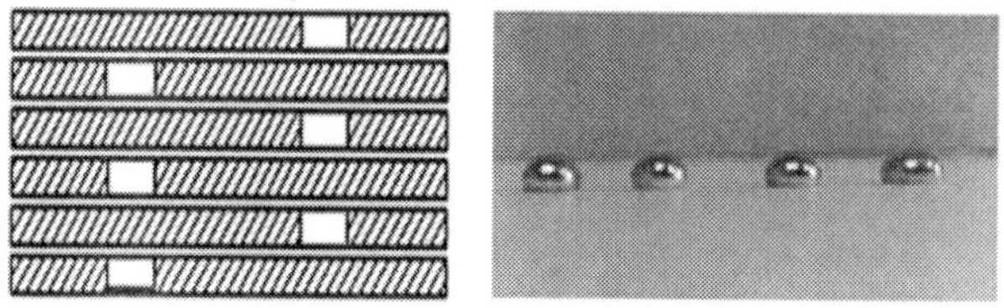

a b

Figure 1. a) the layout of the plates after the "cut"; b) gallium droplets of a metered size obtained by the "cut" method

The liquid disks crystallize (sometimes in liquid nitrogen). In order to the same dosage of drops, each sample was weighed on an analytical balance.

The concentration of each individual sample was determined by contacting with one of the pure components (A or B) (Figure 2) and placed in a thermostat.

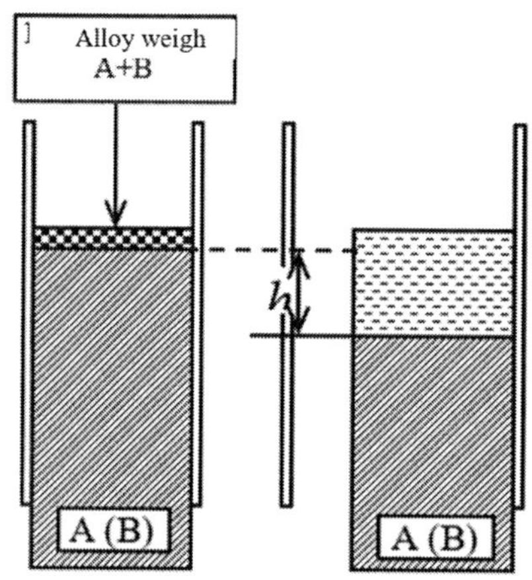

a b

Figure 2. Determination of C (x) by taking liquid samples (determining the concentration of samples): a) solid-phase contacting; b) thermodynamic equilibrium – solid-liquid, at T = const

The thermostat should have a temperature below the melting temperature of the pure component, but above the liquidus temperature of the alloy A + B. At the initial moment, the pure component is in the solid state, and the analyzed sample is in the liquid state.

Since the temperature of the thermostat exceeds the liquidus temperature of the sample, the pure component will dissolve in the liquid sample and the liquid-solid interface begins to move. The movement of the boundary continues until the concentration of the liquid reaches a value corresponding to the liquidus in the melting diagram at the temperature of the thermostat. Upon reaching a thermodynamic equilibrium, the displacement of the liquid – solid boundary ceases.

To reduce the time of equilibrium, a less dense was taken as a substrate (convection develops).

Knowing the mass m of the analyzed sample, the liquid concentration corresponding to the liquidus at a given temperature of the thermostat, and the height h of the molten part of the pure component, we can determine the desired concentration:

$$C_A = \frac{C_L\left(\rho h \frac{\pi d^2}{4} + m\right)}{m},$$

where CA is the desired concentration in weight fractions, CL is the weight fraction A corresponding to the liquidus in the melting diagram at the test temperature, ρ, d, h are density, diameter and height of the melted part of the pure sample, respectively.

The concentration of tablets can be achieved in another way. The concentration of this component lies between the liquidus of the state diagram. The concentration of tablets (disk, drops) can be determined using the state diagram of the system. We place the tablet in a crucible with a deoxidizing liquid (glycerin), raise the temperature and fix the melting point of the tablet (disk), which allows us to determine the concentration (C) based on the fusibility diagram.

The correspondence of the concentration of disks (droplets) is controlled by the x-ray phase and X-ray diffraction methods.

The tablets are contacted with the substrate at various points. The temperature of the system is increased until the tablets melt, turning it into a drop. They are monitored for their wetting and spreading, which makes it possible to determine temporary and concentration dependences of the shape and the diameter of the spreading spot.

Depending on the experimental conditions, two options are possible. The crystallized drop on the inert surface is transferred to the test substrate or the disk (tablet) contacts with the substrate whose temperature is sufficient to transfer the disk into a drop and examine the wetting and spreading of this drop.

2. Results

The methods make it possible to construct concentration distributions – C(x) in contact interlayers [11–13]. If C(x) is known for any system, this dependence can be used to study the dependence of wetting and spreading on concentration.

Figure 3 reflects the dependence of the state diagram and the concentration of spreading on the concentration along transition zones (contact interlayers of variable composition).

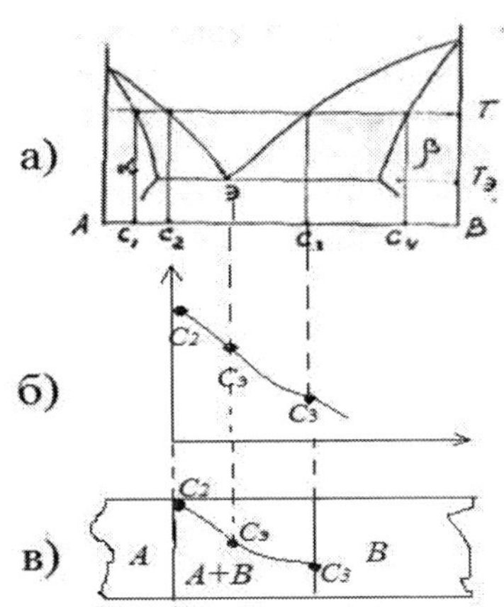

Figure 3. Relationship of the concentration distribution along the contact layer and the state diagram, (a) – state diagram of a two-component system; (b) the concentration distribution in the contact layer; (c) – contact layer in a two-component system

The concentration of liquid melt in contact with solids coincides with the liquidus concentrations on the state diagram of the two-component systems [14–16]. In the plane of the initial contact, the concentration of contact melt always coincides with the eutectic concentration on the state diagram. Contact interlayers obtained in the nonstationary diffusion mode (NDM) have new phases, for example, intermetallic compounds whose concentration is indicated on the state diagram.

When metal melts come into contact with solid surfaces, a spreading phenomenon accompanied by dissolution of the body, diffusion of the melt components into the solid, and chemical reactions leading to the formation of new phases is observed [10].

Figure 4 shows photos of the interaction of gallium at three points on the surface of a zone of variable composition (contact layer) of the Bi-Pb system.

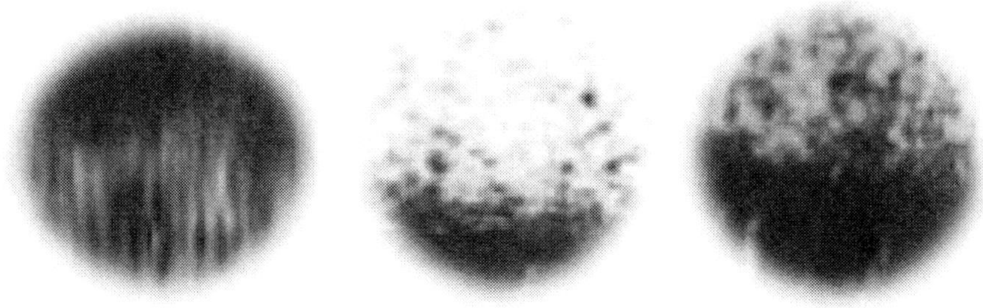

a b c

Figure 4. The microstructure of the diffusion layer during the spreading of gallium droplets over the surface of the contact layer in the bismuth-lead system: a) the bismuth-substrate interface; b) the zone of eutectic concentration; c) the lead-substrate boundary

An analysis of the microstructure patterns of the interaction zone of gallium droplets shows that the depth of dissolution of the substrates depends on the location of droplets on the substrate surface, i.e. on the concentration of the substrate at a given point of contact. An increase in the concentration of bismuth leads to the rearrangement of a dendritic growth due to changes in the nature of diffusion at the interphase boundary – the structure is crushed.

Thus, knowledge of the concentration distribution in contact interlayers (substrates of variable composition), the approaches to the formation and application of metered droplets on the surface of substrates allow us to study the temperature and concentration dependences of the surface characteristics of melts during wetting and spreading. The state diagram specifies the concentration of components at four points of the contact layer: at two liquid-solid boundaries (two liquidus concentrations of this component); the concentration at the point of the initial contact.

3. Conclusion

1. A method for producing metered drops from pure metals and alloys by the "shear" method under the same conditions has been developed

2. The basis for the preparation of substrates of variable composition and droplets of various concentrations is a contact melting phenomenon related to state diagrams, which allows the study of the concentration dependence of the surface characteristics of interacting metals. The relationship between the state diagram and surface characteristics (the configuration and the size of the diameter of the spreading spot) was identified.

3. The nature of the substrate-drop diffusion varies depending on the location of the droplet (substrate concentration) – as the bismuth concentration increases, the structure is crushed.

Acknowledgment
The study was funded by the Russian Federal Research Fund within project No. 18-32-00572 mol_a.

References

[1] Akhkubekov A.A., Akhkubekova S.N., Bagov A.M et al 2016 Gallium spreading over the surface of the contact layer of variable composition of the Pb – Bi system *Bull. of the Russ. Acad. of Sci. Phys.* **80(6)** 701–4

[2] Contor M 1896 Ueber Capillaritatsconstanten *Annalen der Physik* **283(11)** 299–423

[3] Schrödinger E 1915 Notiz Űber den Kapillardruck in Gasblastn *Annalen der Physik* **351(3)** 413–8

[4] Sugner S 1922 The determination of surface tension from the maximum pressure in bubbles *J. of the Chem.Society* **121(715)** 858–66

[5] Jägen F.M 1917 Űber die Temperaturabhängigkeit der molekularen freien Oberflächenenergie von Flüssigkeiten im Temperaturbereich von – 80 bis +1650 °C *Zeitschrift für anorganische und allgemeine Chemie* **101(1)** 1–214

[6] Rebinder P A 1927 Surface phenomena and adsorption *Journal of Experimental Biology and Medicine* **4(14)** 939–53

[7] Pugachevich P P, Beglyarov E M and Lavygin I M 1982 *Surface phenomena in polymers* (Moscow: Chemistry) 200 p

[8] Alchagirov B B and Khokonov H B 2017 *Methods and instruments for research in the field of physics of interfacial phenomena (Special Physical Workshop)* (Nalchik: Kab.-Balk. Univ.) 152 p

[9] Alchagirov B B and Khokonov H B 1994 Wettability of surfaces of solids with melts of alkali metals and alloys with their participation. Theory and research methods *Thermophys. of high temperat.* **32(4)** 590–626

[10] Vityunin M A 2009 *The spreading of aluminum-based melts over the surface of solids and the microstructure features of crystallized materials* (Chelyabinsk) 124 p

[11] Akhkubekov A A 2001 *Diffusion and electric transport in low melting metal systems during contact melting* (Nalchik: KBSU) 312 p

[12] Akhkubekov A A, Orkvasov T A and Sozaev V V 2008 *Contact melting of metals and related nanostructures* (Moscow: Fizmatlit) 152 p

[13] Bagov A.M., Tamaev T.Kh., Khasanov A.I et al 2018 Phase formation in contacting zinc melting with indium and stannic in the presence of strontium impurity and microhardness of derived alloys *Advances in Engineering Research* vol 177 Int. Symp. Engineer. and Earth Sci.: Appl. and Fund. Res. (ISEES 2018) pp 143–5

[14] Rogov V I 1969 *Research of contact melting of metal systems in a diffusion mode* Thesis (Nalchik: Kabard.-Balkar. State Univer.) 13 p

[15] Nilova N N, Bartenov G M, Borisov V T and Matveev Yu E 1972 Investigation of contact melting in a bismuth lead system *Collection of Materials of the conference on physics* pp 13–21 (Nalchik: KBSU)

[16] Nilova N N., Bartenev G M., Borisov V T and Matveev Yu E 1968 Investigation of contact melting in the gallium zinc system *DAN SSSR* **180(2)** 394–7

On kinetics of transfer of interphase borders during contact melting with an electric transfer

A M Bagov[1], A I Khasanov[2,3], M-A V Zubhadzhiev[2] and L Kh Umarova[2]

[1] Kabardino-Balkarian State University, 173, Chernyshevsky St., Nalchik, 360004, Russia
[2] Chechen State University, 17a, Dudayev Boulevard, Grozny, 364060, Russia
[3] Integrated Research Institute n.a. Kh. I. Ibragimov, Russian Academy of Sciences of Grozny, 21A, Staropromyslovsky avenue, Grozny, 364051, Russia

E-mail: vegros@rambler.ru, aslan2001@rambler.ru, zmagamed-ali@mail.ru, lipa-um@yandex.ru

Abstract. The mathematical laws of kinetics of the liquid zone growth under the contact melting of binary metal systems in the presence of constant electric current were described; the analytical and graphic solutions to this problem explain the experimental data.

1. Introduction

Constant electric current can affect the growth rate of the liquid layer formed during contact melting (CM). Depending on the direction of current, the speed of the process can either increase or decrease. In the second case, the growth of the liquid zone ceases, and the process goes into equilibrium [1–3].

Attempts were made [4] to find out a mathematical law expressing the kinetics of movement of interphase boundaries under the CM with an electric transfer. The solution is fraught with certain mathematical difficulties.

The present article aims to find the mathematical law that describes the movement of the melt – solid phase boundaries during the electric transfer in the melt growing during the CM.

2. Results

In the laboratory reference system, the flow of substance of one of the components (B) in the AB melt in the presence of electric current is as follows (taking into account the possible macroscopic flow [4] similar to the Kirkendall displacement) [5, 6]:

$$J = -D\nabla c + DcFEez_B^* \quad (1)$$

where c is the concentration (the number of atoms (ions)) B per unit volume, D is the coefficient of mutual diffusion, E is the electric field strength, z_B^* is the effective charge in relative units of elementary charge (e), $F = V_A/VkT(\partial ln a_B/\partial ln n)$, V – volume of the solution mole, V_A – partial molar volume A, a_B – thermodynamic activity, n – atomic fractions of component B.

The kinetics of displacement of the boundaries of the melt – solid phase is determined by the difference in the flows of substance on both sides of the boundary:

$$J_{L1} - J_{S1} = -(c_1 - c_{S1})\frac{dl_1}{dt}, \quad (2)$$

$$J_{L2} - J_{S2} = -(c_{S2} - c_2)\frac{dl_2}{dt}; \quad (3)$$

where J_{Li} and J_{Si} – boundary flows from the melt and solid phase i, (i=1,2) respectively, l_i – absolute values of coordinates of interphase boundaries, c_i and c_{Si} – boundary concentrations from the melt and solid phase i, respectively.

Denoting the ratio $\frac{J_{Li}}{J_{Si}}$ as α_i and substituting the expression for J_{Li}, i.e. (1), into (2) – (3), we have:

$$\frac{c_1-c_{S1}}{1-\alpha_1}\cdot\frac{dl_1}{dt} = D(\nabla c)_1 - Dc_1FEez_B^*, \quad (4)$$

$$\frac{c_{S2}-c_2}{1-\alpha_2}\cdot\frac{dl_2}{dt} = D(\nabla c)_2 - Dc_2FEez_B^*; \quad (5)$$

where indices 1 and 2 are the values taken for boundary concentrations in the melt (c_1 and c_2, respectively).

Cross–border concentration gradients can be written as:

$(\nabla c)_i = \gamma|c_0 - c_i|/l_i$,

where c_0 – concentration per initial contact plane, γ_i – dimensionless factor taking into account the degree of deviation of the curve c (x) from the linear course (at the melt–solid phase i boundary). Equations (4) – (5) are reduced to

$$\frac{di}{dt} = \frac{h_i}{l_i} - q, \quad (6)$$

where

$$h_i = \frac{\gamma_i(1-\alpha_i)D|c_0-c_i|}{|c_i-c_{Si}|},$$

$$q_i = \frac{(1-\alpha_i)Dc_iFEez_B^*}{|c_i-c_{Si}|}.$$

In the absence of current $q_i = 0$. The coordinates of the interphase boundaries, as can be seen from (6), will change according to the well–known parabolic law:

$$l_i^2 = 2h_{i0}t = k_i^2t, \quad (7)$$

where k_i^2 – the constant of the displacement velocity of the interphase boundary under the CM in the absence of electric transfer, $h_{i0} = h_i|_{E=0}$.

If the current vector coincides with the coordinate (which is measured from the plane of the initial contact in the direction of boundary 2), E> 0. At $zz_B^* < 0$, we have: $q_i < 0$. As can be seen from (6), the electric transfer increases the speed of movement of the interphase boundaries. A similar effect will be observed at E <0 and $z_B^* > 0$. But if $[Ez_B^* > 0$ $(q_i > 0)$, the process slows down. In this case, the value (6) decreases and the movement of the boundaries stops. Moreover, having equalized (6) to zero, we can obtain limiting values of the absolute values of the coordinates of the interphase boundaries:

$$l_i^M = \frac{h_i}{q_i} \quad (8)$$

Using the auxiliary variable $y = h_i - q_il_i$, we transform equation (6) to the form:

$(h_i - y)dy = -yq_i^2dt$.

Its solution, taking into account the initial condition $(l_i|_{t=0} = 0$ is as follows:

$$l_i(t) + \frac{h_i}{q_i}ln\left|\frac{h_i-q_il_i(t)}{h_i}\right| = q_it \quad (9)$$

Let us analyze features of solution (9) for each mode.

1. Slowing down mode ($q_i > 0$). Given (8), let us write (9) as

$$r_i + ln|1 - r_i| = -\frac{q_i^2t}{h_i}, \quad (10)$$

where $r_i = \frac{l_i(t)}{l_i^M}$. Since $0 \le l_i(t) \le l_i^M$, $0 \le r_i \le 1$. Therefore, the sign of the module in (10) can be omitted. Since $l_i(t) = r_il_i^M$, given (8) and (10), we have:

$$l_i^2 = \frac{r_i^2h_i^2}{q_i^2} = \frac{r_i^2h_it}{\frac{q_i^2t}{h_i}} = \frac{r_i^2h_it}{f(r_i)}, \quad (11)$$

where $f(r_i) = -(r_i + ln(1 - r_i))$.

Setting values of parameter l_i^M and varying $l_i(t)$ from 0 to l_i^M, we have corresponding values of parameter $r_i = \frac{l_i(t)}{l_i^M}$ and function $f(r_i)$. (11) can be used to calculate the graph of the dependence $l_i^2(h_i t)$. Figure 1 shows the curves for l_i^M equal to 2, 3 an 4 mm.

(8) taking into account (6) for determining h_i and q_i, shows that $l_i^M \sim \frac{1}{E}$. Therefore, in case of absent electric transfer (E=0) we have: $l_i^M \to \infty$. $l_i^2 \to t$, see (7).

We can determine the time for $l_i(\tau) = r_{iP} l_i^M$, where r_{iP} – the number of target values of r_i. From (11), given (7) and considering $h_i \approx h_{i0} = \frac{k_i^2}{2}$, we have:

$$\tau = \frac{f(r_{iP}) l_i^2(\tau)}{h_i r_{iP}^2} \cong f(r_{iP}) \frac{\left(l_i^M\right)^2}{h_{io}} = 2f(r_{iP}) \frac{\left(l_i^M\right)^2}{k_i^2}. \qquad (12)$$

Thus, it is possible to determine the time during which the system reaches the state of stationary equilibrium. If $r_{iP} = 0{,}95$, a $l_i^M = 0{,}5$ cm, at $k_i^2 = 2 \cdot 10^{-5} \frac{\text{см}^2}{\text{с}}$ (12) gives: $\tau = 0{,}51 \cdot 10^5 c. = 14{,}2$ h.

Given (8) and considering $h_i \cong \frac{k_i^2}{2}$, equation (6) for the slowing down mode can be written as

$$\frac{dl_i}{dt} = \frac{k_i^2}{2}\left(\frac{1}{l_i} - \frac{1}{l_i^M}\right), \qquad (13)$$

In case of absent electric transfer $\left(l_i^M \to \infty\right)$ instantaneous velocities of movement of the interphase boundaries under the CM are:

$$\frac{k_i^2}{2l_i} = \frac{k_i}{2\sqrt{t}}.$$

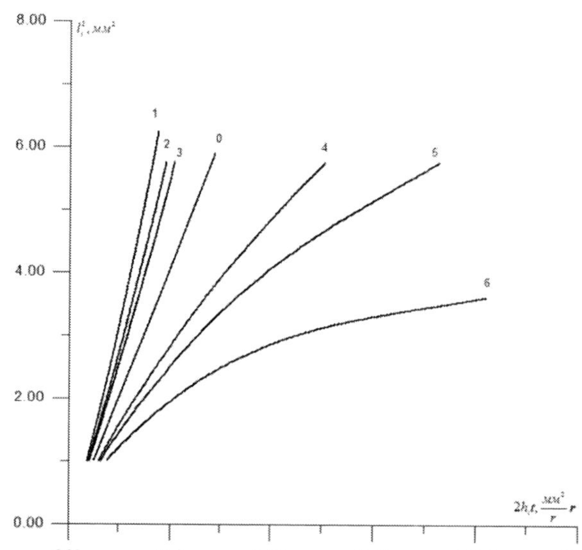

Figure 1. Graph of the growth of the square coordinate of the boundary of the melt – solid phase i (i = 1,2) under the CM with an electric transfer. The numbers of the curves correspond to the following parameters: 1–3: accelerating mode at $l_i^* = 2; 3$ and 4 mm m. 4–6: deceleration mode with $l_i^M = 4; 3$ and 2 mm.; 0 – current mode

2. The accelerating mode ($q_i < 0$). Let us introduce the notation:

$$\frac{h_i}{q_i} = -l_i^* \qquad (14)$$

where l_i^* – a value of l_i. Having divided both parts by l_i^*, we have ($r_i = \frac{l_i(t)}{l_i^*}$):

$$r_i - ln(1 + r_i) = \frac{q_i^2(t)}{h_i} \tag{15}$$

As far as $l_i(t) = r_i l_i^*$, given (14) and (15), we have:

$$l_i^2 = \frac{r_i^2 h_i^2}{q_i^2} = \frac{r_i^2 h_i t}{\frac{q_i^2 t}{h_i}} = \frac{r_i^2 h_i t}{g(r_i)}, \tag{16}$$

where $g(r_i) = r_i - ln(1 + r_i)$. Setting the values of l_i^* and varying $l_i(t)$, we have corresponding values of parameter $r_i = \frac{l_i(t)}{l_i^*}$ and function $g(r_i)$. (16) can be used to build the dependence graph $l_i^2(h_i t)$.

Figure 1 (left part) shows the curves for $l_i^* = 2; 3\ and\ 4$ mm.

Given the values of h_i and q_i, ratio (14) shows that $l_i^* \sim \frac{1}{E}$. $l_i^* \to \infty$ corresponds to the special case (E=0). The parabolic dependence (7) is valid.

Given (14) and considering $h_i \approx h_{i0} = \frac{k_i^2}{2}$, equation (6) for the accelerating mode can be written as

$$\frac{dl_i}{dt} = \frac{k_i^2}{2}\left(\frac{1}{l_i} + \frac{1}{l_i^*}\right). \tag{17}$$

Given that in the absence of electric transfer $q_i = 0$ и $l_i^0 \to \infty$, the following conclusion can be drawn from (17). Value l_i^* has a certain physical meaning: at $l_i = l_i^*$ the instantaneous boundary velocity (melt – solid vase i) will be more than the value of this parameter in the absence of current (in the same size l_i) two times. At $l_i = l_i^*$, $r_i = 1$. $g(r_i) = 1 - ln2 \approx 0{,}307$. From (16), we obtain $h_i \approx \frac{k_i^2}{2}; \frac{l_i^2}{t} = \frac{k_i^2}{2g(1)} = 1{,}63 k_i^2$, while at $E = 0$, $\frac{l_i^2}{t} = k_i^2$ (see (7)).

3. Conclusion

The curves in Figure 1 are similar to the corresponding experimental graphs published in [1, 3]. Moreover, the curves reflect the general process and are valid for any binary system (where the KP process with an electric transfer is possible) and for any experimental temperature, since possible values of the parameters h_i. and l_i^M (or l_i^*) must correspond to the specific experimental conditions.

Acknowledgment

The study was funded by the RFFR within project No. 18–32–00572 mol_a.

References

[1] Akhkubekov A A, Savintsev S P and Bagov A M 2005 Kinetics of the formation of a melt as a result of contact melting, under the conditions of the slowing down action of electric transfer *Melts* **1** 54–8

[2] Bagov A M 2016 *The effect of electrotransfer on mutual diffusion and macroscopic flow of a melt formed during contact melting* Thesis (Nalchik: Kabardino-Balkarian State Univer.)

[3] Akhkubekov A A, Savintsev S P and Bagov A M 2004 The effect of electrotransfer on crystallization and macroscopic flow during contact melting of metals *Metals* **4** 33–6

[4] Bagov A M, Tamaev T Kh, Khasanov A I et al 2018 Phase formation in contacting zinc melting with indium and stannic in the presence of strontium impurity and microhardness of derived alloys *Advances in Engineering Research* vol 177 Int. Symp. Engineer. and Earth Sci.: Appl. and Fund. Res. (ISEES 2018) pp 143–5

[5] Akhkubekov A A and Bagov A M 2009 On the shift of inert labels during contact melting of binary metal systems in the presence of electrotransfer *Proc. of the Russ. Acad. of Sci. Ser. is Phys.* **73(7)** 961–3

[6] Akhkubekov A A 2001 *Diffusion and electric transport in low–melting metal systems during contact melting* Thesis (Nalchik) 312 p

IOP Conf. Series: Materials Science and Engineering **905** (2020) 012006 doi:10.1088/1757-899X/905/1/012006

Ways to improve properties of high-strength building polymer composites

Yu M Bazhenov[1], D K-S Bataev[2], S-A U Murtazaev[3,2], Z Z Alarkhanova[4,2], R U Goitemirov[4], Kh N Mazhiev[3,5], Z S Khasbulatova[4,2] and P D Bataeva[3]

[1] National Research Moscow State University of Civil Engineering, 26, Yaroslavskoye Shosse, Moscow, Russia

[2] Integrated Research Institute named after Kh.I. Ibragimov, Russian Academy of Sciences of the Chechen Republic, 21A, Staropromyslovsky ave., Grozny, 364051, Russia

[3] Grozny State Oil Technical University named after M.D. Millionshchikov, korp. 1,100, prosp. Husejna Isaeva, Grozny, 364068 Russia

[4] Chechen State Pedagogical University, 62, prosp. Husejna Isaeva, Grozny, 364068 Russia

[5] Academy of Sciences of the Chechen Republic, 13, prospect named after M. Esambayev,Grozny, Russia

E-mail: zavkaf@list.ru, kniiran@mail.ru, smurtazaev@mail.ru, alarh2000@mail.ru, kniiran@mail.ru, seismofund@mail.ru, hasbulatova@list.ru, bataeva_ggntu@mail.ru

Abstract. Ways to improve quality of composite materials, in particular, physical and physico-mechanical properties of nanostructured building polymer composites are described. Applications of polymer additives and superplasticizers for improving technological properties of composites are described. The main tasks of improving polymer composites are determined. The effect of surfactants on the properties of high-strength composites is identified. The dependence of the strength of cement stone on porosity is identified.

1. Introduction

The main building composites are concrete and mortar. In recent years, the success of the cement industry and improvement of the concrete and mortar production technology have improved their quality. M800 – M1200 is used in construction; hardness of concrete and mortar improved. However, further development of the construction industry requires harder building materials. It is necessary to reduce porosity of concrete, apply higher strength components, and increase internal cohesion and adhesion [1].

However, using the traditional concrete production technology, it is difficult to reduce its porosity, since it is necessary to add an excess amount of water. By its nature, cement stone is a porous material, and the volume of neoplasms under normal conditions is insufficient to fill the initial voidness of the solid phase.

2. Methods and materials

Hard cement stone and concrete produced in the laboratory of metals, alloys and composite materials of the IRI n.a. Ibragimov cannot be used [5, 7, 10].

It is difficult to increase the tensile strength of concrete t due to the fact that this material has a very heterogeneous structure with a large number of defects, as well as a relatively low adhesion between the components composing the structure, and their low tensile strength.

Reducing the porosity of the concrete composite increases the strength of the materials (Fig. 1).

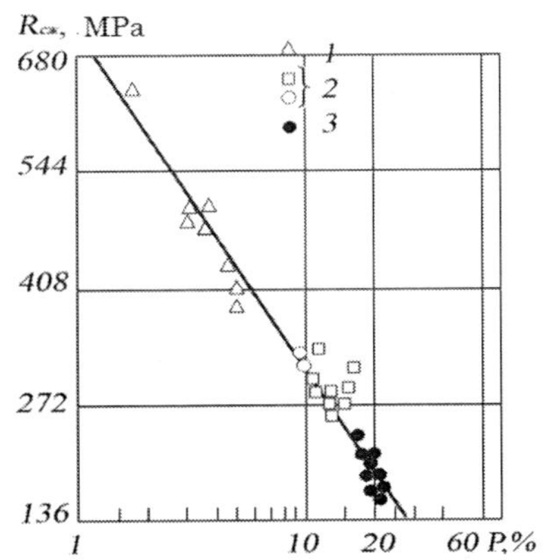

Figure 1. The dependence of the strength of cement stone on its porosity: 1 – samples of hot pressing; 2 – samples pressed at normal temperature; 3 – vibrated samples.

Overcoming these difficulties makes it possible to improve quality of composite materials. The latest achievements of the chemical industry can be used. There are various ways of improving the structure and properties of concrete: introduction of new nanostructured components into the concrete composition, the impact on traditional components, a composite mixture, etc.

There are many materials that combine organic and inorganic components, and these materials have high strength and lightness (for example, shells of insects or animal bones). The combination of the organic skeleton of organic nature and astringent inorganic nature into one system and vice versa allows one to produce new building composite materials that have better properties than traditional building composites. Research has already begun at the IRI n.a. H.I. Ibragimov, RAS [2–5, 7, 9, 11].

In the 1980s, polymer concrete was studied by various researchers. Even the term "P-concrete" uniting various varieties of concrete in which polymers are used was introduced. It is necessary to expand varieties of polymer concrete by introducing new terms "nanostructured building polymer composite" or "quasinanostructured composite", etc.

Currently, there are several ways to improve properties of building composites using polymers [1]:

- introduction of a small amount of polymer additives into the concrete mix (concrete with polymer additives, including superplasticizers, and cement-polymer concrete);
- use of a polymer binder (polymer composites);
- impregnation of finished concrete and reinforced concrete products with special polymer compositions or monomers with their subsequent polymerization in the body of the construction material;
- reinforcement using polymer fibers and microfibres;
- the use of polymeric lightweight aggregates or aggregates modified with polymers;
- introduction of polymer microfillings;
- introduction of polymer ultrafine fillers;
- introduction of composite polymer nano-additives.

Each direction has its own advantages.

IOP Conf. Series: Materials Science and Engineering **905** (2020) 012006 doi:10.1088/1757-899X/905/1/012006

The use of various types of polymer additives in concrete is expanding [6]. Superplasticizers are promising. Superplasticizers based on melamine resin, naphthaleisulfonic acids and modified lignosulfonates are widely used.

Superplasticizers are introduced into the concrete mixture in an amount of 0.2–1 % of the mass of cement. Superplasticizers thin the concrete mix while maintaining the proper bonding of the concrete mix and kinetics of cement hydration (Fig. 2). They are a powerful means of changing rheological properties of a mineral binder. It is most advisable to use superplasticizers to produce high-strength concrete or improve the concrete production technology through the use of cast concrete mixtures. In our experiments, superplasticizers were used to produce M500-M600 grade 400 cement and M600-M700 grade 500 cement. The use of superplasticizers in combination with other methods of modifying concrete with polymers, in particular, impregnating hardened concrete with polymeric materials, can increase the compressive strength up to 70 MPa and more.

By introducing additives in the amount of more than 2 % of the cement mass, cement-polymer concrete can be produced [3]. The presence of polymer in concrete affects properties of the material, especially the tensile and bending strength, and the behavior of concrete under operating conditions. Various latexes and some types of resins are used as additives. The effect of polymer additives is measured in tens of percent compared to concrete without additives, however, such additives are promising for mass application in monolithic and precast concrete and reinforced concrete.

Polymer additives and superplasticizers can be used together with other chemical additives as part of complex additives that change properties of composites.

Polymer composites are a separate large group of new composites. There are more than 30 types of polymer concrete. The most widely used are FAM, PN-1 and UKS.

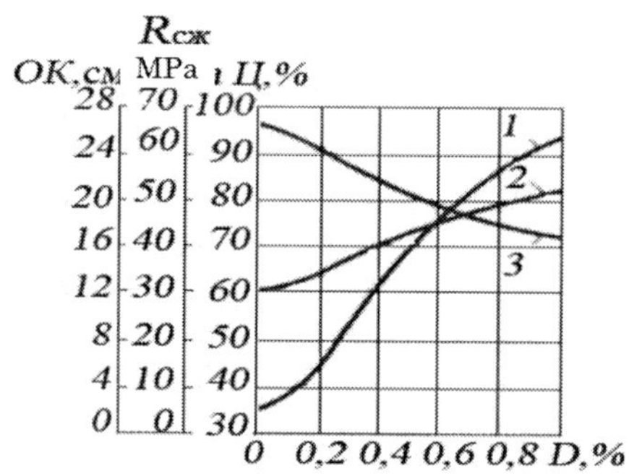

Figure 2. The effect of the content of D superplasticizer on the properties of concrete and concrete mix: 1 – change in the mobility of the mixture when introducing superplasticizers and constant consumption of cement and water; 2 – increased strength of concrete with constant mobility of the mixture and cement consumption by reducing water consumption; 3 – reduced consumption

Heavy polymer concrete has a compressive strength of 60–120 MPa, a tensile strength of 5–10 MPa, several times lower permeability, high chemical resistance and a number of other positive properties. Light polymer concrete with an average density of 1500–1800 kg/m^3 can have a strength of 40–80 MPa, a density of 400–600 kg/m^3, and a strength of 20–30 MPa.

Polymer composites have been studied since 1958 under the guidance of Professor S.S. Davydov [1]. There are several plants producing polymer-concrete and steel-polymer-concrete products used in the construction industry, in the chemical industry, non-ferrous metallurgy, hydraulic engineering, etc.

Important tasks are to improve their properties, reduce creep and shrinkage, increase hardness and reduce costs [4].

Polymer fine fibers are used as reinforcement. They are added to cellular, lightweight and gypsum concrete. Increased deformability of concrete contributes to the joint work with polymer fibers which have a very high level of deformability. As a result, the strength of the material increases.

The use of polymeric lightweight aggregates, especially in combination with a polymeric binder, makes it possible to produce light heat-insulating materials that are resistant to moisture and frost and allow the creation of lightweight building structures.

Polymeric materials are used to modify aggregates, for example, to reduce the water absorption of light aggregates by creating a polymer film on their surface or partially impregnating the aggregate with a high molecular weight technogenic compound. These studies are carried out in the laboratory of metals, alloys and composite materials of the IRI n.a. Ibragimov, RAS. Polymeric materials can be used to improve adhesion between the aggregate and the binder, including by grafting the polymer onto the aggregate. However, methods for modifying aggregates with polymers are not used in the construction industry, since the achieved effect often does not compensate for an increase in the cost of the material and the complex technology. In addition, polymeric materials are still scarce and their use is rational only if they can be economically efficient.

The most noticeable change in the properties of a building composite is observed when it is impregnated with special compounds, i.e. in polymer composites. These are new building materials. Treatment of concrete and reinforced concrete products includes drying, impregnation and sometimes radiation.

Impregnation of concrete, asbestos cement and other capillary-porous materials in order to increase their hardness and resistance has been used for a long time. However, due to the small depth of impregnation and insufficient strength of the impregnating composition, it is not possible to change the strength of the composite.

As a result of special treatment of the composite with polymers, the strength of the material increases several times, its hardness and resistance sharply increase when exposed to aggressive media. Table 1 shows some ways to improve properties of the composite when it is impregnated with methyl methacrylate [1].

By changing the structure of the impregnated composite, the impregnating material and the treatment technology, it is possible to produce polymer composites with a wide variety of properties. In particular, it is possible to produce durable light polymer composites, electrical composites, and a number of other special materials.

Table 1. Properties of construction composites and polymercomposites

Parameter	Composite polymer	Source composite
Tensile strength, MPa: under compression	100–200	30–50
under tension	6–19	2–3
under bending	14–28	5–6
The elastic modulus under compression, MPa	$3.5\cdot10^4$–$5\cdot10^4$	$2.5\cdot10^4$–$3.5\cdot10^4$
Ultimate compressive strain	0,002	0,001
Adhesion to reinforcement, MPa	10–18	1–2
Dynamic tensile strength (τ = 103 MPa·s)	30	7
Shrinkage strain	0–$5\cdot10^{-5}$	$50\cdot10^{-5}$
Creep strain	$6\cdot10^{-5}$–$8\cdot10^{-5}$	$40\cdot10^{-5}$–$60\cdot10^{-5}$
Electric resistance, Ohm	10^{14}	10^5
Water absorption, %	1	3–5
Frost resistance, std. cycles	5000	200
Corrosion Resistance to Sulphates and Acids	High	Insufficient

Table 2 shows properties of polymer composites [1, 9]. For comparison, the results of polymer impregnation and other building materials are shown. These materials can be combined with polymer composites. The data in Table 2 do not exhaust the variety of materials and properties that can be obtained by combining various mineral structures with polymers.

A composite with increased strength and hardness, as well as other improved properties is promising for producing new durable building structures. Figure 3 shows the change in the deformative properties of the polymer composite produced by impregnating the building composite with mixtures of methyl methacrylate with n-butyl acrylate.

Table 2. Properties of various concrete polymers

Material	Weight gain, %	Compressive strength, MPa		Water absorption, %	
High strength concrete	4	76	2100	3.8	0.2
Expanded clay concrete with an average density, kg / m^3:					
750	68	2.9	33.3	23.5	2.5
1400	19	15.2	90.3	13	0.9
Perlite concrete	58	5.5	42	36	7.5
Gypsum concrete	46	22	92	36.5	3
Gypsum cement	13	27	52	16.3	2.5
Arbolite	18	9	48	17.5	2.6
Aerated concrete	75	7.7	75.5	30	1.5
Silicate materials (brick)	12.8	19	120	11	1.1
Ceramics (brick)	15.5	12.5	78	18	1.8
Asbestos cement:		Bending strength			
-standard	6	2	42	-	-
-pressed	6.2	3	70	-	-

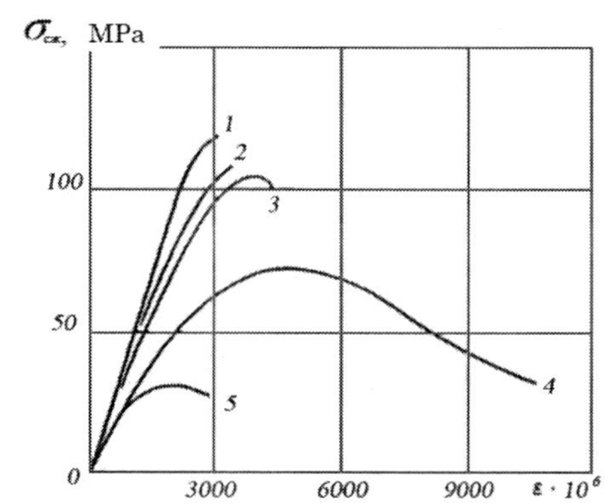

Figure 3. The effect of the impregnating composition on the stress-strain curves: 1 – methyl methacrylate (MMA); 2 – 90 % MMA + 10 % butyl acrylate (BA); 3 – 70 % MMA + 30 % BA; 4 – 50 % MMA + 50 % BA; 5 – concrete (control samples)

Various initial composites or capillary-porous building materials, conventional concrete or reinforced concrete structures and products produced using traditional methods (from cast mixtures without vibration) can undergo subsequent special treatment. It is possible to treat either the entire volume of concrete, or only its surface or individual zones of the product in order to give them required properties.

There are heavy polymer composites produced on the basis of various types of heavy concrete and having an average density of more than 1800 kg/m^3, and light polymer composites produced on the basis of various varieties of light concrete (with an average density of less than 1800 kg/m^3).

Depending on the type of impregnated materials, there are polymer composites based on artificial glassy polymers (styrene, methyl methacrylate, etc.); polymer composites impregnated with viscous organic materials such as bitumen, paraffin, etc.; polymer composites impregnated with sulfur, and polymer composites impregnated with water glass and some other special compounds. The concrete impregnation technology has its own peculiarities, often materials impregnated with various compositions are called "polymer composites".

By the nature of the polymerization, polymer composites are divided into materials produced by thermocatalytic or radiation polymerization.

According to the degree of filling of concrete pores with polymer, there are polymer composites with almost complete impregnation of concrete and polymer composites with zone treatment.

It is possible to produce complex polymer composites with a polymer skeleton is combined with fibrous polymers introduced in the composite, or with polymer fillers or microfillers [8].

In order to increase the efficiency of using polymers, surface or zone impregnation can be used. With a small consumption of monomer (1.5–3 kg/m^2 of concrete), this treatment increases frost resistance, corrosion resistance and improves other properties of structures and products, including those made of a monolithic and lightweight building composite (concrete and reinforced concrete).

The use of secondary products of petrochemicals and oil refining or composite impregnating compositions can reduce the cost of polymer composites.

3. Conclusion

Polymer composites are the most effective materials which lack disadvantages of polymer materials (high cost, aging and creep). Polymer composites do not manifest creep, since the polymer network is combined with a rigid mineral skeleton with low deformability under loading. In polymer composites, the polymer is located inside the material; it is protected from the effects of solar radiation, atmospheric oxygen, and other atmospheric factors, which helps increase its durability.

An advantage of products and structures made of polymer composites is the possibility of gluing and welding them to produce a high strength weld. Some polymer composites can be machined, which is of particular importance for some special applications.

Each material has drawbacks. The disadvantage of polymer composites is their limited heat resistance. By using more heat-resistant polymers, the heat resistance of the polymer composite can be increased, however, these materials must be used for structures that do not have special requirements for heat resistance.

Impregnation of the material with more heat-resistant compositions, for example, liquid metals or other high-temperature melts, can be used to produce more heat-resistant composites.

To ensure the heat resistance of a high-strength composite, high-temperature additives are of great importance. These studies are carried out by the Department of Cementing and Concrete Technology of the Institute of Construction and Architecture of the National Research Moscow State University of Civil Engineering and the Laboratory of High-Molecular Compounds of the IRI named after H.I. Ibragimov of the Russian Academy of Sciences (Grozny).

References

[1] Bazhenov Yu M 1983 *Concrete Polymers* (Moscow: Stroyizdat) 472 p

[2] Bataev D K-S, Goytemirov G U and Muradov M T 2013 The wear mechanism of polymer materials used in the restoration of building structures *Vest. AN CR* **1(18)** 86–9

[3] Mintsaev M Sh, Bataev D K-S, Mazhiev K Kh et al 2018 Prospects for Using 3D-Printing Technologies in Construction of Buildings in Seismic Areas *Advances in Engineering Research* vol 177 Int. Symp. on Engineer. and Earth Sci. (ISEES 2018) pp 311–5

[4] Bazhenov Y, Bataev D, Mazhiev H and Bataeva P 2016 Prospects for the Development of Nanostructured Polymer Composites in the Chechen Republic *Int. J. of Environ. & Sci. Ed.* **11(18)** 12687–97

[5] Abdullaev A and Abdullaev M 2018 High-Quality and High Mobility Concrete with Complex Nanostructured Additives *Advances in Engineering Research* vol. 177 Int. Symp. on Engineer. and Earth Sci. (ISEES 2018) pp 228–31

[6] Salamanova M, Khubaev M, Saidumov M and Murtazayeva T 2017 Self-Consolidating Concretes with Materials of the Chechen Republic and Neighboring Regions *Int. J. Of Environ. & Sci. Ed.* 11(18) 12719–24

[7] Bataev D K-S, Mazhiev Kh N, Goitemirov P U and Umarov M U 2018 Evaluation of Fatigue Wear and the Nature of the Destruction of Polymeric Materials *Advances in Engineering Research* vol 177 Int. Symp. on Engineer. and Earth Sci. (ISEES 2018) pp 218–22

[8] Mashkov Yu K 2004 *Polymer composite materials in tribotechnology*, (Moscow : Khimiya) 261 p

[9] Bazhenov Y, Bataev D, Mazhiev H and Bataeva P 2016 Prospects for the Development of Nanostructured Polymer Composites in the Chechen Republic *Int. J. of Environ. & Sci. Ed.* **11(18)** 12687–97

[10] Abdullaev A, Dadashev R, Abdullaev M et al 2018 Surface Characteristics of Nanosized Bentonite Suspensions as a Modifying Component of Cement Composites *Advances in Engineering Research* vol 177 Int. Symp. on Engineer. and Earth Sci. (ISEES 2018) pp 223–7

[11] Bataev D K-S, Uzaeva A A, Uzaeva S A et al 2018 The Study of Shrinking Deformations of Repair Compositions on Barkhan Sands *Advances in Engineering Research* vol 177 Int. Symp. on Engineer. and Earth Sci. (ISEES 2018) pp 254–7

IOP Conf. Series: Materials Science and Engineering **905** (2020) 012007 doi:10.1088/1757-899X/905/1/012007

The study of application of Maslyanites in pumps designed for pumping aggressive media

D K-S Bataev[1,2,3], R U Goitemirov[1,4] and M A Dzhamaldinova[4]

[1] Integrated Research Institute named after Kh.I. Ibragimov, Russian Academy of Sciences of the Chechen Republic, 21A, Staropromyslovsky ave., Grozny, 364051, Russia
[2] Academy of Sciences of the Chechen Republic, 13, prospect named after M. Esambayev, Grozny, Russia
[3] Grozny State Oil Technical University named after M.D. Millionshchikov, korp. 1,100, prosp. Husejna Isaeva, Grozny, 364068 Russia
[4] Chechen State Pedagogical University, 62, prosp. Husejna Isaeva, Grozny, 364068 Russia

E-mail: kniiran@mail.ru, groznymuh@mail.ru, otd60@mail.ru

Abstract. For experimental studies, chemically active aggressive media pumped by single-screw metering pumps (SSMP) were used. The effect of these media on the destructive compressive stress, tensile strength, elongation, hardness, degree of swelling, antifriction properties and wear resistance of the polymer composite material Maslyanite D (S-1) was studied. It was established that hardness and antifriction properties of Maslyanite D (S-1) are better than those of SKN-26-based rubber 8470.

1. Introduction

Single-screw pumps are applied in the modern construction industry for pumping foam concrete, mortars, paints, adhesives, dibutyl phthalate, carbon tetrachloride, oleic acid nitrile and other aggressive media. The use of oil-petrol-resistant rubbers (grades 3626 and 6470) is limited due to their significant swelling in the aggressive media and irreversible physical and mechanical changes. When dosing media containing a high percentage of abrasive particles, there is a lot of wear of the clip and screw materials.

It is necessary to find and test materials with a low coefficient of friction, high wear resistance and good reciprocal compatibility with pumped fluids, which can be used both in the production and repair of single-screw metering pumps.

The impact of the working medium on polymers can decrease and increase their strength characteristics [1]. The manifestation of this effect is determined by physicochemical properties of the medium and the polymer, the nature and value of the load application, its concentration, temperature and pressure, the ratio of fracture kinetics and diffusion of the medium [2–4].

2. Methods and materials

The study of strength characteristics of the oil group materials was carried out in carbon tetrachloride and dibutyl phthalate in an unstressed state of samples.

Carbon tetrachloride is one of the strongest solvents, and dibutyl phthalate has the greatest plasticizing effect [5].

Materials were tested in an unstressed state by the methods developed on the basis of GOST 4651-2014 (compression), GOST 11262-2017 (tension) and GOST 24622-91 (determination of hardness). [6–8]

Strength characteristics of the material were determined after every 170 hours in the media. The total test duration was 850 hours.

The measuring instrument, preparation techniques, and conditions were the same as in determining the degree of swelling of the materials. The tests were carried out using the UME-10 TM, providing load measurement with an error of not more than 1.0 % of the measured value.

Samples had a cylindric shape with a height of 15 ± 0.5 mm and a diameter of 10 ± 0.5 mm 3–5 samples were placed in each medium; at least 5 samples were tested. Before the test, the dimensions of the samples were measured with an accuracy of 0.01 mm.

The breaking stress in compression was calculated using the known formulas.

Tensile samples had a shape of tongs (GOST 11262-76) and a smooth surface without cracks, shells and scratches. Before the tests, the thickness and width of that part of the sample that is tearing with an accuracy of 0.01 mm in at least 3 places were measured, and the cross-sectional area was determined. The sample was mounted in the clamps so that the direction of the tensile force coincided with the longitudinal axis of the sample.

The tests were carried out using the ALA-LLP at a gripping speed of 111 mm/h.

Hardness was determined using a hardness tester AS-III. A steel ball with a diameter of 5 mm was pressed into the test sample with an initially given force, which was increasing for 30 seconds. The duration of exposure to such a load was 60 seconds. Hardness measurements were made immediately after determining the degree of swelling.

3. Results

The test results (Figure 1) showed that a decrease in the breaking stress of compression of samples made from Maslyanite D (S-1) in carbon tetrachloride is 15.3–16.5 %; in dibutyl phthalate, a decrease in strength in all samples does not exceed 13.8 %.

A decrease in the tensile stress of Maslyanite D (S-1) with a prolonged exposure to carbon tetrachloride and dibutyl phthalate in the media is 27.1–31.7 % (Figure 2). Therefore, the material retains characteristics which are required for the media under study [3].

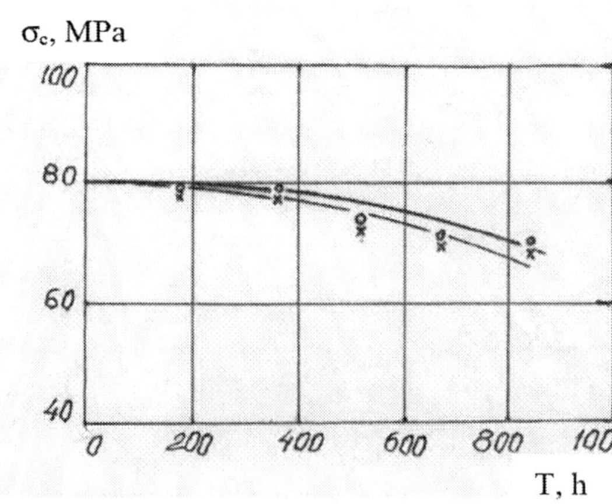

Figure 1. Destructive compressive stress of Maslyanite D (C-1) depending on the exposure time in dibutyl phthalate (o) and carbon tetrachloride (x)

IOP Conf. Series: Materials Science and Engineering **905** (2020) 012007 doi:10.1088/1757-899X/905/1/012007

Several compositions of Maslyanites D (S-1), K-15 and K-35 were tested. The relative elongation of the samples differs from each other (within 15 %). The test results are presented in Table 1 and Figure 3.

To determine the hardness of Maslyanite D (S-1) depending on the time spent in the media, the same samples were used. Hardness was measured immediately after determining the degree of swelling.

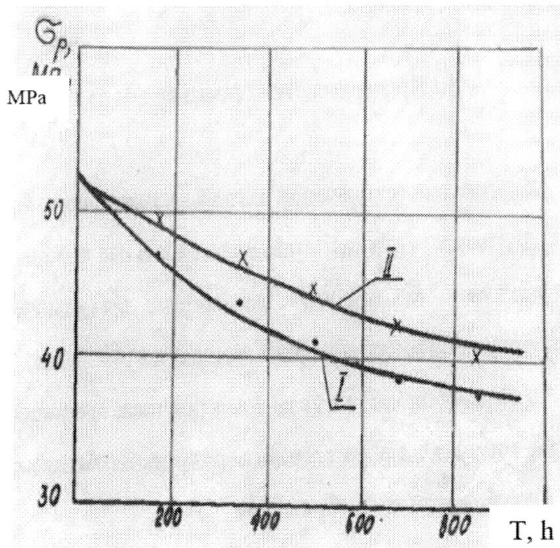

Figure 2. Tensile strength of Maslyanite D (C-1) depending on the exposure time in carbon tetrachloride and (I) dibutyl phthalate (II)

Table 1. Relative elongation of samples from Maslyanite D (C-1) depending on the time spent in the media

Medium	Relative elongation (%) depending on the time spent in the medium (h)					
	0	170	340	510	680	850
CCl4	5,5	5,5	53	5,2	4,8	4,1
DBP	5,5	4,8	4,3	3,8	3,4	3,1

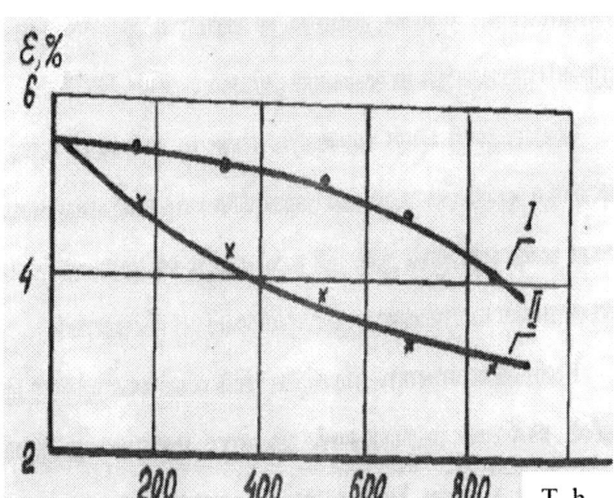

Figure 3. Relative elongation at during the break of Maslyanite D (C-1) samples depending on the time spent in carbon tetrachloride (I) and dibutyl phthalate (II)

The surfaces did not have cracks, shells, or burrs; they were even, smooth, without visible treatment traces. For each test, 5 samples were used; 6 measurements were carried out on each sample. The value of hardness (HV) determined by the depth of indentation of the ball at a given basic force in kgf, was calculated by formula [8]

$$\mathbf{HB=F.0.1/\pi \cdot 0.5 \cdot h} \ (MN/m^2)$$

where F – value of the main indentation force, kgf;
0.5 – ball diameter, cm;
h – maximum indentation depth, cm;
0.1 – coefficient of hardness transfer from the GHS system to the ISS system.

As studies have shown (Table 2, Figure 4), the hardness of Maslyanite D (S-1) when it is in the carbon tetrachloride and dibutyl phthalate for a long time decreases by 20–25 %, which meets the requirements for the material working in these media for a long time.

Table 2. Hardness of Maslyanite D (S-1) depending on the time spent in carbon tetrachloride (CCl_4) and dibutyl phthalate (DBP)

Medium	Hardness of the material (MN / m^2) depending on the time spent in the medium					
	0	170	340	510	680	850
CCl_4	0.9	0.85	0.82	0.78	0.74	0.68
DBP	0.9	0.86	0.83	0.80	0.77	0.69

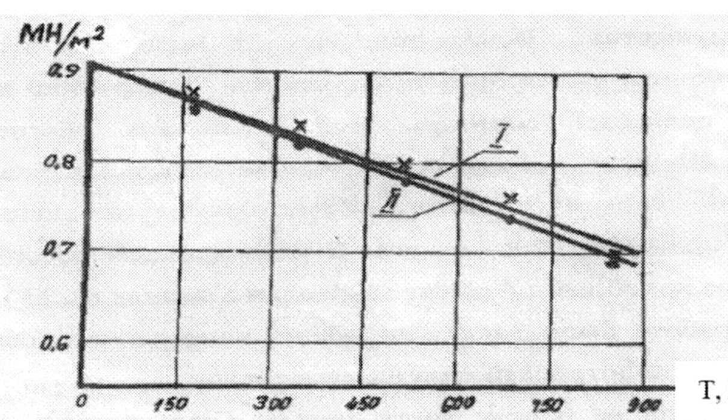

Figure 4. Hardness of Maslyanite D (S-1) depending on the time spent in four-carbon (II) chloride and dibutyl phthalate (I)

By the degree of swelling and modification of physical and mechanical properties of the material depending on the time spent in the medium for a given period of time, it is possible to determine the performance of the material.

The main criterion for the performance of products made from polymer materials is the degree of swelling of the material in an unstressed state. However, studies [12] show that the degree of swelling of a polymer material depends on the applied load. Externally, mechanical loads and internal stresses can change the rate of diffusion penetration of the medium into the polymer, the nature of the effect of the medium itself, or accelerate or slow down the destruction of the material.

There are many methods used to assess strength properties of polymeric materials in contact with liquid media under the influence of static and dynamic loads. However, there are no standard methods for investigating the long-term strength of the fatigue properties of polymer materials working in liquid media [9, 10]. Therefore, to study the chemical resistance and change in the physicomechanical properties of Maslyanite D (S-1), a simple device was designed. Test samples were made in the form of a shovel according to GOST 11262-2017 and had the same parameters and technical requirements as for tensile test samples. The tests were carried out at 20 °C. Before the tests, the base length of the samples was measured, and the cross-sectional area was calculated. The weight of the sample was measured with an accuracy of 0.0001 g. The test sample was loaded into the medium, secured with screw clamps, and loaded through the block with a constant tensile load of 13 MPa (0.35σp).

This load was chosen on the basis of a five-fold margin of safety from the maximum pressure on the material under given operating modes of the screw pump. During testing, under the simultaneous influence of an external load and a liquid medium, the degree of swelling and a change in the linear dimensions of the samples were measured, and the strength properties of the material were evaluated.

When testing samples of Maslyanite D (C-1) in oleic acid nitrile (Tables 3 and 4), no significant effect on the change in the mass of the samples was found. The maximum deviation of the mass of the samples was only 0.15 %. The change in the length of the samples of Maslyanite D (S-1) which spent in the medium for 192 hours was only 0.15 %, and the length changed only in the first 100–120 hours, after which the sizes of the samples were stabilized. The tensile strength did not change, the deviation was within the accuracy of measuring the load by the UME-TM (1 %).

When the material is in the media for a long time, the swelling is insignificant, the residual strength is quite high when compared with stresses developed during the operation of a single-screw pump [11].

Table 3. Change in the mass of samples made from Maslyanite D (S-1) in oleic acid nitrile under tensile loading (%)

No	Load, MPa	Exposure time, h						Note
		24	48	96	120	144	192	
1	13	0.10	0.09	0.10	0.11	0.11	0.11	
2	13	0.10	0.09	0.10	0.12	0.12	0.12	
3	–	0.08	0.06	0.06	0.07	0.07	0.07	Control samples

Table 4. Change in the length of samples made from Maslyanite D (S-1) in nitrile oleic acid under tensile loading (%)

No	Load, MPa	Exposure time, h						Note
		24	48	96	120	144	192	
1	13	0	0	0.07	0.14	0.14	0.14	
2	13	0	0	0.07	0.15	0.15	0.15	
3	–	0	0	0	0	0	0	Control samples

The tests for compression, tension, hardness, and swelling allow us to conclude that high physical and mechanical properties of Maslyanite D (S-1) are preserved in the aggressive media.

Determination of tribotechnical characteristics (the wear rate and the friction coefficient of composite materials working in a friction pair) is an obligatory criterion for assessing their operational resource. [14] For this purpose, tests were carried out on rubber brand R-8470 and Maslyanite D (C-1) using the end friction machine at a speed of V = 0.2 m/s and pressure P = 0.5 MPa The counterbody material was steel 45.

Table 5. Tribological characteristics of composite materials

Material	Wear, g/h	Friction ration
Rubber R-8470	0.008	0.54
Maslyanite D (S-1)	0.002	0.30

The test results (Table 5) showed that the wear of Maslyanite D (S-1) is significantly lower (4 times) that that of rubber R-8470. By the anti-friction properties, Maslaynite D (C-1) surpasses rubber by almost 2 times.

When testing in the abrasive medium, we obtained [13] positive results for many Maslyanite compositions.

4. Conclusion

Comparative tests of the physicomechanical and tribotechnical properties of the materials allows us to conclude that it is possible to use Maslyanite D (S-1) as clips for repairing single-screw metering pumps.

References

[1] Louis-Etienne P 2017 Improving the efficiency and resource of screw pump systems in conditions of high gas content *Engineer. Pract.* **11**. 20–22

[2] Bratu C 2005 Progressing Cavity Pump (PCP) behavior in multiphase conditions *SPE 95272, SPE Annual Technical Conference and Exhibition* (Dallas, Texas, USA, 9–12 October)

[3] Timashev E O and Yamaliev V U 2005 Analysis of causes of destruction of screw pump holders elastomers *Oil and Gas Busin.* **2** Retrieved from: http://ogbus.ru/files/ogbus/authors/Timashev/Timashev_1.pdf.

[4] Latypov B M, Topolnikov A S, Komkov A G and Ismagilov R R Impact of Exploitation Conditions upon the Pump down Time of the Progressive Sucker Rod Pumping Units *Petroleum Eng.*

[5] Gamboa J, Olivet A and Espin S 2003 New *Approach for Modeling Progressive Cavity Pumps Performance: SPE 84137*

[6] GOST 11262-2017 (ISO 527-2: 2012) *Plastics. Tensile test method*

[7] GOST 4651-2014 (ISO 604: 2002) *Plastics. Compression Test Method*

[8] GOST 24622-91 (ISO 2039 / 2-87) *Plastics. Rockwell hardness test*

[9] Shaydakov V V 2002 *Properties and testing of rubbers* (Moscow: Chemistry) 235 p

[10] Urazakov K R, Topolnikov A S and Abramova E V 2011 The area of effective use of screw pumps for oil production *Territory Neftegaz* **2** 16–21

[11] Pyatov I S and Vasilyeva S N 1999 Combined method for modifying the frictional properties of rubbers *Rubber and Rubber* **5**

[12] Korotkikh N I and Gavrilova E I 2003 Development of new elastomers for stator screw downhole motors and screw pumps *Construction of oil and gas wells on land and at sea* no 9 (Moscow: VNIIO-ENG)

[13] Bataev D K-S, Goytemirov S-A and Murtazaev Yu 2015 Out-dressing of polymer composite materials in the presence of an abrasive *March 24–26, 2015) [Modern building materials, technologies and structures* In 2 volumes vol 1 Mater. of the Int. sci.c-pract. Conf. dedicated to the 95th anniversary of CSTU named after Acad. M.D. Millionschikov (Grozny: Grozny Worker) pp 160–166

[14] Urazakov K R, Latypov B M and Ismagilov R R 2015 Experimental Studies of the Friction Coefficient for the Elements of the Progressing Cavity Pumps *Oil and Gas Busin.* **4**

IOP Conf. Series: Materials Science and Engineering **905** (2020) 012008 doi:10.1088/1757-899X/905/1/012008

Stable Tetrahedron NaCl – KCl – PbCl2 – PbWO4 in the quaternary reciprocal system Na, K, Pb // Cl, WO4

Zh A Kochkarov[1], A A Baysangurova[2], Z S Khasbulatova[3], M V Khubaeva[3], Z Z Alarkhanova[4,3], D K-S Bataev[4,5] and D Z Maglaev[4,5]

[1] Institute of Chemistry and Biology, Kabardino-Balkarian State University, Nalchik, Russia
[2] Faculty of Biology and Chemistry, Chechen State University, Grozny, Russia
[3] Faculty of Natural Sciences, Chechen state Pedagogical University, Grozny, Russia
[4] Integrated Research Institute named after H.I. Ibragimova RAS,
[5] Institute of Construction, Architecture and Design, Grozny State Oil Technical University named after M.D. Millionshikova, Grozny, Russia

E-mail: chemest2@rambler.ru, bajsangurova95.95@mail.ru, hasbulatova@list.ru, medina-76@mail.ru, alarh2000@mail.ru, kniiran@mail.ru, maglaev.d@mail.ru

Abstract. The phase diagram of the stable tetrahedron NaCl – KCl – PbCl2 – PbWO4 of the quaternary reciprocal system Na, K, Pb // Cl, WO4 was first studied using methods of differential thermal analysis. Its phase diagram was triangulated and stable triangulating internal sections of NaCl – PbWO4 – KCl.2PbCl2, NaCl – PbWO4–2KCl.PbCl2 and obtained tetrahedra NaCl – PbCl2 – PbWO4 – KCl.2PbCl2, NaCl – KCl – PbWP4, and NaCl – KCl – PbWP4. There are coordinates of three quadruple invariant points revealed.

1. Introduction

The tungstates of d- and some p-elements *(CdWO4, ZnWO4, PbWO4, etc.)* are widely used as heavy scintillators in new-generation electromagnetic calorimeters capable of working in the field of intense radiation [1]. Among them, it is lead tungstate that has the most optimal characteristics, close to the requirements for heavy scintillation materials [2]. Single crystals of lead tungstate have a high density and radiation resistance [2, 3]. That is why *PbWO4* is used in the electromagnetic calorimeter of the Large Hadron Collider (LHC) [4, 5] and the PANDA detector [6–10].

Thus, the development of a fundamentally new technology for the ungstate synthesis and the single crystals development in low-temperature melts of multicomponent reciprocal systems is considered a great theoretical and practical interest [11, 12].

The aim of the research is to study the phase diagram of the quaternary system *NaCl – KCl – PbCl2 – PbWO4,* which is a stable tetrahedron of the quaternary reciprocal system *Na, K, Pb // Cl, WO4.*

The objectives of the study are the following:

1) the triangulation of the phase diagram of the quaternary system *NaCl – KCl – PbCl2 – PbWO4;*

2) identification of the crystallization tree of the quaternary system NaCl – KCl – PbCl2 – PbWO4;

3) calculation of coordinates of the quadruple invariant points of the quaternary system NaCl – KCl – PbCl2 – PbWO4;

4) determination of the coordinates of the quadruple invariant points of the quaternary system NaCl – KCl – PbC12 – PbWO4 by differential thermal analysis (DTA).

References in the text are made in square brackets in the following form: [5, 6, 10] or [5–9] or [1, 3, 9–11] or [6]

2. Methods and materials

All bicomponent systems, with the exception of the KCl – PbC12 system, are simple eutectic. Only the KCl – PbC12 system has the interaction of components with the formation of two compounds: KCl.2PbC12 (D1) – congruent melting and 2KCl. PbC12 (D2) – congruent melting. The KCl – PbWO4 system (Figure 1, Table 1) is a stable diagonal of the three-component reciprocal system K, Pb // Cl, WO4, studied and characterized in the research for the first time [11].

The NaCl – PbWO4 system (Figure 1) is a stable diagonal of the three-component reciprocal system Na, Pb // Cl, WO4, also studied and characterized in the research for the first time [12] (Table 1).

Table 1. Double and triple invariant points of the NaCl – KCl – PbC12 – PbWO4 system

Nonvariant point	Composition in %				t, °C
	NaCl	KCl	$PbCl_2$	$PbWO_4$	
TP (e_1)	50	50	–	–	645
e_2	75	–	–	25	692
e_3	28.4	–	76.1	–	410
e_4	–	23	77	–	424
e_5	–	48	52	–	410
p	–	63.5	36.5	–	490
e_6	–	80	–	20	650
e_7	–	30.5	60.5	9	428
e_8	17	27	56		400
e_9	–	–	91	9	487
E_1	–	21	70	9	410
E_2	–	46	48	6	400
P_1	–	56	34	10	468
E_3	27.5	–	65	7.5	398
E_4	18	11	71		383
E_5	16	36	48		387
P_2	20	46	34		448

The *PbC12 – PbWO4* system (Figure 1) was studied in the research for the first time [11], as well as the coordinates of the double eutectic revealed (Table 1).

The *NaCl – KCl* system (Figure 1) was previously studied repeatedly by different researchers [13–16]; the results show the formation of double solid solutions with a minimum at a component ratio of 1: 1. The *NaCl – PbC12* system (Figure 1) was studied previously studied in different researches as well.

The KCl – PbC12 system (Figure 1) was studied by the method of visual polythermal analysis [13–16], there were revealed compositions of compounds D1, D2 and the coordinates of three invariant points, including two eutectics and one peritectic (Table 1).

The three-component system NaCl – KCl – PbWO4 (Figure 1) was first studied by us; it is an internal section of the four-component reciprocal system Na, K, Pb // Cl, WO4. It was shown that double solid solutions based on sodium and potassium chlorides decompose with the formation of the e6 eutectic (Table 1). The three-component system NaCl – PbC12 – PbWO4 (Figure 1) was studied by us as part of the three-component reciprocal system Na, Pb ‖ Cl, WO4 [12], and a triple eutectic E3 was revealed (Table 1).

The three-component system NaCl – KCl – PbCl2 (Figure 1) was studied earlier [13], the crystallization fields of the components and two compounds D1 and D2, there were revealed coordinates of three invariant points, including two triple eutectics and one triple peritectic (Table 1).

We studied the three-component system KCl – PbCl2 – PbWO4 (Figure 1) as part of the three-component reciprocal system K, Pb ‖ Cl, WO4 [11], revealed the crystallization fields of the components and two compounds D1 and D2, the coordinates of three invariant points, including two triple eutectics and one triple peritectics (Table 1).

We used the method of a priori prediction of the crystallization tree [17, 18], which makes it possible to identify phase complexes forming quadruple invariant points of the system and their relationship based on the development of the tetrahedron (Figure 1).

The method of calculating the coordinates of invariant points using analytical models of the joint crystallization surfaces [17, 18]. In this case, we used the licensed program "Approximation of phase equilibrium states" developed in C ++ Builder 6.0, which allows on the basis of input information (melting points of the initial components, coordinates of double and triple invariant points, coordinates of points on joint crystallization mono-variant lines of two and three phases), approximate all phase equilibria by equations of the first and second order and calculate the coordinates (temperature and concentration) of the desired invariant points.

Used the method of differential thermal analysis (DTA).

During the experimental study of the phase diagram of the *NaCl – KCl – PbCl2 – PbWO4* system, the methods of the projection thermographic method were used [19–21].

3. Results

As can be seen, the NaCl – KCl – PbCl2 – PbWO4 tetrahedron (Figure 1) is triangulated by two internal sections of NaCl – PbWO4 – D1 and NaCl – PbWO4 – D2 into three trimming simplices: NaCl – PbCl2 – PbWO4 – D1, NaCl – KCl – PbWO4 – DbWO4 and NaCl – PbWO4 – D1 – D2.

By the method of a priori forecasting of the crystallization tree, the crystallization tree was revealed (Figure 2). As can be seen from the scattering of the NaCl – KCl – PbCl2 – PbWO4 tetrahedron (Figure 1), the crystallization volume of potassium chloride is limited by the phase equilibria e1E6e6P1pP2 (double, triple invariant points and corresponding monovariant lines), they also form a quadratic peritectic point at NaCl + KCl + D2 + PbWO4 + W phases. Similarly, the crystallization volume of lead chloride is limited by phase equilibria e3E3e9E1e4E4 (double, triple invariant points and corresponding monovariant lines), they also form a quadruple eutectic point in which the phases NaCl + D1 + PbCl2 + PbWO4 are in equilibrium. Finally, there is a quadratic eutectic between two quadruple invariant points (NaCl + PbWO4 + D1 + D2), into which the triple eutectics E2 and E5 are translated (Figure 2).

The identified crystallization tree (Figure 2) calculated the coordinates of the quadruple eutectics and peritectics of the quaternary system NaCl – KCl – PbCl2 – PbWO4 (Table 2).

The results of the coordinate calculating of the quaternary eutectics revealed an experimental study of the tetrahedron *NaCl – KCl – PbCl2 – PbWO4* with a minimum number of polythermal sections.

According to the projection-thermographic method, the crystallization volume of *PbWO4* takes the two-dimensional A – B – C section *(Figs. 1, 3)*. There are central projections of double and triple invariant points obtained. The vertices of the obtained triangle A – B – C are represented by two-component systems and projections of the double eutectics $\bar{e}2$, $\bar{e}6$ *and* $\bar{e}9$, and the sides are represented by three-component systems and projections of the triple invariant points $\bar{E}1$, $\bar{E}2$, $\bar{E}3$ $\bar{E}6$. According to the crystallization set, these triple invariant points are translated into the corresponding quadruple invariant points *(Figure 2)*.

First, a specific composition was expressed through A, B, and C, and then the same composition was found on the matrix, expressed already through the initial components: *Z1 = NaCl, Z2 = PbWO4, Z3 = PbCl2, Z4 = KCl.*

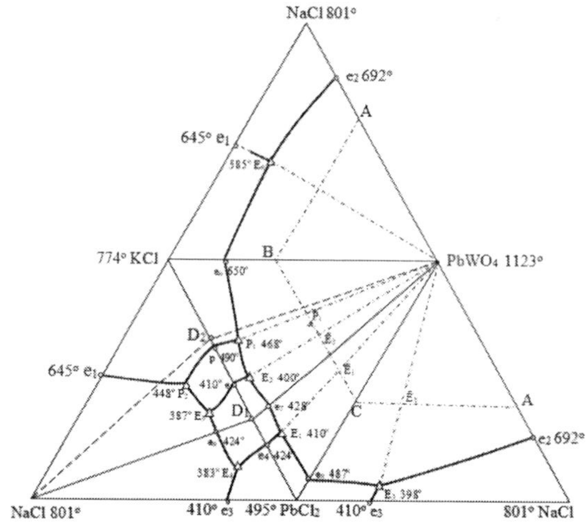

Figure 1. Scanning of the NaCl – KCl – PbCl2 – PbWO4 tetrahedron: 1) dashed and solid lines — triangulation lines; 2) dash-dotted lines show the selected ABC section and the central projections of triple invariant points (Δ) on it; 3) (°) –double invariant points

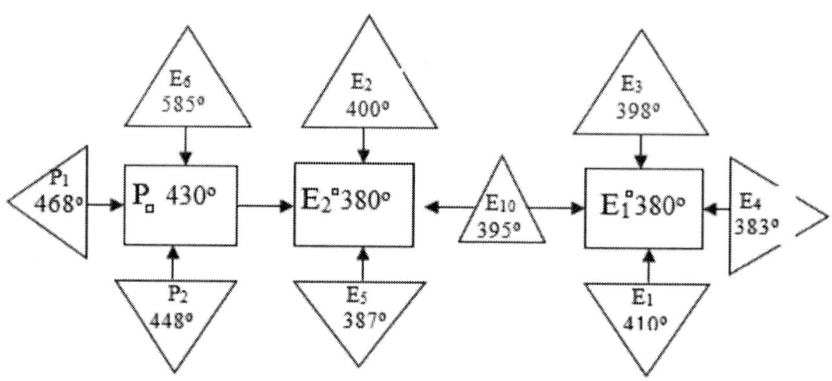

Figure 2. Crystallization set of the four-component system *NaCl – KCl – PbCl2 – PbWO4* and phases crystallizing at invariant points.

At invariant points, phases crystallize the follwing way:
E_1: $PbCl_2$ + $PbWO_4$ + D_1; E_2: $PbWO_4$ + D_1+ D_2;
E_3: NaCl + $PbCl_2$ + $PbWO_4$; E_4: NaCl + $PbCl_2$ + D_1;
E_5: NaCl + D_1 + D_2; E_6: NaCl + KCl + $PbWO_4$;
E_{10}: NaCl + $PbWO_4$+ D_1; P_1: KCl + $PbWO_4$ + D_2;
P_2: NaCl + KCl + D_2; $E_1^{\square}$: NaCl+$PbCl_2$+$PbWO_4$+D_1;
$E_2^{\square}$: NaCl+$PbWO_4$ + D_1+D_2; $P_{\square}$: NaCl+KCl+$PbWO_4$+D_2.

Table 2. The calculated and experimental coordinates of the quadruple invariant points of the NaCl – KCl – PbCl2 – PbWO4 system

System components	Temperature, °C; Composition %			
	Second-order equation		One-dimensional model	DTA
	$\Delta_{ij}^* = 0$	$\Delta_{ij}^* \neq 0$	Platitude	
t ($\bar{P}_{\square}$):	430	435	440	430
NaCl	19	18	17	18.2
KCl	41	43	44	42.8
$PbCl_2$	30	31	28	30
$PbWO_4$	10	08	11	09
t($\bar{E}_1^{\square}$):	385	375	370	380
NaCl	18	17	16	17
KCl	10	09	12	11
$PbCl_2$	62	65	66	65
$PbWO_4$	10	09	06	07
t ($\bar{E}_2^{\square}$):	380	380	380	380
NaCl	15	18	14	17
KCl	32	29	32	30
$PbCl_2$	47	49	48	47
$PbWO_4$	06	04	06	06

Δ_{ij}^* – the liquidus surface deviation from the plane along the line with double eutectics

During the experimental study of the selected two-dimensional cross-section A – B – C, all compositions were determined using the composition matrix:

$$\begin{bmatrix} Z_1 \\ Z_2 \\ Z_3 \\ Z_4 \end{bmatrix} = \begin{bmatrix} 40 & 0 & 0 \\ 60 & 60 & 60 \\ 0 & 0 & 40 \\ 0 & 40 & 0 \end{bmatrix} \begin{bmatrix} A \\ B \\ C \end{bmatrix}$$

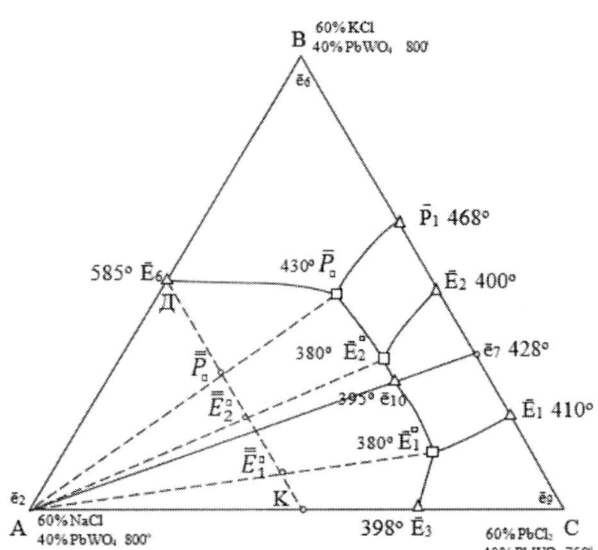

Figure 3. Two-dimensional ABC section: 1) dashed lines – studied polythermal sections; 2) ē2 – ē10 –ē7 – triangulating section; 3) (Δ) and (□) – projections of triple and quadruple invariant points, respectively; ēi – projections of double eutectics

To determine the coordinates of the central projection of the quadruple invariant points on the two-dimensional section A – B – C *(Figure 1.3),* the most informative polythermal section D – K *(Figure 3)* was selected and studied, on the phase diagram of which *(Figure 4)* double projections of quadruple invariant points. The presence of a maximum on the line — tertiary crystallization in the D – K polythermal section at the ratio B: C = 1: 2 (KCl.2PbCl2) indicates that compound D1 and the internal section NaCl – PbWO4 – D1 split the tetrahedron *NaCl – KCl – PbCl2– PbWO4* into two more tetrahedron *NaCl – PbCl2 – PbWO4 – D1 and NaCl – KCl – PbWO4 – D1* with four invariant points.

The coordinates of the double projections of the quadruple invariant points found on the phase diagram of the D – K section determine the direction of the sections A–, A–, and A– . The coordinates of the quadruple invariant projections points is to be revealed *(Figure 3).* Therefore, the polythermal sections A– -, A– -, and A– – were further studied *(Figure 3),* on the phase diagrams of which single projections of quadruple invariant points are reflected and primary crystallization of *PbWO4,* secondary joint crystallization of *NaCl + PbWO4,* quaternary joint crystallizations *NaCl + D1 + PbCl2 + PbWO4 () NaCl + PbWO4 + D1 + D2 and NaCl + KCl + D2 + PbWO4.* During the experimental study of the selected cross sections, all compositions were determined using the composition matrix given above.

The coordinates of the quadruple invariant points projection determine the directions to the quadruple invariant points, and the tetrahedron from the crystallization pole *PbWO4.* Therefore, the polythermal sections *PbWO4– -, PbWO4– – and PbWO4– –* were further studied, the primary crystallization of *PbWO4,* quaternary crystallization lines and quadruple invariant points are reflected on the phase diagrams *(Table 2).*

As seen *(Table 2),* the calculated data on the coordinates of the quadruple eutectics obtained by the equations of the first and second order are agreed with the experimental data obtained by the DTA method.

4. Conclusion

The topology of the quaternary system *NaCl – KCl – PbCl2 – PbWO4* was studied for the first time. There is a phase diagram triangulated into simplexes and a crystallization set revealed.

The study shows the calculated data on the quadruple invariant points coordinates are agreed with experimental data.

Low melting eutectic compositions have been identified to be used in the development of:

1) energy-intensive phase-transition heat-accumulating materials of non-traditional energy sources;
2) the optimal technology for the chemical synthesis of lead tungsten oxide bronzes;
3) methods of electrochemical synthesis of metallic tungsten and lead tungstate.

References

[1] Kachanov V A and Korzhik M V 2008 *J. news and probl. of fund. Phys.* **1** 1
[2] Borisevich A, Dormenev A and Houzvicka J 2016 *IEEE Transact. on Nuclear Sci.* **63** 569
[3] Nikl M 2000 *Phys. Stat. Sol. A* **178** 595
[4] Annenkov A A, Korzhik M V, Lecoq P et al 2002 *Nucl. Instr. Meth. Phys. Res. A* **490** 30
[5] Breskin A and Voss R 2009 *The CERN Large Hadron Collider: Accelerator and Experiments* vol 1-2 (Geneva: CERN)
[6] Chatrchyan S, Khachatryan V and Sirunyan A M 2012 CMS Collaboration *Phys. Lett. B* **710** 403
[7] Novotny R W, Bremer D and Dormenev V 2010 *IEEE Trans. Nucl. Sci.* **57(3)** 1441
[8] L. Bardelli, M. Bini, P.G. Bizzeti, Nucl. Instr. Meth. A., vol. 584, 129 p., 2008.
[9] Korzhik M V, Borisevich A E and Fedorov A A 2008 *IEEE Nucl. Sci. Symp Conffer. Record* **40** 2847
[10] Moreau J, Gladyshevskii R, Galez Ph 1999 *J. Alloys Comp.* **284** 104
[11] Kochkarov Zh A and Sokurova Z A 2018 *J. non-organ. Chem.* **61(7)** 936
[12] Kochkarov Zh A and Sokurova Z A 2017 *J. Melts.* **2** 134

IOP Conf. Series: Materials Science and Engineering **905** (2020) 012008 doi:10.1088/1757-899X/905/1/012008

[13] Korshunov B G, Safonov V V and Drobot D V 1972 *Chloride melting diagrams* (Leningrad: Publ. house "Chem.") 384 p
[14] Mokhosoev M V, Alekseev F P and Lutsyk V I 1978 *The state diagram of molybdate and tungstate systems* (Novosibirsk: Sci.) 319 p
[15] Voskresenskaya N K 1961 *A guide to the fusibility of systems of anhydrous inorganic salts* vol 1, 2 (Moscow: Publ. House of the USSR Acad. of Sci.) 845 p
[16] Posypayko V M and Alekseeva E A 1979 *Diagrams of fusibility of salt systems* part 1–3 (Moscow: Metallurgy) 416 p
[17] Kochkarov Zh A, Khubaeva M V and Khakulov Z L 2011 *J. non-organ. Chemi.* **56(2)** 321
[18] Kochkarov Zh A and Zhizhuev R A 2016 *J. non-organ. Chem.* **61(7)**
[19] Posypayko V I, Trunin A S and Kosmynin A S 1976 *Dokl. USSR Acad. of Sci.* **228**, 911
[20] Posypayko V I and Trunin A S 1984 *Prediction of chemical interactions in systems of many components* (Moscow: Sci.) 216 p
[21] Posypayko V I 1978 *Research methods for multicomponent salt systems* (Moscow: Sci.) 255 p

IOP Conf. Series: Materials Science and Engineering **905** (2020) 012009 doi:10.1088/1757-899X/905/1/012009

Quaternary reciprocal system Li, K, Pb // Cl, WO4

Zh A Kochkarov[1], A A Baysangurova[2], Z S Khasbulatova[3], M V Khubaeva[3], Z Sh Abubakarova[3] and D Z Maglaev[4]

[1] Institute of Chemistry and Biology, Kabardino-Balkarian State University, Nalchik, Russia
[2] Faculty of Biology and Chemistry, Chechen State University, Grozny, Russia
[3] Faculty of Natural Sciences, Chechen state Pedagogical University, Grozny, Russia
[4] Institute of Construction, Architecture and Design, Grozny State Oil Technical University named after M.D. Millionshikova, Grozny, Russia

E-mail: chemest2@rambler.ru, bajsangurova95.95@mail.ru, hasbulatova@list.ru, medina-76@mail.ru, kniiran@mail.ru, maglaev.d@mail.ru

Abstract. There is a topology of the four-component reciprocal system Li, K, Pb || Cl, WO4, its phase diagram was triangulated and stable tetrahedra revealed, the dominant chemical reactions of mutual exchange and complexation were determined. There are chemical synthesis of lead tungstate and lead oxide tungsten bronzes in ionic melts of the Li, K, Pb I Cl, WO4 system carried out. A synthesis technology for lead tungstate in ionic melts is proposed based on the results.

1. Introduction

The lead tungstate single crystal *PbWO4* is used as a scintillator since it has a high density and radiation resistance meeting the basic requirements for scintillators [1–10]. At the same time, existing methods for the synthesis of lead tungstate are not that perfect. That is why the role of state diagrams in solving problems of searching for optimal conditions for the synthesis of substances is extremely large [11–14]. In this regard, there is a great interst in mutual singular systems with mutual exchange reactions.

The study focuses on the phase diagram of the four-component reciprocal system *Li, K, Pb || Cl, WO4,* chemical synthesis of lead tungstate and lead oxide tungsten bronzes in ionic melts of the *Li, K, Pb || Cl, WO4* system.

The aim of the study is to reveal the phase diagram of the four-component reciprocal system Li, K, Pb || Cl, WO4, to develop methods for the synthesis of lead tungstate and lead oxide tungsten bronzes in ionic melts.

The research objectives are the following:

1. Triangulation of the phase diagram of the four-component reciprocal system *Li, K, Pb || Cl, WO4*
2. Identification of chemical reactions of mutual exchange and complexation in the quaternary reciprocal system *Li, K, Pb || Cl, WO4*
3. Studying the phase diagram of the four-component system *LiCl – KCl – Li2WO4 – PbWO4* and determining the coordinates of the low-melting quaternary eutectic.

IOP Conf. Series: Materials Science and Engineering **905** (2020) 012009 doi:10.1088/1757-899X/905/1/012009

4. Chemical synthesis of lead tungstate in ionic melts of the quaternary system *Li, K, Pb || Cl, WO4*.

5. Chemical synthesis of lead tungsten oxide bronzes in ionic melts of the four-component system *LiCl – KCl – Li2WO4 – PbWO4*.

2. Methods and materials

All two-component systems, with the exception of the *KCl – PbCl2* system, are simple eutectic. Only in the KCl – PbCl2 system did the interaction of components with the formation of two new compounds D1 and D2 be found.

The study is conducted using the method of a prediction of the crystallization set [11–14]. This identifies phase complexes forming quadruple invariant points of the system and their reaction based on the development of the tetrahedron (Figure 1).

Another method used in the study is the method of calculating the coordinates of invariant points using analytical models of joint crystallization surfaces of two phases [11–14]. In this case, there is a licensed program "Approximation of Phase Equilibrium States" developed in C ++ Builder 6.0 used to approximate all phase equilibria using first and second order equations and calculate the coordinates (temperature and concentration) of the desired invariant points based on the input information.

An experimental study of the phase equilibria of the elements of the selected systems was carried out by differential thermal analysis (DTA).

3. Results

There are six simplexes of triangulation identified:

The four-component reciprocal system Li, K, Pb // Cl, WO4 (Figure 1) refers to an irreversibly reciprocal diagonal-adiagonal type with developed complexation. It is characterized by the formation of binary compounds D1-D4, including congruently melting – D1, D3, D4 and incongruently melting – D2.

I. $LiCl–KCl–PbWO_4–D_1$

II. $LiCl –PbCl_2–PbWO_4–D_1$

III. $LiCl –KCl–PbWO_4– Li_2WO_4$

IV. $KCl–K_2WO_4–D_3–D_2$

V. $KCl– Li_2WO_4–D_3–D_2$

VI. $KCl–PbWO_4– Li_2WO_4–D_3$

Chemical reactions of mutual exchange and complexation in ternary Li, K // Cl, WO_4; K, Pb // Cl, WO_4; Li, Pb // Cl, WO_4 and the four-component Li, K, Pb // Cl, WO_4 reciprocal systems

The conversion method [12–14] by finding the points and lines of intersection of reactive associations (two-dimensional metastable cross sections) with stable two-dimensional cross sections in the four-component reciprocal system Li, K, Pb // Cl, WO4 (Figure 1) revealed the following mutual exchange reactions:

1. System Li,Pb//Cl,WO4:

$Li_2WO_4 + PbCl_2 = PbWO_4 + 2LiCl$

2. System K,Pb//Cl,WO4:

$K_2WO_4 + PbCl_2 = 2KCl + PbWO_4$

$4K_2WO_4 + 2PbCl_2 = 2[K_2WO_4.PbWO_4] + 4KCl$

$K_2WO_4 + 5PbCl_2 = PbWO_4 + 2[KCl.2PbCl_2]$

$K_2WO_4 + 3PbCl_2 = PbWO_4 + 2[KCl.PbCl_2]$

3. System Li,K//Cl,WO4:

$2LiCl + K_2WO_4 = Li_2WO_4 + 2KCl$

$2LiCl + 2K_2WO_4 = [Li_2WO_4.K_2WO_4] + 2KCl$

4. System Li,K,Pb//Cl,WO_4:

1. $Li_2WO_4 + K_2WO_4 + 2PbCl_2 = 2PbWO_4 + 2LiCl + 2KCl$

2. $Li_2WO_4 + K_2WO_4 + 6PbCl_2 = 2PbWO_4 + 2LiCl + 2[KCl2PbCl_2]$

3. $2LiCl + 3K_2WO_4 + 2PbCl_2 = Li_2WO_4 + 6KCl + 2PbWO_4$
4. $6LiCl+12K_2WO_4+3PbCl_2 = 12KCl+3[Li_2WO_4K_2WO]+3[K_2WO_4PbWO_4]$
5. $2LiCl + 3K_2WO_4 + PbCl_2 = 4KCl + PbWO_4 + [Li_2WO_4K_2WO]$
6. $2LiCl + 3K_2WO_4 + PbCl_2 = 4KCl + [K_2WO_4PbWO_4] + Li_2WO_4$

Thermodynamic substantiation of mutual exchange reactions in ternary Li, K // Cl, WO_4; K, Pb // Cl, WO_4; Li, Pb // Cl, WO_4 and the quaternary reciprocal systems Li, K, Pb // Cl, WO4 (Figure 1, Tables 1–3).

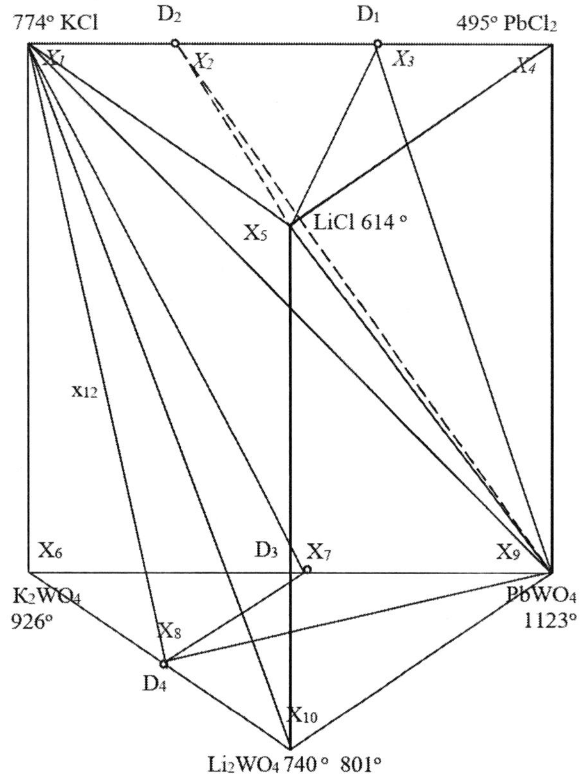

Figure 1. The prism of the quaternary reciprocal system *Li, K, Pb // Cl, WO4*

The isobaric potentials of the mutual exchange reactions were calculated by the Temkin-Schwartzman method. The mathematical processing of the kinetic measurement data was calculated for temperatures of 298, 720, 820, 920 °C according to the equations of Erofeev-Kolmogorov, Sakovich, Arenius. Entropy is calculated by the formula A.F. Kapustinsky and K.B. Yatsemirsky.

Table 1. Metabolic reactions

Metabolic reactions	ΔH°298, kJ/mol	S°298, kJ/mol	Cp° 298, kJ/mol
$K_2WO_4 + 2LiCl = Li_2WO_4 + 2KCl$	–76.9	–15.02	6.6
$K_2WO_4 + PbCl_2 = 2KCl + PbWO_4$	–51.58	27.68	–5.86
$Li_2WO_4 + PbCl_2 = PbWO_4 + 2LiCl$	25.32	39.66	–12.46
$2LiCl + 2PbCl_2 + 3K_2WO_4 = 6KCl +Li_2WO_4 + 2 PbWO_4$	–180.06	43.26	–4.52
$2PbCl_2+Li_2WO_4 + K_2WO_4 = 2PbWO_4 + LiCl + 2KCl$	–26.26	67.3	–18.32

The obtained thermodynamic data indicate that all the presented exchange reactions are fundamentally possible and can be used for the synthesis of lead tungstate in the solid phase or in melts.

Table 2. ΔCp°MoT different temperature values

Reactions	T,k	ΔCp°MoT
$K_2WO_4 + 2LiCl = Li_2WO_4 + 2KCl$	298	0.26546
	720	1.40517
	820	2.03132
	920	2.76786
$K_2WO_4 + PbCl_2 = 2KCl + PbWO_4$	298	0.2357
	720	1.24762
	820	1.8033
	920	2.43089
$Li_2WO_4 + PbCl_2 = PbWO_4 + 2LiCl$	298	0.50115
	720	2.65278
	820	3.83452
	920	5.16877
$2LiCl + 2PbCl_2 + 3K_2WO_4 = 6KCl + Li_2WO_4 + 2 PbWO_4$	298	0.1818
	720	0.92326
	820	1.39101
	920	*1.87023*
$2 PbCl_2 + Li_2WO_4 + K_2WO_4 = 2 PbWO_4 + LiCl + 2KCl$	298	0.73685
	720	3.9004
	820	5.64392
	920	7.59649

Table 3. Isobaric potentials ΔGT° reactions in systems

Reaction	ΔrGT°= φ(T)	T,K	ΔrGT°
$K_2WO_4 + 2LiCl = Li_2WO_4 + 2KCl$	ΔrGT°= –76.9–(–0.01202)T+ΔCp°MoT	298	–73.053
		720	–66.012
		820	–65.012
		920	–63.104
$K_2WO_4 + PbCl_2 = 2KCl + PbWO_4$	ΔrGT°= –51.58–0.02768T+ΔCp°MoT	298	–59.593
		720	–70.262
		820	–72.474
		920	–74.615
$Li_2WO_4 + PbCl_2 = PbWO_4 + 2LiCl$	ΔrGT°=25.32–0.03966T+ΔCp°MoT	298	14.002
		720	0.582
		820	–3.367
		920	–5.998
$2LiCl + 2PbCl_2 + 3K_2WO_4 =$ $=6KCl + Li_2WO_4 + 2PbWO_4$	ΔrGT°= –180.06–0.04326T+ΔCp°MoT	298	–192.77
		720	–210.284
		820	–214.142
		920	–217.984
$2 PbCl_2 + Li_2WO_4 + K_2WO_4 =$ $=2PbWO_4 + LiCl + 2KCl$	ΔrGT°= –26.26–0.0673T+ΔCp°MoT	298	–45.579
		720	–70.816
		820	–75.802
		920	–80.58

The study revealed the phase diagram of one of its stable tetrahedra LiCl – KCl – Li2WO4 – PbWO4 and only one quadruple eutectic in the quaternary reciprocal system Li, K, Pb || Cl, WO4, (Table 4).

The results of calculating the coordinates of the quadruple eutectic made it possible to plan an experimental study of the *LiCl – KCl – Li2WO4 – PbWO4* tetrahedron with a minimum number of polythermal sections.

3.1. Synthesis of lead tungstate

In accordance with the identified metabolic reaction equation:

1) System *Li, Pb // Cl, WO4*: $Li_2WO_4 + PbCl_2 = PbWO_4 + 2LiCl$,
2) System *K, Pb // Cl, WO4*: $K_2WO_4 + PbCl_2 = 2KCl + PbWO_4$,
3) System *Li, K, Pb // Cl, WO4* $Li_2WO_4 + K_2WO_4 + 2PbCl_2 = 2PbWO_4 + 2LiCl + 2KCl$.

Table 4. The calculated and experimental coordinates of the quadruple eutectic of the four-component system

System composition	Temperature °C; Molecular composition %			
	Second order-type equation		Linear model	DTA
	$\Delta_{ij}^* = 0$	$\Delta_{ij}^* \neq 0$	Platitude	
$t(E_1)$:	317	310	315	310
LiCl	50	47	48	48
KCl	34	36	37	35
Li_2WO_4	9	10	8	9
$PbWO_4$	7	7	7	8

Δij * – deviation of the liquidus surface from the plane along the line connecting the double eutectics. Chemical synthesis of lead tungstate and lead tungsten oxide bronzes in ionic melts of ternary *Li, Pb // Cl, WO4; K, Pb // Cl, WO4* and the four-component *Li, K, Pb // Cl, WO4* reciprocal systems.

The initial components taken in equivalent amounts were mixed and put in a mortar, loaded into a platinum crucible and lowered into a shaft furnace. The temperature in it was gradually raised up to 650 °C and held for 30 minutes. The melt was poured onto a steel substrate and, after cooling, it was carefully ground in a mortar and *PbWO4* was washed from slag (sulfates) in hot water for 20 minutes, then it was filtered by rinsing with hot water. The obtained *PbWO4* powder was dried at 300 °C, and then calcined at 500 °C. It is very important that in this case too, soluble salts are formed in the reaction products – potassium and sodium chlorides, lead tungstate from which is easily washed.

X-ray phase analysis of the obtained samples showed that *PbWO4* does not contain impurities (*Figure 2*). The yield of *PbWO4* was 98–99.3 %.

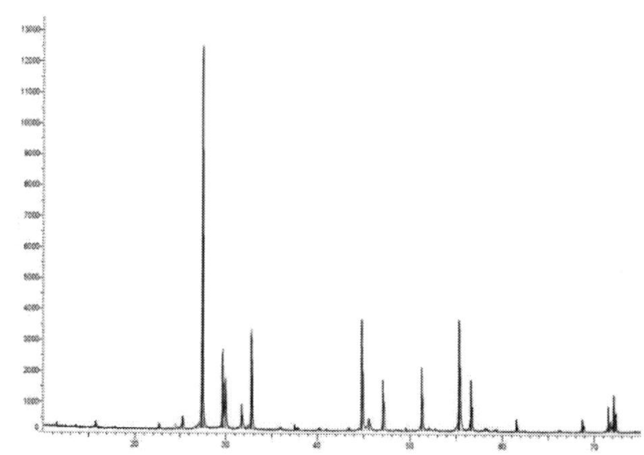

Figure 2. Diffraction pattern of the PbWO4 synthesis at temperature 650 °C according to the exchange reaction

IOP Conf. Series: Materials Science and Engineering **905** (2020) 012009 doi:10.1088/1757-899X/905/1/012009

3.2. Chemical synthesis of lead tungsten oxide bronzes

The initial sample of the eutectic composition of the quaternary system LiCl – KCl – Li2WO4 – PbWO4 introduced the calculated amounts of tungsten powder, tungsten oxide (VI) in accordance with the reaction equation:

$$3xMe2WO4 + (6\text{-}4x)MeO3 + xMe = 6MexWO3 \quad \text{or} \quad 3PbWO4 + 2WO3 + W = 6Pb0.5WO3$$

The resulting mixture was thoroughly mixed in a mortar, then dried at a temperature of 150–200 °C. Next, the charge was transferred to a crucible, lowered into a shaft furnace and heated to the melting temperature. The melt was held at this temperature for up to 30–45 minutes. Then, the melt was poured into a stainless steel cuvette, and after cooling, it was carefully put in a mortar and transferred to boiling distilled water to wash the bronzes from salts. After separation from the filtrate, the bronzes were dried at 100 °C, weighed, and the product yield was determined.

A brown multiphase mixture was obtained with the identified XRD phases *(Figure 3): Pb0.5WO3, Pb0.3WO3, Pb0.45WO3, Pb0.35WO3, Pb0.46WO3, Pb0.47WO3.*

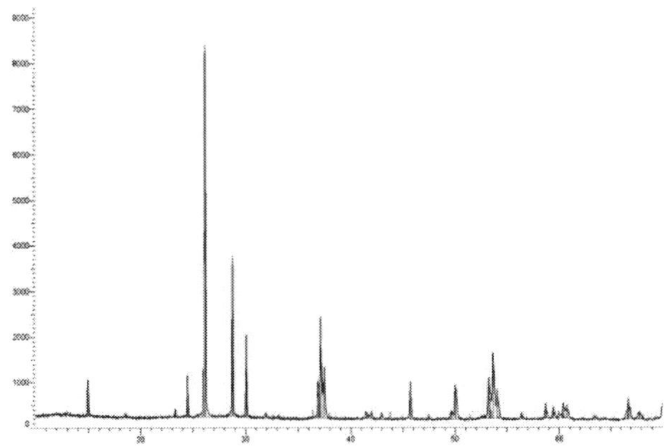

Figure 3. Diffraction pattern of the synthesis product of oxide tungsten bronze lead at a temperature of 650 °C

4. Conclusion

1. The study is focused on the topology of the quaternary reciprocal system *Li, K, Pb // Cl, WO4.* Its phase diagram was triangulated, stable tetrahedron revealed, dominant chemical reactions of mutual exchange and complexation were determined.

2. The phase diagram of the stable *LiCl – KCl – Li2WO4 – PbWO4* tetrahedron of the four-component reciprocal system *Li, K, Pb // Cl, WO4* was studied for the first time, its phase diagram was triangulated into simplexes, the crystallization tree and coordinates of the four eutectic point revealed.

It is shown that the calculated data on the coordinates of the quadruple eutectic point are agreed with the experimental data.

3. The identified low-melting eutectic composition can be used in the development of: 1) energy-intensive phase-transition heat-accumulating materials of non-traditional energy sources; 2) the optimal technology for the chemical synthesis of lead tungsten oxide bronzes; 3) methods of electrochemical synthesis of metallic tungsten and lead tungstate; 4) technologies for growing single crystals of lead tungstate.

4. The chemical synthesis of lead tungstate and lead oxide tungsten bronzes in ionic melts of the quaternary reciprocal system *Li, K, Pb // Cl, WO4* was carried out.

5. Based on the results obtained, a synthesis technology for lead tungstate is proposed, which can be introduced at the Bogoroditsky Plant of Technochemical Products, at OJSC North Crystals (Apatity, Murmansk Region), at the Kurchatov Institute.

References

[1] Borisevich A, Dormenev V, Houzvicka J et al 2016 New Start of Lead Tungstate Crystal Production for High-Energy Physics Experiments *IEEE Transact. on Nuclear Sci.* **63** 569–73

[2] Nikl M 2000 Wide Band Gap Scintillation Materials: Progress in the Technology and Material Understanding *Phys. Stat. Sol. A* **178** 595–620

[3] Annenkov A A, Korzhik M V and Lecoq P 2002 Lead tungstate scintillation material *Nucl. Instr. Meth. Phys. Res. A* **490** 30–50

[4] Breskin A, Voss R. (eds) 2009 The CERN *Large Hadron Collider: Accelerator and Experiments* vol 1–2 (Geneva: CERN)

[5] Novotny R W, Bremer D, Dormenev V et al 2010 The PANDA electromagnetic calorimeter – a high-resolution detector based on PWO-II *IEEE Transact. on Nuclear Sci.* **57(3)** 1441–6

[6] Bardelli L, Bini M, Bizzeti P G, Danevich F A et al 2008 Pulse-shape discrimination with PbWO4 crystal scintillators *Nucl. Instr. Meth. A* **584** 129–34

[7] Korzhik M V, Borisevich A E, Fedorov A A and Missevitch O V 2008 Lead tungstate scintillator for the future particle physics detectors IEEE *Nucle. Sci. Symp. Conf. Record* **40(5)** 2847–9

[8] Deych R and Dolazza E 2006 New trends in X-ray CT imaging In *Radiation Detectors for Medical Applications* (Springer) pp 15–35

[9] Mikhailik V B and Kraus H 2010 Performance of scintillation materials at cryogenic temperatures *Phys. Stat. Sol. B* **247(7)** 1583–99

[10] Danevich F A, Chernyak D M, Dubovik A M, Grinyov B V et al 2009 MgWO4 – A new crystal scintillator *Nucl. Inst. Meth. Phys. Res. A* **608(1)** 107–15

[11] Kochkarov Zh A and Sokurova Z A 2016 Synthesis of lead molybdate in melts of many component systems *Izv. Doug. state ped un-that. Eating. Sci.* **10(2)** 5–9

[12] Kochkarov Zh A and Sokurova Z A 2016 Synthesis of tungstate and oxide tungsten bronzes of lead in melts of multicomponent systems *Izv. Doug. state ped un-that. Eating. Sci.* **10(3)** 20–6

[13] Kochkarov Zh A, Sokurova Z A and Babaeva L Z 2016 Phase equilibria and synthesis in four-component reciprocal systems Na, K, Pb // Cl, WO4; Na, K, Pb // Cl, MoO4 *Izv. Doug. state ped. un-that. Eating. Sci.* **10(4)** 12–8

[14] Kochkarov Zh A, Sokurova Z A, Balkizova A Yu and Tokhov M R 2018 Three-component system KCl-K2WO4-PbWO4 *J. Sci. and Techn. Bull. of the Volga Reg.* **5** 38–40

IOP Conf. Series: Materials Science and Engineering **905** (2020) 012010 doi:10.1088/1757-899X/905/1/012010

Preparation and stereochemistry of nickel phosphine complexes

R A Bisergaeva, Z S Hasbulatova and Y N Sirieva

Faculty of Biology and Chemistry, Chechen State University, Grozny, Russia

E-mail: Radimaa-2009@mail.ru

Abstract. The Nickel (II) ion forms many stable complexes as predicted by the Irving Williams series. This experiment is about preparation and stereochemistry of nickel phosphine complexes. The nickel phosphine complexes, which were prepared are $NiBr_2(PPh_3)_2$ and $NiBr_2(PCy_3)_2$. The first part of the experiment is to prepare the nickel complexes. After preparing nickel phosphine complexes $NiBr_2$ $(PPh_3)_2$ and $NiBr_2$ $(PCy_3)_2$,the complexes have been analyzed for structural determination. The complexes were analyzed by the Magnetic Moments of solids-The Johnson Matthey Balance, UV-visible (electronic spectra), Infrared spectra (normal range) and Far-infrared spectra (which were supplied in the practical booklet).

1. Introduction

Nickel is a chemical element with the chemical symbol **Ni** and atomic number 28. It is a silvery-white lustrous metal with a slight golden tinge. Nickel belongs to the transition metals and is hard and ductile.In its compounds nickel exhibits oxidation states of −1, 0, +1, +2, +3, and +4, though the +2 state is by far the most common. Nickel (II) commonly forms a large number of complexes with three different geometries. These are octahedral, tetrahedral $(PPh_3)_2$, and square planar. Some five-coordinate complexes are known but are rare. Nickel (II) is a $3d^8$ system, so octahedral and tetrahedral complexes will have 2 unpaired electrons and square planar complexes usually will have none. Square planar nickel (II) complexes usually have no unpaired electrons, are diamagnetic and thus have a magnetic moment of zero. Nickel (II) with 2 unpaired electrons. Nickel compounds are suspected carcinogens and phosphines are irritants [1, 2].

Tricyclohexylphosphine is the tertiary phosphine with the formula $P(C_6H_{11})_3$. Commonly used as a ligand in organometallic chemistry, it is often abbreviated to PCy_3, where Cy stands for cyclohexyl. Itis not air stable, so the preparation was carried out under nitrogen.

$NiBr_2$ is a metal, however $(PPh_3)_2$ and $(PCy_3)_2$ are ligands. Nickel complexes can be used as catalysts. They are very efficient in the polymerisation of ethene for the formation of polythene (plastics) and in hydrosilation reactions (the adding of a silane across a double bond).

The general reaction for this experiment:

$$NiBr_2+2P \rightarrow [\ NiBr_2(P)_2],$$

where $P=PPh_3$ or PCy_3 (Ph=phenyl,C_6H_5; Cy= cyclohexyl,C_6H_{11}).(1)

2. Methods and materials

2.1. The preparation of the [$NiBr_2$ (PPh_3)$_2$]

1.4 g of $NiBr_2.3H_2O$, which is dark yellow colour, was dissolved in 15ml of dried absolute ethanol in a small conical flask. As the yellow solution heated in the flask it gradually became green solution. Then, using the reflux apparatus 2.8 g of PPh_3, which is white crystals/flakes was dissolved in 30ml of colourless propan-2-ol. Added few bumping chips to avoid the overflow and heated the mixture, by heating the mixture it gave clear solution. When the phosphine was dissolved turned off the heat and poured the warm nickel bromide solution down the condenser. The precipitated formed immediately. The reaction was fast and changed the colour from green to dark green. Allowed the flask to cool down and then placed it in an ice bath for around 20 minutes to complete precipitation. Sat the suction filtration and washed the filter paper with ethanol to avoid the product to become wet with water. Filtered of the solid product and washed by ethanol and then diethyl ether. Allowed the product to dry and weighed the product and it was 2.63 g. The product was stored in a bottle with a label. The melting point of the product was determined. The melting point range was 190–194 °C.[(1)]

2.2. The preparation of $NiBr_2(PCy_3)_2$

1.4g of $NiBr_2.3H_2O$, which is dark yellow colour was dissolved in 15 ml dried absolute ethanol in a small conical flask and heated. The reflux/nitrogen apparatus has been flushed with nitrogen gas and then dissolved 1.5g of PCy_3, which looks like white sugar powder in 20ml of dried absolute ethanol. Added few bumping chips to avoid the overflow and heated the mixture. PCy_3 is more phosphoric than PPh_3, solt dissolved faster. While the mixture was refluxing poured, the warm nickel bromide solution down the condenser against a positive stream of nitrogen and the precipitate formed immediately. The precipitate was dark green colour. Allowed the flask to cool down and then placed it in an ice bath for around 20 minutes to complete the precipitation. Sat the suction filtration and washed the filter paper with ethanol to avoid the product to become wet with water. Filtered of the solid products and washed by ethanol and then diethyl ether. Allow the products to dry. The final product was black colour. Weighed the product and it was 0.15 g. The product was stored in a bottle with a label. The melting point of the product was determined. The melting point range was 145–167 °C [2].

2.3. Analysis of the Nickel complexes-Structural determination

a) Measurement of the Magnetic Moments of Solids – The Johnson Matthey (JMC) Balance:

$NiBr_2(PPh_3)_2$:Weighed the empty sample tube which was 0.8275g. The balance was switched on with the RANGE knob sat to x1 scale. The apparatus has been switched on and sat by a laboratory supervisor. Pushed the empty sample tube into the apparatus and took the reading Ro,– 029. It is negative since the glass is diamagnetic. Crushed the sample to a fine powder and inserted into sample tube to 2.9 cm. Weighed the sample tube with sample. The weight was 0.9501g. Pushed the tube into the apparatus as before and took the reading, R, 187. The temperature in the room was 21 °C and the calibration constant, C is 1.1260 [3].

$NiBr_2(PCy_3)_2$: Followed the same procedure above.

The empty sample tube weight 0.8170 g. Ro was – 028 the length of the sample was 2.85c m. Weight of the sample tube with sample was 0.9087 g. And R was – 035. The temperature in the room was 21 °C and the calibration constant, C is 1.1260.

b) UV/VIS Spectra

Dissolved 11.1 mg of $NiBr_2(PPh_3)_2$in 10 cm^3 of dichloromethane using a small volumetric flask. Ran the spectrum between 700–400 nm in a quartz cell.

Dissolved 15.6 mg of $NiBr_2(PCy_3)_2$in 10 cm^3 of dichloromethane using a small volumetric flask. Ran the spectrum between 700–400 nm in a quartz cell [4, 5].

c) Infrared Spectra (4000 to 600 cm^{-1}): Done by a laboratory supervisor.

d) Far Infrared Spectra: These are given in the laboratory book.

3. Results

3.1. Measurement of the Magnetic Moments of Solids – The Johnson Matthey (JMC) Balance

(a) Use $\chi_w = \dfrac{Cl(R-Ro)}{10^9 m}$ to evaluate χw

The mass susceptibility for $NiBr_2(PPh_3)_2$ is calculated as followed,
Where, Ro is the reading of the empty tube.
R is the reading of the tube plus sample.
m is the mass of the sample in grams
l is the length of the sample in centimetres
C is the calibration constant

$$X_w = C \text{ x } l \text{ x } (R - R_o) / 10^9 \text{ x m}$$

$$X_w = 1.1260 \quad \text{x } 2.9\text{cm x } (187 - (-029) / 10^9 \text{ x } (0.9501-0.8275)$$

$$X_w = 1.1260 \quad \text{x } 2.9\text{cm x } 216 / 10^9 \text{ x } 0.1226$$

$$X_w = 5.753\text{x}10^{-6}$$

The reading R is negative for $NiBr_2(PCy_3)_2$. So did not proceed any further with the calculation.

(b) Evaluate χ_m where $\chi_m = \chi_w M$ and M = the relative molecular mass of the complex.

$$\chi_m = \chi_w M$$

$$\chi_m = (5.753\text{x}10^{-6})\text{x } 743.075$$

[The relative molecular mass of $NiBr_2(PPh_3)_2$=743.075 g/mol]

$$\chi_m = 4.275\text{x}10^{-3}$$

(c) We now separate χ_m^{para}, the molar susceptibility for the metal ion, from χ_m^{dia} which is the contribution from the ligands.

$$\chi_m^{para} = \chi_m - \chi_m^{dia}$$

$$\chi_m^{para} = 4.275\text{x}10^{-3} - (-173.9)$$

[χ_m^{dia}value is taken from the table 2 which is given in the practical book.]

$$\chi_m^{para} = 173.904275$$

(d) Calculate the magnetic moment from $\mu = 2.83\sqrt{\ }(\chi_m^{para} T)$, Where T is in K.

$$\mu = 2.83\sqrt{\ }(\chi_m^{para} T)$$

$$[21°C= 294.15 \text{ k}]$$

$$\mu = 2.83\sqrt{\ }(173.904275 \text{ x } 294.15)$$

$$\mu = 2.83 \text{ x } 226.172$$

$$\mu = 640.067$$

3.2. UV/VIS Spectra

Calculate the molar absorption coefficient (ε), with units, for any bands using the Beer-Lambert law (A = εcl)

Where A = absorbance;
c = concentration in mol dm^{-3};

l = path length of cell in cm;
Bands where ε is between 0–5000 are most likely to be d–d transitions

$$A = \varepsilon c l$$

$$\varepsilon = A / C \times l$$

$$\varepsilon = 0.348 / 1.49\times10^{-3}\ \text{mol dm}^{-3} \times 1\text{cm}$$

$$[c=(n\times1000)/v=1.49\times10^{-3}\ \text{mol dm}^{-3}]$$

$$(n=0.0111/743.075=1.495\times1^{-5}\ \text{mol})$$

$$\varepsilon = 233.56^{(1)(2)}$$

4. Conclusion

Looking at the infrared spectra (IR) for part 1 it shows thatthe Ni-Br stretch for $NiBr_2(PPh_3)_2$ was observed at 3053 cm^{-1}, and looking at the infrared spectra for part 2 it shows that the Ni-Br stretches for $NiBr_2(PCy_3)_2$ is between 3000–2500 cm^{-1}. However, looking at the Far infrared spectra in a practical book it shows that the Ni-Br stretch for $NiBr_2(PPh_3)_2$ is between 250–300 cm^{-1} and for $NiBr_2(PCy_3)_2$ is between 200–250 cm^{-1}. The mass susceptibility for $NiBr_2(PPh_3)_2$ is $X_w = 5.753\times10^{-6}$. The reading R is negative for $NiBr_2(PCy_3)_2$. So did not proceed any further with the calculation because this will show that the compound is diamagnetic and the magnetic moment is 0.

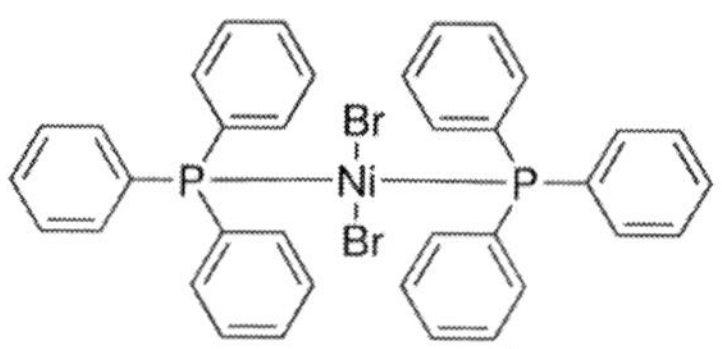

$NiBr_2(PPh_3)_2$ (4)

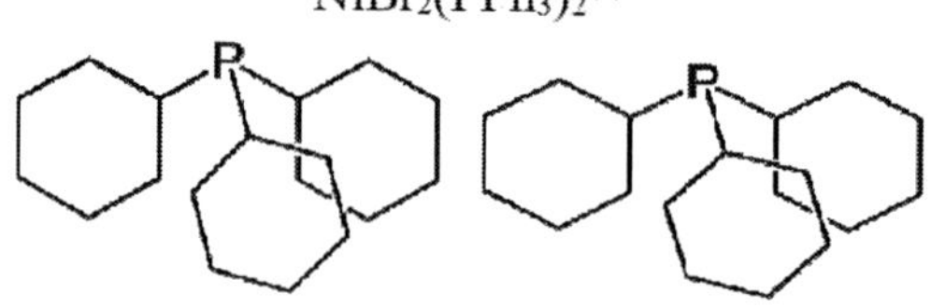

$NiBr_2(PCy_3)_2$ (4)

References

[1] Owen G 2013–2014 *Preparation and Stereochemistry of Nickel Phosphine Complexes* Coordination and organometallic chemistry practical manual PH2S19 (Univer. of South Wales) pp 2–10

[2] Bisergaeva R 2014 *Preparation and Stereochemistry of Nickel Phosphine Complexes* Labartory Book Manual pp 10–5

[3] Lancashire R J 2000 *The Jahn-Teller Theorem* (Jamaica: Univer. of the West Indies) Retrieved from: http://wwwchem.uwimona.edu.jm/courses/JahnTeller.html.

[4] Abd El-Wahab H, Abd El-Fattah M, Ahmed A H et al 2015 Synthesis and characterization of some arylhydrazone ligand and its metal complexes and their potential application as flame retardant and antimicrobial additives in polyurethane for surface coating *J. of Organomet. Chem.* **791** 99–106

[5] *Official website of the University of Toronto Scarborough* Retrieved from: http://webapps.utsc.utoronto.ca/chemistryonline /IR_images/ir_table.png

IOP Conf. Series: Materials Science and Engineering **905** (2020) 012011 doi:10.1088/1757-899X/905/1/012011

The use of carbon nanotubes in the fire extinguishing of oil and oil products

F A Dali[1], G L Shidlovsky[1], M S Khasikhanov[2], R U Zalaev[2], P R Tagirova[2], S S Saidulaev[2], L M Masaeva[2] and R S Erzhapova[2]

[1] Department of Fire Safety of Buildings and Automated Fire-Fighting Systems, St. Petersburg University, State Fire Service of EMERCOM of Russia, St. Petersburg, Russia
[2] Department of Life Safety, Grozny State Oil Technical University named after academician M.D. Millionschikova, Grozny, Russia

E-mail: dalee@igps.ru, shidlovsky.g@igps.ru, ipkggntu@mail.ru, zalaev-rezvan@mail.ru, t-petimat@mail.ru, saydulaev68@mail.ru, lizamasaeva@mail.ru, alikhan1201@mail.ru

Abstract. The process of quenching a flame of liquid hydrocarbons by suspensions of water with carbon nanostructures is studied. It is shown that the dispersion of carbon nanostructures in water intensifies heat transfer in liquids, which leads to their more rapid heating to boiling temperature under conditions of thermal exposure to a flame. During the experiment, it was found that the quenching time of liquid hydrocarbons by nanosuspensions is on average 3.5–5.0 times less than the time of quenching of a liquid with finely divided water.

1. Introduction

When extinguishing fires of liquid hydrocarbons by sprayed water, spraying occurs, convective entrainment of water droplets, and its penetration into the subsurface layer of the burning liquid. The low efficiency of using water as a fire extinguishing composition is associated with insufficient heat removal in the combustion zone [1].

Wetting agents and additives increase the heat sink intensity, create a film on the surface of a burning oil product. Carbon nanotubes increase thermal conductivity and change the rheological properties of liquids at low concentrations (0.01...1.00 vol. %) [2].

The aim of the study was to determine the fire extinguishing characteristics of water suspensions with carbon nanostructures when extinguishing a flame of a flammable liquid.

2. Methods and materials

Functionalized multi-walled carbon nanotubes (MWCNT) and astralen (Astr) in distilled water (DW) were used as an object of study (Fig. 1).

MWCNTs were obtained by catalytic pyrolysis [3]. Parameters of nanoparticles: d = 25…180 nm, l = 1…3 μm. MWCNT was purified from synthesis by-products by the method described in [4]. MWCNT was placed in concentrated nitric acid (HNO_3) 65 % (1 g MWCNT per 50 ml HNO_3) and 35 % sulfuric acid (H_2SO_4). After boiling, the oxidized MWCNTs were filtered off and washed with distilled water to a neutral pH filtrate, and dried at a temperature of 70–80 °C.

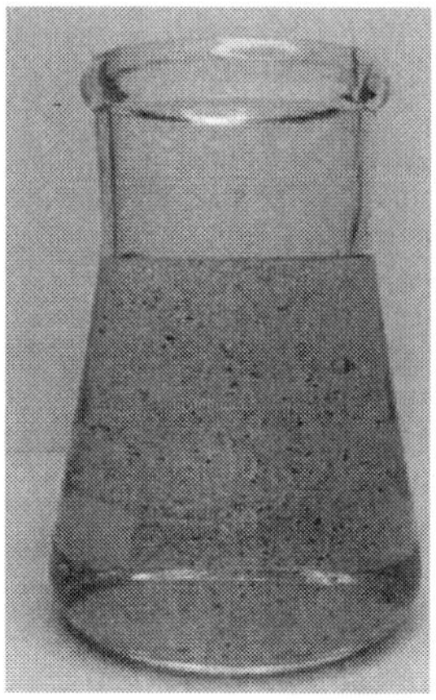

Figure 1. Suspension with carbon nanostructures

Astralenes were obtained by evaporation of graphite anodes in an electric arc discharge [5]. They represent a structure of curved graphite layers with a diameter of 10...150 nm, a distance between graphene layers of 0.336 nm, an average pore size of 20...60 nm. Astralenes are characterized by high thermal stability and in powder form are large agglomerates measuring about 0.5...3 microns in size [6].

Suspensions were prepared by dispersing carbon nanostructures with a volume concentration of 0.05–1.60 % vol. in DW when exposed to an ultrasound source (power 1.2 kW, frequency 50–60 Hz, processing time 30 min). Suspensions DW with a low concentration of MWCNT (φ = 0.01 vol. %). They were used as a control sample.

The following methods were used in the work: atomic force microscopy (AFM) [7]; measuring the surface tension coefficient by the droplet separation method [8]; studies of the heating rate of suspensions to a boiling point [9]; measuring the specific heat of vaporization of suspensions [10]; measuring the extinguishing time of model fires of a class "B" fire described in [11].

2.1. The study of nanostructures in suspension by atomic force microscopy

On the AFM scan of the solid residue, MWCNTs are extended structures with a diameter of ~ 110 nm and a length of 1–3 μm (Fig. 1, a). During the study of the topology of the Astr solid residue, clusters of particles with a diameter of 200–300 nm are observed (Fig. 2, b).

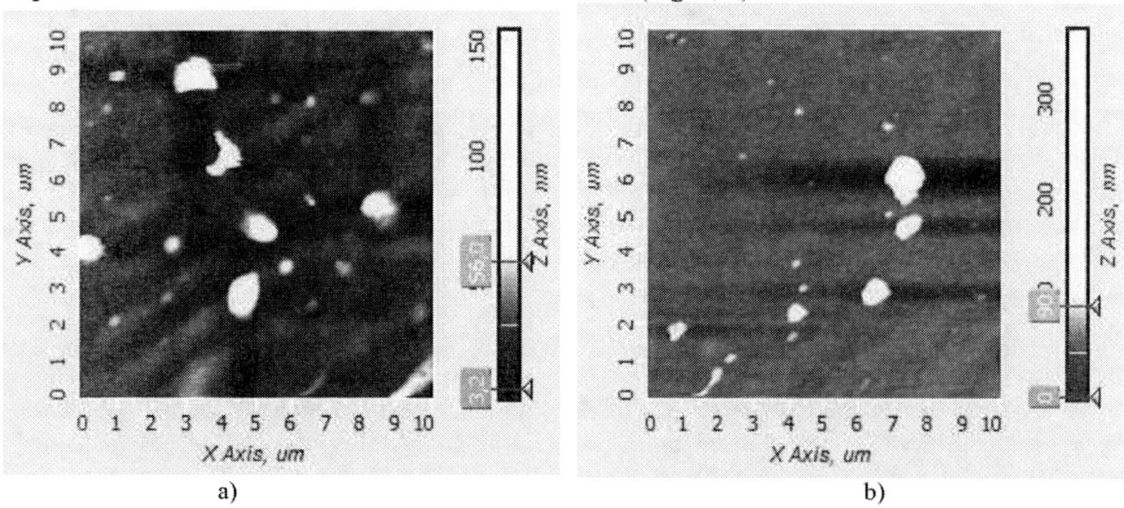

Figure 2. AFM-scans of agglomerations of carbon nanostructures during dispersion in the DW: a – DW+MWCNT; b – DW+Astr.

2.2. Measurement of the surface tension of suspensions

The experimental results showed (Fig. 2) that the surface tension of the DW + MWCNT suspension increases linearly with increasing MWCNT concentration from 0.4 and at a concentration of $\varphi = 1.6$ vol. % reaches 30 %. For the suspension DW + Astr, the logarithmic nature of the change in function is observed. At the initial stage, a sharp increase in the surface tension coefficient occurs at concentrations Astr $\varphi = (0.05...0.5)$ vol. % relative to the suspension DW + MWCNT, but with a further increase in concentration reaches 25 % in comparison with the control sample. It was noted earlier in [12] that the surface tension of water-based nanofluids increases with increasing concentration of carbon nanostructures.

2.3. Investigation of the thermophysical properties of suspensions

During the study of thermophysical properties, it was revealed (Fig. 3) that with an increase in the MWCNT concentration to 1.0 vol. % there is a slight increase in the specific heat of vaporization of the suspension DW + MWCNT by 10–15 %, but at the same time, the heating rate to a boiling point increases linearly to 50 %. An increase in the specific heat of vaporization of the suspension DW + Astr with an increase in the concentration of Astr to 0.5 vol. % reaches 20 % in comparison with the control sample, but with a further increase in concentration it sharply decreases by 30–40 %. In this case, the kinetics of heating to the boiling point increases over the entire observed range of concentrations and reaches 50–70 % at an Astr concentration of 1.0 vol. % Thus, an increase in the specific heat of vaporization is caused by an increase in the surface tension coefficient of water suspensions with carbon nanostructures [13].

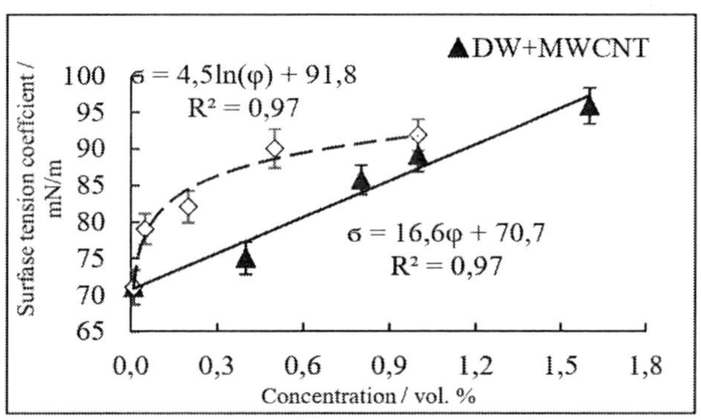

Figure 3. Surface tension suspensions.

2.4. Measurement of the extinguishing time of a model fire of class "B" fire

Evaluation of the fire extinguishing ability of suspensions during the elimination of the combustion of a flammable liquid was carried out in a laboratory setup described in [11].

The free burning time of a flammable liquid was at least 60 s. The extinguishing of a model fire site (pan diameter 450 mm, the combustible mixture is motor gasoline with an octane rating of 95 and water in a ratio of 7: 3) was carried out by a sprayed stream of suspension with a droplet diameter of 100–300 μm and a pressure of 1.0–1.5 MPa. Extinguishing time was recorded at the time of complete elimination of combustion. For each type of extinguishing agent, 5 tests were carried out, the data were averaged.

During the study of fire extinguishing characteristics, it was revealed (Fig. 4) that with an increase in the MWCNT concentration to 1.0 vol. % in suspension DW + MWCNT there was a decrease in quenching time by 70 % compared with the control sample. Extinguishing time with DW + Astr suspension at concentrations up to 0.5 vol. % decreased to 80-90 %. The increase in quenching time at concentrations φ>0.5 (for DW + Astr) and φ>1.0 (for DW + MWCNT) is associated with intensive aggregation of nanostructures [3] and leads to a decrease in the fire extinguishing efficiency of suspensions.

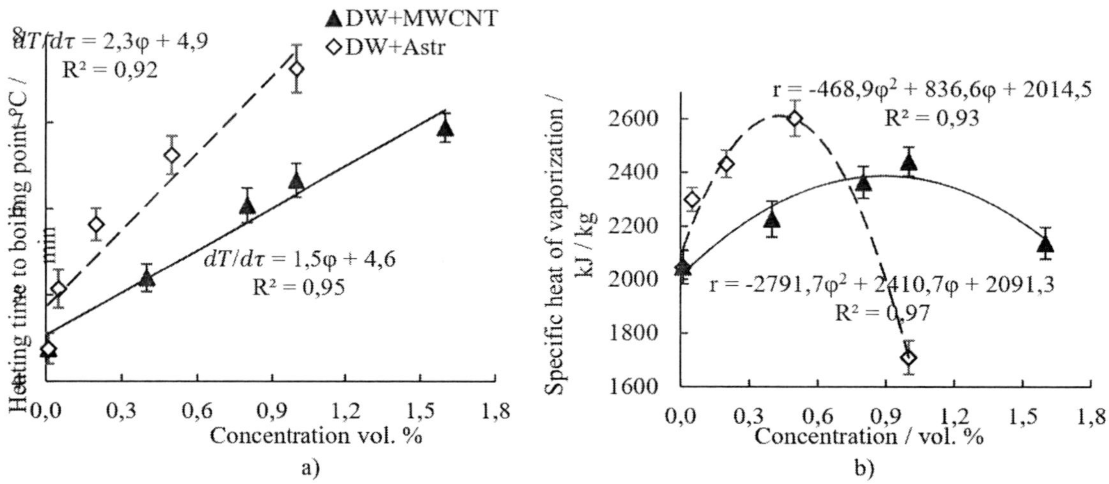

Figure 4. Thermophysical properties of suspensions

It was noted that the specific consumption of extinguishing agent during fire fighting with a suspension of DW + MWCNT decreased by 3.5 times, and for a suspension of DW + Astr by 4.5–5 times.

When quenching with suspensions of water with carbon nanostructures, more intense vaporization in the combustion zone is observed in comparison with the control sample.

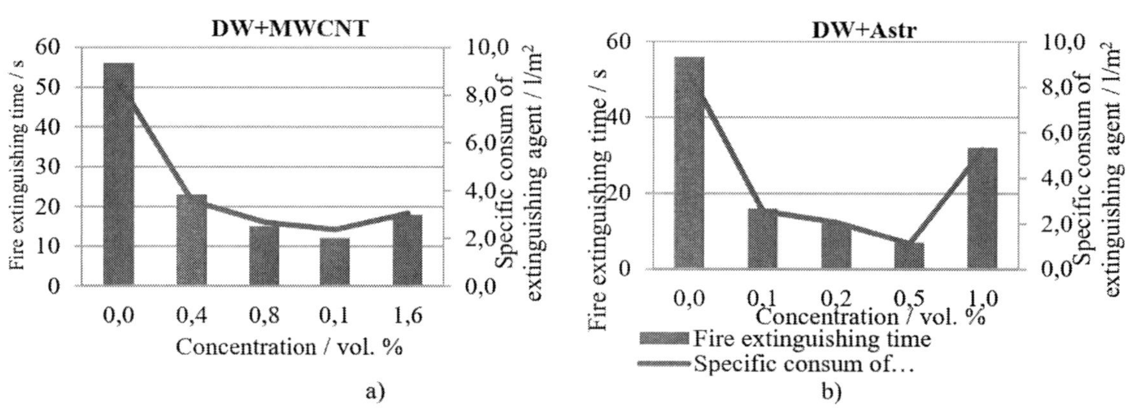

Figure 5. Extinguishing characteristics suspensions

3. Results

The mechanism for extinguishing liquid hydrocarbon fires by sprayed water suspensions with CNS is based on lowering the temperature in the combustion zone to the extinction temperature, at which there is no release of a sufficient amount of hydrocarbon vapor necessary for further continued combustion. A significant factor in changing the properties of nanostructures is the giant resonances of electromagnetic fields on the surface of nanoparticles [14], which determines a significant change in the properties of nanomaterials with a low concentration of CNS due to van der Waals interactions [15].

The mechanism of heat transfer in suspensions with CNS is based on the influence of the Brownian motion of carbon nanoparticles and the formation of a highly thermally conductive liquid layer at the "liquid – solid particle" interface. The process of boiling nanosuspensions depends on the properties of the base liquid, the type and concentration of nanoparticles contained in it, which largely determines the nature of heat and mass transfer with phase transformations, the heat transfer process on the surface of evaporating drops, and the nature of boiling (film or bubble) in the volume of the liquid [16]. The phenomenon of vaporization depends on the surface tension forces in the liquid during the breaking of the bond between neighboring liquid molecules and the movement of molecules into the gaseous medium [12].

It can be concluded that suspensions based on water with CNS are extinguishing agents with a predominantly cooling and diluting effect. When droplets of suspensions enter the combustion zone, intense heating occurs to the boiling point, followed by evaporation and cooling of the combustion zone. With a sufficient amount of water vapor in the combustion zone, flame extinction is observed. In this case, an increase in the specific heat of vaporization entails an increase in the amount of selected heat energy from the combustion zone. Increase Astr concentration to 0.5 vol. % and MWCNT up to 1.0 vol. % in suspensions significantly increases the extinguishing efficiency of Extinguishing Agents. A further increase in the concentration of nanoparticles leads to aggregation of CNS, which reduces the effective thermal conductivity of the suspensions and the specific heat of vaporization.

4. Conclusion

The dependence of the specific consumption and the quenching time of the flame of burning liquid hydrocarbons on the concentration of CNS (MWCNT, astralen) is obtained. The results indicate that suspensions are effective extinguishing agents for extinguishing burning liquid hydrocarbons at a relatively low concentration of carbon nanostructures. Dispersion of CNS intensifies the heat transfer in sprayed droplets of extinguishing agents, which leads to their more rapid heating in conditions of heat exposure to the flame. The effectiveness of controlling the properties of suspensions depends on the physicochemical properties of the base liquid and nanomaterials, as well as on the parameters of the external action.

References

[1] Nolan D P 2014 *Handbook of fire and explosion protection engineering principles: for oil, gas, chemical and related facilities* (William Andrew)
[2] Yu W and Xie H 2012 A review on nanofluids: preparation, stability mechanisms, and applications *J. of Nanomater.* **1** 1 DOI: 10.1155/2012/435873
[3] Zhao Q, Jiang T, Li C and Yin H 2011 Synthesis of multi-wall carbon nanotubes by Ni-substituted (loading) MCM-41 mesoporous molecular sieve catalyzed pyrolysis of ethanol *J. of Indust. and Engineer. Chem.* **17(2)** 218–22 DOI:10.1016/j.jiec.2011.02.009
[4] Sun Y P, Fu K, Lin Y and Huang W 2002 Functionalized carbon nanotubes: properties and applications *Accounts of chem. Res.* **35(12)** 1096–104 DOI: 10.1021/ar010160v
[5] Shames A I, Katz E A, Panich A M et al 2009 Structural and magnetic resonance study of astralen nanoparticles *Diamond and Related Mater.* **8(2–3)** 505–10 DOI: 10.1016/j.diamond.2008.10.056
[6] Ponomarev A and Iudovich M 2015 *Multi-layered carbon nanoparticles of the fulleroid type* US Patent No 9,090,752. 28 July.
[7] Voigtländer B 2016 *Scanning probe microscopy* (Berlin: Springer-Verlag AN) DOI 10.1007/978-3-662-45240-0
[8] Ding Y, Alias H, Wen D and Williams R A 2006 Heat transfer of aqueous suspensions of carbon nanotubes (CNT nanofluids) *Int. J. of Heat and Mass Transfer* **49(1–2)** 240–50 DOI: 10.1016/j.ijheatmasstransfer.2005.07.009
[9] Khaleduzzaman S S, Mahbubul I M, Shahrul I M and Saidur R 2013 Effect of particle concentration, temperature and surfactant on surface tension of nanofluids *Int. Communicat. in Heat and Mass Transfer* **49** 110–4 DOI: 10.1016/j.icheatmasstransfer.2013.10.010

[10] Suriyawong A and Wongwises S 2010 Nucleate pool boiling heat transfer characteristics of TiO2-water nanofluids at very low concentration *Experim. Thermal and Fluid Scince* **34(8)** 992–9 DOI: 10.1016/j.expthermflusci.2010.03.002

[11] Tanvir S and Qiao L 2012 *Nanoscale res. Lett.* **7(1)** 226–36 DOI: 10.1186/1556-276X-7-226

[12] Amiri A., Shanbedi M., Amiri H et al 2014 *Appl. Therm Eng.* **71(1)** 450–9 DOI: 10.1016/j.applthermaleng.2014.06.064

[13] Ponomarev A N, Yudovitch M E, Gruzdev M V and Yudovitch V M 2008 A nonmetallic nanoparticles in a superficial electromagnetic field. Topological factor of mesostructures interference. Voprosy materialovedeniya *Mater. issues* **60(4)** 59–64 Retrieved from: http://www.crism-prometey.ru/science/editions (in Russian).

[14] Dzyaloshinsky I E, Lifshits E M and Pitaevsky L P 1961 The General Theory of the Van der Waals Forces *Uspekhi Fizich. Nauk: Successes of phys. Sci.* **123(3)** 381–422

[15] Das S K, Choi S U S, Yu W and Pradeep T 2007 *Nanofluids: science and technology* (John Wiley & Sons) 397 p DOI: 10.1002/9780470180693.ch1

IOP Conf. Series: Materials Science and Engineering **905** (2020) 012012 doi:10.1088/1757-899X/905/1/012012

Starogroznensky oil and gas field: historical background and current status

A A Daykaev[1], Z G Aydamirova[2], E A Abubakarova[1,2], A A Shaipov[2], A M Movlaeva[2], T Kh Bachaeva[1,2], Kh I Zakriev[2] and R Z Dzharnagaliev[2]

[1] Department of the Issues of Fuel and Energy Complex, Complex Research Institute, RAS, Grozny, Russia
[2] Grozny State Oil Technical University named after Academician M.D. Millionshchikov, Grozny, Russia

E-mail: daykaev@mail.ru, lizi_68@list.ru, eliza_ggni@mail.ru, a.shaipov@gmail.com, amovlaeva@bk.ru, bachaeva@bk.ru, gpf-pg1920@mail.ru, dzharnagaliev@mail.ru

Abstract. The paper is devoted to one of the largest deposits of the North Caucasus, i.e. Starogroznenskoye field. There are identified the main development stages of this field, i.e. well production, development of the Miocene and Mesozoic deposits, and the current stage. Well production continued almost until the end of the 19th century. At the Miocene stage, the production was carried out from thrust and subthrust sand formations of Karagan and Chokrak. The paper also mentions the experimental adit on Mamakayevskaya gill. Which was headed by mining engineers A.A. Klushin and A.E. Yeronin. The beginning of the Mesozoic stage dates to the end of the 1940s. The main production was associated with the Upper Cretaceous sediments. At the end, an assessment of the current state of the structure of proven reserves is given. In conclusion, an assessment of the opportunities aimed at further development of the field associated with the additional exploration and additional development of the Miocene and Lower Cretaceous deposits is given.

1. Introduction

Starogroznenskoye gas and oil field is located within the Grozny ridge in the area of front ranges of the Eastern Fore-Caucasus on the territory of Staropromyslovsky district of Grozny. There are villages of Neftemaysk, Ivanovo, Katayama, etc. near the field the establishment of which is directly related to the development of Starogroznenskoye field. The northern slope of the Grozny ridge is crossed by ravines and gills, the most significant of which are Solenaya, Mamakaevskaya and Groznenskaya. The River Sunzha flows to the south-east of the field.

2. History of development of miocene deposits

Well oil production on the territory of the Grozny region began in the historical period of the late 18th century. The first information about oil sources near Grozny in the region of Mamakaevskaya and Grozny groups was presented by R.I. German at the beginning of the 19th century [1, 5].

The first geological studies date back to 1865 when the work of academician G.V. Abrikh was published. By the mid-1970s, there were more than 100 wells, and 10 years later the oil wells had increased 2 times. After that, the geologists F.G. Koshkul (1879), Andrusov N.I studied geological setting of the area. In 1888, a geological section of Karagan horizon was designed.

A new stage in a detailed geological study of the Grozny region began in the early 1892 when A.M. Konshin proved the fact of opportunities regarding this territory [2].

In the summer of 1893, the company Akhverdov & K° sank a steam well No. 1 in the Alkhan-Yurt section, in which a powerful oil fountain was obtained from the second layer of the Karagan horizon from a depth of 134 m. The well was in operation until 1902; the total oil production amounted to 80.7 thousand tons.

According to Yushkina E.M. in the period from 1893 to 1894 three more flowing wells with small flow rates were drilled. The next well – 7/977 was sank in February 1895, which revealed the oil layer II of Karagansk horizon at a depth of 141 m, which produced a powerful oil gush with a flow rate of 16 thousand tons per day. The well had been flowing for three years and produced more than 710 thousand tons of oil.

Such results made the Grozny region attractive for large oil producers. Besides, Starogroznensky oil-bearing area has become known all over the world. After the flowing in the well – 7/977 there were many published works, which expressed completely different opinions about further development of Grozny oil industry [4].

Since 1901, the Geological Committee of Russia initiated the study of the geological structure and oil and gas potential in connection with unsuccessful drilling results and ever-increasing demand for oil. The work was carried out by a number of famous geologists under the supervision of N.A. Sokolova (G.P. Mikhailovsky, K.P. Kalitsky, I.N. Strizhov, E.M. Yushkin, D.V. Golubyatnikov, etc.) [6]. Studying the structure of Starogroznensky fold I.N. Strizhov and E.M. Yushkin, revealed the presence of a longitudinal gap along the northern wing, overturned anticline to the north, which was confirmed by subsequent detailed exploration. K.P. Kalitsky and E.M. Yushkin performed work in Grozny oil-bearing region [3,7]. Strizhov I.N. (1904) made up a normal section of Karagan horizon.

In 1905, he also performed the first calculation of oil reserves in Starogroznenskoye field on the basis of geological analogies (comparative analysis with oil already produced). The initial balance reserves of oil in the field amounted to 23.3 million tons. Starogroznenskoye oil field workers were among the first in the world to apply various methods of stimulating the formation. So, I.N. Strizhov proposed the use of air injection in Karagan formation in order to increase oil recovery.

In 1906, Kalitsky K.P. compiled the first geological map of Starogroznenskaya area. The works of geologists of that period were not very regular with the exception of the works of Kalitsky and Strizhov.

In 1918, the balance reserves of oil were calculated by S.I. Charnotsky using volumetric and statistical methods, which amounted to 40.2 million tons. However, based on the revaluation of reserves performed in 1923 by V.A. Selskiy the balance reserves in the field amounted to only 20.1 million tons. [11].

During 1921–1922. A.P. Shapovalov compiled a generalized section of the productive strata for Starogroznensky district for the first time in the Grozny region. This work was so thorough that it had been used for 15 years.

At the end of the restoration work in the field, a small amount of exploration work was initiated, which resulted in the discovery of new high-yield deposits in Solenaya gill in Starogroznenskaya area in May 1922. It is also specific that geological conditions that caused the existence of many high-yield deposits in the periclinal part of the anticline remained unclear due to the imperfection of geological documentation of that time. It was found out that the deposits of Solenaya gill had a water pressure regime, while those located directly to the west in the uplifted pool of Starogroznensky anticline were characterized by the dissolved gas regime.

On the example of conditions for the development of oil deposits of Solenaya gill and their comparison with the deposits of the rest of Starogroznenskaya area, N.T. Lidtrop defined the concept of a water pressure mode. His research on this issue played a very important role in the development of further production not only in the Grozny region, but also in other oil and gas fields.

The geologist Lednev N.M. (1925, 1926, 1927), with the participation of Vossoevich in 1928, began a detailed study of the geological structure of Makhachkalinskaya areae and Starogroznensky anticline.

IOP Conf. Series: Materials Science and Engineering **905** (2020) 012012 doi:10.1088/1757-899X/905/1/012012

As a result of studying the structure of Starogroznensky anticline N.A. Kudryavtsev in 1925 made the assumption that Tashkalinsky site is an independent anticlinal fold located echelon-like in relation to Starogroznensky brahianticlinal. At the same time A.P. Shapovalov and (1924) and, somewhat later, I.N. Strizhov (1926) suggested that the northern wing of Starogroznensky fold is complicated by longitudinal thrust.

A detailed study of the structure of Starogroznensky fold begins in the late 1920–1930s. During that period the wells were sunk in Maykop formation (15 exploratory wells) to a depth of 2000–2400 m, none of them discovered Maykop deposits, which, as it became known, lie much lower. At the same time, the drilling results proved the presence of oil in the sub-thrust beds of Karagan. In 1932, the first industrial oil was obtained from well – 5–79 from a depth of 1245 m with a flow rate of 5–10 tons/day. As soon as the next year, dozens of wells have been sunk for sub-thrust.

In 1932, an *experimental adit* was sunk in the section 975 of the old field in *Mamakayevskaya gill*. The work was headed by mining engineers A.A. Klushin and A.E. Yeronin. The adit horizontally passed through bedrock more than 500 m, revealing productive sandstones of the Tertiary age. The adit was intended to carry out research work aimed at creating an industrial oil mine for the development of the upper layers of spaniodontella formation using secondary methods of oil production. The research and experimental production works were carried out in this mine until 1936 [8–10].

In a well – 24/40, drilled in 1934 in *Starogroznenskaya fold (sub-thrust)* an oil gush was obtained from a depth of 1200 m with a flow rate of 40 tons per day from Karagan (stratum III–IV) layer. Thus, a high productivity of these formations was finally proved.

The difficulties of studying the structure of the thrust part of Starogroznensky fold were noted by engineer V.S. Starzhinsky: All the wells in the sub-thrust are either sandstone or clay, without crossing the strata, which is one of the reasons that impede normal lighting of the sub-thrust structure. Wells drilled with oriented north-south curvature would have a significant advantage. In this case, several layers can pass through one well and a marking horizon can be detected.

The first graphical constructions for sub-thrust layers were performed by G.M. Sukharev and V.A. Tilyupo in 1935. In 1937, the geologists N.V. Frolov and V.A. Golyakov developed a special methodology of structural constructions. Since the beginning of the 1930s a great attention was paid to the issues of oil recovery increase and rational development of the Miocene deposits. The importance of using secondary operating methods was proved.

In 1944, due to the problems of development of sub-thrust layers, a special group for well development was created. In the same year, drilling of directional exploratory wells began on the XII and XVI sub-thrust beds of Starogroznenskoye field, and in 1945 they resumed this work on Tashkalinsky site. In the same year, the gushs of oil were produced in several wells drilled within Tashkalinskaya area. This was the largest achievement of geological exploration during the war on the territory of the Chechen Republic. The discovery of Tashkalinskoye field contributed to an increase in oil production and provision of hydrocarbon refineries with raw materials. In general, from 1943 to 1945, approximately 2.5 million tons of oil were produced. [11]

In 1948, in order to select the area for the mine field, an attempt was made to drill 7 appraisal wells in Starogroznensky site in the sections 62, 56, 53, 52, 50, 48 and 146. Based on the analysis of the production materials of the appraisal wells, the eastern field was recommended as more promising in relation to the possibility of its mine method. The total oil area of this field is 82.1 ha. The depth of $2^{в}$ formation is 65–200 m. The initial oil reserves in the 2nd reservoir – recoverable – make up 475.1 thousand tons, geological reserves constitute 1218 thousand tons, and 3b make up recoverable reserves, 426.6 thousand tons, and geological reserves make up 969.5 thousand tons. An exploration project for the selected mine field was drafted in accordance with the temporary instruction for the assessment and exploration of oil fields for open pit mining, in relation to fields with disturbed discharges and well-lit production wells.

The object for development was the main $2^{в}$ layer of Karagansky horizon, as well as $3^{в}$ and I layers. The distance between the exploration profiles is 350 m and between the individual wells in the profile is 125 m. In total, it was planned to drill 21 Craelius wells. In total, about 5000 wells were drilled – the Miocene deposits, Starogroznenskoye field – and cumulative production amounted to about 52.3 million tons of oil. Karagan-Chokrak deposits are at the final stage of development; the degree of development is 99.8 %. Tashkalinskiy site has been fully developed.

3. History of development of mesozoic deposits

In the postwar period (1950–1960) Groznefterazvedka – the geological exploration organization – had worked extensively in the area of geological prospecting and exploration throughout the entire Eastern Fore-Caucasus, including Sunzhensky district. An experienced geologists A. Ragulin compiled geological exploration projects for chalk deposits where he encompassed such oil fields as Starogroznenskoye, Oktyabrskoye and others. An exploration of the Upper Cretaceous reservoir was carried out according to the project since 1958, but the first two wells were liquidated for technical reasons without opening the project horizon.

In total, 37 wells were drilled in the Upper Cretaceous deposit. 27 wells were the productive deposits. The industrial influx of oil from the Upper Cretaceous limestones on the Starogroznenskaya Square was first obtained in 1963 with an open flowing of well – 641.

The structure along the roof of the Upper Cretaceous is an asymmetric fold of the brachiform type of sub-latitudinal strike. The materials of the geological interpretation of detailed gravimetric surveys made it possible to identify a number of features of the gravity field. Sunzhenskaya linear zone includes Starogroznensky maximum, which corresponds to the anticlinal zone within which the Starogroznensky uplift is distinguished. Starogroznensky maximum stretching east to the city of Grozny in the form of a narrow strip is characterized by a restriction from the north with a highly gradient zone and from the south by a less gradient zone of the anomalous gravitational field.

The dimensions of the structure along minus 4250 m enclosing isogypsum are 29.3×3.2 km with an amplitude of 650 m. The Upper Cretaceous section is composed of carbonate deposits; the foraminiferous layers are represented by fractured marls; and the Upper Cretaceous rocks are represented by fractured limestone. The Upper Cretaceous reservoir is classified as a cavern-fractured type. Due to the cracks dissecting the thick strata of the foraminiferous and Upper Cretaceous deposits, the reservoir represents a single hydrodynamic system. The deposit refers to a massive reservoir type. The size of the Upper Cretaceous deposits is 30x3.2 km; the oil floor is 791 m.

An oil deposit in *Aptian deposits* was found in 1973 when they were testing the well – 701. In subsequent years, the Aptian deposits were tested in 16 more wells in which inflows of anhydrous oil were obtained. According to the Lower Cretaceous (Aptian) sediments, a rise corresponding to the shape of the Upper Cretaceous fold was established. The Aptian layer-uplifted deposit. The size of the deposit is 29.7 x 3.1 km with the height of 825.

The Lower Cretaceous Aptian section is represented by terrigenous formations, i.e. sandstones, siltstones and clays. A distinctive feature of the productive Aptian reservoir is the presence of two types of reservoirs, i.e. cavernous-fractured and porous.

In 1989, a hydrocarbon deposit of *Valanginsk deposits* was found (well – 728). The presence of hydrocarbons by testing Valanginsk deposits in well – 729 where gas inflows with a high content of hydrogen sulfide were obtained from the interval 5405–5535 m, and therefore the reservoir was mothballed [12]. A free gas deposit of Valanginsk reservoir lies at a depth of 4900–5300 m; the dimensions are 17.0 x 1.0 km; the gas content floor is 67 m.

The study of the structure of the field by geophysical methods. On the territory of the area under study the materials of detailed gravimetric surveys were processed and their geological interpretation was performed. The analysis of gravimetric maps within the described territory made it possible to identify a number of features of the gravity field. Sunzhenskaya linear zone includes Starogroznensky maximum, which corresponds to the anticlinal zone within which Starogroznensky uplift is distinguished. Starogroznensky maximum stretching east to the city of Grozny in the form of a narrow strip is characterized by a limited gradient zone from the north and a less gradient zone from the south.

Table 1. State of the raw material base of starogroznenskoye field

Deposit	Stratigraphic horizon	Residual reserves of category A + B1, thousand tons		Reserves depletion	Production, thousand tons	
		Geological	Recoverable	%	Accumulated	Annual
Starogroznenskoye field	Miocene	49584	91	99.8	52284	12
	Maikop	–	–	–	8	0
	Up. Cretaceous	16846	0	100	32702	0
	L. Cretaceous	21335	273	92.8	13825	42

4. Current state of raw material base and opportunities

The state of the structure of hydrocarbon reserves and resources is an important element in the exploration planning. Short- and medium-term oil and gas exploration planning is usually based on the information from previously estimated reserves and resources of C2 and C3 categories, and long-term planning is based on the total volume of unexplored part of the ultimately recoverable oil resources along the analyzed area. The table shows the figures for current oil reserves state (01/01/2018), as well as for the depletion and production of Starogroznenskoye field.

5. Conclusion

Future prospects are associated with the additional development of the Lower Cretaceous deposits of Starogroznenskoye field, with the widespread application of modern technologies aimed at maximum extraction of oil and gas from the interior of the earth; additional extraction of oil from Karagan and Chokrak deposits by drilling additional production and exploration wells or by a shaft method. Certain prospects are associated with additional exploration of the Miocene deposits. The resumption of additional exploration of these layers is due to a number of favorable prerequisites for high efficiency of this direction, i.e. favorable hypsometric position of Karagan and Chokrak horizons relative to the lower Upper Cretaceous deposits and higher than Sarmatian deposits, which demonstrated favorable signs of oil and gas deposits; high reservoir properties of sand formations, etc.

One of the topical issues of our time is considered the problem of renewability of oil and gas resources. There are various factors in the renewability of resources. One of the important factors is the excess of the total volume of oil produced over the approved reserves. At the very beginning of the industrial development of the field, mining engineer L.I. Baskakov noted that the volume of trap for Karagan reservoir was exceeded. The first estimation of reserves was done at the end of the 19th century. Subsequently, the oil reserves of the Miocene deposits were calculated repeatedly. There are known facts of the resumption of oil manifestations from the annulus at a number of wells mothballed in the 1990s. Such processes were also observed in other deposits of deeper horizons. There are several ways to determine the beginning of the formation of oil and gas deposit, i.e. paleogeological, historical, geological, geochemical, lithological, gelargon, etc. However, all of them do not record the completion time of the formation of hydrocarbon deposits, i.e. it can be assumed that they are formed at the present stage, which is established by the following criteria: by the potentiometric method, trap occupancy, gas saturation of formation water (for gas deposits). The potentiometric method proposed by N.A. Eremenko, I.M. Mikhailov in 1972, suggests that the energy balance in the reservoir under a nonequilibrium hydrogeological situation can exist only in the presence of hydrocarbon flows, and, consequently, the formation of their clusters. Trap filled with hydrocarbon oil: If the trap is completely filled with oil or gas (full-layer), it is believed that its formation continues at the present time. In 1968, gas saturation of formation water was justified for gas deposits by V.I. Korzenstein. It is assumed that gas deposits surrounded by formation water with maximum gas saturation are in a state of growth at the present stage. Some other criteria also testify to the replenishment of reserves, i.e. the presence of an abnormally high pore pressure in the deposits, the pulsating nature of operation of the production wells, high gas factor, etc. Based on everything stated above, it can be concluded that the oil and gas potential of Starogroznenskoye field and other fields of the Tersko-Caspian trough is significant.

The development of Starogroznenskoye field is an example of a comprehensive study of the features of geological structure for the purpose of prospecting and exploration of complex oil and gas deposits, i.e. subthrust, tectonically shielded, structural and lithological, etc.

References

[1] German R I 1858 Large layer of asphalt in Little Chechnya *Bull. of Natural Sci.* **6** 189–92
[2] Konshin A M 1892 Geological description of Grozny oil-bearing area and oil fields of the Terek region and the Caspian coast *Mater. for the geology of the Caucasus, ser.* **2, B. 6**
[3] Yushkin E M 1909 *The beginning of the Grozny oil industry in essays* (Ekaterinodar) p 50
[4] Kolosov L N 1953 *The emergence of Grozny oil industry (1893–1903)* (Moscow)
[5] Ibragimova Z H 2009 *Royal past of Chechens: politics and economics* (Moscow: Probel-2000) 872 p
[6] Brod I O 1937 *Grozny oil field, Oil fields of the Eastern Fore-Caucasus*
[7] Kalitsky K P 1906 Grozny oil-bearing region *Proc. of Geolcom, new. ser.* **24**
[8] Klushin A A 1934 Turning the experimental adit of Starogroznenskiy district into an experimental mine *Grozny oilman* **2-3** 61–5
[9] Yeronin M A 2000 Starogroznenskaya adit *Oil industry* **3** 78–80
[10] Ershov G A, Tonkonogov P M and Ermolenko A P 1971 *A wonderful source, Essays on the oil industry of Chechen-Ingushetia* (Grozny: Chechingizdat) 327 p
[11] Broda I.O 1958 Geology and oil and gas potential of the Eastern Fore-Caucasus, *Proceedings of Integrated Southern Geological Expedition,* iss. 1 (Leningrad: Gostoptekhizdat) 621 p
[12] Daukaev A A et al 2019 Development of oil and gas research and education institutes in Grozny in 1920–1980 European Proceedings of Social & Behavioural Sciences EpSBS Conference *Social and Cultural Transformations in the Context of Modern Globalism* (SCTCGM 2018) pp 366–75

Strength and deformation characteristics of fiber concrete modified with a peat additive

O V Demyanenko[1], N P Gorlenko[1], Yu S Sarkisov[1], N O Kopanitsa[1], G L Gorynin[2] and K V Ikonnikova[3]

[1] Tomsk State University of Architecture and Building, 2 Solyanaya sq., Tomsk, 634003, Russia
[2] Surgut State University, 1, Lenin Ave., Surgut, 628403, Russia
[3] National Research Tomsk Polytechnic University, 30, Lenin Ave., Tomsk, 634050, Russia

E-mail: Demyanenko.olga.v@gmail.com, gorlen52@mail.ru, sarkisov@tsuab.ru, kopanitsa@mail.ru, gorlen52@mail.ru, ikonna@yandex.ru

Abstract. Dry mixtures are widely used in modern construction. Development of new modifying agents for improving of dry mixtures performance characteristics and increasing its economic efficiency is an important problem. The paper presents study results of the effect of additives based on thermo modified peat on properties of cement mortars. It is established that introduction of additives in the mortars compositions provides improving of their strength and deformation properties.

1. Introduction

The applying of dry construction mixtures in the implementation of masonry and installation kinds of work is able to significantly improve the quality and productivity of labor, to reduce of transporting materials expenses, to provide the presence of the necessary grade and destination mixture in the required amount on the construction site. This way contributes to achieve a more rational organization of production.

Modern technologies of dry building mixtures production allow to obtain compositions with determined rheological and operational characteristics and to regulate their basic properties by introducing modifying additives for various purposes [1].

Plasticizers, stabilizers, water-retaining and water-reducing additives are most often used for regulating dry building mixtures properties, in addition, additives regulating the setting time, antifreezing, hydrophobizing and air-entraining modifiers are applying if necessary [2, 3]. Using of additives of foreign manufacture in the dry building mixtures composition provides a significant increase in the cost of the final product.

One of the methods of solving this problem is to organize the production of modifying additives based on local raw materials. The raw material base of the Siberian region allows the creation of modifying additives based on peat for regulating the properties of dry building mixtures. The material composition of peat is represented by a variety of organic and mineral compounds. The presence of chemically active organomineral functional groups in the raw material provides the opportunity to use various methods of peat modifying: thermal, chemical, mechanical and a combination of several types of impact for the production of various purposes products, including effective modificators for dry building mixes.

The study results of the peat additives effect on the hardened cement paste properties are presented in the scientific works [7, 8]. It has been established that the introduction of the MT-600 additive, obtained at a temperature of 600 °C under conditions of limited air access, leads to significantly increasing of the hardened cement paste compressive strength by 43 % and to reducing the water absorption by 32.5 %. Therefore, the obtained additive can be recommended for using in the composition of mortars with improved strength and hydrophysical characteristics.

2. Methods and materials

Investigations of the mortars strength characteristics of were carried out in accordance with requirements of All Union State Standard (GOST) 5802-86. Cubic samples of cement-sand mortars of size 70x70x70 mm were used for measuring the compressive strength. Flexural strength investigation were carried out on prisms 40x40x160 mm.

In order to study the deformation properties of material, compressive strenght of cement-sand mortar samples with a size of 40x40x40 mm were determined by the testing machine "Instron 3382" at a loading rate of 2 mm per min. The influence evaluation of modifying additive on the nature of deformation fields formation of a cement-sand mortar samples was carried out using a digital optical measurement system "Vic 3D". The system allows to obtain information on the displacement of microvolumes on the sample surface along three orthogonal axes during its deformation [7, 8].

The degree of modifying additives effect on the strength and character of the formation of the deformation of the mortars samples was determined by comparing stereoscopic images of the deformed surface of basic and modified samples.

For testing basic and modified cement-sand mortars were prepared, the binder: sand ratio was 1:3. The MT-600 additive content was 0.5 % by cement mass in the modified mortar. A mixture of sand fractions 1.25–2.5 and 0.16-0.31 in the ratio 70:30 was used as an aggregate for masonry mortar. The using of two or more aggregate fractions in the cement-sand mortar allows to reduce cement consumption due to the reduction in the voidness of the aggregate without impairing the construction and technical characteristics of the material. The flowability of the compositions was the same and was 6-8 cm. As a base sample, a mixture of binder and aggregate was prepared in a ratio of 1:3.

For the study of cement mortar portland cement CEM I 42.5N from "Topkinsky Cement" plant was used as a binder. Cement complies with requirements of All Union State Standard (GOST) 30515-2013 "Cements. General specifications"

The silica sand from Kudrovskoe field of the Tomsk region was used sa an aggregate. The applying sand satisfies the requirements of All Union State Standard (GOST) 8736-2014 "Sand for construction works. Specifications"

In the modified samples, the MT-600 additive based on the low peat of the Gusevskoe field was used. The modifying agent was obtained by pyrolysis of peat at a temperature of 600 °C in a laboratory installation. The elemental composition of the additive represents predominantly silicon, aluminum, carbon, and calcium. The phase composition of the peat additive is presented in Table 1.

Table 1. Phase composition of modifying additive mt-600

Phase	Size, [nm]	Content, [%]
SiO_2	>400	43.81
$CaCO_3$	180	47.99
C (graphite)	15	7.77
$CaAL_2Si_2O_8$	100	0.31
C60	10-20	0.12

The powder of pyrolysed peat has a high dispersion: the average particle size is (42 ± 3) μm, and the specific surface area is 600 m^2/kg.

Electron microscopy analysis allowed to stablish that the additive contains nano-sized elements as various forms of nanocarbon, such as fullerenes, ranging in size from 5 nanometers, nanotubes and nanofibers, whose dimensions reach 200 nanometers or more, with a transverse dimension of 20 nanometers or less.

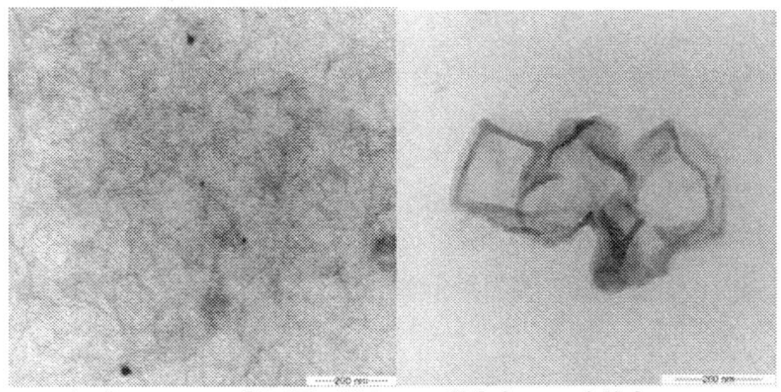

Figure 1. Electron microscopic images of MT-600 modifying additive

To justify the increase in strength, electron microscopic studies of the microstructure of cement stone modified by the addition of TM 600 to the surface of the samples were carried out. Figures 2 and 3 show electron microscopic images of a control and modified cement stone with a magnification of 5,000 times.

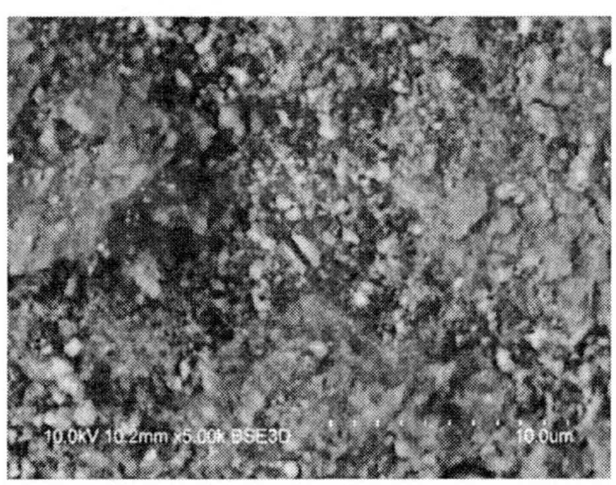

Figure 2. Electron microscopic image of cement stone without additives, magnification x5000

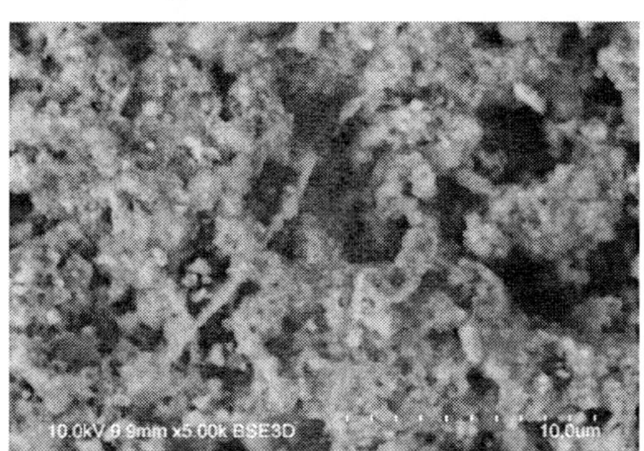

Figure 3. Electron microscopic image of a cement stone with MT-600 modifying additive, magnification x5000.

In the micrographs (Figures 2, 3) it can be seen that the volume of neoplasms in comparison with the control cement stone increases, which is associated with a large number of crystallization centers. The size and number of micropores are also increased. In addition, needle-shaped or fibrous crystals were found in the images of modified cement stone, which are more clearly visible at a lower magnification – 1800 and 1000 times.

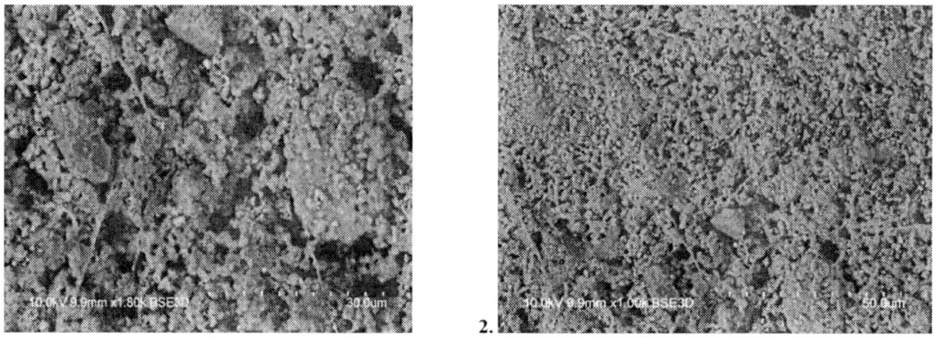

Figure 4. Electron microscopy images of a modified cement stone, magnification x1800 (1) and x1000 (2)

3. Results

Mechanical strength is the main performance characteristic of hardened mortars.

Bending tensile strength and deformation characteristics do not relate to the main quality parameters of mortars, according to All Union State Standard (GOST) 4.233-86), however, the working conditions of the masonry and assembly hardened mortars suppose the possibility of tensile stresses occurrence during exploitation. In this connection, the paper compares the investigation results of the compressive strength, flexural strength characteristics and elastic modulus of solidified mortars.

The study results of the strength characteristics of cement-sand mortars, presented in Figure 5, show that in the early time of hardening (7 days), the introducing of the MT-600 additive increased the compressive strength by 44 % compared to the basic samples. And at 28 days of hardening the strength of samples modified by MT-600 is up to 20 % higher than basic ones strength.

At the same time, the average flexural strength of samples modified with additives MT-600 is 8.05 MPa, and the basic samples flexural strength is 6.996 MPa. Thus, the flexural strength of the modified hardened mortar by 15 % exceeds the same characteristic of the basic samples.

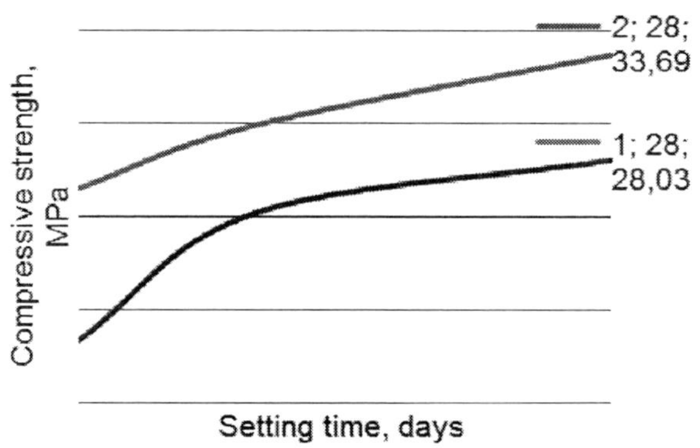

Figure 5. Influence of additive based on thermally modified peat on the compressive strength of cement-sand mortars; 1 – control cement-sand mortar; 2 –modified cement-sand mortar

The deformation curves presented in Figures 6 and 7 show that the introduction of additives based on peat in cement compositions leads to increasing of the elasticity modulus of the modified mortar to 25 % comparing to the basic mortar.

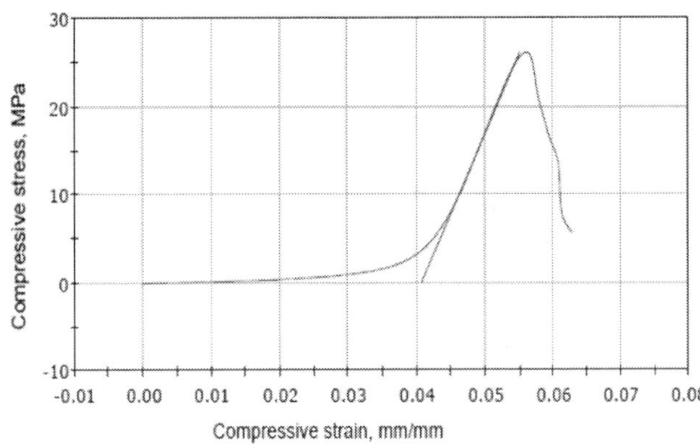

Figure 6. Diagram of the stress-strain state of a cement-sand mortar basic sample

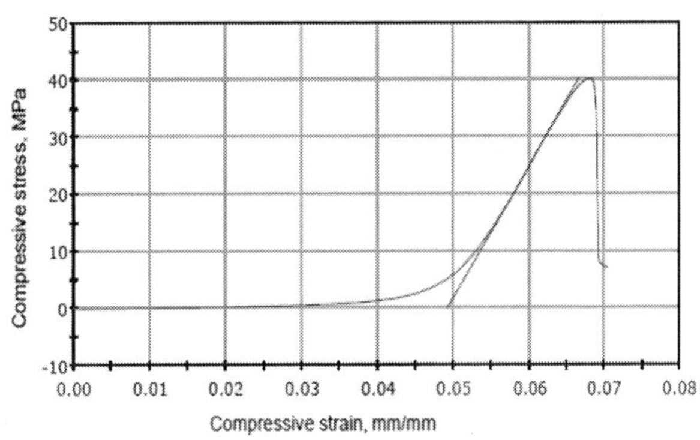

Figure 7. Diagram of the stress-strain state of a sample of cement-sand mortar with the MT-600 additive

The research results showed that the modified cement-sand mortar can work elastically in a wider range of loads, that ensures high crack resistance and deformability of mortar mixtures based on modified cement. The increase in the modulus of elasticity of cement stone with the introduction of additives based on peat can be associated with the formation of hardening structures that work under load, without the appearance of zones with plastic deformations. Let us consider the results of testing the specimens for compression. Four series of samples were tested, two of which are test samples, modified additive MT-600 and two series of control samples of cement-sand mortar, made without the use of additives.

Figure 5 shows the compression test results of the samples.

The results are shown in Figure 8. Three images of the deformed surface correspond to each sample. The first picture was taken at loads equal to 1/2 from the destructive value, the second was made with destructive loads, the last photograph characterizes the stage of destruction of the samples.

In this case, the development of transverse deformations, which are the cause of the destruction of the samples, is considered. To the right of each image of the deformed surface, a scale of values of transverse strains is given in accordance with the color scale. The development of elastic deformations in samples with additives is characterized by the formation of a sealing zone in the middle part of the cube and gradual decompression of the medium in the vicinity of the free faces of the cubes. It is possible to note the linear dependence of the growth of transverse deformations from stresses to the level of average loads. The destruction of the investigated cubes began with the appearance of microcracks in the peripheral areas. At the same time, under loads preceding the destructive values, the rate of transverse deformations of the cubes decreased.

IOP Conf. Series: Materials Science and Engineering **905** (2020) 012013 doi:10.1088/1757-899X/905/1/012013

Sample 1 of a cement-sand mortar with the addition of MT-600

Sample 2 of a cement-sand mortar with the addition of MT-600

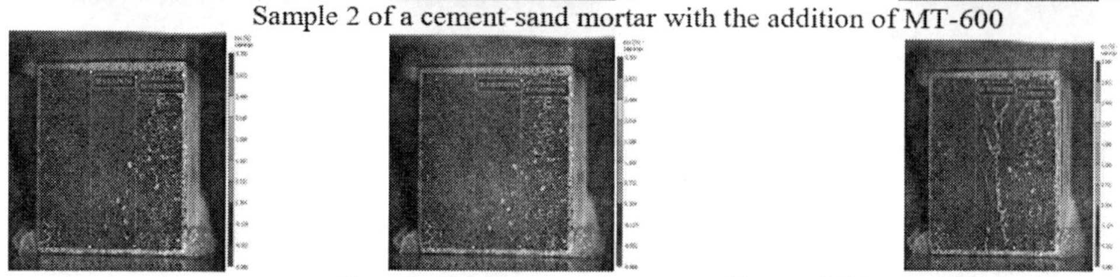

Sample 1 cement-sand mortar without additives

Sample 2 cement-sand mortar without additives

Figure 8. Images of the deformed surface of samples of cement stone

Comparison of patterns of deformations of samples with additives and deformation patterns of control samples showed their practical coincidence. Unlike control samples, the destruction of samples with additives occurred somewhat later, meanwhile, in time the destruction occurs faster and is characterized by a large decrease in the descending section of the strain diagram. This is confirmed by a comparison of the deformation diagrams.

4. Conclusion

Strength characteristics of solidified solutions were investigated. It was found that the introduction of a peat-based modifier into the composition of the solution mixture makes it possible to increase the compressive strength of the solution by 20 % and the bending strength by 15 %. The increase in compressive strength is explained by the presence in the additive of particles with a size of approximately 100 times smaller than the cement grains. In [5] Yu.M. Bazhenov calls such particles "seals", as they fill the voids between the particles of the binder, providing a more dense packing of particles. In addition, the effectiveness of sealing additives is due to the formation of additional crystallization centers and high surface energy, which contributes to the acceleration of hardening and increase the strength of cement stone. The higher bending strength is due to the presence of fibrous inclusions, which were found on micrographs of peat additives. It is assumed that such inclusions provide the effect of reinforcing neoplasms of cement stone at nano- and microlevels [1–4].

It is known that the modulus of elasticity of a material largely depends on its compressive strength. The conducted studies showed that for the compositions under study the directly proportional dependence of these quantities is preserved.

Acknowledgment
The article is supported by the grand of the Russian Foundation for Basic Research, project № 18-47-860015\18. Mathematical simulation of mechanical properties of constructions made of fibre concrete with metallic and nonmetallic fillers was carried out with the use of the super computer and additive technologies.

References
[1] Hyun-Oh Sh, Seung-Jung L and Doo-Yeol Y 2018 Bond Behavior of Pretensioned Strand Embedded in Ultra-High-Performance Fiber-Reinforced Concrete *Int. J. of concrete struct. and mater.* **12(1)** 1–13 DOI: 10.1186/s40069-018-0249-4
[2] Bicer K, Yalciner H and Balks A P 2018 Effect of corrosion on flexural strength of reinforced concrete beams with polypropylene fibers *Construct. and build. mater.* **185** 574–88
[3] Abu-Obaida A, El-Ariss B and El-Maaddawy T 2018 Behavior of Short-Span Concrete Members Internally Reinforced with Glass Fiber-Reinforced Polymer Bars *J. of composit. for construct.* **22(5)** 04018038
[4] Xie J, Fang Ch and Lu Zh 2018 Effects of the addition of silica fume and rubber particles on the compressive behavior of recycled aggregate concrete with steel fibres *J. of cleaner product.* **197(1)** 656–67
[5] Ferrara L, Park Y-D and ShahSurendra P 2007 A method for mix-design of fiber-reinforced self-compacting concrete *Cement and concrete res.* **37(6)** 957–71
[6] Rybin V A, Utkin A V and Baklanova N I 2016 Corrosion of uncoated and oxide-coated basalt fibre in different alkaline media *Corros. Sci.* **102** 503–9
[7] Demyanenko O.V., Kopanitsa N.O., Sarkisov Y.S et al 2017 Quantitative phase analysis of modified hardened cement paste *IOP Conf. Ser. Earth and Environ. Sci.* **87(9)** 092008 DOI: 10.1088/1755-1315/87/9/092008
[8] Ionov V.V., Larionov S.A., Sarkisov Y.S et al 2017 Tribological properties of hydraulic fluids modified by peat-based additives *IOP Conf. Ser. Mater. Sci. and Engineer.* **177(1)** 012108 DOI: 10.1088/1757-899X/177/1/012108
[9] State Standard (GOST) 8736-2014 *Sand for construction works. Specifications* 12 p
[10] State Standard 8736-2014 *Sand for construction works. Specification* 11 p

Solution of multidimensional system with XY differentiation of probability density equations for identification of silver nanoparticles on fibers

V M Emelyanov, T A Dobrovolskaya and V V Emelyanov

Department of Standardization, Metrology, Quality Management, Technology and Design, South-West State University, Kursk, Russia

E-mail: vmemelianov@yandex.ru, dobtatiana74@mail.ru, emelianov@nm.ru

Abstract. The article presents the results of compiling and solving a system of multidimensional differential correlation equations of probability densities for identifying colloidal silver nanoparticles on polyester fibers with multidimensional correlation components of Raman polarization spectra. A method is proposed for increasing the accuracy and speed of identification of silver nanoparticles on polyester fibers, taking into account the longitudinal and transverse polarization of laser radiation over the entire range of the Raman spectrum with the analysis of two peaks sequentially and in order at the same time along the X-side and along the Y-side of the fibers. To implement the method, a program was developed in the Mathcad software. When solving the system using the nonlinear quadratic and differential equations with respect to XY, the probability density equations of distribution ellipses obtained very high accuracy p0 and p1 up to 10^{-16}.

Keywords: polyester fiber, silver nanoparticles, Raman spectra, system of multidimensional equations, recognition accuracy, probability of intersection of distribution ellipses, accuracy of nanoparticle identification

1. Introduction

The current development of nanotechnology allows elaborating textile materials with improved hygienic and bactericidal properties. In the production of such products, the urgent task is to control the deposition of nanoparticles on the surface of materials, which in turn is impossible without the precise identification of nanoparticles [1–6]. To study nanostructured materials, microscopy and spectroscopy methods are used [7–8]. Multidimensional mathematical methods are used to recognize nanoparticles on the surface of textile materials. Development of methods for detecting nanoparticles on the surface of textile materials is impossible without the use of modern computer programs and the development of algorithms for the implementation of mathematical methods for identifying nanoparticles in applied mathematical programs. When solving the system of differential correlation equations, an uncertainty was revealed in the fact that when checking the solution for the graphic intersection of distribution ellipses without nanoparticles and with nanoparticles, two intersection points appeared. In this case, the probability densities of the intersection of ellipses, found when solving nonlinear quadratic equations, were not equal. The difference between these values was significant, and the accuracy was 23 %. Such accuracy for the identification of silver nanoparticles is unacceptable. The methods developed by the authors for identifying nanoparticles on the surface of textile materials with an estimate of the probability density of the intersection of the distribution ellipses of the peaks of the Raman polarization spectrum are described in [9–10]. Moreover, in the proposed methods there are elements of manual selection of some components in the equations. The analytical expression of the differential correlation equation is obtained by differentiating the equation with respect to X. The accuracy is quite high at 10^{-15}, but depends on the selection of additives in the equation.

To increase accuracy, it is necessary to develop a method for automatically determining the reliability of identification of nanoparticles.

2. Methods and materials

For the mathematical and software implementation of the method of identification of nanoparticles, Raman spectra were used, which were obtained at the regional center of nanotechnology of the South-Western State University in the study of polyester fibers (PE) upon deposition of silver nanoparticles on them from a colloidal solution of silver nanoparticles AgBion (TU 2499-003-44471019-2006, concern "Nanoindustry"). Using a scanning probe microscope (SPM) with a confocal Raman and OmegaScope™ fluorescence spectrometer, measurements were made at 5 points on each sample with polarization of the laser beam across X- and along Y-fibers. For analysis, 9 main peaks of Raman spectrograms were selected. Data processing was performed using the MathCad 15.

In this paper, we consider a possible increase in the number of equations for solving the system by using differentiation with respect to X and Y at the same time when estimating the probability density of the intersection of the ellipses of the peak height distribution of the Raman polarization spectrum when identifying silver nanoparticles on the surface of polyester fibers.

To compile a system of differential correlation equations, we use the p–densities of intersection probabilities of distribution ellipses in vector-matrix analytical expressions along the coordinates of the intersection points. In this paper, we consider a system of only two vector-matrix analytical expressions of probability densities $\ln(p)=X^T \cdot R^{-1} \cdot X$ and $\ln(p_{Ag})=X_{Ag}{}^T \cdot R_{Ag}{}^{-1} \cdot X_{Ag}$ for p and p_{Ag} [14].

For the analytical estimation of probability densities (p and pAg) from the intersection points of distribution ellipses, in accordance with the method developed by the authors, vector-matrix expressions 1–5 are used.

$$R_{Ag} := \begin{pmatrix} 1 & rXYAg_i \\ rXYAg_i & 1 \end{pmatrix}; \quad R := \begin{pmatrix} 1 & rXY_i \\ rXY1_i & 1 \end{pmatrix}. \tag{1}$$

$$
\begin{cases}
f(x,y) := \left[\ln\left[\dfrac{1}{(2\pi)\cdot\left[\left(\left|R_{Ag}\right|\right)^{0.5}\right]} \right] + \dfrac{-1}{2}\cdot\left[\begin{pmatrix} \dfrac{x-M_{XAg_i}}{\sigma_{XAg_i}} & \dfrac{y-M_{YAg_j}}{\sigma_{YAg_j}} \end{pmatrix} \cdot {R_{Ag}}^{-1} \cdot \begin{pmatrix} \dfrac{x-M_{XAg_i}}{\sigma_{X_i}} \\ \dfrac{y-M_{YAg_j}}{\sigma_{YAg_j}} \end{pmatrix} \right] \right] - \\
- \left[\ln\left[\dfrac{1}{(2\pi)\cdot\left[\left(\left|R\right|\right)^{0.5}\right]} \right] + \dfrac{-1}{2}\cdot\left[\begin{pmatrix} \dfrac{x-M_{X_i}}{\sigma_{X_i}} & \dfrac{y-M_{Y_j}}{\sigma_{Y_j}} \end{pmatrix} \cdot R^{-1} \cdot \begin{pmatrix} \dfrac{x-M_{X_i}}{\sigma_{X_i}} \\ \dfrac{y-M_{Y_j}}{\sigma_{Y_j}} \end{pmatrix} \right] \right] \\
g(x,y) := \dfrac{d}{dx}\left[\dfrac{-1}{2}\cdot\left[\begin{pmatrix} \dfrac{x-M_{XAg_i}}{\sigma_{XAg_i}} & \dfrac{x-M_{YAg_j}}{\sigma_{YAg_j}} \end{pmatrix} \cdot {R_{Ag}}^{-1} \cdot \begin{pmatrix} \dfrac{x-M_{XAg_i}}{\sigma_{XAg_i}} \\ \dfrac{x-M_{YAg_j}}{\sigma_{YAg_j}} \end{pmatrix} \right] \right] \cdot \\
\cdot \left[\dfrac{d}{dy}\left[\dfrac{-1}{2}\cdot\left[\begin{pmatrix} \dfrac{x-M_{X_i}}{\sigma_{X_i}} & \dfrac{x-M_{Y_j}}{\sigma_{Y_j}} \end{pmatrix} \cdot R^{-1} \cdot \begin{pmatrix} \dfrac{x-M_{X_i}}{\sigma_{X_i}} \\ \dfrac{x-M_{X_i}}{\sigma_{X_i}} \end{pmatrix} \right] \right] \right] - \\
- \left[\dfrac{d}{dx}\left[\dfrac{-1}{2}\cdot\left[\begin{pmatrix} \dfrac{x-M_{X_i}}{\sigma_{X_i}} & \dfrac{x-M_{Y_j}}{\sigma_{Y_j}} \end{pmatrix} \cdot R^{-1} \cdot \begin{pmatrix} \dfrac{x-M_{X_i}}{\sigma_{X_i}} \\ \dfrac{x-M_{X_i}}{\sigma_{X_i}} \end{pmatrix} \right] \right] \right] \cdot \\
\cdot \dfrac{d}{dy}\left[\dfrac{-1}{2}\cdot\left[\begin{pmatrix} \dfrac{x-M_{XAg_i}}{\sigma_{XAg_i}} & \dfrac{x-M_{YAg_j}}{\sigma_{YAg_j}} \end{pmatrix} \cdot {R_{Ag}}^{-1} \cdot \begin{pmatrix} \dfrac{x-M_{XAg_i}}{\sigma_{XAg_i}} \\ \dfrac{x-M_{XAg_i}}{\sigma_{XAg_i}} \end{pmatrix} \right] \right]
\end{cases}
\tag{2}
$$

The following notation was used: rXY, rXYAg are correlation matrices of the Raman polarization spectra of polyester fibers without silver nanoparticles and silver nanoparticles, respectively; M_X, M_{XAg} are the mathematical expectation vectors of peak intensities of spectra with polarization across X fibers without silver nanoparticles and silver nanoparticles, respectively; M_Y, M_{YAg} are the mathematical expectation vector of peak intensities of spectra with polarization along Y-fibers without silver nanoparticles and with silver nanoparticles, respectively; σ_X, σ_{XAg} are the root-mean-square deviations of the peak intensities of the spectra with polarization across the X-fibers without silver nanoparticles and with silver nanoparticles, respectively; σ_Y, σ_{YAg} are the root-mean-square deviations of the peak intensities of the spectra with polarization along the Y-fibers without silver nanoparticles and with silver nanoparticles, respectively; i, j are the numbers of the peaks of the spectrograms, the suppression of which is studied. Moreover, in the system of equations (2), the expression g(x, y) is the Jacobian double derivative (with respect to dx and with respect to dy), and there is no differentiation in the expression f(x, y). The analytical expression of the differential correlation equation g(x, y) is obtained by differentiating the equation f(x, y) with respect to X and Y. In this case, the solution occurs automatically without manually selecting the values of the parameter p.

3. Results

To solve the system of equations (2) using vector-matrix analysis, an algorithm was developed that was implemented using the built-in functions of the MathCad Enterprise Edition 15:

Given

$$f(x,y) = 0$$

$$g(x,y) = 0$$

$$v := \mathrm{Find}(x,y), \tag{3}$$

where v is the coordinate vector of the intersection of distribution ellipses with and without nanoparticles.

At the next stage, a software implementation of the solution of the system of equations (2) was carried out. So, when examining the intersection of 4 and 1 peaks:

$$v_{(4-1)} = \begin{pmatrix} 398.439542 \\ 371.170082 \end{pmatrix}$$

$$f_{4-1}\left(v_{(4-1)_0}, v_{(4-1)_1}\right) = -2.6645352591003757 \times 10^{-15}$$

$$g_{4-1}\left(v_{(4-1)_0}, v_{(4-1)_1}\right) = 1.0842021724855044 \times 10^{-17}$$

To determine the recognition accuracy of nanoparticles, expressions (4–5) were proposed, where p is the identification accuracy for the equation without nanoparticles and p_{Ag} is the identification accuracy for the equation with silver nanoparticles.

$$p_{Ag} := \frac{1}{(2\cdot\pi)\cdot\left[\left(\left|R_{Ag}\right|\right)^{0.5}\right]}\cdot e^{\frac{-1}{2}\left[\left(\frac{v_0 - M_{XAg_i}}{\sigma_{XAg_i}} \quad \frac{v_1 - M_{YAg_j}}{\sigma_{YAg_j}}\right)\cdot R_{Ag}^{-1}\cdot\begin{pmatrix}\frac{v_0 - M_{XAg_i}}{\sigma_{XAg_i}} \\ \frac{v_1 - M_{YAg_j}}{\sigma_{YAg_j}}\end{pmatrix}\right]} \tag{4}$$

$$p := \frac{1}{(2\cdot\pi)\cdot\left[\left(\left|R\right|\right)^{0.5}\right]}\cdot e^{\frac{-1}{2}\left[\left(\frac{v_0 - M_{X_i}}{\sigma_{X_i}} \quad \frac{v_1 - M_{Y_j}}{\sigma_{Y_j}}\right)\cdot R^{-1}\cdot\begin{pmatrix}\frac{v_0 - M_{X_i}}{\sigma_{X_i}} \\ \frac{v_1 - M_{Y_j}}{\sigma_{Y_j}}\end{pmatrix}\right]} \tag{5}$$

Reliability values p and p_{Ag} calculated from expressions (4–5) under the condition of simultaneous differentiation with respect to X and Y for the study of the intersection of peaks 4 and 1:

$$p_{Ag_{(4-1)}} = 0.00082199604570955$$

$$p_{(4-1)} = 0.00082199604570954$$

The analysis of the obtained values shows that p and p_{Ag} for the corresponding peaks are almost equal.

A graphical estimate of the intersection of distribution ellipses for various differentiation options is shown in Figures 1 and 2. If we solve the usual system of equations for the same peaks of the Raman spectrum, then the probability density will be:

$$p_{(4-1)}=0.00082188904657987$$

$$p_{Ag(4-1)}=0.0008221540489625.$$

When solving a system of equations with differentiation with respect to X, we obtain for the same peaks of the Raman spectrum the probability density:

IOP Conf. Series: Materials Science and Engineering **905** (2020) 012014 doi:10.1088/1757-899X/905/1/012014

$$p_{(4-1)}=0.00082199573528761$$

$$p_{Ag(4-1)}=0.00082199573528757.$$

By solving the system of equations with differentiation with respect to Y, we obtain the probability density for the peaks of the Raman spectrum:

$$p_{(4-1)}= 0.00082199573106$$

$$p_{Ag(4-1)}= 0.0008219957310$$

Based on the solution of the system of equations (2), it is possible to determine the radii of the ellipses of the distribution of the radiation intensity of the Raman spectra from expressions 6–7. In this case, R is the radius of the ellipse for Raman spectra without nanoparticles; R_{Ag} is the ellipse radius for Raman spectra with silver nanoparticles.

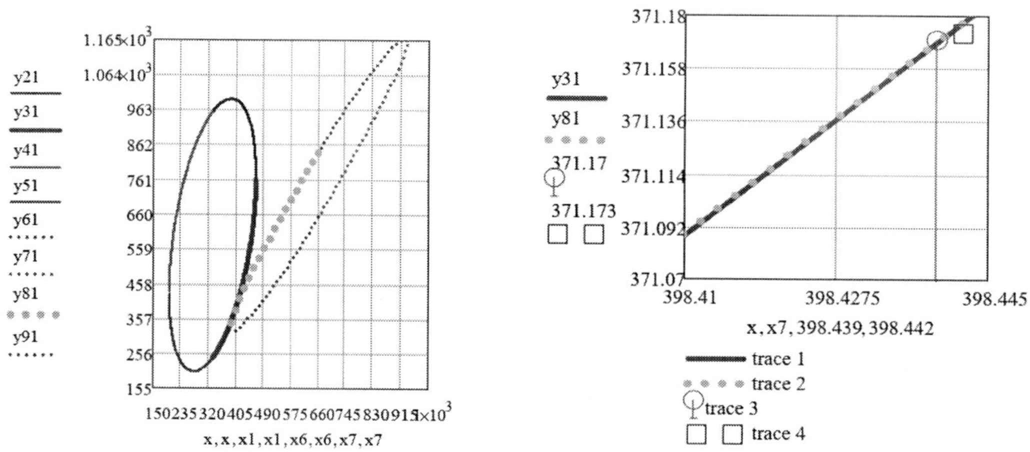

a) b)

Figure 1. A graphical estimate of the intersection of the ellipses of the distribution of Raman spectra and the results of solving the system of equations for identifying nanoparticles with polarization Y-along and X-across the fibers: a – general view; b – differentiation with respect to XY – ○ and general solution of the system of equations with respect to XY – □

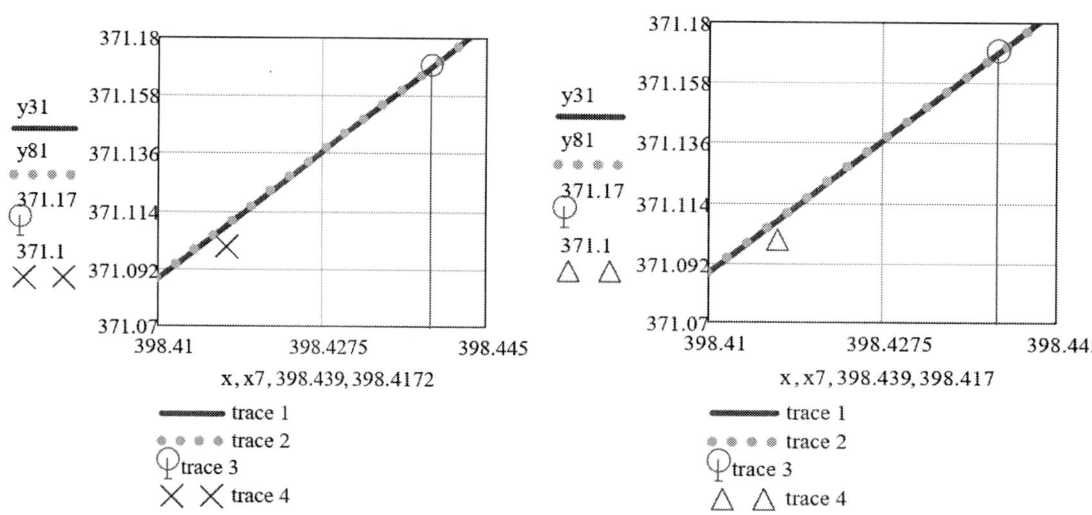

a) b)

Figure 2. Graphical estimate of the intersection of the ellipses of the distribution of Raman spectra and the results of solving the system of equations for identifying nanoparticles with polarization Y-along and X-across the fibers: a – differentiation with respect to XY – ○ and along X – ×; b – differentiation with respect to XY – ○ and with respect to Y – Δ

$$R_{Ag} := \left[\left(\frac{v_0 - M_{XAg_i}}{\sigma_{XAg_i}} \quad \frac{v_1 - M_{YAg_j}}{\sigma_{YAg_j}} \right) \cdot R_{Ag}^{-1} \cdot \left(\begin{array}{c} \frac{v_0 - M_{XAg_i}}{\sigma_{XAg_i}} \\ \frac{v_1 - M_{YAg_j}}{\sigma_{YAg_j}} \end{array} \right) \right] \cdot \tag{6}$$

$$R := \left[\left(\frac{v_0 - M_{X_i}}{\sigma_{X_i}} \quad \frac{v_1 - M_{Y_j}}{\sigma_{Y_j}} \right) \cdot R^{-1} \cdot \left(\begin{array}{c} \frac{v_0 - M_{X_i}}{\sigma_{X_i}} \\ \frac{v_1 - M_{Y_j}}{\sigma_{Y_j}} \end{array} \right) \right] \cdot \tag{7}$$

The values of R_{Ag} and R for the distribution ellipses of the Raman spectra at the intersection of 4 and 1 peaks resulted in the following:

$$R_{Ag_{(4-1)}} = 3.680765923137155$$

$$R_{(4-1)} = 3.275528794713935$$

4. Conclusion

It should be noted that when solving the system of differential equations (2), high accuracy was revealed in the results with automatic numerical solution in a large range of initial parameters $x = 50 \div 6000$ and $y = 50 \div 5000$ with stable solution parameters.

The solution on the values of the coordinates of the intersection of distribution ellipses $X = 398.4395417708532$, $Y = 371.1700822350267$ has high accuracy and does not depend on the initial value of the parameters x and y in the specified range.

The practically required accuracy of obtaining the results depends on the necessary resolution (sensitivity) of the proposed method for identifying silver nanoparticles.

In this paper, the accuracy of the solution by the equation $f(v_0, v_1)= -4.440892098500626\cdot 10^{-15}$ and by the equation $g(v_0, v_1)= -3.9031278209478155\cdot 10^{-18}$ for the initial given parameters $x = 50$, $y = 400$ is revealed. With the initial given parameters $x = 6000$, $y = 5000$, the accuracy of the solution is according to the equation: $f(v_0, v_1)= -0.00000000000002398$ and according to the equation $g(v_0, v_1)= 2.6020852139652106\cdot 10^{-17}$.

The accuracy of obtaining the output parameters: $p = 0.00082199604570954$ and $p_{Ag} = 0.00082199604570955$ can be estimated up to 10^{-16}. Higher accuracy cannot be given by any of the investigated methods.

The objectives of further research are as follows. construct and solve a system of differential equations when differentiating along X and Y when compiling a system with the number of unknown parameters more than two, evaluate the increase in the reliability of increasing the sensitivity of identification of silver nanoparticles on polyester fibers using multidimensional differentiation by X and Y simultaneously with the study of two or more peaks spectrum.

References

[1] Hajipour J M, Fromm M K, Ashkarran A A et al October 2012 Antibacterial properties of nanoparticles *Trendsin Biotechnol.* 30(10) 499–511

[2] Carina I C C and Barrosb M T November 2017 Polymeric nanoparticles: A study on the preparation variables and characterization methods *Mater. Sci. and Engineer. C* 80 771–84

[3] Gajanan K and Tijare S N 2018 Applications of nanomaterials *Materialstoiday: proc.* 5(1) part 1 1093–6

[4] Kobrakov K I, Kuznetsov D N, Rodionov V I et al 2017 Influence of silver nanoparticles inclusions in textile materials on their thermophysical properties *Izv. Vysshikh Uchebnykh Zavedenii Ser. Teknol. Tekstil. Promyshl.* 371(5) 52–6

[5] Khammatova V V 2016 Manufacture of experimental samples of nanomodified textile materials affect the strength of special-purpose clothing *Izv. Vysshikh Uchebnykh Zavedenii Ser. Teknol. Tekstil. Promyshl.* 362(2) 59–65

[6] A Ya Kolpakov et al 2019 Nanostructured coatings based on amorphous carbon and gold nanoparticles obtained by the pulsed vacuum-arc method J. of Nano- and Electr. Phisics. 11(4) 04019-1–04019-7

[7] Khammatova V V and Razumeyev K E 2016 Researches of the microstructure of samples of the nanomodified textile materials for special clothes by microscopy methods *Izv. Vysshikh Uchebnykh Zavedenii Ser. Teknol. Tekstil. Promyshl.* 5(365) 84–9

[8] Kovalenko A V, Plakhtii Ye G and Khmelenko O V 2019 Research of Photoluminescence Spectra of ZnSXSe1 – X:Mn Nanocrystals Obtained by Method of Self-propagating High-temperature Synthesis *J. of Nano- and Electr. Phisics* 11(4) 04031-1–04031-5.

[9] Emelyanov V M, Dobrovolskaya T A and Emelyanov V V 2018 Multidimensional system of equations with XY-Y differentiation of probability densities p0-p1 for identification of gold nanoparticles *Biointerface Res. in Appl. Chem.* 8**(6)** 3652–6

[10] Emelyanov V M, Dobrovolskaya T A and Emelyanov V V 2017 Computer simulation of Raman spectra of textile materials when silver and gold nanoparticles are applied to them *Fund. Study* 10(1) 7–12

IOP Conf. Series: Materials Science and Engineering **905** (2020) 012015 doi:10.1088/1757-899X/905/1/012015

Morphodynamic conditions of particle size distribution of beach sediments of Coastal Area of Karadag Natural Reserve

A V Drygval[1], E V Stanis[2], P V Drygval[2], E A Latushkina[2] and N S Buldovich[2]

[1] Geomatics Research Center, A.O. Kovalevsky Institute of Biology of the Southern Seas of RAS, Sevastopol, Russia
[2] Department of Geoecology, Peoples Friendship University of Russia (RUDN University), Moscow, Russia

E-mail: drygval95@mail.ru, estanis@mail.ru, any-poly@mail.ru, elena_latushkina@mail.ru, nbuldovich91@gmail.com

Abstract. The article presents the results of studying particle size distribution of sediments in the beach areas of the Karadag coastal area of the Crimean peninsula. In the framework of this study, this area is considered as a lithodynamic system. The paper describes the connection between the granulometric structure of sediments and external processes that affect this system, which is an important indicator of response of the condition of a coastline to dynamic environmental conditions. It is established that the wave processes, exogenous geological processes (mainly of a gravitational nature), abrasion of cliff and bench result in the transportation and accumulation of beach sediments along the coast, as well as determine the peculiarities of their distribution according to particle size distribution. Based on the particle size analysis, the differentiation of beach sediments along the coastline of Karadag is revealed. When calculating the static parameters of the particle size data, the morphodynamic conditions of distribution of sediments in the coastal area are characterized.

1. Introduction

The coastal area of the Karadag coast (the zone of contact between land, sea and atmosphere) with a total length of indentation of the coastline of 8 is an active lithodynamic system. Any activity that takes place here requires a complete study of the current state of the coast, its trends and dynamics in order to predict changes in the coastline.

A number of works with a very wide geography are devoted to the study of the modern lithodynamic state of sediments of seacoasts and rivers. For example, such studies are conducted in Ghana (the Volta River) [11], in New Zealand in the Canterbury Bay [10], as well as on the Pacific coast near the city of Timaru [5]. The distribution features of beach material, both sandy and pebbled, are studied in China (Beibu Bay beaches) [14] and in Canada (the Bay of Fundy) [16]. The relationship between distribution of sediment size, their volumes and the study of beach erosion is carried out in the USA (Plum Island and the Merrimak River) [9]. It should be noted that during these studies, particular attention is paid to the specific features of the redistribution of solid sediments of the coasts by particle size.

Domestically the study of the lithodynamic processes of seacoasts was carried o t by such researchers as G.A. Safyanov [15], P.K. Bozhich [4], V.P. Zenkovich [17], N.A. Belov [2], whose works were devoted to the issues of geomorphology of seacoasts and particle size differentiation of sediments. The features of modern lithodynamics, particularly in the coastal area of the Black Sea, are reflected in the works of A.A. Klyukina [12], I.V. Agarkova-Lyakh [1], A.S. Boguslavsky et al. [3].

2. Methods and materials

As part of the study of differentiation processes of sediments on the beaches of the Karadag coastal area in the summer of 2018, sampling of solid sediment material was carried out, particle size analysis was made, as well as statistical processing with the main results represented in graphs.

Figure 1. Points of beach sediment profiles in the bays of the coastline of the Karadag natural reserve

The main object of the study is the deposits of the beach area, represented by coarse and sandy material.

The sediment material was collected in the beach areas of 15 bays: Karadag Bay, bay near the Kuzmichev Kamen cliff, Razboinichya Bay, Putstsolanovaya Bay, Pogranichnaya Bay (western part), Lvinaya Bay, Barakhta Bay, Yuzhnaya Serdolikovaya Bay, Srednyaya Serdolikovaya Bay, Severnaya Serdolovaya Bay, Livadia Bay, Graviinaya Bay, Lyagushachya Bay, Cape Malchin Bay, Koktebel Bay. In all the bays, the particle size analysis was carried out along the transverse profile in the widest part of the beach. However, due to the development of an extended beach area, the number of transverse profiles in the Karadag and Koktebel Bays amounts to 2 and 4, respectively. The sampling of solid material of beach sediments, represented by coarse and sandy material, was carried out along 19 profiles at certain points with the range from 2 to 3 meters, depending on the width of the beach. The morphological parts of it were also covered, i.e. the wave run-up area, coastal shafts, the shallow surface of the beach and the area at the foot of the adjacent slope.

At the selected points, the soil samples were collected from the surface of the beach sediments to a depth of 30 cm. A at the interval of 2 or 3 meters cm. A coarse material was measured using a caliper. The size of the sand fractions was determined by the sieve method in accordance with the current GOST 25100-2011 standard – "Soils. Classification "[8]

In order to process and interpret the results of the obtained data, the following statistical particle-size parameters were calculated for each profile: roof-mean-square or standard deviation S_0, (used as sorting factor); average or median diameter (median) M_d; kurtosis (E_k) and graphic average diameter (GAD) [7].

1) Sorting factor S_0 is calculated using the following formula:

$$S_0 = \sqrt{\frac{Q3}{Q1}},$$

where Q1 and Q3 are the first (smaller) and the third (large) quartiles in the graph of the cumulative curve of material distribution of beach sediment by size.

The first quartile is such a grain size (i.e., such an abscissa) with respect to which 25 % of the grains are smaller and 75 % are larger; the third quartile, on the contrary, is a size relative to which 25 % of the particles are larger and 75 % are smaller. The quartiles on the distribution graph are determined by drawing from the ordinates corresponding to the content of 25 and 75 % of the data, the normal lines on the abscissa axis, which corresponds to the particle size and describes the quartiles sought [6]

The smaller the difference in grain sizes, the closer the quartiles are to each other and to the average diameter M_d, the closer S_0 to the unit, i.e. the better the sorting. In a well-sorted material $S_0<2$; in a medium-sorted material $S_0 = 2–3$; in a poorly sorted material $S_0> 3$ [6].

The degree of sorting indicates the duration of the dynamic processing of material, during which the beach sediments of the clastic material tend to become homogeneous (or monofractional). If you look at the Gaussian distribution curve of such a material, then it would be more compressed and higher. Conversely, a bi- or polymodal distribution curve will indicate a mixture of material from different sources or a lack of dynamic environmental forces necessary for processing and sorting.

2) The average, or median, diameter M_d (mm) is the size relative to which half of the particles are smaller and the other half is larger (this is the second quartile Q2). This is an important indicator of the particle size structure of the rock and a genetic trait indicating the strength or speed of the current that has deposited this material. It is most informative for relatively well-sorted rocks. By the size of the debris (in particular, by M_d value), one can navigate in the strength or velocity of the current that processes the beach material [6].

3) The graphic average diameter – GAD (mm) – is determined not by one point (50 % of Q2 ordinate), but by the biggest part of the particle size spectrum. This indicator is used for classification purposes when evaluating particle size distribution in a sample:

$$\mathrm{GAD} = \frac{Q4+Md+Q5}{3},$$

where Q4 and Q5 are the particle sizes, with respect to which 16 and 84 % of the particles in the sample are smaller.

4) Kurtosis (E_k) is a measure of peaks or "steepness" of the distribution curve:

$$\mathrm{E}_k = \frac{Q3-Q1}{2(P90-P10)},$$

where P_{90} and P_{10} are percentiles (or deciles), i.e. the diameters of the grains by which the tails of the total curve are determined with $P_{90}>P_{10}$. Moreover, there is the following gradation: <0.67 – strongly flat peak distribution curves; 0.67–0.90 – flat topped; 0.90–1.11 – normal (correspond to the normal distribution); 1.11–1.50 – weakly compressed (weakly pointed); 1.50–3.0 – medium (or highly) compressed (mid-peak) and> 3.0 – extremely compressed (highly peaked) [6]. High variability of values indicates that the modal and related classes are characterized by a high percentage of the material.

In accordance with the excess in statistics [13], the normal distribution has a value of the indicator equal to 0 (or tending to it), positive values characterize the peaks of the distribution curve, and negative values indicate the flatness of the curve. Linking this information with the kurtosis interpretation indices during particle size analysis, one may conclude that the calculated kurtosis values > 1.0 indicate stable active hydrodynamics, the stability of processing and re-sorting of clastic material, or the effect of active forces on the system for a long time. A significant positive excess indicates that strong hydrodynamic processes acted in a short period of time, or weak processes acted for a long period, and also the speed of dynamic processing (sorting) of the brought clastic material exceeded the intensity of its receipt. The kurtosis value that is <1.0 indicates the instability of external environmental forces (wave-breaking process within the lithodynamic system) or the lack of activity, as well as massive, often avalanche, flow of debris material that is not able to process dynamically the

IOP Conf. Series: Materials Science and Engineering **905** (2020) 012015 doi:10.1088/1757-899X/905/1/012015

forces of the medium. For example, the bringing of clastic material and its deposition in dynamically stagnant conditions [6].

The final diagrams of the calculated particle size distribution were compiled in ArcGis 10.5. The numbers on the diagrams correspond to the following transverse profiles of bays:

1) Karadag Bay (profile1);
2) Karadag Bay (profile 2);
3) Bay near the Kuzmichev Kamen cliff;
4) Razboinichya Bay;
5) Putstsolanovaya Bay;
6) Pogranichnaya Bay;
7) Lvinaya Bay;
8) Barakhta Bay;
9) Yuzhnaya Serdolikovaya Bay;
10) Srednyaya Serdolikovaya Bay;
11) Severnaya Serdolovaya Bay;
12) Livadia Bay;
13) Graviinaya Bay;
14) Lyagushachya Bay;
15) Cape Malchin Bay;
16) Koktebel Bay (профиль 1);
17) Koktebel Bay (profile1 2);
18) Koktebel Bay (profile1 3);
19) Koktebel Bay (profile1 4).

3. Results

The analysis of the particle size composition of beach sediments makes it possible to establish a number of patterns of their distribution along the entire coastline of the Karadag reserve. The material in the beach sediments under the influence of the wave-breaking process is distinguished differently. The decrease in particle size in the northeastern part of the coast is explained by more yielding rocks and an actively eroded cliff. On the contrary, sediment enlargement is the source of terrigenous material and is observed on coastal areas composed of more solid rocks, as well as in places where the impact of disturbance is limited

To analyze the coastal variability of statistical parameters of the grain size distribution of sediments, distribution diagrams of sorting values (S_0), median (Md), and graphic average diameter (GAD) were compiled.

When analyzing the distribution of sorting values (S_0), it was found that well-sorted beach sediments prevail within the considered area. The sorted sediments are found on the periphery of the coastal area, usually at the foot of a slope or in the center of the beach. Poorly sorted material is present in the Karadag Bay, in the upper part of the beach. This indicates the inflow of foreign material associated with active exogenous geological processes on the front of the coastal ledge, as well as with the heterogeneous composition of the flyschoid strata

Having carried out the analysis of the median particle size along the coast it was revealed that larger sediment particles are located in the northeastern part of the coastline, starting from the central part of the beach to the sea edge (Bay at Cape Malchin, Koktebel Bay profile 1 and profile 2). In the eastern and western parts of the coastline, the median size decreases sharply. The consolidation of sediments was noted in the bay near the Kuzmichev Kamen cliff at the beginning and at the end of the transverse profile (at the point at the foot of the adjacent slope and in the zone of the wave run, respectively), as well as at the edge of the sea in Putstsolanovaya Bay. The consolidation of sediments can be explained by the local, avalanche-like inflow of coarse clastic material and the composition of the rocks.

When passing to the central part of the beaches of the coastline, the redistribution of sediment material occurs in such a way that their median size is relatively uniform, and is not represented by large particles (sizes to small pebbles).

When analyzing the distribution of the values of the graphic average diameter, we can conclude that the central part of the coastal section under consideration is characterized by a minimal coarseness of deposits. In the eastern part of the coastal area, sediment consolidation is observed in the bay near the Kuzmichev Kamen cliff in the upper and lower parts of the beach. In this transverse profile, landslide processes are observed at the adjacent slope. In the eastern part of the coast, the GAD indicator has the maximum value in the coastal section under consideration.

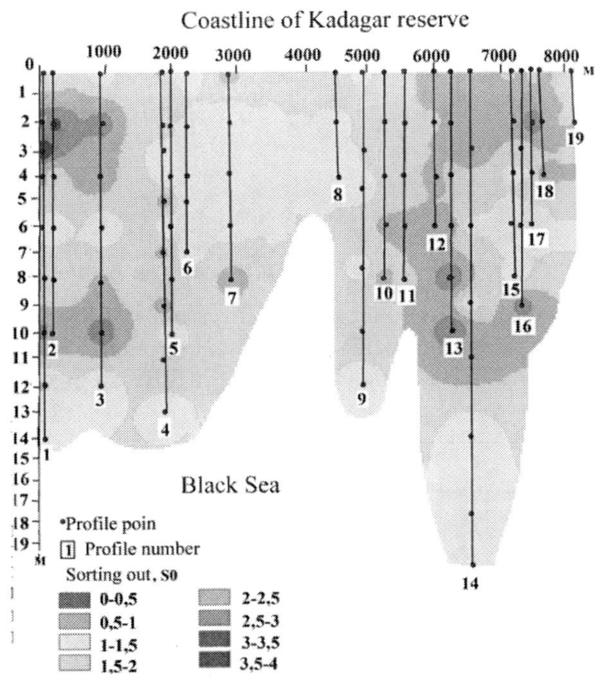

Figure 2. Distribution of sorting values (S_0) of beach sediments of the Karadag coastal area. *Notes*: • – sediment sampling point; $S_0<2$ – well-sorted material; $2<S_0<3$ – medium sorted material; $S_0>3$ – poorly sorted material. Horizontal direction is a distance along the coastline, m; vertical direction is a distance perpendicular to the coastline (beach profile), m

The enlargement of sediments is also explained by the local inflow of coarse clastic material and is observed in the areas where the influence of waves is limited. What is observed here is an active inflow of coarse clastic material into the lithodynamic system due to the active abrasion activity influencing the yielding rocks. In general, one may say that the beach sediments of the Karadag coastal area are composed of small and medium pebbles.

The kurtosis value along the entire coastline of the Karadag reserve (<1.0) indicates the instability of external environmental forces (wave-surf process within the lithodynamic system) or their weakness., as well as massive, often avalanche inflow of clastic material, which can not be processed by the external forces of lithodynamic system due to casting and deposition of clastic material into dynamically stagnant zones, where the activity of the beach's lithodynamic system is minimal.

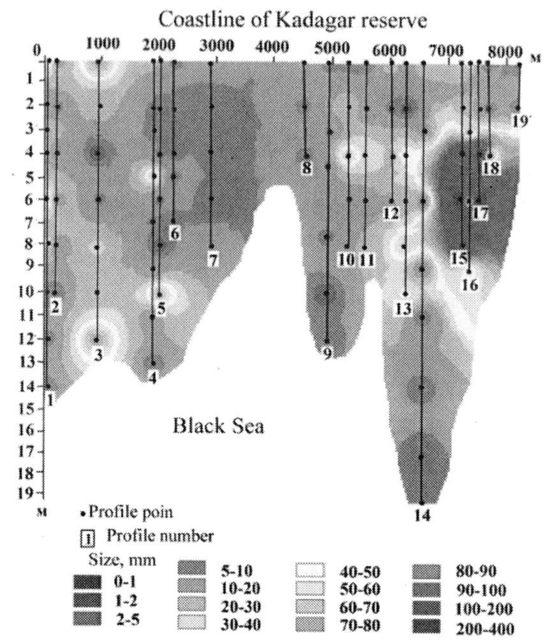

Figure 3. Distribution of the values of median particle size (mm) of beach sediments of the Karadag coastal area. *Notes*: • – sediment sampling point. Horizontal direction is a distance along the coastline, m; vertical direction is a distance perpendicular to the coastline (beach profile), m

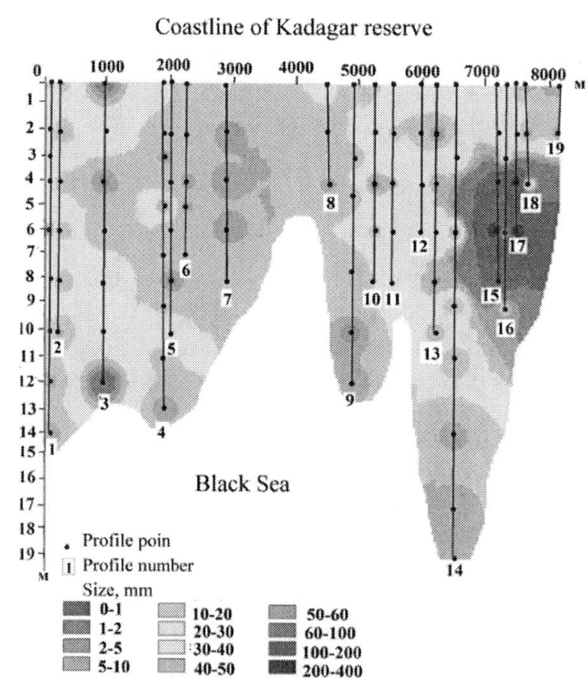

Figure 4. Distribution of values of graphic average particle diameter (mm) of beach sediments of the Karadag coastal area. *Notes*: • – sediment sampling point. Horizontal direction is a distance along the coastline, m; vertical direction is a distance perpendicular to the coastline (beach profile), m

4. Conclusion

There is a connection between the particle size differentiation of beach sediments and the dynamic processes effecting the system (beach). Testing the material of beach sediments and revealing spatial variability of the particle size structure gives general characteristic of various processes that manifest themselves within the lithodynamic system. When conducting particle size analysis and identifying sediment differentiation along the coast, some conclusions can be drawn about the lithodynamic processes of coastal zone functioning.

In general, sediments in the bays of the Karadag coast are mainly well-sorted. Sorted sediments are present on the periphery of the coastal area, and, as a rule, at the foot of the slope or in the center of the beach. Poorly sorted material is present in the Karadag Bay, in the upper part of the beach, which indicates the inflow of foreign material into this area. The heterogeneous material of small-plate structures entering the beach area as a result of exogenous processes on the frontal part of the coastal ledge is associated with geological structure represented by the flyschoid strata.

It was revealed that, according to the median particle size, larger sediment sizes are located in the northeastern part of the coastline, starting from the central part of the beach to the sea edge (Bay at Cape Malchin, Koktebel Bay profile 1 and profile 2). In the eastern and western parts of the coastline, the median size decreases sharply. The enlargement of sediments was noted in the bay near the Kuzmichev Kamen cliff at the beginning and at the end of the transverse profile (at the point at the foot of the adjacent slope and at the edge of the sea, respectively), as well as at the point near the sea of the Putstsolanovaya Bay. The enlargement of sediments is explained by the local inflow of coarse clastic material and is observed in the areas where the impact of waves is limited.

Thus, it was found that in the particle size composition of beach sediments at the foot of the slope along the entire coast of Karadag dominated by medium-size and small-size pebbles, except for the southwestern part of the coast where large pebbles prevail. In the last 2 profiles of the Koktebel Bay there is a sharp decrease in the size of sediments to fine gravel. In the northeastern part of the coast, the size of sediment particles also sharply decreases compared with the central bays – pebbles and fine sand.

Pebbles prevail in the center of the beach area of the coast, with the exception of some bays. In the bay at the Kuzmichev Kamen cliff in the center of the beach as a part of sediment coarse sand prevails. In the eastern part of the coast, there is a tendency for increased size of sediments material up to small and medium boulders (Bay at Cape Malchin, Koktebel Bay, profile 2). In the East, the prevailing sediment size is reduced to pebbles and fine sand.

Along the entire coast the deposits of the beach area are represented by pebbles of various sizes in the bays where the 1st coastal ramp stands out, with the exception of Yuzhnaya Serdolikovaya Bay, where the 1st coastal rampart consists mainly of fine gravel. In the zone of the wave run-up, the composition of beach sediments includes mainly medium and small pebbles, with the rare exception of some bays. In the Putstsolanovaya Bay fine gravel prevails and in the north-eastern part of the coast there are medium boulders in the bay.

The decrease in particle size distribution in the northeastern part of the coast can be explained by less durable types of sediment rocks and an actively eroded cliff.

Throughout the coastline of the Karadag reserve, instability or inadequate activity of external forces acting on the coastal area was observed in combination with the massive inflow of clastic material due to exogenous geological processes in the coastal area, as well as the inflow of clastic material into dynamically inactive areas.

Acknowledgment

The work is a result of the following state order: Study of spatial and temporal organization of water and land ecosystems with the aim of developing an on-line monitoring system based on remote sensing data and GIS technologies. The registration number is AAAA-A19-119061190081-9.

References

[1] Agarkova-Lyah I V 2015 Natural reserves of shoreline areaof the South coast of Crimea Sci. notes of Taurida national Univer. V I Vernadsky Geogr. Geol. **1(67)** 42–58

[2] Belov N A 1938 About the movement of pebblesin Lemenskaya Bay *Sci. notes of Moscow State Univer. Geogr.* 19 249–69

[3] Boguslavsky A S 2016 Formation of pebble beaches of costal area of mountain Crimea Environmental safety of coastal and offshore areas of the sea pp 47–55

[4] Bozhich P K 1930 About movement of sea sediments (Moscow) 96 p

[5] Chen B and Stephenson W November 2015 Measuring pebble abrasion on a mixed sand and gravel beach using abrasion baskets *Geomorphol.* 248 24–32

[6] Frolov V T 1993 Litologiya vol 2 (Moscow: MSU) 432 p

[7] Frolov V T 1964 Guideline for laboratory studies on petrography of sedimentary rocks (Moscow: MSU) 310 p

[8] GOST 25100-2011 *Grunty. Classification*

[9] Hein C J., Fallon A R., Rosen P et al May 2019 *Shoreline Dynamics Along a Developed River Mouth Barrier Island: Multi-Decadal Cycles of Erosion and Event-Driven Mitigation* vol 7 103 p

[10] Hemmingsen M, Eikaas H and Marsden D September 2019 A GIS approach to sediment displacement in mixed sand and gravel beach environments J. of Environ. Manag. **249(109083)**

[11] Jayson-Quashigaha P, Addo K, Amisigo B and Wiafe G September 2019 Assessment of short-term beach sediment change in the Volta Delta coast in Ghana using data from Unmanned Aerial Vehicles (Drone) Ocean & Coastal Management in press

[12] Klyukin A A 2004 Balansed sedimentation in the coastal area of the Black Sea near Karadaga Sci. notes of Taurida national Univer.V.I. Vernadsky **17(56)** 82–90

[13] Eleseeva I I *Momenty raspredeleniya i pokazateli ego formy* Retrieved from: http://www.stathelp.ru/ots/g5p9.html.

[14] Pang W, Dai Z, Ge Z et al February 2019 Near-bed cross-shore suspended sediment transport over a meso-macro tidal beach under varied wave conditions *Estuar., Coastal and Shelf Sci.* 217 69–80

[15] Safyanov G A and Kovalenko M A 2015 Morphodynamic conditions of particle size differenciation of sediments of coastal line of Tersky Coast oft he White Sea *Geomorfolo.* 2 88–9

[16] Stark N and Hay A E December 2016 Pebble and cobble transport on a steep, mega-tidal, mixed sand and gravel beach *Marine Geol.* 382 210–23

[17] Zenkovich V P 1958 *Morphology and dynamics of soviet seashores of the Black Sea* ed O K Leon'ev (Moscow: AN USSR) 188 p

IOP Conf. Series: Materials Science and Engineering **905** (2020) 012016 doi:10.1088/1757-899X/905/1/012016

Advanced method for measuring slurry surface tension

R Dzhambulatov[1], V Savvin[2], D Bataev[1], A Khasanov[3], R Uspazhiev[4,1], Z Khasbulatova[3] and A Bagov[5]

[1] Kh.I. Ibragimov Complex Research Institute, Russian Academy of Sciences, Grozny, Russia
[2]National Research Nuclear University MEPhI, Institute of Nuclear Physics and Engineering, Obninsk, Russia
[3]Chechen State University, Grozny, Russia
[4] Grozny State Oil Technical University named after M.D. Millioshikov. Grozny, Russia
[5] Kabardino-Balkarian State University, Nalchik, Russia

E-mail: asldzam@mail.ru, Savvin-vs@yandex.ru, denabataev61@mail.ru, aslan2001@rambler.ru, russlan-chr@mail.ru, hasbulatova@list.ru, vegros@rambler.ru

Abstract. The work is devoted to a method that makes it possible to level the influence of sedimentation on the surface tension of slurry. To do so the methods for measuring surface tension are used, where the sedimentation process is multidirectional relative to the surface under study. An analysis of the results shows that the proposed method can virtually eliminate the influence of sedimentation on the surface tension of dispersed systems with low kinetic stability. Using the proposed method, the concentration dependence of the surface tension of aqueous slurries of bentonite in the range of 0–20 mass% of the solid phase is studied.

1. Introduction

Bentonite is a clay mineral used in various fields of human activity. The range of bentonite use is very wide and extends from the food industry to the strengthening of hydraulic structures and preparation of drilling mud.

The analysis of the state of research of finely dispersed bentonite shows that there is a considerable amount of experimental and theoretical information on the crystal structure and the influence of lattice defects on a number of practically important properties [1–5].

In many cases, the surface properties of bentonite slurries are used. The methods of physical and chemical analysis are aimed at measuring and studying the surface properties of homogeneous substances. The existing practical request to study the surface properties of disperse systems remains unsatisfied. In particular, there is no reliable method for measuring the surface tension (ST) of slurries, which allows one to take into account the influence of sedimentation of particles on ST. Therefore, the development of reliable methods and techniques for measuring the ST of suspensions remains an urgent task.

2. Theoretical part

When developing a method that allows one to obtain reproducible ST magnitude, it is necessary to take into account that bentonite slurries differ in their complex chemical composition and structure, which makes it difficult to study a number of practically important properties [6–8].

In suspension, the following interfaces can be distinguished: solid particle – liquid, solid particle – air, liquid – air. The liquid – air interface is called the surface of the suspension, and it is the object of our study.

Bentonite slurry is a polydisperse system in the process of formation of which sedimentation is observed. As a result, there is a change in the content of particles of the solid phase in the surface layer, which has a significant influence on the magnitude of ST suspension.

It is generally accepted that solid phase particles do not significantly affect the ST magnitude of bentonite slurry. Therefore, the features of bentonite slurries on ST isotherms are usually associated with the composition of liquid phase, the presence of excess charges on the surface [9], and the influence of electrolytes [10]. However, the available experimental results indicate a significant influence of particles of bentonite solid phase on the ST magnitude [11–12].

Based on the features of surface formation characteristic of each method, the methods for measuring the ST of slurry can be divided into two groups. The first group includes methods in which the process of measuring ST is accompanied by sedimentation of solid phase particles on the surface (hanging drop technique, stalagmometric method). Indeed, when using these methods, coarse particles contained in a drop of slurry move to the surface.

The second group includes methods in which particles of the solid phase, as a result of sedimentation, leave the surface, i.e. the surface is depleted over time by coarsely dispersed particles (lying-drop techniques, maximum pressure in a gas bubble or in a drop, capillary method of surface tension measurement, and so on).

Therefore, when using any of the above methods for measuring ST, the researcher will encounter problems caused by the dynamic nature of the composition of the surface layer, which leads to the dependence of ST on the time of formation of droplet surface. Therefore, before proceeding with the development of a methodology for measuring the ST of slurries, it is necessary to find out what this dependence is and how to take it into account when measuring the ST of slurry.

3. Methods and materials

Samples of slurry of bentonite were prepared in distilled water by gravimetric method. To achieve the optimum degree of dispersion and reduce the measurement accuracy caused by sedimentation, after thorough mechanical grinding, bentonite was transferred to a dispersion medium and was subjected to stirring for 1–1.5 hours (magnetic stirrer; slow heating). After that, the prepared slurry was placed into storage for a certain time. As experience shows, to achieve a sufficient degree of dispersion required for measuring the bentonite ST, two days of exposure of a solid phase in the dispersion medium at room temperature should be sufficient. The measurements were carried out using the methods of a lying and a hanging drop on DSA-100 Analyzer [13].

4. Results

Using these two methods, the time dependence of the ST of bentonite slurry was investigated. The measurement results are presented in Figure 1 and 2.

As is clear from these Figures, when measuring the ST by the lying-drop technique, an increase in ST is observed over time (Figure 1), and in the hanging drop technique (Figure 2), the ST decreases with time. With an increase in the concentration of the solid phase, the linear time dependence of the ST remains to be unchanged; however, the angular coefficient (positive for lying-drop technique and negative for the hanging drop technique) grows in absolute magnitude. Such a change in the magnitude of ST is due to the fact that in the lying-drop technique, the concentration of solid phase particles in the surface decreases, which leads to an increase in ST approaching the magnitude of ST of water. As a result of sedimentation, in the hanging drop technique the surface, on the contrary, is enriched with particles of the solid phase and a decrease in the magnitude of ST is observed.

There is a question of how to level the influence of sedimentation on the results of measuring the ST of slurry. How to extrapolate the obtained results to the results that could be obtained in zero gravity? The analysis and generalization of the results allowed us to develop the following methodology.

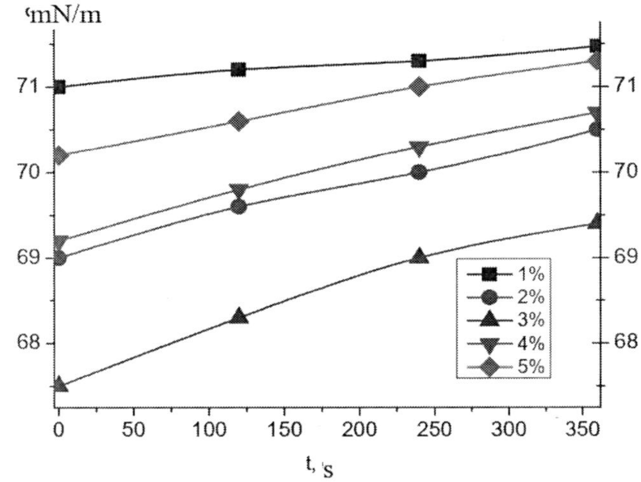

Figure 1. Time dependence of ST of bentonite slurry with different content of solid phase at 298 K (lying-drop technique)

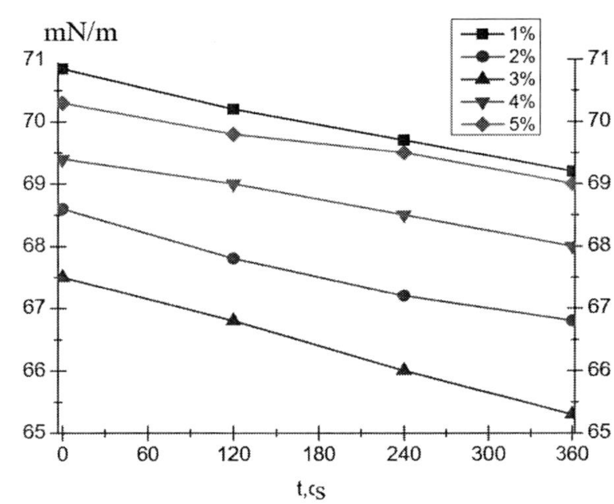

Figure 2. Time dependence of ST of bentonite slurry with different content of solid phase at 298 K (hanging drop technique)

The essence of the proposed methodology for measuring the ST of slurry is that the ST is measured in two ways, i.e. by the technique of hanging and a lying drop [14]. The process of sedimentation when using these two measurement techniques occurs in the opposite directions. In the first case, the surface of the droplet is enriched with particles of the solid phase and in the second it is depleted. As a result, the time dependences of ST in these two cases show opposite changes: when using the hanging drop technique, a linear (within the limits of measurement accuracy) decrease in ST is observed (Figure 2), the ST of the lying droplet increases linearly with time (Figure 1). Moreover, the time dependences of ST are the same in absolute magnitude.

The superposition of the ST results obtained using indicated measurement techniques, where the time dependences of ST are equal in absolute magnitude but are differently directed relative to the surface under study, allows one to minimize the influence of sedimentation. Figure 3 shows the ST of bentonite slurry obtained by averaging the results of measuring the ST by the techniques of lying and hanging drop in the indicated time interval. Insignificant deviations of the ST isotherms from the horizontal line are within the experimental limits (0.1–0.3 mN/m). The constancy of ST (within the measurement accuracy) indicates that the method of superposition of measurement results obtained for hanging and lying drops makes it possible to find ST characterizing a suspension with particles uniformly distributed over the volume.

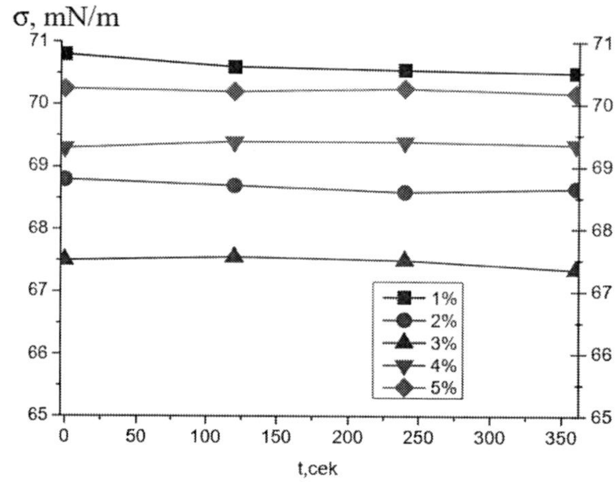

Figure 3. The time dependence of the ST of t, s ite slurry with different solid phase contents (superposition of the results of measuring the surface tension by the methods of lying and hanging drop), at 298 K.

The absence of the influence of substrate on the test liquid (a standard, fluoroplastic plate recommended by the manufacturer of the measuring equipment is used as the substrate) on the ST magnitude can be judged by comparing the measurement results using lying drop that is located on the substrate and hanging drop when there is no substrate.

Figure 4 shows the ST isotherms obtained by two indicated methods, as well as the isotherm calculated by the superposition method at a concentration of 3 % of the mass of solid phase. As one can see from the isotherms 2 and 3, the angle of inclination of both ST isotherms relative to the composition axis is almost symmetric which indicates the absence of significant influence of the substrate on measured ST magnitude.

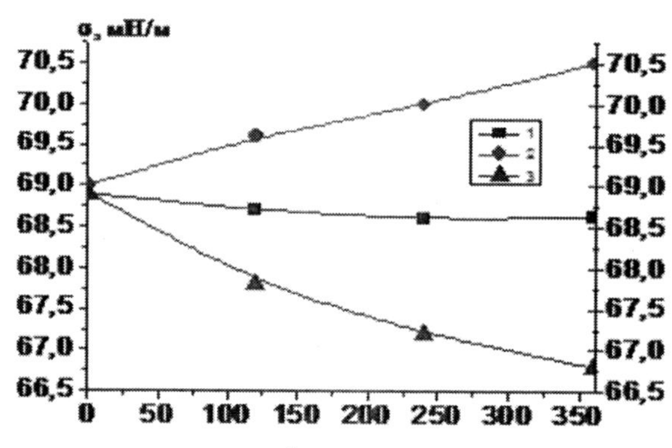

Figure 4. Time dependence of surface tension of bentonite slurry at 3 mass. % of solid phase: 1 – averaged values; 2 – lying-drop technique; 3 – hanging drop technique, at 298 K.

Figure 5 shows the concentration dependence of the ST of an aqueous slurry of bentonite. The measurements were carried out using the proposed method of superposition (at a confidence level of 0.95, the error in the measurement of ST does not exceed ± 0.6 mN/m).

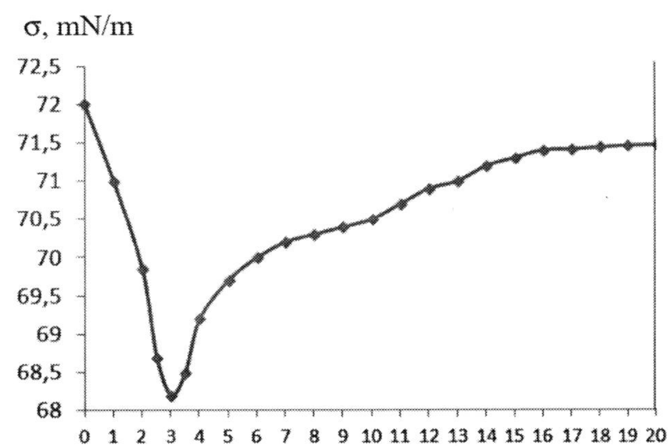

Figure 5. Dependence of ST on the concentration of solid phase in an aqueous slurry of bentonite at 298 K. (hanging drop technique)

Figure 5 shows the ST isotherm of an aqueous slurry of bentonite at 298 K. In the concentration range of 3 mass. % there is a local minimum. With an increase in the concentration (up to 14–15 %) an increase in ST occurs. A further increase in the concentration of the solid phase (starting from a concentration of 16 %) does not lead to a significant change in ST.

5. Conclusion

Since the sedimentation rate with the techniques of hanging and lying drop is the same, but differently directed relative to the surface, the concentration of solid particles has the same influence on the magnitude of ST in time, regardless of the technique applied. Therefore, a superposition of measuring results of the ST of slurry by the indicated methods eliminates the influence of sedimentation on the value of slurry.

In addition, the analysis of the averaged ST isotherms (Figure 3) makes it possible to conclude that it is advisable to measure the concentration dependence of ST by the techniques of lying and hanging drop at a surface age of at least 120 s, since with a shorter drop formation time, there is observed an increased scatter of experimental ST magnitudes. The increased scatter of the measured values or magnitudes in the initial period of droplet formation is explained by the fact that at the beginning (regardless of the measurement method) there are an intense processes of redistribution of solid phase particles between the surface layer and the bulk phase in the droplet due to residual liquid perturbations that occur at the moment of surface formation. After the specified time (120 s), the system approaches a quasi-equilibrium state, which leads to a decrease in the intensity of redistribution of particles, and, as a consequence, to the stability of results.

References

[1] Mezhidov V Kh, Viskhanov S S and Daudova A L 2013 Chemical composition of bentonite (deposits in the Chechen Republic) *Bull. of the Acad. of Sci. of the Chechen Republic. Grozny* **1(18)** 13–9

[2] Ponomarev V V, Bakun V G, Kononenko S A and Savostyanov A P 2008 Study of the structure and adsorption properties of natural and modified bentonites Izv. of HEIs North Caucasus Reg. Techn Sci Ser. **3** 94–7

[3] Vakalova T V, Khabas T A, Pogrebenkov V M and Vereshchagin V I 2005 *Clays. Structure, properties and research methods* (Tomsk: TPU Publ. House) 248 p

[4] Van Olphen H 2008 Forces between suspended bentonite particles Shell Development Company". Exploration and prediction research division *Huston* **62** 204–24

[5] Uryev N B 2011 Rheological and thixotropic properties of an aqueous slurry of bentonite clay, previously subjected to electrodynamic effects *Colloid J.* **73(1)** 90–6

[6] Nanda K and Bulk K 2005 Cohesive energy and surface tension from the size-dependent evaporation study of nanoparticle *Appl. Phys. Letters* **87** 21–9

[7] Uryev N B 2017 Dynamic, aggregative and structural stability of highly concentrated colloidal dispersed systems *Phys. Chem. of surface and protect. of mater.* **53(1)** 103–12

[8] Uryev N B 2018 *Highly concentrated dispersed systems and materials* 2rd ed add. (Moscow: Techn. Polygraph Center) 407 p

[9] Kochurova N N, Rusanov A I and Myrzakhmetova N O 1991 The Jones-Ray Effect and Surface Electrification *J. the Reports of the USSR Acad. of Sci.* **316(6)** 1425–7

[10] Shantarin V D and Voitenko V S 1990 *Physical Chemistry of Dispersed Systems* (Moscow: Nedra) 315 p

[11] Dadashev R Kh, Dzhambulatov R S and Elimkhanov D Z 2015 Features of the concentration dependence of the surface tension of aqueous slurries of bentonite *J. of Phys. Chem.* **89(8)** 1338–40

[12] Dadashev R H., Dzhambulatov R S and Elimkhanov D Z 2018 Effect of Nanoscale Particles on the Surface Properties of Aqueous Suspensions of Bentonites *Bull. of the Russ. Acad. of Sci. Phys.* **82(7)** 902–4

[13] Dadashev R Kh, Dzhambulatov R S and Elimkhanov D Z 2012 Measurement of surface tension by hanging drop method on a DSA-100 tensiometer *Coll. Proc. of the CRI RAS* **5** 3–7 (Grozny)

[14] Rusanov A I and Prokhorov V A 1994 *Interfacial tensometry* (St. Petersburg: Chemistry) 397 p

Obtaining SiC single crystals, and comparing characteristics of its solid solutions, films, Schottky diodes

R H Dadashev[1], V I Altukhov[2], A V Sankin[2], D K Sysoev[2], D Z Elimkhanov[3], M-A A Gudaev[4] and R T Uspazhiev[5,6]

[1] Center for Materials Science, Academy of Sciences of the Chechen Republic, Grozny, Russia
[2] North Caucasus Federal University (branch), Pyatigorsk, Russia
[3] Faculty of Physics and Mathematics, Chechen State University, Grozny, Russia
[4] Faculty of Natural Sciences, Chechen State Pedagogical University, Grozny, Russia
[5] Department of Physics. Grozny State Oil Technical University named after M.D. Millionschikova, Grozny, Russia
[6] Integrated Research Institute named after Kh.I. Ibragimova of the Russian Academy of Sciences, Grozny, Russia

E-mail: raykom50@mail.ru, altukhovv@mail.ru, zam-id@pfncfu.ru, omrc@pfncfu.ru, edzhabrail@mail.ru, gudaevm@mail.ru, russlan-chr@mail.ru

Abstract. The main methods for producing 4H-SiC single crystals and films of its solid solutions are considered. A new technique for producing single crystals of SiC polytypes is described. A new nonlinear model of the Schottky barrier height (BS) has been developed, and the current-voltage characteristics of Schottky diodes have been obtained, their comparison with the I–V characteristics of silicon-based diodes has been made.

1. Introduction

According to the conclusion of CREE [1–9], silicon carbide is the promising primary material for power electronics. We note a number of the most well-known technologies for producing SiC single crystals.

(a) Previously, direct ohmic heating technology was used to produce SiC. In this case, a SiC single crystal is grown from the gas phase in a growth chamber (crucible), where a field of operating temperatures of the order of 2500 °C is created [7].

(b) SiC sublimation technologies are also applied, where the temperature of the single seed crystal is kept below the temperature of the source [10–13].

(c) The technology for growing SiCc single crystals using special thermal insulation is also of interest [14].

The main disadvantage of the above technologies is the lack of reliable control over the values of the working temperature gradients. The lack of reliable control leads to the appearance of a large number of cracks and impurities in the single crystal.

This article discusses a device that allows for the production of 4H-SiC polytypes by the method of controlled induction heating [9, 10]. In the device under consideration, additional oscillatory heating circuits are introduced for precise control of temperature field gradients.

The structure of the device for producing perfect silicon carbide single crystals is shown in Figure 1.

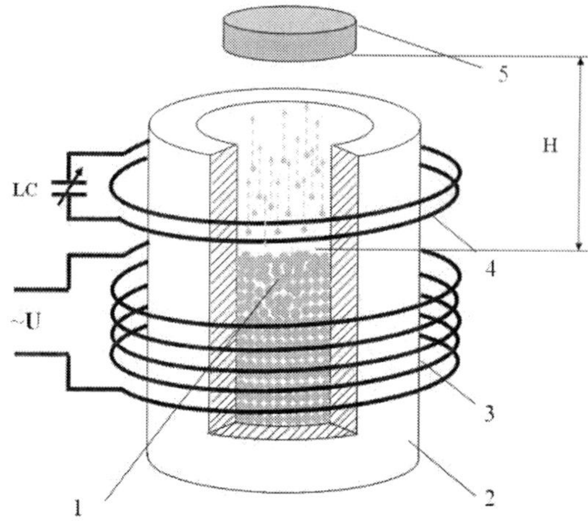

Figure 1. Scheme of a device for producing perfect single crystals of silicon carbide (SiC) using high-temperature induction heating (1 – source with a mixture of silicon carbide; 2 – the body of the growth chamber (crucible); 3 – inductor (induction coil); 4 – additional oscillatory circuit; 5 – seed single crystal)

An additional oscillating circuit (4) allows maintaining the necessary temperature gradients with high accuracy due to trim capacitors, providing a resonant mode with the inductor (3) – primary oscillatory circuit. The stability of the sublimation mode is maintained by changing the parameters of additional controlled oscillatory circuits.

In the growth zone, stable working temperature fields are created with an axial gradient in the direction from the seed crystal (5) to the source of silicon carbide (1). A stable field of operating temperatures ensures uniform evaporation of the source of silicon carbide and its crystallization from the vapor phase on the growth surface of the seed crystal. Uniform crystal growth is achieved by sublimation of SiC vapor from the gas phase in the growth chamber of the crucible by maintaining stable temperature gradients in the growth chamber and constant temperature values near the surface of the seed crystal. The crucible (2) of the proposed device contains housing and a lid with a pedestal, on which the SiC seed is fixed. At the base of the crucible, the body is a source of SiC (1) in the form of "pure" polycrystals or silicon carbide powder.

2. Theoretical part

2.1. Modified Schottky barrier height model for metal-semiconductor structures

It is known that in the contact region of structures, there are localized states of defects. In this situation, we consider the Schottky barrier height model [19, 21], but in a second-order approximation nonlinear in the concentration of defects. $N_i = c \cdot 10^{13}$ cm$^{-2} \cdot$eV^{-1}; c=0–30, where c is a convenient "concentration factor" in units of 10^{13} cm$^{-2} \cdot$eV^{-1}.

In this approach, along with the concentration of N_i defects, it is convenient to introduce the occupation numbers $n_x(c)$. These numbers are determined by the form of the Hamiltonian of the system. In this approach, we obtain higher values of the Schottky barrier, which for small N_i ($N_i < 10^{13}$ cm$^{-2} \cdot$eV^{-1}) leads to better agreement with the experimental data than in the classical Bardin – Schottky – Mott model. In the model under consideration, the barrier height Φ_B^x is described by the formula [19, 21]:

$$\Phi_B^x(c) = \Phi_m - \chi + \Delta\Phi_x(c) \quad (1)$$

$$\Delta\Phi_x(c) = 4\pi\left(e^2 / 4\pi\varepsilon_0\varepsilon\right)\lambda N_i n_x(c) \quad (2)$$

where χ – the electron affinity; Φ_m – the work function of the metal; $\Delta\Phi_x(c)$ – the potential barrier due to the tunneling of carriers (electrons) between localized quasilevels (E_i states) and the metal; λ – the thickness of the contact region with permittivity $\varepsilon_0\varepsilon$; N_i – surface density of localized states of defects, $n_x(c)$ – the filling number of the quasilevel of the defect E_i with half-width $\Gamma=\pi\rho U^2$ (U is the hybridization energy of localized states and metal); ρ – the (constant) density of states of the metal; E_F is the Fermi level energy.

In this case, the metal in contact with the semiconductor is described by defective surface states $|d>c$ energy E_i located in the forbidden zone. The metal in connection with the level $|d>c$, can be described by the Andersen's Hamiltonian. In the simplest case, it has the following form [19]:

$$H = \sum_k \varepsilon_k c_k^+ c_k + E_i d_i^+ d_i + U \cdot \sum_k (c_k^+ d_i + h.c.). \quad (3)$$

where ε_k – the electron energy; U – value of the matrix element of hybridization of the metallic |k> bond and the defective |d> state; c^+_k – the operator of the electron production with momentum $|k>$; d_i^+ – the electron creation operator in |d>. If we assume that the Gaussian defective state E_i overlaps with a wide metal conduction band, i.e., in the range $|E_0 – E| \leq \Gamma \approx E_0$ ($E_0 = E_i – E_F$). Then one can find the state occupation numbers by the formula $n_d = n_x(c)$:

$$n_d(c) = n_x(c) = \int_{-\infty}^{E_F} \rho_d(E) dE, \quad (4)$$

$$\rho_x(E) = \frac{1}{\pi} \cdot \frac{\Gamma}{(E - E_i)^2 + \Gamma^2}, \quad (5)$$

$$n_x(c) = \frac{1}{\pi} \cdot \int_{-\infty}^{E_F} \frac{\Gamma dE}{(E - E_i)^2 + \Gamma^2} = \frac{1}{\pi} \cdot arc\cot\left(\frac{E_i - E_F}{\Gamma}\right). \quad (6)$$

The forbidden zone of a semiconductor in a solid solution of silicon has the form

$$E_g^x \equiv E_g - ax + bx^2, \quad (7)$$

where E_g=3.3 eV, a=0.56, b=3.86.

Then the position of EF relative to the ceiling of the valence band is determined by the expression:

$$E_F = \chi + E_g^x - \Phi_m - \Delta\Phi_x(c). \quad (8)$$

Next, we introduce the notation

$$(E_i - E_F) / \Gamma \equiv \delta_x(c) \quad (9)$$

$$E_i = E_g \xi_i \quad (10)$$

Moreover, for the state, ξi = 0.3 (0.5; 0.7), taking into account (9), we obtain:

$$\Gamma\delta_x(c) = (\Phi_m - \chi) - (1 - \xi_i)E_g^x + \Delta\Phi_x(c) \quad (11)$$

For the characteristic barrier width $\lambda = 3\eta$Å (η=0.5–2.0), the potential barrier $\Delta\ \Phi x$ has the form:

$$\Delta\Phi_x(c) = k\eta c 2 n_x(c) \quad (12)$$

where the coefficient k = 0.272 eV. For justified values of $E_F \approx E_g/2$, for $E_i = E_F$ (Γ=0.5–2.0 eV), the quantity $2n_{x0}(c) \approx 1$.This data is in accordance with calculations in and gives the best agreement with the experimental data for the quantity $\Phi^x_B(c)$.

The Schottky barrier model nonlinear in the concentration of surface states is presented in [12, 21]. The approach given in [12, 21] shows higher values of the Schottky barrier. For low N_i defect concentrations in the contact region (N_i<10^{13} sm^{-2}·eV^{-1}; N_i=c·10^{13}sm^{-2}.eV^{-1}), this leads to better agreement with the experimental data presented in [12, 13] than the formula Bardina, Schottky-Mott. Moreover, in our model, for the height of the potential Schottky barrier Φ^x_B and the occupation numbers $n_x(c)$, we obtain [12, 13, 21]:

$$\Phi_B^x(c) = p + k\eta c 2n_x(c)\,, \tag{13}$$

$$\Gamma\delta_x(c) = p - (1-\xi_i)E_g^x + k\eta c(1 - c\nu), \tag{14}$$

$$n_x(c) = (1/\pi)\cdot \mathrm{arccth}\,\delta_x(c). \tag{15}$$

where k = 0.272 eV, $p=\Phi_m-\chi$; energy of surface states $\xi_i \cdot E_g{}^x$; ξ_i = 0.3 (0.5; 0.7); barrier width λ=3η Å, η=0.5–2.0.

3. Methods and materials

The microcurrents in Si diodes and Schottky barrier diodes based on SiC and Si–Cr were measured by the method [18]. The measurement circuit is shown in Figure 2. Similar schemes are used to measure ultra-low currents.

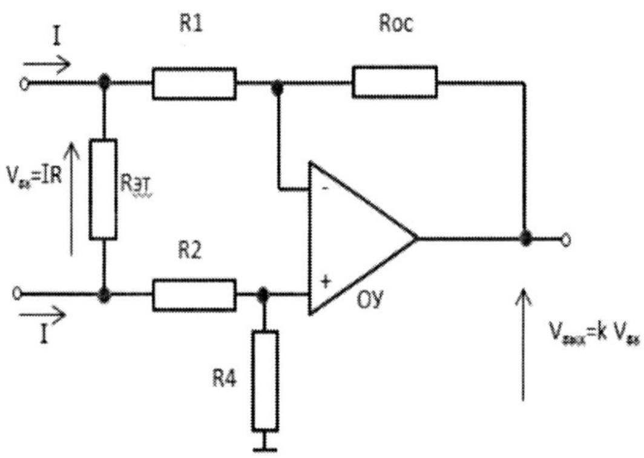

Figure 2. Production of single crystals, silicon carbide films, analysis of current-voltage characteristics of conventional silicon diodes and Schottky barrier diodes.

4. Results

The Schottky barrier model nonlinear in the concentration of surface states is presented in [12, 19]. The approach proposed in [12] gives higher values of the Schottky barrier. At the low defect, concentrations lead to better agreement with the experimental data than the classical Bardin – Schottky – Mott formula [12, 13].

It was also shown in [13, 19, 22] that "ordinary" heterojunctions based on SiC solid solutions of n-SiC/p-(SiC)1-x(AlN)x (n-p junction) type have similar characteristics and similar volt-ampere characteristics with diodes based on SiC with a Schottky barrier Me/(SiC)1-x(AlN)x (Me=Al; Ti;Cr; Ni). The calculated and experimental heights of the Schottky barriers are given in Table 1, respectively.

Table 1. Schottky Barrier Heights for different metals

Metal	Al (eV)	Ti (eV)	Cr (eV)	Ni (eV)
Theory	1,74	1,90	1,78	2,24
Experiment	1,78	1,85	1,98	2,16

The calculated values are consistent with experimental data [12].

The value of the Schottky barrier of SiC-based diodes can also be experimentally obtained using the regimes of small signals on alternating current when removing the current volt-ampere characteristics according to a hybrid equivalent circuit [18].

A comparison of the volt-ampere characteristics of various types of diodes is shown in Figure 3. An essential difference between Schottky diodes is revealed here (Figure 3a). The diode type is 1N5822, with a current of 2A at a voltage of 0.4V and 1N5819 at a voltage of 0.6V from conventional Si–p-n junctions. For silicon diodes of type 1N5408, we have a current of 2A at a voltage of 0.85V and for 1N4004 at a voltage of 1.2V. It can be seen that Schottky diodes at the same operating currents have a lower voltage drop at the junction in the direction of direct connection, as a result of which their temperature heating, power loss, and reliability are increased.

Figure 3 (a) shows that the two upper branches belong to ordinary diodes, the two lower ones to diodes with a Schottky barrier [18]. Figure 3 (b) shows the experimental I–V characteristics of Schottky diodes with voltages up to 4V, where different charge transfer mechanisms appear (direct (a) and (b) – the theory for different ideality coefficients). The calculated straight lines in the range 01–0.5 V correspond to the direct volt-ampere characteristics of the CREE Schottky diodes of the CREE type in the range 0.5–0.9 V obtained at different temperatures (Figure 3c) [20].

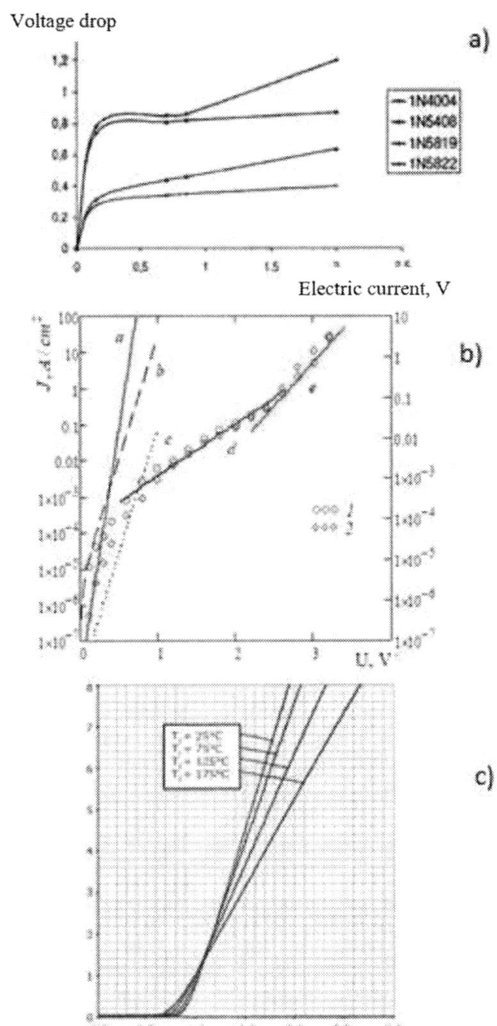

Figure 3. Analysis of the volt-ampere characteristics of conventional silicon diodes and Schottky barrier diodes

5. Conclusion
Thus, the experimental data are in work, and the volt-ampere characteristics of ordinary diodes and diodes with Schottky barriers are calculated. The mechanisms of charge transfer at different ideality coefficients are revealed. The results obtained are of great practical importance in modern power, extreme micro-nano electronics.

References
[1] Luchinin V and Tairov Yu 2010 Silicon carbide is a diamond-like material with controlled nanostructure-dependent properties *Nanoindustry (Nanomater.)* **1** 129–34
[2] Ramazanov Sh M 2014 *Preparation and study of epitaxial films of solid solutions (SiC)1-x(AlN)x, and surface-barrier structures based on them* (Makhachkala) 21 p
[3] Lebedev A O 2013 *Growth and defect formation of single crystals of semiconductor silicon carbide grown by the LETI method* (St. Petersburg) 35 p
[4] Shakhnovich I 2014 If the state does not have electronics, it has no future *Electr.: sci., technol., busin.* **7** 80–90
[5] Kargin I I, Safaraliev G K, Kharlamov N A et al 2013 Kinetic features of obtaining films of solid solution (SiC)1-x(AlN)x by ion sputtering. University News. North Caucasus region *Techn. Sci.* **6** 118–21
[6] Luchinin V and Tairov Yu 2000 Domestic semiconductor silicon carbide: a step towards parity *Modern electr.* **7** 12–5
[7] Kirillov B A, Bakin A S, Solnyshkin S N and Tairov Yu M 1997 Modeling of heat and mass transfer during the growth of silicon carbide single crystals *FTP* **31(7)** 794–8
[8] Mills G and Cohn G 2010 Silicon Carbide for Russian Industry: CREE Comes to Russia *Electron.* **8** 4–6
[9] Sankin A V , Altukhov V I , Kazarov B A et al *A device for producing perfect single crystals of silicon carbide with additional control circuits for induction heating* Patent No 173041 February 20, 2017
[10] Altukhov V I, Sankin A V and Kazarov B A 2017 *Calculation of the new properties of wide-gap semiconductor materials* (Dusseldorf, Saarbrucken: Lambert Acad. Publ.; Germany: SKFU) 149 p
[11] Gnesin G G 1977 *Silicon carbide materials* (Moscow)
[12] Safaraliev G.K., Bilalov B.A., Kurbanov M.K et al 2015 Calculation of the height of the Schottky barrier at the contact of a metal with a semiconductor solid solution (SiC)1–x(AlN)x *Microelectr., Elements of integrated circuits* **44(6)** 453–8
[13] Altukhov V I, Kasyanenko I S, Sankin A V et al 2016 Calculation of the Schottky barrier and current – voltage characteristics of metal – solid solutions based on silicon carbide *Phys. and technol. of semiconductors* **50(9)** 1190–4
[14] *Method for growing low defect density siliconcarbide* US Patent No 6428621 August 6, 2002
[15] Gusev A S, Ryndya S M, Kargin N I and Bondarenko E A 2010 Low-temperature synthesis of thin films of silicon carbide by vacuum laser ablation and the study of their properties *Surface. X-ray, synchrotron and neutron stud.* **5** 18–22
[16] Safaraliev G K 2011 *Solid solutions based on silicon carbide* (Moscow) 296 p
[17] Kargin N I, Gromo V D, Kuznetsov G D and Grekhov M M 2014 The effect of irradiation on the instrument characteristics of transistor structures based on AlGaN/GaN *Bull. of the National Res. Nuclear Univer. MEPhI* **3(1)** 68–70
[18] Afonsky A A and Dyakonov V P 2011 *Electronic Measurements in Nanotechnology and Microelectronics* ed V P Dyakonova (Moscow: DMK Press)
[19] Altukhov V I, Sankin A V, Sigov A S et al 2018 The Schottky barrier model nonlinear in the concentration of surface states and the calculation of the current – voltage characteristics of

IOP Conf. Series: Materials Science and Engineering **905** (2020) 012017 doi:10.1088/1757-899X/905/1/012017

SiC-based diodes and its solid solutions in a composite current transfer model *Semiconductor Phys. and Technol.* **52(3)** 366–70

[20] *Datasheet.su. Electrotechnical portal* Retrieved from: http:// datasheet.su/datasheet/CREE/c3d04060f (accessed: 10/20/2019).

[21] Davydov S Yu, Lebedev A A and Tikhonov S K 1997 On the Schottky barrier at the contact of a metal with silicon carbide *Phys. and Technol. of Semiconductors* **31** 597–9

[22] Dadashev R 2008 *Thermodynamics of Surface Phenomena* (Cambridge: Int. Sci. Publ.) 281 p

IOP Conf. Series: Materials Science and Engineering **905** (2020) 012018 doi:10.1088/1757-899X/905/1/012018

Influence of various factors on the spreading kinetics

V S Novosadov[1], R Kh Dadashev[2], R A Kutuev[3], D Z Elimkhanov[3], Z I Dadasheva[3] and Kh S Talkhigova[3]

[1] Moscow State University of Food Production, 11, Volokolamskoe sh., Moscow, 125080, Russia

[2] Academy of Sciences of the Chechen Republic, 13, prospect named after M. Esambayev, Grozny, Russia

[3] Chechen State University, 17a, Dudayev Boulevard, Grozny, 364060, Russia

E-mail: novosadov@list.ru, raykom50@mail.ru, kra-07@mail.ru, edzhabrail@mail.ru, zarema-69-69@mail.ru, talhigova@mail.ru

Abstract. It was found that at the stage of fast spreading ($\tau \leq 0{,}02c$) at constant driving force within the wetting perimeter, the following exerts the main impact on the spreading kinetics: inertial forces that prevent spreading and change their sign when the speed slows, viscous resistance forces, the influence of which increases in proportion to the growth of the thickness of the boundary layer and vertical forces, which begin to appear when the spreading velocity slows down.

1. Introduction

A large number of works is devoted to wetting and spreading problems [1–20]. It is known that the driving force of wetting and spreading is the reduction of free energy of the system mainly by reducing the free surface energy in the process of replacing two solid and liquid surfaces with a common interfacial boundary.

The Young's equations are usually used in wetting and spreading analysis. In [5–9] it was shown that the Young's equation obtained within the framework of the molecular theory of capillarity can be correctly used only under equilibrium conditions on a non-deformable surface in the absence of wetting [3]. Under irreversible wetting and spreading conditions, when $\theta < 90^{\circ}$ and the adhesion energy $W_A > W_K$ is the work of cohesion, the Young's equation does not work since $\cos\theta$ cannot be greater than 1. The equation also does not take into account the vertical component of the total vector resulting in deformation of the substrate. Hence, the kinetic regularities of the stage of fast drop spreading over the solid surface are considered, and the influence of various factors (inertial forces, viscosity and vertical forces) on the spreading process is studied.

2. Methods and materials

Features of the mechanism and kinetics of the spreading process are discussed in [1–20, etc.]. Two main cases should be considered: 1) Absence of wetting and spreading or reversible (limited [1]) wetting and spreading, when the liquid drop can be easily removed from the surface (wetting angle $\theta > 90^{\circ}$). In reversible processes, liquid adhesion to the solid phase results from intermolecular interaction (e.g. Van der Waals forces). 2) Wetting and spreading of liquid with different intensity on solid and liquid surface or irreversible spreading at $\theta < 90^{\circ}$. In irreversible spreading, adhesion results from the formation of chemical bonds. The interaction energy at irreversible adhesion is two orders of magnitude greater and makes $\sim 10^2$ kJ/mol. The drop can be partially removed from the surface only during its cohesive fracture in the area of the interfacial boundary. Whereas, a thin layer of liquid phase remains on the surface. In this case, as a result of chemisorption of liquid clusters by the solid phase atoms, the elementary wetting act ends with the formation of a common interfacial boundary. An important conclusion follows from this definition. In systems far from chemical equilibrium, mass transfer processes will occur across a common interfacial boundary (dissolution and diffusion of liquid atoms into the solid phase). In the first case, the absence of wetting and, accordingly, dissolution enables efficient selection of materials for storage and transport of reactive substances and food media, and the lower the adhesion, the lower the resistance to liquid flow through channels and capillaries. In the second case, wetting enables dissolution and saturation of the surface of the solid phase with liquid atoms. Mass transfer processes in the area of interfacial boundary significantly change rheological and physicochemical properties of contacting solid and liquid phases and their compounds.

The driving force of wetting and spreading is the reduction of free energy of the system mainly by reducing the free surface energy in the process of replacing two solid and liquid surfaces with a common interfacial boundary. Figure 1 shows the balance of forces at the edge of the lying drop.

The Young's equations are commonly used in wetting and spreading analysis: $\cos\theta = (\sigma_{тг} - \sigma_{тж})/\sigma_{жг}$ and $\cos\theta = 2W_A/W_K - 1$. In [5-9] it was shown that the Young's equation obtained within the framework of the molecular theory of capillarity can be correctly used only under equilibrium conditions on a non-deformable surface in the absence of wetting [3]: $\theta > 90^\circ$ and $W_A < W_K$. Under irreversible wetting and spreading conditions, when $\theta < 90^\circ$ and the adhesion energy $W_A > W_K$ is the work of cohesion, the Young's equation does not work since $\cos\theta$ cannot be greater than 1. Spontaneous spreading corresponds to a positive value of the spreading coefficient $S = \sigma_{тг} - (\sigma_{жг} + \sigma_{тж}) > 0$. In this case, $\sigma_{тг} - \sigma_{тж} > \sigma_{жг}$ and in the Young's equation, the value $\cos\theta > 1$. The equation does not take into account the vertical component of the total vector leading to deformation of the substrate, the peculiarities of the solid phase and liquid structure in the area of the interfacial boundary. A new record of Young's equation is proposed: $\cos\theta = N^d(\sigma^{|}_{тг} - \sigma^{|}_{тж})/\sigma^{|}_{жг}$, where $N^d \le 1$, $\sigma^{|}_{ij}$ – dynamic tension at interfacial boundaries, N^d – function that includes an unknown scalar equation that takes into account the peculiarities of non-equilibrium spreading. At $N^d = 1$ we get the known Young's equation [5].

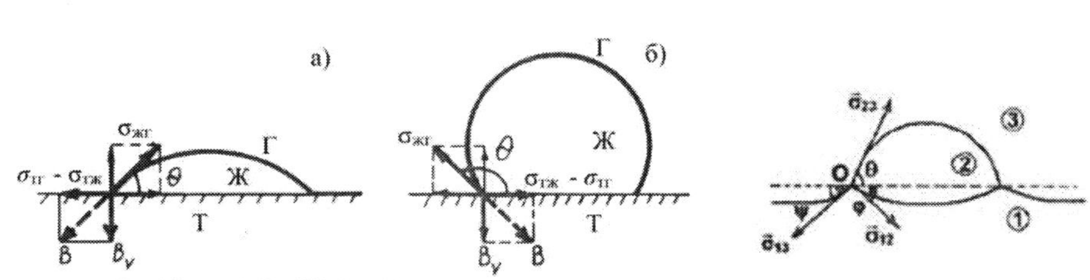

Figure 1. Equilibrium of forces at the edge of the lying drop on a solid undeformed (a, b) and deformable (c) substrate: a, c) at $\theta < 90^\circ$; b) at $\theta > 90^\circ$; c) solid body equilibrium profiles, drop – pairs: $\vec{\sigma}_{12} \equiv \sigma_{тг}$, $\vec{\sigma}_{13} \equiv \sigma_{тж}$, $\vec{\sigma}_{23} \equiv \sigma_{жг}$ [10].

On the role of vertical forces in the spreading process. Young's formulas are derived as a projection on the OX axis of the total surface tension vector σ_{ij} under equilibrium conditions and do not reflect the vertical component B_v (Fig. 1) of this vector. The presence of the vertical component of the solid body (elastic and plastic deformation) causes the contact surfaces of thin substrates to bend. Baily [11] found that a thin layer of mica (2...3 μm) with a mercury drop experiences a characteristic bend. Gegusin Ya.E. [12] shows that when a thin (up to 5 mm) polymer film is wetted with water, the drop begins to bend the film and at the final stage of spreading wraps into it under the vertical force. Figure 1a shows that three surface tensions in no form can the drops come into equilibrium, since σ_{mz} and $\sigma_{mж}$ lie in one plane and are directed against each other, and the third one $\sigma_{жz}$ is located at an angle. Therefore, there is an uncompensated force applied to the outline of the drop and directed vertically upward. The convex surface of the drop creates Laplacian capillary pressure that presses it to the flat surface. On flat surface this pressure is zero ($R \to \infty$), and at convex it is equal to $P = 2\sigma_{жz} / R_{жz}$. In order for capillary pressure to be balanced over the entire drop-limiting surface, it is necessary to form a well beneath it so that all three forces along the contour are compensated, i.e. $2\sigma_{mz} / R_{mz} = 2\sigma_{mж} / R_{mж}$. Thus, strictly speaking, the Young's equation applies to the non-equilibrium case.

The well is found in the absence of wetting, as well as in experiments on the interaction of mutually insoluble metals (Fe+Pb, Mo+Sn, etc.) and low-soluble metals, as well as in iron wetting of oxide materials and borosilicate melts of tungsten and rhenium [4, 12–16]. It was found that the surface of the solid phase is deformed at high temperatures and over more than 2…10 hours under the influence of unbalanced surface forces. A rib-shaped cuff (Fig. 1 c) is thus formed, the height of which increases with time and can reach 10 μm at high temperatures with a width of up to 5 μm [15]. Vertical forces $F_0 - \sigma_{жz} \sin\theta$ increase as the value θ approaches 90°. A stretched state region occurs along the wetting perimeter. The transport of the substance (viscous flow mechanisms, surface diffusion, evaporation – condensation) will be carried into this area, leading to the formation of a cuff. A well also occurs during dissolution of solid phase atoms in melt [16]. Thus, the measured value θ depends not only on the nature of the interfacial interaction (ratio of σ_{mz}, $\sigma_{mж}$ and $\sigma_{жz}$), but also on the intensity of dissolution processes. In this case, when a well is formed (Fig. 1c), the measured value θ is less than the true edge angle $\theta = \theta + \varphi$.

Figure 1b shows a profile of the equilibrium state of the system according to the second Neumann-Gibbs capillary equation $\sum \sigma_j \bar{t}_j = 0$ (σ_j – interfacial tensions, t_j – unit vectors tangent to the phase boundaries), which in the projection on the vertical and horizontal axes gives: $\sigma_{mz} \cos\psi = \sigma_{жz} \cos\theta + \sigma_{mж} \cos\varphi$ and $\sigma_{жz} \sin\theta = \sigma_{mz} \sin\psi + \sigma_{mж} \sin\varphi$. If you put the value of the angle $\psi = 0$ [4, 13–14] due to its small value or difficulty to determine, it is possible to measure σ_{mz} and $\sigma_{mж}$ in one experiment by measuring the value $\sigma_{жz}$ separately using equations: $\sigma_{mz} = \sigma_{жz} \cos\theta + \sigma_{mж}$ and $\cos\varphi \sigma_{жz} \sin\theta = \sigma_{mж} \sin\varphi$. The use of equations at $\psi = 0$ requires special analysis. In [13] at the equilibrium of the Pb drop on the Fe surface at 920 °C a reasonable value σТГ for Fe that equals 1645 mJ/m2 was determined. On the other hand, from the experimental data given in [14] we see that the values σ_{mz} and $\sigma_{mж}$ in the melt drop Cd+ system (Cd-Sn) decrease with time within 8 hours in argon and 2 hours in glycerol, which is contradicts the physics of the process. Obviously, after achieving the chemical equilibrium, the system's tendency to achieve mechanical equilibrium, resulting in surface warping, cannot lead to the decrease in σ_{mz} and $\sigma_{mж}$ [7–9].

In general, the force leading to non-equilibrium spreading looks as follows:

$$F_p = F_{дв}^d\left(\sigma_{mz}^d - \sigma_{mж}^d\right) - 1/N^d \sigma_{жz}^d \cos\theta \pm$$
$$F_i(\tau) - F_v(\tau) - P_Л(\tau) - P_B(\tau) + P_g \geq 0,$$

where $F_{дв}^d\left(\sigma_{mz}^d - \sigma_{mж}^d\right)$ – true driving force determined by the decrease in free surface energy during the spreading process when replacing solid and liquid surfaces with a common interfacial boundary. At $\sigma_{mж}(\tau) = -\sigma_{mж}$, $F_{дв}^{\max} = \sigma_{mz} + \sigma_{mж}$, $W_A^{\max} = \sigma_{mz} + \sigma_{mж} + \sigma_{mж}$ $1/N^d \sigma_{жz}^d \cos\theta$ – the contribution of this spreading retardant component is determined by $\sigma_{жz}^d$ that varies from the initial $\sigma_{жz}^0$ to equilibrium $\sigma_{жz}^p$ and an increase in $\cos\theta$ as the decrease θ, $F_i(\tau)$ – inertial force preventing spreading with a maximum value at the start of spreading, acts until the start of the second spreading stage. As the speed of movement slows down, the inertia force changes its sign, facilitating the movement of the liquid.

$F_v(\tau)$ – viscous resistance force, which increases in proportion to the growth of the thickness of the boundary layer $\delta=\sqrt{\nu\tau}$. The process of dissolution on the perimeter of wetting will also increase the strength of the flow resistance while increasing the viscosity of the boundary layer fluid.

$P_Л(\tau)$ – Laplacian capillary pressure, which presses a drop having a convex surface with a radius of $R_{жг}$. $P_Л(\tau)=2\sigma_{жг}/R_{жг}(\tau)$. Pressure $P_Л(\tau)$ prevents leakage and its contribution decreases over time.

$P_B(\tau)$ – vertical force deforming the surface of the solid phase, the action of which begins to manifest itself as the contribution of inertia forces $F_i(\tau)$ decreases.

P_g – gravity field will contribute to spreading.

According to experimental data, it is common to distinguish between: 1) a fast spreading stage (up to 0.02 seconds) and 2) a slow spreading stage from 0.03 sec to 10 minutes or more. The slow spreading stage is typical for systems with strong chemical interaction resulting in the formation of chemical compounds in the surface layer of the solid phase. This stage is determined by the kinetics of diffusion processes and the liquid spreading over a layer of chemical compounds (intermetallides), as well as in the case of oxide films and organic contaminants on the surface. In the vast majority of cases, upon contact of clean surfaces, the first stage is completed within $\tau\sim 0{,}02$ seconds. At further 10 minutes or more, the values of the edge angle do not change [8, 17–20].

Kinetic patterns of fast spreading stage and the role of viscous resistance forces. Let us consider the fast spreading stage that is typical for all systems and independent of the chemical activity of the interacting phases. At this stage, the drop flows over the surfaces, occupying its main area. The spreading rate, like that of any process, is determined by the ratio of a driving force to resistance forces, mostly inertial and viscous forces. Based on the analysis of numerous experimental studies [3–9, 15–20] it was found that in the first stage of fast spreading ($t_1=0...5\cdot 10^{-3}$ sec) the radius of the drop r increases from time τ according to the linear law $r=K_1\tau$, in the second stage $t_2=5\cdot 10^{-3}$ to $1{,}4\cdot 10^{-2}$ sec there is a parabolic dependence of spreading $r^2=K_2\tau$. The spreading rate decreases on average from 1.8 to 0.4...0.6 m/s (Fig. 2–3). The linear dependence r^2 on τ at this stage confirms the constancy of radial consumption and, hence, the constancy of driving forces. Since fresh layers of liquid flow onto a clean solid surface, $F^d_{дв}$ is constant. The driving force at the periphery of the drop has a maximum value $F^d_{дв}=\sigma_{mг}-\sigma_{mж}(\tau)=const$ and is determined mainly by dynamic values of the interfacial tension $\sigma^{|}_{mж}(\tau)$. At this stage, $\sigma^{|}_{mж}(\tau)$ has the minimum value and can make $\sigma^{|}_{mж}(\tau)<0$ [5-9]. If drop movement is significantly slowed down and equilibrium values of concentration are reached on its periphery during mass transfer, the driving force of spreading is significantly reduced [5–8]: $F^P_{дв}=\sigma^P_{mг}-\sigma^P_{mж}<<F^d_{дв}$. As the thickness of the flowing liquid layer decreases, its thickness h becomes commensurate with the thickness δ of the boundary layer, which at the surface has a maximum viscosity gradient. The velocity of the drop slows down as the driving force becomes insufficient to overcome the viscous resistance and friction at the solid surface, and the third stage ($t_2=1{,}4\cdot 10^{-2}$ up to $(2...9)\cdot 1{,}4\cdot 10^{-2}$ sec) occurs at which $r^n=K\tau$. The spreading rate and driving force decrease by 1...2 orders, and n increases to 3...4. At the end of this period, in the process of mutual dissolution, the diffusion zone reaches a triple point on the periphery of the drop. Interfacial tensions $\sigma_{mж}$ and $\sigma_{жг}$ reach equilibrium values and the fourth stage of slow spreading $t_4=(2...9)\cdot 10^{-2}$ sec to $1...30$ sec and more $1...10$ min; $r^n=K_3\tau$, $n=8$). Thus, where a variable viscosity boundary layer is formed at the surface of the solid phase, the liquid droplet cannot flow into the liquid phase film of minimum

thickness in the 3…4 molecular layers (6...12Å) and generally $\theta_{\min} \geq 3...4^{\circ}$. To confirm this, the works [5–8] carried out estimates to determine the thickness of the boundary layer.

On the role of inertial forces in non-equilibrium spreading. At the initial point of contact, the spreading rate $V = 0$, and then under the action $F^{d}_{\partial e}$ having the maximum value at this stage, the drop begins to flow rapidly. High Reynolds number values ~ 4500 at initial spreading rates and drop oscillations after flowing onto the substrate indicate that the spreading occurs in inertial resistance mode. This is further confirmed by the fact that the effect of temperature decreases as the spreading rate increases. Besides, the calculation of speed constants – K coefficients in equations $r^{n} = K\tau$ at the initial stage according to Ya.I. Frenkel's equation for viscous flow, gives velocity values different by more than an order from experimental [16–20].

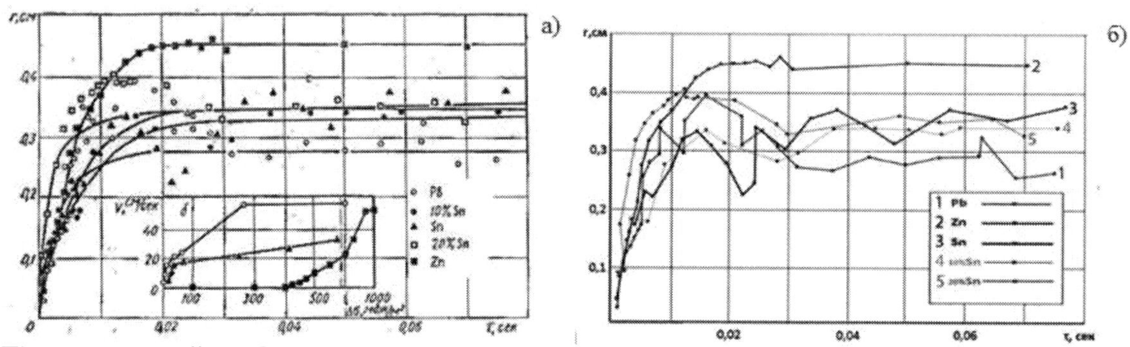

Figure 2. Spreading of Zn, Pb, Sn and Pb alloys with Sn on iron surface [17]. The dependence of the drop base radius r on time τ (a) and velocity V on $\Delta\sigma(\sigma)$. Same as Figure 2a, experimental points are connected by lines (b)

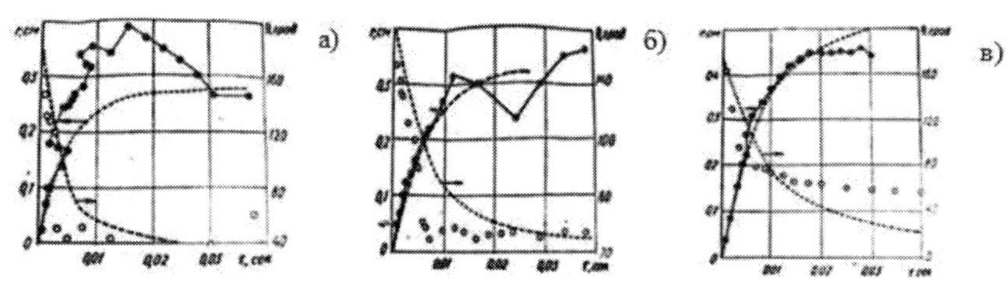

Figure 3. Changes in the drop base radius r, contact angle and spreading rate $V = K\Delta\sigma$: a – lead on iron at 350 °C; b – tin on iron at 235 °C; c – zinc on iron at 450 °C. Points – experimental values connected by lines, and dashed lines – calculated values [18]

Experimental data [17–18] given in Fig. 2 and 3 confirm the change of inertial forces action sign. With careful consideration of the escperimental points, relatively large deviations from the middle curves can be seen. If you connect the experimental points with lines (Fig. 2b and 3), the wave movement of the drop front due to the periodic change of the sign of inertial forces becomes obvious. When the motion of the drop is slowed by viscous forces, inertial forces begin to act in the opposite direction resulting in vibration of the drop.

3. Results

Thus, the kinetics of drop spreading is determined by a variety of factors. The analysis of the driving forces shows that when there is a constant driving force on the peripheries of the drop, the growing force of the viscous resistance acts on it. The increase in the thickness of the boundary layer by law $\delta = \sqrt{\nu\tau}$ decreases the spreading rate and stops the movement. Where it is possible to dissolve the solid phase in the liquid, there will be an increase in the viscosity of the boundary layer. When a drop of equilibrium concentrations is reached in the process of mass transfer, the driving force of spreading is significantly reduced. At high flow rates at the first and second stages, in addition to viscous resistance forces, inertial forces play a determining role, which change their sign over time.

The critical values of edge wetting angles are considered in [5, 8]. The edge angles θ cannot be equal 0.09 and 180°. The value θ cannot be zero, since the angle at the edge of the *equilibrium* liquid phase film of minimum thickness in 3...4 molecular layers (6...12Å) is not zero and minimum values $\theta_{min} \geq 3...4^{\circ}$. The values θ on the solid and liquid surfaces are always less than 180°, since equality $\theta = 180^{\circ}$ implies a vapor layer separating the drop from the surface. The drop or soft particle of spherical shape approaching the surface of the solid body falls under the influence of far-acting attraction forces and then short-acting Brownian repulsion forces, which will lead to tensions in it. Calculations of molecular forces acting at a distance of 0.1 μm show that these tensions cause local deformation relative to the soft bodies prior to their contact, i.e. prior to the moment of action of the Brownian repulsion forces. Thus, the soft particle loses its original spherical shape and the point Hookian contact is excluded. The value θ also cannot be 90°. At $\theta > 90°$ there is no wetting, dispersion forces are in effect and the drop can be separated from the surface. At $\theta < 90°$, chemical bonding occurs in the contact zone, accompanied by irreversible drop spreading.

4. Conclusion

1. The driving forces of the separation are analyzed and kinetic regularities of the stage of fast separation of the drop on the solid surface are considered.

2. It was found that the deceleration of the spreading rate at constant driving force is caused by the increase in the thickness of the viscous boundary layer according to the law $\delta = \sqrt{\nu\tau}$, which decreases the spreading rate and stops the movement. At high flow rates at the first and second stages, in addition to viscous resistance forces, inertial forces play a determining role, which change their sign over time. The analysis of effect of vertical forces on the spreading kinetics was performed and critical values of edge angles were determined.

References

[1] Rawlinson J and Widom B 1986 *Molecular theory of capillarity* Transl. from Engl. (Moscow: Mir) 379 p

[2] Zadumkin S N and Khokonov Kh B 1977 General conditions of equilibrium of interphase boundaries and capillary equations In *Methods of research and properties of interphase boundaries of contacting phases* (Kiev: Naukova Dumka) pp 163–75

[3] Naidich Yu V 1972 *Contact phenomena in metal melts* (Kiev: Naukova Dumka) 196 p

[4] Bykhovsky A I 1983 *Spreading* (Kiev: Naukova Dumka) 191 p

[5] Novosadov V S September 2017 Physical and chemical features of surface wetting with small-sized drops in metal systems *Works of the Int. interdisciplinary symp. Physics of interphase boundaries and phase transitions* iss 7 vol 1 pp 100–5

[6] Novosadov V S and Khokonov Kh B 2011 Correctness of molecular capillarity equations in irreversible interaction processes of solid and liquid phases *Works of the Int. interdisciplinary symp. Physics of interphase boundaries and phase transitions* pp 115–20 (19–23 September)

[7] Novosadov V S 2012 Driving forces of spreading. Change of dynamic values of interfacial tension In *Soldering in various industries. Seminar materials* Coll. 2 (Moscow: CRDZ) pp 41–56

[8] Novosadov V S September 2018 Physicochemical and rheological patterns of wetting and spreading in metal systems *Soldering* Coll. of mater. of the Int. scie. and techn. Conf. pp 101–20

[9] Novosadov V S 2012 Wetting in non-equilibrium conditions. Passing through minimum dynamic values *Works of the Int. interdisciplinary symp. Physics of interphase boundaries and phase transitions* pp 92–5

[10] Popel S I 1994 *Surface phenomena in melts* (Moscow) pp 250–4
[11] Baily J 1957 II Intern. *Congress Surface activity* III Discussion (London: Butter worths scient publ) 189 p
[12] Geguzin Ya E. 1977 *Drop* (Moscow: Nauka) 175 p
[13] Geguzin Ya E and Ovcharenko N N 1962 Surface energy and processes on the surface of solid bodies *Successes of Phys. Sci.* 76(2) 283–328
[14] Khokonov Kh B, Shebzukhov A A and Savintsev P A 1977 Measurement of interfacial characteristics in eutectic systems In *Physical chemistry of melt surface* (Tbilisi: Metsniyerab publ. house) pp 267–72
[15] Khlynov V V, Ishimov V I and Kutiin A B 1972 Peculiarities of iron wetting of solid and liquid oxides In *Wettability and surface properties of melts and solid bodies* (Kiev: Naukova Dumka) pp 88–91
[16] Boxer E L, Khlynov V V et al 1972 *About adhesion of molten enamels to hard iron oxide* (Kiev: Naukova Dumka) pp 304–6
[17] Popel S I, Zakharova T V and Pavlov V V 1974 Spreading of lead-tin melts on iron surface In *Adhesion of melts* (Kiev: Naukova Dumka) pp 53–8
[18] Popel S I, Pavlov V V and Zakharova T V 1974 Kinetic features of liquid metals spreading on solid surface In *Adhesion of melts* (Kiev: Naukova Dumka) pp 7–11
[19] Sorokin Yu V, Khlynov V V and Esin O A 1965 Spresding rate of slag on oxide materials In *Surface phenomena in melts and solid phases* (Nalchik: KBSU publ. house) pp 558–62
[20] Eremenko V N., Lesnik N D and Ivanova T S 1977 Kinetics of flow and contact interactions in metal systems with intermediate phases In *Methods of research and properties of interface boundaries of contacting phases* (Kiev: Naukova Dumka) pp 51–70

IOP Conf. Series: Materials Science and Engineering **905** (2020) 012019 doi:10.1088/1757-899X/905/1/012019

Self-consistent model of fluid structure in volume and boundary layer

V S Novosadov[1], R Kh Dadashev[2], R A Kutuev[3], D Z Elimkhanov[3], Z I Dadasheva[3], and Kh S Talkhigova[3]

[1] Moscow State University of Food Production, 11, Volokolamskoe sh., Moscow, 125080, Russia
2 Academy of Sciences of the Chechen Republic, 13, prospect named after M. Esambayev, Grozny, Russia
3 Chechen State University, 17a, Dudayev Boulevard, Grozny, 364060, Russia

E-mail: novosadov@list.ru, raykom50@mail.ru, kra-07@mail.ru, edzhabrail@mail.ru, zarema-69-69@mail.ru, talhigova@mail.ru

Abstract. The study proposes the self-consistent theory that combines an activation model, a free volume model and a cluster model. The theory makes it possible to explain why the flow of fluid in slip conditions requires extremely minimal tangential stresses and describes the real structure of the fluid in the boundary layer, as well as to explain why the fluid is almost not squeezed out of the capillary gap.

1. Introduction

At present, various fluid models have been developed. In general, these models can be divided into two groups: in one of them monoatom is taken as a structural unit, and in the other – clusters. Each of these models has its own advantages and disadvantages. In certain approximations these models describe some properties and processes that occur at the interfacial boundary. However, despite a relatibely good coincidence of the theory with the experiment for simple polyatomic fluids, the existing models are only the first approximation to describe viscosity and diffusion coefficients in solutions.

It seems interesting to develop a self-consistent fluid model that combines an activation model, a free volume model and a cluster model. The paper analyzes the main models of fluid structure: monoatomic, hole theory, activation models, free volume models and a cluster model. The cluster fluid model is proposed, which explains peculiarities of fluid structure in volume and boundary layer.

2. Methods and materials

In order to understand and describe the processes of wetting, spreading and capillary flow of fluid, the impregnation of porous materials, let us consider the structure and diffusion permeability of fluid using the model of microgroups (clusters) [1–5]. In general, the existing fluid models are divided into two groups: in one of them the monoatom is taken as a structural unit, and in the other – microgroups. From the analysis of the fluid structure and methods of its stabilization (homogenization) it is possible to significantly improve the quality and properties of the fluid, in particular metal melts and solders.

Self-consistent fluid theory shall combine an activation model, a free volume model and a cluster model. It should answer the following questions: why for fluid flow under slip conditions the required

IOP Conf. Series: Materials Science and Engineering **905** (2020) 012019 doi:10.1088/1757-899X/905/1/012019

threshold minimum shear tangents τ do not exceed 10^{-6} N/m^2, comparable to measurement error; how the fluid structure changes over the thickness of the boundary layer, and why it is almost impossible to extrude the fluid from the gap. Tangential stresses for the movement of the fluid particle at the free flow rate U decrease from the wall deep into the fluid according to the law of change of the thickness of the boundary layer. This is obviously due to particular structure of the fluid in the boundary layer, which prevents it from being extruded out of the gap. Based on the analysis of the main fluid models [1–14] the following conclusions can be drawn about its structure. The numerical simulation [6] within the monatomic model shows that the mean square displacement d of the diffusing particle makes ~ 1.5, i.e. less than 0.5 of the average distance between the fluid atoms. In other models [14] the value d for self-diffusion in fluid metals is 0.4...1.1, i.e. 10...40 % of the interatomic distance. Unfortunately, based on the analysis of radiograms of the radial distribution of the monatomic fluid, no definite conclusions can be drawn regarding the structure (especially for polyatomic fluids). The hole models [7–10] (Frenkel Ya.I., Glesston S.) represent a relatively rough approximation of the real movement of the fluid molecule. Ignoring the features of interaction between molecules, the Stokes-Einstein equation makes it possible to establish a link between diffusion coefficient D and viscosity η. The model seems fair for fluid metals in case of ionic radius of monovalent ion used as a particle radius. Regardless of the fluid model, the temperature dependence $D(T)$ and $\eta(T)$ is determined by the exponential dependence with the corresponding values of the activation Q_D and Q_η, which in reality is confirmed within a limited temperature range. Within the framework of the hole theory of fluid (Frenkel Ya.I., Eyring D. et al.) it is assumed that cavities ("holes"), which have a certain equilibrium concentration, arise and burst in fluid as a result of fluctuation in various places. Frenkel estimates the radius of holes making less than ~ 10 nm, which in order of magnitude corresponds to the size of the cluster. These calculations can only be considered as estimates, as it is illegal to use surface tension in the calculation of macroscopic concepts. As already noted, numerical modelling of particle paths indicates that there is no stick-slip phenomenon of particles and thus the basis for introducing the concept of activated state disappears. The activated and primary states become indistinguishable.

Within the framework of the non-activation models in order to calculate the diffusion coefficient Cohen and Turnbull [11] applied the "free volume model" (the difference between the volume of fluid and the self-volume of molecules), which is continuously redistributed between different areas of the solution. The particle moves when a cavity of more than some critical volume $V_{кр}$ and the redistribution of the free volume does not require work. However, here too, the diffusion motion of particles essentially remains uneven, so that this model is not quite consistent. The best correspondence of theory with experiment in all models is obtained by using ionic radii as diffusion units. To describe the temperature dependence of viscosity, the simple exponential dependence is often used. Free volume theories and activation theories are also used in viscosity modeling. Despite a fairly good coincidence of the theory with the experiment for simple polyatomic fluids, the existing models can serve only as the first approximation to describe viscosity and diffusion coefficients in solutions. According to the data given in [1–3], the cluster model quite well describes the self-diffusion and dynamic viscosity coefficients. The models proposed in [1–3] allow determining the dimensions of cluster radii: r = 0.46 (Pb)...2.3 (Fe)...6.2 (Li) nm. Inter-cluster breaks at melting temperature occupy a small volume and vary within 1...6 %. Inter-cluster breaks result in a large number of activated atoms having at least one broken bond on the "surface" of the cluster and serve as a measure of fluid disordering. The concentration C_a – activated atoms for different metals at $T_{пл}$ is within 10...40 % indicating that a significant proportion of is are activated. The cluster has a lifetime $t_{ж}^{кл}$ of about 10-7...10-8 s and can only be conditionally isolated over a period of time greater than its thermal oscillation period t_Q^{am} near some equilibrium position (t_Q^{am} ~ 10-11...10-10 s). According to the cluster model [1...3], both individual atomic groups (clusters) and activated atoms in inter-cluster breaks will participate in the elementary act of diffusion in the fluid. Thus, the resulting mass transfer coefficient D_0 is composed of members: D_{cl} – cluster diffusion coefficient and D_a – diffusion coefficient by inter-cluster mechanism in the zone of inter-cluster breaks

$$D_0 = D_{cl} + D_a C_a \tag{1}$$

However, there is a question of justifying the form of clusters. The authors use a simple cubic shape or sphere inscribed in a cube, with inter-cluster breaks creating a free volume of 1...6 %. It should be noted that when the spherical particles are laid, the least dense cubic laying and the denser hexagonal laying are distinguished. Porosity of such medium does not depend on the radius of spheres and makes: at cubic laying 47.64 % and at hexagonal – 25.94 %, which is almost an order of magnitude more than free volume in fluid 1...6 % (on average we accept 3 %). The amount of movement of cluster d is greater than 2 $r_{кл}$, which conflicts with the main fluid models. Besides, the theory does not consider the possibility of a vacancy in the cluster, although it is postulated that the size of the cluster is determined by the sphere of stress in the vacancy area. The absence of vacancies in the cluster of fluids excludes the possibility of activation mechanism of diffusion "on the body of the cluster", and therefore in the volume of fluids.

3. Results

On the role of stress in the interfacial boundary area. The surface of the solid phase, like macrodefect, represents a powerful stimulating factor that changes the structure of the fluid at the interfacial boundary. Wetting in the process of formation of the structured layer of clusters in the area of interfacial boundary [4] creates stresses [5].

$$\sigma = \sigma_1 + \sigma_2\left(\frac{\Delta a}{a}, MCP\right) + \sigma_2^d, \tag{2}$$

where σ_1 – stress in the surface layer of the solid phase (surface stress) due to surface layer rearrangement; σ_2 – takes into account the contribution of dilatations due to the crystallographic mismatch between the parameters of crystal lattices of cluster atoms and the solid substrate. This contribution can be as high as about $5 \cdot 10^4$ MPa [15]. Obviously, this value is overestimated and will be reduced by stress relaxation when the dislocation structure of the surface layer is rearranged. It was experimentally established that the dilation effect does not allow explaining from the state diagram the influence of coatings on the solid substrate at the level of stress in the contact zone. These stresses will also be determined by the difference in metallochemical properties (MCP), which are determined by the features of the electronic structure of the contacting phases. σ_2^D – defines the diffusion and concentration stresses arising during mass transfer, which are caused by different mobility of atoms crossing the interfacial boundary of solid and fluid phases (diffusion) and the appearance in the solid phase of atoms of a different variety (concentration). The length of stress field in the area of interfacial boundary in metal systems exceeds tens of microns. The effect of contact strengthening of solder joints depending on the gap proves the correct estimates. If the thickness of the soft layer of solder is less than 50...70 μm, it is possible to obtain equivalent-strength joints. For organic compounds, the length of this field is reduced by an order of magnitude.

Features of cluster structure of fluid in volume and boundary layer [5]. Let us consider the velocity field in laminar fluid movement along a capillary of size $d=h<1.5$ mm in which convective flows can be neglected. The distribution of speeds (current lines) along the section near the entrance is almost uniform (Fig. 1). Then, under the influence of viscosity forces, the speeds are redistributed along the section. The layers of fluid at the wall are slowed and a boundary layer is formed, and the central part of the flow (core), where a uniform distribution of speeds is still maintained, moves faster, which is due to the need for a certain flow rate of fluid to pass through the capillary section. Outside the defined initial section ($l_{нач}$), the two boundary layers are closed and then a uniform flow is established at some distance greater than the $l_{нач}$. Thus, the uniform flow in the flat capillary consists of 2 "uniform" boundary layers.

Viscosity is the property of a fluid (opposite to fluidity) to resist shear (sliding of its layers), and tangential stresses occur in the fluid. Fluid braking at the wall is caused by viscosity and changes directly in proportion to the velocity gradient U:

$$\tau = \eta du/dy.$$

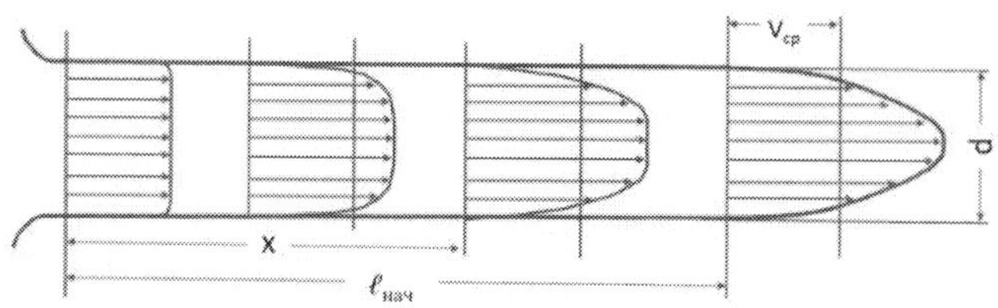

Figure 1. Formation of velocity profile at the initial section in capillary

Speed profile diagram (Fig. 1) resembles a deck of cards during slanting. It is reasonable to assume that the current lines consist of a layer of clusters (the map itself), and the gap between the layers represents inter-cluster breaks, most of which are parallel to the flow. When calculating the viscosity of simple monatomic fluids Andrade [12] uses the concept of average distance between layers, but does not determine its absolute value. According to [16], a gap of 0.15 nm (1.5 $\AA$) remains between the liophobic walls of the quartz capillaries at the flow between mercury and the walls, which provides a slip effect. As already noted, numerical modeling clearly indicates that the displacement of particles in the fluid occurs at distances less than 0.5 times the interatomic distance, about 1.5$\AA$. Obviously, in the case of inter-cluster interaction, this distance h should be smaller and commensurate with the amount of inter-cluster discontinuities. In cluster models [1...3] this distance is not defined.

The change of volume observed in experiments when melting (ΔV) can be presented as the sum of three factors: $\Delta V = \Delta Vr - \Delta Vc + \Delta Vg$, where ΔVr – increase in volume due to increase of distance between clusters; ΔVc – reduction of volume due to their regrouping. The value $\Delta Vc = (n`/2n)K_y$, where $K_y$ – consolidation coefficient, which changes depending on the type of a crystal lattice in a firm state: $n`$ – number of atoms on a surface of a cluster, n – total number of atoms in a cluster. Therefore, K_y at the capillary wall will be substantially higher than in volume. The third factor ΔVg – increase in volume due to emergence and closure of voids as a result of fluctuation. According to Ya.I. Frenkel's calculations, the dimensions of these voids are less than 10 nm, which is commensurate with the dimensions of clusters. Repackaging of clusters becomes possible in the process of completion of inter-cluster breaks after melting. Obviously, under the influence of the stress field, this factor should lead to a dense hexagonal packing of clusters regardless of the type of packing of atoms within the clusters. The effect of rearrangement is zero for dense-packed metals and becomes more prominent with the structure of the metal becoming more "loose" in a solid state, thus decreasing the melt volume. The equation for assessment of ΔVr and ΔVc for metals with various type of packing is given in work [3].

For closely packed metals $Ky=0$ and $\Delta Vc=0$ the calculation models are in a relatively good compliance with the experiment. Thus, the most energetically advantageous form of clusters in the area adjacent to the surface should be the hexagonal form. While moving away from the surface and approaching the "normal" fluid at the edge of the boundary layer, the shape of the clusters may change. A simple cubic shape is preferable, which corresponds to the model [3] for fluid in volume. The number of holes and their dimensions will also increase when removed from the surface.

Non-vacancy diffusion mechanism in fluid. Similar to the non-vacancy diffusion mechanism in substitutional solid solutions [15] let us consider possible mechanisms for moving clusters in the fluid volume. We consider two versions of the theory of activated diffusion of clusters: 1) the presence of empty space ("holes") is not assumed and 2) its presence is justified (Fig. 2). The cluster may be displaced either because of the simple exchange of places with neighbors (Fig. 2-1) or due to collective rotation of the group of atoms (Fig. 2-2) in the neighborhood. If there is a void, it is possible to move the cluster in the same way as to move it to the vacancy (Fig. 2-3), etc. According to the crowdion mechanism (crowdion – one-dimensional compression of atoms on certain sections of a row), cluster movement is a consequence of a sequence of small movements of crowdion in the direction of its orientation. This mechanism provides a significant speed of movement with little activation energy. In accordance with the model shown in Figure 3, it is reasonable to present a boundary layer of variable thickness and viscosity clusters in order to provide a shape for changing the velocity profile. The thickness of these layers and the concentration of voids will naturally increase when approaching the boundary layer. In the solid body, the pores may have a different shape [17]: spherical, cubic or polyhedral, which is limited to eight planes of type (111) and six of type (100), etc. It is highly likely that the equilibrium shape of the hole will also change as the cluster shape changes. Let us consider the mechanism of pulse transmission in the fluid outside the boundary layer. The dashed line in Figure 3 shows current lines. Let us arbitrarily select any layer and consider the interaction of clusters located in the central layer with nearby layers.

Combining the models of Andrade [12] and G.M Panchinkov [13], it is possible to present the following scenario of pulse transmission from one cluster layer to another. Pulse transmission is accomplished by those clusters whose energy reaches some optimal value and when the cluster is at a distance greater than the equilibrium distance. Pulse transmission will be connected with temporal association of two clusters from adjacent layers. The ultimate energy ε constitutes a certain fraction of fluid evaporation heat (E_{eva}), namely $\varepsilon = 2E_{eva} / z$, where z – coordination number. Let us assume that the middle layer cluster has transmitted a pulse of more energy to the layer above. Depending on the pulse energy, the associated cluster will move in the direction of the total pulse. In the next step, its dissociation will occur by creating an inter-cluster rupture to achieve equilibrium size and shape. Further migration of dissociated clusters will be consistent with the non-vacancy mechanisms presented in Figure 3, which is typical of all layers.

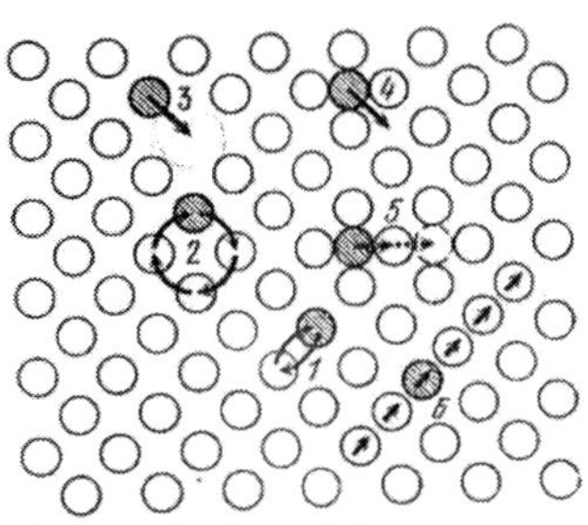

Figure 2. Possible non-vacancy models of cluster diffusion mechanism (free volume model): 1) simple exchange; 2) cyclic exchange or ring exchange; 3) movement at the presence of hole void; 4) simple inter-cluster mechanism; 5) inter-cluster displacement mechanism; 6) crowdion mechanism

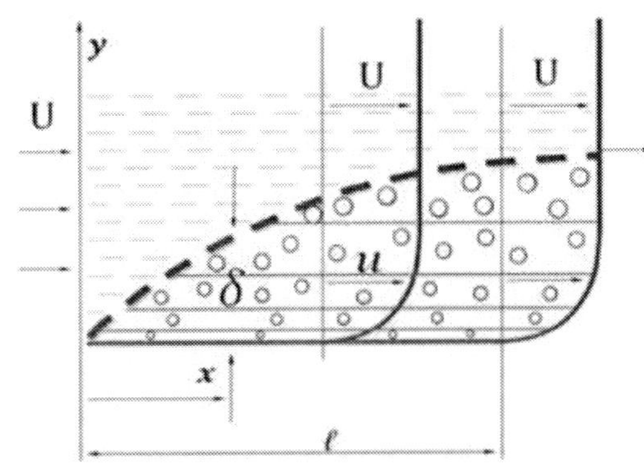

Figure 3. Boundary layer and its cluster structure

In order to explain the ultra-small tangential stresses at the flow of fluid in free volume, it is reasonable to assume that the probability of directing most of the inter-cluster breaks between the layers will be planar oriented, i.e. parallel to the action of applied forces. Such arrangement is energetically more advantageous and allows easy sliding of layers. In the boundary layer, the stress field in the interfacial boundary region will orient the inter-cluster breaks orthotropically, i.e. in the direction normal to the surface. While moving from the surface and decreasing the stress field, the orientation of the inter-cluster breaks will gradually approach (at the periphery of the boundary layer) their location in the fluid volume, allowing easy sliding of the layers. Abnormally high viscosity at the solid surface is caused by the association of clusters in the fluid into agglomerations. Figure 3 shows a diagram of the multilayer distribution of associated clusters in the direction of decreasing the intensity of the association process. This model makes it possible to explain the nature of the change in velocity field at the wall. Figure 3 also shows a diagram of the increase in voids and their dimensions towards the boundary layer.

When the fluid interlayer is deformed to a value of ~ 10 μm [5], the surfaces approach with increasing density and, accordingly, viscosity of the fluid. In this case, a quasi-solid body structure with high yield strength values is formed in the gap. Inter-cluster fractures normal to the surface make it very difficult for clusters of this layer to slide in the direction of tangential stresses. The yield

IOP Conf. Series: Materials Science and Engineering **905** (2020) 012019 doi:10.1088/1757-899X/905/1/012019

strength of this layer approaches these values for the solid phase at high temperatures. This is confirmed by the proximity of the pressure values required to extrude fluids and to provide welded joints during pressure diffusion welding. The yield strength of this layer approaches these values for the solid phase at high temperatures.

4. Conclusion

1. The main models of fluid structure were analyzed: monoatomic, hole theory, activation models, free volume models and cluster model. The cluster model of fluid is proposed, which explains peculiarities of fluid structure in volume and boundary layer.

2. On the basis of the model, the reasons are determined, which ensure easy sliding of fluid in the volume, and make its movement in the boundary layer difficult.

3. It is shown that the difficulty of extrusion of fluid interlayer from gap under pressure is caused by special associative structure of clusters and arrangement of inter-cluster breaks along normal to surface, i.e. formation of quasi-solid body structure in a gap with yield stress close to solid phase at high temperatures.

References

[1] Ershov G S and Chernyakov V A 1978 *Structure and properties of fluid metals* (Moscow: Metallurgy) pp 248
[2] Ershov G S and Bychkov Yu B 1982 *Physical and chemical bases of rational alloying of steel and alloys* (Moscow: Metallurgy) 376 p
[3] Ershov G S and Poznyak L A 1985 *Microinhomogeneity of metals and alloys* (Moscow: Metallurgy) 214 p
[4] Novosadov V S 2010 Features of structure and properties of structured fluid layer near interfacial boundary *Analysis of current state, discussion, prospects and achievements in the field of soldering* Mater. Seminar Soldering 2010 Coll. 1 CRDZ (Moscow) pp 10–6
[5] Novosadov V S 2018 Cluster model of metal melts during metal soldering. Review *Coll. of mater. of Int. sci. and techn. Conf. Soldering 2018* (Toglyatti, 11–14 September) (TSU publ. house) pp 73–101
[6] Rahman A 1964 *Phys. Rev. A* 136, 405
[7] Frenkel Ya I 1945 *Kinetic theory of fluids* (Moscow; Leningrad: USSR Acad. of Sci.)
[8] Glesston S et al 1948 *Theory of absolute reaction rates* Transl. from Engl. (Moscow: Foreign literat. Publ. house)
[9] Belaschenko D K 1970 *Transport phenomena in fluid metals and semiconductors* (Moscow: Atomizdat) 397 p
[10] Frenkel Ya I 1945 *Kinetic theory of fluids* (Moscow; Leningad: USSR Acad. of Sci.)
[11] Cohen M N and Turnbull D S 1959 *Chem. Phys.* **31** 1164
[12] Andrade E N da C 1934 *Philos. Mag.* **17** 497, 698
[13] Panchenkov G M 1947 *Viscosity theory of fluids* (Moscow; Leningrad: Gostekhizdat)
[14] Lepinskikh B M, Kaibichev A V and Savelev Yu A 1974 *Diffusion of elements in fluid metals of iron group* (Moscow: Nauka) 191 p
[15] Geguzin Ya E 1979 *Diffusion zone* (Moscow: Nauka) 344 p
[16] *Gas-solid interfacial boundary* Transl. from Engl. ed E. Flada 1984 (Moscow: Nauka) 160 p
[17] Geguzin Ya E 1967 *Physics of sintering* (Moscow: Nauka) 360 p

Geoecological monitoring of the Khankal field based on thermal infrared imaging

E V Elsunkaeva[1], T B Ezirbaev[2] and T Kh Ozdieva[1]

[1] Grozny State Oil Technical University named after Acad. M.D. Millionschikov, 100, Ordzhenikidze sq., Grozny, 364051, Russia
[2] Integrated Research Institute n.a. Kh.I. Ibragimov, Russian Academy of Sciences, 21A, Staropromyslovsky ave., Grozny, 364051, Russia

E-mail: elina.elsunkaeva@mail.ru, timersno@mail.ru, tami.93.93@mail.ru

Abstract. The aim of the work is to review technology for assessing and modeling quality of the environment in industrial zones, the emission of pollutants into the atmosphere based on geographic information systems and environmental modeling taking into account environmental monitoring data. The article presents results obtained by using Geoscan 201 for simultaneous shooting with a Thermophrame-M thermal imager and a visible range camera.

1. Introduction

The Khankala thermal water deposit has been in operation since the 1970. To date, some of the wells have been abandoned or are not in operation. The surviving well stock is used for heating greenhouses.

The field is located 10 km southeast of Grozny, the Chechen Republic (Fig. 1).

The thermal groundwater of the Khankala deposit is concentrated in the Middle Miocene Karagan-Chokrak deposits, composed of sandstones and clay interlayers and lenses. In total, there are up to 22 productive formations with a thickness of several to 50–60 m [4, 5].

In 2015, a pilot geothermal station was put into operation on the territory of the Khankal field. It was based on the circulation scheme for using the Earth deep heat (DCS).

The station with a capacity of 8 MW has been in operation since the beginning of 2016. It is based on the circulation system consisting of injection and production wells; recirculation involves the operation of the station with complete injection of water back into the reservoir; the greenhouse system acts as a heat consumer (Fig. 1) [6].

During the operation of thermal groundwater deposits using doublet technology, the main environmental threat is leakage and spills of fluid from the primary circuit. To solve this problem, it is necessary to monitor the field [3].

2. Methods and materials

Environmental monitoring is carried out by field and remote methods. Field (contact) methods include monitoring carried out at stationary observation posts and ground route studies. Remote methods are photo and video shooting, as well as lidar sensing (thermal, infrared, and other types of surveys) Remote methods are widely used to monitor the environment [1, 2, 4].

IOP Conf. Series: Materials Science and Engineering **905** (2020) 012020 doi:10.1088/1757-899X/905/1/012020

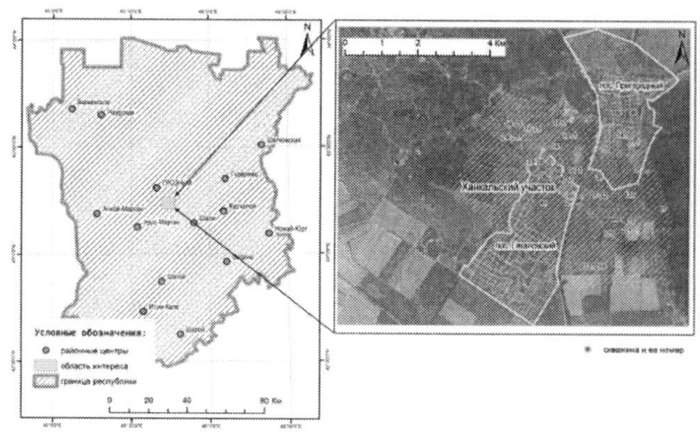

Figure 1. The map of the Khankala license area

The dynamic development of unmanned aerial and ground technical equipment and geo-information technologies contributed to the development of alternative methods for monitoring vast territories. One of these methods is the use of unmanned aerial vehicles (UAVs) for aerial photography and creation of digital terrain models [8, 9]. In 2016, monitoring using an UAV equipped with a digital camera and a thermal imager was carried out at the Khankala thermal groundwater deposit and the adjacent territory. Results of photographic, thermal and infrared shooting and their analysis are presented below.

To monitor the Khankala field of thermal underground water, the UAV Geoskan 201 designed for shooting area and extended objects was used. This UAV can be offline for 3 hours, with an average flight speed of 80 km/h. Such parameters allow for observation of a large area. The launch of the drone is carried out using a catapult, the device lands on a parachute. The system is automatic, the ground control panel controls the process at each flight stage.

The UAV can carry up to 1 kg of load.

As a payload, a set consisting of a gyrostabilized optoelectronic system and a thermal imager of 640 × 480, 17 μm, 25 Hz was used.

The paramaters are as follows: image expansion is MPEG or H.264, 2-axis stabilization, accuracy of 250 mrad, rotation range in azimuth, unlimited elevation + 10–100°.

Overall dimensions: the diameter of 110 × 144 mm, the weight of 620 g

The Thermophrame-M thermal imager was used with the visible range camera; the shooting was synchronous. The shooting was carried out in accordance with the algorithms of the GeoScan Planner program, which is part of the UAV ground control station. The images were stitched together using the Agisoft PhotoScan program by the block editing algorithm. The total area of the monitoring site was 3710 m^2. Along with photographing, the thermal imaging survey of the area was carried out. It took two flights of 2.5 hours to capture the entire area.

3. Results

As a result of heat recording data processing, more than a dozen thermal anomalies were revealed (Fig. 2). Interpretation of the data (forms and temperatures) on the identified anomalies made it possible to determine their sources. [6].

Cameral works were followed by the field study of the identified anomalies. The coordinates were determined using the Garmin GPSMAP 64ST GPS navigator. Assessment of reliability of the monitoring results using UAVs was presented below. The results were compared with the results of field studies at the site (Fig. 3).

The above anomaly is due to the spill of hot water. It has a temperature of 50–55 (conventional temperature units) and an isometric shape of 50 × 40 pixels [7].

IOP Conf. Series: Materials Science and Engineering **905** (2020) 012020 doi:10.1088/1757-899X/905/1/012020

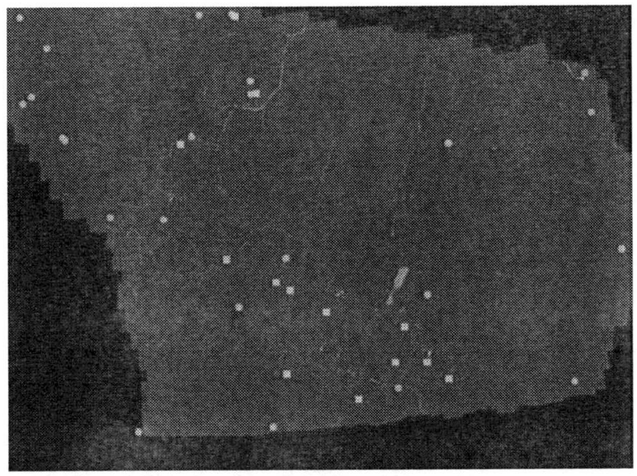

Figure 2. The photo of the area obtained as a result of heat shooting. Circles mark wells, squares – heat anomalies

a) Anomaly recorded using the UAV

b) The area where the anomaly was recorded

c) Color gradation of temperature changes

Figure 3. Comparative analysis of thermal survey results

4. Conclusion

Studies using UAVs conducted at the Khankal field identified 13 anomalies. The results of the experiment showed the possibility of monitoring exploited deposits of thermal groundwater using an unmanned aerial vehicle and a thermal imager to detect anomalies caused by leaks of old wells and pipelines, draining of used fluid, etc. In addition, the monitoring will allow for timely repair and cleaning works increasing the life of the systems.

Environmental monitoring of the area with operated wells of heat energy water using the UAV Geoskan 201 and a thermal imager showed that it is possible to sort out anomalies and determine the source by shapes and temperatures of the anomaly.

The development of thermal water deposits causes micro-earthquakes, release of hydrothermal steam, spills of thermal water and subsidence. Continuous monitoring of geothermal water deposits and the development of methods for timely assessment of the environment of industrial facilities using remote monitoring data can prevent possible environmental and other risks or eliminate them.

References

[1] Ayroyan Z A, Korkishko O A and Sukharev G V 2016 Monitoring of oil trunk pipelines using unmanned aerial vehicles *Don Engineer. Bull.* 4 Retrieved from: ivdon.ru/ru/magazine/archive/n4y2016/3898

[2] Gryadunov D A, Mitrofanov E V and Bubnenkov D I 2012 On the use of unmanned aerial vehicle complexes for multi-level environmental monitoring *Bull. of Moscow State Reg. Univer. Natural Sci.* **4** 95–9

[3] Labazanov M M, Shaipov A A, Damzaev Z M E and Ozdieva T Kh 2017 *Thermal waters of the North Caucasus* All-Russian Congress of Ecologists. Collection of abstracts dedicated to the Year of Ecology in Russia (Grozny) pp 39–49

[4] Zinchenko O N 2011 *Unmanned aerial vehicles: application for aerial photography for mapping* (Moscow: Racurs) Retrieved from: www.racurs.ru/www_download/articles/UAV_1.pdf

[5] Kerimov I A, Daukaev A A, Moiseenko N A et al 2009 *Minerals of the Chechen Republic* Reference book (Grozny) 245 p

[6] Farkhutdinov A M, Cherkasov S V, Shaipov A A et al 2017 Monitoring of thermal groundwater using an unmanned aerial vehicle on the example of the Khankal field *Vest. of the Acad. of Sci. of the RB* **23((2)86)** 75–82

[7] Farkhutdinov A M, Fuke de S, Mintsaev M Sh and Cherkasov S V 2015 The use of geostatistics to analyze the prospects for the exploitation of the Khankal deposit of heat energy water *Geoinform.* **1** 60–8

[8] Carrivick J L, Smith M W, Quincey D J and Carver S J 2013 Developments in budget remote sensing for the geosciences *Geology Today* **29(4)** 138–43

[9] Jordan B R 2015 A Bird's-eye view of geology: "The use of micro drones. UAVs in Geologic Fieldwork and Education" *GSA Today* **25** 42–3

IOP Conf. Series: Materials Science and Engineering **905** (2020) 012021 doi:10.1088/1757-899X/905/1/012021

Method of complex interpretation of results well logging (TABC Gintel) for the terrigenous rocks sections Tersko-Sunzhensky petroleum region

T B Ezirbayev[1], M A Khasanov[2] and S S-A Gatsaeva[2]

[1] Kh. Ibragimov Complex Institute of the Russian Academy of Sciences (CI RAS), Grozny State Oil Technical University, Grozny, Russia
[2] Grozny State Oil Technical University, Grozny, Russia

E-mail: timersno@mail.ru, geofizikggni@mail.ru, geofizikggni@mail.ru

Abstract. A brief overview of the proposed method of processing and interpretation of well logging data in Terrigenous deposits. And also presents results of data well survery with its application using the example terrigenous deposits Tersko-Sunzhensky petroleum region. The findings suggest that the interpretation of the results of the GIS, as described TABC gives accurate information to determine geological characteristics and hydrocarbon saturation Lower Cretaceous terrigenous deposits of the Tersko-Sunzhensky petroleum region.

1. Introduction

As the problems solved by well logging methods become more complex, the problem of studying complex collectors also increases.

Special attention is paid to manifolds with polymineral matrix composition, complex structure of capacitive space and multicomponent composition of fluid, as well as low-permeability clay manifolds. The identification of such collectors within the productive horizon is very important in the development of oil and gas fields.

The non-optimality of well logging complex and research technology, imperfections of metrological support, methods of individual and complex interpretation of well logging data are added to the objective difficulties of solving the set tasks.

Well logging methods are very diverse and use all kinds of physical fields (electric, electromagnetic, nuclear radiation, gravitational, mechanical stresses, thermal). No geological problem can be solved by any of the well logging methods alone. Hence the need for integrated application of well logging methods and interpretation of the results.

2. Methods and materials

The current state of the industry makes it possible to create an information and computing system, within the framework of which it is possible to implement a generalized algorithms mic approach to the complex interpretation of well logging data, which implements the entire arsenal of currently existing methodological approaches in Russia and abroad with the use of various systems of petrophysical models and connections. Developing this direction, the Gintel system developed the technology of integrated processing and interpretation of well logging ESKS-TABC data, within the

IOP Conf. Series: Materials Science and Engineering **905** (2020) 012021 doi:10.1088/1757-899X/905/1/012021

framework of which there is a "Technique of automated restoration of rock properties in terrigenic section according to well logging data in the Gintel system."

The TABC methodology uses various algorithm schemes for processing and interpreting well logging data based on the use of petrophysical model systems, stochastic connections, quantitative and qualitative criteria and limitations. At the same time, it considers the use of collector rock properties both all known petrophysical models based on single and two-component rock models (sandstones, clay sandstones containing wet clay) and the use of a system of petrophysical models describing a three-component model of sand-aleurite-clay rock, which allows to restore the full volume of structural and thermic model of the well logging according to the data of the well logging . Optimal selection of computational circuits and systems of petrophysical models and connections is determined in the process of adaptation of TABC procedure to specific geological conditions of studied section. When processing well logging data in the computer system, the method of interpretation modeling is used, which provides multiple determination of rock properties with assessment of reliability of found geological characteristics of the section on the basis of automated use of core data, results of tests, materials of trial or current operation on the date of logging in each well.

Results of processing of well logging data by wells shall be prepared in accordance with the requirements of regulatory documents of State Commission for Mineral Reserves. In addition, they can be exported in various accepted formats for input to software complexes that perform geological modelling of oil and gas deposits and calculation of hydrocarbon reserves therein, for example, to the currently used Pertel (Schlumberger) and Irap RMS (Roxar) packages in oil companies. [6, 7].

3. Results

This article discusses the principles of using the TABC technique to determine the calculated parameters of oil-bearing formations of lower-Cretaceous deposits of the Tersko-Sunzhensky oil-and-gas region (TSOGR).

In order to determine the physical properties of such rocks as terrigenic deposits of lower chalk according to well logging data (on the date of logging), it is necessary to use a system of petrophysical models, which should take into account the peculiarities of change of properties of terrigenic deposits and their oil saturation. Which can be different in values depending on the geographical location [10]. Justification of the petrophysical support system for sediments lower chalk in various fields of the Tersko-Sunzhensky oil and gas region was carried out taking into account a number of factors:

1. In the vast majority of wells there are data of electrometrics, acoustic and radioactive methods, cavernometry necessary for study of rock properties. All logging data are recorded by equipment developed in the 1970s.

2. Core studies are often performed by determining a standard set of physical properties of rocks: porosity, absolute air permeability, bound water saturation by centrifugation, bulk density, specific electrical resistance measured under atmospheric and formation conditions under full and partial water saturation of samples with water with mineralization of about 80 g/l. Granulometric analysis is performed on some part of cores. There is a description of a small collection of grinding. Sand, aleurite and clay fractions were formed on a special core collection, and natural gamma activity and cationic exchange capacity were measured on these fractions. There is no data from the capillarimetry study.

Laboratory analysis was carried out on kerns from wells of various deposits in small volumes (Carabulak-Achaluki, Starogroznenskoe, Oktyabrsky, Akhlovsky, Malgobek- Voznesenskoe, Eldarovsky, Hayan-Kortovskoe, Yastrebynny, Bragunskoye). The depth of core sampling is not related to well logging data – only the slipping intervals within which the core was sampled for testing are known. Therefore, the generalization of the core data can only be carried out in an integrated form. In justifying the petrophysical support, the core and well logging data were independently linked by the authors.

A comparative analysis of the values of the parameters on the core taken from different deposits showed that in the graphs of the comparison of the parameters, for example porosity, permeability and bound water saturation (Figure 1), or relative resistance and porosity (Figure 2), while maintaining a uniform pattern of the parameters relationship, there is a wide variation of points, which makes the establishment of stochastic bonds, which are usually used in traditional approaches to the interpretation of well logging data, uncertain.

The location of the points on the correlation connections reflects the natural and significant changes in the properties of the lower chalk rocks lying at different depths within the Tersko-Sunzhensky oil and gas bearing region. Therefore, it was concluded that in justifying petrophysical models of the interpretation of well logging data, it is necessary to use a different approach to the determination of relationships from the construction of stochastic dependencies of the type "core-core," core-well logging".

This approach should be based on the application of generalized petrophysical models that establish physically sound relationships between the parameters, and the verification of the execution of such linkages on the entire collection of core samples, regardless of where they are taken within the lower chalk thickness.

This approach was developed and applied by the authors of the TABC methodology [6] and is considered as adaptation of the set of accepted petrophysical connections to the studied rocks based on the use of mathematical modeling methods. In the process of adaptation, two problems are solved: 1) the applicability of the system of petrophysical models to describe the properties of the studied rocks is proved; and 2) the values of empirical coefficients, which are included in the formulas and adjust them for use in determining the properties of rocks in section, are established.

Therefore, in justifying the petrophysical support of the technique of interpretation of well logging data, the approach to the study of petrophysical patterns and evaluation of parameters of petrophysical connections was applied. It consists in mathematical modeling using actual core data on Lower Cretaceous sediments of known generalized petrophysical patterns inherent in terrigenic rocks and identified by petrophysics to date in the study of many types of sand-aleurite-clay rocks of different mineral composition.

The system of petrophysical models developed by V.S. Afansyev, [7] was used as the basis for the research. To date, this model system has already been successfully adapted to many geological sections of Western Siberia (mesose), Timano-Pechor Province (paleose), West Caucasus (kynose), Egypt (mesose and kynose) and other geological areas.

This system is based on a three-component model of sand-aleurite-clay rock. It follows from this model that the rock consists of a structural framework and pore space. The structural framework is folded with sand, aleurite and clay fractions, carbonate cement, primary rock debris and various other minerals.

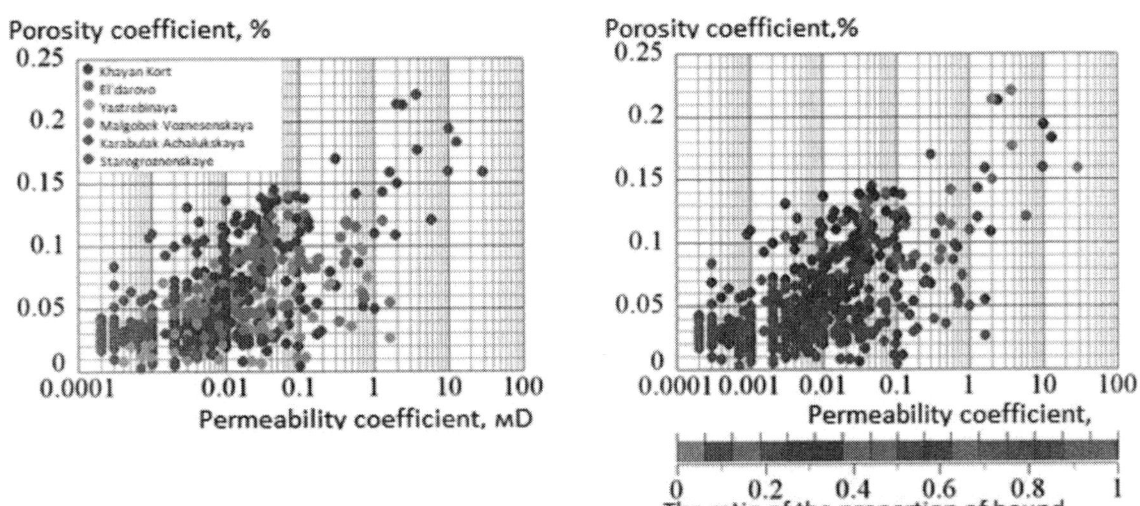

Figure 1. Comparison of porosity of absolute permeability of rocks: a – change of comparable values at different deposits, b – dependence Kpp = f (Kp, Sq.sv)

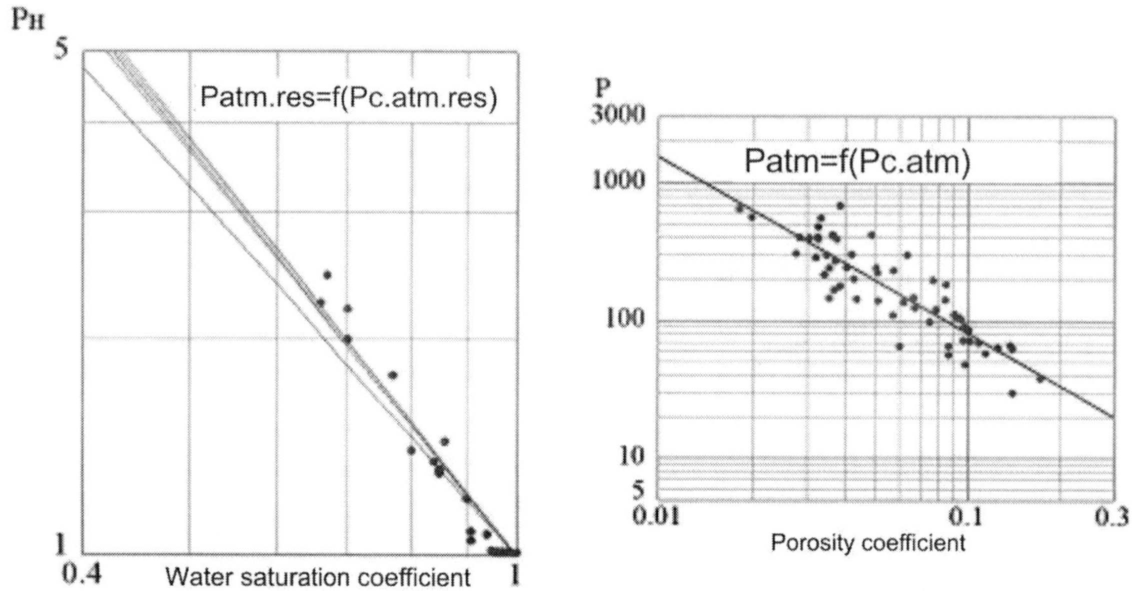

Figure 2. Dependency graphs P = f (Kp) (a) and Ph = f (Kb) (b)

The proposed model takes into account the influence of aleurite material on the physical properties of rock along with the clay component. This model also takes into account the change in the physical properties of the structural framework of the rock as a function of the ratio of its fractional composition, the change in the charge of the forming fractions of the particles, the mineralization of the produced water, the oil and gas saturation and the reflection of these changes in the adsorption processes and physical properties, which are determined by the well logging data.

An important step in the interpretation of well logging data for wells is the assessment of the reliability of certain geological properties of rocks. One example of the evaluation of the reliability of the developed methodology (comparison of the results of interpretation using the proposed methodology with the results of well testing) is shown in Figure 3.

For the three-component model of terrigenic rock a system of petrophysical equations is justified, describing physical properties of rocks and readings of well logging methods as a function of petrophysical characteristics of rocks and their fluid saturation taking into account thermobaric conditions of rock formation in section.

Underlying the creation of a system of generalized petrophysical models is the notion that terrigenic sediments are generally characterized by common basic petrophysical patterns. They form physical properties of specific sand-aleurite-clay deposits depending on actual structural-mineralogical composition of rock skeleton, hydrogeological and thermodynamic conditions of rock formation and actual oil and gas saturation of rock collectors.

In the development of petrophysical models for the interpretation of well logging data for terrigenic rocks, a comprehensive analysis of theoretical and experimental petrophysical studies carried out by many scientists in Russia and other countries was carried out and special studies were carried out.

The author together with the developer of the proposed system S.V. Afanasyev carried out studies on adaptation of the system of petrophysical models in geological conditions of alb-pharmacy rocks of lower chalk and application of the method of TABC for interpretation of well logging data at fields of Tersko-Sunzhensky oil and gas bearing region [3, 5].

To determine effect of adsorption effect on physical properties of rock, parameter is used, which determines electric charge of pore channels, formed by particles, which form rock skeleton and is expressed in form of capacity of cationic exchange of rock Q, mol/l, determined in units equivalent to value of concentration of ions in electrolyte, which saturates pore space of rock.

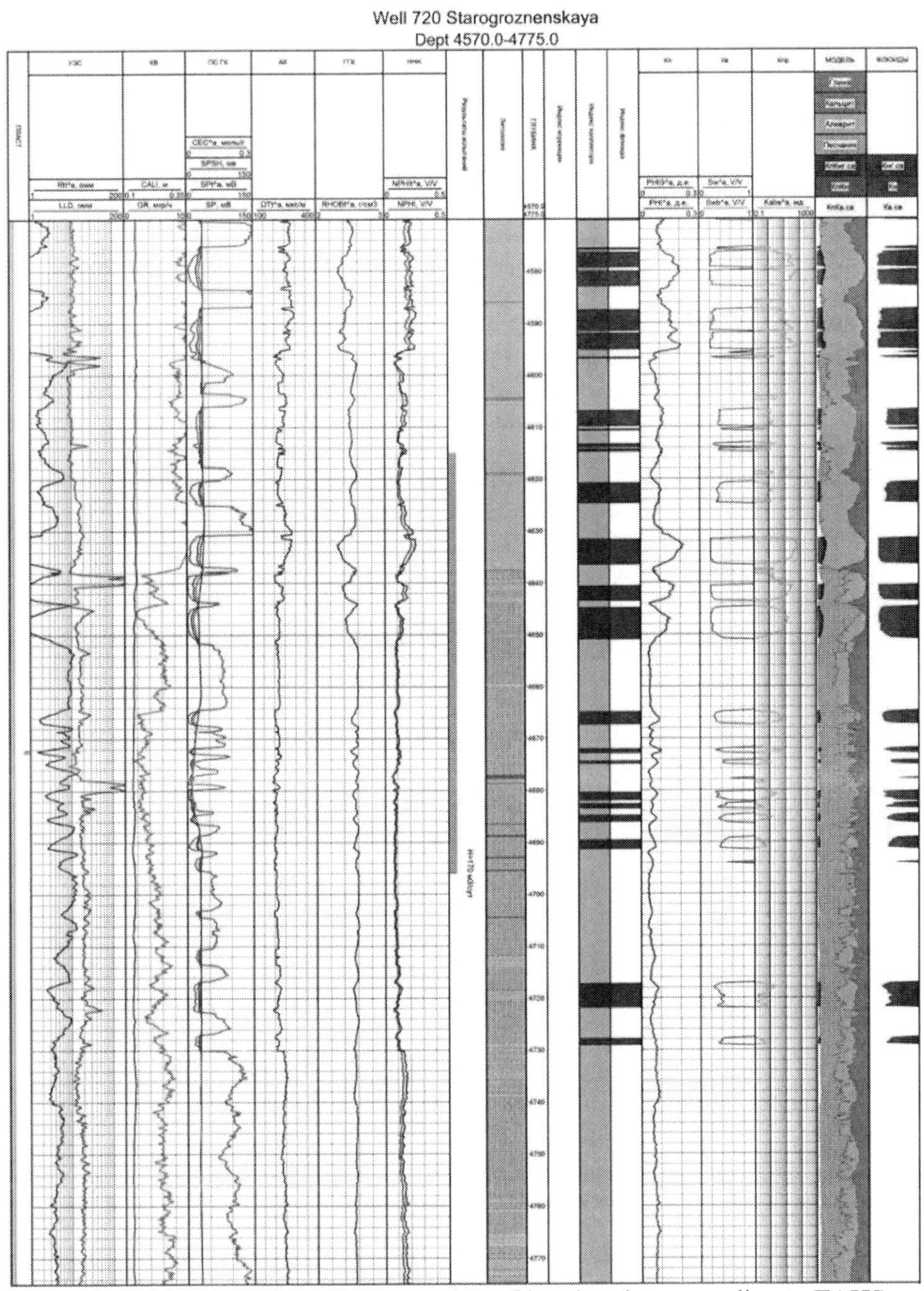

Figure 3. Tablet of results of interpretation of logging data according to TAHS method. Well № 712 Starogroznenskoe deposit Tersko-Sunzhenskiy oil and gas bearing region

The natural radioactivity model takes into account the contribution to the GC curve readings of the sand, aleurite and clay structural components of the rock skeleton.

On the basis of the synthesis of petrophysical survey data for many geological objects, a model has been developed to estimate the content of residual water in the rock. This model determines the total content of two components: the volume of water of the double electric layers formed as a result of integral adsorption processes in the rock pore system, and the volume of molecular bound water controlled by the content of fine-grained (aleurite) structural component in the rock [1, 2].

The absolute permeability of terrigenic rock is calculated from the theoretical model Crc =/[Kp, Kwhv) developed using Kozeni's equation and expressing it taking into account the proportion of bound water in Kwsv rock presented by V.N. Kobranova [8, 9].

References

[1] Ezirbaev T B, Eljayev A S, Hasanov M A and Gatsaeva S S Determining Reservoir Type in Lower Cretaceous Terrigenic Sediments of Tersko-Sunzhensky Oil and Gas Bearing Region
[2] Uze O et al 2008 *Analysis of dynamic flow* (KAPPA)
[3] Shnurman I G 2003 Studies of terrigenic reservoirs in the submountain region of Caucasus from the results of well surveys (Krasnodar: Prosveshcheniye-Yug)
[4] Hasanov M A and Ezirbaev T B 2009 Petrophysical features of terrigenic Lower Cretaceous productive reservoirs of Tersko-Sunzhensky Oil and gas bearing region as a basis for GIS data interpretation *NTV Karotazhnik* **8(185)** 3–9
[5] Hasanov M.A, Aleksandrov B.L., Eljayev A.S et al 2017 Forecasting cracked collectors on anticlinal type structures at late stage of exploration in oil and gas area *IOP Conf. Ser. Earth and Environ. Sci.* **87** 1–6
[6] Afanasyev S V 2005 Technology of a complex reinterpreting of geophysical well logging data when creating a 3D geological model of a deposit in prolonged development *Oil Industry* **2** 12–7
[7] Afanasyev V S and Afanasyev S V 2003 Gintel, a system of automated visual interpretation of log survey data. Description and User Manual, OOO *Geoinform. Technol. and Syst.* 910
[8] Gachaev A M 2011 On fractal structure of oil and gas deposits *Engineer. Annals of the Don* **1** 1–10
[9] Kozhevnikov D A 2005 Interpretation issues with NTZH *Inform. and Space* **1** 29–34
[10] Batashov M V 2013 Peculiarities of petrophysical features of Lower Cretaceous deposits of the Tersko-Sunzhensky oil and gas bearing region from geophysical data *Geol. in the Develop. World* **1** 134–6

Geophysical prospecting and well reliability indicators

N G Fedorova, Yu K Dimitriadi, I V Muradkhanov, S A Kaverzin, A S Akopov, K I Chernenko, N A Ryapolov and E D Voropaeva

North-Caucasus Federal University, 1, Pushkin Str., Stavropol, 355017, Russia

E-mail: fng8@mail.ru, SNGS@mail.ru, dvo125@yandex.ru, ars.akopov@gmail.com, kabiso@rambler.ru

Abstract. The paper considers the issues of well reliability indicators (strength, durability, safety) determined using the results of geophysical prospecting of the technical condition of the support as initial data. It is shown that it is possible to establish (recognize) the technical condition of bearing elements of well support, if according to the results of instrument studies, the types of damages are identified, their dimensions are established and actual strength parameters are calculated. Forecasting of service life is connected with the identification of a support element limiting service life of wells and determination of dominant mechanism of its damage for predicted operation period. The definition of well safe operation life is connected with the development of well reliability system. An example of elements included in the well reliability system is given, each of which is assigned a corresponding significance factor. The evaluation algorithm of the impact of properties of reliability system elements on its safe operation life is shown.

1. Introduction

The strength, durability and tightness of a well as a capital construction object are determined by the corresponding properties of the support.

Therefore, it is difficult to overestimate the importance of geophysical prospecting in order to determine the technical condition of support elements, in particular, well bearing structures – casing strings.

It is possible to determine (recognize) the technical condition of the casing if:

- as a result of instrument studies the types of damages are identified and their characteristic dimensions are determined;
- according to the obtained data, the actual strength parameters of the studied casing strings are calculated (values of excessive internal and external pressures, at which maximum stresses arising in the areas of damage are equal to the yield point of the material).

The solution of diagnostic problems is always connected with reliability forecasting for the nearest period of operation [1].

It is possible to predict the service life of wells if there is statistics related to the technical condition of the main bearing element of the support – production casing and the dominant mechanism of its damage for the predicted period of operation is established.

Gamma-percentile (or standard) service life is established when data on actual well service life are accumulated [2].

Leakage of the support (casing strings and cement sheath) affects the safety of well operation, can provoke fluid migration and lead to well liquidation [3].

2. Definition of actual strength indices of load-bearing elements

The load-bearing elements of the support, which receive the working loads during construction and operation of wells, include casing strings (intermediate and operational, respectively).

It should be noted that as the object of calculation the casing strings (except for very thick-walled ones) meet the engineering criterion of fineness [4] and belong to the class of thin-walled roller shells.

The analytical calculation of cylindrical roller shells having an arbitrary cross-sectional configuration is extremely complex [5] and is generally related to the division of the cross-sectional contour into regions having a constant radius of curvature. By way of example Figure 1 shows the cross section of a string with deformed cross section.

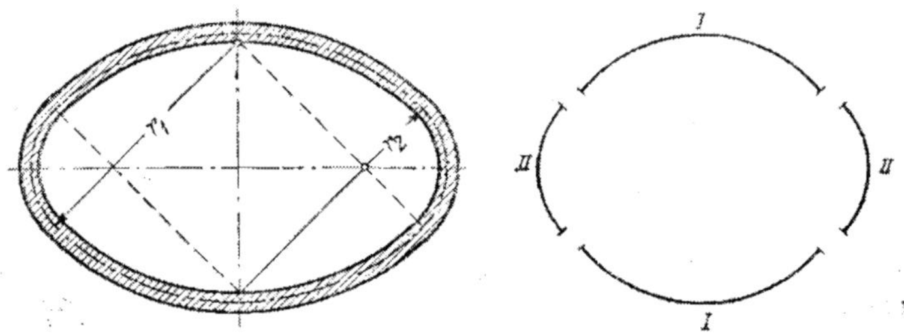

Figure 1. Oval cross-section in the form of joint of cylindrical plates of different radii.

As can be seen, the casing is presented in the form of articulated cylindrical plates I and II (Figure 1) having radii r_1 and r_2 respectively. The solutions for plates are built together with boundary conditions of deformation continuity in places of their connection.

As can be seen, the analytical calculation of casings with cross-section of an arbitrary configuration is extremely difficult.

Therefore, prior to the development of software based, in particular, on the finite element method, there were no methods to calculate the strength of damaged strings. The exception is temporary guidelines for calculation of casing strings worn out by tool joints of a drill string [6].

However, in the calculation formula, the trough-like working lost clear lines and a string with increased values of ovality and wall variation (relative to their normalized value) became the calculation scheme of the damaged string.

The development and implementation of strength calculations of software based, in particular, on the finite-element method eliminated all difficulties related to modeling of the actual configuration of damaged casing strings.

At the moment there is a set of methods for strength calculation of damaged casing strings by data of geophysical study (GIS) of technical condition [7]. At the same time, each particular damage required the individual procedure for calculating casing strings.

Typical types of damages were determined as a result of the GIS analysis of the technical condition of casing [8].

The calculations of actual strength parameters of casing strings according to GIS data "crossed the T's" in the issue of recognizing the technical condition of bearing elements of the support. In practice, this made it possible to focus on the selection of safe loads for casing strings both during construction and operation of wells.

The design support allowed the engineering staff to be oriented in determining safe loads for casing strings.

At the same time, naturally, there was a question about the value of error of calculations (for strength of damaged pipes) and its connection with instrumental errors at GIS of casing strings. The value of relative error of calculations is determined based on the following [9].

The results of calculations in accordance with procedures [7] are excessive external P'кр and internal P'т pressure (MPa), at which maximum stresses in damaged strings are equal to the yield point of the material. The limit pressure values are calculated according to the following formulas:

IOP Conf. Series: Materials Science and Engineering **905** (2020) 012022 doi:10.1088/1757-899X/905/1/012022

$$P'_{кр} = P_{кр} \cdot K_1 \quad и \quad P'_{т} = P_{т} \cdot K_2 \tag{1}$$

where $P_{кр}$ and $P_{т}$ – nominal values of limit overpressures for pipes of the type size under consideration (MPa); K_1 and K_2 – dimensionless factors of reduction of load capacity of damaged pipes to excessive external and internal pressures, respectively.

It shall be noted that:

- obtaining expressions for calculating K1 and K2 as a function of nominal wall thickness δ (mm) and characteristic damage size u (mm) made the basis of methods for calculating the strength of damaged casing strings;
- expressions for calculation of K1 and K2 are individual for each diameter D of casing pipes and their type of damage.

It is understood that in such a case, the relative error in calculating P'кр and P'т arises precisely from the error in determining K1, 2.

Let us consider the algorithm for calculating the relative error.

The expressions for calculating the coefficients K1, 2 were derived from the statistical processing of arrays of their discrete values calculated on finite-element models of damaged pipes.

The basic provisions for obtaining discrete coefficient values were as follows:

- load reduction factors were taken to be equivalent to the effective stress concentration factors (which, under loads constant over time, show how many times the presence of a defect reduces the load capacity of the structure);
- effective stress concentration coefficients were taken equal to inverse values of theoretical stress concentration coefficients.
- theoretical coefficients of stress concentration were calculated observing the geometric similarity of finite-element models and damaged casing strings.

Thus, with regard to the theory, the expressions for calculations of load capacity reduction factors met the requirements for strength calculations of damaged structures. The theoretical aspect did not introduce errors in determining coefficients K1, 2.

Let us look at functional dependencies for calculating the coefficients K1, 2.

The general view of expressions is as follows:

$$K_{1,2} = C_1 \cdot \exp(C_2 \cdot u) \tag{2}$$

where C1 and C2 – coefficients defined by expressions:

$$C_1 = c \cdot \delta + d$$
$$C_2 = a \cdot \delta - b \tag{3}$$

where u (mm) – characteristic damage size; a, b, c, d – numerical coefficients (d – dimensionless, a, b, c – dimensional); δ (mm) – thickness of the pipe wall.

From formula (2), it can be seen that the parameter specifying the magnitude of the error in determining the coefficients K1, 2 is the extent of damage "u".

It shall be noted that "u" is:

- established according to the data of geophysical surveys of the technical condition of casing strings;
- related to the actual (minimum) thickness of pipe walls in the area of damage, except for dimensions of pipe cross-section deformation (radial deflection).

The instruments, according to which "u" is determined, have different values of absolute error, in particular 0.5 mm or 1.5 mm. It is understood that "u" will have the same absolute error Δu.

The relative error ε1(K1,2) in calculating the K1 and K2 coefficients caused by the error in determining "u" was calculated using the first differential technique.

At the same time, it was found that at Δu = 0.5 mm the relative error of calculation of residual strength parameters is as follows:

- for pipes with ratio $\frac{\delta}{D} \geq 0.06$, $\varepsilon 1(K1, 2) = 5\text{–}10$ %;
- for pipes with ratio $0.02 < \frac{\delta}{D} < 0.06$, $\varepsilon 1(K1, 2) = 11\text{–}14$ %.

For engineering calculations, the permissible error value makes 15 % [5].

3. Conclusions

1. Calculations of residual strength help to prevent emergency situations related to non-conformity of operating loads of actual strength of casing strings, which improves safety of well construction and operation.

2. The calculation error of residual strength of damaged pipes will be within the range of permissible values, if the absolute error of pipe wall thickness (or radial deviation of cross section) during geophysical studies of technical condition does not exceed 0.5 mm.

3. The calculations according to GIS data (technical condition of casing strings) place increased requirements for their accuracy and reliability.

4. Determination of well life indicators

4.1. Forecast of residual well life

Forecasting of residual well life is performed based on the following:

- residual well life is determined by the residual life of production string [10];
- the dominant type of damage to production casing strings for the predicted period of operation is the total corrosion wear of the internal surface. This type of damage is identified on the basis of GIS analysis through the study of production well strings with different service life;
- residual life is forecast on the basis of characteristic reflecting the reduction of carrying capacity of string pipes at the increase of corrosion damage depth.

Such characteristic is represented by expressions for calculation of load capacity reduction factors of pipes (to excessive external and internal pressures) at general corrosion damage of internal surface [11].

Let us illustrate the above by the calculation of residual life of pipes with the diameter of 168.3 mm. Let us accept that the operational load of the forecast period of operation is excessive internal pressure Pin.ex.

The characteristic of reducing the carrying capacity of pipes with the diameter of 168.3 mm at the total corrosion of internal surface is as follows:

$$K2 = (0.0235 \cdot \delta f + 0.8038) \cdot \exp((0.0304 \cdot \delta f - 0.5455)u) \quad (4)$$

where K2 – coefficient of reduction of load capacity of damaged pipes to excessive internal pressure; δf – actual thickness of pipe wall, mm; u – depth of corrosion damage, mm. δf and u are set at casing non-destructive testing at the moment of calculation.

The criterion of the maximum condition of the pipes is the reduction of the wall thickness to the maximum permissible value [u]. [u] is calculated from the expression converted with respect to it:

$$[u] = \frac{\ln\{[K_2]/(0{,}0235 \cdot \delta_{\Phi} + 0{,}8038)\}}{0{,}0304 \cdot \delta_{\Phi} - 0{,}5455} \quad (5)$$

[K2] – permissible value of the load reduction factor:

$$[K2] = n2 \cdot Pin.ex. / P_T' \quad (6)$$

where P_T' – excess internal pressure at which maximum stresses in the damaged pipe walls are equal to material yield point, MPa. The value P_T' is calculated, for example, according to [7]; Pin.ex. – operating load for the forecast operation period, MPa. Pressure character – static (re-static); n2 – factor of safety to internal pressure.

Residual life T (years) of casing strings is calculated by formula

$$T = \frac{[u]}{V} \tag{7}$$

where V – corrosion rate, mm/year:

$$V = \Delta\delta/\Delta t \tag{8}$$

where $\Delta\,\delta$ – changes of thickness of a wall by Δt time period (year) for which it occurred (mm).

5. Conclusions

1. The residual service life of operational casing is forecast taking into account its technical condition and individual reserve of load capacity to operational load.

2. This is rational since individual forecasting of operational life of objects usually leads to the increase of their average life, as well as to inter-repair periods of operation [12].

3. However, if another dominant mechanism for damage to production strings is installed according to GIS data, for example, a combination of total corrosion damage to the inner and outer surfaces of the string pipes, this will require the development of an appropriate methodology for calculating their residual life.

5.1. Determination of gamma-percentile life

The gamma-percentile life is determined according to the statistics on their actual service life. Such data, in particular, can be the results of Safety Expert Review (SER) of wells.

Table 1. Distribution of the number of wells by extended service life

Extended operation life	Number of wells, % Year of SER				
	2007	2008	2009	2010	2011
full	57.8	55.2	90.3	81.1	91.1
refuced	38.4	41.0	5.1	16.2	5.5
not extended	3.8	3.8	4.6	2.7	3.4

Let us consider one example. Table 1 shows the SER results related to the extension of safe operation life of wells at underground gas storage facility from 2007 to 2011 [13].

Non-extension of service life is considered as a failure (event when operational condition of an object is violated [4].

Comment to Table 1:

- starting from 2009, the number of wells with the maximum service life is dramatically increasing. Explanation – in 2007–2008, a significant number of SERs was carried out for wells with service life of more than 30 years (40, 50 and more). Such wells extended their service life at the second SER of wells;
- the number of wells, the service life of which is not extended, is almost stable.

From the latter it is concluded that the probability of failure-free operation of wells γ at service life from 30 to 50 years is on average 96 %.

It shall be noted that the method of statistical data processing is the simplest, while the conclusions are essential. In accordance with the obtained value of gamma-percentile life, the service life of underground gas storage facility was adjusted. In this case, the gamma-percentile life is an operational indicator of well reliability.

Note – Operational reliability index – reliability indicator, point or interval estimation of which is determined by operation data.

Let us note the representativeness of sampling comprising eight hundred (800) wells, as well as different reasons of well failures, namely:

- leakage of production string;
- intercasing pressure;
- fluid migration, etc.

Despite various reasons, the wells were classified as one group – failed. In the practice of engineering reliability analysis this is possible on the basis of the principle of the same consequences, when random events (failures, damages, etc.) can be included in the common classification group leading to the same consequences [14].

6. Conclusions

1. The determination of actual gamma-percentile life requires the accumulation of statistics on actual well life.

2. It appears that such data should be generated for wells of each field separately.

7. Well reliability

From the classical perspectives of the reliability theory, an object (technical system) is divided into elements interacting with each other according to some logical schemes – a reliability structure diagram is designed. The reliability of the system is dependent on the probability of failure-free operation of all components.

It is essential that the elements of technical systems are mass-produced products for which reliable statistical estimates of reliability indicators are possible [13].

Since wells are not objects of mass production, it is understood that the system theory is not suitable for studying their reliability.

In practice, the applied methods are used to study and forecast well reliability [15], the main components of which are as follows:

1. strength and durability calculations of well bearing structures;
2. determination of safe operation life.

When solving the issue of extension of the safe operation life of wells, the following measures are taken:

1. residual well life is calculated;
2. well reliability system is created, consisting of i elements (for example, production string, annular space, wellhead area). Each element is assigned a significance factor δi;
3. impact of reliability system components properties on safe operation life is assessed.

Quantitative estimates of the state of the above objects are obtained using the methodology proposed in [15].

It is essential that the reliability of elements is evaluated by the characteristics of their state, for example, as shown in Table 2, where the code a_i – numerical value of the characteristic.

Table 2. Performance and reliability of the production string

State characteristic	Degree of reliability	Code, a_i
During operation, the casing leakage along the threaded joints was noted. Repair works to eliminate leakage of threaded joints of the casing were performed two or more times. Triangular threaded pipe joints	Low	1
During operation, the casing leakage along the threaded joints was noted. Repair works to eliminate leakage of threaded joints of the casing were performed once. String OTTG, OTTM joints	Average	2
During operation, the casing maintained its tightness. String joints include metal-metal hermetic seal	High	3

The integral characteristic of objects within the well reliability system is the K0 coefficient, which is calculated using the following formula:

$$\mathbf{K_0} = \sum_{i=1}^{5} \left(a_i \cdot \delta_i\right) \tag{9}$$

The value of K0 coefficient determines not only the degree of reliability of a well, but also the reduction of its residual life, as shown in Table 3.

The extended safe operation life $T^{Б}_{ост}$ of a well is defined as the product of its residual life and coefficient C:

$$T^{Б}_{ост} = T_{ост} \cdot C \tag{10}$$

As can be seen, the extension of safe operation of wells is caused by the use of qualimetry methods.

Table 3. Reliability levels and factors of well residual life reduction

Well reliability factor K	Well reliability level	Well residual life reduction C
0.8–1.0	Very high	0.9
0.64–0.8	High	0.7
0.37–0.64	Average	0.5
0.2–0.37	Low	0.3
0.0–0.2	Very low	0

It shall be noted that formally the safe operation terms are forecast taking into account the operating conditions, as well as the technical condition of both the bearing structure and the objects of the well reliability system.

8. Conclusion

1. Geophysical studies of technical condition are closely connected with the definition of reliability indicators (strength and durability) of wells, but without design tracking of well indicators the GIS results are not informative.

2. The use of GIS results as initial information for calculations is a powerful incentive to improve methods and means of diagnostics of the technical conditions of wells.

References

[1] Birger I A 1978 *Technical Diagnostics* (Moscow: Mechanical Engineering) 234 p

[2] GOST 27.002-15 *Equipment reliability. Terms and definitions* 2016 (Moscow: Standartinform) 41 p

[3] *Safety rules for oil and gas industry* 2013 Federal norms and rules of industrial safety series 08 iss 19 (Moscow: ZAO "Sci. and Techn. Center for Res. on Industr. Safety Probl.") 288 p

[4] Novozhilov V V 1951 *Thin shell theory* (Moscow: State publ. of shipbuild. Liter.) pp 344

[5] Boyarshinov S V 1973 *Basics of machine construction mechanics: "Safety rules for oil and gas industry"* Federal norms and rules of industrial safety series 08, iss 19 (Moscow: Mechan. Engineer.) 456 p

[6] Goncharenko A M 1975 *Temporary guidelines on calculation of worn-out casing strings* Approved by Deputy General Director Grozneft: Valid from 18.11.1975 (Grozny: SevKavNIPIneft) 10 p

[7] *Instructions for calculation of damaged casing strings and casing strings under special operating conditions* 2007 STO Gazprom 2-2.3-117-2007 Approved by Deputy Chairman of the Board of OAO Gazprom A G Ananenkov on 22.11.2007 (Moscow: RPI Gazprom) 71 p

[8] Klimov V V 2001 Technical condition of well case in fields and underground gas storage facilities: problems and their solution, Review In *Drilling of gas and gas condensate wells* (Moscow: RPI Gazprom) p 65

[9] Tolpaev V A, Fedorov N G and Kolesnikov V V 2007 On calculation error of residual strength of casing strings according to GIS data *Onshore and offshore construction of oil and gas wells* **9** 32–4 (Moscow)

[10] Fedorov N G 2017 Cement shells and well durability *Oil facilities* **8** 86–8 (Moscow)

[11] *Instructions for calculation of well life and residual life: STO Gazprom 2-3.2-346-2009* 2009 (Moscow: Gazprom Expo LLC) 36 p

[12] Bolotin V V 1984 *Forecasting the operational life of machines and structures* (Moscow: Mechan. Engineer.) 321 p

[13] Shamshin V I, Fedorov N G and Dubenko V E 2014 On definition of terms of safe operation of oil and gas wells *Onshore and offshore construct. of oil and gas wells* **3** 30–2

[14] Efremov L V 2008 *Practice of probabilistic analysis of reliability of computer-based technologies* (St. Petersburg: Nauka) 216 p

[15] *Methodological recommendations for assessment of risk of accidents of hydraulic structures of reservoirs and industrial waste accumulators* 2001 Approved by VODGEO Research Institute, 01.01.2001; coordinated by Emercom of Russia, order no 9-4/09-644 of 14.08.2001 (Moscow)

IOP Conf. Series: Materials Science and Engineering **905** (2020) 012023 doi:10.1088/1757-899X/905/1/012023

Studies of vertical position of building walls via surveying method

I G Gairabekov[1], I Yu Pimshin[2], M Sh Mintsaev[1], M-B I Gairabekov[1], A I Gayrabekova[1], A T Mishieva[1] and E I Ibragimova[1]

[1] Grozny State Oil Technical University named after M.D. Millionshchikov, korp. 1,100, prosp. Husejna Isaeva, Grozny, 364068 Russia
[2] Don State Technical University, 1, Gagarin square, Rostov-on-Don, 344000, Russia

E-mail: ibragim.ggni@mail.ru, ivan.pimschin@yandex.ru, gayrabekovbashir@gmail.com, Gayrabekovamina@mail.ru, zuba-866@yandex.ru, lina.leams@mail.ru

Abstract. It is impossible to get a truthful overview of the strain-stress state of any object using the methods recommended by the current regulatory documents to determine the roll of buildings and structures. This paper implements the proposed method of determining the set of roll values by vertical sections of the object. Such a method allows obtaining a detailed picture of deformations, on the basis of which it is possible to develop a qualitative project to restore its design geometry. The paper presents the implementation of methods for determining the complex of values of building rolls with both classical and hinged ventilated facades. The rolls were determined according to 17 vertical sections (9 measurement cycles in total) for a building with a classic facade and 53 vertical sections (53 measurement cycles in total) for a building with a hinged ventilated facade. The results are presented in graphical and tabular form.

1. Introduction

The methods of studying building deformations recommended by the current regulatory instruments require further improvement. Thus, it is recommended to use generalized roll values to determine deformations of buildings and structures in regulatory documents, which cannot sufficiently reflect the real picture of deformed state of buildings and structures [1–6]. Based on such results, it is not possible to make a viable project to restore its design geometry.

In the proposed method, a much larger number of roll values is determined – with a certain pitch, by vertical sections of an object, which characterizes the deformed state of the building in more detail. This method makes it possible to detect stress zones caused by deformations of engineering structures, which is necessary to restore the spatial geometry of the building without additional deformations and faults [7–11].

2. Methods and materials

Roll measurement was performed according to the following procedure. The device on the tripod was installed at a distance approximately equal to the height of the building under control. The coordinate system of the electronic total station was oriented parallel to the controlled plane (building wall). Then, in a single coordinate system, the device coordinated the points under study located in characteristic places of the building. The measurement results were automatically recorded by an electronic total station in the selected tool file. This file was translated to a computer after the measurements were made. After that, it was imported into the Credo_Dat program intended for camera processing of field engineering and geodetic measurements. In this program, the polar coordinate system of the electronic total station was converted into a rectangular Cartesian coordinate system. Then the resulting 3D point coordinates were exported to the *.dxf (AutoCAD) format. The resulting file was then opened in AutoCAD, where drawing, measurement and analysis of the basic geometry of the building was performed.

The rolls of the structure can be expressed in relative and absolute measures. Besides, there are private $\mathbf{q_x}$, $\mathbf{q_y}$ (for some coordinate plane) and absolute Q (full roll) for both individual elements and the structure as a whole.

In determining the rolls of the studied building, the method of coordination was used. This method defines the rectangular coordinates of the top ($x_в$, $y_в$, $z_в$) and bottom point ($x_н$, $y_н$, $z_{вн}$) of the construction structure in various ways. It is more rational to implement this method using a modern electronic total station (Figure 1).

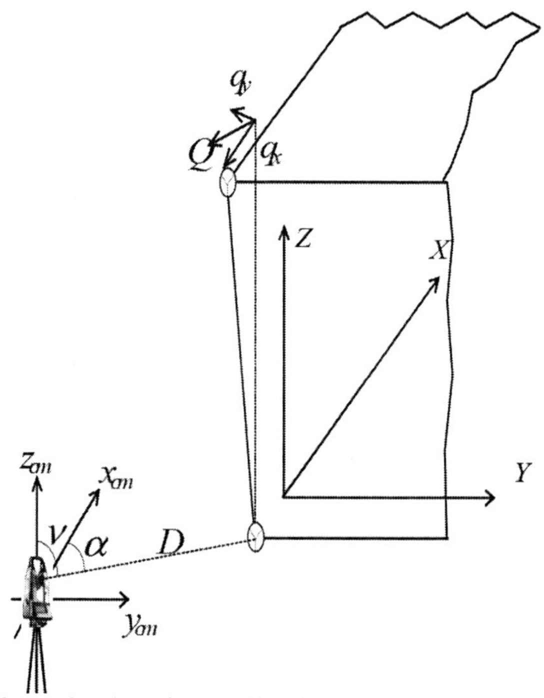

Figure 1. Diagram of roll determination via coordination method

The electronic total station measures a horizontal angle α, a vertical angle ν and an inclined distance s (Figure 1). The following working formulas are used to convert the polar coordinate system to a rectangular Cartesian coordinate system:

$$h = s \cdot \sin\nu;\ x = s \cdot \cos\nu \cdot \cos\alpha;\ y = s \cdot \cos\nu \cdot \sin\alpha\,.$$

Private rolls were determined by the following formulas:

$$q_x = x_1 - x_2;\ q_y = y_1 - y_2.$$

The processing of field materials when defining the structure rolls includes the calculation of absolute (complete) roll by the following formula:

$$Q = \sqrt{q_x^2 + q_y^2}\ ,$$

where q_x and q_y – private rolls along X and Y, respectively.

3. Results

Let us analyze the results of defining the roll complex for the building with classic facade. Figure 1 shows the layout of controlled sections and stations of the device relative to the object under study. The geometry of the building was controlled via 5 stations.

In accordance with Section 5, GOST 24846-81 (Soils. Measurement methods of base deformations of buildings and structures), the limiting errors of roll measurement depending on height H should not exceed 0.0001·H for civil buildings and structures. In our case (at H = 19 m) the limiting error of roll measurement makes 1.9 mm. This measurement accuracy is ensured by the device used (*Trimble 3602 DR* electronic total station).

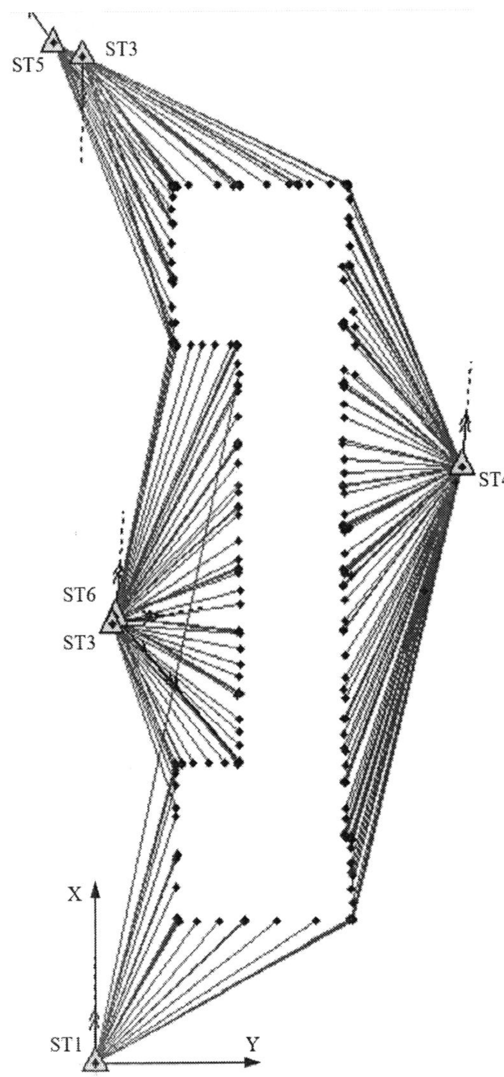

Figure 2. Diagram of device standing stations relative to monitored points located on the studied building

The roll not exceeding 0.005·H is possible. Therefore, if the height is H = 19 m, the permissible roll of the building is 95 mm.

Table 1 shows the results of building vertical monitoring performed during different observation periods.

Rule of roll signs. The rolls directed towards the increase in the size of a construction (from the building) have the sign (+), and the rolls directed towards the reduction of the size (inside the building) have the sign (–).

Table 1. Control of vertical position of the building

Date	02.04.2005		28.07.2005		05.03.2009		04.02.2017		05.05.2017		02.11.2017		04.02.2018		03.03.2018	
Frame	Rolls in mm		Rolls in mm		Rolls in mm		Rolls in mm		Rolls in mm		Rolls in mm		Rolls in mm		Rolls in mm	
No.	q_x	q_y	q_x	q_y	q_x	q_y	q_x	q_y	q_x	q_y	q_x	q_y	q_x	q_y	q_x	q_y
1	+60	–101	–40	–15	+46	–88	+12	–89	+11	–88	+7	–90	+7	–87	+5	–89
2	+38	0	–	–	–	–28	+40	–5	+37	+2	+37	0	+39	1	+36	–3
3	–	–	–	–	–	+51	–	+49	–	+47	–	+48	–	+44	–	+46
4	–	–	–	–	–	+39	–	+42	–	+44	–	+50	–	+47	–	+49
5	–	–	–	–	–	+27	–	+53	–	+55	–	+60	–	+58	–	+61
6	–	–	–	–	–	+56	–	+66	–	+69	–	+70	–	+75	–	+71
7	–45	+42	–	–	–	+61	–46	+57	–48	+60	–46	+56	–45	+59	–49	+57
8	–97	+26	–66	+50	–43	+40	–47	+41	–48	+38	–53	+37	–51	+40	–49	+39
9	+20	–69	+70	–20	+72	–16	+66	–15	+62	–18	+62	–19	+60	–20	+64	–17
10	–38	+28	+20	+98	–35	–48	–46	–24	–47	–25	–45	–21	–49	–22	–47	–20
11	–67	–55	–35	–79	–44	–76	–69	–71	–72	–69	–69	–68	–67	–67	–65	–65
12	+40	+80	–45	–69	+29	+57	+31	+56	+35	+58	+30	+56	+28	+54	+26	+52
13	–	–	–	–	–	+63	–	+62	–	+61	–	+59	–	+63	–	+60
14	–94	+60	–36	+36	–78	+38	–82	+46	–85	+48	–86	+49	–87	+51	–90	+52
15	–43	+80	–34	+46	–27	+59	–35	+45	–37	+47	–36	+46	–39	+50	–37	+48
16	0	–39	+52	–26	+2	–18	+3	–27	+5	–26	+4	–28	+3	–29	+6	–32
17	+11	–38	+40	–30	+6	–28	+12	–34	+10	–37	+15	–35	+13	–36	+11	–40

Figure 3 shows the location and numbering of vertical control sections.

Then, let us analyze the results of defining the roll complex for the building with a hinged ventilated facade. As mentioned above in accordance with Section 5, GOST 24846-81 (Soils. Measurement methods of base deformations of buildings and structures), the limiting errors of roll measurement depending on height H should not exceed 0.0001·H for civil buildings and structures. In our case (at H = 19 m) the limiting error of roll measurement makes 2 mm. This measurement accuracy is ensured by the above mentioned device (GTS-102N electronic total station).

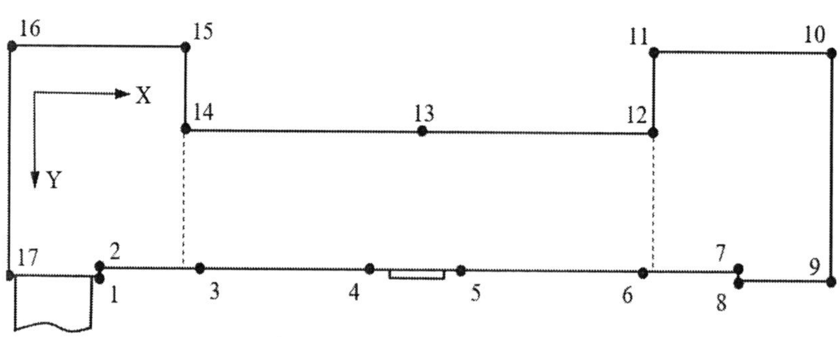

Figure 3. Arrangement and numbering of vertical control sections

The roll size which is not exceeding value 0.005·H is permissible (Annex 3). The height of the building under study is 19 m. Therefore, the permissible roll makes 95 mm.

Next, let us analyze the results of defining the roll complex for the building with a classic facade. Due to the fact that the building is covered on all sides with lining, it was decided to control the verticality by the middle of the frames (windows) of the building. Figure 2 shows the diagram of controlled points and standing stations of the device relative to the studied object. Figures 3, 4 and 5 show the results of the window verticality control.

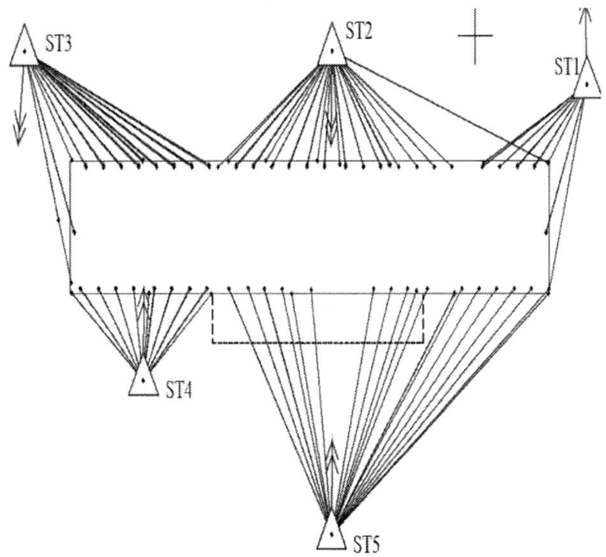

Figure 4. Diagram of controlled points and standing stations of the device (tacheometer) relative to the studied object

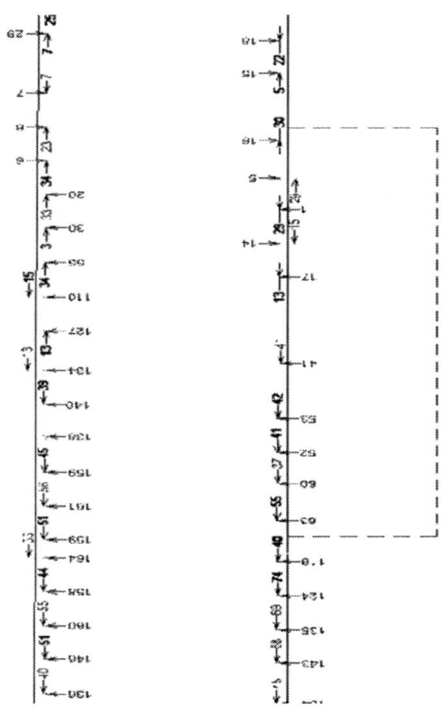

Figure 5. Window row verticality control (1 measurement cycle)

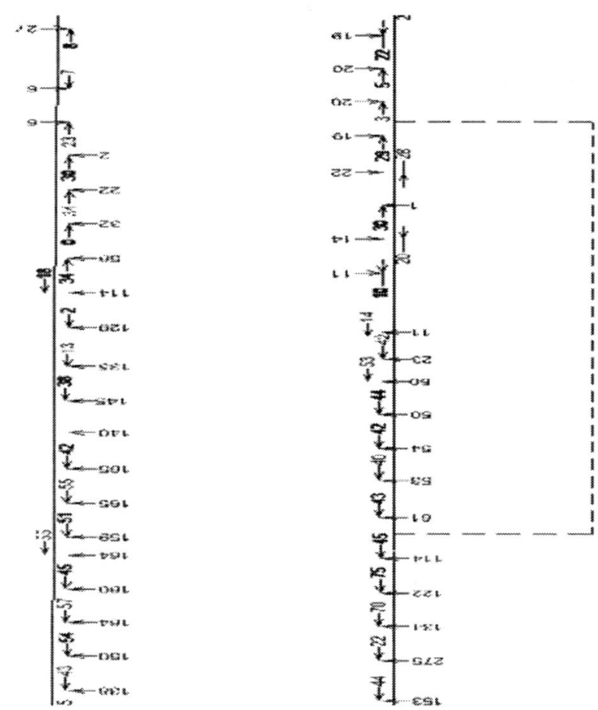

Figure 6. Window row verticality control (3rd measurement cycle 12.09.2009)

4. Conclusion

The analysis of results on defining the complex of rolls of the building with a classic facade showed the following:

1. No dynamics of verticality change of the main building structures were revealed. These values are within the measurement error range.

2. We see from the verticality control diagram that the building has different-directional rolls. This characterizes the significant accumulated deformation and disruption of the integrity of the internal geometry of the object under study.

3. Maximum wall rolls are fixed on the elevation facing yard of the building.

The analysis of results on defining the complex of rolls of the building with a ventilated facade showed the following:

1. During the period of the studies the recorded vertical displacements of sedimentary grades are within the limits of errors of performed measurements.

2. During the period of the studies the recorded changes of the building roll are within the limits of errors of performed measurements.

The dynamics of deformation of the building is not fixed.

It is recommended to continue observations of deformation dynamics of the studied building.

The measurement results can be used for measures to restore operational reliability of the studied buildings.

References

[1] GOST 24846-2012 *Soils. Measurement methods of base deformations of buildings and structures* 2012 (Moscow: Ministry of Reg. Development of the Russ. Fed.)

[2] SP 22.13330.2011 *Foundations of buildings and structures. Updated version of SNiP 2.02.01-83* 2011 (Moscow: Ministry of Reg. Development of the Russ. Fed.)

[3] GOST P 53778-2010 *Buildings and constructions. Rules of technical survey and monitoring* 2010 (Moscow. Standartinform)

[4] *Manual on geodetic works in construction (SNiP 3.01.03-84)* approved by the Order of SiM Networks no 147 of 10 July 1985

[5] RD 34.21.322-94 *Methodology guidelines for observations of precipitation and deformation of buildings and structures of thermal power plants under construction and operation*

[6] *Guidelines for observations of deformations of bases and foundations of buildings and structures* 1975 (Moscow: Stroyizdat)

[7] Zotov V D and Zotov M V 2005 Lifting and leveling of buildings using flat jacks Coll. *Reconstruction of cities and geotechnical construction* (Rostov-on-Don) pp 156–67

[8] Boyko M D 1975 *Diagnosis of damages and methods of restoration of operational qualities of buildings* (Leningrad: Stroyizdat) 335 p

[9] Gairabekov I G, Gairabekov M-B I, Pimshin I Y et al August 2019 Geodetic Control of the Geometry of the Levelled Building *Atlantis Highlights in Material Sciences and Technology* Proc. of the Int. Symp. Engineer. and Earth Sci.: Appl. and Fund. Res., dedicated to the 85th anniversary of H.I. Ibragimov (ISEES 2019) pp 416–20 Available Online

[10] Gairabekov I G, Gairabekov M-B I, Pimshin I Y et al August 2019 Accuracy of Determining Buidings' Geometric Characteristics and Structures by Modern Means of Geodetic Measurements *Atlantis Highlights in Material Sciences and Technology* Proc. of the Int. Symp. Engineer. and Earth Sci.: Appl. and Fund. Res., dedicated to the 85th anniversary of H.I. Ibragimov (ISEES 2019) pp 421–5 Available Online

[11] Gairabekov I G and Gagaeva H R August 2019 Distortions Introduced by the Cartographic Projection of Gauss at a Distance of the Construction Site from the Axial Meridian *Atlantis Highlights in Material Sciences and Technology* Proc. of the Int. Symp. Engineer. and Earth Sci.: Appl. and Fund. Res., dedicated to the 85th anniversary of H.I. Ibragimov (ISEES 2019) pp 583–7 Available Online

IOP Conf. Series: Materials Science and Engineering **905** (2020) 012024 doi:10.1088/1757-899X/905/1/012024

Study of static and dynamic geometric characteristics of buildings

I G Gairabekov[1], I Yu Pimshin[2], M Sh Mintsaev[1], M-B I Gairabekov[1], A I Gayrabekova[1], A T Mishieva[1] and E I Ibragimova[1]

[1] Grozny State Oil Technical University named after M.D. Millionshchikov, korp. 1,100, prosp. Husejna Isaeva, Grozny, 364068 Russia
2 Don State Technical University, 1, Gagarin square, Rostov-on-Don, 344000, Russia

E-mail: ibragim.ggni@mail.ru, ivan.pimschin@yandex.ru, gayrabekovbashir@gmail.com, Gayrabekovamina@mail.ru, zuba-866@yandex.ru, lina.leams@mail.ru

Abstract. It is impossible to get a truthful overview of the strain-stress state of any object using the methods recommended by the current regulatory documents for determining deformations. This paper implements the proposed method of determining the spatial geometry of an object. Such a method allows obtaining a detailed picture of deformations, on the basis of which it is possible to develop a qualitative project to restore its operational reliability. The rolls were determined according to 38 vertical sections (9 measurement cycles in total) for a building with a classic facade and 53 vertical sections (2 measurement cycles in total). Besides, five cycles of determination the settlement of structure were performed. The trigonometric leveling was used to control the horizontal position of different level building structures (top of window aperture of the 2nd floor). Elevations of horizontal building structures were determined in different conditional systems of heights. Elevations of horizontal building structures were determined in a single conditional system of heights. The results of the method are presented in graphical and tabular form.

1. Introduction

The methods of studying building deformations recommended by the current regulatory instruments require further improvement. Thus, it is recommended to use generalized roll values, settlements of the building foundation, and in some cases deflections, to determine deformations of buildings and structures in regulatory documents, which cannot sufficiently reflect the real picture of deformed state of buildings and structures [1–6]. Based on such results, it is impossible to determine a real picture of the stress-strain state and to make a viable project to restore its design geometry.

The proposed method – the method of determining the spatial geometry of an object – defines a larger number of geometric parameters – settlement, horizontality of horizontally oriented structures, a set of rolls on separate vertical sections, characterizing in full and in detail the deformed state of the building. This method makes it possible to detect stress zones caused by deformations of engineering structures, which is necessary to restore the spatial geometry of the building without additional deformations and faults [6–11].

IOP Conf. Series: Materials Science and Engineering **905** (2020) 012024 doi:10.1088/1757-899X/905/1/012024

2. Methods and materials

Static and dynamic deformation characteristics were studied on the example of the shopping center building with built-in fitness club located at 133/177 Krasnoarmeyskaya Str., Moscow.

The work was carried out using the latest geodetic measurement tools, such as:

- total station instrument GTS-102N;
- high-accuracy electronic geodetic level TrimbleDINI 0.3;
- invar levelling staff LD11 NedoTrimble;
- invar levelling staff LD12 NedoTrimble;
- metal measuring reel Geobox PK2-50 No. 7;
- sliding calipers (0-300) mm No. GK11275363;
- metal measuring reel 1000 mm No. 21051;
- ultrasonic tester UK 1401M No. 4011698;
- pachometer IPA-MG4 No. 2317.

These tools have verification or calibration certificates.

The office study of geodetic measurements was performed in the Credo_Dat program. Analysis and design of geometric parameters of objects is performed in ZWCAD software.

The purpose of preliminary inspection was to establish compliance of the layout and structural diagrams of existing structures with the requirements of technical documentation. During the inspection, the most damaged sections of the structures were revealed, as well as bearing elements under the most unfavorable conditions of operation. The general condition of the structures was visually assessed: presence of wetted concrete sections, condition of protective coating, presence of corrosion, etc. Thus, the preliminary inspection made it possible to collect the information, which allowed clarifying the program and scope of instrumental works.

No defects of both the entire building and its individual elements were detected from the visual inspection.

Observations of settlement points are made in the system of heights close to Baltic from wall benchmarks.

Observations of settlement points are made in a conditional system of heights. The levelling line is laid in forward and reverse directions. The determination of settlement points in each cycle is made with the accuracy characterizing the mean quadratic error of determination of settlement points in a weak place not more than 1 mm (the most remote mark from original benchmarks), which is provided by the method of high-accuracy geometric leveling of the II class in accordance with the current regulatory documents. The settlement points are determined in the Baltic system of heights through geometrical levelling in forward and reverse directions. The average length of the directional ray was 25–30 m for the main traverse and 2–25 m – for levelling of settlement points. This made it possible to reduce the time of observations at the station, to improve the reliability, quality and speed of monitoring the measurement process.

The height of the directional ray above the underlying surface of the earth was allowed not less than 0.5 m, shoulder inequality at the station – not more than 1 m, and accumulation of shoulder inequality in the section – not more than 2 m. According to Paragraph 3, GOST 24846-2012 (Soils. Measurement methods of base deformations of buildings and structures) the geometric leveling is used as the main method to measure vertical displacements (settlement). Basic technical characteristics and tolerances for geometric leveling were adopted in accordance with regulatory requirements.

The defects of the closed levelling loop did not exceed the permissible value determined by the formula:

$$f_{доп} = \pm 5мм\sqrt{L} \text{ km}, \quad (1)$$

where L – levelling line length in km.

Permissible misalignment of levelling line between settlement points did not exceed the following value

$$\mathrm{Fh(add)} \leq \pm 0.3 \text{ mm} \sqrt{n}\,, \tag{2}$$

where n – number of levelling stations.

As a result of the levelling, cycle and accumulated vertical movements of settlement points were determined.

In geometrical sense, settlement is expressed by vertical sections lowered from initial to subsequent position of the settlement point. In cases where these lines for different brands are the same, the settlements are called even and when the lines are not equal, they are called uneven. The principle of processing repeated measurements is that there are differences in elevation of settlement points between measurement cycles. At the same time the following is defined:

- total settlement $S_{i,(полн.)}$, the value of which is equal to the difference of elevations of the same point of the current $H_{i,(послед.цикл)}$ and initial cycle of observations $H_{i,(1цикл)}$;

$$S_{i,(полн)} = H_{i,(1цикл)} - H_{i,(последцикл)} \tag{3}$$

- settlement between adjacent cycles $S_{i,с.ц.}$, the value of which is determined as the difference of elevations of the same point of the current $H_{i-1,(предыд.цикл)}$ and the previous $H_{i,(послед.цикл)}$ of the observation cycle; $H_{i,(послед.цикл)}$ цикла наблюдений;

$$S_{i,с.ц.} = H_{i,(последцикл)} - H_{i-1,(предыдцикл)} \tag{4}$$

- uniform settlement $S_{(рав).}$, the value of which is equal to the minimum total settlement $S_{min,(полн.)}$ of the total settlement points;

$$S_{(рав.)} = S_{\min,(полн)} \tag{5}$$

- uneven settlement $S_{(нерав.)}$, the value of which is defined as the difference of the total settlement $S_{i,(полн.)}$ and the uniform settlement $S_{рав.}$;

$$S_{(нерав)} = S_{i,(полн)} - S_{(рав.)} \tag{6}$$

The average settlement is calculated by the following formula:

$$S_{ср.} = \frac{\sum S_i}{n} \tag{7}$$

On the basis of calculated values of total settlement points the average monthly settlement rates υ_i are defined as follows:

$$\upsilon_i = \frac{\Delta S_i}{t}, \tag{8}$$

where $\Delta S = S_{i+1} - S_i$ – inter-cycle settlement;

t – inter-cycle time period (usually expressed in months).

In the physical sense the velocities characterize the settlement dynamics.

Acceleration (attenuation) of settlement is calculated by the following formula:

$$a_i = \frac{\upsilon_i - \upsilon_{i-1}}{t} \tag{9}$$

This parameter characterizes the change in the dynamics of settlement (attenuation, acceleration).

Roll measurement was performed according to the following procedure. The device on the tripod was installed at a distance approximately equal to the height of the building under control. The coordinate system of the electronic total station was oriented parallel to the controlled plane (building wall). Then, in a single coordinate system, the device coordinated the points under study located in characteristic places of the building. The measurement results were automatically recorded by an electronic total station in the selected tool file. This file was translated to a computer after the measurements were made. After that, it was imported into the Credo_Dat program intended for camera processing of field engineering and geodetic measurements. In this program, the polar coordinate system of the electronic total station was converted into a rectangular Cartesian coordinate system. Then the resulting 3D point coordinates were exported to the *.dxf (AutoCAD) format. The resulting file was then opened in AutoCAD, where drawing, measurement and analysis of the basic geometry of the building was performed.

The rolls of the structure can be expressed in relative and absolute measures. Besides, there are private **qx, qy** (for some coordinate plane) and absolute **Q** (full roll) for both individual elements and the structure as a whole.

In determining the rolls of the studied building, the method of coordination was used. This method defines the rectangular coordinates of the top ($x_в$, $y_в$, $z_в$) and bottom point ($x_н$, $y_н$, $z_{вн}$) of the construction structure in various ways. It is more rational to implement this method using a modern electronic total station).

The electronic total station measures a horizontal angle α, a vertical angle ν and an inclined distance s. The following working formulas are used to convert the polar coordinate system to a rectangular Cartesian coordinate system:

$$h = s \cdot \sin \nu ;\; x = s \cdot \cos \nu \cdot \cos \alpha \quad (10)$$

$$y = s \cdot \cos \nu \cdot \sin \alpha \,. \quad (11)$$

Private rolls were determined by the following formulas:

$$q_x = x_1 - x_2;\; q_y = y_1 - y_2. \quad (12)$$

The absolute (total) roll Q is calculated by the following formula:

$$Q = \sqrt{(}\, q_x^2 + q_y^2, \quad (13)$$

where q_x and q_y – private rolls along X and Y, respectively

Absolute roll orientation angle θ *is* calculated by the following formula:

$$\theta = \text{arctg}\,(q_x/q_y) \quad (14)$$

The geometry of the building was controlled from 6 stations.

The trigonometric leveling was used to control the horizontal position of different level building structures (top of window aperture of the 2nd floor). Elevations of horizontal building structures were determined in different conditional systems of heights. Elevations of horizontal building structures were determined in a single conditional system of heights. The point of the horizontal construction structure having the highest conditional point was taken as "0", hence, all other points have a negative sign.

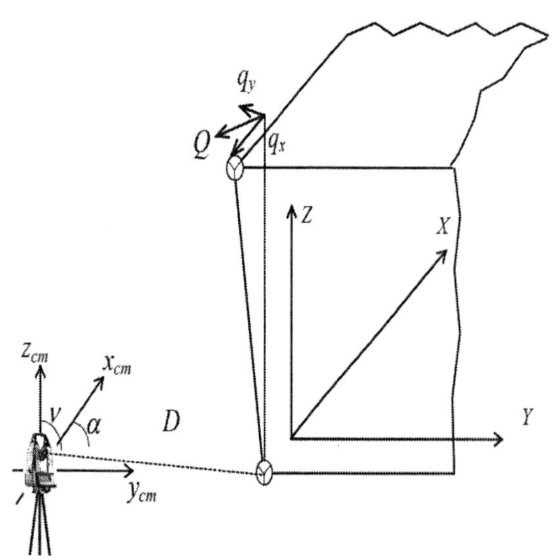

Figure 1. Diagram of roll determination via coordination method

IOP Conf. Series: Materials Science and Engineering **905** (2020) 012024 doi:10.1088/1757-899X/905/1/012024

3. Results

Figure 2 (Curves of settlement points) shows the settlement curves. This figure shows that the average settlement of the whole building, for the current observation period (from May to December) made 1.0 mm, and for the whole year (2016) it made 1.1 mm. The average accumulated settlement of the entire building was 4.2 mm. The largest accumulated settlement (since the beginning of observations) is recorded for point No. 9 (4.5 mm), No. 10 (4.5 mm), No. 23 (5.2 mm) and No. 24 (2.7 mm). The largest settlement in 2016 was recorded at points No. 15, 16. 17 and amounted to about 2 mm.

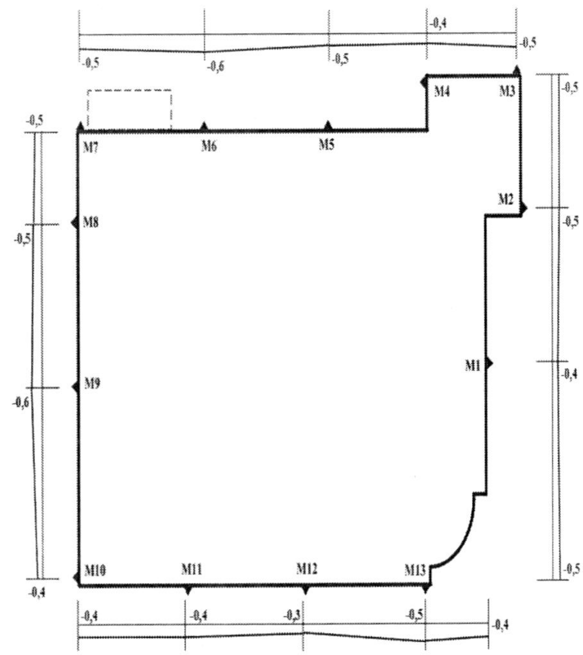

Figure 2. Settlement point curves

Figure 3 (Building roll diagram as of 16.12.2015) and Figure 4 (Building roll diagram as of 07.11.2016) graphically presents the results of the building rolls as of 16.12.2015 and 07.11.2016. These figures show the total and private rolls of all controlled building structures (wall edges and planes) in mm, as well as the complete roll of the whole building.

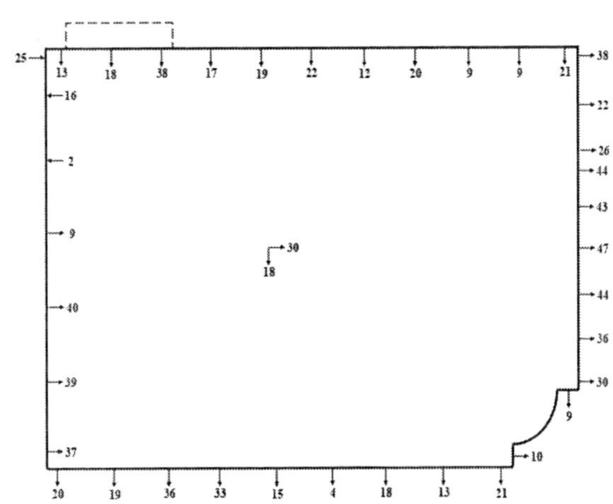

Figure 3. Building roll diagram as of 16.12.2015

This figure shows that the complete roll of the whole building makes 37 mm and is placed in the south-east direction.

In accordance with Section 5, GOST 24846-2012 (Soils. Measurement methods of base deformations of buildings and structures), the limiting errors of roll measurement depending on height *H* should not exceed 0.0001·H for civil buildings and structures. In our case (at $H \approx 25$ m) the limiting error of roll measurement makes 2.5 mm. This measurement accuracy is ensured by the device used (electronic total station).

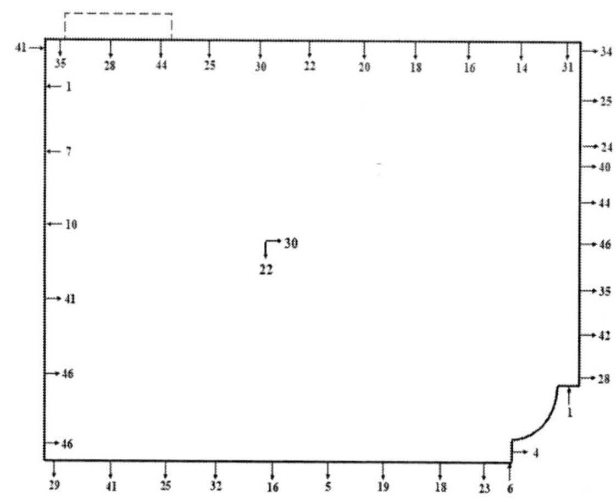

Figure 4. Building roll diagram as of 07.11.2016

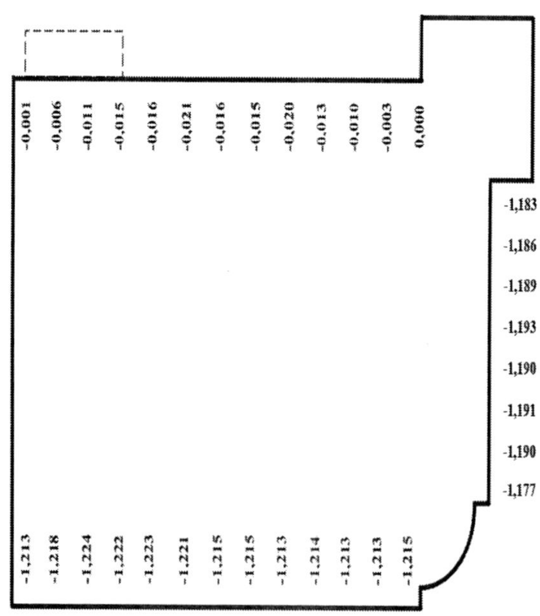

Figure 5. Results of horizontal position control of building structures

According to SNiP 2.02.01-83 (Foundations of buildings and structures), the permissible value is the roll value, which does not exceed 0.005·*H*. Therefore with a height of the building of $H \approx 25$ m, the admissible roll of the building is equal to 125 mm. All rolls of building structures do not exceed this tolerance.

The results of horizontality control of building structures are schematically shown in Fig. 5. The results of conditional elevations are given in meters.

This figure shows that along the north facade of the building, the top of the window aperture of the 2nd floor bent by about 20 mm. For all other elevations, the elevations of the 2nd floor window aperture change within the measurement errors.

4. Conclusion

The study made it possible to conclude the following:

1. The visual inspection of the whole building and its individual elements did not reveal any deformation defects.
2. the recorded results of settlement points are practically within the limits of instrumental errors of observations. The maximum accumulated settlement is recorded in points No. 6 and No. 9 (0.6 mm). The minimum settlement is recorded in point No. 12 (0.3 mm). All settlement points do not exceed the permissible limit.
3. Full roll of the entire building equals 37 mm and is placed in the southeast direction. All fixed rolls of building structures do not exceed the permissible limit of 125 mm established by regulatory documents.
4. According to the results of the horizontal control of the building structures, it was determined that on the northern facade of the building, the top of the window aperture of the 2nd floor bent by about 20 mm. For all other elevations, the elevations of the 2nd floor window aperture change within the measurement errors.
5. The building is in the permissible design operating mode.
6. More complete analysis of the geometric characteristics of the building structures will be carried out based on the results of the next measurement cycle.

References

[1] GOST 24846-2012 *Soils. Measurement methods of base deformations of buildings and structures* 2012 (Moscow: Ministry of Reg. Development of the Russ. Fed.)

[2] SP 22.13330.2011 *Foundations of buildings and structures. Updated version of SNiP 2.02.01-83* 2011 (Moscow: Ministry of Reg. Development of the Russ. Fed.)

[3] GOST P 53778-2010 *Buildings and constructions. Rules of technical survey and monitoring* 2010 (Moscow. Standartinform)

[4] *Manual on geodetic works in construction (SNiP 3.01.03-84)* approved by the Order of SiM Networks no 147 of 10 July 1985

[5] RD 34.21.322-94 *Methodology guidelines for observations of precipitation and deformation of buildings and structures of thermal power plants under construction and operation*

[6] *Guidelines for observations of deformations of bases and foundations of buildings and structures* 1975 (Moscow: Stroyizdat)

[7] Zotov V D and Zotov M V 2005 Lifting and leveling of buildings using flat jacks Coll. *Reconstruction of cities and geotechnical construction* (Rostov-on-Don) pp 156–67

[8] Boyko M D 1975 *Diagnosis of damages and methods of restoration of operational qualities of buildings* (Leningrad: Stroyizdat) 335 p

[9] Gairabekov I G, Gairabekov M-B I, Pimshin I Y et al August 2019 Geodetic Control of the Geometry of the Levelled Building *Atlantis Highlights in Material Sciences and Technology* Proc. of the Int. Symp. Engineer. and Earth Sci.: Appl. and Fund. Res., dedicated to the 85th anniversary of H.I. Ibragimov (ISEES 2019) pp 416–20 Available Online

[10] Gairabekov I G, Gairabekov M-B I, Pimshin I Y et al August 2019 Accuracy of Determining Buidings' Geometric Characteristics and Structures by Modern Means of Geodetic Measurements *Atlantis Highlights in Material Sciences and Technology* Proc. of the Int.

IOP Conf. Series: Materials Science and Engineering **905** (2020) 012024 doi:10.1088/1757-899X/905/1/012024

Symp. Engineer. and Earth Sci.: Appl. and Fund. Res., dedicated to the 85th anniversary of H.I. Ibragimov (ISEES 2019) pp 421–5 Available Online

[11] Gairabekov I G and Gagaeva H R August 2019 Distortions Introduced by the Cartographic Projection of Gauss at a Distance of the Construction Site from the Axial Meridian *Atlantis Highlights in Material Sciences and Technology* Proc. of the Int. Symp. Engineer. and Earth Sci.: Appl. and Fund. Res., dedicated to the 85th anniversary of H.I. Ibragimov (ISEES 2019) pp 583–7 Available Online

IOP Conf. Series: Materials Science and Engineering **905** (2020) 012025 doi:10.1088/1757-899X/905/1/012025

Development of a digital surface model and a digital terrain model based on ERS data

I G Gairabekov, A I Hamzatov, A T Mishieva, E I Ibragimova, M-B I Gairabekov and A I Gayrabekova

Grozny State Oil Technical University named after M.D. Millionshchikov, korp. 1,100, prosp. Husejna Isaeva, Grozny, 364068 Russia

E-mail: ibragim.ggni@mail.ru, khamzatov.akhmed@mail.ru, zuba-866@yandex.ru, lina.leams@mail.ru, gayrabekovbashir@gmail.com, Gayrabekovamina@mail.ru

Abstract. The paper considers the development methods of a digital surface model and a digital terrain model of the *Dead City* territory. The work utilizes the Phantom 4 PRO UAV with high flight stability and excellent shooting quality (camera 20Mg/px) with 4K resolution that allows creating qualitative DSM, DTM with a possibility of storage, processing, transfer on various platforms in different formats.

1. Introduction

The development of a digital surface model and a digital terrain model of the Dead City territory began with the collection of photographic material.

Before the beginning of aerial photography, the determining factor is correct selection of the required type, mode of aerial photography as well as competent preparation of the flight mission. There are many varieties of aerial photography. The type of aerial photography is chosen depending on the final objective and the software that will be used for further processing of the material.

We used high-altitude (height more than 40 m) general coverage photography, since the Dead City represents a mountainous terrain with frequent wind changes. The use of UAVs in conditions of variable winds and variable relief is safe at such altitudes.

Leveling survey (90°) makes it possible to chosen see the objects: their adjacent sides, inclination angles, vegetation, forms of relief, to minimize deviations of UAV from the selected flight trajectory.

The flight mission was to compile the general scheme of UAV flight, in which all factors of future flight are taken into account with high accuracy. Competent flight mission is one of the most important factors of aerial photography, which is directly related to the quality of thus obtained material.

The next step is to select the pilot mode. This work is based on “autonomous” mode of UAV flight, it provides accurate and high-quality shooting, as well as a sufficient percentage of overlap between photographs of the area.

2. Methods and materials

Flight mission and flight mode was compiled in SKYCATCH program. When you start the program, the “NEW MISSION” window appears. Here we define a new flight mission plan and select the flight mode, additional parameters are specified later.

After selecting the "NEW MISSION", the program should give access to your geodatabase, i.e. to the operator's location. This is done to more accurately link the model to the orthophotomap, and to allow the UAV navigating space based on the geographic coordinates of the terrain in real time. Orthophotomap is a photographic terrain plan derived from the processing of remote sensing data of the Earth's surface using UAVs.

The second step is to "Find Yourself on the Map". After we clicked "NEW MISSION", a satellite map pops up, and the access to mobile device geodatabase allows the satellite to find your location automatically.

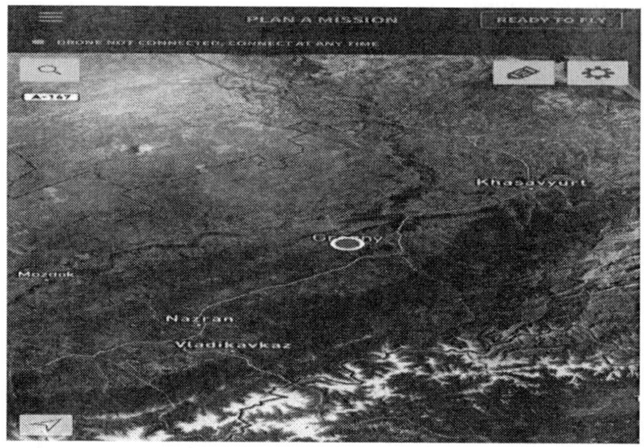

Figure 1. Software Dialog Box

After you start the map, the blue dot automatically detects and displays your location in the spatial coordinate system.

Next, it is necessary to set the flight parameters and the mode of UAV take-off and landing.

We click on the pencil icon in the main window of the program and the "Settings Window" appears. With a finger we highlight the terrain section that we want to shoot on the map, and the program automatically lays out the optimal route of shooting, which will require the minimum flight time, will allow covering the required objects as much as possible and creating the required viewing angle during aerial photography.

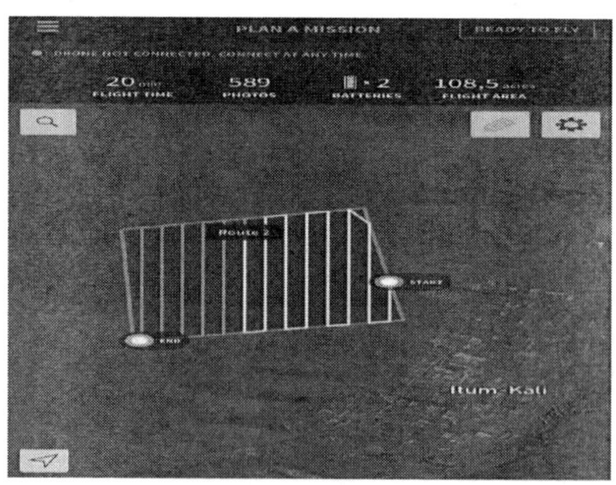

Figure 2. Software Dialog Box

After the route is laid, we proceed to add the main parameters of the survey: UAV flight altitude, descent angle, camera rotation angle. On the main window of the program interface press the icon “Flight Settings” and in the program window the settings window appears, where we mechanically enter the required flight altitude and survey angles.

After setting the flight mode and entry of precise parameters of UAV flight, press “READY TO FLY”. Then the UAV starts the flight, it will move along the trajectory determined by the flight mission, collect photographic material with the specified shooting resolution and with the optimal percentage of overlap between photos. For safety, we can set the UAV flight to a manual mode at any moment and start control from the console. The landing accuracy makes ± 1 meter from the start point.

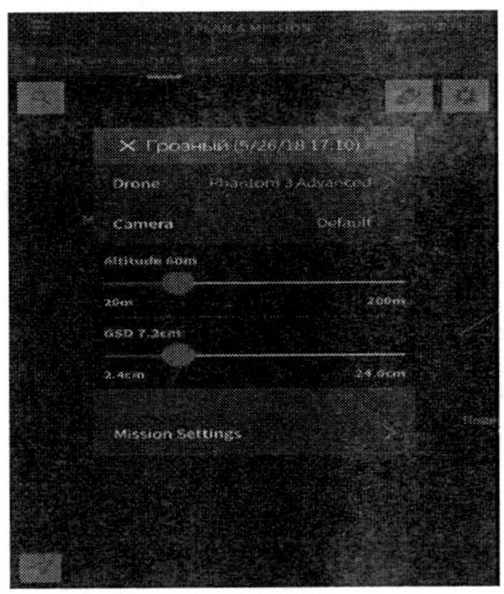

Figure 3. Software Dialog Box

After the completion of the UAV flight and the collection of the necessary data of remote sensing of the Earth’s surface (based on aerial photography), the stage of processing of the obtained data begins, on the platform of which a digital surface model and a digital terrain model will be built.

Agisoft Photoscan Pro was used to process the resulting material. It is a modern automated GIS platform designed to collect, store, process and transmit remote sensing data of the Earth’s surface.

The interface of the program includes hundreds of functions in the sphere of creation of digital models of a relief, an area, processing of space pictures, creationsof orthophotomaps, georeferencing of maps and creation of relief maps.

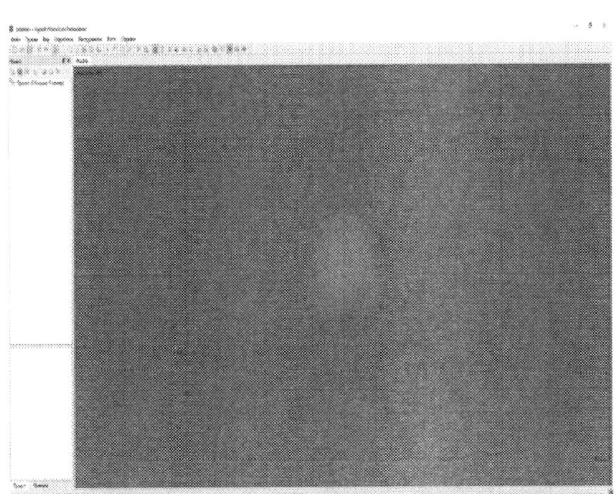

Figure 4. Agisoft Photoscan Pro interface

When you start the program, a "blue ball" immediately pops up in the center of the program window. It is an auxiliary element, if you move the cursor and scroll the "blue ball", you immediately get the impression of a three-dimensional space.

The processing of received data begins with the fact that we load the filtered data into the program itself. This is done to run the program itself. Next, in the "Processing" section, click "Download Photos", then the program window appears and the "Desktop" opens. After that, we select the folder we want and upload all the photos to the program.

After a short processing, the program loads the material we want. At the stage of loading the material into the program and before processing, it is important to keep track of what we upload into the program. Next, we check whether there are any photos with unsatisfactory quality in the general array.

After loading, all selected material is displayed in the program itself. This is done to make it easier for the user to adjust image quality and edit material as data processing proceeds.

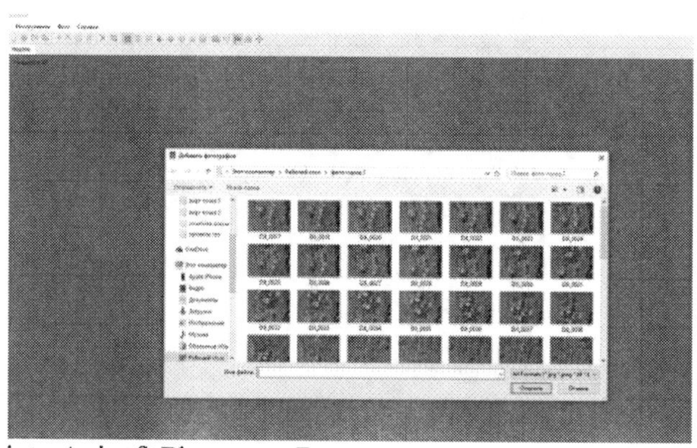

Figure 5. Loading data into Agisoft Photoscan Pro

After downloading the data, the next step is "Processing", which includes "Photo Alignment". This is the first stage in the algorithm for building a digital surface model or a digital terrain model.

In the "Photo Alignment" step, the program begins to calculate snapshot points using algebraic and mathematical algorithms.

Depending on the area of the snapshot and the number of pixels on its area, the program calculates the places from where the UAV produced the snapshot and begins to collect and align the points from which subsequent snapshots were taken into one flat plane. This process is called photo alignment.

The program principle is inextricably linked to the PC control point, namely the processor. As we know the PC has two control points: basic and mathematical.

The main is set by default, and when processing data, the program starts to recognize only it, and when using the program, it is recommended to use the mathematical processor of the PC. This approach greatly accelerates PC operations and processes.

When we do high quality ERS processing, it takes a lot of computer time, and the mathematical processor accelerates material processing. To do this, the program has its own algorithm. The mathematical processor of PC control in the program includes the following actions:

- "tools";
- "settings";
- "these models";
- "GPU" (we tick off).

After the "Photo Alignment" stage, a weak dot sketch for the future model is created in the additional control window, i.e. the program begins to find common areas of "overlap" between paired photos at this stage and prepares the frame for their further bonding.

Aerial photography in autonomous mode with the help of the flight mission is the main step that provides for qualitative observation of the percentage of overlap required for model creation.

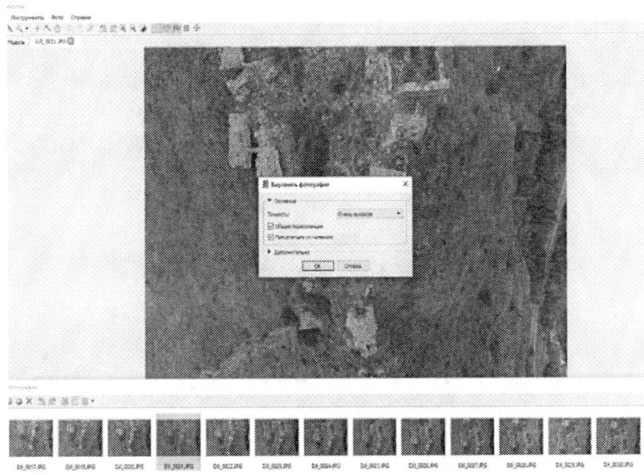

Figure 6. Alignment of photos

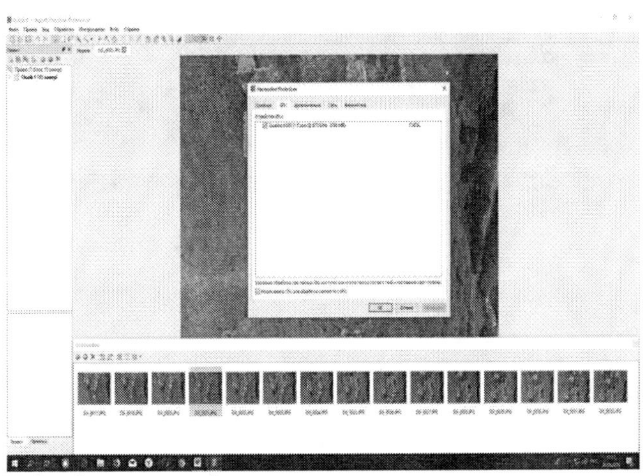

Figure 7. Enabling GPU for data processing

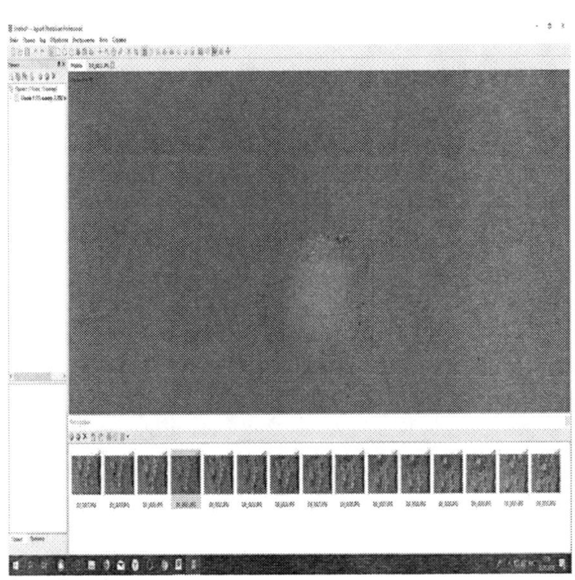

Figure 8. Creation of a point cloud

After the point cloud is built, the second stage of material processing begins to create a digital surface model and a digital terrain model. The second stage of processing in Agisoft Photoscan Pro is called "Building a Dense Point Cloud". It is based on the principle of "mosaic". As in the first case, the PC recognizes all algorithms that the user sets as mathematical operations, so it is not recommended to turn off the graphics (mathematical processor).

To create a dense point cloud, click "Processing" – "Create Dense Point Cloud".

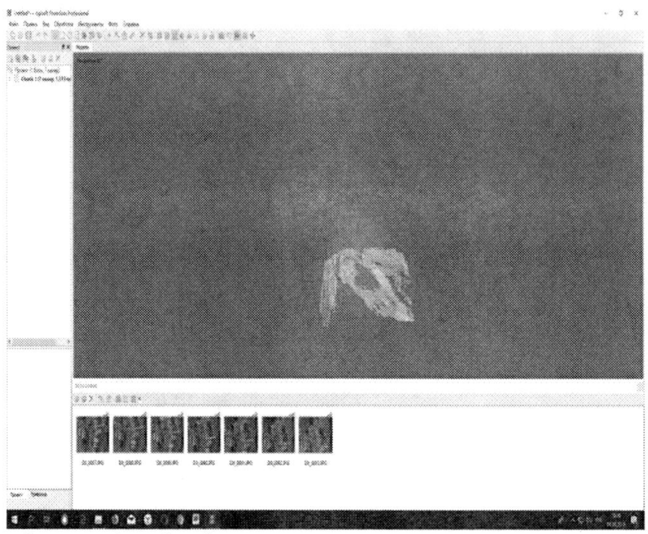

Figure 9. Creation of a dense point cloud

Building a dense point cloud is the process of the program binding common points on various images obtained from the UAV. After completing the "Create Dense Point Cloud" process, the model gets a pronounced contour and gray color.

In general, during the whole processing process, we can view the model on different spectra by switching modes on the panel window. This is done so that each download and processing step can be monitored.

After completing the "Create Dense Point Cloud", we proceed to the next processing step. It is called "Build Model".

The "Build Model" phase is the main step of processing. That is why it takes a long time. To start building a model, you must refer to the "Processing" menu and click "Build Model", where we specify the required quality.

Depending on the specified quality, the download time will vary following the principle of "higher quality – more time".

At the stage of model construction, the future digital surface model or the digital terrain model becomes complete.

To do this, you must create a texture and specify the geometric dimensions and shape. Immediately we indicate in the program whether we want to get a further map of heights.

After the "Build Model" phase, the texture phase begins.

Texture is necessary to give the digital surface model or the digital terrain model the greatest possible real appearance. The texture also provides an opportunity to express contours and outlines that allow you to better see small areas and contours. They, in turn, give detail and contrast.

To begin the texture process, open the "Processing" window and click "Draw Texture". It does not take a long time to build a texture, as we do the main steps with the PZU graphics processor.

Texture construction is the final step of processing remote sensing data using UAVs in Agisoft Pfotoscan PRO.

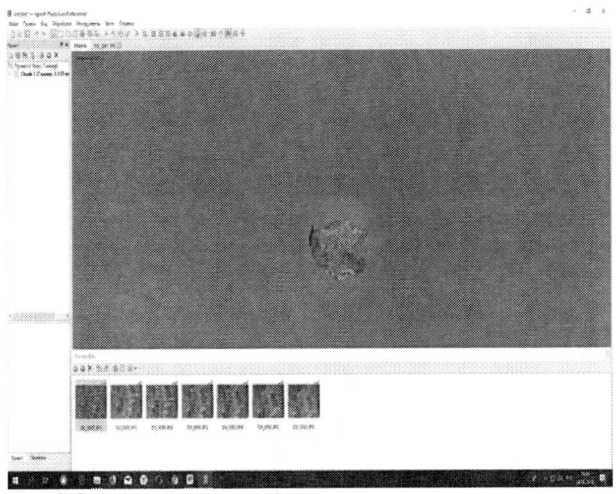

Figure 10. Creation of a model

In the program itself, you can view the results of any stage of model creation. To do this, the program interface offers very fast and convenient switches that allow you to instantly turn on the "View Model" in different spectra and layers.

Next, we can start doing additional operations and actions on models depending on what we want to get.

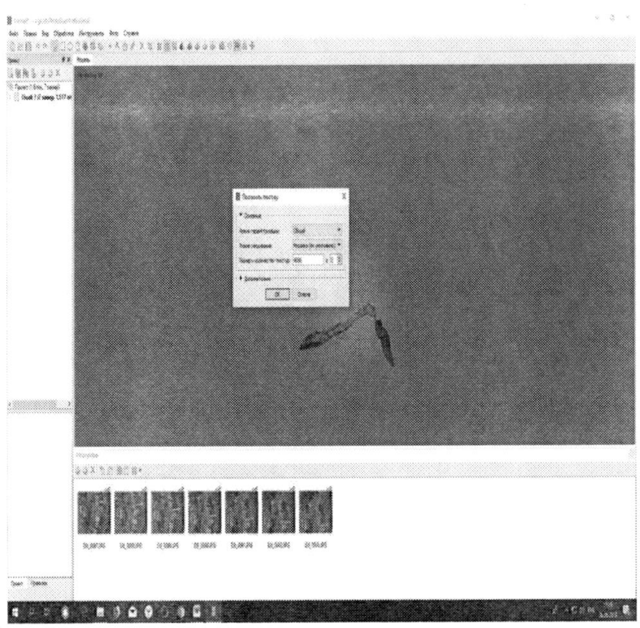

Figure 11. Creation of texture

The most common use is to "Create a Height Map", which allows obtaining the elevation of any section of the model.

The accuracy of elevations depends on the accuracy with which previous actions were carried out on the Earth's remote sensing data.

Building a height map is a long process that is divided into several stages. Each stage has its own meaning and duration within a time frame.

The first stage – the model shall be saved in STL format. It is done so that we can make a model leader and create a backup (for security reasons) for external media.

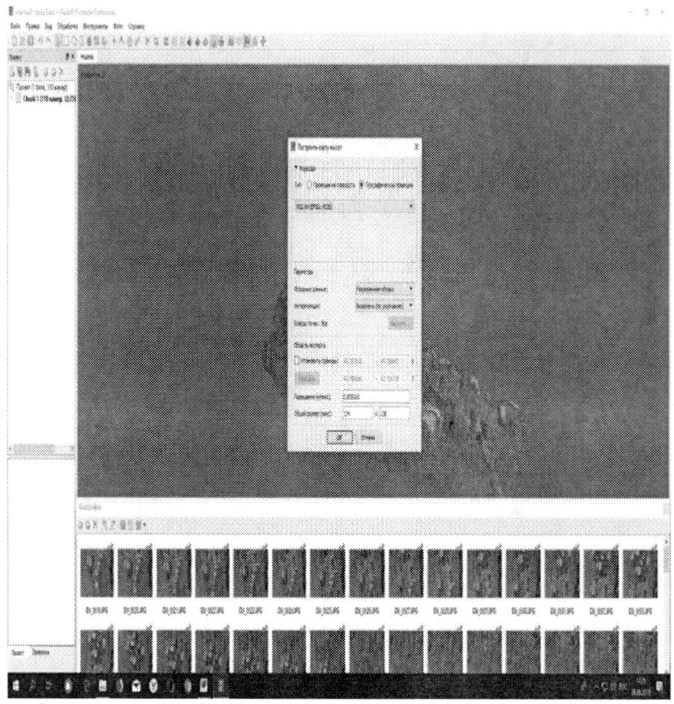

Figure 12. Height mapping

3. Results

The software provides a wide range of functions for exporting and importing data. After building a height map, we can export the acquired data in different formats and to different media regarding the fact where we want to use them and what we want to get from these exported data.

A height map is a platform target map that provides a characteristic of the excess of a certain territory or relative to the elevation of an object.

As mentioned above, you can export spot elevations after you create a map. To do this, open the "File" window and click "Export Height Map". Next, a dialog box appears in which you need to specify the format in which we want to store our data. Without a height map, it is impossible to obtain orthophotomap in the software.

One of the most important and irreplaceable features of the AGISOFT PFOTOSCAN PRO is that it enables the creation of high-quality orthophotomaps based on created digital surface models or digital terrain models.

The most important high-precision engineering tasks in the field of geodesy, photogrammetry, mine surveying, digital modeling and construction are solved on the basis of orthophotomaps.

The creation of orthophotomaps is the main platform for further geodetic justification, as georeferencing is done on the map using accurate and qualitative orthophotomap.

The AGISOFT PFOTOSCAN PRO creates the orthophotomap based on the built digital surface model or digital terrain model, which, in turn, are created on the basis of aerial photography.

To start the creation of the orthophotomap, you must access the "File" menu in the program window and the "Build Orthophotomap" dialog box. After clicking "Build Orthophotomap" the data processing begins in a green loading format that does not take long.

The created orthophotomap can be unloaded in the formats we need. We recommend using the STL format, which allows taking the file to external sources. Then the orthophotomap can be opened without the initial remote sensing data of the Earth's surface.

Figure 13. Orthophotomap of the area obtained on the basis of DTM

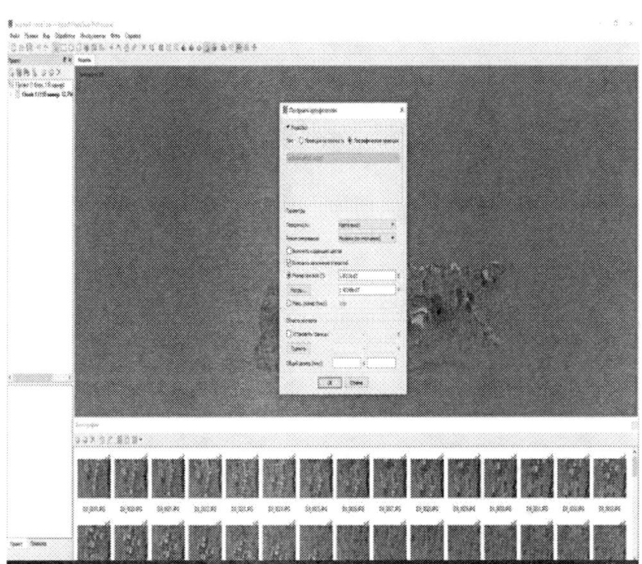

Figure 14. Dialog box of orthophotomap drawing

4. Conclusion

This paper explores the processes of creating a digital surface model and a digital terrain model. The analysis of the crowdsource technology in the creation of digital maps was carried out.

The functional model is the basis for building a data model. In fact, without data about what kind of information is involved in creating a digital terrain model, it is pointless to build a data model.

Based on the analysis, a model of processes for a digital terrain model was created, which is the source for the data model. The collaborative modeling method linked entities and attributes in process and data models.

Digital modeling technologies make revolutionary changes in almost all spheres of human activity: construction, medicine, geodesy and mine surveying, etc.

Thus, the use of the latest digital technologies in engineering geodesy, construction and monitoring of buildings, structures and the Earth's surface seems quite relevant.

References

[1] Shechovtsov G A and Novikov V M 2015 *Basics of photogrammetry* Information letter no 336-95 (N. Novhorod: Nizhny Novgorod RI)

[2] Movsesyan R A and Taplashvili I A 2009 New tools and methods for automation of digital modeling measurements *Problems of astronomy and geodesy* 6th Congr. of All-Union society of astronomy and geodesy (Yerevan; Moscow) pp 88–9

[3] Ustavich G A, Oleynik A M and Shalygina E L 2014 Development of the basics of modeling and visualization *Geodesy and aerial photogr.* **1** 40–53

[4] Gayrabekov I G 2011 *Development of geodetic support technology for engineering protection of buildings and structures* Thesis 271

[5] Korenea B G and Rabinovich I M (ed) 1972 *Guide to photogrammetry* (Moscow: Stroizdat) 511 p

[6] SP-13-102-2003 *DTM-based survey rules* 2003 (Moscow: DPC)

[7] Ibragimova E.I., Gayrabekov I.G., Mishiyeva A.T et al *Solution to the Problem of Lanslide-Prone Territories Using Geodesic Methods*

IOP Conf. Series: Materials Science and Engineering **905** (2020) 012026 doi:10.1088/1757-899X/905/1/012026

Geotechnical monitoring

I G Gairabekov, Sh S-E Kaimov, A T Mishieva, E I Ibragimova, M-B I Gairabekov and A I Gayrabekova

Grozny State Oil Technical University named after M.D. Millionshchikov, korp. 1,100, prosp. Husejna Isaeva, Grozny, 364068 Russia

E-mail: ibragim.ggni@mail.ru, kaimov.shamil@bk.ru, zuba-866@yandex.ru, lina.leams@mail.ru, gayrabekovbashir@gmail.com, Gayrabekovamina@mail.ru

Abstract. The paper considers methods of observation over foundation settlement of buildings and structures on the southern slope of the Veduchi Ski Resort and its accompanying infrastructure. The study was carried out from control stations via geometric levelling of the II class. Based on the results of two observation cycles, the settlement values were obtained, all buildings and structures settled within 0.4–2.2 mm. Based on the data on the limit permissible values presented in the paper it can be concluded that buildings and structures on the southern slope of the Veduchi Ski Resort and its accompanying infrastructure are within the permissible settlement values.

1. Introduction

Geotechnical survey of buildings (structures) of surrounding facilities of I and II degrees of responsibility, including underground internal communications, is required in case of their location within the boundaries of the influence area of newly erected or reconstructed facilities.

Geotechnical monitoring was carried out on the southern slope of the Veduchi Ski Resort and its accompanying infrastructure.

Geotechnical monitoring includes:

- laying of geodetic and observation networks;
- horizontal and vertical control;
- observations of horizontal displacement of buildings and structures;
- observations of foundation settlement of buildings and structures;
- analysis and calculation of horizontal and vertical displacements of buildings and structures [1].

2. Methods and materials

In order to create an observation network and ensure the quality of works, 6 soil benchmarks were laid down. Places for laying of soil benchmarks were selected taking into account slope stability, convenience of observation of deformation marks [2–3]. The layout is shown in Figure 1.

Ground benchmarks under the drill were laid as follows: a well of the required depth was drilled (tubular benchmark at the depth from 3 meters). The benchmark was laid into a drilled well with a diameter of 95 mm and a depth of up to 3 m. The benchmark represents a metal pipe with a diameter of 89 mm with a wall thickness of 5 mm, with a steel tip at the end with three anchor discs. Cement slurry was poured into the well in the amount of cement and sand 1:3 with a layer sufficient for unobstructed immersion of the metal pipe; the upper part of the pipe was fixed in a vertical position via braces and a plumb; the well was filled with soil through layer-by-layer trampling. The well was filled with soil before setting of cement slurry in the anchor earth cut, i.e. not later than 30 minutes after the pipe installation.

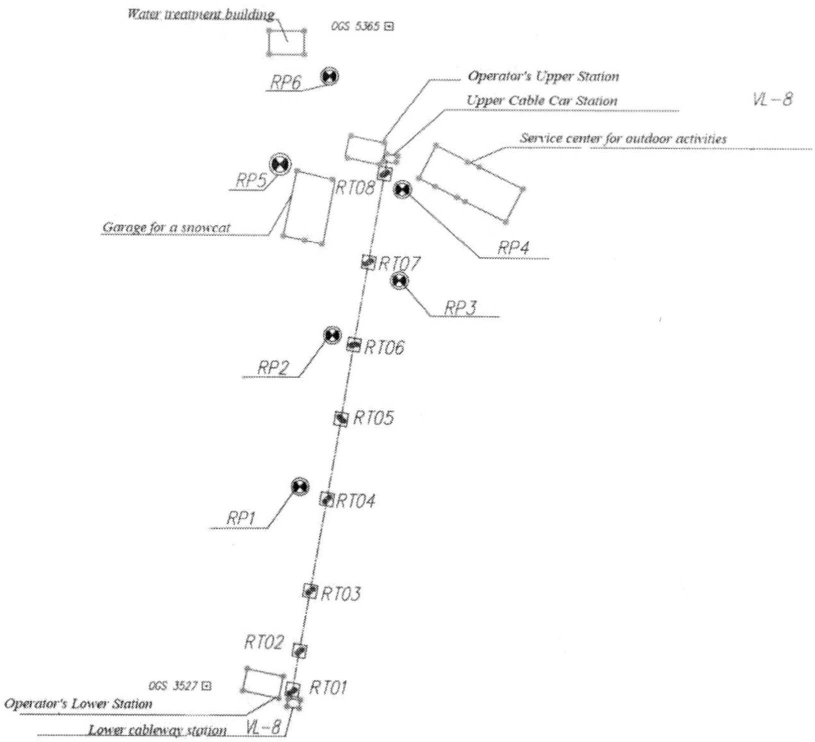

Figure 1. Diagram of geodetic and observation networks

A mark with the identification number was embedded in the upper face of the benchmark pipe. A keyline was made for each laid down item.

The number of ground benchmarks, their location and the method of fixation ensure the required accuracy of deformations of foundations and structures. In order to optimize the scope of works and unify the method of geodetic measurements, the benchmarks intended for observation of vertical and horizontal displacements were combined.

The structures (method of attachment) of ground benchmarks complied with the requirements of regulatory and technical documentation and were determined during preparatory works on the basis of the results of reconnaissance of the area of work. Design of signs installed for points of geodetic thickening networks – tubular signs with an anchor of 160 op type (Figure 2–3).

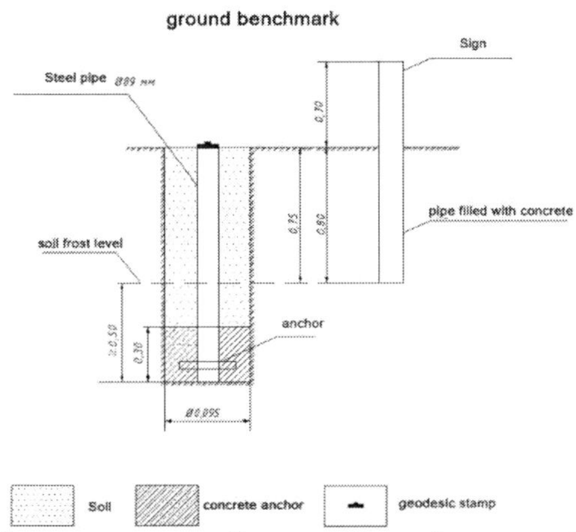

Figure 2. Tubular sign with armature according to 160 op sign type with the identification column

The required marks were made of a 32mm diameter iron pipe with a corrosion resistant steel reference head (Figure 4) (Type 1) and installed by the end of March 2019 (Figure 5).

In the 0 cycle, the settlement points were installed on the foundations: service-center of active activities, garage, water treatment building, lower station of VL-8 ropeway, control room of the lower station, upper station of VL-8 ropeway, control room of the upper station. Besides, these settlement points were painted with oil paint and numbered.

Figure 3. Ground benchmark No. 1

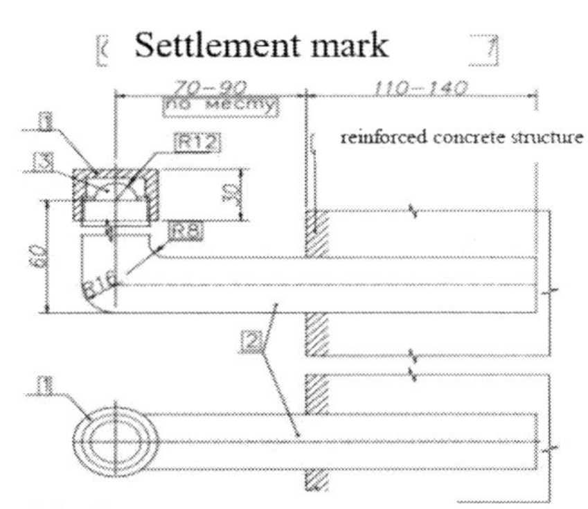

Figure 4. Settlement point design

Figure 5. Type 1

Two diagonal anchor bolts were used on VL-8 ropeway supports to observe deformation. For identification, anchor bolts were marked with oil paint, where their numbers were written.

The settlement points were transferred for safety observation under the Act on Delivery of Geodetic Benchmarks for Safety Observation to JSC Resorts of the North Caucasus.

In order to transfer the planned and altitude elevations, a horizontal and vertical control (HVC) was carried out from the HVC stations Rp1 and Rp2 previously laid by LLC MarGeo. Closed traverses of I order and traverse leveling of IV class were laid from these stations. The works were carried out by electronic total station *Leica FlexLine TS09 plus R1000 1*. Horizontal and vertical angles were measured by reciprocal observation with root mean squared error amounting to not more than 5. The line length was measured by two techniques in forward and reverse directions by pointing to the prism of a beam deflector. The line length was measured in the horizontal mode [4].

3. Results

Determination of points heights of geodetic control was performed by trigonometric leveling using electronic total station *Leica FlexLine TS09 plus R1000 1* in forward and reverse directions, two directions per reflector. The height of the instrument and reflector above the center mark was measured to an accuracy of 1 mm. The measurements were monitored by trigonometric leveling using the method "from the middle" with equal shoulders.

Permissible measurement misalignment in geodetic moves did not exceed the boundary values 1, 2, limit angular misalignment [5]: $f\beta_{\partial on.} = 10''\sqrt{n}$, where n – number of angles;

limit relative misalignment: $\frac{1}{N} = \frac{[S]}{Fs} \leq \frac{1}{10000}$;

The horizontal and vertical control diagram is shown in Figure 6.

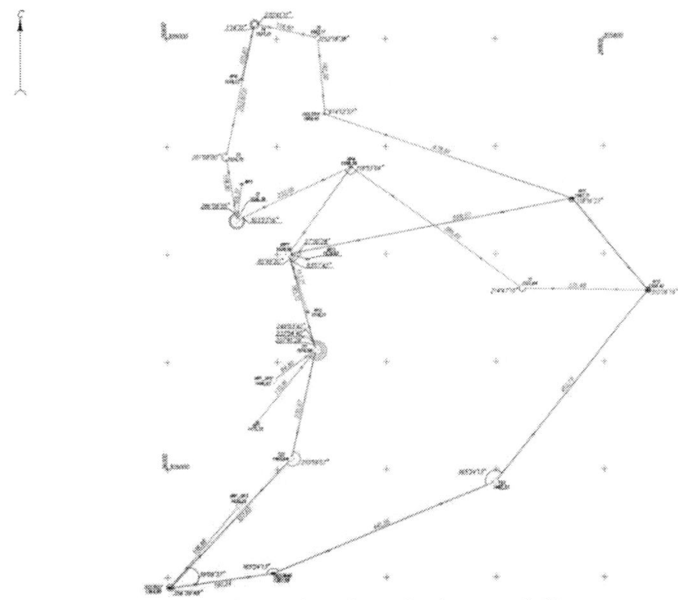

Figure 6. Horizontal and vertical control diagram

Processing of transit traverse of 1 order was carried out in the CredoDat program, upon completion of processing the program generates progress sheets, the example of which is shown in Table 1.

4. Conclusion

From the obtained coordinates and heights of the benchmarks of the two measurement cycles, which are given in Tables 3, 4, it is possible to determine the stability of the benchmarks.

Table 5 shows the benchmark heights in two observation cycles and the settlement of these benchmarks.

Table 2 shows the traverse characteristics. When studying the characteristics, we see that Fb does not actually get out of Fb permissible defects of the 1 order traverse. In this way the 1 order traverse moves can be used for horizontal and vertical control [6–10].

The tables show that the benchmarks settled to a minor value, but to equalize for the most accuracy were used the initial points Rp1 and Rp2 laid by LLC MarGeo. These benchmarks remained stable and thus we concluded that in the second cycle of observations it is possible to use our ground benchmarks, as they equalized from the initial points, which are considered stable.

Table 1. Sheet of transit traverses

Paragraph	Measured angle	Directional angle	Measured distance	Equalized distance	X	y
1	2	3	4	5	6	7
T22						
		222°45'46"				
3527	39°09'37"				25780.7795	305003
		81°55'73"	1907397	190.2407		
OGS3640	165°2442"				25S07.50S4	305192
		67°19'37"	441.5489	441.5501		
	151°04'05"				25977.7135	305599
T33						
		38°23'43"	452.1851	452.1872		
RP2					26332.113	30588
OGS3640		"				
		261°55'73				
3527	320°5024"				25780.7795	305003
		42°45'46"	327.6057	327.6047		
T22	149°40'03"				26021.2976	305226
T11	26°13'41"	12°25'49"	205.9721	205.9715		
		218°39'00"			26222.441	305270
NRP1						
RP1		140°38'78"				
RP2	310°26'20"				26332.113	305880
		271°04'55"	231,4744	231,4756		
T1						
	214°4740"				26336,4841	305648
RP4	118°53'03"	305°52'05"	385,6575	385,6575		
		244°45'40"			26562,4483	305336
T2						
RP4		244°45'40"				
T2	286°08'54"				26463,4729	305126
T3	201°07'58"	350°54'07"	118,7963	118,7958	26580,7742	305107
		12°02'09"	252.S749	252,8731		
T4	270°41'07"	102°43'73"			26828,0885	305160
T5						
RP1					26502,6700	305740

Table 2. Characteristics of transit transverses

Travel points	Length move	N	Nb	Fb fact..	Fb aux..	Discrepancy before adjustment				Discrepancies in equalized directional angles			
						Fx	*Fy*	*Fs*	*[SJ/Fs]*	*Fx*	*Fy*	*Fs*	*[S]/Fs*
3	4	5	6	7	8	9	10	11	12	13	14	15	16
3527, OGS3640 , RP2	1083,9737	4	3	0°00'04"	0°00’7"	-0.0089	0.0121	0.015	72229	-0.0022	-0.0034	0.0041	265568
3527, T22, Til	533,5778	3	3	-0°00'01"	0°00’7"	-0.0004	0.0028	0.0029	185744	0.0013	0.0008	0.0015	355281
RP2, Tl, RP4	617,132	3	3	-0°00'00"	0°00’7"	-0.0087	0.0075	0.0115	53607	0	0.0012	0.0012	510796
T2, T3, T4	371,6712	3	3	0°00’4"	0°00’7"	-0.0108	0.0042	0.0116	32080	0.0022	0.0003	0.0022	165628
RP1, OGS3314 , T4	739,4116	4	3	-0°00’7"	0°00’7"	-0.0092	0.0205	0.0224	32941	0.0019	-0.0086	0.0088	84206

As a result of geodetic works carried out on the southern slope of the Veduchi Ski Resort and its accompanying infrastructure 50 deformation marks of the observation network and 6 deep benchmarks of the support network were installed.

In order to transfer the planned and altitude elevations, the horizontal and vertical control was carried out from horizontal and vertical control points Rp1 and Rp2 earlier laid by LLC MarGeo. In the first cycle, the horizontal and vertical control showed no unacceptable misalignment of the moves, thus the moves were consistent with the accuracy of 1 order in plan and IV class in height.

In the second geotechnical monitoring cycle, the benchmark network was also equalized from earlier laid points. The result of the equalization showed that the Rp5 remained stable, Rp2 and Rp6 were set at 2 mm, the rest were set at 1 mm, thus we see that the whole benchmark network did not exceed the permissible settlement values in the second measurement cycle.

Deformation control benchmarks and anchor bolts were used to observe horizontal displacements.

The horizontal movement of the elements of the observation network was recorded on the southern slope of the Veduchi Ski Resort and its accompanying infrastructure from the points of the support network via the radiation method.

As a result of the observation of horizontal displacements in two cycles, we determined the displacements of buildings and structures. All buildings and structures were displaced within 0.6–2.5 mm, which fully satisfies the permissible values of displacements of buildings and structures.

Table 3. Index of coordinates

N	*Item name*	*X*	*Y*	*mXY*	*H*	*mH*
1	*2*	*3*	*4*	*5*	*5*	*5*
			Estimated attitude			
1	3527	25780.78	305003.71	0.01905	1364.9042	0.02554
2	RP1	26088.962	305163.73	0.0137	1473.5474	0.02123
3	RP2	26293.264	305256.33	0.01035	1519.2141	0.01753
4	OGS3314	26660.439	305288.7021	0.0117	1569.4378	0.0237
5	OGS3640	25807.508	305192.06	0.01631	1387.5789	0.02599
6	ISHRP1	26502.67	305740.46		1587.705	
7	ISHRP2	26332.113	305880.34		1559.424	
8	RP3	26391.377	305254.59	0.00984	1529.6169	0.01732
9	RP4	26562.448	305336.38	0.00827	1558.3509	0.0176
10	RP5	26531.151	305136.17	0.01294	1528.0021	0.02124
11	RP6	26725.309	305136.18	0.01523	1518.1299	0.02385
12	T1	26336.484	305648.9	0.00464	1551.6439	0.01397
13	T2	26463.473	305126.49	0.01188	1526.3853	0.02117
14	T3	26580.774	305107.71	0.01301	1524.7542	0.02234
15	T4	26828.089	305160.44	0.01492	1527.2051	0.0237
16	T5	26801.903	305276.41	0.01298	1562.1117	0.02387
17	Til	26222.441	305270.48	0.00993	1514.5754	0.02074
18	T22	26021.298	305226.14	0.01298	1460.3441	0.02331
19	T33	25977.714	305599.49	0.01078	1482.3305	0.02482
20	VRP1	26401.06	305221.74	0.00866	1528.5582	0.0173
21	VRP_OP3	25934.399	305130.01	0.01681	1420.2541	0.02607
22	VRP_OP5	26166.451	305193.9748	0.01251	1496.8294	0.02092

Table 4. Index of coordinates

N	*Item name*	*X*	*Y*	*mXY*	*H*	*mH*
1	*2*	*3*	*4*	*5*	*5*	*5*
			Estimated attitude			
1	RP1	26088.9705	305163.7253	0.00843	1473.5463	0.00482
2	RP2	26293.2511	305256.3302	0.00792	1519.2122	0.00315
3	OGS3314	26660.4454	305288.7044	0.00348	1569.4474	0.00195
4	OGS3527	25780.7915	305003.7016	0.00560	1364.8992	0.00210
5	OGS3640	25807.5195	305192.0568	0.00480	1387.5794	0.00213
6	ISHRP1	26502.6700	305740.4560		1587.7050	
7	ISHRP2	26332.1130	305880.3350		1559.4240	
8	RP3	26391.3840	305254.5894	0.00255	1529.6164	0.00148
9	RP4	26562.4536	305336.3772	0.00256	1558.3505	0.00146
10	RP5	26531.1590	305136.1717	0.00804	1528.0017	0.00241
11	RP6	26725.3166	305136.1842	0.00866	1518.1282	0.00328
12	T1	26323.6802	305658.0388	0.00138	1551.6431	0.00113
13	T2	26463.4808	305126.4897	0.00351	1526.3895	0.00175
14	T3	26580.7828	305107.7064	0.00386	1524.7490	0.00184
15	T4	26828.0962	305160.4396	0.00443	1527.2083	0.00195
16	T5	26801.9089	305276.4160	0.00386	1562.1079	0.00196
17	T11	26222.4479	305270.4732	0.00293	1514.5678	0.00171
18	T22	26021.3061	305226.1359	0.00381	1460.3384	0.00192

Table 5. Difference in ground benchmark heights

n/n	Ground benchmark	Cycle 1. April 2019. Ground benchmark heights, m	Cycle 2. May 2019. Ground benchmark heights, m	Difference in ground benchmark heights Cycle 1 – Cycle 2, mm
1	2	3	4	5
1	RP1	1473.547	1473.546	1
2	RP2	1519.214	1519.212	2
3	RP3	1529.617	1529.616	1
4	RP4	1558.351	1558.350	1
5	RP5	1528.002	1528.002	0
6	RP6	1518.130	1518.128	2

References

[1] Shekhovtsov G A and Shekhovtsova R P 2009 *Modern geodetic methods. Definitions of deformations of engineering structures* Monograph 156 p
[2] Bolshakova V D and Levchuk G P *Reference guide to engineering and geodetic works edited in 1980* 785 p
[3] Klushin E B, Kiselev M I and Mikhelev D Sh 2001 *Engineering Geodesy* 464 p
[4] Bolgov I F 1984 *Geodetic works during construction and testing of large structures* 144 p
[5] Gairabekov I G 2011 *Development of geodetic support technology for engineering protection of buildings and structures* Thesis 271 p

[6] Korenev B G and Rabinovich I M (ed) 1972 *Guide to structure dynamics* (Moscow: Stroizdat) 511 p
[7] Aime K A 2005 Monitoring of buildings and pits Part 2 *Constr. Mater., equipment, technol. of the century* **11** 37–9
[8] *Temporary design norms and rules of multifunctional high-rise buildings and complexes in Moscow* 2005 MGSN 4.19-2005 (Moscow) 129 p
[9] GOST R 53778-2010 *Buildings and structures. Rules for survey and monitoring of technical condition*
[10] *Technical regulations on safety of buildings and structures introduced by the Federal Law of the Russian Federation* no 384-FZ of 30.12.2009

Cable robot design method using the NX CAD / CAM / CAE system

E V Gaponenko S I Anciferov and V S Kuzmina

BSTU named after V.G. Shukhov, 46, Kostyukova street, Belgorod, 308012, Russia

E-mail: gaponenkobel@gmail.com, gaponenko.ev@bstu.ru, anciferov.sergey@gmail.com, anciferov.si@bstu.ru, viktorija.kuzmina@yandex.ru

Abstract. The article describes a method for designing a cable robot used for moving goods under special conditions. It is a parallel structure mechanism using the NX CAD / CAM / CAE system controlled by the Teamcenter PLM system. The mechanism is a rigidly fixed frame connected by tensioned ropes to a platform containing an axial displacement device. The diagram of the device is presented. To develop an electronic-digital model of the robot, the control structure that allowed for designing using the "Top-Down" method was built. The control structure reduces time required to develop a model. The spatial parameters of the device were determined. The product concept was developed.

1. Introduction

Currently, when developing a new product, it is necessary to ensure its high quality, low economic costs of manufacturing, high strength characteristics, compliance with customer requirements, etc. To solve engineering problems and achieve required quality, the NX CAD / CAM / CAE system controlled by the Teamcenter PLM system which allows for achieving the set goal and maximum efficiency is used [1, 2].

The use of the NX CAD / CAM / CAE-system can reduce the cost of the product, improve quality and reduce the time of launching it into the market. The Teamcenter PLM-system ensures interaction of all participants in the workflow in a single information space that provides each user with real-time access to product development. The integration of the NX CAD / CAM / CAE and PLM-systems allows you to cover all stages of the product life cycle [3, 4].

2. Development of a cable robot

Based on the terms of reference and outline documents, the electronic structure of the cable robot of a parallel structure was developed in the Teamcenter PLM system application "Structure Manager".

Cable robots are a special kinematic structure, consisting of a working body connected to a fixed base platform using cables [5–8]. Unlike the known parallel robots [9–10], cable robots use cables instead of extension rods to control the position of the output link. In these mechanisms, the position of the output link is controlled by changing the length of the cables. Cables are usually wound around coils attached to the base and driven by a rotating motor.

The kinematic structure of the robot is shown in Figure 1. The cable robot is designed to move goods in special conditions and contains electric motors 1, 2, 3, 4, each of which includes a device for laying cables and a gearbox. The electric motors are rotated by control device 5. The electric motor reducers 1, 2, 3, 4 are connected to cables 6, 7, 8, 9 to move the load in the horizontal plane through

IOP Conf. Series: Materials Science and Engineering **905** (2020) 012027 doi:10.1088/1757-899X/905/1/012027

rollers 10, 11, 12, 13. Rollers 10, 11, 12, 13 are located on frame 14. The gripping device is movable platform 15 connected to cables 6, 7, 8, 9. On movable platform 15, there is output link 18, on one end of which actuator 17 is located. It can perform axial movements. Drive 17 is a direct current motor kinematically connected with a gear rack. The gear rack is at the end of the output link. At the opposite end of the output link, the drive is rigidly fixed. Its output shaft is equipped with adjustable grip 16.

When motors 1, 2, 3, 4 rotate with cable laying devices and gearboxes, the length of cables 6, 7, 8, 9 changes. As a result, the gripper moves to the desired point in space. It is assumed that cables 6, 7, 8, 9 are able to move the gripper due to output link 18 which can perform axial movements by means of axial displacement actuator 17 connected to the rack mounted on platform 15 in a horizontal plane; the drive consisting of an electric motor and a kinematic transition node ensures the conversion of rotational motion into reciprocating and adjusts the size of capture 16.

The Teamcenter "Structure Manager" application provides an ability to work with the product throughout the entire life cycle. It also allows you to download revisions and manage variant conditions. The structure manager works with revisions of elements. When creating the structure of the load moving device, a mixed approach is used to develop the structure, that is, the design engineer develops the structure of the top-level product elements, and then the product composition can be supplemented from the NX CAD / CAM / CAE system [1] (Figure 2)

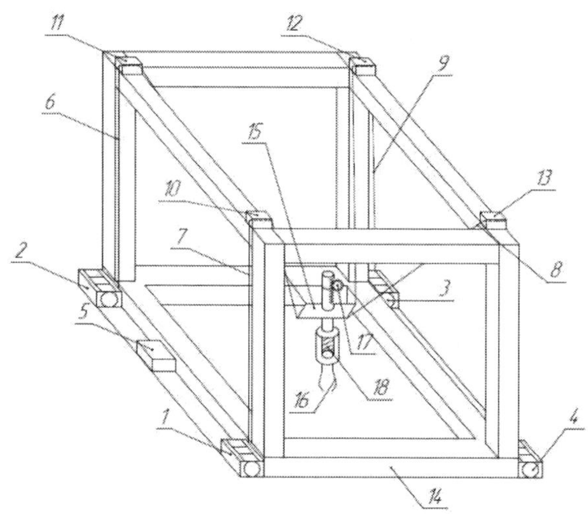

Figure 1. The kinematic diagram of a cable robot for moving goods

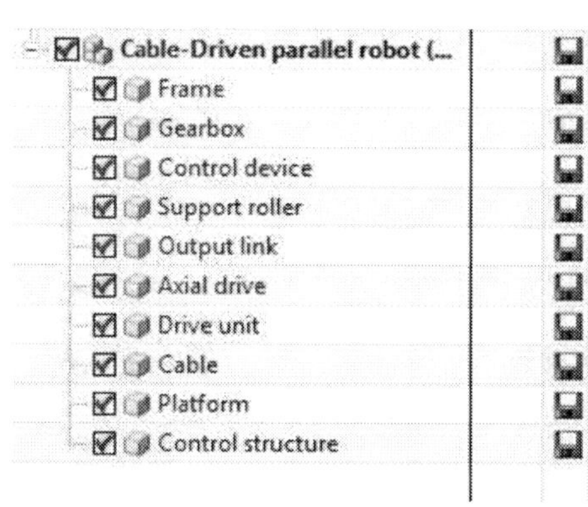

Figure 2. The structure of the device for moving load.

After saving the composition of the new product to fill the datasets with geometry, you need to select the top assembly level and click the button "Run / Open in NX".

NX offers to select a template file in which the geometry will be created: "Assembly unit". Only the top level of the assembly is reflected in the assembly navigator. To add datasets, you need to right-click the top assembly level and select "Manage components in standby mode (Teamcenter)."

The Top-Down method was used. To facilitate the work after the composition of the product is formed, a control structure of the assembly is created in the NX CAD / CAM / CAE-system "Modeling". The control structure of the product is a special approach to the design of a new product, which allows for collective work, reduces the time of designing and making changes. The control structure is a set of geometric objects – axes, coordinate planes located relative to each other.

In the NX CAD / CAM / CAE-system "Modeling", using the "Expressions" command located on the "Tools" toolbar, the variables are created for constructing and managing the control structure (Figure 3). The variables entered in the future will be associated with dimensions of the control structure and used to simplify the editing process of the assembly. This is due to the fact that the geometry and the variable of the control structure will not be responsible for the location of the components.

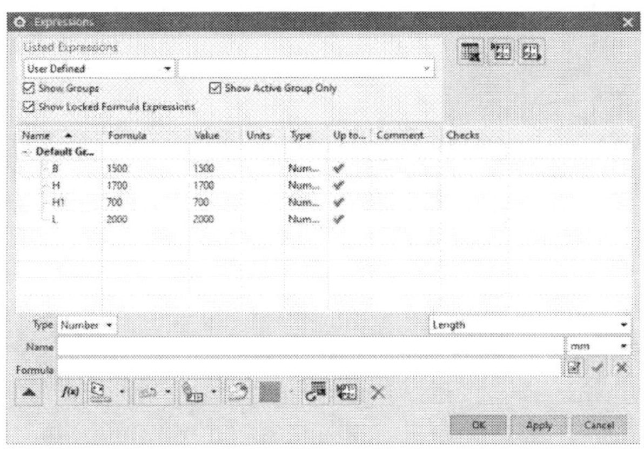

Figure 3. Dialog Window "Expressions".

Using the "Coordinate plane" command, the support tower base plane is created parallel to the XC-YC plane (Figure 4).

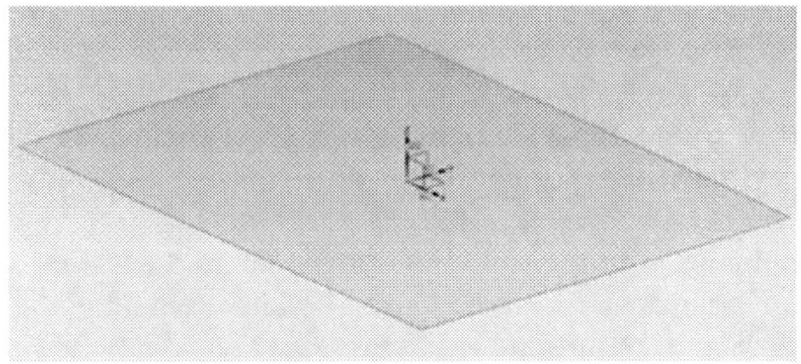

Figure 4. The base plane of the support towers.

The planes are created parallel to the ZC-XC and ZC-YC planes that limit the length and width of the product being developed (Figure 5). The platform plane of the load handling device is created. Ultimately, the control structure includes 3 horizontal planes and 4 vertical ones. The plane of the platform is located at a distance of 700 mm from the plane of the base of the support towers, and the plane limiting the overall size in height is located at a distance of 1700 mm. The planes parallel to the

ZC-YC planes are located at a distance of 2000 mm from each other, and the planes parallel to the ZC-XC planes are located at a distance of 1500 mm from each other. Figure 6 shows the control structure.

When the control structure is ready, the geometry of the model is created. The frame consists of several elements: a vertical strut, lower profiles 1 and 2, upper profiles 1 and 2, OB-4545L-BP-1655. Profile 1 is the first one to be designed.

Using the "Sketch" command, the section of the profile located on a plane parallel to the ZC-YC plane is constructed, then the sketch is completely limited (Figure 7).

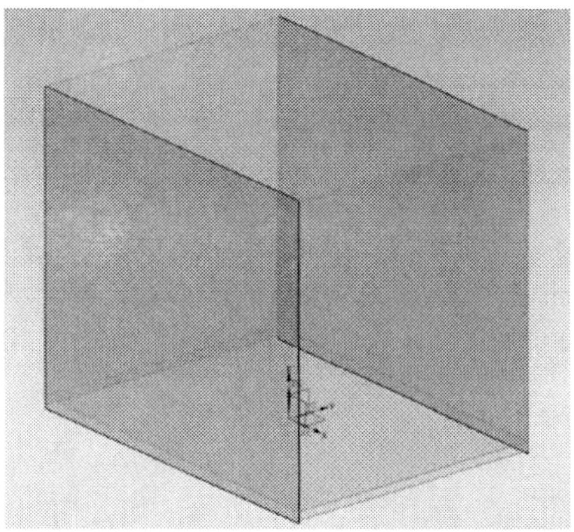

Figure 5. Building of planes parallel to the ZC-XC plane

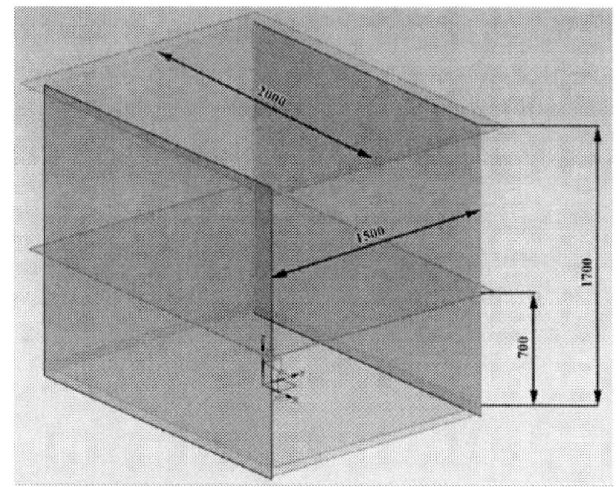

Figure 6. The control structure

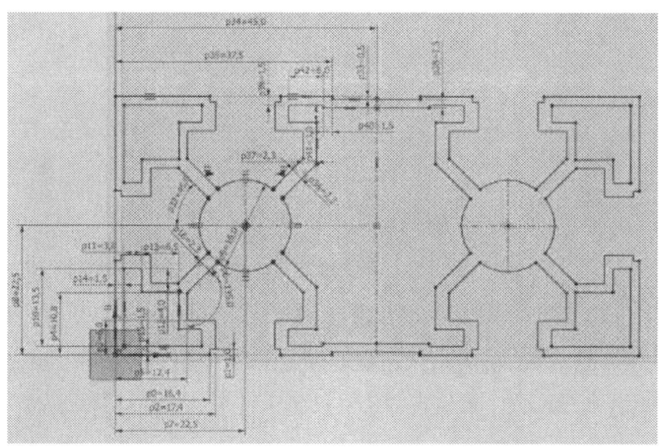

Figure 7. Vertical Rack Profile Sketch

When the profile sketch is completed, a solid body is built using the "Pull" command (Figure 8). The cut is made in the cylindrical sections of the profile using the commands "Sketch" and "Pull" (Figure 9).

Using the "Slope" command, trim the right end face at an angle of 45 degrees (Figure 10).

The left end trim is constructed in the same way. The last element is the rounding of the ribs of the vertical strut using the command "Rib rounding " (Figure 11).

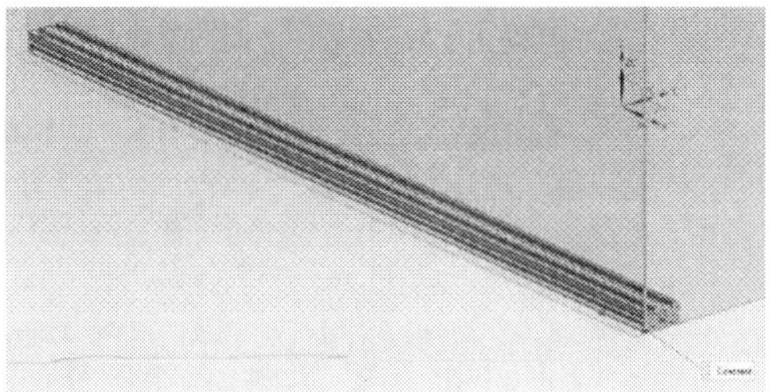

Figure 8. Construction of the geometry of the vertical rack

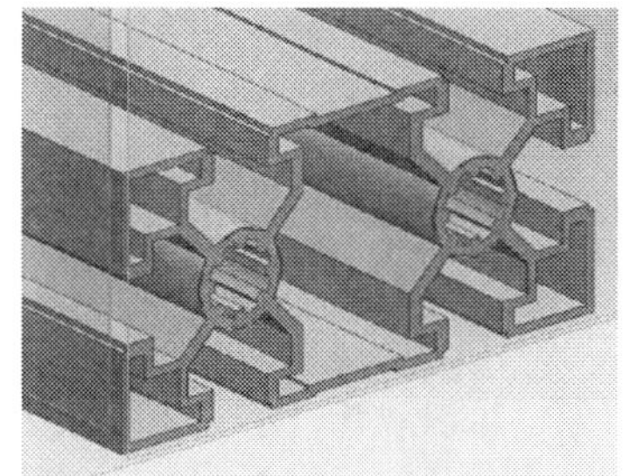

Figure 9. Cut-out

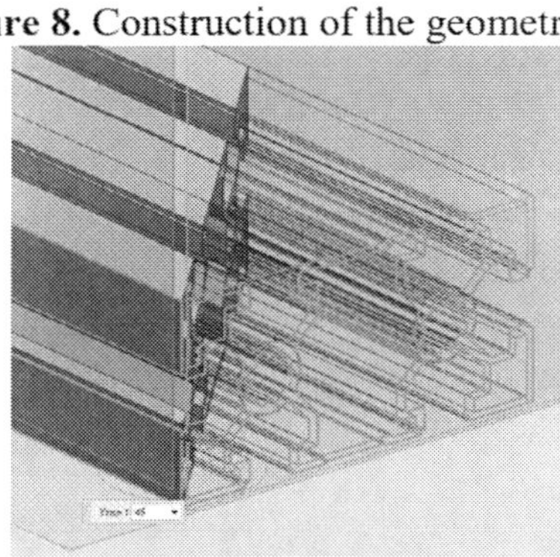

Figure 10. Right end trim

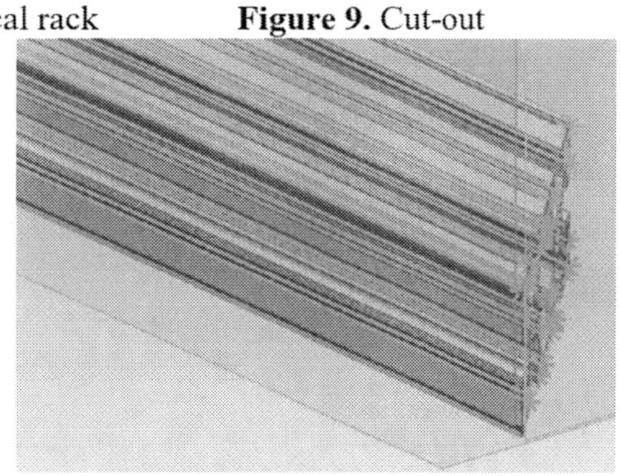

Figure 11. Rib rounding

After constructing the lower profile, the remaining frame elements are constructed. Using the commands located on the "Assemblies" toolbar, the frame assembly model is created (Figure 12). When designing products using the control structure, the main components of the assembly are mated t with elements of the control structure. In this case, using the "Mates" command, the lower profile is mated with planes parallel to the ZC-XC and ZC-YC planes (Figure 13). Then, the remaining components of the top-level assembly are added to the upper assembly level.

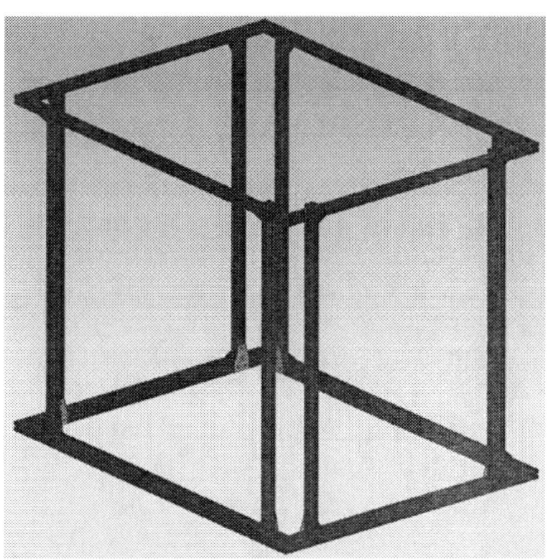

Figure 12. Assembly Frame

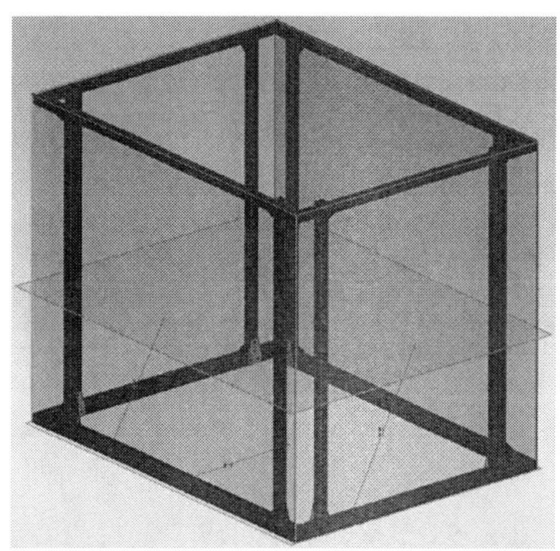

Figure 13. Mating the frame with the control structure

The geometric model of the platform on which the axial displacement drive and the output link with the gripper will be located is created (Figure 14).

After all the components have been built, fasteners are added to the frame assembly. To do this, the Teamcenter Classifier PLM system is used or fasteners from the NX CAD / CAM / CAE reuse library are added.

In the assembly of the cable robot, the cables are constructed in a simplified representation supporting the platform to provide information about their position in the assembly. Figure 15 shows the assembly model of a cable robot with a movable gripper for moving goods in special conditions.

3. Conclusion

Thus, using the NX CAD / CAM / CAE system controlled by the Teamcenter PLM system, it is possible to design new products and carry out outline designing. The use of the control structure ensures accurate control of the assembly, as a result of which the change in the size of the control structure allows you to increase or decrease the size of the working area.

Figure 14. Building of the platform

Figure 15. Cable robot for moving goods

Based on the preliminary design, a detailed study of the components and details of the cable robot can be carried out. The CAD / CAM / CAE-system NX controlled by the PLM-system Teamcenter reduces the time for implementing the outline design and changing the structure.

Acknowledgments

The research was funded by RFBR within the research project № 18-31-20060, using equipment of High Technology Center of BSTU named after V G Shukhov.

References

[1] Goncharov P S, Eltsov M Yu, Korshikov S B et al 2010 *NX for the designer-mechanical engineer* (MQM Press) 503 p
[2] Ovsyannikov V E, Shpitko G N and Vasiliev V I 2014 *Technical and engineering-psychological principles of machine design* (Kurgan: Publ. house of Kurgan State. un-ta) 115 p
[3] Torop D N and Terlikov V V 2011 *Teamcenter. Beginning of work* (Ed. DMK Press) 280 p
[4] Eltsov M Yu, Kozlov A A, Sedoykin A V and Shirokov L Yu 2011 *Designing in NX under Teamcenter control: studies Manual* (Belgorod: BSTU publ. house) 781 p
[5] Merlet J P and Daney D A 2010 A portable, modular parallel wire crane for rescue operations *IEEE Int. Conf. Robot. and Automat. (ICRA)* pp 2834–9
[6] Gosselin C 2013 *Cable-Driven Parallel Robots* (Berlin; Heidelberg: Springer) pp 3–22

[7] Duan Q J and Xuechao D 2014 *Workspace Classification and Quantification Calculations of Cable-Driven Parallel Robots* pp 1–9

[8] Gaponenko E V, Fisherman L A and Kholoshevskaya L R 2019 Structural analysis and classification of robotic systems with drive mechanisms based on cable elements *Bull. of BSTU named after V.G. Shukhov* **99** 126–36

[9] Aleshin A K, Glazunov V A, Rashoyan G V and Shai O 2016 Analysis of kinematic screws that determine the topology of singular zones of parallel-structure robots *J. of Mach. Manufact. and Reliability* **45(4)** 291–6

[10] Rybak L A, Khalapyan S Yu, Kholoshevskaya L R and Malyshev D I 2018 Planning the optimal trajectory of the robot-machine parallel structure *AIP Conf. Proc.* **2029** no 020066

Geodynamic features of the territory of Grozny city

M Ya Gaysumov[1,2], I A Kerimov,[2,3] S V Badaev[1,2], O M Gaysumov[4] and A A Doduev[3]

[1] Kh. I. Ibragimov Complex Research Institute, RAS, 21A, Staropromyslovskoe highway, Grozny, 364051, Russia
[2] Academy of Sciences of the Chechen Republic, 13, prospect named after M. Esambayev, Grozny, Russia
[3] Grozny State Oil Technical University named after M.D. Millionshchikov, korp. 1,100, prosp. Husejna Isaeva, Grozny, 364068 Russia
[4] Public Autonomous Institution, State Expertise of the Chechen Republic, Grozny, Russia

E-mail: gro_ss@bk.ru, ibragim_kerimov@mail.ru, badaev_sv@mail.ru, g1uck@mail.ru, ayub_doduev@mail.ru

Abstract. The work contains data on tectonic structure, modern geodynamic activity and materials on the study of negative exogenous processes obtained as a result of the development of oil fields located on the urban land. The analysis of geodynamics of the Tersko-Sunzhen structure located in the central part of the Tersko-Caspian deflection was performed. The history of the formation of this tectonic zone is long and diverse and is determined by its fault-block tectonics. At the present stage, the territory is also defined as geodynamically active, consisting of blocks bounded by faults and fractures, which are characterized by intense movements of the earth's surface. Stress unloading is expressed in the deformation of rock massif and in case of rapid (stepwise) unloading of tectonic stresses – through earthquakes. A special condition is that Starogroznenskoe and Oktyabrskoe oil fields are located on the territory of the city. To date, the cumulative production amounted to more than 150 million tons of oil. Long-term development of deposits led to equilibrium disturbances of the geological environment with significant change of formation conditions and accordingly to the development of hazardous geodynamic processes.

1. Introduction

The Tersko-Sunzhen structural and tectonic zone, where the urban territory of the Grozny city is located, is in the central part of the Tersko-Caspian Deflection (TCD) and is presented in a relief by the ridges of the same name. According to many researchers, the zone was formed simultaneously with the geological development of the Tersko-Caspian deflection as a whole [2, 4, 5, etc.].

It is assumed that the TCD has been formed since the Jurassic Era, when against the background of calm tectonic conditions, the placement of existing raises and depressions in the Paleozoic foundation was determined by the Tersko-Khasautsky and the eastern end of the Pshekish-Tyrnyauz faults. The tectonic activation of the folding structure of the Greater Caucasus at the turn of the pont-Akchagylian also had a significant geodynamic influence on TCD formation.

Against the background of the general meridian compression of the Greater Caucasus, and possibly the regional stress field, there is an intensification of faults and as a consequence "explosive" in geological time formation of Tersk and Sunzhen folds [2]. Later in Late Akchagylian and Apsheron periods the growth of ridges stopped and against the background of the immersion of the area a strong precipitation thickness, which under the influence of internal stresses and gravitational forces was unloaded by the formation of folds, has accumulated. The dramatic activation of submeridional compression at the turn of the Pleistocene-Hologene the formation of the Terco-Caspian tectonic zone continued. There is a renewal of old and the appearance of new faults of north-western and north-eastern extensions on already existing folds, the complication of relief forms and the growth of folds in general.

The breakdown of the deflection with the appearance of the Tersko-Sunzhen elevating system (zone) refers to the beginning of the late orogenic stage and is characterized by the reduction of the capacity of the Sarmatian stage, the appearance of sedimentation interruptions and some facies changes (for example, the erosion in the base of the Upper Sarmatian on the western extension of the Sunzhen elevation in the Argudan anticline). The relative growth of elevations is recorded for the Meotian and Pontian periods. In the Cimmerian age there is a significant denudation of anticlines of the Tersky and Sunzhen raises. In the Upper Pliocene, these raises are involved in the active deflection of the edge deflection and the Akchagylian deposits on these anticlines lie within different horizons of the Miocene and Oligocene periods. The beginning of the orographic growth of modern range-raises refers to the early anthropogenic and continues to the present.

2. Problem statement

In geological perspective, the territory of the Grozny city has a complex tectonic structure and is accordingly affected by geodynamic phenomena of both endogenous and exogenous nature. This complexity is determined by the following factors: complexity of tectonic forms, high seismicity of the area, strong thickness of sediments with high relative and total soil subsidence, as well as high water table.

All these factors contribute to negative geodynamic processes in the territory of the city. The study and subsequent consideration of geodynamic features are necessary to assess, forecast and prevent the development of negative geodynamic effects.

3. Study of structural-tectonic features of the territory

At present, at the surface the Tersko-Sunzhen zone is represented by Tersky and Sunzhen ridges-raises separated by a chain of depressions. In the latter, from west to east, the following can be identified: Upper Kurpskaya Dam (south-west of Malgobek), Alkhanchurtskaya Valley, Pervomayskaya Dam and Grozny Valley. At the same time, the structure of raises differs significantly at different stratigraphic levels.

In the structure of the Cretaceous and Paleocene-Eocene and later deposits, the Terskaya range of raises represents a box fold with an amplitude of 1.5–2 km, a width of 6–15 km with a flat (width of 3–6 km) vault and an inclination of wings up to 30–45 degrees. In clay Early Orogenic (Maykopian series) molasses with a capacity of 1–2 km, the lifting is transformed, and in the structure of medium and upper Miocene, more competent carbonate-terrigenous deposits with a capacity of 1.5–2 km, 1–3 of link anticlines are identified in different places, sometimes tipped, complicated by overlap faults with amplitudes of up to 1 km and separated by synclines [1, 18].

Sunzhen arc-shaped elevation is also heterogeneous at different stratigraphic levels. It is assumed that in its base there is a fault zone of the foundation, which serves as an extension of the Pshekish-Tyrnyauz suture zone. Its fractured nature is confirmed by geophysical data [1]. In sedimentary mantle it is represented by local anticline structures such as Starogroznenskaya, Oktyabrskaya, North-Oktyabrskaya, etc. On the Upper Cretaceous and Paleocene-Eocene deposits in the core of lifting the anticlinal is fixed, in some places box-shaped with the fall of wings at angles of about 20–40 degrees. In plastic early-orogenic clay molasses (capacity 1–1.5 km) and deposits of middle and upper Miocene (up to 2–3 km) the nucleus of anticlinal is squeezed and it acquires a comb-shaped or diapir appearance, and in places it turns to the south and is complicated by unloading, uplifts and surges. The north wing of the main anticline is

IOP Conf. Series: Materials Science and Engineering **905** (2020) 012028 doi:10.1088/1757-899X/905/1/012028

complicated in places by smaller narrow folds of 10–15 km length. In the east, the rise splits to form large Novogroznenskaya and Starogroznenskaya anticlines.

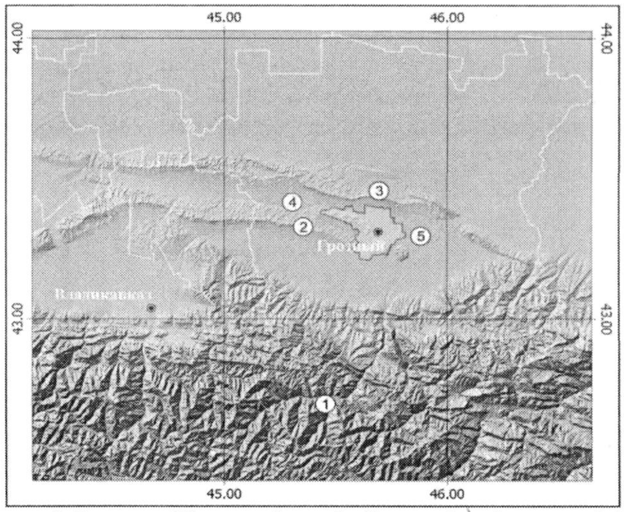

Figure 1. Location map of Grozny. The numbers indicate: 1 – horst-anticlinorium of the Greater Caucasus; 2 – Sunzhen elevation; 3 – Terskoe elevation; 4 – Alkhanchurtskaya valley; 5 – Grozny valley.

The *Alkhanchurtskaya valley* is the most studied within the chain of valleys dividing Terskoye and Sunzhen rise. Milanovsky E.E. and other researchers consider this structure as a relatively passive block of the edge deflection, where the top of Cretaceous decreases to the east in the interval of depths minus 4–6 km (i.e. located 2–4 km lower than in adjacent rises), and the bottom of the upper Sarmatian is identified at depths minus 2–3 km. The cavity is almost horizontal with molasses and is characterized by increased capacities of Upper Sarmatian sublayer (up to 1 km), presence of Meotian blurred in rises and powerful (up to 1 km) Upper Pliocene. However, the low capacity of quaternary deposits obviously indicates the "retraction" of this territory [2]. The Upper Kurpskaya dam limiting the Alkhanchurtskaya depression from the west is seen as an independent brachyanticline with exits in the core of the upper Pliocene complicated by low-amplitude uplifts.

The frowziness of the folds of these ridges is expressed by the "wedging" of folds: Terskaya to the north and Sunzhen to the south, with the formation of rupture disorders both in the vault and on the wings of the folds. The results of paleotectonic analysis implies stage (tectonic) formation of folds [4, 14].

The *Grozny valley* is located between the eastern parts of the Tersky and Sunzhen rises, i.e. it has steep north-eastern and south-western sides.

The Grozny city is located on the junction of the Grozny valley with the branching edges of the Sunzhen rise.

4. Modern geodynamics

In 1971 and 1986, the maps of modern vertical movements of the Earth's surface were designed for the territory of the European part of the USSR and Eastern Europe. In general, for the mountain areas of Caucasus the speeds make up to +10 mm/year and more and in deflections, including TCD, the speed of vertical movements makes $0 \div (\pm 2)$ mm/year, but the nature of speed distribution is different.

It should be noted that according to the data of long-term measurements, the oscillatory-wave nature of vertical tectonic movements of the Earth's crust within the entire Caucasus region was established [5, 10]. The most intense local anomalies of vertical and horizontal movements of the Earth's surface are confined to zones of deep faults and currently active tectonic structures of different order. For such areas it was established that these modern movements are high-amplitude (over 50 mm/year), short-period (from 0.1 years to the first years), spatially localized (from 0.1 km to the first tens of km) and have pulse-coupled and sign-variable direction [8].

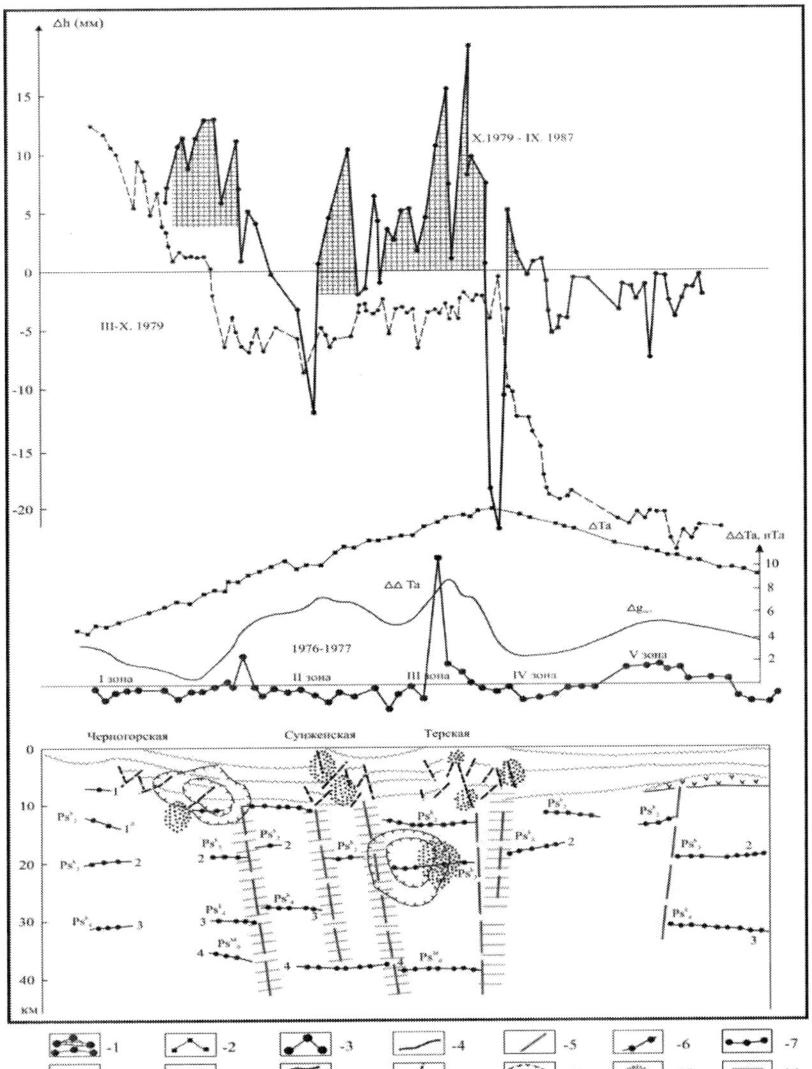

Figure 2. Results of data comparison of SVDZP and geophysical fields by regional profile II-II combined with ECWM profile. (According to Kuznetsova V.G. et al., 1991; Sidorov V.A. et al., 1989; Kadurina L.S. et al., 2000; and additions by authors). Legends: 1 – SVDZP curves over various observation cycles; 2 – magnetic field curve; 3 – residual gravity anomaly curve; 4 – magnetic field variation curve; 5 – faults according to geological and geophysical data; 6–11 – seismic boundaries of sections according to ECWM data; b – surface of crystalline foundation; 7 – granite layer surface; 8 – basalt layer surface; 9 – upper mantle roof; 10 – effusion rocks; 11 – faults in consolidated part of crust; 12 – areas of rock decompression in the Earth's crust according to the data of interpretation of gravity anomalies; 13 – zones of modern softening (increased fracturing) of medium rocks according to the data of tectonophysical interpretation of SVDZP; 14 – potential subvertical kinematic fluid permeable zones. Earth surface subsidence caused by oil and gas fields development

More detailed study was carried out in 1974-88. The purpose of these works was to identify the comparative characteristic of modern vertical movements of the Earth's surface for various structural-tectonic deflection zones. In total, 4 cycles of repeated measurements were performed for the studied territory. Observations were performed along a series of local and regional geological profiles located transversely to the extension of the main geostructural deflection elements. The accepted technique of observations and the used tools allowed conducting repeated leveling with an average quadratic error ± 0.50–0.95 mm/km [11].

The studies made it possible to establish high modern tectonic activity of the studied territory step-wise increasing towards the Greater Caucasus.

The interpretation of profile observations shows that the considered territory is characterized by heterogeneity in the distribution of modern geodynamic activity. This differentiation is of a block nature, the boundaries between the areas of modern elevation and descent of the Earth's surface are expressed by the gradient change in the speeds of the SVDZP. The width of the blocks is 2–3 km or more and the value of gradients is about 5–8 mm/km, the background values being (directly within descending and rising blocks) 1–2 mm/km.

It is surely possible to distinguish a single block corresponding to the Tersky and Sunzhen anticlines – the Alkhanchurtskaya Synclinal separating them (Figure 2). This block is limited by faults with high geodynamic activity; in the east the width of the fault is 8–10 km, the value of relative vertical displacements – 25–28 mm/year, in the west the width of the fault is 6–8 km with the value of displacements – 8–12 mm/year. The length of the block is 30–35 km and the width is 6–7 km.

The comparative analysis of geodynamic survey data and seismic survey materials makes it possible to clearly establish that the gradient zones reflect the location and activity of foundation faults on the Earth's surface and to assume that the occurrence of these zones is caused by high geodynamic activity of the corresponding deep faults. Most likely all established gradient interblock zones are compared with the main deep faults identified via the earthquake converted-wave method (ECWM) [9, 11, 12].

5. Subsidence of the earth's surface caused by the development of oil and gas fields

At the initial stage of development, rock pressure (lithostatic pressure) on the top of the formation containing hydrocarbons is balanced by stresses in the matrix and formation pore pressure. During development of hydrocarbons with the reduction of pore pressure their compression occurs in deposit volume with subsequent deformation of rocks and accordingly lowering of surface above the developed formation. According to literary data, the rates of descent of the Earth's surface during hydrocarbon extraction usually make 1–2 cm per year, the total values of subsidence do not exceed the first tens of centimeters. However, there are also cases of significant losses – up to several meters, accompanied by deformations of ground structures, collapse of casing strings and other damages [6, 8, 15, et al.].

Depending on the geology and mechanical characteristics of the reservoir and cap rocks, as well as field operation modes, the deformations occurring in the reservoir can spread into the top of the reservoir, the layers covering it, etc. up to the surface where they appear mainly in the form of vertical displacements. Thus, various cases are possible:

- if the reservoir is represented by very hard rocks (hard sandstone, hard dolomites, limestone), its deformation will be small and close to elastic. There will be little vertical compression of the formation without apparent consequences on the surface;
- if the reservoir is represented by weak easily compactible rocks (such as loose sandstones, sands, chalk) or pure aleurites, the deformation of the reservoir will be more significant, far beyond the limits of elasticity.

The relevance of the study was increased further by the end of our century due to the intensification of human engineering activities. Despite considerable research by scientists of different countries, it should be noted that there is no single methodology to determine the value of vertical displacements of the Earth's surface depending on the development intensity of oil and gas fields.

Geostatic equilibrium is broken during field development. Productive formation is subject to compressive pressure, which is equal to the difference of mountain and formation pressure. Formation pressure during development decreases, except for short periods of seismodynamic activity [8, 13, 17, et al.].

Elastic and plastic models, environment models with consolidation also the so-called "tent" model which reflects the main features of deformation of porous saturated environments under loading are most frequently used as the mechanical model of reservoir behavior under loading. There are known works by Chin L.Y., Boade R.R. et al. who implemented this model for the conditions of the Ekofisk reservoir in the North Sea, where the settlement of the sea floor reached 6-7 m [18]. Later, this model was finalized by Kashnikov Yu.A. et al., and successfully used for settlement calculations at oil and gas fields of the Volga region and Siberia [6].

It is known that the relative deformations of liquid-saturated collectors (deformation of matrix, pore space and solid phase) under comprehensive compression are caused by two types of stresses: effective stress $(\sigma - P)$ and pressure of liquid in pores. The relative total deformation of the reservoir is determined by the following ratio:

$$\frac{dV}{V} = \beta_{c\kappa} \cdot d(\sigma - P) + \beta_{m\epsilon} \cdot dP \,, \quad (1)$$

where $\beta_{c\kappa}$ – volume compressibility coefficient of the matrix;

$\beta_{m\epsilon}$ – volume compressibility coefficient of the solid phase;

V – sedimentary rock matrix volume (m^3);

dV – change of sedimentary rock matrix volume (m^3);

P – formation pressure (Pa);

$(\sigma - P)$ – effective pressure (Pa).

In order to estimate the subsidence of the oil-bearing horizon top, let us shift from the volumetric deformation to the vertical compression of the formation. The uniaxial compression may be presented as follows:

$$dh = h(\beta \cdot d(\sigma - P) + \beta_{m\epsilon} \cdot dP) \,, \quad (2)$$

where dh – vertical compression of the formation;

h – average oil-saturated formation thickness;

$d(\sigma - P)$ – change of effective stress during field operation;

dP – change of fluid pressure in pores during operation.

This equation allows estimating the subsidence physically based on laboratory measurements of mechanical properties of rock samples and field-geological study at oil fields.

Starogrosnenskoe and Oktyabrskoe oil fields are located in a large part on the built-up territory of Grozny. These are multilayer fields, are in long-term development, the operation mode – elastic-water-driven is characterized by a significant drop of formation pressure and accordingly changes in the stress state of both the reservoir and the adjacent rocks. For these fields, the authors calculated the amount of subsidence of the formation top using formula (2).

Further, the transfer of formation deformations to the daylight surface was calculated using Young's modulus and Poisson's coefficient for transfer of elastic deformations to the surface for shortest distance from the top of the productive formation to the surface. Values of elastic modules for cover mass of sand-clay deposits are accepted according to the reference data.

During the period under study, the longest time in operation belongs to the Miocene deposits in Karagano-Chokrak formation and the subsidence of the Earth's surface caused by development are insignificant. This is caused by relatively small drop in formation pressure as well as the elastic properties of the formation. The period of development of the Miocene deposits of the Tashkalinsky section of the Starogroznenskoe field amounts to 72 years, the formation pressure has fallen by 3.6 MPa, the total subsidence of the formation top is more than 400 mm, the surface – 94 mm, and the average subsidence per year is about 1 mm/year. Although other Miocene deposits are characterized by a large drop in formation pressure, their subsidence rate is less than 1 mm/year. The greatest drops in formation pressures occurred in Cretaceous deposits of both fields. The deposits are currently in the final stages of their development. Against the background of intensive oil extraction, the formation pressures decreased from 69–72 to 41–43 MPa. The total subsidence of the formation top and, accordingly, the surface for the whole period of operation is quite large and make about 345–422 mm, and the average annual subsidence rate – 6–8 mm.

The greatest influence on the total subsidence of the formation top is the amount of change of effective stress during the deposit operation. This indicator is the largest in Starogrosnenskoe field – 10.9 (Table 1).

Besides natural seismicity, during long-term development of hydrocarbon deposits the equilibrium conditions in the formation are disturbed, which can cause critical redistribution of stresses in the developed formation and adjacent rocks and leads to the technogenic seismicity.

Earthquakes with proven technogenic nature occurred in Starogroznenskoe and Oktyabrskoe oil fields. Weak earthquakes in Starogroznenskoe district occurred quite often, both before and during various periods of operation. Against the background of weak earthquakes in 1938 and 1963 there were significant earthquakes with the magnitude M = 3.8–4.3, excited by various factors of oil field development. The Starogroznenskoe earthquake of 1971 with M = 4.3, h = 2.8 km, I_0 = 7–8 points, which caused destruction of residential and industrial buildings of the city, occurred against the background of a sharp drop in pressure, exorbitant oil extraction and the beginning of water injection into the upper cretaceous deposit to intensify oil extraction [15].

Table 1. Calculation of oil field top subsidence and forecast of earth surface displacements above them

n/n	Deposits	T, Year of development	dt, years	$P_{нач}$, MPa	$P_{тек}$, MPa	dP, MPa	$d(\sigma - P)$ MPa	$\beta_{ск\ min}$, MPa^{-1}	$\beta_{ск\ max}$, MPa^{-1}	β, MPa^{-1}	H, m	h_{cp}, m	dh_{top}, mm	$dh_{surface}$, mm	$V_{cp.\ settlement}$, mm/year
1	2	3	4	5	6	7	8	9	10	11	12	13	14	15	16
	Starogroznenskoe field														
1	Cretaceous (upper and lower Cretaceous)	1964	54	72.1	42.7	29.4	10.9	0.35	0.57	0,2	4100	272 +168	3252–3725	–345	–6
2	Miocene (karagan – chokrak deposit)	1893	125	10.6	6.5	4.1	2.4	1.2	1.46	1.06	140	27	195–212	–55	–0.5
3	Tashkalinsky area (karagan – chokrak deposit)	1946	72	17.9	3.6	14.3	7.20	1.4	1.77	1.73	320	12	417–450	–94	–1.3
	Oktyabrskoe field														
4	Cretaceous (upper Cretaceous)	1966	52	69.2	41.2	28.0	1.8	0.32	0.43	0.3	3844	275	2475–2522	– 422	–8.1
5	Miocene (karagan – chokrak deposit)	1913	105	7.0	4.5	2.5	1.50	1.6	2.29	1.42	1102	71	422–496	–97	–0.9

6. Conclusion

The analysis of geodynamics of the Tersko-Sunzhen structure located in the central part of the Tersko-Caspian deflection was performed. The history of the formation of this tectonic zone is long and diverse and is determined by its fault-block tectonics. The formation of advanced deflection structures took place in the environment of horizontal compression. However, the question of the source of force remains unclear. Thus, many see the source of compression in the transformation of vertical forces into horizontal ones. Most likely, such a source could be the growth of the Caucasus mountain structure at the orogenic stage of development, starting with the Sarmatian, which created large additional stresses in the lithosphere, or the horizontal component of gravity could arise due to the own weight of rocks when they are tilted.

At the present stage, the territory is also defined as geodynamically active, consisting of blocks bounded by faults and fractures, which are characterized by intense movements of the earth's surface. Stress unloading is expressed in the deformation of rock massif and in case of rapid (stepwise) unloading of tectonic stresses – through earthquakes.

Long-term development of deposits led to equilibrium disturbances of the geological environment with significant change of formation conditions and accordingly to the development of dangerous geodynamic processes. Earthquakes with proven technogenic nature occur at Starogroznenskoe and Oktyabrskoe fields. In general, the obtained data show that the territory of the Grozny city is located in a highly active geodynamic zone and as a result is subject to the development of intensive geodynamic processes of both natural and technogenic nature with negative consequences for residential buildings and infrastructure.

Acknowledgment
The study was performed under the grant of the Russian Foundation for Basic Research r_a 18-45-200001.

References

[1] Gaysumov M Ya and Badaev S V 2018 Depth structure of the Tersko-Caspian deflection by geophysical data In I A Kerimov, V B Zalishvili and V I Chernashina (ed) *Modern problems of geology, geophysics and geoecology of the North Caucasus* vol VIII (Moscow: IIET RAS) pp 252–60

[2] Koronovsky N V, Guschin A I, Nikitin M Yu et al 1990 Geological development and formation of the modern structure of the Tersko-Caspian foretrough In *Tectonics of orogenic structures of the Caucasus and central Asia* (Moscow: Nauka) pp 4–35

[3] Kerimov I A, Borisenko Z G, Daukaev A A, Gaisumov M Ya et al 2010 *Geology of oil fields of Tersko-Sunzhenskiy oil and gas bearing region* Reference book (Grozny: Acad. of Sci. of the Chechen Republic) 254 p

[4] Zhemerichko M I 1981 Development of morphostructures of Advanced Ranges of the Chechen-Ingush ASSR in Pliocene-Quaternary Time In *Nature and natural resources of the central and eastern part of the North Caucasus* (Ordzhonikidze: NOSU) pp 3–8

[5] Kasyanova N A 1994 Features of modern geodynamics of the central part of the Tersko-Caspian deflection *Geotectonics* **5** 85–90

[6] Kashnikov Yu A and Ashihmin S G 2007 *Rock mechanics during development of hydrocarbon raw deposits* (Moscow: Nedra) 467 p

[7] Kerimov I A., Gaysumov M Ya and Abubakarova E A 2009 Geophysical fields and fracture tectonics of the Tersko-Caspian deflection *Geodynamics. Deep structure. Thermal field of Earth. Interpretation of geophysical fields. Fifth scientific readings in the memory of Yu P Bulashevich* (Yekaterinburg, 06–10 July 2009) pp 226–30

[8] Kuzmin Yu O 2004 Modern geodynamics of fault zones Phys. of the Earth **10** 95–112

[9] Kuznetsova V G and Maksimchuk V E 1991 Results of tectonic magnetic studies of the structure and modern geodynamics of the lithosphere of the Tersko-Caspian deflection *Geophys. J.* **13(6)** 47–55

[10] Nikonov A A 2006 *Modern Earth crust movements* 2nd ed (Moscow: Komkniga) 192 p

[11] Sidorov V A, Bagdasarova M V, Atanasyan S V et al 1989 *Modern geodynamics and oil and gas content* (Moscow: Nauka) 199 p

[12] Kadurina L S, Grigorieva N V, Ishhanova B H et al 2000 Structure of the earth crust of the Tersk-Caspian deflection along the Tersk profile according to seismic data *Geology and mineral raw material base of the North Caucasus* Mater. of the IX Int. sci. and pract. Geol. Conf. (Essentuki) pp 262–3

[13] Barbour A J and Wyatt F K 2014 Modeling strain and pore pressure тassociated with fluid extraction: The Pathfinder Ranch experiment *J. Geophys. Res. Solid Earth* **119** 5254–73 DOI: 10.1002/2014JB011169

[14] Gaysumov M.Ya., Badaev S.V., Akhmatkhanov R.S et al Study of the Paleozoic Base of the Terek-Caspian Trough Based on the Complex of Geological and Geophysical Data *Advan. in Engineer. Res., Atlantis Press* **182** 106–10 DOI: 10.2991/ciggg-18.2019.60
[15] Kerimov I A., Gaysumov M Ya and Akhmatkhanov R S 2012 Technogenic seismicity in gas and oil fields *Geol. and Geophys. of South Russ.* **1** 22–45
[16] Hauksson E, Gobel T, Ampuero J-P and Cochran E A June 2015 *Century of oil-field operations and earthquakes in the greater Los Angeles Basin, southern California. The Leading Edge* SEG. Downloaded 06/03/15 to 130.118.44.150 pp 650–6
[17] McGarr A, Simpson D and Seeber L 2002 Case Histories of Induced and Triggered Seismicity *Int. Geophys.* **81A** 647–61
[18] Zoback M and Zinke J 2002 Production-induced normal faulting in the Valhall and Ekofisk oil fields *Pure and Appl. Geophys.* **159** 403–20

IOP Conf. Series: Materials Science and Engineering **905** (2020) 012029 doi:10.1088/1757-899X/905/1/012029

Average magnetization of quasi-binary solid solution of magnetic and non-magnetic ions

A Yu Gufan[1], Yu M Gufan[1], E N Klimova[2], O V Naskalova[3] and I A Osipenko[2]

[1] Research Institute of Physics, Southern Federal University, Rostov-On-Don, Russia
[2] Don State Technical University, Rostov-On-Don, Russia
[3] Faculty of Physics and Mathematics, Chechen State University, Grozny, Russia

E-mail: alexander.gufan@gmail.com, yu.gufan@gmail.com, javor05@mail.ru, naskalova_o@mail.ru, iaosip@mail.ru

Abstract. The paper describes features of calculating the magnetization of solid solution based on statistical approach on the example of a one-dimensional crystal. Calculations are made in approximation of Weiss effective field. The result shows the error of the third Gilleo hypothesis, which is still widely used to forecast the magnetic properties of solid solutions.

1. Introduction

The need to study the properties of solid solutions in order to obtain new materials with electrodynamic characteristics capable of solving applied problem seems quite relevant. For a specific case of complex oxides with perovskite structure simultaneously exhibiting nonlinear properties of ferroelectrics and magnetically ordered media, this need was justified by Smolensky et al in [1]. However, the theory of magnetic properties of solid solutions of magnetic[1] and non-magnetic atoms on the nodes of one sub-lattice was built even earlier in the works Gilleo and Goodenough [2–4]. In this case, the need to develop such theory is associated with the construction within molecular engineering of multicomponent ferrite spinels and garnet ferrites widely used in communication means operating within the centimeter wavelength range. In work [2], three hypotheses were advanced when calculating the temperatures dependence of magnetic phase transition (PT) on the concentration of solid solution components. The first one assumed that only those magnetic atoms (ions) whose "closest neighbors" have at least two magnetic atoms (ions) contribute to the formation of the long-range magnetic order in crystalline solid solutions of magnetic and non-magnetic ions. In this basic postulate of Gilleo and Goodenough's theory there are some uncertainties that manifest themselves when applying the theory to both ideal and real crystals. Ferrites are multicomponent oxides containing both magnetic ions of 3-d and 4-f metals and non-magnetic ions (Zn, Si, Al...) as structure-forming agents. Elementary cells of ferrites have quite complex structure, and due to proximity of ion radii different ions with a certain probability are often placed in the structure according to the same system of correct points of a degenerate structure [5]. It is these features of the structure of ferrites that give ferrites the name *solid solutions*. Besides, magnetic ions in oxides are positively charged and their closest neighbors in the equilibrium structure of the dielectric or semiconductor are always negative

[1] For short let us call the transition metal atoms, the exchange interactions between which are responsible for phase transitions to magnetically ordered states simply "magnetic atoms". The atoms of other elements placed on the same sub-lattices in a crystal as magnetic atoms are referred to as "non-magnetic atoms".

ions. It is the negative ions that provide the indirect exchange interaction between magnetic ions, i.e. determine the structure of the existing magnetic order. Moreover, the exchange interaction through the p-electrons of oxygen (p-ion) is most severe if the center of the oxygen ion is located on a straight line connecting the centers of magnetic ions. Thus, the indirect exchange provided by oxygen ions is sharply anisotropic, so the number of closest neighbors also depends on the structure of the ligand. Therefore, the above words “nearest neighbors of magnetic ions” are taken in quotation marks and require clarification. It shall be noted that the first Gilleo hypothesis for use without reservations should be reformulated. For example in the form of a statement that the highest magnetic PT temperatures are to be expected in substances with a paramagnetic phase structure collected from ligands with highly symmetrical (weakly distorted cubic) structure and the maximum possible magnetic ion density in the structure. In this formulation, the hypothesis is well supported by the experiment [6, 7].

The second hypothesis of Gilleo and Goodenough’s theory is that all magnetic atoms, among which the closest neighbors are 2, 3, 4, etc. magnetic atoms, equally contribute to the formation of the long-range magnetic order. This hypothesis does not seem very convincing. Indeed, for general reasons, it is clear that the more bonds supporting ferro- or ferrimagnetic ordering of magnetic moments of neighboring atoms, the stronger the effective field must be, and accordingly the transition temperature to the magnetically ordered state is higher. However, the phenomenological nature of Gilleo and Goodenough’s theory offset all advantages in predictions of a theory based on the notion of the number of real connections between magnetic atoms. The questionable claims of the second hypothesis and its obvious disadvantages would necessarily manifest themselves if a real microscopic theory of physical properties of the condensed state of matter, in particular perfect crystals, were built. Such theory should take into account polyatomic interactions and abandon the approximation of a self-consistent field. In the approximation of the self-agreed or effective field [6, 8], the adjustable parameters are used to predict the magnetic transition temperature, the values of which are selected so that the responses of the theory correspond to the experiment. The disadvantages of Gilleo and Goodenough’s theory are being addressed by modern researchers using quantum chemistry models [9–11]. In particular, recent works rely on the approximation of DFT in the form of “ab initio” [7]. Two provisions should be taken into account when assessing the results thus obtained. First, the DFT method relies on the approximation of a self-consistent field when selecting electronic wave functions. Thus, the use of DFT approximation does not remove objections to disadvantages of the approximation of a self-consistent field, and the closely related extremely coarse monatomic approximation. Apart from the above noted loss of differences between the number of links or the multivariate of the task, the approximation of an effective field with necessity leads to the loss of differences between predicted ordered structures. The work [12] gives an example where the crystal structure of the ordered phase predicted in the approximation of the self-agreed field is hexagonal, while the cubic structure is predicted in the more complete consideration of the relation dimension. The fact that the example relates to PT ordering rather than magnetic transitions is not essential to approximate a self-consistent field. Indeed, in the approximation of a self-consistent field, there is an unambiguous relationship between these two types of orders [6, 8]. It should also be borne in mind that the true Hamiltonian of interatomic interaction leading to PT is always unknown. Thus, the work [13] discusses in detail the relationship between the interatomic exchange interactions and the formed exchange interaction of the respective magnetic sub-lattices. Even from such a simple example we see a complex dependence and ambiguity of such connection. In [12], a specific example discusses the problem that arises when predicting Hamiltonian-based properties limited by the condition of accounting for not more than tetratomic interactions.

The second hypothesis of Gilleo and Goodenough’s theory proved to be the main obstacle to the application of the theory to solid solutions (alloys). There are few pure metals that have the properties necessary to solve the applied problems. The difficulty arises from the fact that the uniformity of the solid solution with necessity is broken on an atomic scale. Thus, in an ideal binary solid solution, each lattice node can only be occupied by one atom – A or B. This means that each lattice node has its own individual surroundings, which may be described using statistical methods. However, the statistics of the distribution of atoms in the solid metal solution is unknown. All statistical methods are based on certain assumptions about interatomic interactions. If atoms do not interact at all, their distribution on lattice nodes in a binary solid solution is determined by binomial distribution. The presence of interatomic interactions results in a difficult to control change of distribution. Therefore, all general statistical methods prove powerless in the study of properties of specific solid solutions. We can either justify the intended distribution or try to find basic distributions by which any pre-determined distribution in the ensemble of interacting particles could be decomposed. This way was proposed in the work by Kikuchi [14]. In [15], this method was considered for more accurate definition of the distribution of atoms in the solid solution of ions Fe^{2+}–Nb^{5+} in ferroelectromagnet $PbFe_{1/3}Nb_{2/3}O_3$.

2. Methods and materials

1. This paper describes an example of a solid solution model of magnetic and non-magnetic atoms distributed across the nodes of a one-dimensional chain. It is shown that the third Gilleo hypothesis does not correspond to the relationship T_c and c obtained within the framework of approximation of the self-agreed field under any modification of hypotheses expressed in [2]. The work also illustrates some features of the theory of spontaneous magnetization and magnetic PT in solid solutions. It is noted what type of experiments would help to establish a link between the conditions of occurrence of the long-range magnetic order and the distribution of magnetic and non-magnetic ions at the nodes of a crystal lattice for specific solid solutions, including partially ordered solutions. The following calculations are made assuming the applicability of the self-agreed Weiss field approximation [6, 8].

It is known that self-consistent field models are not applicable to: 1) the description of PT between different magnetically ordered states [6, 8]; 2) the description of stable existence of complex magnetic structures [12]. The listed prohibitions on the applicability of the approach of a self-consistent field are due to the fact that this approach only takes into account paired interatomic interactions. The more nonlinear components of the non-equilibrium potential F are related to the configuration entropy [12] and are characterized by a symmetry group, the order of which is $r!$. Here r – number of nodes of a primitive lattice of the magnetically ordered phase of a crystal, where magnetic atoms can also be placed. Typically, these m nodes are equivalent in the paramagnetic phase of a crystal if the solid solution is not characterized by a distant order in the spatial distribution of magnetic and non-magnetic atoms.

Let the distant order in the spatial distribution of the different grades of atoms that make up the solid solution also exist in the paramagnetic phase. Then m is the total number of nodes belonging to different sub-lattices available for settlement by magnetic atoms in the paramagnetic phase that entered the primitive cell of the magnetically ordered phase. Let, for example, PT in magnetically ordered state be accompanied by multiplication of crystallochemical primitive cell by 6, 8, etc. Let us also suppose that a primitive cell in a disordered state contains just one node available for filling with a magnetic and non-magnetic atom. Then the symmetry of nonlinear components of the non-equilibrium potential in the approximation of the self-consistent field is described by permutation groups $G6$ and $G8$, the order of which is $6! = 720$ and $8! = 40320$, respectively. It is clear that the set of stabilizers of groups L determining the symmetry and structure of ordered phases with full consideration of symmetry of crystal and order parameter does not coincide with the set of stabilizers of the group of symmetry of non-equilibrium potential in approximation of self-consistent field at "complex" orders [12]. In this regard, we will assume that as a result of magnetic PT in a given one-dimensional crystal one of two magnetically ordered states occurs: ferromagnetic or simplest antiferromagnetic.

The ferromagnetic state of a solid solution implies that the interatomic exchange interaction between two magnetic atoms placed on adjacent nodes of a chain favors the parallel direction of their magnetic moments. The simplest antiferromagnetic ordering implies that the exchange interaction between the nearest neighbors makes the antiparallel direction of their magnetic moments the most probable. We neglect the interaction with more distant neighbors in the consideration below, which reduces neither the commonality of consideration nor the interesting results for our purposes obtained in the approximation of a self-agreed field.

We consider the crystal structure in the form of a one-dimensional chain to make some claims clearer without using bulky formulas. Let us emphasize that it is the approximation of the self-agreed field that allows considering magnetic ordering on a one-dimensional chain of a solid solution of magnetic and non-magnetic atoms while taking into account the interaction of only the nearest neighbors. It is known that the exact solution of the statistical problem in this case does not allow any ordering [16]. In the approximation of a self-consistent field, the results for crystal lattices of different dimensions differ only quantifiably. For purposes of illustrating the error of the Gilleo hypothesis, this difference is insignificant.

2. An arbitrary linear chain assembly filled with magnetic and non-magnetic atoms can be present in six structurally different configurations. In three of them this unit is filled with the magnetic ion, and in three – non-magnetic ion. Configurations differ in the number of closest neighbors creating an efficient magnetic field on this node. There may be be 0, 1, or 2 of such neighbors. The second variant (one magnetic neighbor) turns out to be symmetrically twice degenerate.

In the considered model, only two of the six configurations in the distribution of the three atoms on adjacent lattice nodes can contribute to the ensemble average magnetization of an arbitrary lattice node. These are configurations in which a magnetic atom sits on a node, and at least one adjacent node is occupied by a magnetic atom. Let us denote the probability of filling an arbitrary node of a one-dimensional chain with non-magnetic atoms p. Then the probability that one magnetic and one non-magnetic atom may be in the chain next to a node filled with a magnetic atom is $W_2 = 2p(1-p)^2$. The probability that all three atoms will turn out to be magnetic is $W_3 = (1-p)^3$. Let the spin of magnetic atoms S = 1/2 and, accordingly, magnetic moment $(m_i = \pm\mu_B S_z^i)$. Then, considering the possible magnetic interaction, the three considered atoms may be in eight energy-distinct configurations. Indeed, given that the effective magnetic field on the side of one magnetic neighbor in the approximation of the self-matched field $H_{eff}^{(1)} = \chi^{-1}m$, it can be written that the probability P of the configuration is one nearest neighbor – the magnetic atom and the magnetic moment is directed parallel to $H_{eff}^{(1)}$

$$P = W_2 \cdot \exp\left\{-\beta\chi^{-1}m\right\}/Z_{m2}, \tag{1}$$

here, χ^{-1} – inverse atomic magnetic susceptibility that can be written through Lande factor [18] to approximate a free atom or through crystal field operators in the case of an atom in a crystal lattice [18]. The "magnetic" part of the statistical sum of the considered structural configuration in the approximation of the self-consistent field:

$$Z_{m2} = 2ch(\beta\chi^{-1}m). \tag{2}$$

Finally, we get the average magnetic moment of the magnetic ion

$$m = 2p\ (1-p)^2 th(\beta\chi^{-1}m) + (1-p)^3 th(2\beta\chi^{-1}m). \tag{3}$$

Accordingly, in (3) the temperature of the second PT type in magnetically ordered state is determined by the ratio:

$$1/\beta = k_B T_c = 2(1-p)^2\chi^{-1}. \tag{4}$$

3. Let us discuss the consequences (2)–(3). First of all, let us note that the result (2)–(3) is recorded in the form of dependence $m(p_i)$, where p_i – probability of location of non-magnetic ions in places of nearest neighbors. This representation of the result of the calculations makes it possible to determine the dependence of the average magnetization on one crystal unit and the transition temperature to magnetically ordered state on the degree of ordering of arrangement of magnetic and non-magnetic ions, and shows the way of generalization of this result in case of three-dimensional crystals. Thus, in

the case of completely disordered distribution of magnetic and non-magnetic atoms along the nodes of the chain, the calculations make it possible to obtain a ratio that can be tested experimentally: $T_c \sim (1-c)^2$, where c – concentration of non-magnetic ions of atoms.

Let us suppose by analogy with the first Gilleo hypothesis that only those magnetic atoms, which inner circle has two magnetic atoms, contribute to the establishment of the magnetic order. Then, from (3)–(4) we get

$$m^* = (1-p)^3 th(2\beta\chi^{-1}m); \quad (5)$$

$$k_B T_c^* = 2(1-p)^3 \chi^{-1}. \quad (6)$$

Therefore, Gilleo hypothesis of the number of closest magnetic neighbors that make the magnetic ion an active participant in the formation of the magnetic order significantly affects the dependence $m(p, T)$ and $T_c(p, T)$ obtained in the approximation of the self-agreed field. Thus, if the distribution of magnetic and non-magnetic ions is completely disordered and there must be at least two magnetic neighbors, then in the model under consideration we obtain $T_c \sim (1-c)^3$, which is qualitatively different from the previous result $T_c \sim (1-c)^2$.

To illustrate the delicacy of the question on the necessary number of connections to neighboring magnetic ions so that an arbitrary lattice assembly filled with a magnetic atom can belong to an infinite magnetic-ordered region, let us give one result of the flow theory (Fig. 1, 2) [19]. Let lattice nodes filled with magnetic atoms be randomly distributed in space. For example, this is possible in solid solutions, where chaotic displacements of atoms are caused by stochasticity in the distribution of solution components to nodes of the "middle lattice" or otherwise to the nodes of the "virtual crystal" with idealized structure. Let the radius of exchange interaction of magnetic atoms with other atoms be equal R.

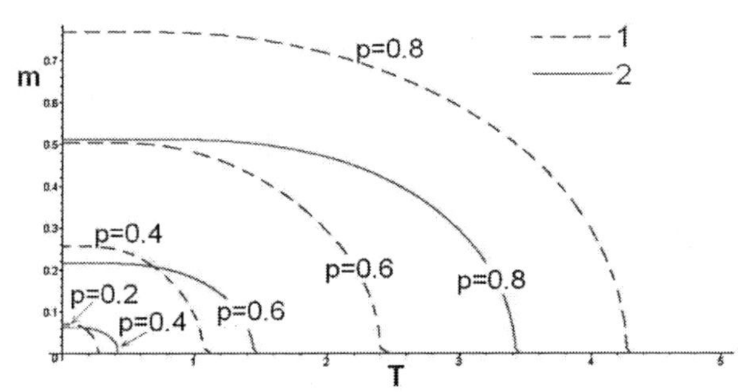

Figure 1. Dependence of m (T) in disordered quasi-binary solid solution of magnetic and non-magnetic ions of $Fe_{1-p}(Nb, Sn)_p$ type for cases where the establishment of a magnetic order in a chain involves: 1 – all magnetic ions, 2 – only those magnetic ions that have two magnetic neighbors

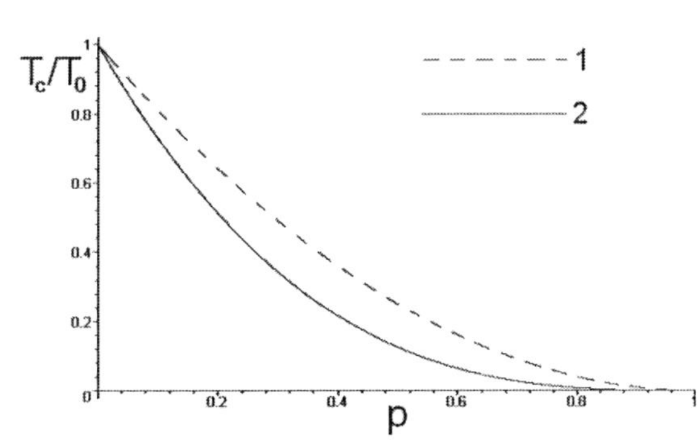

Figure 2. Dependence of magnetic transition temperature on probability of non-magnetic neighbor of magnetic ion

Due to the mechanism of indirect exchange through electrons the conductivity R can significantly exceed the distance a to the nearest neighbors in the correct lattice. Under these assumptions, an arbitrary magnetic atom can belong to an infinite cluster formed of magnetic atoms only if the average number of nodes (k) occupied by magnetic atoms and falling into radius sphere R is greater than 2.5–3 (greater accuracy at numerical experiments on the final number of nodes cannot be obtained [19]).

This result is interesting because as part of the approximation of a self-consistent field or any other model approximation when solving the dependence problem $T_c(p)$ it is impossible to establish how many exchange magnetic connections between a given ion and its closest neighbors must be broken by non-magnetic atoms so that the magnetic atom does not participate in collective effect of establishing magnetic order. However, the answer to this question can be obtained experimentally by comparing the measured dependencies $m(T,p), T_c(p)$ and $(dT_c/dp)\,|=\Psi(p_0)$ with those obtained from theories implying a different minimum number of broken bonds.

3. Conclusion

Finally, in two and three-dimensional crystals, apart from the number of bonds of a given magnetic atom with magnetic atoms located in the nearest environment, it is necessary to further consider the conditions of propagation ("flow") of exchange interactions to macroscopic regions of a crystal. This circumstance will make it possible to establish a lower limit of magnetic ion concentration, at which it is still possible to transition to a magnetically ordered state.

Acknowledgment

The study was carried out within the framework of the basic part of the State Task of the Russian Ministry of Education and Science 3.6439.2017/8; topic: No. 3.6386.2017/BC and 3.6439.2017/8.

References

[1] Smolensky G A et al 1961 On the coexistence of ferroelectric and ferrimagnetic States *Bull. of the Acad. of Sci. USSR: Phys.* **25(11)** 1333–9

[2] Gilleo M A 1958 Superexchange Interaction Energy for Fe^{3+}-O_2-Fe^{3+} Linkages *Phys. Rev.* **109(3)** 777–81

[3] Goodenough J B, Wickham D G and Croft W J 1958 Some magnetic and crystallographic properties of the system LixNi1-xNixO *J. Phys. Chem.* **5** 107

[4] Gilleo M A and Geller S 1958 Magnetic and Crystallographic Properties of Substituted Yttrium-Iron Garnet *Phys. Rev.* **110(1)** 73–8

[5] Pearson W B 1972 *The crystal Chemistryand Physics of metals and Alloys* vol I (New York: Willey-interscience) pp 34–43

[6] Brout P 1965 Phase *Transitions* (New York) pp 24–62

[7] Saltas V et al 2017 Modelling solid solutions with cluster expansion, special quasirandom structures, and thermodynamic approaches *Appl. Phys. Rev.* **4(4)** 041301

[8] Smart J S 1966 *Effective Field Theories of Magnetism* (Saunders)

[9] D Connolly J W and Williams A R 1983 Density-functional theory applied to phase transformations in transition-metal alloys *Phys. Rev. B* **27(8)** 5169–72

[10] Baettig P and Spaldin N A 2005 Ab initio prediction of a multiferroic with large polarization and magnetization *Appl. Phys. Letters* **86(1)** 012505 (1–3)

[11] Qu W et al 2006 Room temperature magnetoelectric multiferroism through cation ordering in complex perovskite solid solutions *J. Phys. Condens Matter* **18** 8935–42

[12] Gufan Yu M and Moshchenko I N 1991 Symmetric degeneration of statistical models of complex ordered alloys *Solid state phys.* **33(4)** 1166–72

[13] Gufan Yu M and Kalita V M 1987 Molecular theory of exchange mettamagnets *Sow. Phys. Solid State* **29(11)** 1893–6

[14] Kikuchi R A 1951 Theory of Cooperative Phenomena *Phys. Rev.* **81(6)** 988–1003

[15] Gufan K Yu, Petrov K Yu and Stryukov D V 2009 Model of local structure of multicomponent oxides with hexagonal unit cell *Bull. of the Russ. Acad. of Sci.: Phys.* **73(8)** 1225–6
[16] Landau L D and Lifshits E M 1995 *Statistical physics* (Moscow)
[17] Landau L D and Lifshits E M 2001 *Quantum mechanics* (Moscow)
[18] Ballhausen C J 1962 *Introduction to ligand field theory* (McGraw Hill)
[19] Shklovsky B I and Efros A L 1979 *Electronic properties of doped semiconductors* (Moscow)

IOP Conf. Series: Materials Science and Engineering **905** (2020) 012030 doi:10.1088/1757-899X/905/1/012030

Synthesis of group compressed-air sources for marine seismics

V I Gulenko[1], E I Zakharchenko[1], A V Rudakov[2] and Yu I Zakharchenko[1]

[1] Kuban State University, 149, Stavropolskaya Street, Krasnodar, 350040, Russia
[2] JSC Yuzhmogeologia, 20, Krymskaya St., Gelendzhik, 353461, Russia

E-mail: v_gul@mail.ru, evgenia-zax@yandex.ru, rudakov_ymg@mail.ru, ofis-2010@yandex.ru

Abstract. The improvement of geological efficiency of marine seismic survey implies continuous improvement of the entire technological complex used in water areas. This is typically accomplished by using complex linear or spatial non-uniform compressed-air groups. In the development of such sources, in addition to solving a number of purely technical problems related to their placement on board the ship and safe operation, one of the most difficult tasks is to determine the initial requirements for selection of the main parameters of the group, such as the volumes of working chambers sources and geometry of their grouping, as well as other factors determining amplitude, time and frequency of emitted acoustic signals. This study considers the main principles of grouping the compressed-air sources for marine seismic survey. It is shown that the most important factors determining the choice of source parameters and specific research technology include the nature of geological tasks and seismic conditions, which primarily depend on the sea depth in the study area. A simple method of calculating the configuration of linear subgroups with fluctuation suppression is described, as well as examples of building linear and spatial groups based on them with pronounced directional properties are considered.

1. Introduction

The improvement of geological efficiency of marine seismic survey implies continuous improvement of the entire technological complex used in water areas. At the same time, along with the use of modern geophysical vessels, satellite positioning systems and digital seismic telemetry systems, it is necessary to use sufficiently powerful sources that excite high-amplitude and broadband signals, which are not complicated by oscillations and solve complex geological tasks in different seismogeological conditions.

This objective is usually achieved by the use of complex linear or spatial non-uniform compressed-air groups, which sometimes contain dozens of pneumatic radiators combined in a number of close subgroups – clusters [1, 9, 11]. In the development of such sources, in addition to solving a number of purely technical problems related to their placement on board the ship and safe operation, one of the most difficult tasks is to determine the initial requirements for selection of the main parameters of the group, such as the volumes of working chambers sources and geometry of their grouping, as well as other factors determining amplitude, time and frequency of emitted acoustic signals [1, 9, 11].

This paper considers some principles of construction of linear and spatial groups of compressed-air sources for marine seismic survey.

2. Tasks of selecting source characteristics during marine seismic survey

Similar to ground seismic survey, during marine seismic survey the selection of source characteristics is fully determined by the geological task, for which the on-board seismic survey equipment is created, as well as by technological peculiarities of seismic survey operations in different seismogeological conditions and capacities of the used vessels.

Thus, relatively big research geophysical vessels on a displacement of several thousand tons equipped with the powerful compressor and specialized lifting equipment necessary for exploitation of large areal compressed-air groups consisting of several (from 4 to 8)identical linear subgroups with the total volume of working chambers from 40–50 dm^3 to 100 dm^3 (and sometimes more) are most often used to solve the geological tasks related to *search and exploration of deep-seated hydrocarbon fields* in a shelf zone with the sea depths of over 10–20 m. In recent decades, third-generation research vessels – the so-called "ramforms" with advanced seismic equipment, which allows simultaneous registration of seismic information from several thousand channels located in several dozens of simultaneously towed streamers, have been used for 3D marine seismic survey. At the same time the most powerful areal groups consisting of 6–8 simultaneously towed streamers with the total number of pneumatic radiators reaching hundreds are used.

At a towing speed of 4–6 knots, which is typical for operations with floating piezo-streamers, it takes 8–12 s for such a vessel to pass the interval between excitation points of 25 m. The capacity of the compressor station should ensure guaranteed filling of all working chambers of group radiators with compressed air at the operating pressure of 12–15 MPa, with all compressed air energy stored per exposure may reach several (2–4) megajoules.

Thus, on the one hand the *total volume of the group is determined by the capacity of the compressor station* and, on the other hand, *by the need to ensure such a density of acoustic energy flow in a given frequency band and in given spatial angles that provide the required signal-to-interference ratio at the input of the towed receiver, i.e. the necessary noise immunity of the group thus ensuring successful solution of the geological task.*

In case of *detailed seismic survey* of objects limited in area lying at medium depths, specialized geophysical vessels of the second generation with displacement of up to one and a half thousand tons are used more often during profile *2D works*, which have significantly less energy capacity and allow simultaneous use from 2 to 4 linear subgroups with the total number of radiators of up to 20–30, with the total volume of chambers of up to 30–40 dm^3. A good example in such conditions are twin-hulled ships, such as the Searcher [8], which, at low settlement (about 1.5 m) ensure the operation of similar radiating systems.

Until the late 1980s, geophysical vessels often used *compact compressed-air groups* with grouped radiators placed on towed metal frames. Compared to linear and even more spatial sources, compact groups have a significantly higher level of mutual acoustic influence, have weak directional properties and with the same total volume are usually inferior to linear groups in all acoustic indicators [1, 4].

Relatively small geophysical vessels with the displacement of up to 600–800 t are usually used to solve *engineering-geological tasks* during the study of shallow lying thicknesses and the top part of a section. As part of a technological complex that has recently operated in JSC Yuzhmorgeologia, in CJSC ROMONA, etc. [7], along with popular seismoacoustics, side-scan sonar and other methods, high-resolution high-frequency seismic survey is often used with the registration in the frequency band from 20–30 to 300–400 Hz. At the same time, the best results were obtained when relatively small compact compressed-air groups containing 3–5 radiators (1–2 clusters) with a total volume of up to 2–3 dm^3, and towed at a depth of 2–3 m are used as excitation sources [7].

At *seismic exploration in the conditions of shallow water* if the sea depth is from 2 m to 8–10 m, relatively shallow low-tonnage vessels, which specific power and loading capacity allow operating only very small compressed-air groups (8–16 radiators) with the total volume of 10–20 dm^3, are usually used [1, 4]. Until recently, in similar seismogeological conditions the so-called "yo-yo" technology with periodic uncoiling – streamer sampling at continuous movement of a vessel are used. In recent decades, with the advent of modern digital telemetry systems such as BOX, ARAM ARIES and others, spatial surveillance systems with fixed bottom receivers and mobile explosion point [6] have become common on the shallow sea with the "yo-yo" technology no longer being applied.

In the most complex seismogeological conditions of the *limit shallow water and in the transit zone* spatial observation systems with a fixed receiving device (bottom telemetry, bottom streamer, or geophones installed on metal pins) and a mobile explosion point moving along a given network of excitation points are widely used in recent years [4]. Relatively small groups of pneumatic radiators installed on small-size floating facility (pontoon, barge, etc.) and towed on floats with suspension at a depth of 1.0–1.5 m, or moved by dragging along the bottom at a sea depth of less than 1.0 m are usually used as sources (Figure 1). In the latter case, the required information may be accumulation.

Figure 1. Towed pontoon with compressed-air sources [1, 4]

In the *transit zone* at the sea depth of less than 0.5–1.0 m where excitement in a water layer becomes inefficient, the best results are achieved with the use of *submersible compressed-air sources* exciting signals in small wells and usually working in the accumulation mode. At the same time the whole equipment set, including seismic station, compressor and auger drill or hydraulic monitor, is placed on amphibious class vehicles with increased cross-country capacity (Figure 2).

Figure 2. Explosion point with auger drill and compressed-air source PIC-3 on the basis of floating caterpillar transporter (works of JSC Yuzhmorgeologia in the North Caspian Sea) [4]

Thus, considering the variety of geological tasks, the used technologies, as well as the difference in the energy and load capacity of vehicles used in different seismogeological conditions, it seems most relevant to develop technical means for excitation of elastic waves on the basis of the *modular principle*.

In due course, this approach was justified by A.V. Kalinin with regard to the development of an electric spark source for marine seismoacoustic survey [2, 3]. *With regard to the development of group compressed-air sources, the modular principle of equipment implies the creation of such a set of technical means that with a minimum nomenclature variety of devices used would ensure the maximum variety of characteristics necessary to solve a wide range of tasks in different seismogeological conditions.*

With modular design of the compressor station and the control system, this task is essentially reduced to the creation of several universal linear subgroups with different total volumes, high reliability and good acoustic characteristics on the basis of a number of similar radiators (Bolt, Sleeve Gun, G.Gun, Pulse, Baby or others). The volumes of radiator chambers in a linear subgroup, their number, base of grouping and depth of immersion are chosen so that at simultaneous operation of the group the first pressure peaks of signals of separate radiators are summarized inphase while repeated shocks (taking into account reflection from "water-air" surface) are developed in antiphase.

3. Calculation of synchronous non-uniform compressed-air groups

Monograph [1] describes different methods of synthesis of *synchronous non-uniform compressed-air groups*, which imply the generation of a sum signal with fluctuation suppression. One of the simplest methods is as follows.

A linear subgroup is created from several clusters, the distances between which are chosen so that their mutual acoustic influence is absent or minimal. The distance between adjacent radiators in each cluster is selected so that their air bubbles during exhaust merge into a single volume. The pulsation period of the signal emitted by the cluster is about the same as the single total volume radiator, but the amplitude characteristics are much better: the amplitude of the first peak of the sum cluster signal is significantly higher and almost equal to the sum of the amplitudes of the signals of individual radiators. The amplitude of the second peak (i.e. pulsation) is significantly lower than that of a single total volume radiator.

The volumes of working chambers of radiators (clusters) V_i are chosen so that all pulsations (second peaks) do not develop inphase, but form the sequence of impulses with approximately constant difference of pulsation periods by ΔT (Figure 3, b). The ΔT is chosen so that the reflection of repeated peak of *(i-1)* signal from "water-air" surface coincided with repeated peak of i signal and ensured its quenching. At the same time ΔT shall be equal to wave transit time from the radiator to "water-air" surface and back: $\Delta T = 2h/c$, where h – depth of immersion of the radiator, c – speed of distribution of elastic waves in water $c = 1500$ *m/s*.

According to the known Rayleigh-Willis formula [1–3, 9, 10], at fixed operating pressure of compressed air and equal depth of radiation immersion, the pulsation periods of T_i signals and volumes of working chambers V_i are bound by the ratio:

$$T_i = k\, V_i^{1/3}, \tag{1}$$

where k – coefficient depending on the radiator design; at the same time

$$T_{i+1} = T_i + \Delta T. \tag{2}$$

Substituting in (2) the expression (1) for the pulsation period T_i, after small transformations we get a recurrent formula for calculating the volumes of working chambers of compressed-air radiators (clusters) in a linear subgroup:

$$V_i = V_1[1+2h(i-1)/(cT_1)]^3, \quad i = 2,\dots,n \tag{3}$$

where V_1 and T_1 – volume of the working chamber and the pulsation period of the signal of the 1st radiator of the linear subgroup having the minimum volume, n – number of clusters in the line. The final configuration setting can be performed after an experimental study of the acoustic characteristics of the group.

As an example, Figure 3 shows the calculation of configuration of a linear subgroup of 4 clusters (9 radiators) with the total volume of 10 dm^3 located on the basis of 12 m.

The calculation of cluster signals, sum signal of linear group and its amplitude spectrum is performed in MathCAD on the basis of algorithms similar to those implemented in known Gundalf™ [10].

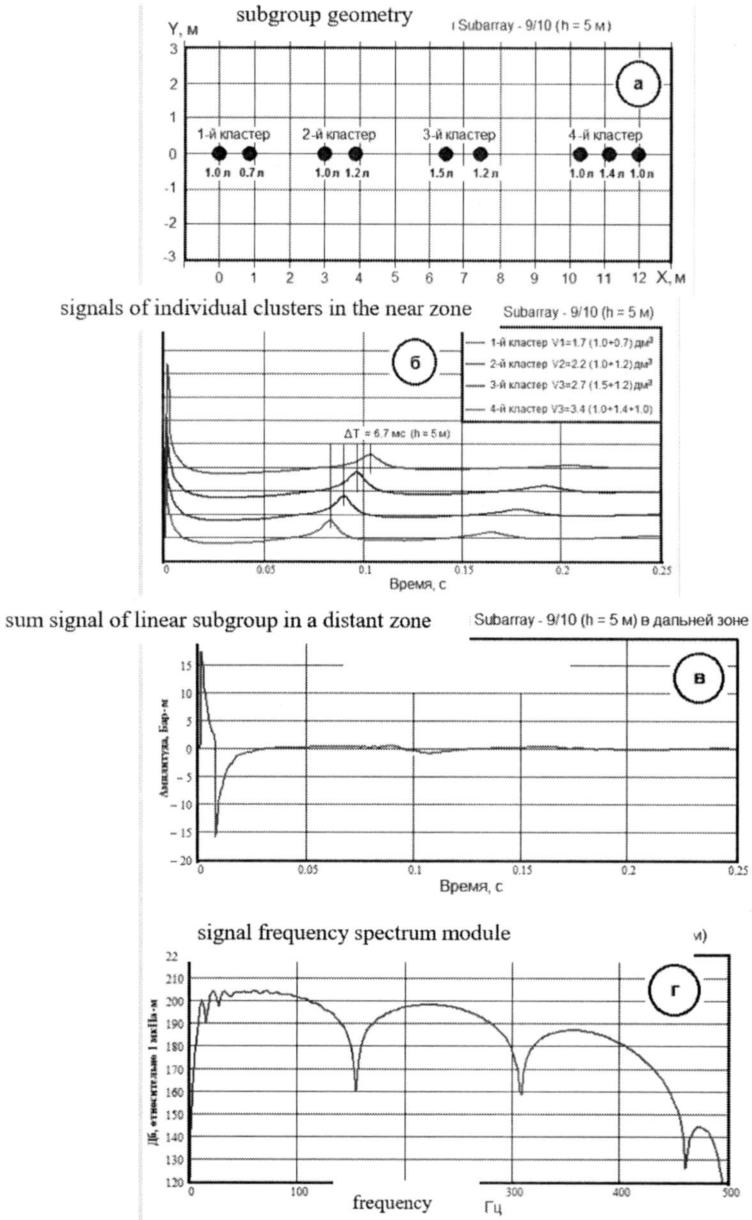

Figure 3. Calculation of Subarray linear subgroup configuration ($N = 9$, $V_\Sigma = 10$ dm^3, $h = 5$ m).

4. Implementation of interference systems with adjustable directional characteristic

The use of compressed-air groups with a large total volume is only possible on specialized geophysical vessels equipped with both complex lifting equipment necessary for round-trip operations and for towing large compressed-air groups, as well as improved and powerful compressor equipment and corresponding control systems.

However, the use of large non-uniform groups is justified because along with the formation of simple in shape and short in duration signals with a wideband spectrum, it is also possible to significantly increase the amplitude and energy of radiation and thereby significantly increase the noise immunity and depth of marine seismic surveys.

Another important circumstance is that large groups are generally located on a base that is commensurate with wavelength allowing such groups to be used as directional interference systems as well. At synchronous actuation such linear or spatial groups have directional characteristic with the maximum in vertical direction. By introducing time delays at the moments of actuation of adjacent radiators or adjacent subgroups, it is possible to implement an interference system with a controlled directional characteristic, which at the excitation stage will generate waves at predetermined apparent speeds.

Such an interference system can be created by simply combining identical linear subgroups into a corresponding areal system. Due to such association amplitude of a sum signal in the direction of an acoustic axis of the α_0 group increases approximately in proportion to the number of the united subgroups *K*, and the density of acoustic energy flow – in proportion to K^2. In addition to the change in amplitude and energy indices, there is a significant change in directional characteristics of the group.

As an example Figure 4, a and b shows the modules of the generalized interferential characteristics, respectively, in longitudinal (XOZ) and cross (YOZ) planes of linear subgroup Subarray-9/10 with the immersion depth of $h = 5\ m$ calculated at $\alpha_0 = 0°$ (synchronous operation), on Y axis – angle down. The configuration of the line group Subarray-9/10 is shown in Figure 3, a.

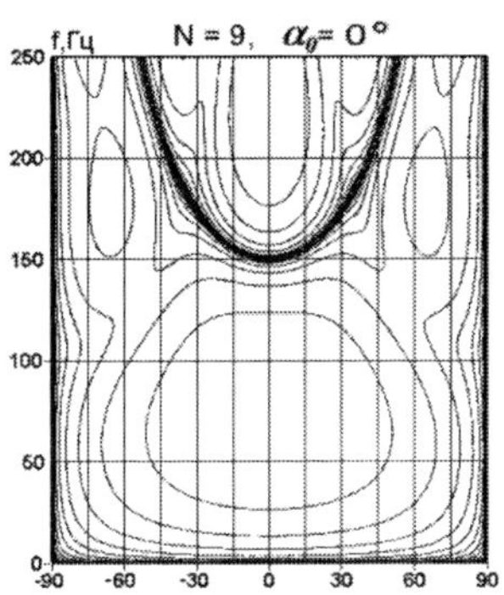

a – characteristic in the *XOZ* plane, one line Subarray-9/10

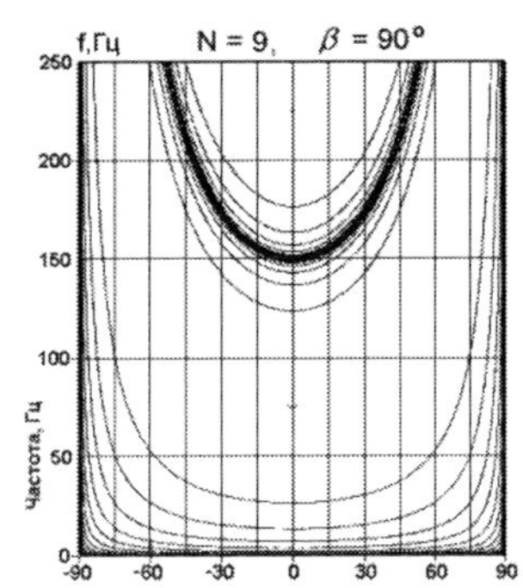

b – characteristic in the *YOZ* plane, one line

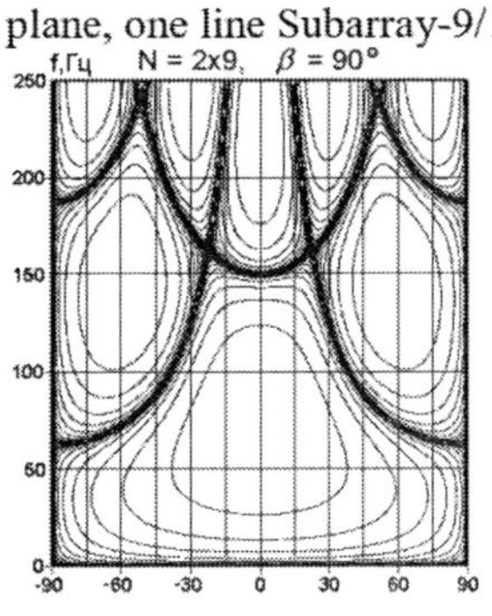

c – characteristic in the *YOZ* plane, two lines Subarray-9/10 in parallel with a cross span $\Delta Y = 6\ m$

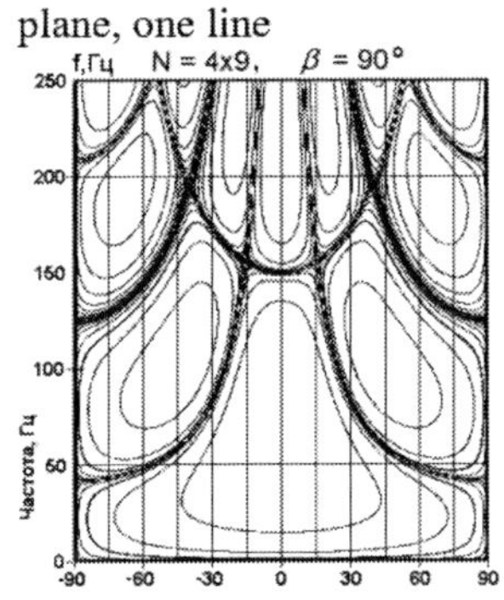

d – characteristic in the *YOZ* plane, four lines in parallel with a cross span between the lines $\Delta Y = 6\ m$

Figure 4. Modules of generalized characteristics of linear compressed-air group at $h = 5\ m$

Figure 3, c shows that one such linear subgroup has good acoustic parameters: the calculated amplitude of *P-P* signal is 32.7 bar·m, the degree of pulsation quenching *P/B* – 16.7, however, as Figure 4 shows, the group length of $D = 12\ m$ is characterized by relatively low direction even in the longitudinal *XOZ* plane ($\beta = 0°$). =In the cross *YOZ* plane ($\beta = 90°$) the direction of any one linear group corresponds to the direction of a point source.

To compress the directional characteristic of a group in the YOZ plane, a transverse grouping of lines is used, i.e. their combination into a single areal group. For example, Figure 4, c and d show the generalized characteristics in the YOZ plane of two areal groups consisting of two and four identical linear subgroups (Subarray-9/10, $h = 5\ m$), respectively, towed in parallel with the spacing between the lines $\Delta Y = 6.0$ m. From comparison of Figure 4, b, c and d we see that there is very effective cross compression of the directional characteristic of the spatial group. However, in the longitudinal plane XOZ, the directional characteristics of all these variants of the groups still look as shown in Figure 4, a.

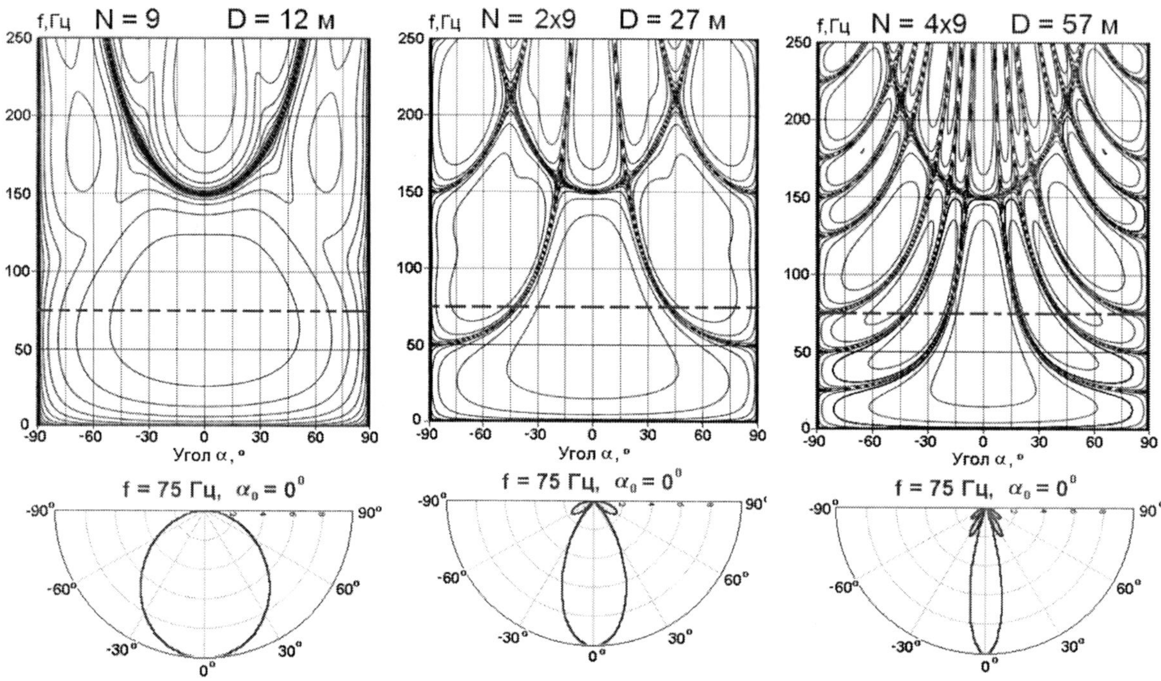

a – one line Subarray-9/10, N = 9

b – two lines Subarray-9/10 (N = 18) with longitudinal span ΔX = 15 m

c – four lines Subarray-9/10 (N = 36) with longitudinal span ΔX = 15 m

Figure 5. Modules of generalized characteristics in *XOZ* plane and directional diagram at frequency $f_0 = 75$ Hz of linear compressed-air group at $h = 5\ m$

Clearly, in the longitudinal plane *XOZ*, further compression of the main characteristic can also be achieved by increasing the longitudinal grouping base *D*. Considering that the length of each line is structurally fixed and is determined by the length of the scaffold on the deck of a vessel, on which all linear subgroups in transport situation are placed, the longitudinal base of the group is most often increased only due to longitudinal span ΔX between subgroups.

To illustrate the above, Figure 5 shows a series of *XOZ* plane characteristics and patterns calculated for one (a), two (b), and four subgroups (c) Subarray-9/10 at the depth of immersion $h = 5\ m$, $\alpha_0 = 0°$ and the longitudinal span between subgroups $\Delta X = 15\ m$, D – common base of the group. The angle of setting of all diagrams $\alpha_0 = 0°$ (synchronous operation of radiators).

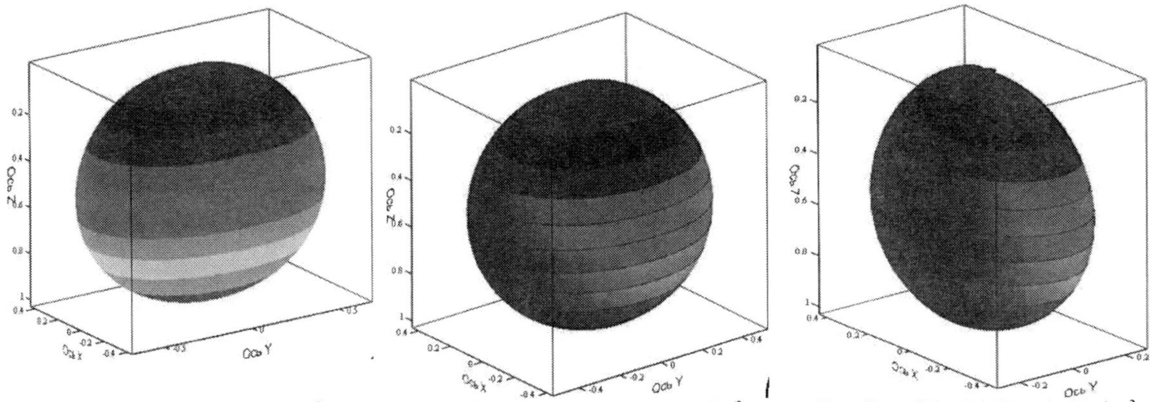

a – one line (N = 9, V = 10 dm^3, h = 5 m) b – two lines (N = 18, V = 2×10 dm^3, h = 5 m, ΔY = 6 m) c – four lines (N = 36, V = 4×10 dm^3, h = 5 m, ΔY = 6 m)

Figure 6. Free-space diagrams of areal groups at f_0=75 Hz

Figure 5 shows that the main radiation lobe created in the vertical direction with the increase of span *ΔX* is considerably narrowed on axis *X*.

Figure 6 shows spatial directional patterns of three options of the areal groups calculated at a frequency of f_0 = 75 Hz for one (a), two (b) and four parallel subgroups (c) of Subarray-9/10 at the immersion depth of $h = 5\ m$, $\alpha_0 = 0°$ and the cross span between subgroups $\Delta Y = 6\ m$.

Figure 6 shows that the main radiation lobe created in the vertical direction with the increase in the group base on *Y* axis (from 0 to 3*ΔY*) is also considerably narrowed on *Y* axis.

Thus, *choosing the corresponding longitudinal and cross spans between subgroups ΔX and ΔY, as well as setting a certain operation law of radiators for each subgroup via delays* $\{\Delta t_{ij}\}$ *it is possible to concentrate the emitted energy rather efficiently in any direction of the lower half-space within any set space angle.*

However, the practice shows that the compression of the directional characteristic of the group in the longitudinal plane XOZ should be confined to certain limits. Thus, for subsequent processing of information according to CDP algorithms, it is desirable that target reflections on seismograms are maintained in shape and amplitude along the whole length of the hodograph. This allows maximizing the effect of the subsequent summation.

Figure 7 shows that for this purpose it is necessary that when grouping radiators the main radiation lobe is created in the α_0 direction, and the width of this lobe on level 0.707 is determined by the angles α_{min} and α_{max} corresponding to the exit of seismic beams to final points X_{min} and X_{max} of the end-on receiver.

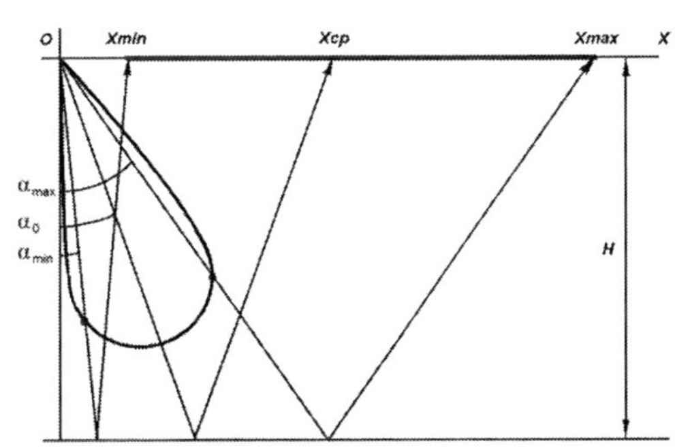

Figure 7. Choice of group radiation lobe width: X_{min} – offset of the receiving device; X_{max} – far end coordinate of the receiver; $X_{max} = X_{min} + L$; *L* – length of the receiver; $X_{cp} = X_{min} + L/2$ – streamer midpoint; *H* – reflector depth

Depending on the length of the receiving device L and the reflector depth H, the values α_0, α_{min} and α_{max} as a first approximation can be presented by the following obvious expressions:

$$\alpha_{\min} = arctg\left(\frac{X_{\min}}{2H}\right);$$

$$\alpha_0 = arctg\left(\frac{X_{cp}}{2H}\right);$$

$$\alpha_{\max} = arctg\left(\frac{X_{\max}}{2H}\right)$$

Figure 8 shows the dependences $\alpha_0(H)$, $\alpha_{\min}(H)$ and $\alpha_{\max}(H)$ calculated for receiving devices of different length and allowing estimating approximately the necessary width of the main reflection lobe of the compressed-air group in the longitudinal XOZ plane.

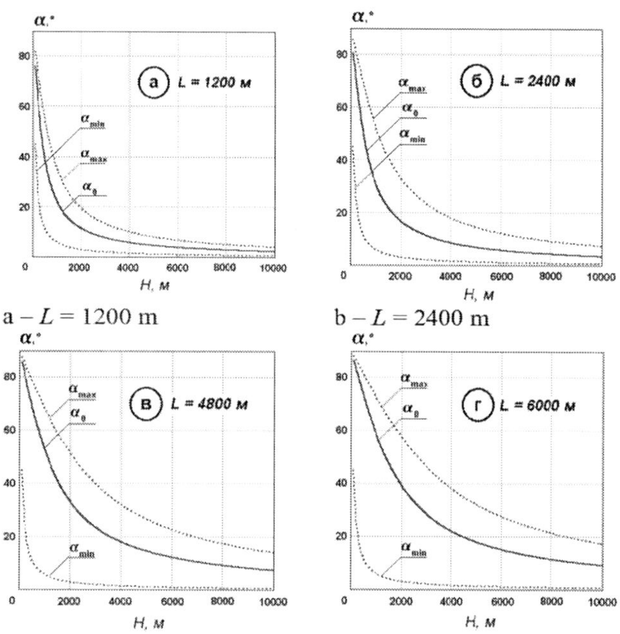

a – L = 1200 m
b – L = 2400 m
c – L = 4800 m
d – L = 6000 m

Figure 8. Dependences $\alpha_0(H)$, $\alpha_{min}(H)$ and $\alpha_{max}(H)$ calculated for receiving devices of different length

More precisely, taking into account the speeds of elastic waves in the environment and complex trajectories of beams, the values α0, αmin and αmax can be defined, for example, by the VOLNA-M software package for the modeling of seismic wave fields, which is specially developed for the correct solution of such tasks [5]. The modeling of beam diagrams shows that the interval of angles αmin – αmax restricting the limits of a beam tube in the longitudinal XOZ plane getting for receiving setting can significantly depend on high-speed differentiation of a section (Figure 9).

For example, the comparison of Figure 9, a and Figure 9, b shows that in order to study the borders under a high-speed layer, the energy radiated by a source shall be concentrated in narrower interval of angles α_{min} – α_{max} that will require the increase in the base of group D in the longitudinal XOZ plane.

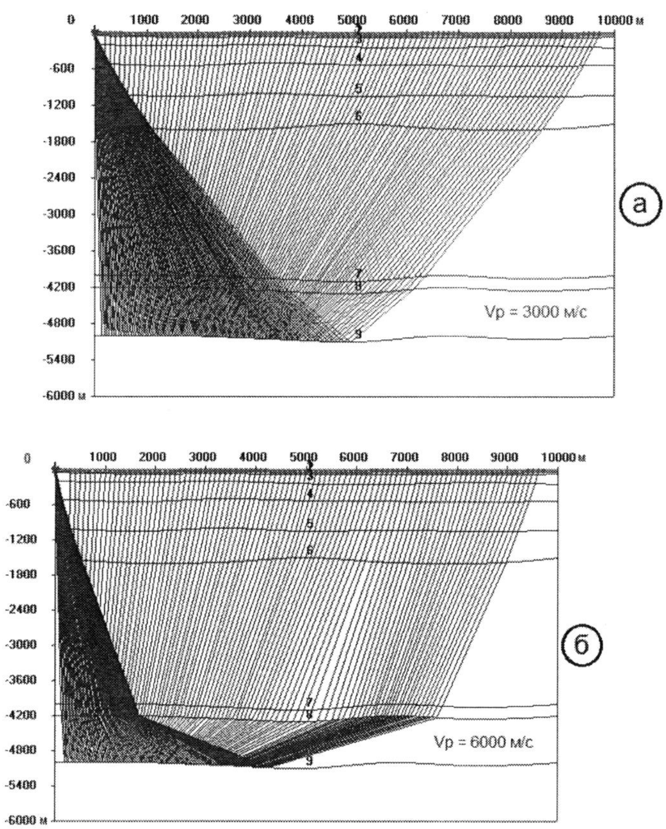

Figure 9. Beam patterns of reflected waves obtained for two model sections with different speed *Vp* in the ninth layer

5. Conclusion

This paper briefly discusses the basic principles of construction of group compressed-air sources for marine seismic survey. It is shown that the most important factors determining the choice of source parameters and specific research technology are the nature of geological tasks and seismogeological conditions, which primarily depend on the sea depth in the study area.

A simple method of calculating the configuration of linear subgroups with fluctuation suppression is described, as well as examples of building linear and spatial groups based on them with pronounced directional properties are considered. For more accurate and reasonable selection of seismic survey methods, including parameters of compressed-air sources grouping, it is recommended to use modern software tools for simulation of seismic wave fields and calculation of characteristics of any linear and spatial interference systems on the basis of pneumatic radiators.

References

[1] Gulenko V I 2003 *Compressed-air sources of elastic waves for marine seismics* Monograph (Krasnodar: KubSU) 313 p

[2] Kalinin A V, Azimi Sh A, Kalinin V V et al 1972 of immersion depth of pressure sparker source in seismoacoustic study in water areas In *Applied geophysics* iss. 65 (Moscow: Nedra) pp 84–95

[3] Kalinin A V, Kalinin V V and Pivovarov B L 1983 *Seismoacoustic studies in water areas* (Moscow: Nedra) 203 p

[4] Gulenko V I and Shumsky B V 2007 *Marine seismic survey technology in shallow water limit and transit zone Monograph* (Krasnodar: KubSU) 111 p

[5] Gontarenko I A 2012 Development of software tools for rapid simulation of seismic wave fields *Georesources* **1(43)** 15–8

[6] Shumsky B V 2008 *Technology of seismic studies of 2D MOV OGT in the limit shallow water and transit zones* Cand. Dissertation (Krasnodar: KubSU) 28 p

[7] Samsonov E A 2012 Comparative characteristic of high-resolution seismic survey (HRS) and ultra-HRS during engineering surveying in water areas *Proc. of the 39 session of the Int. Sci. Seminar named after Uspensky* (Voronezh, 30 January – 02 February) 246 p

[8] Shallow seismic vessel NIS Seeker-5 *Materials of JSC Sevmorneftegeophysics* Retrieved from: http://www.smnggeophysics. com/i/iskatel5-rus.pdf

[9] Johnston R C, Reed D H and Desler J F 1988 Special report on the SEG Technical Standards Committee: SEG Standard specifying marine seismic energy sources *Geophys.* **53(4)** 566–75

[10] Hatton L *Gundalf™ – an airgun array modelling package* Retrieved from: http://www.gundalf.com

[11] Hatton L 2008 The broadband acoustic output of marine seismic airgun sources *Proceedings of the Institute of Acoustics, Underwater Noise* (Southampton, England)

IOP Conf. Series: Materials Science and Engineering **905** (2020) 012031 doi:10.1088/1757-899X/905/1/012031

Resource-saving technology of ceramic brick production with nickel slags of Orenburg Region

V A Gurieva and A A Ilina

Modern Construction Materials and Technologies, Orenburg State University, Orenburg, Russia

E-Mail: victoria-gurieva@rambler.ru, ilyina.a06@gmail.com

Abstract. The paper considers the problem of using technogenic products in the production of construction materials thus obtaining ceramic products with specified properties both in Russia and abroad. It provides data on chemical and mineralogical composition of initial raw materials, defines pre-calcination and calcination properties of clay and nickel slag. Compositions with different percentage of clay/slag are developed and two-factor experiment is carried out, where the first factor is the percentage of slag in a sample, while the second factor is the sintering temperature of samples. The diagrams show the influence of material composition of clay/slag charge mixture on compression strength, water absorption and density of test samples after firing at a temperature of 900, 950, 1000 and 1050 °C. The obtained results allowed identifying the need for sintering of masses from clay of Khalilovsky deposit with nickel slags of OAO Yuzhuralnickel of Orenburg Region in the production of ceramic brick.

1. Introduction

Every day, huge amounts of raw materials are extracted worldwide and are used for processing and construction of industrial and residential buildings and structures thus spending kilowatts of electricity and non-renewable resources. Resource saving in all areas of human activity was one of the most key tasks of the modern industry. The construction industry, as one of the most material-intensive, needs resource-saving technologies. The paper studies the development of basic compositions of charge with nickel slags of Orenburg Region in the technology of ceramic brick production. These developments are necessary for the production of quality materials thus reducing the costs of extraction, transportation, processing of primary mineral raw materials, regulating their properties by introducing additives – various by-products (wastes) of industry.

Orenburg Region has the richest mineral reserves and fertile lands. The region has large deposits of gas, oil, ores, ferrous and non-ferrous metals, construction and decorative stone, ceramic clays, carbonate and silica rocks, brick and refractory clays, etc. Considerable reserves of mineral raw materials and fuel resources triggered the development of mining and processing industries [1]. Most of all non-ferrous metals in Russia are extracted in the Urals. The geographical location of Orenburg Region links it with developed non-ferrous metallurgy having large deposits of copper, nickel, aluminum, etc. The enterprises that develop such deposits besides non-ferrous metals produce tons of waste that are stored occupying huge areas and causing damage to the environment. Therefore, the problem of developing methods of recycling and reutilization of industrial waste is relevant for Orenburg Region. As a result of the processing of unused slags of nonferrous metallurgy into the production of construction materials the development of waste disposal technologies will increase the profitability of processing plants and provide the construction industry of Orenburg Region with affordable quality materials. The disposal of these wastes will minimize the impact of slag deposits on the environment, solve social and economic problems. The use of non-ferrous metallurgy wastes is one of the efficient ways of saving natural materials.

The lack of primary resources such as high-quality clays forces scientists around the world to develop ceramic brick production technologies using accumulated waste. The reduction of high-quality natural raw material reserves at the accumulation of technogenic products makes the problem of using wastes as secondary raw materials in production of construction ceramics relevant [2]. Scientists of different countries studied the use of copper-nickel ores, copper-nickel production slags, etc. as additives to ferro-chromium slag charge mixture, zeolite of apatite-nepheline ores mill tailings [3–10].

A great contribution to the development of technological principles for improving the quality of ceramic products was made by the following domestic scientists: A.I. Avgustinik, P.P. Budnikov, I.I. Moroz, R.Z. Rakhimov, L.L. Maslennikova, I.F. Schlegel, P.I. Bozhenov, etc.

V.Z. Abdrakhimov and his co-authors described the technology of ceramic brick production with the addition of phosphorus slag to the charge [3, 4].

Turkish researchers from Bartin University [5] studied the influence of ferro-chromium slag and zeolite and their various combinations on physical, mechanical properties and microstructure of bricks.

2. Methods and materials

Clays of Khalilovsky deposit and slags of OAO Yuzhuralnickel Plant (Kempirsaisky massif, Orenburg Region) were selected for the study.

The experiments were based on the standard methods to study compositions and technological properties of raw materials. Mineralogical and chemical analyses were performed to study compositions of clay raw materials and slag wastes. The reparation of raw materials, development of prepared mixtures, drying and burning of samples were carried out according to standard enterprise methods.

3. Results

Earlier studies on the production of ceramic bricks from clays of Alimsaisky deposit and nickel slags showed the prospect of using this kind of slag for the production of ceramic bricks [11].

Based on the results of the study the clay of Khalilovsky deposit was chosen for this experiment [11]. At the same time, the condition of haulage was assumed. The Alimsaisky clay deposit is located at a distance of 300 km from the dumps of nickel slags, in turn the distance to Khalilovsky deposit is only 40 km. Closer territorial location to slag dumps in case of good technological properties of clays will make it possible to significantly reduce the transportation costs for the delivery of raw materials in the future.

The Khalilovsky group of iron ore deposits is located in Orenburg Region on the eastern slopes of the Southern Urals (Figure 1). The total reserves of iron ores with 30–40 % Fe content exceed 310 million tons. Khalilovsky deposit is characterized by quite thick capping and adjacent rocks represented by clay rocks. Khalilovsky clay deposit is promising for use in ceramic brick production.

Until now, clay is the main component in the raw material charge for the production of ceramic products, so it is extremely important to assess its compositions and processing properties. The assessment of clay quality and its suitability for ceramic production in the course of the study was carried out on the basis of laboratory tests. At the first stage, chemical and mineralogical compositions of Khalilovsky clay deposit were analyzed. Figure 2 shows the chemical composition.

In terms of alumina content (Al_2O_3) the clay of Khalilovsky deposit belongs to the group of acid argillous raw materials. According to the content of iron oxides ($Fe_2O_3 > 3$ %) – to a group with high content of coloring oxides. The largest component in the clay under study is quartz, which is present in clay in bound and free form. SiO_2 share – 49.5 %.

The Okhotin diagram was used to define the type of clay of Khalilovsky deposit – loam (Figure 3).

Figure 1. Khalilovsky group of deposits, Gaisky district, Orenburg Region, Southern Ural, Russia

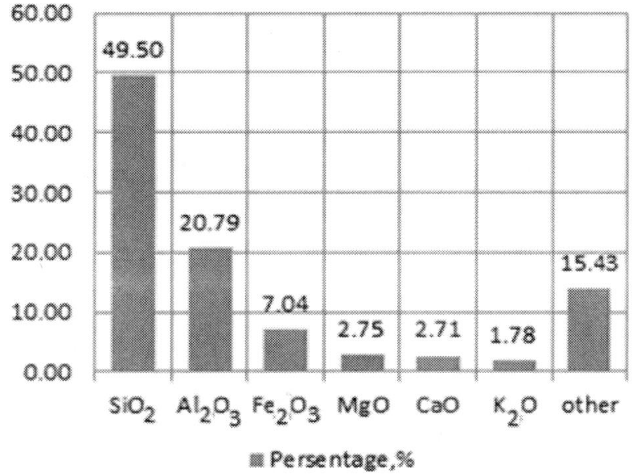

Figure 2. Chemical composition of Khalilovsky clay deposit

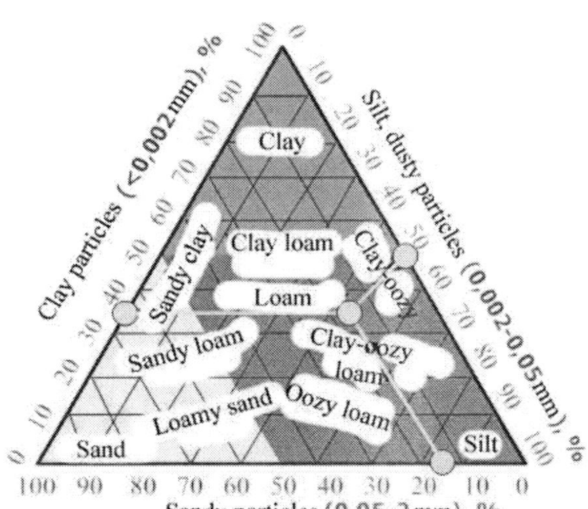

Figure 3. Okhotin diagram for clay of Khalilovsky deposit

The mineral composition of clay rocks was defined via X-ray diffraction method. The X-ray phase analysis of clay test samples of Khalilovsky deposit allowed establishing that the clay is polymineral, chlorite-illite-kaolinite with admixed quartz.

The samples were made from the clay of Khalilovsky deposit, and pre-calcination and calcination properties were studied [12]. The results are shown in Table 1.

Table 1. Pre-calcination properties of clay raw material

Pre-calcination properties of clay raw material	Relative molding-moisture content, %	Plasticity	Drying sensibility factor
	6.25	П=27 (high plasticity)	0.23 (low plasticity)
Calcination properties of clay raw material	Water absorption, %	Potsherd density, g/cm^3	Compression strength, MPa
	9.5	1.73	10.5

The use of technogenic dumps – unused non-crystalline nickel-containing slags of the South Ural Nickel Plant (OAO Yuzhuralnickel Plant) was studied as the second component of the charge. It is a large Russian non-ferrous metallurgy enterprise, whose share in the world nickel production is more than 1 %. It is located in Orsk, Orenburg Region.

Figure 3 shows the chemical composition of slag samples taken at OAO Yuzhuralnickel Plant.

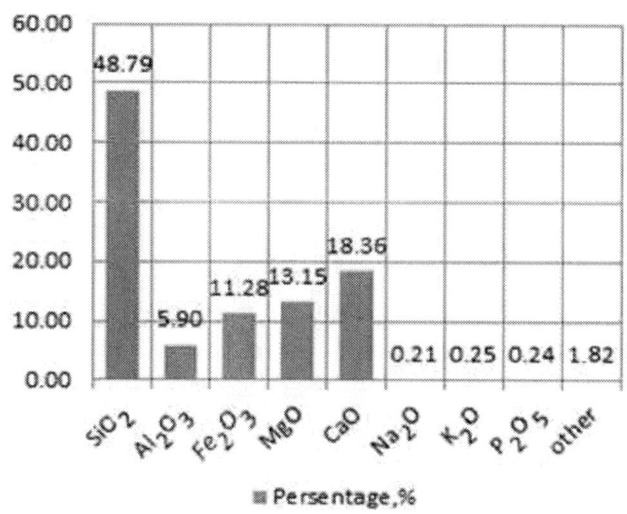

Figure 4. Chemical composition of slags of OAO Yuzhuralnikel Plant

The method of mathematical multi-factor analysis was used in the planning of the experiment. Clay composition from Alimsaisky deposit and nickel slags of OAO Yuzhuralnickel Plant were used as samples. At the first stage, the main factors influencing the properties of ceramic item are selected: X_1 – slag content in a two-component charge in the amount of 5–40 % and X_2 – firing temperature, 900–1050 °C.

Based on clay plasticity (P = 27) the method of sample forming was adopted – plastic (Figure 4).

According to the design matrix 4 compositions were developed: composition 1 (slag – 5 %, clay – 95 %, burning temperature – 900 °C), composition 2 (slag – 40 %, clay – 60 %, burning temperature – 900 °C), composition 3 (slag – 5 %, clay – 95 %, burning temperature – 1050 °C) composition 4 (slag – 40 %, clay – 60 %, burning temperature – 1050 °C).

The tests were carried out on samples measuring 2x2 cm, 6 samples of each composition in a series of tests.

In order to carry out the experiment and process the obtained data, a mathematical model was obtained in the form of a second order polynomial for each characteristic of an item.

$$Y = a_0 + a_1X_1 + a_2X_2 + a_3X_1^2 + a_4X_2^2 + a_5X_1X_2, \quad (1)$$

where a_0; a_1; a_2; a_3; a_4; a_5 – beta coefficients; X_1 – slag content in charge, %;. X_2 – firing temperature, °C.

a)

b)

Figure 5: a) as-formed samples; b) samples dried for 48 hours at 20 % relative humidity and 22 °C

Compression strength, MPa, water absorption, %, average density, g/cm³ are used as optimization parameters. The selection of these parameters is based on their significant influence on structure formation of ceramic products and operational properties, which ensures the duration of ceramic operation under fixed conditions [13].

After processing the results, the beta coefficients were determined (Formulas 2–4) and the mathematical models of the optimization parameters were constructed (Figures 6–8).

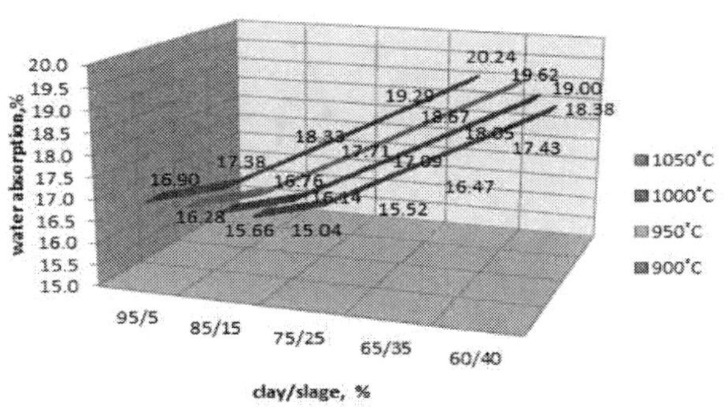

$W=17.4-1.43x_1-0.94x_2$(2)

Figure 6. Change of water content depending on clay/slag ratio

When calculating the compression strength of the samples by the volume of standard size brick, we took 60 % of the volume specified on the diagram (Figure 8).

4. Conclusion

The analysis of data in the diagrams (Figures 6–8) shows that the increase in the slag content of the charge leads to the increase in water absorption. At the lowest studied temperature equal to 900 °C, water absorption increases with the growth of slag fraction in a charge from 16.9 to 20.24 %.

At a temperature equal to 1050 °C the slag content in a charge increases from 15.04 to 18.38 %. In general, water absorption in experimental samples exceeds the recommended technological interval, which for common brick makes 12–14 %, for ashlar brick – 8–10 %.

Besides, the increase of slag content in a charge contributes to the reduction of the following parameters: compression strength, average density. This is caused by the fact that at a firing temperature of 900–1050 °C the fracture density of slag increases thus increasing its effective porosity. Structural changes in slag composition lead to the fact that with the increase in the amount of technogenic raw material (nickel slag) in a charge regardless of the clay component properties the effective porosity and water absorption of a ceramic product as a whole increase.

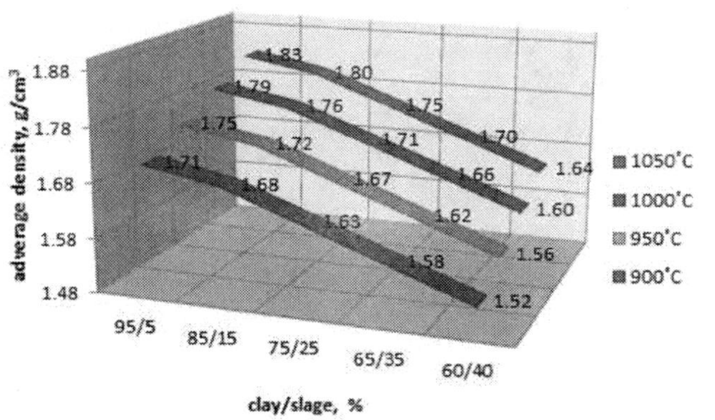

$p=1.69+0.08x_1+0.06x_2$(3)

Figure 7. Change in average density depending on clay/slag ratio

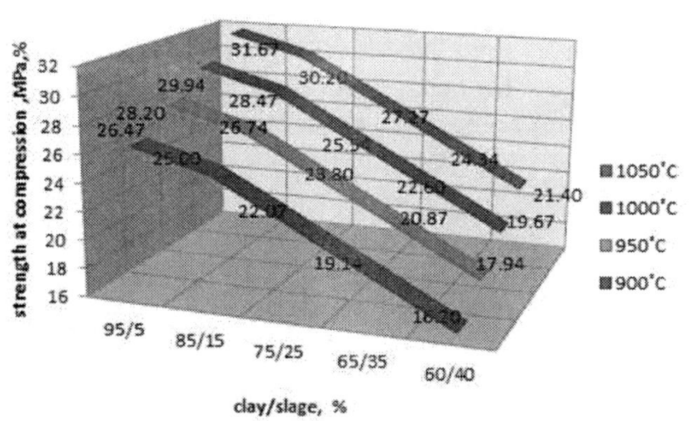

$Rs=24.67+4.4x_1+2.6x_2$(4)

Figure 8. Change in compression strength depending on clay/slag ratio

These studies serve as the basis for further development of ceramic brick production technology with the use of clay raw materials and technogenic wastes – nickel slag, and ensure the production of a rational charge with directional structure formation of ceramic brick.

References

[1] Turchaninov V I 2017 Building materials from technogenic raw materials: a tutorial (Orenburg: OGU) 207 p
[2] Kolodina O A 2003 Metallurgyofthe Orenburg Region *Geography* **43** 27–30
[3] Abdrakhimov V Z and Abdrakhimova E S 2014 The use of slag from coal combustion in the Kansk-Achinsk basin in the production of ceramic materials based on inter-shale clay *Ecol. and Industry of Russ.* 36–9

[4] Abdrakhimov V Z 2008 Ecological, theoretical and technological principles of the use of phosphoric slag and slag material in the production of high-quality ceramic bricks (Novosibirsk) pp 20–4
[5] Gencel O, Sutcu M, Erdogmus E and Koc V 2016 Cay Properties of bricks with waste ferrochromium slag and zeolite *J. of Cleaner Product.* 112–3
[6] Prokofieva V V 1986 The *Use of Associated Iron Ore Enrichment Products in Construction in the North* (Leningrad) 176 p
[7] Bozhenov P I 1994 *Complex use of mineral raw materials and ecology* (Moscow: Publ. House of the Associat. of construct. Univer.) 263 p
[8] Monteiro Fontes S N and Vieira C M 2014 On the production of fired clay bricks from waste materials: A critical update *Construct. and Build. Mater.* **68** 599–610
[9] Welington L F, Reis E L and Lima R M F 2015 Incorporation of residues from the minero-metallurgical industry in the production of clayelime brick *J. of Cleaner Product.* 505–10
[10] Augustinik A I 1974 *Ceramics* (Leningrad: Stroizdat) 592 p
[11] Gurieva V A and Ilyina A A August 2019 The Manufacturing of Structural Ceramics with the Addition of Non-Ferrous Slags *Atlantis Highlights in Material Sciences and Technology* vol 1 Proc. of the Int. Symp. Engineer. and Earth Sci.: Appl. and Fund. Res. Dedicat. to the 85th anniversary of H I Ibragimov (ISEES 2019) pp 174–78
[12] Pavlov V F 1977 Physico-chemical bases of firing of products of construction ceramics (Moscow: Stroizdat) 240 p
[13] Svatovskaya L B, Maslennikova L L, Slavina A M et al *The mixture for the manufacture of ceramic bricks* RF Patent No 241213 Byul no 5, 2011
[14] Kulchitsky L I and Usyarov O G 1981 *Physico-chemical bases of formation of properties of clay rocks* Monograph (Moscow: Nedra) 178 p ISBN: 5-262-00159-9

IOP Conf. Series: Materials Science and Engineering **905** (2020) 012032 doi:10.1088/1757-899X/905/1/012032

Analysis of carbon fiber destruction by lighting current

I A Guschin and T N Erina

Chuvash State University, 15, Moscow ave., Cheboxary, 428015, Russia

E-mail: elpardon@gmail.com

Abstract. The paper considers the unique properties of conducting carbon fiber composite: specific resistivity at different directionality of layers and critical energy of layering and destruction on the basis of experimental data. High mechanical strength and low specific gravity are noted, allowing its use in aircraft engineering to improve aircraft tactical and technical data of aerial vehicles. Strong anisotropy of electrically-conductive properties affects current spreading through carbon fiber and leads to significant destruction of the material. Theoretical analysis of lightning current spreading along anisotropic-conducting material in the form of extended plate was carried out on the basis of two proposed models of carbon fiber destruction by lightning currents. For the first continuum model an exact analytical solution to the Laplace equation with Neumann boundary conditions via the Greenberg method was found. The criteria for composite destruction was found, and the radius and depth of destruction of a conductive material by lightning current were calculated. Layer-by-layer destruction is considered on the basis of carbon fiber equivalent circuit. The strong anisotropy of the material resulting in the release of full energy in the first layer is taken into account. It is shown that the destruction of the top layer changes the distribution of currents over the remaining layers. The results of numerical modeling of layer-by-layer destruction of carbon fiber for different number of layers are given. The conclusion on the application of criteria to forecast the lightning effects and optimize lightning at the stage of aircraft design is made.

1. Introduction

Carbon fiber as a conducting composite material is widely used in many industries, including aircraft engineering. It is caused by its unique properties – high mechanical strength and low specific weight considerably improving flight characteristics of aerial vehicles. The application of nanotechnologies allows influencing hardness and strength, thermal conductivity and electrical conductivity of materials. However, with increasing use of carbon fibers in aircraft engineering, they become more vulnerable to lightning than solid metal structures and suffer extensive damage [1–4].

In order to increase flight safety and lightning resistance of carbon fiber products there is a need to conduct experimental studies of their electrical conductivity affected by pulse currents simulating lightning discharge, analysis and assessment of material destruction, creation of models of lightning current spreading along conducting material and development of criteria for its destruction. The solution of these tasks will make it possible to forecast lightning effects at the stage of aircraft engineering and to choose lightning protection methods.

The electrical conductivity of carbon fiber and its destruction are sufficiently studied. A wide variety of carbon fibers resulted in a wide spread of electrophysical characteristics, but on average the main properties for the construction of destruction models and selection of structural material were determined. Such characteristics include longitudinal and cross resistance of carbon fiber, critical energy of layering and destruction [5]. Table 1 shows the values of these parameters.

Table 1. Electrophysical characteristics of carbon fiber

Longitudinal resistance Characteristic		Value
Specific resistivity, Ohm·cm	Unidirectional fibers, ρ_0	$(1–2)10^{-3}$
	at an angle of 45°, ρ_{45}	$(2–4)10^{-3}$
	at an angle of 90°, ρ_{90}	$(2–4)10^{-3}$
	cross, ρ_{tr}	10–100
Critical energy of layering, W_{cr1}, J/cm³		**$1.2\cdot10^3$**
Critical energy of layering, W_{cr2}, J/cm³		**$2\cdot10^3$**

It was found that the greatest contribution to the destruction is made by the pulse component of the lightning current determined by the integral action of the current. Getting into carbon fiber the Joule heating destroys it. The aggressive heating of the binder forms gases with pressures capable of causing destruction of a material. On the basis of the available data on conductive properties of carbon fibers and energies leading to layering and destruction of a material it is possible to build models of lightning currents spreading along the conductive composite material, find criteria of material destruction, compare models and choose the most promising for further theoretical analysis of carbon fiber destruction and increase of lightning resistance of products from this material.

2. Methods and materials

For the analysis of lightning current spreading along carbon fiber two models are considered: continuum model and layered structure model. The first – continuum model – takes into account the current flow in longitudinal and cross direction in case of strong anisotropy of electrically-conductive properties of carbon fiber. The model of such a medium represents an extended plate of R radius from carbon fiber with thickness H, which includes a lightning current of the channel radius r_{ch}, spreading along the plate to the outer ring electrode at a distance R from the center of the plate. The distribution of current density within a circumference was considered as uniform.

For this medium, it is valid to consider the Laplace equation with Neumann boundary conditions, which has an exact analytical solution. It was found via the Greenberg method and included the Bessel functions [6]. Similar problem is solved for coaxial arrangement of elements opposite to each other. The case of $R\rightarrow\infty$ considers infinitely extended plate with thickness H. The densities of cross and longitudinal currents $j_r=-\varphi_r/\rho_r$ and $j_z=-\varphi_z/\rho_z$, allowing analyzing the distribution of currents in the anisotropic conductive medium, are found from potential distribution φ (r, z). The correlation with the experiment is achieved by defining the longitudinal $Rr = [\varphi(r_{ch},0) - \varphi(R, 0)]/I_l$ and cross resistance $Rz = [\varphi(r_{ch},0) - \varphi(r_{ch}, H)]/I_l$ of passing lightning current I_l.

In order to understand the complete picture of destruction, it is necessary to consider the model taking into account the layered structure of the real carbon fiber [7]. This can be done using a series-parallel composite equivalent circuit (Figure 1).

The resulting system of equations was solved by the Gauss method. At the same time the quantity of elementary longitudinal ΔRr and cross resistance ΔRz was chosen final and equal accordingly to N and n = nlay – 1 with the number of layers of the material nlay. From nN equations longitudinal and cross currents were found, from which the analysis of lightning current spreading was carried out. At ΔRr, ΔRz tend to zero, the results of calculations for both models coincided and are given in Section III.

For practical calculations, the concept of equivalent fracture depth Ze (similar to the depth of electromagnetic radiation penetration into the conductive medium) is introduced to determine the degree of material destruction. It is found from the equality of the energy released in the equivalent layer of thickness Ze at the maximum current density in the first layer and the total energy released in all layers through which the unevenly distributed current flows:

$$z_э/d = \sum_{i=1}^{n} \frac{j^2{}_r(i)\rho_r + j^2{}_z(i)\rho_z}{j^2{}_r(1)\rho_r + j^2{}_z(1)\rho_z}$$

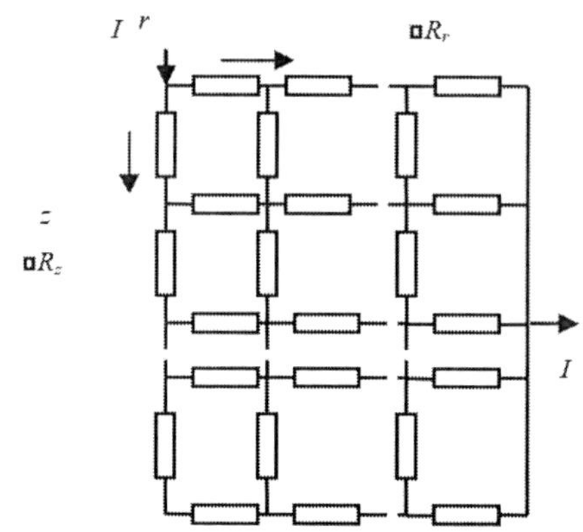

Figure 1. Carbon fiber equivalent circuit with discrete conductivity

Then we face the matter of choosing a particular model for specific calculations and ways to increase the lightning resistance of a composite material at the stage of aeronautical design engineering. The necessary comparative analysis of calculations for the two proposed models is carried out in the next section.

For experimental confirmation of the chosen models of destruction the carbon fiber samples were produced and longitudinal and cross resistance at lightning current transmission were defined. Cylindrical electrodes were brought in contact in the middle part of a sample transversely to the layers on both sides. The simulation of discharge channel of rch radius took into account one electrode, and the second – structural conducting object under carbon fiber skin. The experiment used cylindrical electrodes of diameter D from 4 to 50 mm. The general parameters for the two models (radius of the lighting channel rch, thickness (number of layers) H (nlay), specific resistance ratio ρr/ρz) fully corresponded to the experiment. At the same time, the error of calculation with the experiment did not exceed 15 % (Table 2).

Table 2. Experimental validation of calculations

D, mm	R, Ohm	Longitudinal resistance Experiment, R_{av}, Ohm	Number of samples	Variation, %
4	0.052	0.050	4	15
20	0.024	0.023	4	12
		Cross resistance		
4	0.202	0.185	4	10
50	0.173	0.178	4	10

3. Results

Calculations of current density distributions make it possible to draw the conclusion on the nature of current spreading in a plate with anisotropy of conductivity. Thus, with a large relationship between cross and longitudinal specific resistivity ρz/ρr, the current will mainly flow in the narrow surface layer, and the resulting current density in the vicinity of the channel radius will be determined by the radial component of the current density jr. Figure 2 shows dependencies of reduced current density jr/jm from parameter r/R at different ratios z/H (H/R = 0.01). The data clearly show that the maximum current densities are located near the radius of the channel. With the increase of r/R and z/H, the distribution of the radial component of the current density is equalized.

Figure 3 shows the dependence of *jz/jm* on *z/H* at different *r/R*. It follows from the comparison of the figures that in calculating the resulting current density, the cross component *jz* can be neglected. The current density *jz* decreases rapidly with increasing thickness of a composite, resulting in highly uneven current distribution across the material near the discharge channel.

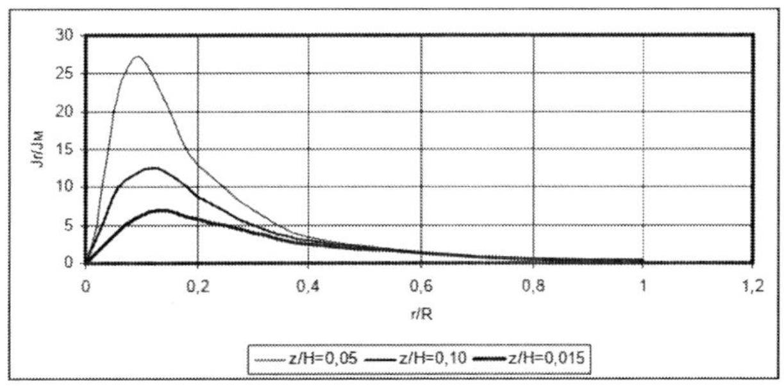

Figure 2. Distribution of *jr/jm* along radius *r/R* at different ratios *z/H*.

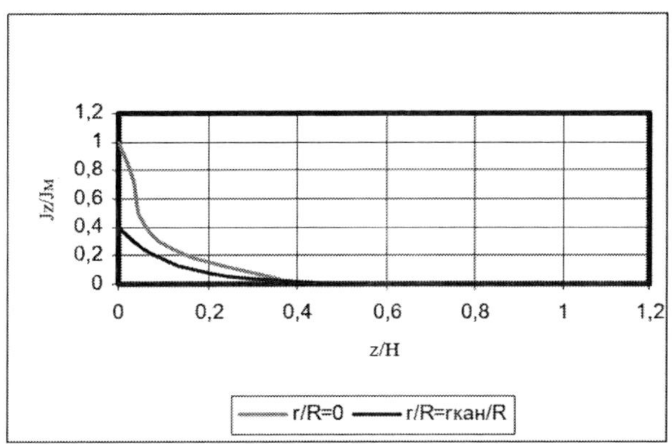

Figure 3. Distribution *jz/jm* along thickness *z/H* at different *r/R* (r_{ch} / R = 0.1).

Based on the calculation of current densities in solid anisotropic-conducting medium with thickness H, it is possible to calculate the radius and depth of material destruction. The destruction criterion is the specific energy input per unit of time, for which the following expression is true:

$$W_{rel} = (j_r^2\, \rho_z + j_z^2\, \rho_r) \cdot A/(2 \cdot I_m^2),$$

where A – integral of lightning current action; I_m – maximum current in a pulse.

Hence, the current densities resulting in the release of a given specific energy W_{rel} lie on the ellipse. Let us estimate the depth and radius of composite destruction by constructing isoenergetic curves in coordinates (r, z) at preset values of lightning channel radius r_{ch}, A, ρ_z/ρ_r (Figure 4).

Figure 4 shows the results of calculation of isoenergic curves for carbon fiber at r_{ch} = 1 cm, A = $0.6 \cdot 10^6$ A^{2}c and two values ρ_z/ρ_r. At $\rho_z/\rho_r = 10^3$, the destruction radius and depth will make respectively $r_p = 2.5\, r_{ch}$ and $z_p = H/3$, and at $\rho_z/\rho_r = 10^4$ $r_p = 3\, r_{ch}$ and $z_p = H/4$. Hence, the non-uniformity of current spreading due to severe anisotropy results in painful destruction of the outer layers.

The calculations taking into account the layered structure of carbon fiber showed that the distribution of current density weakly depends on the number of layers, which is caused by the strong anisotomy of the conductive properties of carbon fiber, namely at ρ_z / ρ_r >>1. For the case of strong anisotropy ρ_z / ρ_r >>10^4 it can be considered that the energy release near the channel occurs only in the first layer and there is a layer-by-layer breakdown of the material. The distribution of equivalent depth of destruction at various ρ_z/ρ_r is shown in Figure 5 and 6.

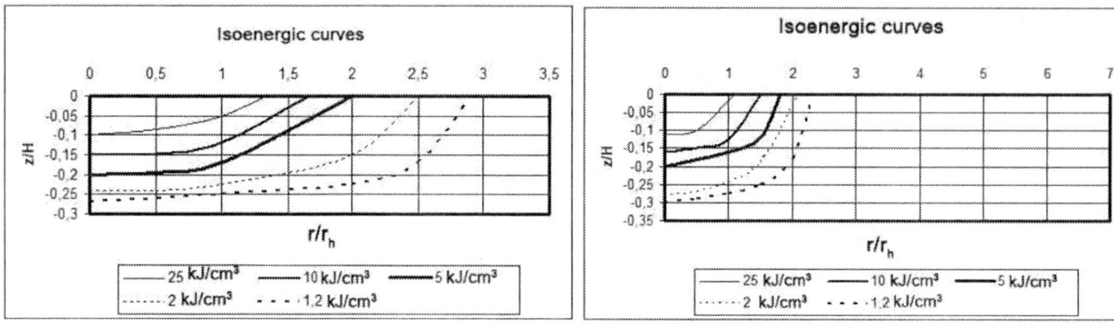

Figure 4. Isoenergic curves at $A = 0{,}6\cdot 10^6 A^2c$ and $\rho_z/\rho_r = 10^4$(a) and $\rho_z/\rho_r = 1\ 0^3$(b)

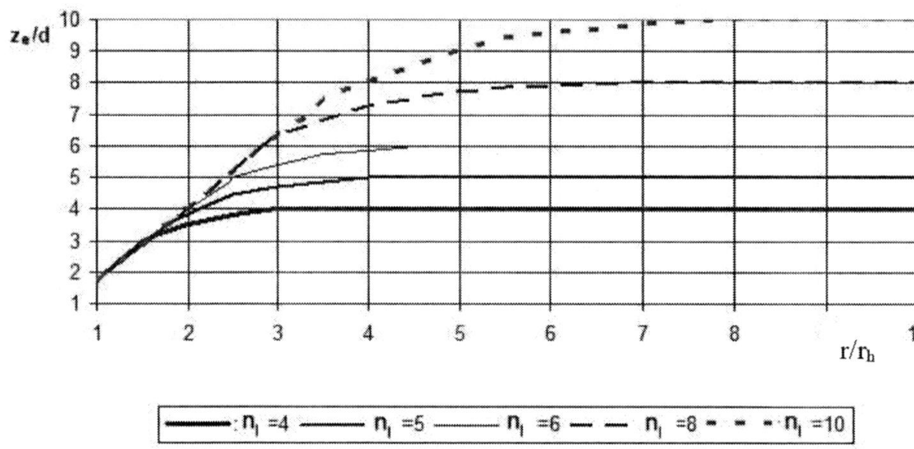

Figure 5. Distribution of relative depth of energy density z_e/d along radius r/r_{ch} for different number of layers at $\rho_z / \rho_r = 10^3$

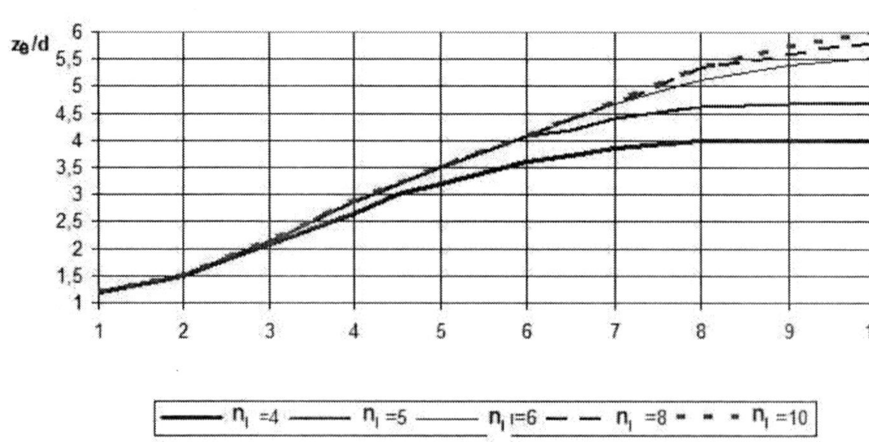

Figure 6. Distribution of relative depth of energy density z_e/d along radius r/r_{ch} for different number of layers at $\rho_z / \rho_r = 10^4$

For example, Figure 5 shows that the relative radius at which the current density is aligned with the layers, i.e. $z_e = n_{lay}$ d, can be expressed as follows:

$$r_e/r_{ch} = n_{lay} - 1$$

It follows from the continuum model that in the real material the current density across the layer thickness is uneven. In other words, each individual layer has anisotropy properties typical for the entire material.

In a model taking into account the layered structure of the material, the current density was considered to be uniformly distributed within each layer. At a small layer thickness (0.01cm) this assumption is quite justified. Figure 6 shows the comparison of calculation results regarding both models, in particular for j_r (r) and j_z(r) at n_{lay}=10, $\rho_z/\rho = 10^4$, r_{ch}=1 cm, R=10 cm. The solid line shows the graphs for the continuum, and the dash-and-dot line is used for the layered medium.

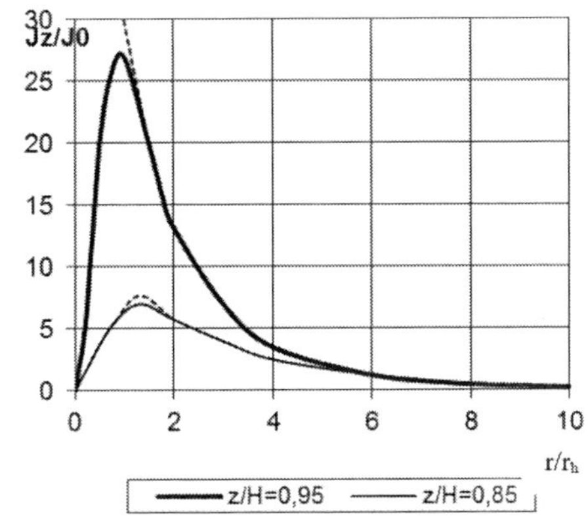

Figure 7. Distribution of radial current density j_r/j_m by radius r/r_{ch} at different ratios z/H

4. Conclusion

Thus, both models made it possible to construct the destruction curves and forecast the degree of destruction, to obtain expressions for destruction characteristics (radius, depth, and integral of destructive current action), but the layered model proved to be more consistent with the layer-by-layer destruction mechanism. Thus, it becomes possible to consider the dynamics of destruction over time, which does not allow making the static models. The proposed model of current spreading through the conducting composite material taking into account its layered structure makes it possible to forecast the degree of destruction of the real layered carbon fiber material without the experiment and reasonable selection of the necessary structural material.

Thus, the results of the calculations for two independent models of current spreading along carbon fiber fully coincided with the experiment with the same initial data. This confirms the correctness of theoretical conclusions and the possibility of its application to forecast the consequences of carbon fiber destruction by lightning currents. The simplicity of analysis of the numerical method for the layered structure of the material compared to the continuum makes this model more cost-efficient and suitable for the analysis of material destruction. Of all tested models, the layer-by-layer destruction mechanism most reflects the real processes that occur in carbon fiber when it is exposed to lightning channel. The dynamics of layer-by-layer destruction over time will be considered in the following works.

References

[1] Chemartin L, Lalande P, Peyrou B et al December 2012 Direct Effects of Lightning on Aircraft Structure: Analysis of the Thermal, Electrical and Mechanical Constraints *J. Aerospace Lab.* **5** 1–15

[2] Chemartin L, Lalande P, Montreuil E et al Feb. 2009 Three Dimensional Simulation of a DC Free Burning Arc. Application to Lightning Physics *Atmospheric Res.* **91** 371–80
[3] Madsen S F and Carloni L June 2015 Lightning exposure of Carbon Fiber Composites in wind turbine blades *J. of Lightn. Res. (24th Nordic Insulation Symposium on Materials, Components and Diagnostics*) pp 105–10
[4] Madsen S F 2012 Proposal of New Zoning Concept Considering Lightning Protection of Wind Turbine Blades *J. of Lightn. Res.* **4** 108–17
[5] Vishnyakov L R, Kokhanyi V A, Chernyavsky I I, Kokhanaya I N et al 2017 Impact of nanomodification and surface reinforcement of carbon fiber by bonded-soldered grids on lightning resistance of composite *Technol. Syst.* **4** 29–39
[6] Guschin I A 2016 Analytical model of current flow by conducting composite material *J. of Chuvash Univer.* **3** 30–5
[7] Guschin I A 2017 Accountin of the layered structure of the composite in the model of lightning current flow *J. of Chuvash Univer.* **3** 23–9

Aggregation properties of p- and n-functionalized calix[4]resorcine in water/oil emulsion in the presence of sulfate reducing desulfobacter bacteria

E Guseva[1], M R Ziganshina[1], D R Isakov[2] and V A Mendelson[1]

[1] Kazan National Research Technological University, 10, K.Marx St., Kazan, Tatarstan 420111, Russia
[2] Kazan Federal University, Kazan, 18, Kremlyovskaya str., Kazan, 420008, Russia

E-mail: leylaha@mail.ru, zigmay4@mail.ru, geofak@kpfu.ru, vallery78@mail.ru

Abstract. This paper presents a comparative study of the aggregation properties of P- and N-functionalized calix[4] resorcines in "water/oil emulsion + Desulfobacter culture". It has been shown that both compounds under study displayed aggregation properties in low concentration range. Nevertheless, their structural features affect the aggregation process. Moreover, the beginning of aggregation of N-functionalized calix[4]resorcine with petroleum hydrocarbons molecules and microorganism cells shifted to lower concentration zones than the one of P-functionalized calix[4]resorcine. However, P-functionalized calix[4]resorcine was characterized by a wider interval at which the structuring of the water/oil emulsion is observed.

1. Introduction

It is well known that interfacial phenomena are essential in petroleum extraction and recovery. The combination of hydrophilic and hydrophobic groups in surfactants lead to widely regulate their properties [1–3]. In general terms, Calix[4]resorcines can be defined as macrocyclic cavitary systems with different interesting properties [4–6]. Thus, the presence of hydrophilic and hydrophobic groups in their composition, as well as the possibility of functionalization their lower and/or upper rims with functional fragments may allow them to acquire surface-active properties (surfactants) [7], which may be of great importance in oil biodegradation processes for controlling the reactivity of substances. Previously, the aggregation properties of P- and N-functionalized calix[4]resorcines I-II (Figure 1) were studied on model systems that mimic the composition of hydrocarbons that are found in larger quantities in oil [7, 8]. The purpose of this work is to compare the characterization of the aggregation properties of P- and N-functionalized calix[4]resorcines I–II in a system of "50 % OFWW (oil field wastewater) – 50 % oil – 10 % of a Desulfobacter culture medium".

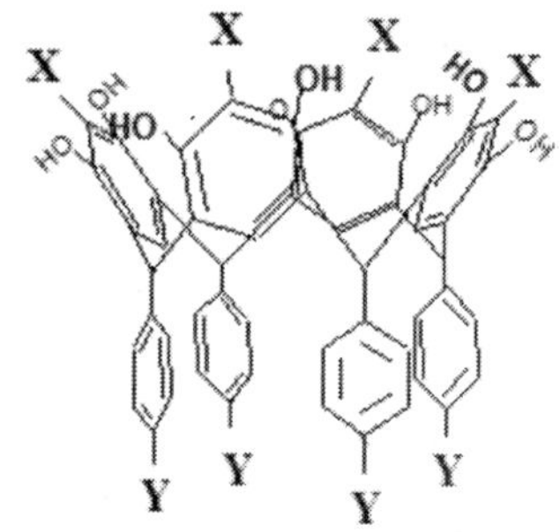

compound I: X = CH_2-$N(CH_3)_2$; Y = CH_3.

compound II: X = H; Y = P(O)(OEt)$_2$

Figure 1. Structural formulas of compounds **I-II**

2. Methodology

In order to compare the aggregation properties characteristics, N-functionalized calix[4]resorcine, substituted on the upper rim of the molecule by aminomethyl fragments [9] and P-functionalized calix[4]resorcine, substituted on the lower rim of the molecule by aryldiethoxyphosphoryl fragments [10] were taken as objects of study.

During the research process we created a system of "50 % OFWW (oil field wastewater) –50 % oil –10 % culture medium Desulfobacter" in which compounds I or II were introduced in various concentrations. In parallel, control experiments were carried out without the addition of compounds I and II. Further, the system "50 % OFWW (oil field wastewater) – 50 % oil – 10 % culture medium Desulfobacter" is designated as "water/oil emulsion- Desulfobacter".

The aggregation ability of P- and N-functionalized calix[4] resorcines was determined according to the composition-properties diagrams using two independent methods:

- conductometric method based on measuring electrical conductivity of the studied compounds solutions G = f(C);
- spectrophotometry method based on measuring the optical density of the studied compounds solutions A = f (C).

All measurements were taken at 25 °C. The temperature was maintained using a thermostat with an accuracy of ±0.1°C. The concentration of the studied compounds varied within 10^{-3}–10^{-6} mol/l. Solutions of the studied compounds were prepared immediately prior to studies using the dilution method. The electrical conductivity of the studied compounds solutions was measured on an LM-301 conductometer (standard cell LM-3000). The measurement of the optical density of the compounds solutions was carried out on an SF-16 instrument at wavelengths of λ = 280 nm and λ = 288 nm, which characterize the presence of a calix-resorcinal structure [11].

3. Results and discussions

The change in electrical conductivity of "water/oil emulsion-Desulfobacter" system when adding compounds I–II d (Figure 2–3) shows that the aggregation of both compounds occurs in 2 stages. At CI = $0.012 \cdot 10^{-3}$ mol/l and CII= $0.023 \cdot 10^{-3}$ mol/l (0.03 g/l), the dependences G=f (C) show a clear bend, indicating a change in the physicochemical properties of the systems and the beginning of their structuring process. The intense bending next point of the graph corresponds to CI = $0.093 \cdot 10^{-3}$ mol/l (0.1 g/l) and CII = $0.184 \cdot 10^{-3}$ mol/l (0.25 g/l), at which form new bonds with micellar system. The change in optical density in the "water/oil emulsion-Desulfobacter" system when adding compound I also indicates that the structure of the solution changes at 2 points (Figure 4–5). Thus, compound I-II exhibits aggregation properties at low concentrations.

Both compounds can be attributed to ionic surfactants, in which already at low concentrations an intense increase in electrical conductivity occurs, apparently due to an increase in the number of protonated calixresorcines molecules to a bend corresponding to the critical concentration of association. In the area of these concentrations, spontaneously form associates by the head-to-tail type of electrostatic mechanism, most likely, in the form of ionic micelles. After this, the mobility of the ions decreases and, consequently, the conductivity will decrease.

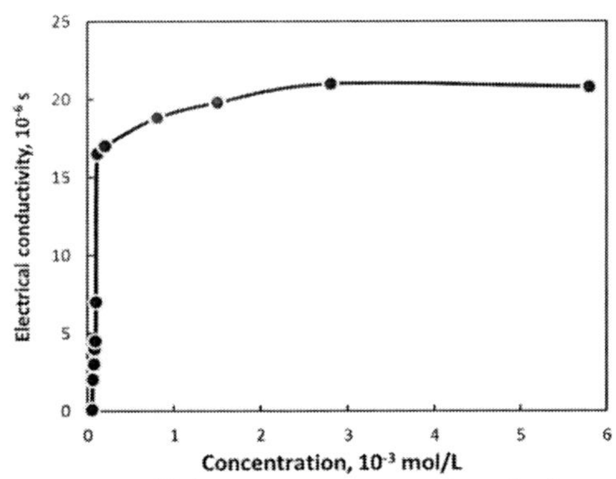

Figure 2. "water/oil emulsion – Desulfobacter" system electrical conductivity dependence on the concentration of the compound **I**.

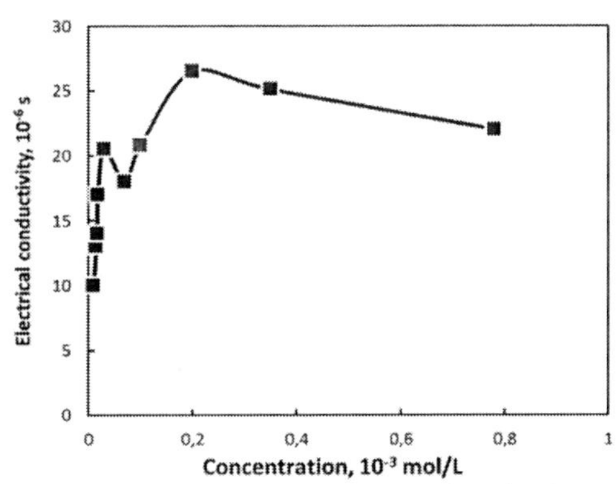

Figure 3. "water/oil emulsion – Desulfobacter" system electrical conductivity dependence on the concentration of the compound **II**.

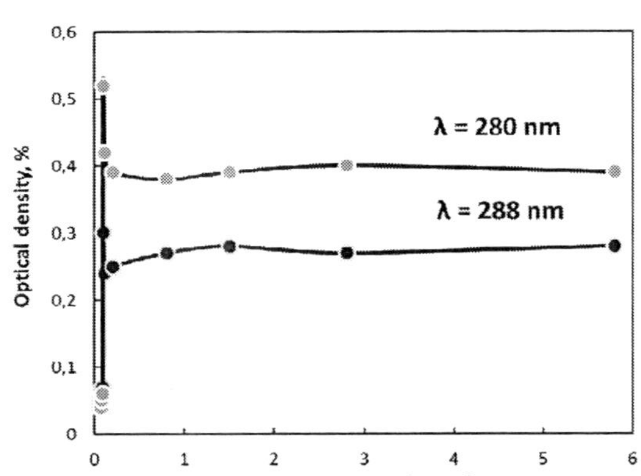

Figure 4. "water/oil emulsion – Desulfobacter" system optical density dependence on the concentration of the compound I.

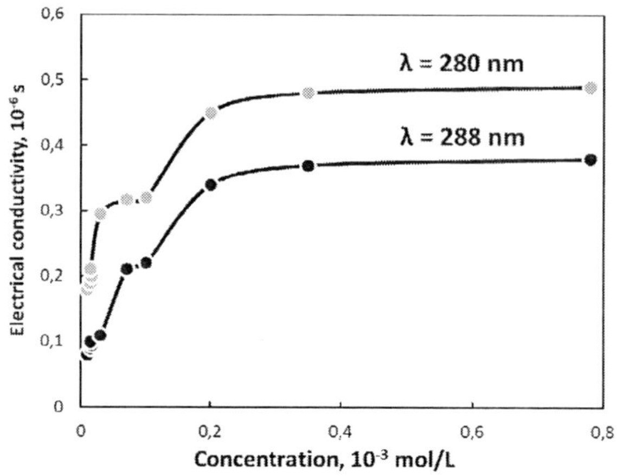

Figure 5. water/oil emulsion – Desulfobacter" system optical density dependence on the concentration of the compound II.

However, for compound I, the process of changing the physicochemical characteristics of the solution occurs at lower concentrations. Obviously, the presence of a lone electron pair in nitrogen atom of the amino groups will contribute to the formation of hydrogen bonds with hydroxyl groups, turning compound I into a zwitterion [12]. Protonated forms of amino groups lead to the formation of a charged lower rim, which facilitates association. In addition, the presence of a lone electron pair in nitrogen atom of the amino groups will contribute to the formation of bonds with the constituent parts of oil that carry acceptor groups, as well as with negatively charged cells of microorganisms. Obviously, therefore, the concentration at which the structuring of the solution begins in the presence of compound I should be lower.

In compound II, the localization of a positive charge in phosphorus atom transforms it into a cation-active surfactant, which will bind the negative charge of sections of the molecules of petroleum hydrocarbons. The upper hydrophobic resorcinal part of molecule rim will encapsulate those parts of the oil molecules or individual molecules of oil fractions that are compatible with it in size, as well as the cells of microorganisms. Obviously, the encapsulation process requires higher concentrations of compound II. In this case, hydroxyl groups can also interact via hydrogen bonds.

4. Conclusion

We have presented a comparative study of the aggregation properties of P- and N-functionalized calix[4]resorcines in the "water/oil emulsion + Desulfobacter culture" and have demonstrated that both compounds reveal aggregation properties in low concentration range. However, the concentration at which the structuring of the solution begins in the presence of compound I is lower, which is explained by structural features and different ways of interacting with molecules of petroleum hydrocarbons and microbial cells of compounds I–II.

Acknowledgment

The work is performed according to the Russian Government Program of Competitive Growth of Kazan Federal University and for the state assignment in the sphere of scientific activities.

References

[1] Busu T N Z T M, Norasikin S, Mohtar S S, Noor A M M et al 2019 An evaluation of lignocellulosic solutions from OPEFB pulping process as demulsifiers for crude oil emulsion demulsification *Petrol. Sci. and Technol.* **37** 1–8

[2] Pu W, Chao S, Shishi P et al 2019 Viscosity reduction of acidic heavy oil through emulsification: effects of salinity and pH *Petrol. Sci. and Technol.* **37** 402–8

[3] Sainath K and Ghosh P 2013 Stabilization of silicone oil-in-water emulsions by ionic surfactant and electrolytes: the role of adsorption and electric charge at the interface *Industr. and Engineer. Chem. Res.* **52** 15808–16
[4] Guseva E V, Morozov V I, Zinkicheva T T et al 2012 Complexes of Rhodium(III) with N-Functionalized Calix[4]Resorcinolol Russian *J. of General Chem.* **82** 1323–33
[5] Guseva E V, Buslaeva T M, Grishin E I and Zinkicheva T T 2015 Influence of Solvents on Protolytic and Aggregation Properties of Aminomethylated Calix[4]resorcine *Macroheterocycles* **8** 415–23
[6] Guseva E V, Buslaeva T M and Polovnyak V K 2016 Rhodium complexation with phosphoryl-containing calix[4]resorcine *Russ. J.of Inorganic Chem.* **61** 1436–44
[7] Kutlahmetova A R and Guseva E V 2015 The qualities of N-fuctionalized Calix[4]resorcines *Bull. of the Technol. Univer.* **18** 7–11
[8] Guseva E V and Kutlahmetova A R 2014 The qualities of p-fuctionalized Calix[4]resorcines. Part 1. Aggregation qualities *Bull. of Kazan Technol. Univer.* **17** 7–10
[9] Shatalova N I, Sidorov N A, Gavrilova E L and Krasilnikova E V 2007 Synthesis of Aminometilorized Calix[4]resorcines *Bull. of Kazan Technol. Univer.* **3** 40–3
[10] Gavrilova E L, Naumova A A, Shatalova N I et al 2008 The New Type of Calix Resorcines Bearing Phosphonates and Phosphonium Fragments at The Lower Rim *Phosphorus. Sulfur. and Silicon* **183** 561–5
[11] Fegtle F and Weber E 1988 *Chemistry Complexes of Host-Guest* (Moscow: Publ. House MIR)
[12] Yanilkin V V, Ryzhkina I S, Nastapova N V et al Single-electron oxidation and nucleophilicity of aminomethylated calix[4] resorcinarenes *Russ. Chem. Bull.* **52** 1142–9

IOP Conf. Series: Materials Science and Engineering **905** (2020) 012034 doi:10.1088/1757-899X/905/1/012034

Optimization of the regeneration of the sodium-cation ion-exchange filter

M Yu Kalinichenko[1], N I Stojanov[1], D V Abornev[1] and S V Ovchinnikova[2]

[1] North-Caucasus Federal University, 1, Pushkin Str., Stavropol, 355017, Russia
[2] Kuban State Agrarian University, 13, Kalinina Str., Krasnodar, 350044, Russia

E-mail: mix2710@mail.ru, cnjzyjd1949@mail.ru, a.deniss@mail.ru, svetlana.swetlana-ov@yandex.ru

Abstract. A new methodology has been developed for determining the volumetric coefficients of internal mass transfer according to the results of an experimental study. The values of volumetric coefficients of internal mass transfer were calculated and their dependence on the technical characteristics of the regeneration process of the sodium-cation exchange filter under dynamic conditions was established. A computer model of the process of regeneration of the sodium-cation exchange filter in nonequilibrium and non-stationary conditions in the form of a computer software package ("Regeneration-1", "Regeneration-2"), recommended for scientific, design, commissioning and operating organizations, relevant industries. For the first time, the task of optimizing the ion-exchange nonequilibrium regeneration of the sodium-cation exchange filter as a separate stage of the filter cycle was posed and solved. The analytical dependence of optimizing the specific consumption of NaCl salt for nonequilibrium regeneration of the sodium-cation exchange filter as a separate stage of the filter cycle is obtained.

1. Introduction

Over the twentieth century, the world's population has increased almost four times. At the same time, the consumption of fresh water increased nine times, including for drinking and communal needs – 13 times. As a result, water resources in various regions of the world became catastrophically short of water [1–5]. In the next 20 years, taking into account current trends in population growth and the global economy, an increase in the demand for fresh water by at least 100 km^3 per year is expected. According to the UN [1, 2], more than 1.2 billion people live in conditions of constant shortage of fresh water, about 2 billion people suffer from it regularly (in the dry season, etc.).

The existing shortage of fresh water can be compensated for by the use of natural highly saline waters (water of the seas and oceans, salt lakes, groundwaters) using ion-exchange technologies.

Ion-exchange technology is widely used in industry [5–11]. Study their patterns is one of the main problems of modern science. The technological process of ion-exchange solution processing includes two main stages:

- removing from a solution passed through an ion-exchange material, the desired ion (or group of ions) of a certain type in exchange for the ion contained in the ion-exchange material. In this case, the solution is purified from ions of a certain type, and the ion-exchange material is saturated with these ions, the number of ions capable of such an exchange in it decreases and the ion-exchange material is "depleted", the stage of sorption or absorption of ions of a given type;
- introducing into the "depleted" ion-exchange material ions capable of being exchanged for an

ion of a given type by passing through it a special solution containing an ion capable of being exchanged, usually in a higher concentration, the stage of regeneration or restoration of the ability of an ion-exchange material to absorb ions of a given type from processed solution.

Ion exchange softening of natural, waste and sea waters to produce suitable for feeding steam and hot water heat generators and heat desalination plants is one of the most relevant areas of application of ion exchange technologies. The process of ion-exchange water softening consists in replacing the Ca2 + and Mg2 + cations, which are the basis of poorly soluble scale-forming hardness salts (CaCO3, CaSO4, Mg (OH) 2), with a Na+ ion, as a result of which only well-soluble sodium salts remain in the water. Regeneration of the ion-exchange material is carried out with a NaCl solution with a concentration of from 6 to 12 %. This process is called sodium-cation exchange water softening and is carried out using ion-exchange materials capable of exchanging cations (sulfonated coal, KU-2, KU-4, etc.).

The dynamics of ion-exchange sorption is the subject of numerous works by domestic and foreign scientists [6, 10, 11]. However, along with significant achievements in this area, the results obtained are not enough to solve some practical problems. The scientific and mathematical description of real sorption systems operating under dynamic conditions remains a very difficult issue, due to insufficient knowledge of many processes and phenomena that occur during ion-exchange.

It is important to take into account that most of the works on the calculation of dynamic ion-exchange systems consider dilute solutions whose concentration does not exceed 0.1 H, where the external diffusion kinetics is the limiting stage of ion exchange. Nevertheless, the use of alternative sources of natural and industrial wastewater involves waters with high salinity up to 0.5 H. Since real ion-exchange processes are accompanied by a mass of additional external and internal factors that influence the course and quality of the ion exchange process itself under dynamic conditions, there are empirical dependencies cannot fully claim scientific validity. However, one must not underestimate the accumulated practical experience in operating ion-exchange systems at various enterprises – it needs to be studied and generalized with efforts to develop a mathematical apparatus that describes ion-exchange processes.

Currently, the characteristics of the process of softening fresh water of drinking quality (total salinity up to 2000 mg / l) have been most fully studied. The characteristics of the cation exchanger regeneration process (salt content of the regenerating NaCl solution from 50 g / l to 120 g / l) have not been completely studied.

The regeneration process was studied mainly experimentally. Attempts to quantify the process are few in number and have been undertaken mainly to demonstrate computational methods for the simplified kinetic scheme, while modeling the regeneration process of the sodium-cation exchange filter requires a detailed consideration.

Therefore, the search for a new method for calculating the non-stationary and nonequilibrium dynamic regeneration process, as well as the development of a computer model of the cation exchanger regeneration process for its quantitative description and optimization, is undoubtedly relevant, which was the main objective of this work.

2. Methods and materials

Experimental studies were conducted on the basis of the laboratory of the Department of Heat and Gas Supply and Real Estate Expertise of the North Caucasus Federal University. For the experiment was designed and installed special setup (Fig. 1).

The experimental setup consists of two ion-exchange columns, which are connected in a special way for the supply and removal of solutions through a hose system. In the system there are tanks for the initial and regeneration solution, pumps for supplying the initial and regeneration solution to the ion-exchange columns. While the regeneration process is taking place in one column, the other column undergoes depletion in order to save time. Before the regeneration process, it is necessary to loosen the cation exchanger, and after the regeneration process it is necessary to wash the cation exchanger, for this purpose a separate pump is provided (Fig. 2).

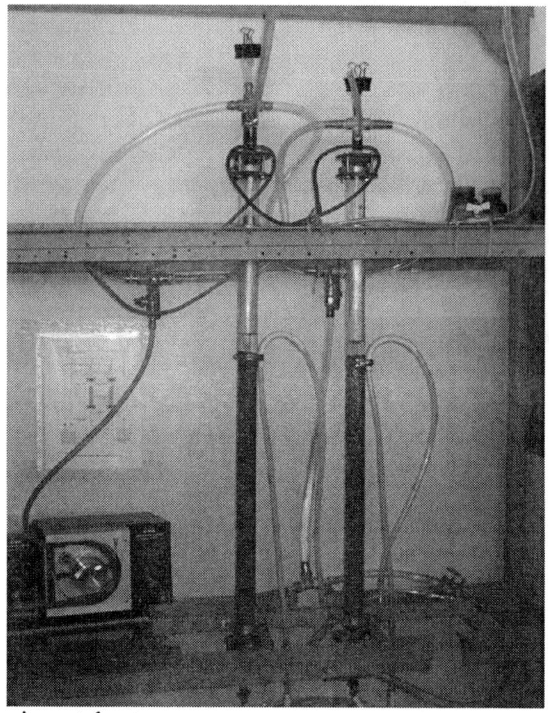

Figure 1. Photo of the experimental setup

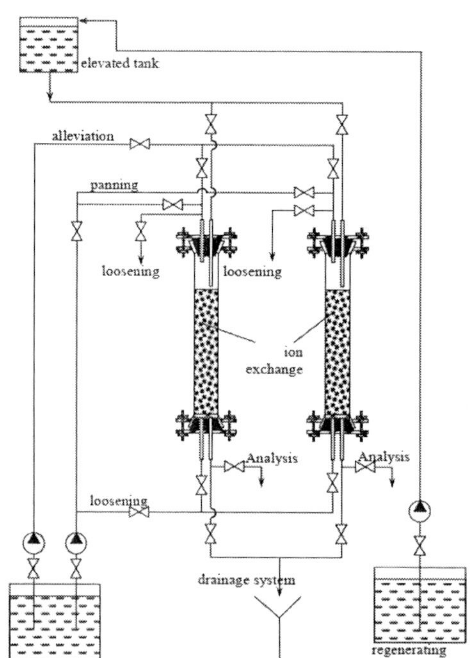

Figure 2. Schematic diagram of the experimental setup

Ion-exchange columns with a diameter of 32 mm and a height of 1060 mm are made of quartz glass. The columns are filled with swollen cation exchanger КУ-2×8 with a layer height of 600 mm. The remaining volume of the column above the cation exchange resin is filled with an appropriate solution, depending on the process. Bottom layer of cation exchanger in the ion exchange column is limited by a special drainage device, which is covered with nylon fabric, the cell diameter of which is about 0.2 mm, which prevents cation exchanger from flowing.

In this study, the analysis of the samples was carried out according to the titrimetric method, as the most accessible and simple method for determining the quantitative content of ions Ca^{2+}, Mg^{2+} и Cl^{-}. This method provides sufficient measurement accuracy.

The titrimetric method of stiffness analysis is based on the use of reactions of complexing metal ions with complexones, accompanied by the formation of stable, slightly dissociated, water-soluble intramuscular salts.

The experimental research plan is designed to determine the numerical values of the volumetric coefficients of the external and internal mass transfer (γ_1 и γ_2, β_1 и β_2) when comparing the experimental output curves of changes in the total concentration and concentration of the absorbed ion with the calculated ones according to the developed programs "Regeneration-1" and "Regeneration-2".

The "Regeneration-1" program is designed to solve the system of equations for the dynamics of regeneration of the sodium-cation exchange filter by changing the total concentration of the regeneration solution using the finite difference method, and the "Regeneration-2" program is used to solve the system of equations of the regeneration dynamics of the sodium-cation exchange filter according to the change in the concentration of the displaced ion in the regeneration solution by the finite difference method.

38 experiments were conducted to study the regeneration process in pair exchange Na^{+}–Ca^{2+} and 31 experiments in pair exchange Na^{+}–Mg^{2+}.

In experiments on the study of the regeneration process in pair exchange Na^{+}–Mg^{2+} the following values of the main initial characteristics in different combinations were taken:

- $C_{PP} = 1680 \div 1920$ milligram-equivalent/l;
- $W_P = 0.338 \div 5.72$ m / h;
- $C_{PH} = 100 \div 515$ milligram-equivalent/l;
- $C_{1H} = 40 \div 99.5$ milligram-equivalent/l;

where C_{PP} – concentration of regeneration solution supplied to the upper section of the ion exchanger layer in the filter, milligram-equivalent/ l; C_{PH} – total concentration of softening solution in the process of sodium-cation exchange softening prior to the regeneration process, milligram-equivalent / l; C_{1H} – initial concentration of ion absorbed by softening (Ca^{2+} or Mg^{2+}) in softened solution preceding the regeneration process, milligram-equivalent/l; W_P – transmission rate of the regeneration solution through the ion exchange column, m / h.

In experiments on the study of the regeneration process during the pair exchange of Na^{+}–Ca^{2+} the following values of the main initial characteristics were taken:

- $C_{PP} = 1675 \div 1920$ milligram-equivalent/l;
- $W_P = 0.49 \div 1.49$ m / h;
- $C_{PH} = 15 \div 507$ milligram-equivalent/l;
- $C_{1H} = 10 \div 122$ milligram-equivalent/l.

In all experiments, the height of the cation exchanger layer was maintained at 60 cm and the concentration of hardness ions in the initial regeneration solution, C_{1PP}, fed to the upper section of the cation exchanger layer in the filter is not higher then 0.267 milligram-equivalent/l.

Particular numerical values of volumetric coefficients of internal mass transfer with a change in the total concentration of the regeneration solution, as well as β_2 when the concentration of the ion displaced from the cation exchanger changes (Ca^{2+} or Mg^{2+}) are determined by comparing the experimental output regeneration curves at the output of the sodium-cation exchange filter and the calculated output regeneration curves determined using the "Regeneration-1" and "Regeneration-2" programs for each experiment. Moreover, the volumetric coefficients of external mass transfer γ_1 with a change in the total concentration of the regeneration solution and β_1 with a change in the concentration of the ion (Ca^{2+} or Mg^{2+}) displaced from the cation exchanger are not determinative, because during the regeneration of the sodium-cation exchanger filter on KU-2 × 8 cation exchanger, which occurs at large concentrations (more than 0.5 H), the intradiffusion region is limited. The values γ_1, β_1, c^{-1}, were set:

- for pair exchange $Na^+–Ca^{2+}$ $\gamma_1 = 0.05$ c^{-1}, $\beta_1 = 0.038$ c^{-1};
- for pair exchange $Na^+–Mg^{2+}$ $\gamma_1 = 0.05$ c^{-1}, $\beta_1 = 0.031$ c^{-1}.

Moreover, the values of the coefficients γ_2 and β_2 were selected in such a way that the calculated output curve of the regeneration of the sodium-cation exchange filter coincided as much as possible with the experimental experimental data. In accordance with the theory of mathematical statistics, the smallest sum of the squared deviations of the points of the experimental and calculated output curves is taken as a selection criterion (Fig. 3).

Having obtained the values of the effective volumetric mass transfer coefficients and for all the experiments carried out, a correlation analysis was carried out, which was performed on a computer in Microsoft Excel. The tightness of the relationship between the obtained dependences was estimated by the correlation relation R.

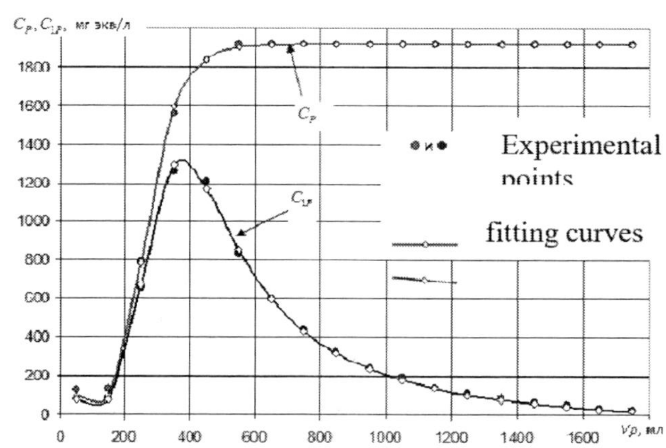

Figure 3. Comparison of calculated and experimental points of output curves.

Analysis of the correlation coefficients allows us to identify significant dependences by internal diffusion characteristics $\beta_2 = f(C_{PP}, C_{PH}, C_{1H})$ and $\gamma_2 = f(C_{PP})$.

The obtained empirical coefficients are averaged between all the experiments carried out and as a result the following dependences are obtained:

- for pair exchange $Na^+–Ca^{2+}$

$$\gamma_2 = 0,0048 \cdot C_{PP} + 1,05 \cdot 10^{-6} \cdot C_{PP}^2, \text{ c}^{-1},$$

$$\beta_2 = 0,00345 \cdot C_{PP} + 0,00562 \cdot \alpha_{PP}, \text{ c}^{-1};$$

- for pair exchange $Na^+–Mg^{2+}$

$$\gamma_2 = 0,0073 \cdot C_{PP} + 7,9 \cdot 10^{-7} \cdot C_{PP}^2, \text{ c}^{-1},$$

$$\beta_2 = 0,00747 \cdot C_{PP} + 0,0067 \cdot \alpha_{PP}, \text{ c}^{-1},$$

where C_{PP} – concentration of regeneration solution, milligram-equivalent/ml; α_{PP} – relative concentration of regeneration solution.

3. Results

Using the experimental values of the volumetric coefficients of the external and internal mass transfer (γ_1, β_1 и γ_2, β_2) and the programs "Regeneration-1" and "Regeneration-2", it is possible to calculate the output curves of changes in the concentration of the regeneration solution and changes in the concentration of the displaced ion (Ca^{2+} или Mg^{2+}) for given initial characteristics of the process of regeneration of sodium-cation exchange filter (C_{PP}, W_P, C_{1H}, α_{PP} etc.) within the range of experimental research data.

Examples of output curves of field experiments and constructed using a computer model are presented in Fig. 4.

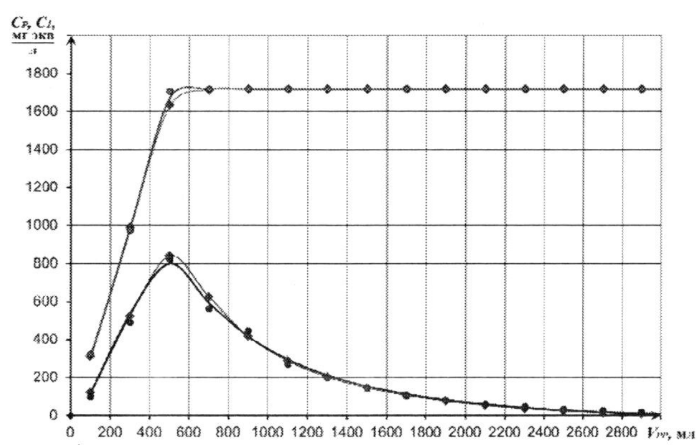

Figure 4. The output curves of the regeneration process of the sodium-cation exchange filter with the Na^+–Ca^{2+} pair exchange according to the conditions of private experience: round dots — experimental data, thick line — calculated curve for individually selected mass transfer coefficients; thin line with rhombic dots – calculated curve according to the derived equations of mass transfer coefficients

The developed numerical model of the regeneration process can be used to assess the effectiveness of this process and its optimization.

The effectiveness of any process is determined by the ratio of the value of the beneficial effect from it to the costs of its implementation in comparable units.

The costs of the regeneration process are determined mainly by the specific consumption of the regenerating agent (for the sodium-cation exchange filter, the Na^+ ion in the form of $NaCl$ salt). The remaining costs: depreciation of equipment, energy costs for pumping solutions, costs for maintenance personnel, etc. are almost constant for various modes of the regeneration process and may not be taken into account when optimizing the process.

The reagent consumption ($NaCl$) for regeneration of the sodium-cation exchange filter is taken into account by the following indicators, the formulas of which are developed for the numerical model:

- specific current consumption of $NaCl$ salt for regeneration, d_{P_j}, milligram-equivalent $NaCl$ /ml,
- degree of regeneration n_{P_j} ,
- current relative specific consumption of $NaCl$ salt for regeneration g_{P_j}, milligram-equivalent / milligram-equivalent.

Fig. 5 shows dependency graphs $n_{P_j} = f\left(d_{P_j}\right)$, $g_{P_j} = f\left(d_{P_j}\right)$ for a special case: regenerated cation exchanger KU-2x8 is initially in equilibrium with the source softened water; relative concentration of

IOP Conf. Series: Materials Science and Engineering **905** (2020) 012034 doi:10.1088/1757-899X/905/1/012034

absorbed cation in the initial state $\alpha_{1PH} = \frac{C_{1H}}{C_P} = \frac{70.7}{326} = 0.22$; вlayer height of KU-2x8 cation exchanger in the sodium-cation exchanger H_K=200 см; total concentration of the regeneration solution entering the upper layer of cation exchanger in the sodium-cation exchanger C_{PP}=1800 milligram-equivalent/l; the rate of passage of the regeneration solution through the cation exchanger layer in the sodium-cation exchanger W_P=1 m/h.

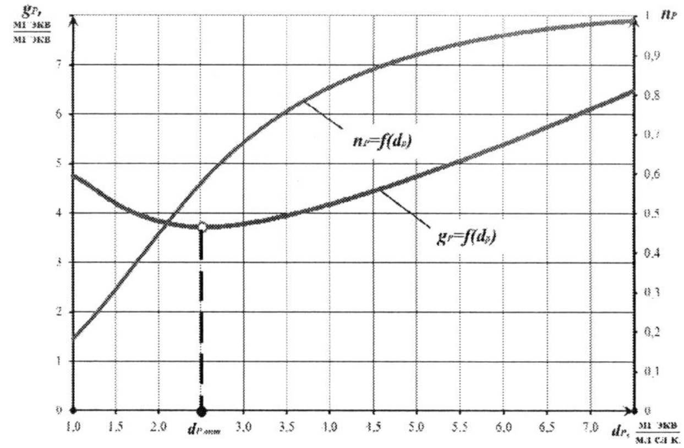

Figure 5. Dependency graphs $n_{P_j} = f\left(d_{P_j}\right)$ и $g_{P_j} = f\left(d_{P_j}\right)$ for a special case.

Analyzing the graphic dependencies of the main characteristics of the regeneration process $n_{P_j} = f\left(d_{P_j}\right)$ and $g_{P_j} = f\left(d_{P_j}\right)$ we can conclude that there are inflection points and extreme points (Fig. 5), i.e. about the possibility of optimizing the nonequilibrium regeneration of the sodium-cation exchange filter as a separate stage of the filter cycle.

Considering the regeneration process as a separate stage of the filter cycle of the sodium-cation exchange filter, it is advisable to take the specific relative consumption of salt as an optimization criterion g_P.

Due to the fact that analytical dependencies are more accessible to engineers and technicians than computer models, the goal of further work is to obtain a regression multivariate dependence of the optimality of reagent consumption based on planning and conducting computational experiments, as well as adding an optimization method to the software package.

The purpose of conducting computational experiments with a computer model of an isolated regeneration process is to obtain a regression multivariate dependence $d_P = f\left(C_{PP}, W_P, C_{PH}, C_{1H}\right)$ at extreme points of relative specific consumption g_P (Fig. 5). In other words, using a computer model, the problem of optimizing an isolated process of nonequilibrium regeneration is solved and the regression dependence of the specific consumption of $NaCl$ salt at extreme points of the relative specific consumption g_P on four factors is determined: the total concentration of the softened solution in the process of sodium-cation exchange softening preceding the regeneration process C_{PH}, the initial concentration of absorbed when softening an ion (Ca^{2+} or Mg^{2+}) in a softened solution C_{1H}, the input rate of the regeneration solution W_P and the concentration of the regeneration solution C_{PP}.

So, for a multifactor computational experiment, a first-order plan for PFE 24 at two levels has been adopted. In order to facilitate practical use for describing the objective function, the regression equation in the form of a linear polynomial with terms of interactions is adopted.

Thus, the obtained equation for determining the optimal specific consumption of *NaCl* salt for the regeneration of the sodium-cation exchange filter, depending on significant factors, is presented below.

$$\begin{aligned} d_P^{on} &= 0,66 + 0,64 \cdot C_{PP} - 0,012 \cdot W_P + 5,61 \cdot C_{PH} - \\ &-6,77 \cdot C_{1H} - 0,033 \cdot C_{PP} \cdot W_P - 0,39 \cdot C_{PP} \cdot C_{PH} + \\ &+3,3 \cdot C_{PP} \cdot C_{1H} - 0,34 \cdot W_P \cdot C_{PH} + 1,14 \cdot W_P \cdot C_{1H} - \\ &-25 \cdot C_{PH} \cdot C_{1H}, \end{aligned} \tag{1}$$

where C_{PP}, C_{PH}, C_{1H} – concentrations in milligram-equivalent /ml, W_P – in m/h, d_P^{on} – milligram-equivalent *NaCl*/ml.

4. Conclusion

Thus, the search for the optimal specific salt consumption can be carried out using both the developed computer numerical model and the analytical optimization equation (1). The equation for optimizing the specific consumption of salt (1) is more acceptable than a computer model, and also accessible and recommended for engineers designing and operating sodium-cation exchange filters.

Application of the above formula makes it possible to develop practical recommendations for improving the process of regeneration of KU-2x8 cation exchanger while softening highly mineralized waters. Maintaining an optimal specific salt flow rate will reduce the operating costs of sodium-cation exchange filters.

The developed numerical model can be used to simulate the technological processes of water treatment of thermal power plants in order to predict and optimize their operation, as well as to quantify and optimize other ion-exchange processes.

References

[1] Danilov-Danilyan V I 2008 The global challenge of freshwater scarcity *The Century of Globalizat.* **1** 45–56

[2] Danilov-Danilyan V I 2005 Fresh water shortage and the world market *Water resources* **32(5)** 625–33

[3] Allan J A 2002 *The Middle East water question: Hydropolitics and the global economy* (London) pp 3–20

[4] Gain A K and Wada Y 2014 Assessment of Future Water Scarcity at Different Spatial and Temporal Scales of the Brahmaputra River Basin *Water Resources Manag.* **28(4)** 999–1012

[5] Zhang Q., Liang X., Fang Z et al 2016 Urban water resources allocation and shortage risk mapping with support vector machine method *Natural Hazard* **81(2)** 1209–28

[6] Mamchenko A V, Kosygina I M and Savchenko O A December 2010 H^+-Na^+ exchange dynamics on the caboxylic cation exchanger in the fluidized bed *J. of Water Chem. and Technol.* **32(6)** 324–8

[7] Feyziev G K, Dzhalilov M F, Kuliev A M and Badalov B Sh April 2007 Good prospects for repeated use of spent regenerative solutions at chemical desalting plants with reduced consumptions of reagents and wastewater *J. of Water Chem. and Technol.* **29(2)** 110–4

[8] Abdullaev K M, Agamaliev M M, Kosmodamianskii V E and Dadashova O O June 2010 Substantiation of flow charts for reverse-osmosis desalting with the ion-exchange softening of sea water and permeate *J. of Water Chem. and Technol.* **32(3)** 187–92

[9] Larin B M, Bushuev E N, Larin A B et al April 2015 Improvement of water treatment at thermal power plants *Thermal Engineer.* **62(4)** 286–92

IOP Conf. Series: Materials Science and Engineering **905** (2020) 012034 doi:10.1088/1757-899X/905/1/012034

[10] Agafonova Yu V, Khamizov R Kh, Tikhonov N A and Venitsianov E V 2002 Patterns of Dynamics of Ion Exchange in Chromatographic Systems of the Ecological Police *Sorpt. and Chromatogr. Proces.* **2(6)** 586–95

[11] Geyvandov I A, Abornev D V, Kalinichenko M Yu and Geybatov R A 2010 Determination of the dynamic characteristics of sodium-cationite softening of mineralized waters from experimental data based on a mathematical model *Bull. of the North Caucasus State Techn. Univer., Physicochem. Ser.* **4(25)** 44–9

Hybrid modeling of wave processes in the scientific justification of hydraulic solutions

I G Kantarji

Moscow State University of Civil Engineering, 26, Yaroslavskoye sh., Moscow, Russia

E-mail: Kantardgi@yandex.ru

Abscract. To determine the wave or ice load on offshore hydraulic structures, calculation methods and modern mathematical models are used, as well as field observations and laboratory studies. Each of these methods has advantages and disadvantages. Sharing existing methods provides an approach called "hybrid or composite modeling." Hybrid modeling provides the most reliable approach for designing safe facilities. However, the interaction of the individual components of hybrid modeling requires further research. This article is devoted to the study of the problems that arise when verifying the results of numerical modeling of wind waves using field and laboratory model measurements. The experience of studying the projects of the port-shore protection complex in Imeretinka, Sochi, the LNG terminal in the Ob Bay and the port in the East Bay of the Sea of Japan is used.

1. Introduction

Currently, marine hydraulic engineering in the world and in Russia is an area of active development. The main external loads on offshore hydraulic structures are determined by surface waves and ice fields, the correct determination of the characteristics of which largely determines the safety of the designed structures. Both computational methods and modern mathematical models are involved, as well as field observations and laboratory studies.

In this case, a rather complex interaction of calculation, numerical and experimental methods arises, requiring special study. The interaction of methods is called a hybrid or, more recently, a composite approach.

For a long time, the author has been leading a group that is engaged in pre-design studies of wave processes to ensure the design and construction of port, shelf and shore protection structures of various projects. In these studies, the problems of the interaction of methods with each other within hybrid modeling arose. Some of these problems and related results are discussed in this article.

2. Hybrid modeling of wave processes

The main regulatory document providing the determination of the design parameters of waves and ice are the construction rules of SP 38.13330.2012 "Loads and impacts on hydraulic structures (wave, ice and from ships)", introduced in 2013. In accordance with them, engineering materials are used hydrometeorological surveys, long-term field observations of hydrometeorological processes in the design area.

In this case, observations of the characteristics of ice fields are used directly for ice, and observations of the fields (speed and direction) of the wind are used for waves. Recommended methods for calculating the elements of waves in deep water from the wind fields, as well as waves at

IOP Conf. Series: Materials Science and Engineering **905** (2020) 012035 doi:10.1088/1757-899X/905/1/012035

the structure. Since the calculation method was supposed to be applicable for the whole variety of natural conditions of the shelf and coastal zone of the seas, some averaged approaches were proposed. And although these methods are supplemented by some empirical corrections, they remain averaged and allow unpredictable errors when applied to the specific conditions of the design area. It is assumed that the errors in the determination of the design parameters of waves and ice will be compensated by the engineering reserves adopted during the design.

However, this approach is poorly applicable for the most critical facilities. Therefore, as can be assumed, these circumstances led to the inclusion in SP 38.13330.2012 of the following requirement: "Loads from waves and ice on hydraulic structures of class I, as well as the calculated wave elements in open and enclosed waters, must be clarified based on field observations and laboratory studies". That is, the elements of waves obtained by calculation must be compared with field observations in the construction area and adjusted according to the results of such a comparison. This norm applies to the most responsible hydraulic structures of class I.

The method for determining the design characteristics of waves in a joint venture is the equivalent wave method [1]. In accordance with the distribution function of wind speed in a given wave-hazardous direction (storm mode), a design storm corresponding to the class of construction is determined. This estimated storm is assumed to be equivalent to the storm regime for the construction area.

However, the estimated storm parameters cannot be directly compared with the data of field measurements of waves in the area of the construction site, even if such data are available. Since the measurement data refer to specific time and hydrometeorological conditions, and the calculated storm is virtual, equivalent in terms of determining the load on the designed structure of a long-term sequence of storms.

Another approach to determining the design parameters of waves on the approach to the construction uses reanalysis to restore the hydrometeorological conditions for the construction area for a sufficiently long previous period (several decades) and determine the characteristics of severe storms for this period [2, 3].

To determine the operational characteristics of waves in the open sea on the approach to the projected object, the data of the analysis of wind fields for the last 30-50 years are used, on the basis of which periods of the most severe storms in the region under consideration are distinguished, the wave characteristics for which are calculated on the basis of the SWAN numerical model. This approach, which has become generally accepted in modern foreign and Russian studies [4–7], allows one to obtain wave characteristics directly in the study area, where there are no permanent stations for observing waves.

When re-analyzing hydrometeorological conditions, the characteristics of specific storms are established, which can be compared with the data of direct measurements of waves in the region and, thus, fulfill the above requirement of the joint venture to refine the calculated characteristics of waves by comparison with the data of field measurements.

The resulting sequence of storms can be used to construct the distribution functions of heights and wave periods. In addition, if the estimated storm with a security of once every hundred years (for Class I facilities) is determined on the basis of data from long-term observations of the wind, then storms close to the calculated characteristics can be distinguished from the sequence of storms obtained by reanalysis.

As an example of the application of the method of reanalysis of hydrometeorological conditions to determine the calculated wave parameters, the data on the coast protection project in Imeretinka (Black Sea, Sochi) are given below..

The wave regime of the coast of the Imereti Lowland is modeled from long-term data of the wind field over the Black Sea, with subsequent calculation of wave transformation in the coastal zone based on equations of approximation of gentle slopes. Such an approach made it possible to calculate the wave parameters of the largest storms of recent decades directly in the coastal zone of the Olympic complex and to evaluate the impact of the designed shore protection structures on changes in wave characteristics in the coastal zone.

To determine the operational characteristics of waves in the open sea on the approach to the Imereti coast, we used data from the analysis of wind fields over the Black Sea in 1980–2010, based on which periods of the most severe storms in the region under consideration were identified, for which the wave characteristics were calculated based on the SWAN model [7]. This approach allows one to obtain wave characteristics directly in the study area, in which there are no permanent stations for observing waves.

To calculate the characteristics of the waves on the deep water border of the Imereti coast, we analyzed the NCEP / NCAR reanalysis data of the wind fields from 1980 to 2010 using data adjusted for satellite observations and wave measurements in the region for corrections.

After 1999, the NCEP "final analysis" meteorological fields were used as the main source of meteorological data [8], in which the meteorological fields were additionally adjusted due to satellite remote observation data.

For calculations by the SWAN wave model in the specified 30-year period, periods of strong winds were selected, from the South to the North-West, with wind speeds of 12 m/s and more, which can lead to high storm waves on the coast near the city of Sochi. Wind was estimated for the entire eastern part of the Black Sea (east of Kerch). In total, in the period 1979–2010, 67 severe storms were revealed by the specified criterion.

The wave field calculations were carried out by the spectral model of SWAN on three regular rectangular grids in spherical coordinates (latitude-longitude). The first grid includes the entire Black Sea region. The square mesh size of the first grid is 1.5 minutes. The second grid is nested in the first and includes the coastal zone of Sochi. The mesh size of this grid is 9 s. The third grid, nested in the second, covers the area of Imereti Bay from the mouth of the river. Mzymta to the river. Psou, and the cell size of this grid is also 10 times smaller than the cell size of the previous (second) grid and is 0.9 s. Bathymetry for computational grids was interpolated at grid nodes according to GEBCO topographic data publicly available on the Internet on a thirty-second grid, and for the coastal part of Imereti Bay, depths measured by an echo sounder were used as part of the development of the shore protection project.

The calculations were carried out by the SWAN spectral model in three stages – on each of the three nested grids with sequential data transmission at the boundary points between them.

Based on the results of calculations of the characteristics of the identified 63 strongest storms in the period 1980–2010, their statistical characteristics were analyzed using the WMO method [6]. To assess the frequency of storm waves in the mode, empirical integral probability distribution functions of wave heights (IFDW) are plotted at the seashore border of the coastal shelf zone in the region of the eastern tip of the port. The transition from IFDW to the values of security P is carried out according to the formula P = 1–IFDW, and, accordingly, the frequency of the storm n (the number of years the value of this parameter can be exceeded) n = 1 / P. The repeatability of wave periods was similarly determined..

The 15 strongest storms in the indicated period were received, ranked by the height of a significant wave, and Table 1 shows the calculated frequency of wave heights and periods in the storm system.

Table 1. Parameters of significant waves of different repeatability in the system at the seaford border of the port imeretinsky

Repeatability every n years	IFDW	H, m	T, s	Storm Dates
1	0	3.1	5.7	
5	0.8	5.5	7.5	
10	0.9	5.8	8.0	1995-01-01, 1998-02-17, 1999-12-29, 2006-03-03, 2007-11-11
25	0.96	6.5	8.4	1989-12-07, 1987-01-03, 1987-01-27
50	0.98	7.0	9.5	1993-01-25

As follows from the presented results, the fifteen most severe storms in the region have a security of 2–20 %. The average direction of the approach of the waves to the coastline along these 15 storms

IOP Conf. Series: Materials Science and Engineering **905** (2020) 012035 doi:10.1088/1757-899X/905/1/012035

is 222 degrees. Moreover, the average direction of approach of the waves of the four strongest storms is 216 degrees with deviations of only 1–2 degrees in this group. Thus, we can conclude that the highest waves approaching the sea border of the Imereti coast, taking into account its exposure, are mainly frontal.

Testing the results of calculations based on field observations is possible only with the use of measurement results in other areas. Directly in the coastal zone of Imeretinka, such measurements were not carried out.

The eastern section of the shore of Imereti Bay from the river. Mzymta to metro Konstantinovsky has an extremely complex relief of the underwater slope due to the presence of underwater canyons (Fig. 1). Canyon the New comes close to the shore, its edge is expressed in relief from depths of 7–8 m. The canyon bed is relatively straightforward. The front of the canyon has a width of about 400 m. According to repeated surveys, the edge of the canyon Novy moves towards the coast with an average speed of 1.0 m/year.

The waves above the underwater canyons can be focused, significantly increasing in height. At the same time, high and steep waves pass along the canyon channel to the shore, producing disastrous effects on structures and beaches. Thus, according to unconfirmed data from BBC News Europe, January 30, 2013, a record high wave with a height of 100 feet (about 30 m) was observed on the coast of Portugal. This wave was observed over the Nazare Canyon. At the same place in November 2011, a record wave of 78 feet (24 m) in height was recorded in the Guinness Book of Records. The effect of focusing waves over canyons is not described by the method of calculating wave refraction based on geometric optics.

Wave fields for the coastal zone of the Imereti Lowland were modeled by the SWAN model on unstructured meshes. The choice of unstructured grids for calculating coastal wave fields was made because of the need for a more detailed resolution of the numerical grid (about 6 m) in the coastline area, which made it possible to more accurately take into account the influence of complex bathymetry of the region and to avoid interpolation errors in further calculations of hydrodynamic models.

Storm scenarios were calculated for 5 years from 1983 to 1987, as well as 3 large storms – December 1989, January 1993 (maximum for a 30-year period) and November 2007. At the sea boundary of the computational grid, wave spectra obtained from calculations on grid No. 2 of Sochi.

The calculated wave characteristics at the outer boundary of the detailed computational grid of the coastal zone were used to calculate wave fields in the coastal zone with a detailed description of the HWAVE refractive-diffraction model, which is based on the monochromatic or spectral version on the hyperbolic approximation of the gentle slope equation [9–15]. Figure 2 shows the wave field off the coast of the Imereti Lowland, calculated respectively by the spectral version of the HWAVE model during periods of maximum storm in December 2009.

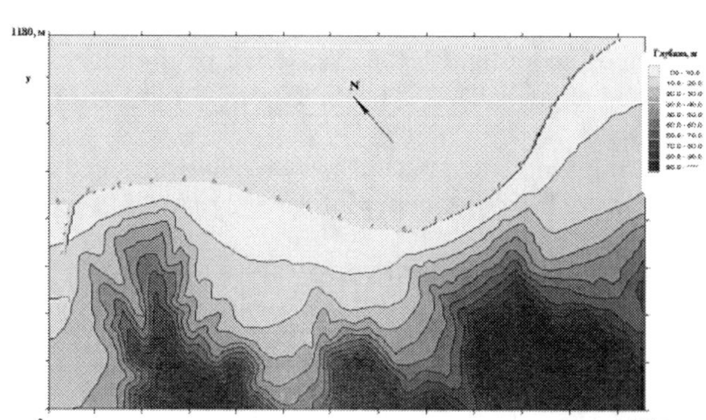

Figure 1. Bathymetric plan of the underwater coastal slope for the eastern section of the coastal zone of Imeretinka (with canyons)

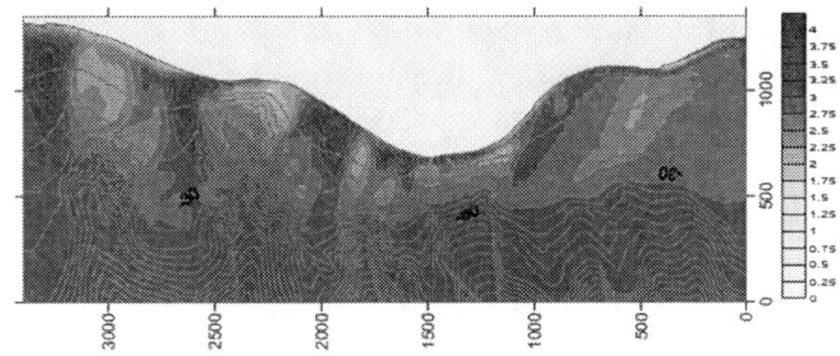

Figure 2. Field of significant wave heights near the Imereti coast during the storm maximum period 12.2009 calculated by the HWAVE_S model (incident wave H = 3.03 m, T = 9.22 s)

The calculation results indicate a more pronounced zone focusing and divergence in the wave field than follow from similar calculations SWAN model due to rugged terrain of the coastal area of Imeretinskaya lowland.

In order to further verify the calculated wave characteristics for the coast of the Imereti Lowland, in the shallow basin of the laboratory of MGSU Research Laboratory Of Offshore Oil And Gas Industrial Facilities, physical modeling of the waves was performed in the selected hazardous areas of the shore protection dam design (Fig. 3). Simulation scale 1: 25.

Высоты волн в процессе экспериментов The wave heights during the experiments were measured at several points along the coastal profile. The measurement results were compared with mathematical modeling of wave transformation.

The transformation of waves on slopes for modeling model studies was simulated using a nonlinear numerical model based on the Boussinesq-Zheleznyak-Pelinovsky (Serre) equations with special additives [10]. An example of calculating the transformation of waves on a model coastal slope is shown in Fig. 4.

Statistical processing of measurement and calculation data was carried out, and a regression dependence of the measured and calculated wave heights was constructed. Correlation coefficient of measured and calculated wave heights: R = 0.985, which indicates a very good agreement between the results of physical and numerical modeling.

Figure 3. A model of a section of the shore of the Imereti Lowland under the influence of waves in the MGSU basin

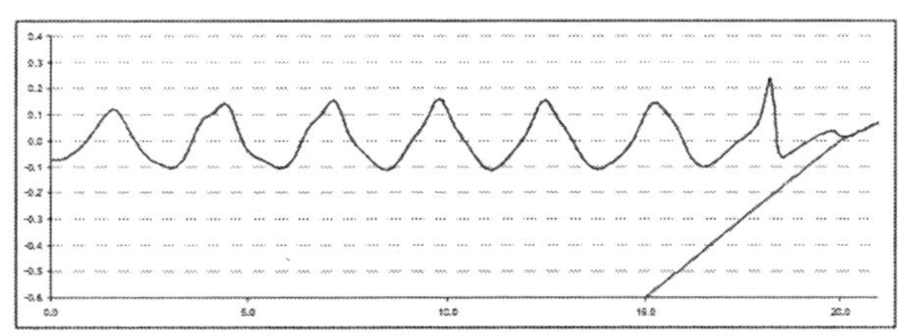

Figure 4. An example of mathematical modeling of wave transformation on the coastal slope of a physical model

3. Verification of the results of numerical simulation

Another important task of hybrid modeling technology is to verify the results of numerical simulations by comparing them with field and laboratory measurements. This issue is considered on the example of modeling the wave regime and lithodynamic processes for the design of port hydraulic structures of the Terminal for liquefied natural gas and stable gas condensate "Utrenny", seaport of Sabetta.

The location of the designed Terminal is the eastern coast of the Ob Bay, south of Cape Halcyneisal, between the mouths of the Haltsanayakh and Nyadayngcho rivers. Figure 5 shows the red layout of the Layout option of the Remote terminal, orange – the boundaries of the simulation area (Projection UTM43, EPSG: 32643).

Modeling was carried out for 3 layout options of the Terminal, characterized by the presence and configuration of the protective structures, as well as the number of berths and the size of the approach channel.

The simulation technology used includes the following steps.

At the initial stage of the work, the calculations were carried out using the SWAN model, version 40.85, then the calculation was based on the later version of SWAN 41. 10AB. Since the beginning of the century, the SWAN model has been used as a tool for calculating coastal wave fields (for example, [7]). The model is based on the equation of the balance of the density of the wave action (or the balance of energy conservation, in the absence of flows) with sources and sinks. The model describes the generation, propagation, refraction, diffraction, reflection and passage of waves through flooded obstacles.

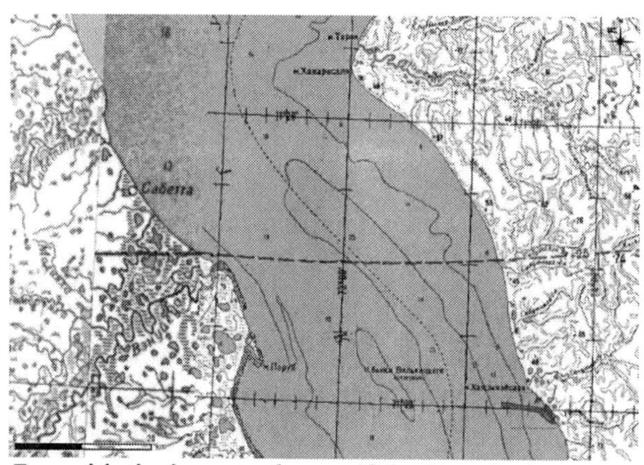

Figure 5. Map of the Ob Bay with the layout scheme of the Terminal

The ARTEMIS model [16, 17] is based on an extended version of the equation for the gentle slopes of the UPS [10], which, along with the originally incorporated capabilities for calculating the transformation of waves in the coastal zone, taking into account refractive and diffraction processes, adds the ability to calculate the influence of dissipation, due to bottom friction and destruction of waves in the coastal zone. In the ARTEMIS code, the equations of gentle slopes are solved numerically by the finite element method using a parallel computing algorithm.

Three time scenarios were calculated for verification – August and September 2012, September 17, 2015 – October 18, 2015, and July 10, 2016 – July 27, 2016. For the first period, the calculated heights and directions of the waves were compared with those measured near the Sabetta and Nord points (Fig. 6-a). For the second and third periods, a comparison was made with measurements of two wave-measuring buoys in the area of the Terminal (Fig. 6-b)

The results of comparing the measured and calculated wave heights show relatively high values of the correlation coefficient of these values in the range 0.87–0.90 and low values of standard deviations (SD) – 0.21–0.25 for Sabetta and Nord stations.

The results of comparing the measured and calculated periods of significant waves give the values of the correlation coefficients for these values of 0.67–0.73 and the values of standard deviations (VSD), about 0.8 sec. Moreover, the calculation results somewhat overestimate the wave periods, the characteristic period of high waves, about 5 seconds.

To verify the ARTEMIS wave model, a numerical simulation of wave fields was performed for the conditions of physical modeling of the port in the East Bay of the Sea of Japan. Physical modeling of wave propagation in the port water area was carried out in the wave pool of the "GS" Research Laboratory for sections in the area of the East and West breakwaters. A schematic representation of the plan of the port model in the wave basin is shown in Fig. 7.

a) b)

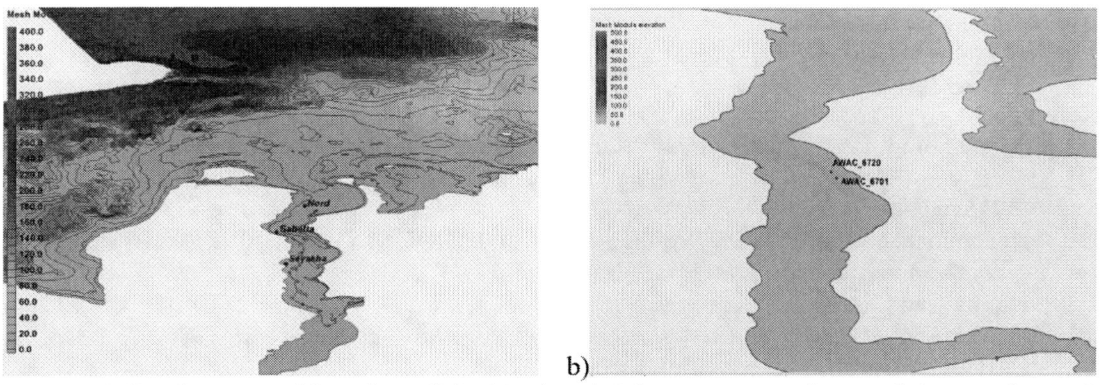

Figure 6. a) Depth map and location of the Nord and Sabetta wave stations and the Seyah weather station b) Location of the AWAC_6701 and AWAC_6720 wave buoys

The results of comparing the measured and calculated values of the wave heights shows a high value of the correlation coefficient of the data (0.94). However, in this case, the values of the waves obtained by numerical simulation were averaged in a circle with a radius of about a quarter of the wavelength centered at the control point. This allowed to reduce the effect of partially standing waves forming in the port water area as a result of multiple reflection [9].

As part of the development of a mathematical model of waves for the three options for arranging port facilities and presetting the model, a 3-directional wave was selected with a frequency of 1 time per 100 years, 1 % coverage, the parameters of which were obtained from the initial statistical processing of a 30-year period of extreme waves. Azimuth of wave 270 °, wave period 7.03 s, wave height 3.57 m.

To obtain wave distributions of varying repeatability and security in the water area of the Terminal, the SWAN spectral model calculates the parameters of extreme wind waves over a 30-year period, on the approach to the Terminal facilities.

a) b)

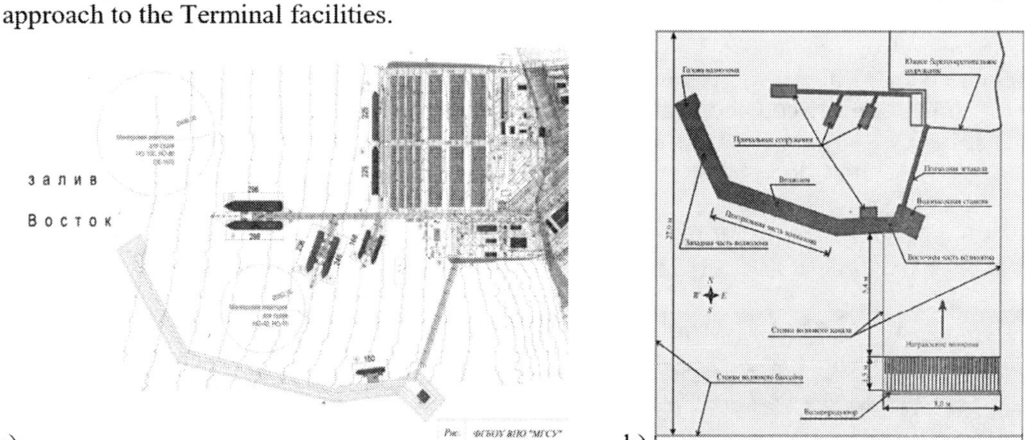

Figure 7. a) Port chart in Vostok Bay, b) Port model

To calculate the wave mode in the water area of the Terminal, the ARTEMIS wave model was chosen [16]. It describes the reflection of waves and allows you to enter different reflection coefficients at the border areas of the computational domain corresponding to port structures of various types.

4. Conclusion

When designing modern offshore hydraulic structures to determine the main external wave load, it is necessary to apply the technology of hybrid modeling. The technology includes the use of numerical wave modeling for the design area, as well as verification of calculation results using data from field observations and experimental laboratory studies. This approach allows us to ensure reliable determination of the characteristics of loads and impacts at the level of modern scientific research and is used in modern specialized scientific centers [18–22]. Details of hybrid modeling technology require further research.

References

[1] Lopatukhin L I 2012 *Wind Waves* (St. Peterburg: St. Peterburg State Univer.) 165 p

[2] Kalnay E, Kanamitsu M et al 1996 The NCEP/NCAR 40-year reanalysis project *Bull. Amer. Meteor. Soc.* 437–47

[3] Kanamitsu M, Ebisuzaki W, Woollen J et al 2002 NCEP-DOE AMIP-II Reanalysis (R-2) *Bull. of the Amer. Meteor. Soc.* **11** 1631–43

[4] Holthuijsen L H 2007 *Waves in oceanic and coastal waters* (Cambridge Univer. Press) 236 p

[5] Holthuijsen L, Booij N, Ris R et al 2009 *SWAN Cycle III version 40.51, User Manual* (Netherlands: Delft Univer. of Technol.) 67 p

[6] *WMO Guide to wave analysis and forecasting, WMO (Series) no 702* 1998 (Geneva: Secretariat of the World Meteorol. Organizat.) 159 p

[7] *SWAN team, Swan cycle III version 41.10AB, Scientific and Technical documentation* 2017 (Delft Univer. of Technol.: Faculty of Civil Engineer. and Geosci.)

[8] Dikiy P V, Dzuba P V, Jelezniyak M I and Sorokin M V 2011 Int. J. for Computat. Civil and Struct. Engineer. **7(2)** 54–63

[9] Kantarzhi I and Zheleznyak M 2016 Laboratory and numerical study of waves in port water body *Civil Engineer. Magazine* **6** 49–59

[10] Demchenko R I and Jeleznyak M I 2002 *Appl. Hydromech.* **4(2)** 23–9

[11] Berkhoff J C W 1972 Computation of combined refraction–diffraction *Proc. of the 13th Int. Conf. on Coastal Engineer.* (Vancouver, Canada: ASCE) pp 471–90

[12] Rogers W E, Maihatu K J, Hsu L et al 2007 Forecasting and hindcasting waves with the SWAN model in the Southern California Bight *Coastal Engineer.* **54(1)** 1–15

[13] Lemkea N, Fontourab J A S, Callaria D F and Fonseca D 2017 Comparative study between modeled (SWAN) and measured (waverider buoy) wave data in Patos Lago RS *Brazil Pan-Amer. J. of Aquatic Sci.* **12(1)** 1–13

[14] Dietrich J C, Zijlema M, Westerink J J et al 2011 Modeling hurricane waves and storm surge using integrally-coupled, scalable computations *Coastal Engineer.* **58(1)** 45–65

[15] Dietrich J C, Tanaka S, Westerink J J et al 2012 Performance of the unstructured-mesh, SWAN+ ADCIRC model in computing hurricane waves and surge *J. of Sci. Comput.* **52(2)** 468–97

[16] Aelbrecht D 1997 ARTEMIS 3.0: A finite element model for predicting wave agitation in coastal areas and harbours including dissipation *WIT Transactions on The Built Environment* p 30

[17] Hervouet J M 1999 TELEMAC, a hydroinformatic system *La Houille Blanche* **3–4** 21–8

[18] Madsen P A and Larsen J 1987 An Efficient Finite-Difference Approach to the Mild-Slope Equation *Coastal Engineer.* **11** 329–51

[19] Guillou N and Chapalain G 2012 Modeling penetration of tide-influenced waves in Le Havre harbor *J. of Coastal Res.* **28(4)** 945–55
[20] Prodanovic P 2014 Numerical simulation of coastal climate at a harbour site in the Great Lakes *21st TELEMAC-MASCARET User Conf.* pp 15–7
[21] Kumar V S, Babu V R, Babu M T et al 2008 Assessment of storm surge disaster potential for the Andaman Islands *J. of Coastal Res.* **24(sp)** 171–7
[22] Reed C W, Erown B M, Sánchez A et al 2011 The coastal modeling system flow model (CMS-Flow): Past and Present *J. of Coastal Res.* 1–6

IOP Conf. Series: Materials Science and Engineering **905** (2020) 012036 doi:10.1088/1757-899X/905/1/012036

A new method for controlling the quality of technological products

A M Karmishin[1], L K Orlik[1] and G S Zhukova[2]

[1] Russian State Social University, Moscow, Russia
[2] Financial University under the Government of the Russian Federation, Moscow, Russia

E-mail: vaxzk8chif@mail.ru, lubov.orlik@gmail.com, galsevzhukova@mail.ru

Abstract. The problem of conformity of products to the requirements of technical control or between products of different batches puts the methods of experimental determination of product properties high on the agenda. The measurement of the parameters of these properties is stochastic. Some product properties can only be determined in specially designed experiments. Thus, in the production of powders, their dispersion can be normalized, the specific charge of condenser powders can increase, the location of magnetic granules of nanoparticles on information carriers can be controlled, the content of physiologically active substances in drugs can be controlled, and the spectrum of mobilities of atmospheric particles can be analyzed. Possible approaches to solving the problem of checking a batch of products for compliance with the requirements of technical specifications (TS) are considered using the example of powder dispersion. For the first time, a method has been proposed for comparing product properties, which are continuous random variables, by determining the joint area under their differential curves. The decision on the conformity of the measured product properties to the requirements of technical specifications is based on a comparative assessment with the directive established joint area under the distribution density curves. The technique of determining the probability of correspondence of normal distribution laws to each other is described. A final analytical solution is obtained for two exponential distribution laws.

1. Introduction

In the production of products in practice, the question arises of its compliance with the requirements of technical specifications or the correspondence between products of various lots. Some product properties can only be determined in specially designed experiments. In the production of powders, their dispersion can be normalized, when information is placed on special magnetic media, the location of magnetic granules of nanoparticles of the same size, shape and orientation can be controlled, in the manufacture of drugs, the content of physiologically active substances can be controlled, while studying natural and anthropogenic factors involved in the formation aerodispersion systems, the spectrum of mobility of atmospheric particles can be analyzed; in the production of tantalum electrolytic x capacitors the specific charge capacitor powders can be increased [1–6]. The experimental determination of product properties should be measured using scientifically justified methods.

Depending on the physical meaning of a product property, on a quantitative level it can be described in two different ways: in the form of a determinate value (the product property must be greater or less than a given value); in the form of a distribution law (for properties that are discrete or continuous random variables). If the controlled property of products varies from batch to batch slightly, then we can talk about high quality, otherwise the quality of the process is low and measures must be taken to improve it.

Despite the fact that some product properties are a deterministic quantity, in practice they are judged by the measurement results, which, as is well known, are random in nature.

In this regard, one can speak of the conditional determinism of these properties.

2. Methods and materials

2.1. Methods for experimental determination of product properties that are discrete random variables

The task of evaluating an unknown parameter of the studied product property with a given reliability $\gamma = 1-\alpha$ is solved using the confidence interval taking into account the additional condition for confidence boundaries, based on the essence of the quantitative characteristics. If the accuracy of the parameter should be no more than a given value, then the additional requirement to the boundaries is usually due to the fact that the probability of deviation of the measured quantity is greater than the specified one, that is, to the left or to the right of the confidence limits should be the same and equal to $(1-\alpha)/2$ [7].

If, according to the condition of the problem, the measured value, for example, the content of the reagent in the sample, should be not less than a certain minimum admissible value C_{min}, then the condition for estimating the confidence probability is as follows:

$$P\{C > C_{min}\} = \alpha$$

If we assume that the measurement errors of the relative content of the substance in the sample are small, then, due to the large number of commensurate random errors affecting the measurement, it can be assumed with sufficient accuracy for practice that the result of measuring the relative content of the active principle in the sample is a continuous random variable distributed over normal law. Then, to estimate the probabilities, one can use an integral function with an unknown mean square deviation s_1, which is estimated from the results of measurements. So, the probability that a random value of the measured quantity, for example, the concentration of the active principle in the sample C will be less than a certain value C_{min} determined by the requirements of TU, can be found as follows:

$$P = 0{,}5\left[1 + erf\left(\frac{C_{min} - \overline{C}}{\sqrt{2}s_1}\right)\right],$$

where $\overline{C}$ is the average value of the measured substance content in the sample.

The probability that the random value of the measured concentration of the active principle in the sample C will be not less than a certain value C_{min} determined by the requirements can be found by the formula:

$$P = 0{,}5\left[1 - erf\left(\frac{C_{min} - \overline{C}}{\sqrt{2}s_1}\right)\right],$$

while the probability that the substance content in the sample is in the range from C_1 to C_2 will be equal to:

$$P = 0{,}5\left[erf\left(\frac{C_2 - \overline{C}}{\sqrt{2}s_1}\right) - erf\left(\frac{C_1 - \overline{C}}{\sqrt{2}s_1}\right)\right].$$

It follows from the foregoing that there will always be some likelihood that the studied product property will not meet the requirements of technical specification. It follows that the requirements of technical specifications should be set based on the guaranteed value of the studied quantitative characteristics of the product, which in practice is not always observed.

2.2. Methods for experimental determination of product properties that are continuous random variables

The most complete characteristic of a continuous random variable is the law of its distribution, which is given as a function of the probability density function, or as a distribution function. Note that the law is considered given if its type and its parameters are known.

So, for example, the distribution of the mass of powder particles or the mass of aerosol particles by their size is described by a logarithmically normal law, which has the following form [8]:

$$\varphi_m(d) = \frac{1}{\sqrt{2\pi}\,\tilde{d}\,\sigma_{ln\,d}} e^{-\frac{(ln\,d - ln\,d_{em})^2}{2\sigma^2_{ln\,d}}};$$

$$P = 0{,}5\left[1 + erf\left(\frac{ln\,d - ln\,d_{em}}{\sqrt{2}\sigma_{ln\,d}}\right)\right],$$

where d is the particle diameter, μm; d_{em} is a law parameter: median particle mass diameter, μm; $\sigma_{ln\,d}$ is law parameter: standard deviation of the natural logarithm of a random particle diameter.

Note that a similar distribution has a random dose of the drug that causes a given therapeutic effect (for this, the particle size d can be replaced by dose D) [9].

In relation to powder, the question arises whether or not the dispersion of the powder differs significantly from batch to batch and whether it meets the requirements of TS. If the differences from the point of view of mathematical statistics are not significant, then the technological process can be considered high-quality and stable. Otherwise, process improvement will be required.

For a physiologically active substance/drug, the following question can be formulated. Does the physiological activity of the drug differ significantly from batch to batch, the requirements of TS, or compared with another drug? If these differences are not significant, then it is considered that the batch of product meets the requirements of TS. Summarizing the above, it can be noted that in practice, according to the results of experimental studies of a continuous random variable, the distribution law and its parameters are determined, so the question arises of the conformity of the laws of distribution of the measured property of a product from batch to batch.

Let us consider possible approaches to solving the problem of checking a batch of products for compliance with the requirements of technical specifications using the example of powder dispersion. Let it be established from the primary measurement data of particle sizes that their mass is located according to a logarithmically normal law and its parameters d^*_{em} and $\sigma^*_{ln\,d}$ are determined.

It is necessary to compare this distribution law with the requirements of TS (with another batch), according to which the log-normal law has parameters d_{em} and $\sigma_{ln\,d}$.

Following the ideas of classical mathematical statistics, one can compare the parameters of the laws in question in pairs. For each of the compared parameters, three incompatible events are possible: one of the considered parameters will be less, equal or greater than the other. Then the total number of possible combinations of incompatible events when comparing the two parameters of the laws of distribution will be 3^2=9:

1) $d^*_{em} < d_{em}$ and $\sigma^*_{ln\,d} < \sigma_{ln\,d}$;
2) $d^*_{em} < d_{em}$ and $\sigma^*_{ln\,d} = \sigma_{ln\,d}$;
3) $d^*_{em} < d_{em}$ and $\sigma^*_{ln\,d} > \sigma_{ln\,d}$;
4) $d^*_{em} = d_{em}$ and $\sigma^*_{ln\,d} < \sigma_{ln\,d}$;
5) $d^*_{em} = d_{em}$ and $\sigma^*_{ln\,d} = \sigma_{ln\,d}$;

6) $d_{em}^{*} = d_{em}$ and $\sigma_{\ln d}^{*} > \sigma_{\ln d}$;

7) $d_{em}^{*} > d_{em}$ and $\sigma_{\ln d}^{*} < \sigma_{\ln d}$;

8) $d_{em}^{*} > d_{em}$ and $\sigma_{\ln d}^{*} = \sigma_{\ln d}$;

9) $d_{em}^{*} > d_{em}$ and $\sigma_{\ln d}^{*} > \sigma_{\ln d}$.

Similar combinations can be made for the parameters of the law of distribution of random doses of a drug that cause a given therapeutic effect.

It should be borne in mind that when comparing two random variables, the following hypotheses can be formulated:

- hypothesis H is true and is accepted according to the selected criterion (correct decision);
- hypothesis H is false and is rejected according to the selected criterion (correct decision);
- hypothesis H is true, but rejected according to the selected criterion (incorrect decision is an error of the first kind);
- hypothesis H is false, but is accepted according to the selected criterion (incorrect decision is an error of the second kind).

Thus, an error of the first kind is associated with the probability of rejection of the null (true) hypothesis, and a mistake of the second kind is associated with the probability of adopting an alternative null (false) hypothesis [10].

Sometimes, a quantile can be specified as the requirements of a technical specification: a guaranteed particle size and a certain mass, concentrated on particles of smaller or larger sizes. Such a specification of requirements does not fully characterize the law of distribution of a random variable. We illustrate this with the following example: let the particles of one batch have a median mass diameter $d_{em} = 4$ µm and $\sigma_{lnd} = 0.175$, while for other particles $d_{em} = 1$ µm and $\sigma_{lnd} = 1.263$.

For each of the considered distributions, we have: F(5) = 0.9 (i.e., 90% of their mass is concentrated on particles up to 5 µm). However, these are completely different distributions, which is clearly illustrated in Figures 1 and 2.

From the given example it follows that for the properties of products described by continuous random variables, the law parameters or a quantile of order p must not be compared, but their most complete characteristics, i.e. distribution laws.

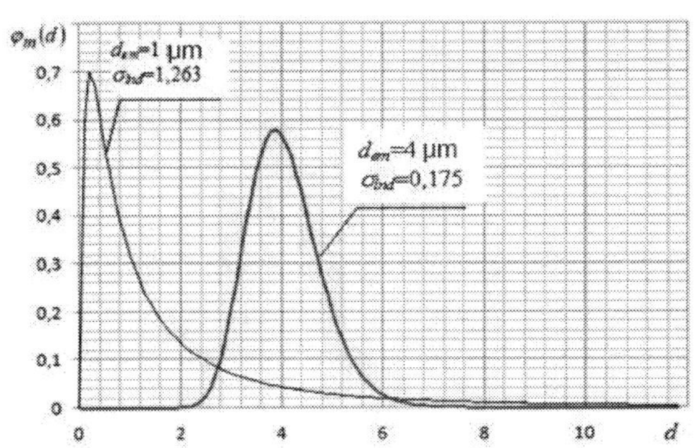

Figure 1. Two probability densities of logarithmically normal laws of mass distribution by particle size

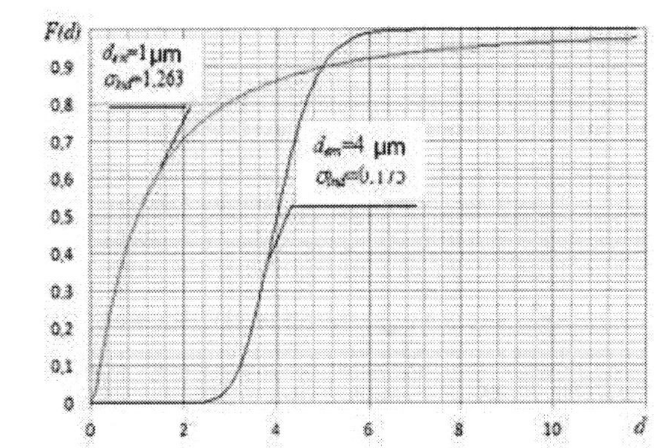

Figure 2. Two distribution functions of the logarithmically normal laws of mass distribution over particle sizes

As far as the authors of this article know, in mathematical statistics such a question is posed for the first time.

3. Results

As a measure of the correspondence of the two distribution laws to each other, we consider the total (combined) area under two differential curves (Figs. 3a and 3b).

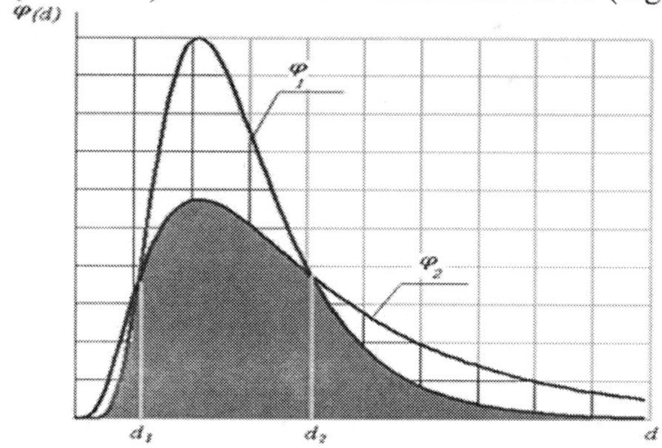

Figure 3. Illustration to a comparison of two distribution laws. Possible configuration a)

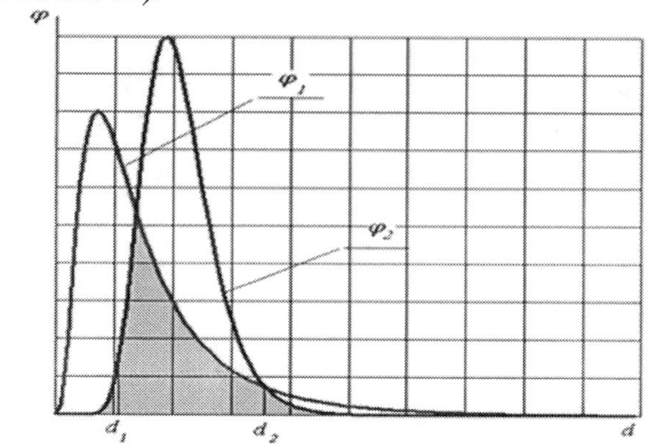

Figure 4. Illustration to a comparison of two distribution laws. Possible configuration b)

Interestingly, the area under the differential curve inherently determines the probability and is normalized to unity. Then, if two differential curves coincide, which is possible only if the parameters of the distribution laws are completely equal, then the laws themselves coincide with a probability equal to unity. If the total area under the differential curves of the distribution laws tends to zero, then they coincide with the probability tending to zero and do not coincide, that is, they are different, with the probability tending to unity.

To find the total area under two differential curves, we need to find the points of their intersection. So, for example, equating two logarithmically normal densities and solving the quadratic equation, we get two roots, respectively, two intersection points:

$$d_{1,2} = e^{\frac{b \pm \sqrt{b^2 - 4ac}}{2a}},$$

where

$$a = \frac{1}{2}\left(\frac{1}{\sigma^{*2}_{ln\,d}} - \frac{1}{\sigma^{2}_{ln\,d}}\right); \quad b = \frac{ln\,d^{*}_{em}}{\sigma^{*2}_{ln\,d}} + \frac{ln\,d_{em}}{\sigma^{2}_{ln\,d}};$$

$$c = \frac{\left(ln\,d^{*}_{em}\right)^2}{\sigma^{*2}_{ln\,d}} - \frac{\left(ln\,d_{em}\right)^2}{\sigma^{2}_{ln\,d}} - 0.5\,ln\frac{\sigma_{ln\,d}}{\sigma^{*}_{ln\,d}}.$$

After the intersection points of two differential curves are found, determining the total area under them is not difficult.

In computer-assisted probability calculations for Figs. 3 and 4, four situations are possible. However, if the distribution laws are arranged in such a way that the parameters of a law which median particle diameter is smaller to be the first to be input (less than the median dose for a physiologically active substance), then only two situations can be considered.

For the case when the parameters of the distribution laws for computer calculations are introduced in exactly this sequence, that is, for $d_{em1} < d_{em2}$, the total area under two differential curves is calculated as follows:

$$\text{if } \sigma_{ln\,d1} > \sigma_{ln\,d2} \left(\sqrt{k_1} < \sqrt{k_2}\right), \text{ then}$$

$$P_s = P_1(d_1) + [P_2(d_2) - P_2(d_1)] + [1 - P_1(d_2)],$$

$$\text{if } \sigma_{ln\,d1} < \sigma_{ln\,d2} \ (\sqrt{k_1} > \sqrt{k_2}), \text{ then}$$

$$P_s = P_2(d_1) + [P_1(d_2) - P_1(d_1)] + [1 - P_2(d_2)].$$

Here $P_1(d)$ and $P_2(d)$ are integral functions of mass distribution over particle sizes, calculated by the following equation:

$$P(d) = \frac{1}{2}\left[1 + erf\left(\sqrt{k}\, ln \frac{d}{d_{em}}\right)\right], \quad \sqrt{k} = \frac{1}{\sqrt{2}\sigma}.$$

We obtain similar equations when comparing the physiological activity of two batches of a product or two different substances, for which it is sufficient to replace the particle size symbol d with the dose symbol D.

For the final decision on the conformity or divergence of the two laws of distribution or the conformity of the measured product properties to the requirements of TS, it is necessary to establish the level of probability of compliance of the two distribution laws. If as a result of the calculations it is established that the joint area under the two distribution curves is not less than the directive established, then the hypothesis of the equality of the two distribution laws is accepted, otherwise the laws do not match.

Figs. 5 and 6 show the correspondence of two distribution laws having the following parameters:

d_{em1}=0.3 μm and $\sqrt{k_1}$=1.0;

d_{em2}=0,5 μm and $\sqrt{k_2}$=1.5 (a);

d_{em1}=0.3 μm and $\sqrt{k_1}$=1.0;

d_{em2}=0.325 μm and $\sqrt{k_2}$=0.94.

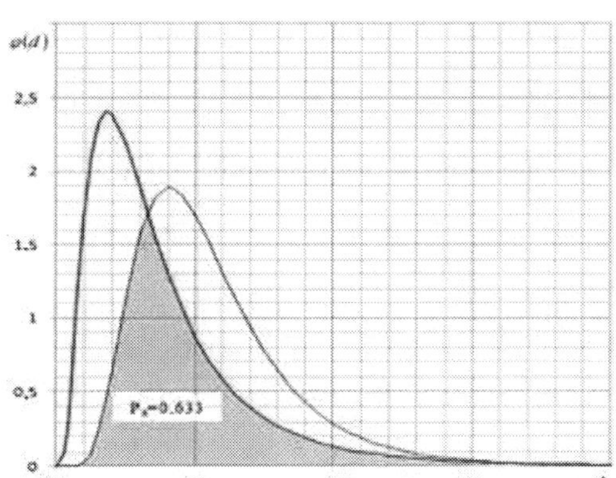

Figure 5. Illustration of the case of a low probability of coincidence of two distribution laws

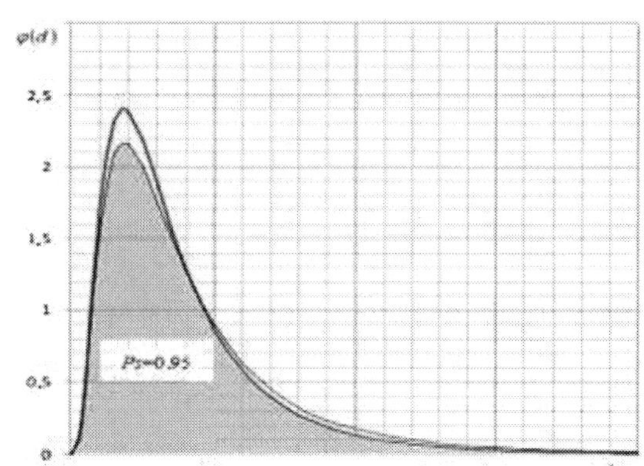

Figure 6. Illustration of the case of high probability of coincidence of two distribution laws

Similar considerations apply to product properties, which are continuous random variables distributed according to the normal law.

The presented idea of determining the probability of correspondence of the exponential laws of the distribution of random variables allowed obtaining the final analytical solution.

Let there be two exponential distribution laws with parameters λ_1 and λ_2. Let us specify $\lambda_1 < \lambda_2$ (Figure 7).

From the analysis of the two densities of the exponential distribution laws presented in Figure 7, it follows that the joint area under two differential curves is found as

$$P_s = F(\lambda_1, x_1) + [1 - F(\lambda_2, x_1)] \tag{1}$$

where x_1 is the intersection point of two densities of exponential distribution laws

$$x_1 = \frac{1}{\lambda_2 - \lambda_1} ln \frac{\lambda_1}{\lambda_2}; \qquad F(\lambda, x) = 1 - e^{-\lambda x}.$$

Substituting x_1 into (1), we find the probability of coincidence of two exponential distribution laws:

$$P_s = 1 - \left(\frac{\lambda_1}{\lambda_2}\right)^{\frac{\lambda_1}{\lambda_2 - \lambda_1}} + \left(\frac{\lambda_1}{\lambda_2}\right)^{\frac{\lambda_2}{\lambda_2 - \lambda_1}}. \tag{2}$$

An analysis of formula (2) shows that it is correct in limiting cases. If $\lambda_1 = \lambda_2$, then P_s=1. Thus, the laws of distribution coincide with a probability of unity. If $\lambda_1 << \lambda_2$, then P_s tends to zero, which means that the laws in question do not coincide.

Let us introduce dimensionless criterion k_{ov} [0, 1]:

$$k_{ov} = \frac{\lambda_1}{\lambda_2}.$$

Then the probability of coincidence of distribution laws (2) through the criterion k_{ov} can be written as

$$P_s = 1 - k_{ov}^{\frac{k_{ov}}{1-k_{ov}}} + k_{ov}^{\frac{1}{1-k_{ov}}}. \tag{3}$$

Table 1 contains calculated by relation (3) probabilities of coincidence of the laws of distribution of two exponentially distributed random variables.

Table 1. Probabilities of coincidence of laws of distribution

Complex k_{ov}	Probability Ps	Complex k_{ov}	Probability Ps
0	0	0.55	0.783
0.05	0.189	0.60	0.814
0.10	0.303	0.65	0.843
0.15	0.392	0.70	0.869
0.20	0.465	0.75	0.895
0.25	0.528	0.80	0.918
0.30	0.582	0.85	0.940
0.35	0.631	0.90	0.961
0.40	0.674	0.95	0.981
0.45	0.714	1.00	1.00
0.50	0.750		

Figure 8 shows a graph of the probability of coincidence of two exponential laws depending on the dimensionless complex k_{ov}.

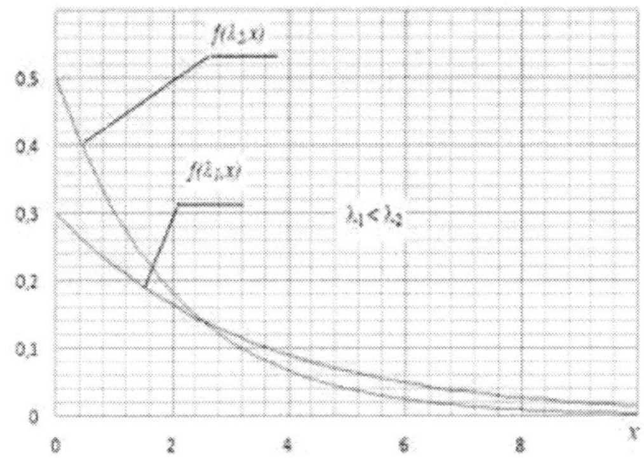

Figure 7. Illustration to comparison of two exponential distribution laws

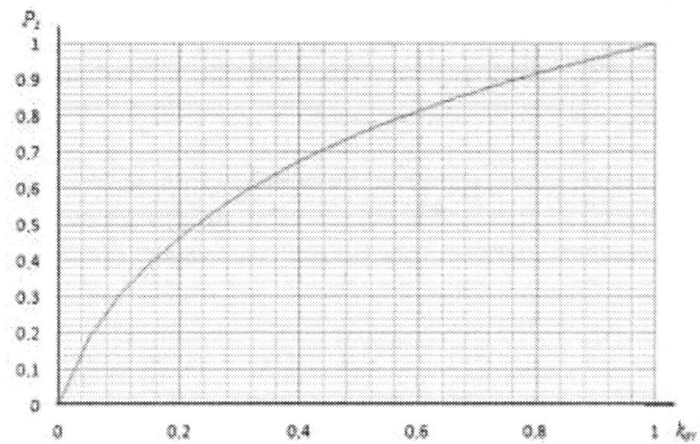

Figure 8. Probability of coincidence of two exponential laws depending on the dimensionless complex k_{ov}

Obviously, similar calculations can be made for other distribution laws, for example, for the law of equal density.

4. Conclusion

The requirements of technical specifications for the level of certain quantitative characteristics should be formulated on the basis of statistical, not deterministic, ideas about these properties of products.

For the first time, a method has been proposed for comparing the properties of products that are continuous random variables by determining the combined area under their differential curves, which, in essence, is nothing more than the probability of coincidence of the laws of distribution of random variables under consideration.

Using the developed method for comparing the two distribution laws, various practically important applied problems can be solved. For example, an analysis of the professional training of specialists performing experimental research. If the results of measuring the properties of products from the same batch, carried out by two different employees, do not differ from the point of view of mathematical statistics, then the level of their professional training is the same, otherwise the level of their professional training is different.

In a similar way, process stability can be compared.

References

[1] Meilichov E Z and Farzetdinova R M 2014 Maximum Density of Magnetic Recording and Distribution of Switching Fields *Solid Body Phys.* **56** 2326–34

[2] Laakso L, Gagne S, Petäja T et al 2007 Detecting charging state of ultra-fine particles: instrumental development and ambient measurements *Atmos. Chem. Phys.* **7** 1333–45

[3] Varekhov A G 2014 Spectral analysis mobility of atmospheric particles *Sci. instrum. Making* **24(4)** 51–8

[4] Edward M C 2000 Tantalum and Niobium A Review of Worldwide Industry Statistics *TIC Bui 102* 2–3

[5] Kolov V N, Miroshnenko M N, Orlov V M and Prokhorov T Y 2005 Kalietermic production of tantalum powders *ZHPC* **78(4)** 545–7

[6] Libenson G A, Lopatin V J and Komarnicki G V 2001 *Powder metallurgy processes. Production of metal powders* Textbook for univer. vol 1 (Moscow: MISIS)

[7] Kramer G 2003 *Mathematical methods of statistics* (Moscow; Izhevsk: Res. Center Regular and stochastic dynamics)

[8] Kobsar A I 2006 *Applied mathematical statistics. For engineers and researchers* (Moscow: Phys. Math. Lit.)

[9] Karmishin A M and Kireev A A 2011 *Mathematical methods of pharmacology, toxicology and radiology* 2rd ed (Moscow: APR) 330 p
[10] Orlik L.K., Semenov S.A., Guseva S.V et al 2009 *Terrorism: problems, models, scenarios* (Moscow)

IOP Conf. Series: Materials Science and Engineering **905** (2020) 012037 doi:10.1088/1757-899X/905/1/012037

Stoichiometric modification of $BiFeO_3$ / REE to create new materials and methods for their design for artificial intelligence systems

S V Khasbulatov[1,2], L A Shilkina[1], V A Aleshin[1], D I Gordienko[1], S N Kallaev[3], Z M Omarov[3], I A Verbenko[1] and L A Reznichenko[1]

[1] Research Institute of Physics, Southern Federal University, 194, Stachki Ave., Rostov-on-don, 344090, Russia
[2] Faculty of physics and mathematics, Chechen State University, 32, Sheripova str., Grozny, 364024, Russia
[3] Dagestan Scientific Center of the Russian Academy of Sciences, 45, M. Gadjieva st., Makhachkala, 367025, Russia

E-mail: said_vahaevich@mail.ru, shilkina@yandex.ru, alyoshin09@ip.rsu.ru, grash002@yandex.ru, kallaev-s@rambler.ru, gadjiev@mail.ru, ilich001@yandex.ru, reznichenko@mail.ru

Abstract. The article discusses the effects of modifying bismuth ferrite $BiFeO_3$ with rare-earth elements, REE, (large-sized, group 1, with $0.94 \leq \bar{R} \leq 1.04$ Å – La, Pr, Nd, Sm, Eu, Gd, medium-sized, small-sized Group 2, with $\bar{R}$ <0.94 Å – Tb, Dy, Ho, Er, Tu, Yb, Lu). The authors describe the study results of the influence of the crystallophysical parameters of stoichiometrically introduced dopants on the type of phase diagrams of $Bi_{1-x}REE_xFeO_3$ systems, the grain structure of ceramics, the dielectric spectra of samples, and the behaviour of their thermophysical characteristics over a wide temperature range. They show the possibility of using new multiferroic materials in artificial intelligence systems.

1. Introduction

At the current stage of development, science and technology in the Russian Federation are becoming a decisive factor that ensures the country's ability to meet big challenges with significant risks, especially for the economy: exhaustion of raw materials, demographic changes in society, an increase in anthropogenic pressures on humans and the environment, the need to increase Russia's competitiveness in world markets, etc. All this determined the contemporary priorities and prospects of the scientific and technological development of the Russian Federation, including in the material science field, which will create a transition to new materials and methods for constructing them for artificial intelligence systems, to environmentally friendly, resource-saving energy [1]. In this regard, we consider relevant research aimed at establishing correlation relationships: composition – phase state diagrams – grained landscape – dielectric/thermophysical properties of environmentally friendly compositions based on solid solutions, SS, $BiFeO_3$ promising spintronics materials [2].

2. Methods and materials

The objects of the study were following SS: Bi_{1-x} REE_xFeO_3 (REE – La, Pr, Na, Sm, Eu, Gd, Tb, Dy, Ho, Er, Tu, Yb, Lu; $0 \leq x \leq 0.5$; $\Delta x = 0.025$; 0.05; 0.10).

The main method for preparing the analyzed SS is double solid-phase synthesis with intermediate grinding and granulation of dispersed crystalline powders, sintering using conventional ceramic technology.

Synthesis and sintering modes: $T_1 = 1063 \div 1183$ K, $\tau_1 = 5 \div 10$ h; $T_2 = 1073 \div 1183$ K, $\tau_2 = 4 \div 10$h; $T_{sintering} = 1143 \div 1273$K, $\tau_{sintering}$=1,5÷2h;

We performed X-ray diffraction studies at room temperature using the powder diffraction method with a DRON-3 diffractometer and vapour pressure apparatus, VPA (Sok$_\alpha$ – radiation; Mn – filter; Bragg – Brentano focusing scheme). We used a VPA-1 automatic diffractometer with a VEB Freiberqer Prazisionsmehaniko goniometer for studying a high-temperature (300–1000) K) study of the crystalline structure of SS. X-ray density ($\rho_{x\text{-ray.}}$) was determined by the formula: $\rho_{x\text{-ray}} = 1.66 * M / V$, where M is the weight of the formula unit in grams, V is the volume of the perovskite cell in Å^3. The measurement errors of structural parameters have the following values: linear

$\Delta a = \Delta b = \Delta c = \pm$ (0.002 ... 0.004) Å; angular $\Delta\alpha = 3$ '; volume $\Delta V = \pm$ 0.05Å^3 ($\Delta V / V * 100$ % = 0.07 %).

We determined the experimental (ρ_{exp}) density of the samples by hydrostatic weighing where we used octane as a liquid medium. Relative density ($\rho_{rel.}$) was calculated by the formula ($\rho_{exp.} / P_{x\text{-ray.}}$) * 100 %.

We studied the grain structure of objects using a Neoophot 21k optical microscope and a Leica DMI-5000M inverted high-precision microscope

We used a special bench constructed by the Scientific Research Institute of Physics of the Southern Federal University with Agilent 4980A, WayneKerr 6500B precision LCR meters for studying the high-temperature dielectric spectra. We performed the measurements in the temperature range (300 ... 900) K and the frequency range (20 ... 10^6) Hz.

Thermophysical studies were carried out at the Kh.I. Amirkhanov Institute of Physics of the Dagestan Scientific Center of the Russian Academy of Sciences. We measured the coefficient of thermal expansion (CTE, α) by a capacitive dilatometer developed at the Institute of Physics, DSC RAS; we determined specific heat capacity (C_p)on a differential scanning calorimeter DSC 204 F1 Phoenix German company NETZSCH; we studied thermal diffusivity (χ) by laser flash using an LFA 457 G Microflash installation of the German company NETZSCH; we calculated thermal conductivity (λ) according to the formula: $\lambda(T) = \chi(T)C_p(T)\rho_{exp.}(T)$[3].

3. Results

Figure 1–2 shows the dependence of the structural characteristics and densities of the 1st group objects on the content (x) of modifiers. All cases of the state diagram contain five concentration regions corresponding to successive structural transformations $Rh \rightarrow Rh+R_1 \rightarrow Rh+R_1+R_2 \rightarrow R_1+R_2 \rightarrow R_2$ with small variations in the phase filling, and Figure 3 illustrates the position dependences of the boundaries of the existence of the pure Rh phase, its mixture with the R_1 – and R_2 – phases, and the differences between R_{Bi} and R_{REE}, V_{exp} and V_{theor} (ΔV) on the radius, $\bar{R}$, REE.

The stabilization of the Rh phase upon the introduction of La, Pr, Nd (with $\bar{R} \geq 0.99$Å) probably results from the creation of the most favourable (dimensional) conditions for the existence of $BiFeO_3$, which, as is known [4], has a boundary position in the perovskite family. At lower $\bar{R}$, the shift of the Rh – boundary toward lower REE concentrations is a consequence of a significant deviation of the conditions from the necessary for the formation of SS substitution ($\Delta\bar{R}$ should be less than 15 % [5]). The latter resulted in the stability loss of the region of coexistence of the Rh– and $R_{1,2}$ – phases due to the greater instability of the structure of multiphase SSs compared to single-phase ones.

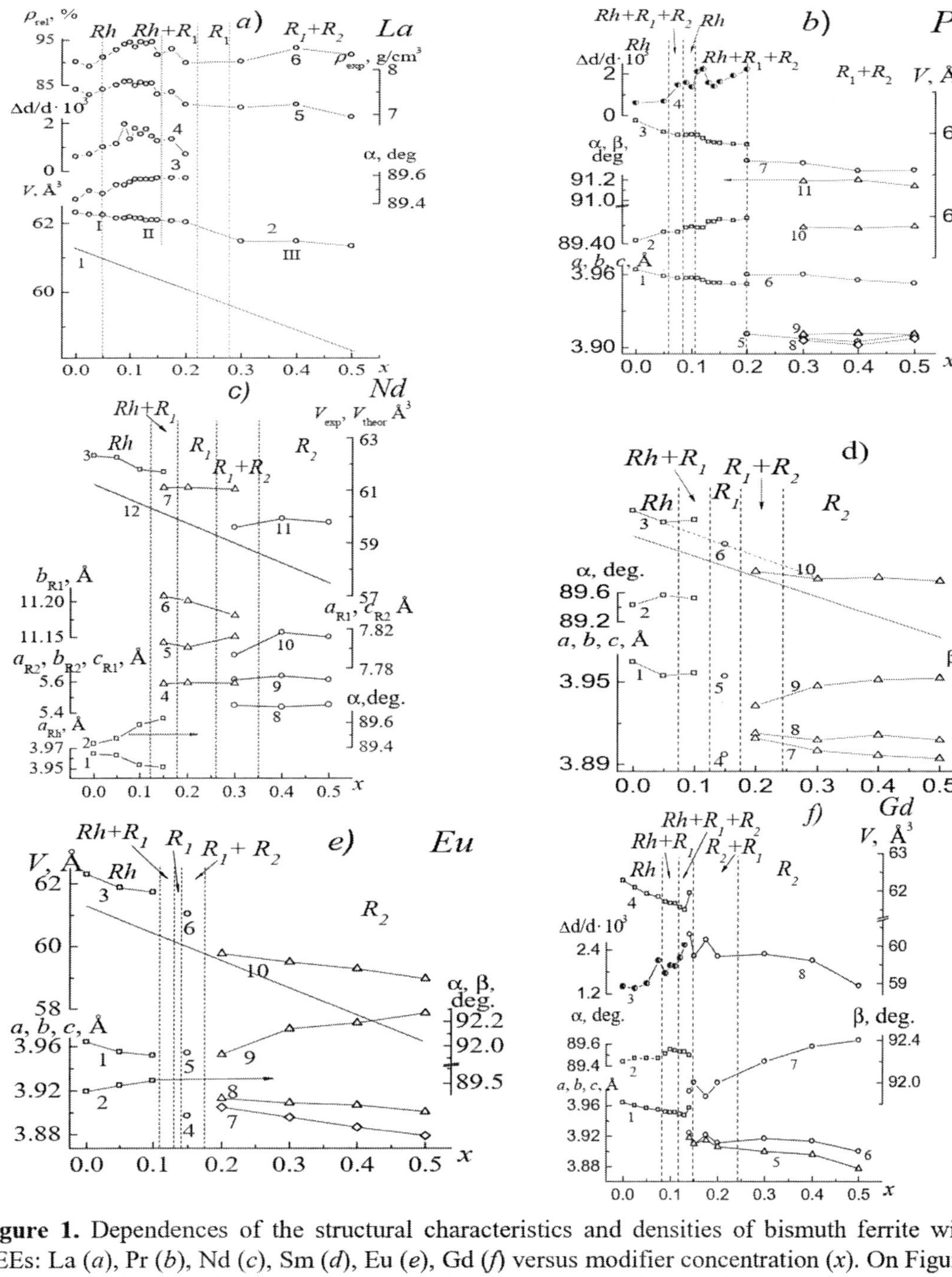

Figure 1. Dependences of the structural characteristics and densities of bismuth ferrite with large REEs: La (*a*), Pr (*b*), Nd (*c*), Sm (*d*), Eu (*e*), Gd (*f*) versus modifier concentration (*x*). On Figure *a)* – 1 – theoretical, 2 – experimental perovskite cell volumes, 3 – *a*, *4* – $\Delta d/d$, *5*-$\rho_{exp.}$, *6*-ρ_{rel}.; on Figure *b)* – 1, 2, 3 – *a*, λ, V_{Rh}, 5, 6, 7 – *a*, *c*,V_{R_1}, 8, 9. 10, 11 – *в*, *a*, V_{R_2}, respectively, 4 – microdeformations, $\Delta d/d$; on Figure *c)* – 1, 2, 3, *a*, λ, V_{Rh}, 4, 5, 6 – *c*, *a*, *в* , 7–V_{R_1}, 8, 9, 10, 11 – *a*, *в*, *c*, V_{R_2},respectively, 12– $V_{Theor.}$ perovskite cell; on fig *d)* 1, 2, 3 – *a*, λ, V_{Rh}, 4, 5, 6 – *c*, *a*, V_{R_1}, 7, 8, 9, 10 – *в*, *a*, *β*, V_{R_1},respectively; on fig *e)* 1, 2, 3 – *a*, *λ*, V_{Rh}; 4, 5, 6 – *c*, *a*, V_{R_1}, 7, 8, 9, 10 – *в*, *a*, *β*, V_{R_2}, respectively; on Figure *f)* 1, 2, 4 – *a*, *λ*, V_{Rh}; 5, 6, 7, 8 -*в*, *a*, *β*, V_{R_2}, 3 – $\Delta d/d$, respectively. Straight lines without experimental points are the dependences of $V_{theor.}(x)$ calculated for SS substitution in *A* – positions

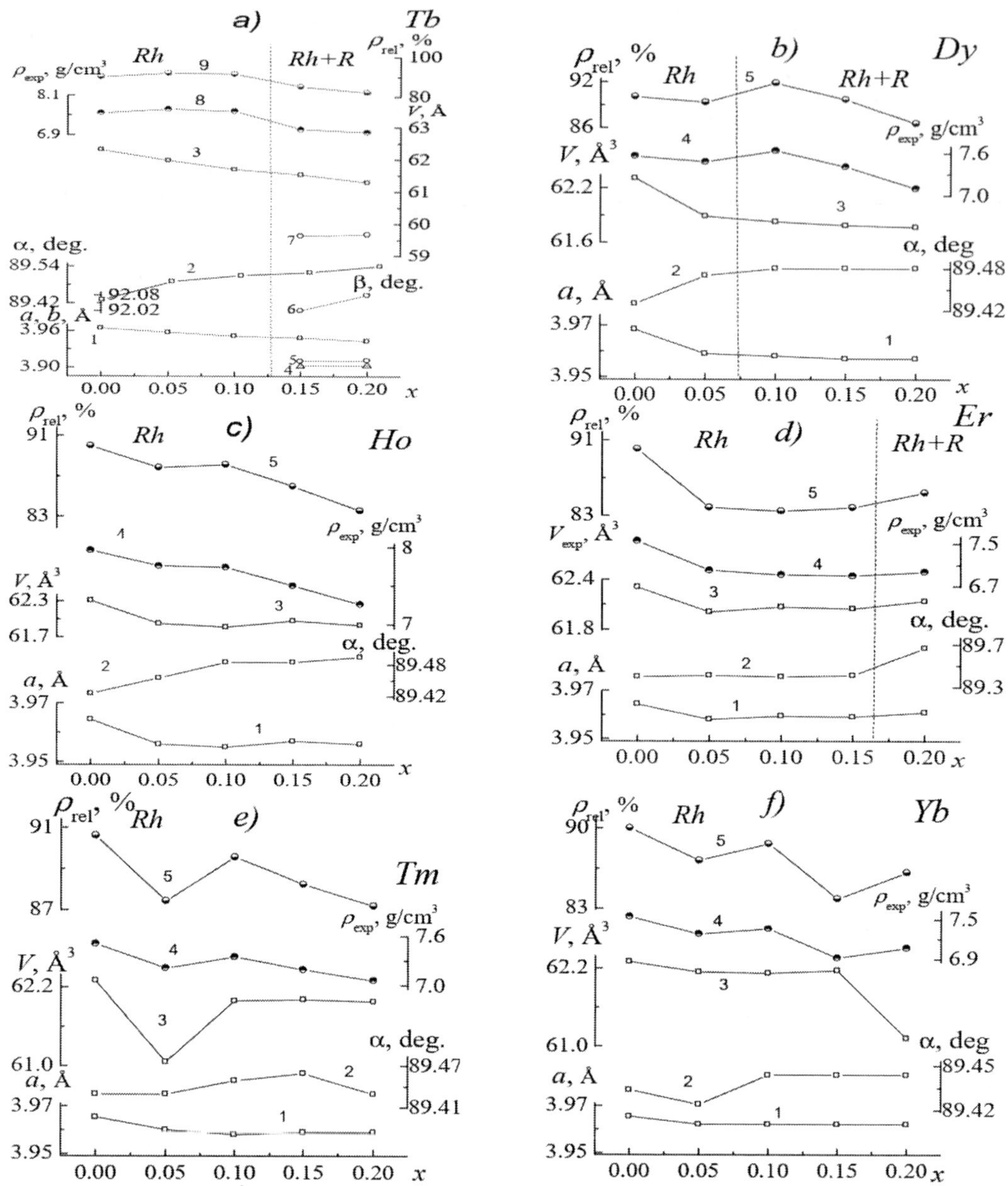

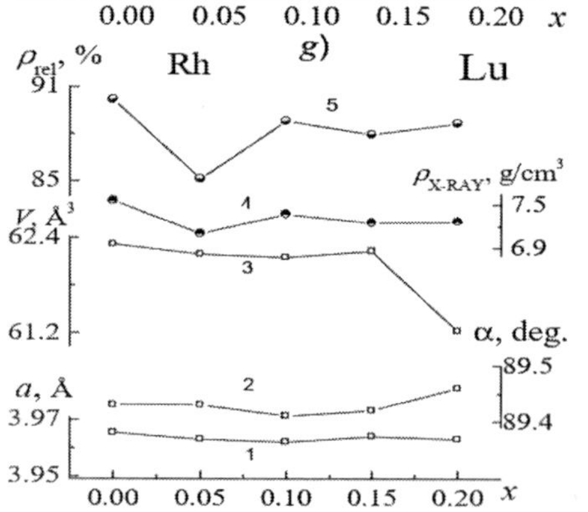

Figure 2. Dependences of the structural ıcteristics and densities of bismuth ferrite with ım-sized: Tb (*a*), Dy (*b*), Ho (*c*) and small-l: Er (*d*), Tm (*e*), Yb (*f*), Lu (g) REEs depend on :oncentration (*x*) of the modifier. On Figure (*a*): 3 – parameter *a*, angle α, volume *V* of the ıbohedral cell, 4, 5, 6, 7 – parameters *b*, *a*, angle of the rhombic cell, respectively. On Figure (B-j): , 2 – α, 3 – *V*, 4 – ρ_{exp}, 5 – ρ_{rel} *Rh*-cells.

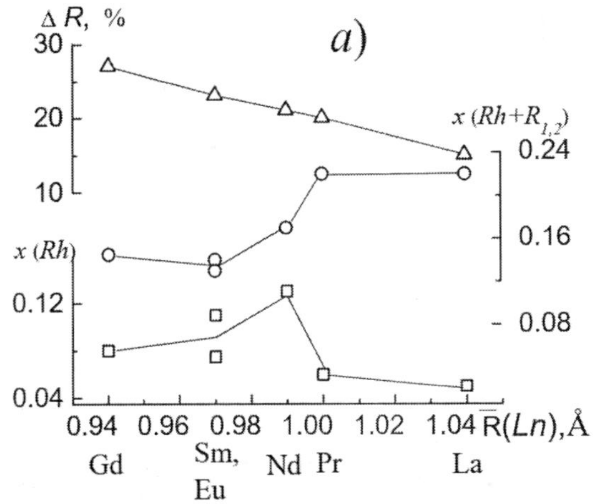

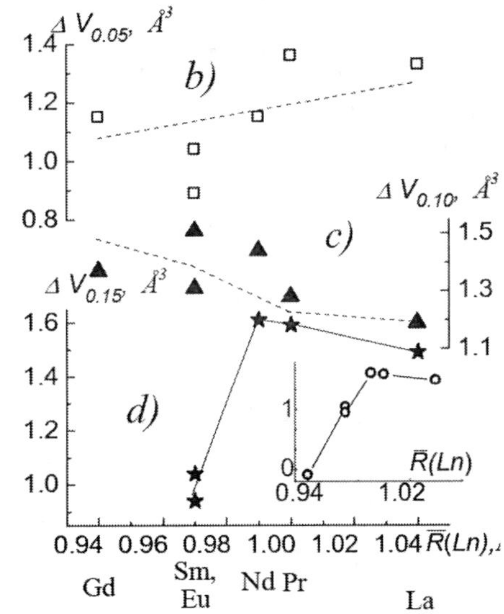

Figure 3. *a* – Dependences of the position of the boundary (x) of the existence of a pure *Rh* – phase, its mixture with $R_{1,2}$ – phases, the relative difference between $\bar{R}_{Bi}$ and $\bar{R}_{REE}$ ($\Delta\bar{R}$) on the radius,$\bar{R}$, REE
b-d – Dependencies of the difference between $V_{exp.}$ and $V_{Theor.}$ (ΔV) from the ionic radius,$\bar{R}$, REE in SS with $x = 0.05$ (b); 0.10 (s); 0.15 (d)

The reasonableness of the above assumption is also confirmed by the behaviour of ΔV (Figure 3, *a*) which characterizes the structural nonstoichiometry of $BiFeO_3$ – compounds with a variable valence ion ($Fe^{3+} \leftrightarrow Fe^{2+}$) and, as a result, it is vacancy-saturated with the defect situation excluded by a crystallographic shift with the development of anion and A – deficiency [6]. At low REE contents (Figure 3, *b-d*), their effect is minimal, and $\Delta V_{0.05}$ is practically independent of$\bar{R}$ REE against the background of a noticeable spread in ΔV values.

At large x, in the region with $\bar{R}$> 0.99 Å, ΔV changes little contributing to the stabilization of the Rh phase, and at $\bar{R}$<0.99Å in the case $x = 0.10$, $\Delta V_{0.10}$, grows naturally with decreasing $\bar{R}$ destabilizing *Rh*, – *Rh* + $R_{1,2}$– phases; in the case of $x = 0.15$, the decrease in structural non-stoichiometry, apparently, resulted not so much from size effects as the specifics of other crystallographic characteristics of the introduced REEs.

In the second group of SSs, the Rh phase characteristic of $BiFeO_3$ either remains in the entire range of REE concentrations (Ho, Tm, Yb, Lu) or coexists with the arising R phase (Tb, Dy, Er). Such a simplification of the phase patterns is due to a large amount of Bi– and Fe-containing impurities usually concomitant with the formation of $BiFeO_3$ and to ballast phases of the non-perovskite structure with the participation of REEs, the amount of which increases with decreasing $\bar{R}$REE. Figure 4 shows it with microphotographs of the grain structures of the studied objects. The accumulation of such phases, grey grains, in the La → Lu ranges leads to loosening of the microstructure, thickening of the crystallite boundaries and deforming the habit of the grains of the main phase.

Figure 5 shows the dependences $\varepsilon' / \varepsilon_0$ (T), $\varepsilon'' / \varepsilon_0$ (T), tg δ (T) on different frequencies of an alternating electric field. The results indicate the formation of a relaxation maximum $\varepsilon' / \varepsilon_0$ in $BiFeO_3$ with all large-sized REEs and some small-sized REEs near (400–450) K and a frequency-dependent maximum tg δ near 600 K. While the position of latter shifts toward lower temperatures in the process of enrichment of $BiFeO_3$ with introduced REEs. With small-sized REEs, the first maximum is localized near (550 ... 600) K, the second is near (750 ... 800) K.

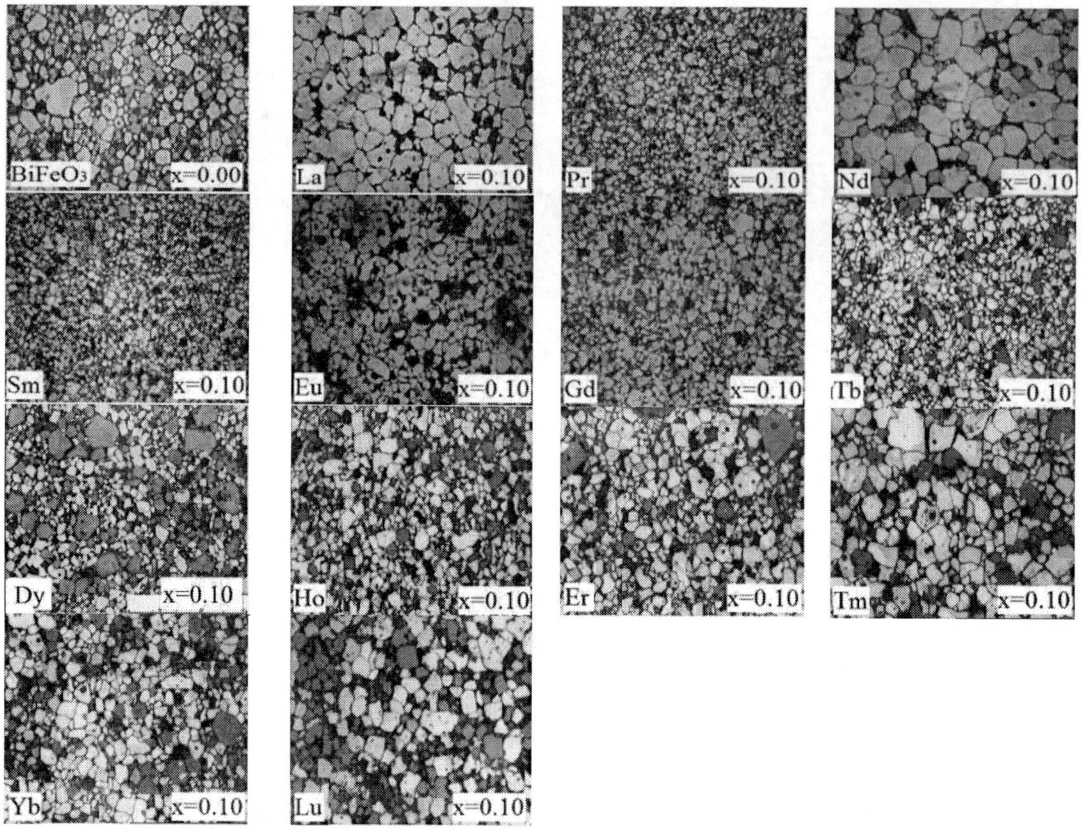

Figure 4. Fragments of the microstructures of ceramics $BiFeO_3$, $Bi_{1-x}REE_xFeO_3$ (x = 0.10). (Magnification: x1050, scale: 2 mm – 1 μm;) scale: 2 mm – 1 μm).

We developed $ln\omega$ ($1 / T_m$) graphs, where T_m is the temperatures of the extrema ε ' / ε_0 measured at a frequency $f = \omega / 2\pi$, to describe the process of low-temperature dielectric relaxation according to the data obtained from the dependences ε ' / ε_0 (ω). The obtained dependences satisfy the Arrhenius law with an activation energy E_a close to 1 eV (in $BiFeO_3$ E_a = 1.007 eV) and an average time to overcome the potential barrier ≈ $1.06 \cdot 10^{-16}$ s, which is typical for Maxwell – Wagner relaxation [7], associated with the accumulation of free charges at the interface of components in spatially inhomogeneous media against the background of interlayer, interphase, and intraphase rearrangements. The reason for its development is the natural composite structure of $BiFeO_3$ and $BiFeO_3$ / REE, which forms based on at least four (except for $BiFeO_3$ itself) Bi, -Fe-containing compounds (Bi_2O_3, Fe_2O_3, $Bi_{25}FeO_{40}$, $Bi_2Fe_4O_9$) that almost always accompany the formation of $BiFeO_3$, remaining in it (in different quantities) in the form of ballast phases and undergoing a series of bifurcations in the above temperature ranges (Figure 6) [8].

Dependences of thermophysical characteristics (Figure 7-9) on temperature are traditional with maxima and minima near T_N t

IOP Conf. Series: Materials Science and Engineering **905** (2020) 012037 doi:10.1088/1757-899X/905/1/012037

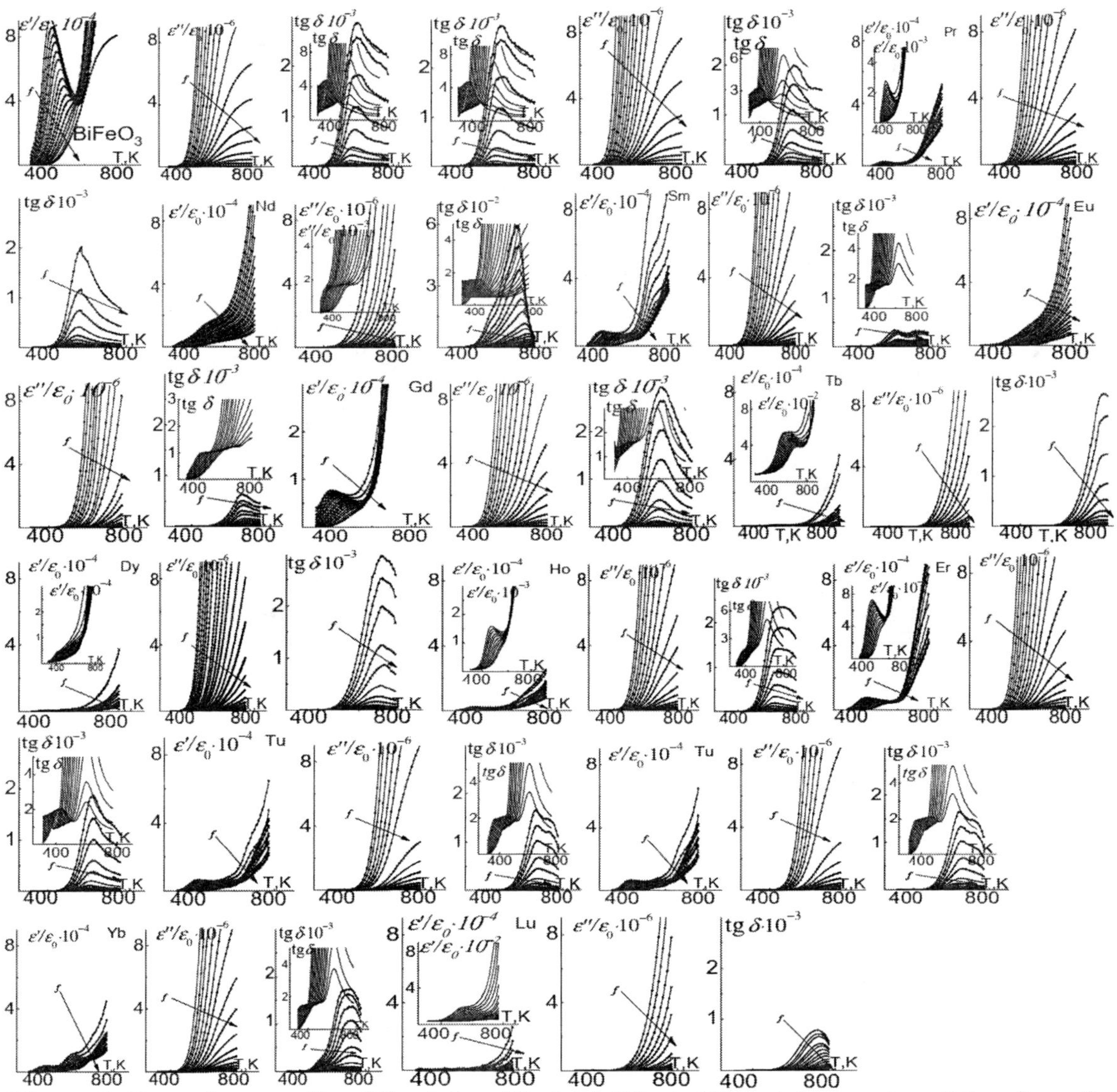

Figure 5. Dependences ε '/ ε_0 (T), ε' '/ ε_0 (T), tg δ (T) at different frequencies of the alternating electric field $BiFeO_3$ (left), $Bi_{0.9}REE_{0.1}FeO_3$ (bottom), in the frequency range (25 ÷ 1, 5 • 10^6) Hz.

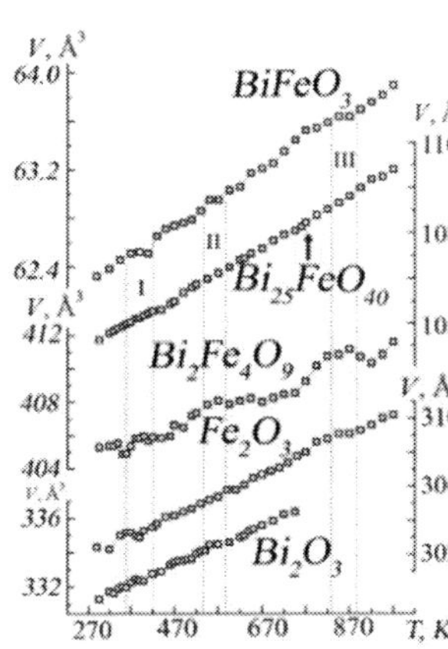

Figure 6. Dependences of the unit cell volumes of compounds – components of the naturally – composite $BiFeO_3$ structure (I – III – regions of the invar effect)

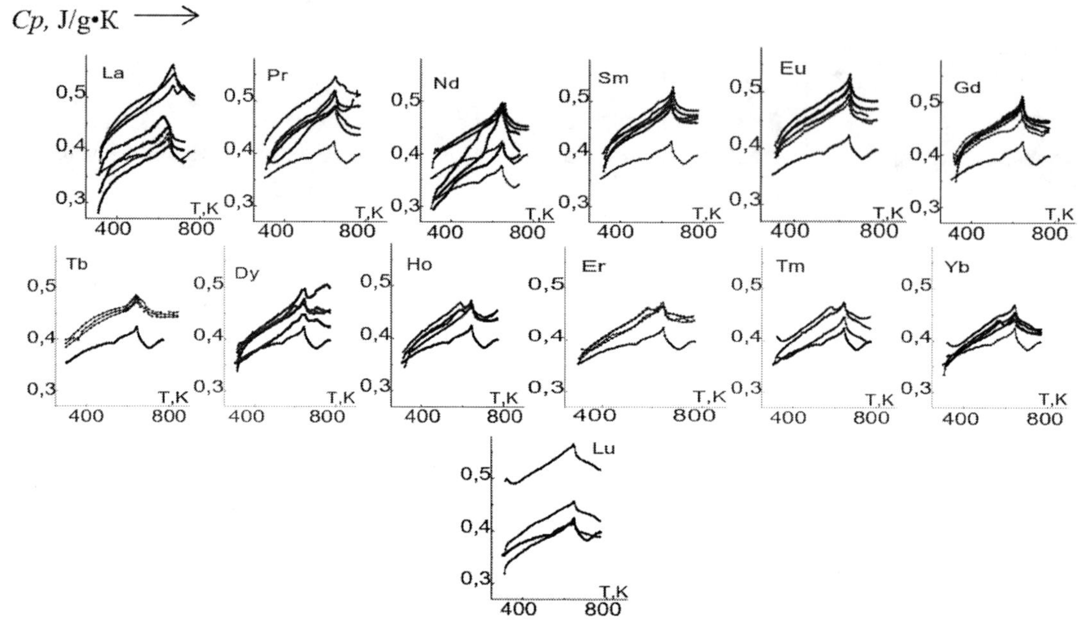

Figure 7. (a). Temperature Dependences of Heat Capacity on $Bi_{1-x}REE_xFeO_3$ (Several curves in each figure correspond to different REE concentrations in $BiFeO_3$)

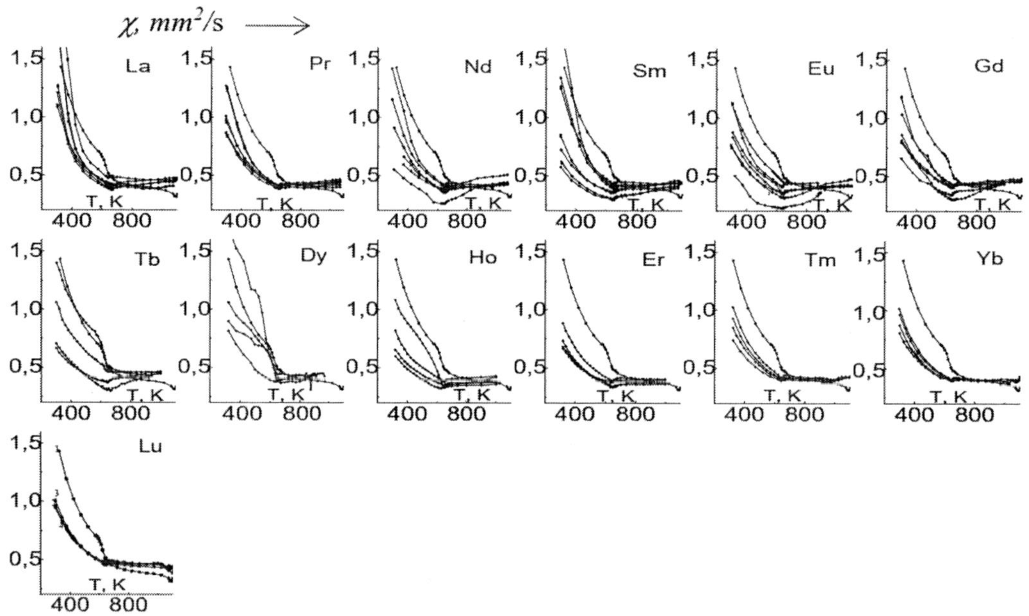

Figure 8. Dependences of thermal diffusivity on the temperature of $Bi_{1-x}REE_xFeO_3$ (Several curves in each figure correspond to different REE concentrations in $BiFeO_3$)

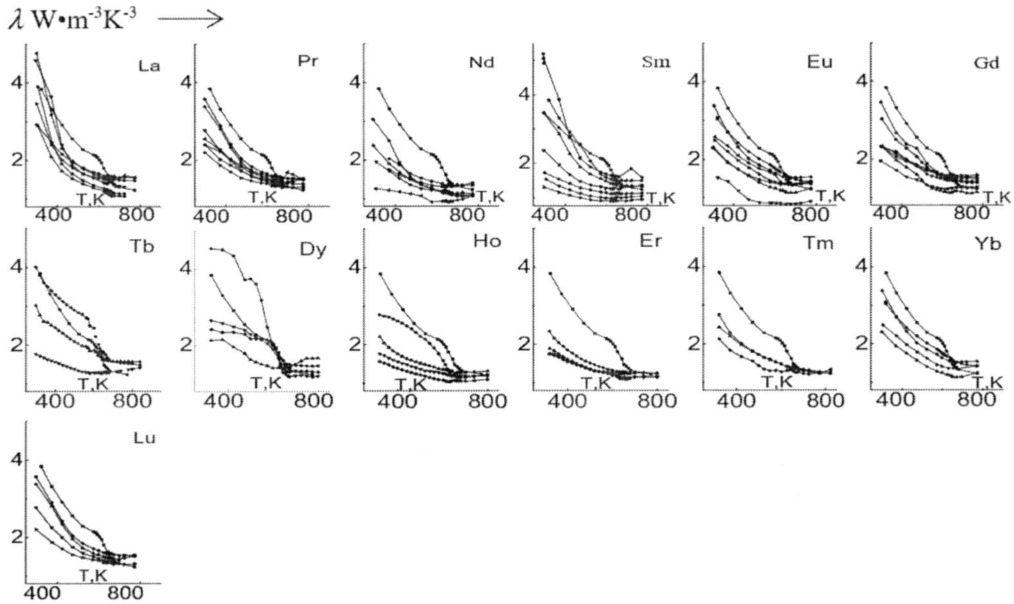

Figure 9. Dependences of thermal conductivity on temperature $Bi_{1-x}REE_xFeO_3$ (Several curves in each figure correspond to different REE concentrations in $BiFeO_3$)

4. Conclusion

The authors showed the possibility of using multiferroic based on $BiFeO_3$ in artificial intelligence systems.

Acknowledgement

Research was financially supported by the Ministry of Science and Higher Education of the Russian Federation (State assignment in the field of scientific activity, Southern Federal University, 2020)"..

References

[1] *Strategy for scientific and technical development of the Russian* Federation Approved by By the Decree of the President of the Russ. Fed. dated 01.12.2016 no 642

[2] Boldyrev N A, Pavlenko A V, Shilkina L A et al 2019 Structure, microstructure, dielectric and piezoelectric properties of (1-x-y)$BiFeO_3$-x$PbFe_{0.5}Nb_{0.5}O_3$-y$PbTiO_3$ ceramics *Ceram. Int.* **45(12)** 14768–74

[3] Abdulagatov I M, Abdulagatova Z Z, Kallaev S N and Bakmaev A G Thermal-Diffusivity_and_Heat Capacity Measurements of Sandstone at High Temperatures Using Laser Flash and DSC Methods *Int. J. Thermophys* DOI: 10.1007 / s10765-014-1829-4

[4] Khasbulatov S.V., Pavelko A.A., Reznichenko L.A et al 2017 Dielectric and Thermal Properties of Multiferroic Bismuth Ferrite Doped with Praseodymium and Neodymium (Ch. 11, pp. 109–124), Advanced Materials – Techniques, Physics, Mechanics and Applications. (Heidelberg; New York, Dordrecht, London: Springer) vol 193 570 p

[5] Urusov V S 1977 *The theory of isomorphic miscibility* (Moscow: Science) 251 p

[6] Rao Ch N R and Gopalakrishnan J 1990 *New Directions in Solid State Chemistry: Structure, Synthesis, Properties, Reactivity, and Materials Design* (Cambridge: Cambridge Univ. Press; Novosibirsk: Nauka)

[7] Turik A V and Radchenko G S 2003 Maxwell-Wagner relaxation of elastic constants in layered polar dielectrics *Solid State Physics* **45(6)** 1013–16

[8] Miller A I, Verbenko I A, Gesev A A et al 2012 Mechanochemical synthesis of $BiFeO_3$ *Designs from compos.mater.* **2** 47–50

IOP Conf. Series: Materials Science and Engineering **905** (2020) 012038 doi:10.1088/1757-899X/905/1/012038

Effects of modification of bismuth ferrite with small-sized rare earth elements

S V Khasbulatov[1,2], L A Shilkina[1], V A Aleshin[1], D I Gordienko[1], I A Verbenko[1] and L A Reznichenko[1]

[1] Southern Federal University, 105/42, Bolshaya Sadovaya Str., Rostov-on-Don, 344006, Russia

[2] Chechen State University, Grozny, 17a, Dudayev Boulevard, Grozny, 364060, Russia

E-mail: said_vahaevich@mail.ru, shilkina@yandex.ru, alyoshin09@ip.rsu.ru, grash002@yandex.ru, ilich001@yandex.ru, reznichenko@mail.ru

Abstract. The phase and grain structures and the dielectric properties of solid solutions (SS) of bismuth ferrite modified with small-sized rare-earth elements (SRE) are studied. The regularities of the appearance of impurities of various compositions depending on the ion radius and SRE concentration in objects are revealed. Based on the analysis of the dielectric spectra of the SSs, the development of Maxwell-Wagner relaxation associated with the accumulation of fSRE charges at the interfaces between the components was established in them. The conclusion is made about the possibility of using the results obtained when creating a new generation of multifunctional materials.

1. Introduction

Recently, there has been a rapid increase in interest in a wide class of substances – multiferroics based on bismuth ferrite, due to the combination of magnetic (TN ~ 643 K), and ferroelectric (Tc ~ 1083 K) orderings [1]. However, it was most often not possible to synthesize thermally stable powder products suitable for producing ceramics [2]. The introduction of rare-earth elements (SRE) makes it possible to stabilize the structure of bismuth ferrite, optimize its properties, and also strengthen the interaction between magnetic and electrical properties by suppressing the spatially modulated magnetic structure [3]. In this case, in most works, the effects of BiFeO3 modification by large-sized SREs with $0{,}94 \le \bar{R} \le 1{,}04$Å (La, Pr, Nd, Sm, Eu, Gd) are studied. Information on the effect of small-sized SREs with $\bar{R} < 0{,}94$Å on the internal structure (crystalline, grain) and the $BiFeO_3$ macroresponses is practically absent in the literature. In this connection, the aim of this work was to establish the features of phase formation, the formation of a polycrystalline (grain) structure and the dielectric properties of ceramic samples of solid solutions (SS) of bismuth ferrite modified with small SREs.

2. Methods and materials

The objects of study were SS binary systems of the form $Bi_{1-x}SRE_xFeO_3$, где SRE=Tb (0.89), Dy (0.88), Ho (0.86), Er (0.85), Tm (0.85), Yb (0.81), Lu (0.80) (in brackets are given the values of ionic radii SRE, R_i, in Å according to [4]), x = 0.05–0.20, Δx = 0.05. Samples were obtained using conventional ceramic technology, including two-stage synthesis from oxides Bi_2O_3, Fe_2O_3, SRE_2O_3

high purity at temperatures T_1 = 1070 K, 10 ч; T_2 = 1070–1120 K, 5 h (depending on the composition, with an increase in T_2 ~ by 10 degSREs for every 5 mol.% SRE), and subsequent sintering without pressure at temperatures $T_{сп}$ = 1140–1200 K, 5h (depending on a composition) [5].

X-ray diffraction studies were performed by powder diffraction using a DRON-3 diffractometer (filtered $Co_{k\alpha}$- radiation, Bragg-Brentano focusing scheme). The calculation of structural parameters (linear parameters of the cell; experimental, V_{exp}, and theoretical, V_{theor}, volumes) was carried out according to the standard method [6]. The experimental ($\rho_{exp.}$) density of the samples was measured by hydrostatic weighing in octane; X-ray density calculation ($\rho_{x\text{-}ray.}$) производили по формуле: $\rho_{x\text{-}ray.}=1.66\times M/V$, where M – is the weight of the formula unit in grams, V – is the perovskite cell volume in Å^3; the relative density ($\rho_{rel.}$) was calculated by the formula ($\rho_{exp.}/\rho_{x\text{-}ray.}$)×100%.

The polycrystalline (grain) structure of multiferroics was studied in reflected light using a Neophot 21 optical microscope and an Leica DMI 5000M inverted high-precision microscope.

The relative permittivity ($\varepsilon/\varepsilon_0$) was studied on a special bench using an Agilent E4980A precision LCR meter in wide temperature ranges (300–800 K) and frequencies (f) of an alternating measuring field (20–2· 10^6 Hz).

3. Results

X-ray phase analysis carried out at room temperature showed (Table 1) that all the studied SSs contain impurity phases $Bi_{25}FeO_{40}$ (1) (S.G. 123, a = 10.181 Å, cubic symmetry), $Bi_2Fe_4O_9$ (2) (S.G. Pbam, a = 7.965 Å, b = 8.44 Å, c = 5.994 Å, rhombic symmetry), usually concomitant formation of $BiFeO_3$ [7], and phases with a garnet-type structure $SRE_3Fe_5O_{12}$ (3) (cubic symmetry, space group Ia3d) [8], the concentration of which varies depending on the type of SRE and their concentration. Thus, compound (1) enriched in bismuth is always formed, and (2) enriched in iron, in cases where $I_{пр}/I_{пер}$ phases (3) less than 10 % (excluding SS composition $Bi_{0.85}Yb_{0.15}O_3$, in which a small amount of compound (2) is fixed in the presence of ~ 40% $I_{пр}/I_{пер}$ phase (3)). The smallest number of ballast phases (1), (2) is characteristic of TPs with larger SRE cations (Tb), and a larger number of TPs with Lu — the smallest of all SREs. At the same time, a minimum of such impurities is contained in TR with average concentrations of introduced modifiers (~ 10–15 %). Going beyond the specified interval leads to an increase in impurities, which is undoubtedly associated with the incomplete entry of small-sized SRE cations into the base structure $BiFeO_3$ with very small and large quantities. That this is so confirmed by the greatest ρ_{rel} ceramic specimen compositions $Bi_{0.90}SRE_{0.10}FeO_3$.

In fig. 1 shows the changes in value $I_{пр}/I_{пер}$ phase (3) depending on the concentration of SRE, and on the insert – the dependence of the minimum SRE concentration at which phase (3) is already formed on the R_i of the SRE cation. It can be seen that the formation of phase (3) begins the earlier, the finer the SRE cation (with the exception of SS with the smallest Lu cation, in which phase (3) does not form at any Lu concentration), and its amount (estimated from $I_{пр}/I_{пер}$) increases with an increase in the SRE content, and at x > 0.15 – it increases with a higher rate. The observed is associated with a greater likelihood of impurities in compositions with a higher SRE concentration and smaller radii of their cations.

The table shows the parameters of the cubic cell of phase (3) and the known values of the SS structural parameters of the $SRE_3Fe_5O_{12}$ compounds. It can be seen that the parameter a phase (3) significantly exceeds the values of parameter a of the $SRE_3Fe_5O_{12}$. This means that during the synthesis of SS $(Bi_{1-x}SRE_x)FeO_3$ apparently, SSs of the form $(SRE_{1-y}Bi_y)_3Fe_5O_{12}$ are formed, which is indirectly confirmed by the identity of the dependences of the cubic cell parameters of the $SRE_3Fe_5O_{12}$ and SS compounds of this type on the ionic radius SRE (Fig. 2). In this case, the stoichiometry of a given composition is violated due to the “departure” of Bi from the base structure. This explains the absence of this impurity in SS with SRE, in which a sufficiently large amount of Bi-containing impurities crystallizes already (1), (2).

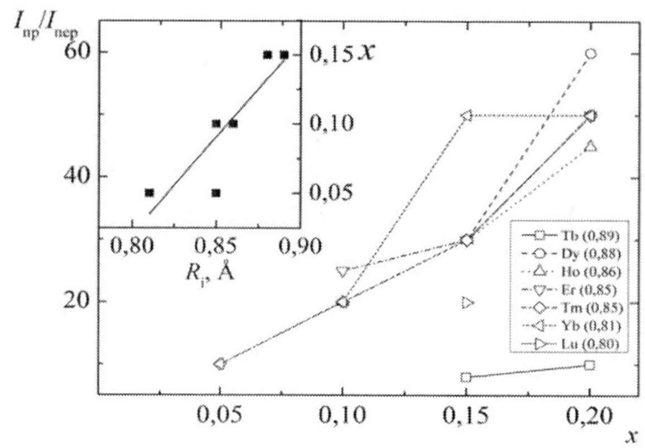

Figure 1. Dependence of $I_{пр}/I_{пер}$ of the impurity phase of the $SRE_3Fe_5O_{12}$ type on the SRE concentration. The inset shows the dependence of the minimum SRE concentration at which phase (3) is already formed on the R_i of the SRE cation

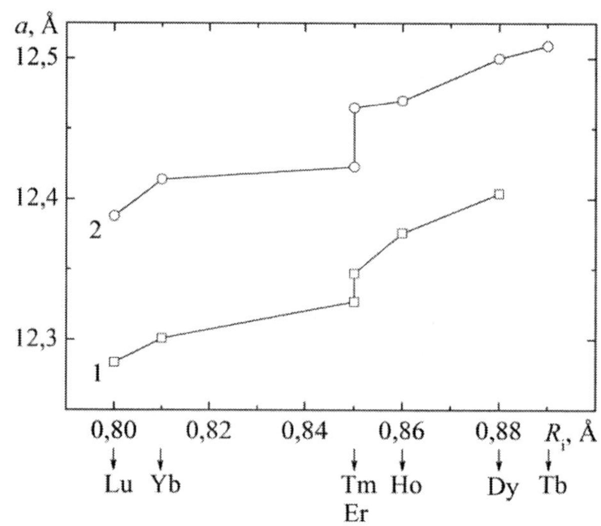

Figure 2. Dependencies of the parameters of cubic connection cells $SRE_3Fe_5O_{12}$ according to the JCPDS database) (1) and SS $(SRE_{1-y}Bi_y)_3Fe_5O_{12}$ (2) from SRE ionic radius

The table shows that SS $(Bi_{1-x}Tb_x)FeO_3$ and $(Bi_{1-x}Dy_x)FeO_3$ undergo a transition from the rhombohedral (Rh) phase to the morphotropic area (MA) containing a mixture of Rh and a rhombic (R) with monoclinic (M) subcell (the latter is of the type of $GdFeO_3$, implemented in $SREFeO_3$ compound). All other SSs belong to the Rh area (except for the SS composition $Bi_{0.8}Er_{0.20}FeO_3$ with "traces" of the R phase). The absence of MA in these SSs may be due to the formation of garnet-like structures, as a result of which the SRE concentration decreases and the Rh → R transition becomes impossible.

In fig. Figure 3 shows the changes in the volumes of rhombohedrally distorted elementary perovskite cells of the studied SS. In almost all SS V cells decrease with increasing SRE, which corresponds to the logic of decreasing the average radius of the A cation, the exception is SS with Er and Lu, in which V increases, and SS with Yb, in which V remains almost unchanged. This, as well as the fact that, at the same concentration (starting from x = 0.10) of the introduced SRE V does not decrease with a decrease in the average radius of the A cation, is a consequence, as noted earlier, of the crystal chemical features of the formed SS: for small and large x, most likely, SS embeddings are formed, with medium SS substitutions or a combined type. R cell options ($a = c$, b, β) SS composition

$Bi_{0.85}Tb_{0.15}FeO_3$, $Bi_{0.80}Tb_{0.20}FeO_3$, and Rh cells (a, α) $Bi_{0.90}Dy_{0.10}FeO_3$ are equal, so, 3.910 Å, 3.902 Å, 92.02°; 3.911 Å, 3.903 Å, 92.08°; 3.954 Å, 89.48°.

Table 1. The relative intensity of the strong line of the impurity phase, (I approx/I first , SS symmetry and density $Bi_{1-x}SRE_xFeO_3$, experimental and tabular values of the cubic cell parameter of the $SRE_3Fe_5O_{12}$ compounds, impurity phases formed during the solid-state synthesis of SS species ($(Bi_{1-x}SRE_x)FeO_3$. ((25-1-40) relates to $Bi_{25}FeO_{40}$; (2-4-9) – $Bi_2Fe_4O_9$.)

Chemical formula	I approx/I first	Symmetry	P_{exp}, г/см^3	P_{rent}, г/см^3	P_{rel}, %	A_{exp}, Å $SRE_3Fe_5O_{12}$	A_{tab}, Å JCPDS
$(Bi_{0.95}Tb_{0.05})FeO_3$	7(25-1-40) 7(2-4-9)	Rh	7.67	8.35	91.88		
$(Bi_{0.9}Tb_{0.1})FeO_3$	8(25-1-40) 7(2-4-9)	Rh +traces R	7.61	8.32	91.49		
$(Bi_{0.85}Tb_{0.15})FeO_3$	5(25-1-40) 3(2-4-9) ~8 $Tb_3Fe_5O_{12}$	~60 % Rh + ~40 % R (SREFeO3)	7.06	8.29	85.18	12.509	
$(Bi_{0.8}Tb_{0.2})FeO_3$	6(25-1-40) ~10 $Tb_3Fe_5O_{12}$	~15% Rh + 85 % R	6.97	8.46	82.40	12.512	–
$(Bi_{0.95}Dy_{0.05})FeO_3$	10(25-1-40) 10(2-4-9)	Rh	7.48	8.36	89.44		
$(Bi_{0.9}Dy_{0.1})FeO_3$	8(25-1-40) 5(2-4-9)	Rh +traces R	7.64	8.31	91.95		
$(Bi_{0.85}Dy_{0.15})FeO_3$	21(25-1-40) ~30$Dy_3Fe_5O_{12}$	Rh +traces R	7.41	8.25	89.79	12.500	12.404 (23-237)
$(Bi_{0.8}Dy_{0.2})FeO_3$	35(25-1-40) ~60$Dy_3Fe_5O_{12}$	Rh +traces R	7.10	8.19	86.67	12.498	
$(Bi_{0.95}Ho_{0.05})FeO_3$	12(25-1-40) 10 (2-4-9)	Rh	7.36	8.36	88.03		
$(Bi_{0.9}Ho_{0.1})FeO_3$	9(25-1-40) ~20$Ho_3Fe_5O_{12}$	Rh	7.34	8.31	88.33	12.470	12.376 (23-282)
$(Bi_{0.85}Ho_{0.15})FeO_3$	17(25-1-40) ~30$Ho_3Fe_5O_{12}$	Rh	7.10	8.24	86.19	12.489	
$(Bi_{0.8}Ho_{0.2})FeO_3$	24(25-1-40) ~45$Ho_3Fe_5O_{12}$	Rh	6.86	8.19	83.78	12.487	
$(Bi_{0.95}Er_{0.05})FeO_3$	9(25-1-40) 3(2-4-9)	Rh		8.35			
$(Bi_{0.9}Er_{0.1})FeO_3$	10(25-1-40) ~25 ($Er_3Fe_5O_{12}$)	Rh		8.29		12.465	12.347 (23-240)
$(Bi_{0.85}Er_{0.15})FeO_3$	17(25-1-40) ~30 ($Er_3Fe_5O_{12}$)	Rh		8.23		12.459	
$(Bi_{0.8}Er_{0.2})FeO_3$	26(25-1-40) ~50 ($Er_3Fe_5O_{12}$)	Rh +traces R		8.16		12.460	
$(Bi_{0.95}Tm_{0.05})FeO_3$	6(25-1-40) ~10 ($Tm_3Fe_5O_{12}$)	Rh	7.26	8.35	86.98	12.423	12.327 (23-591)
$(Bi_{0.9}Tm_{0.1})FeO_3$	11(25-1-40) ~20 ($Tm_3Fe_5O_{12}$)	Rh	7.40	8.30	89.12	12.434	
$(Bi_{0.85}Tm_{0.15})FeO_3$	15(25-1-40) ~30 ($Tm_3Fe_5O_{12}$)	Rh	7.24	8.25	87.79	12.434	
$(Bi_{0.8}Tm_{0.2})FeO_3$	24 (25-1-40) ~50 ($Tm_3Fe_5O_{12}$)	Rh	7.11	8.20	86.74	12.432	
$(Bi_{0.95}Yb_{0.05})FeO_3$	16(25-1-40) 11(2-4-9) ~10 ($Yb_3Fe_5O_{12}$)	Rh	7.29	8.34	87.38	12.414	12.301 (23-730)
$(Bi_{0.9}Yb_{0.1})FeO_3$	17(25-1-40) ~20 ($Yb_3Fe_5O_{12}$)	Rh	7.37	8.30	88.83	12.399	

Table 1. Continuation

$(Bi_{0.85}Yb_{0.15})FeO_3$	54 (25-1-40) 3(2-4-9) ~50 ($Yb_3Fe_5O_{12}$)	Rh	6.93	8.24	84.06	12.414	
$(Bi_{0.8}Yb_{0.2})FeO_3$	37 (25-1-40) ~50 ($Yb_3Fe_5O_{12}$)	Rh	7.08	8.20	86.33	12.410	
$(Bi_{0.95}Lu_{0.05})FeO_3$	39(25-1-40) 34 (2-4-9)	Rh	7.10	8.34	85.12		
$(Bi_{0.9}Lu_{0.1})FeO_3$	24(25-1-40) 21 (2-4-9) traces ($Lu_3Fe_5O_{12}$)	Rh	7.37	8.30	88.79		
$(Bi_{0.85}Lu_{0.15})FeO_3$	22(25-1-40) 16(2-4-9) ~20 ($Lu_3Fe_5O_{12}$)	Rh	7.25	8.24	87.94	12.388	12.284
$(Bi_{0.8}Lu_{0.2})FeO_3$	57(25-1-40) 36(2-4-9)	Rh	7.27	8.20	88.66		

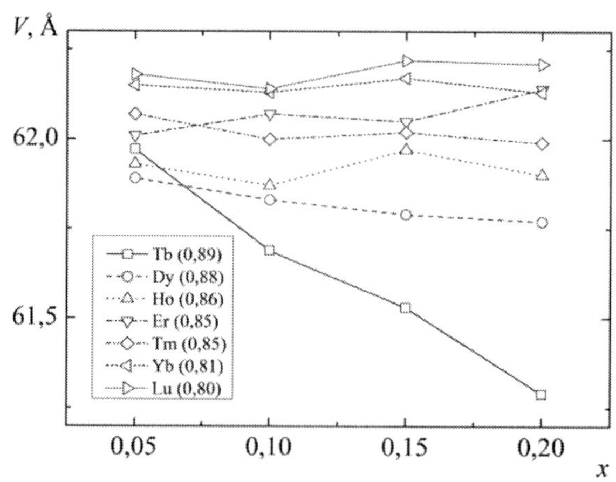

Figure 3. Dependence of the volumes of rhombohedrally distorted elementary perovskite cells SS $Bi_{1-x}SRE_xFeO_3$ on x.

In fig. 4 shows photographs of the microstructure of the studied objects. Black areas of rounded and irregular shape – pores. They are not uniformly distributed on the surface. Etching revealed the boundaries of crystallites: closed dark lines around the lighter areas (grains of ceramics). It was found that the most advanced microstructures of bismuth ferrite are realized with larger SREs. With a decrease in Ri and an increase in x, the amount of the impurity phase ("gray" grains) increases, the structure loosens, intergranular spaces are filled with the liquid phase, grain boundaries become thicker, and their habit is transformed.

The results of studying the temperature dependences $\varepsilon/\varepsilon_0$ of ceramic samples are shown in Fig. 5, 6. As can be seen from the figures, in the areas of 400–500 K two strongly dispersed maxima $\varepsilon/\varepsilon_0$ are formed, the peculiarity of which is a shift to the high temperature region, a decrease and smearing of their peak values with increasing frequency. Similar phenomena, known as Maxwell-Wagner relaxation, were observed earlier in [9–10] and were associated with the accumulation of free charges on the component interfaces on the background of interlayer, interphase, and interphase interplanes. The development of such effects is due to the natural composite with the $BiFeO_3$ state.

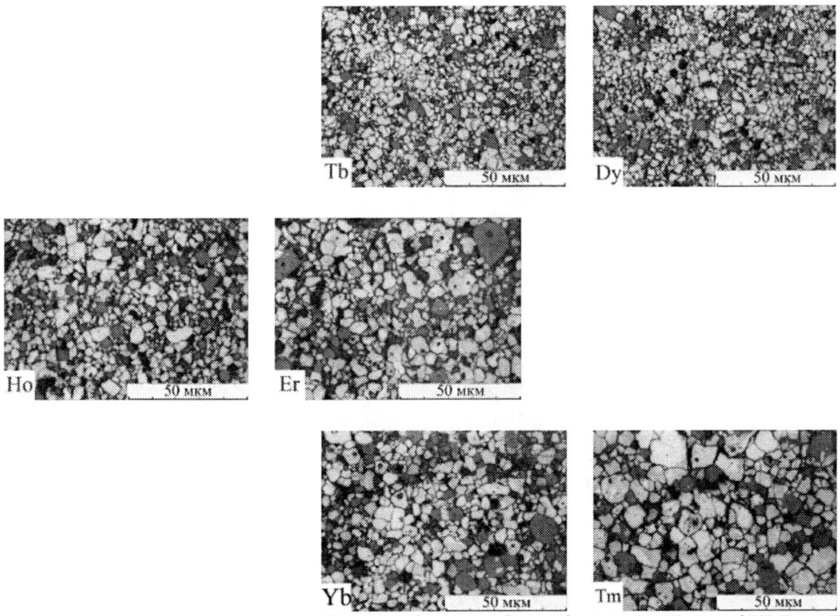

Figure 4. Microstructures of ceramic samples $Bi_{0.9}SRE_{0.1}FeO_3$ (SRE – Tb, Dy, Ho, Er, Yb, Tm) at room temperature. SRE is indicated in the lower left corner of the photos

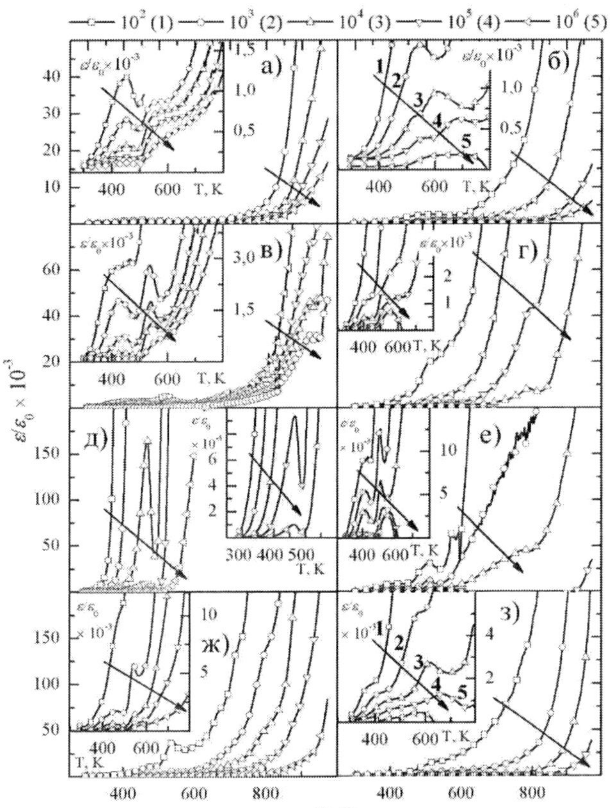

Figure 5. Temperature dependences $\varepsilon/\varepsilon_0$ of $Bi_{1-x}SRE_xFeO_3$ ceramic samples and f at x = 0.05 и 0.20, SRE = Tb (a, b), Dy (в, г), Ho (d, e), Er (f, g), respectively

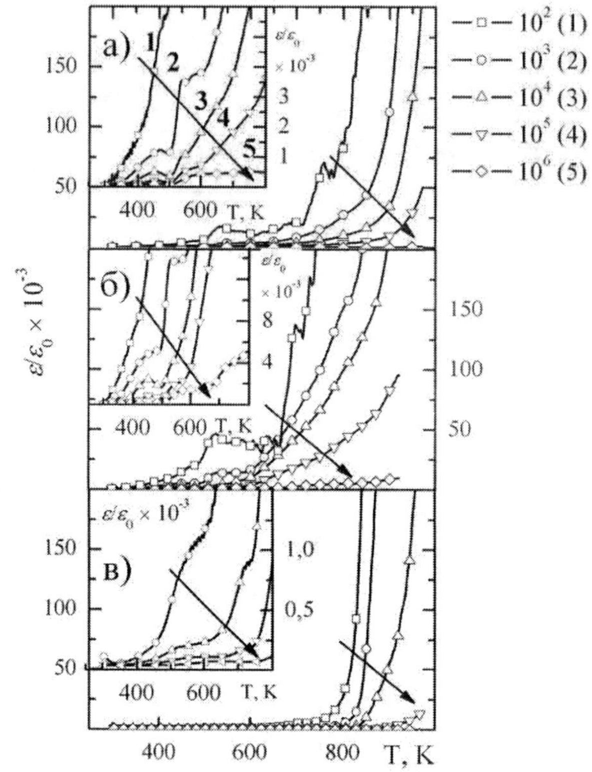

Figure 6. Temperature dependences $\varepsilon/\varepsilon_0$ of $Bi_{0.80}SRE_{0.20}FeO_3$ ceramic samples and *f*, SRE = Tm (a), Yb (b), Lu (c)

4. Conclusion

As a result of the studies, it was found that during the synthesis of all the studied SS, impurity phases are formed, while bismuth enriched ($Bi_{25}FeO_{40}$) are always formed, and iron-enriched ($Bi_2Fe_4O_9$) phases are formed when $I_{пр}/I_{пер}$ phases with a garnet type structure ($SRE_3Fe_5O_{12}$) are less than 10 %. The crystal chemical features of the formed SSs were revealed: at small and large x, most likely SS interstitials are formed, and at medium x SS substitutions or a combined type. The deterioration of the grain structure of ceramics was established with a decrease in the radius of the introduced SRE. It is shown that the dielectric spectra of SS indicate the development of Maxwell-Wagner relaxation in the studied objects.

The results obtained in this paper must be taken into account when creating a new generation of multifunctional (combining ferroelectric, magnetic, and ferroelastic properties) materials that can find promising applications in areas such as sensory technology, diagnostics, processing and storage of information, and many others

Acknowledgment

Research was financially supported by the Ministry of Science and Higher Education of the Russian Federation (State assignment in the field of scientific activity, Southern Federal University, 2020)..

References

[1] Smolensky G A and Chupis I E 1982 Ferroelectromagnets *Physics-Uspekhi* pp 137, 415
[2] Carvalho T T and Tavares P B 2008 Synthesis and Thermodynamic Stability of Multiferroic $BiFeO_3$ *Mater. Letters* **62** 3984

[3] Zalessky A A V, Frolov A A, Khimich T A and Bush A A 2003 Concentration transition of a spin-modulated structure to a homogeneous antiferromagnetic state in the $Bi_{1-x}La_xFeO_3$ system system according to ^{57}Fe NMR data *Phys. of the solid state* **45** 134

[4] Bokiy G B 1954 *Introduction to Crystal Chemistry* (Moscow: Moscow State Univer. Publ. House)

[5] Razumovskaya O N, Verbenko I A, Andryushin K P et al 2009 *Fund. Probl. of Radioelectr. Instrumentat.* **9** 126

[6] Fesenko E G 1972 *The perovskite family and ferroelectricity* (Moscow: Atomizdat) 248 p

[7] Denisov V M, Belousova N V, Foal V P et al 2012 *J. of Siber. Fed. Univer. Chem.* **2** 146

[8] Smith J and Wayne H 1962 *Ferrites* (Moscow: Foreign Liter.) 504 p

[9] Khasbulatov S V, Pavelko A A, Reznichenko L A et al 2017 Dielectric and Thermal Properties of Multiferroic Bismuth Ferrite Doped with Praseodymium and Neodymium *Advanced Materials – Techniques, Physics, Mechanics and Applications* vol 193 Chapter 11 pp 109–124 (Heidelberg; New York; Dordrecht; London: Springer) 570 p

[10] Khasbulatov C C V, Pavelko A A, Shilkina L A et al 2016 Crystal and grain structure, thermophysical and dielectric properties of multiferroic $Bi_{1-x}Dy_xFeO_3$ *Thermophys. and Aeromech.* **23(3)** 461–6

Identification of intermittent point source capacity and location of the initial contamination plume

V S Kochergin and S V Kochergin

Marine Hydrophysical Institue of the Russian Academy of Sciences, 2, Kapitanskaya Str., Sevastopol, 299011Russia

E-mail: vskocher@gmail.com, ko4ep@mail.ru

Abstract. While working on a model of mass impurity transition for the waters of the Sea of Azov and the Kerch Strait, we calculated the capacity of the immediate point source and identified the location of the initial contamination plume. Contamination source capacity identification is based on minimizing the quadratic functional of the forecasting quality and solving associated problems. Associated problem solving is also used in order to identify the location of the initial contamination plume in the body of water under analysis. The first goal was reached for the Gulf of Taganrog area, and the second one – for the Kerch Strait area. It must be noted that both areas have heavy marine traffic and experience the construction of new transport corridors. Simulation results were produced taking into consideration north-western winds that are dominant in the region. Quantitative experiments show that it is possible to identify the capacity of an intermittent point source of contamination using a three-dimensional model of mass impurity migration, as well as to identify the location of the initial contamination plume.

1. Introduction

While solving ecological problems using quantitative dynamic models and those describing the propagation of contaminants [1, 2], it is necessary to improve forecasted simulation fields by identifying the inputs for the models.

The selected quantitative simulation inputs must provide for the concentration field evaluation models to be in optimum compliance with the existing measurement data. One of the methods of such input selection is the variational assimilation method for the measurement data [3–7] based on minimizing forecasting discrepancies and solving some specific problems. The goal of this work is developing and trying-out efficient algorithms to perform the task set. Developing such algorithms is an important and relevant objective due to the growing anthropogenic load on the waters of the Sea of Azov and the Black Sea.

2. Methods and Materials

In this work, we used the adjoined equation method [8-10] in order to identify the intermittent capacity of contamination source for the mass impurity migration model used in the Sea of Azov. Model [1, 2] is used as a dynamic one. The produced model fields of speed and turbulence diffusivity are used as inputs for the integration of the associated problem and the mass impurity migration model. Let us describe the instruments and methods used in more detail.

Considering the migration model, let us present the equation of mass impurity migration in σ coordinates

$$\frac{\partial DC}{\partial t}+\frac{\partial DUC}{\partial x}+\frac{\partial DVC}{\partial y}+\frac{\partial WC}{\partial \sigma}=\frac{\partial}{\partial x}A_H\frac{\partial DC}{\partial x}+\frac{\partial}{\partial y}A_H\frac{\partial DC}{\partial y}+\frac{\partial}{\partial \sigma}\frac{K}{D}\frac{\partial C}{\partial \sigma} \quad (1)$$

with conditions on lateral boundaries

$$B{:}\frac{\partial C}{\partial s}=0 \quad (2)$$

with boundary conditions on the surface and on the floor

$$\sigma=0{:}\quad \frac{\partial C}{\partial \sigma}=Q\quad (t), Q(t)=Q_S(t)\cdot\delta(x-x_0)\cdot\delta(y-y_0)\ \ \sigma=-1{:}\quad \frac{\partial C}{\partial \sigma}=0, \quad (3)$$

and initial data

$$C(x,y,\sigma,0)=C_0(x,y,\sigma) \quad (4)$$

where t is the time; x_0, y_0 are the coordinates of point source; C is the impurity concentration; $Q(t)$ is the point source of intermittent capacity; U, V, W are speed field components; A_H and K are the horizontal and vertical turbulence diffusivities respectively; D is the dynamic depth; n is the normal to a lateral boundary. Let us assume that the point source is on the sea surface. This problem is solved the same way if the source is assumed to be located on the floor or on the lateral boundary, but it requires relevant changes in boundary conditions (2), (3).

Then let us consider the adjoined equation method: we shall multiply (1)–(4) by C^* and integrate them taking into consideration the boundary conditions and a continuity equation counterpart in σ coordinates:

$$\frac{\partial D}{\partial t}+\frac{\partial DU}{\partial x}+\frac{\partial DV}{\partial y}+\frac{\partial W}{\partial \sigma}=0 \quad (5)$$

and select C^* as the solution to the next associated problem:

$$-\frac{\partial DC^*}{\partial t}-\frac{\partial DUC^*}{\partial x}-\frac{\partial DVC^*}{\partial y}-\frac{\partial WC^*}{\partial \sigma}-D\frac{\partial}{\partial x}A_H\frac{\partial C^*}{\partial x}-D\frac{\partial}{\partial y}A_H\frac{\partial C^*}{\partial y}-\frac{\partial}{\partial \sigma}\frac{K}{D}\frac{\partial C^*}{\partial \sigma}=0, \quad (6)$$

$$B{:}\frac{\partial C^*}{\partial s}=0,\quad \sigma=0{:}\frac{\partial C^*}{\partial \sigma}=0,\quad \sigma=-1{:}\frac{\partial C^*}{\partial \sigma}=0, \quad (7)$$

$$t=T{:}C^*=h, \quad (8)$$

We get:

$$\int_M hCdM=\int_M C_0C^*dM \quad (9)$$

Let us select h in the following form:

$$h=\begin{cases}\frac{1}{m(\Omega)}- & in\ field\ \Omega\\ \&\quad 0\ - & out\ of\ field\ \Omega\end{cases} \quad (10)$$

where m is a measure of a domain $\Omega \in M$. At that, on the left side of expression (8), we get the mean concentration $\bar{C}_T$ in Ω at the moment T.

If we choose Ω as a computation mesh cell, we get:

$$\bar{C}_T=\int_M C_0C^*dM\ . \quad (11)$$

Thus, using the solution of the adjoint problem (6) – (8) by formula (11), it is possible to evaluate the concentration of the impurity in the given cell. Let us note that in (9) and (11) C^* is the influence function of the initial data. Its spatial arrangement reflects the impact of initial data on the final solution of the model.

Further on, let us consider the variational identification algorithm. If we assume that the measurement data are present at the end of the time period, the problem of measurement data assimilation for $C_{changed}$ is in searching the minimum next quadratic functional.

$$I_0=\frac{1}{2}\left[P\left(RC-C_{changed}\right),P\left(RC-C_{changed}\right)\right]_M, \quad (12)$$

where M is the integration domain of the model for the time interval of $[0,T]$; P is a zero fulfillment operator for the discrepancy field of the forecast when no measurement data are present; R is a projection operator for the observation point, and the scalar product is calculated using the standard method. Minimization (12) with model limitations (1)–(3) is equivalent to the extremum seeking for the next functional

$$I = I_0 + \left[\frac{\partial DC}{\partial t} + \frac{\partial DUC}{\partial x} + \frac{\partial DVC}{\partial y} + \frac{\partial WC}{\partial \sigma} - - \frac{\partial}{\partial x} A_H \frac{\partial DC}{\partial x} - \frac{\partial}{\partial y} A_H \frac{\partial DC}{\partial y} - \frac{\partial}{\partial \sigma}\frac{K_H}{D}\frac{\partial C}{\partial \sigma}, C^*\right]_{M_t} +$$
$$+ \left(\frac{\partial C}{\partial n}, C^*\right)_{B_t} + (C - C_0, C^*)_M + \left(\frac{\partial C}{\partial \sigma} - Q_S(t) \cdot \delta(x - x_0) \cdot \delta(y - y_0), C^*\right)_{\sigma_t^0}, \quad (13)$$

where $M_t = M \times [0, T]$, $\sigma_t^0 = \sigma^0 \times [0, T]$.

Writing down a variance of functional (13) and integrating it piece by piece taking into consideration the boundary conditions and the continuity equation counterpart in σ coordinates, we get

$$\delta I = (\delta Q_S(t) \cdot \delta(x - x_0) \cdot \delta(y - y_0), C^*)_{\sigma_t^0} \quad (14)$$

where C^* are Langrangian multipliers selected from the solution of the next adjoint problem

$$-\frac{\partial DC^*}{\partial t} - \frac{\partial DUC^*}{\partial x} - \frac{\partial DVC^*}{\partial y} - \frac{\partial WC^*}{\partial \sigma} - D\frac{\partial}{\partial x} A_H \frac{\partial C^*}{\partial x} - D\frac{\partial}{\partial y} A_H \frac{\partial C^*}{\partial y} + \frac{\partial}{\partial \sigma}\frac{K_H}{D}\frac{\partial C^*}{\partial \sigma} = 0$$

$$: \frac{\partial C^*}{\partial s} = 0, \sigma = 0: \frac{\partial C^*}{\partial \sigma} = 0,$$

$$\sigma = -1: \frac{\partial C^*}{\partial \sigma} = 0$$

$$t = T: C^* = -P(RC - C_{changed})$$

Due to the functional stationary state and gradient calculation, we get

$$\nabla_{Q_S(t)} I = C^*(0, x_0, y_0, t) \quad (18)$$

The capacity of the source is calculating using the following iterative formula

$$Q_S^{n+1}(t) = Q_S^n(t) + \tau \cdot \nabla_{Q_S(t)} I \quad (19)$$

where τ is the iterative parameter selected using the following formula

$$\tau = \frac{(P(RC - Cchanged_M), PR\delta C)_M}{(PR\delta C, PR\delta C)_M} \quad (20)$$

where δC is the solution for the respective problem (1) – (4) in its variations

$$\frac{\partial D\delta C}{\partial t} + \frac{\partial DU\delta C}{\partial x} + \frac{\partial DV\delta C}{\partial y} + \frac{\partial W\delta C}{\partial \sigma} = \frac{\partial}{\partial x} A_H \frac{\partial D\delta C}{\partial x} + \frac{\partial}{\partial y} A_H \frac{\partial D\delta C}{\partial y} + \frac{\partial}{\partial \sigma}\frac{K}{D}\frac{\partial \delta C}{\partial \sigma} \quad (21)$$

with conditions on lateral boundaries

$$B: \frac{\partial \delta C}{\partial s} = 0 \quad (22)$$

with boundary conditions on the surface and on the floor

$$\sigma = 0: \quad \frac{\partial \delta C}{\partial \sigma} = C^*(0, x_0, y_0, t)$$
$$\sigma = -1: \quad \frac{\partial \delta C}{\partial \sigma} = 0, \quad (23)$$

with boundary conditions on the surface and on the floor

$$\sigma = 0: \quad \frac{\partial \delta C}{\partial \sigma} = C^*(0, x_0, y_0, t)$$
$$\sigma = -1: \quad \frac{\partial \delta C}{\partial \sigma} = 0, \quad (23)$$

and initial data

$$\delta C(0, x, y, \sigma) = 0. \quad (24)$$

The identification procedure in general consists of the following:

- direct problem (1)–(4) is solved using the initial value of $Q(t)$;
- forecast discrepancies are found and functional (12) is defined;
- adjoint problem (15)–(17) is solved;
- gradient (18) of functional (13) is determined;
- the problem is integrated into variances (21)–(24);
- the iterative parameter is calculated using formula (20);
- the capacity of the source $Q(t)$ is determined iteratively using formula (19).

3. Results

Quantitative experiments were carried out using the model [1, 2] for the waters of the Sea of Azov. In order to identify the intermittent capacity of contamination, we calculated the model field of currents due to constant north-western wind, blowing at 10 m/s. The simulation resulted in the spatial distribution of factors A_H and K. Speed and turbulence diffusivity fields were used as inputs when integrating the model of mass impurity migration for 5 days. Figure 1 shows the model field of sea surface concentration, and the location of the point source of contamination is circled. The initial profile $Q(t)$ was specified as a sinus with the maximum in the middle of the integration term. The total integration term for the model amounts to 1800 steps. The produced model field was used as measurement data for the sea surface both wholly and partially, i.e. Data was assimilated from some of the computation mesh nodes. Calculations with 2.5% noise level showed that in order to achieve the functional minimum, 5 iterations would suffice. Further iterations do not improver the result and they are obviously induced by the uncertainty of the model to the data. Similar results were produced when assimilating the data with 5.0% and 10% random noise levels. Quantitative experiments performed showed that the increase of the measurement data, especially in the domain of the significant concentration values, leads to the improved identification of the parameter in question. Besides, the reduction of measurement errors induced by the random number counter also leads to the increased fidelity of the results. In general, the procedure provides for a proper representation of profile values $Q(t)$.

The initial contamination plume was specified for the strait surface in the area of heavy marine traffic. Besides, in this region, there is a possibility of the anthropogenic impact of other nature. Figure 2 shows a rectangular area on the surface of which initial concentration values were specified. It also shows the simulation results for the contamination spread under the north-western wind. The concentration field on Figure 2 is normalized for the maximum value. The contamination plume at the end of the time period reaches its maximum near the southern coast of the Kerch Peninsula. A circle in Figure 2 marks the location of the maximum concentration value, and the diamonds show the point selected at the periphery of the contamination plume.

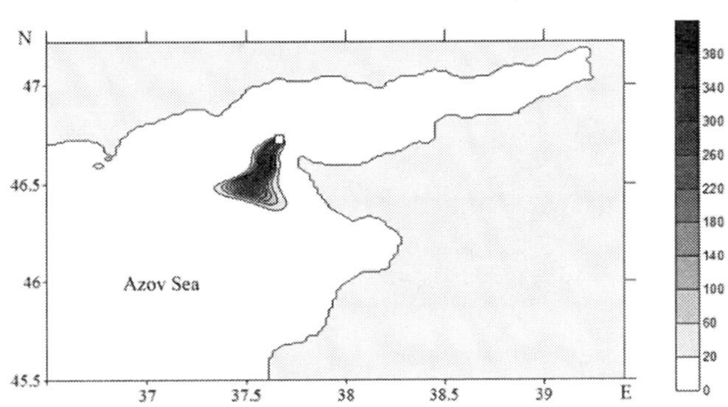

Figure 1. Contamination source location and model field of concentration at the end of the time period.

Within five days of the simulation, the impurity spread along the coast up to the gulf of Feodosia. This contamination spread rate is typical for the region under northern and north-eastern winds. In order to identify the location of the initial contamination plume, we solved a series of adjoint problems of (13)–(15), (17). Diamonds in Figure 2 show the points of initial data specification for adjoint problems.

Figure 3 characterized the possible location of the initial concentration field produced by solving a series of adjoint problems with initial conditions (17) specified in mesh cells at the periphery of the model field (Figure 2). Therefore, solving a series of adjoint problems allows identifying the area (grey in Figure 3) of the possible location of the initial contamination plume. In order to determine its location more accurately, a priori information on the nature and the composition of the impurity can be used. In order to identify the values of the initial concentration field, it is necessary to use the variance method for the identification of migration model inputs, as described above. The problem of seeking the immediate point source of contamination can be solved in the same way.

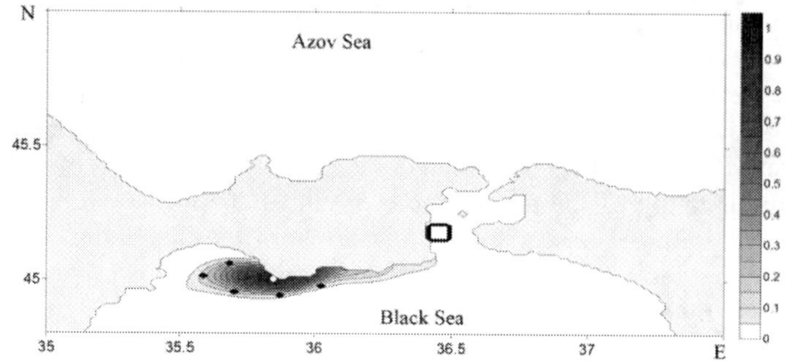

Figure 2. Specification domain for the initial concentration field and the model field at the end of the period

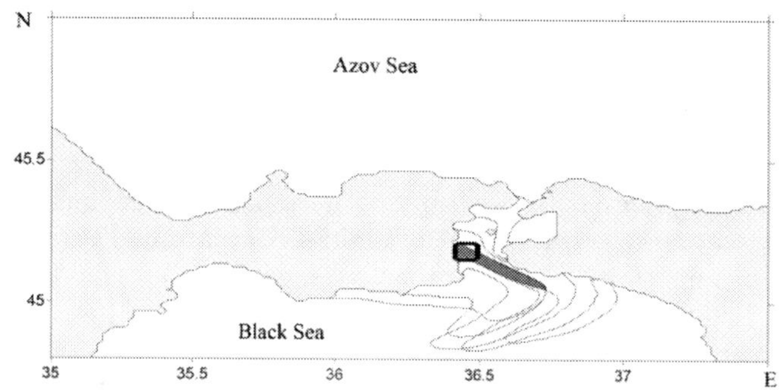

Figure 3. Specification domain for the initial concentration field and the adjoint problem solution transverse (calculations 2–6)

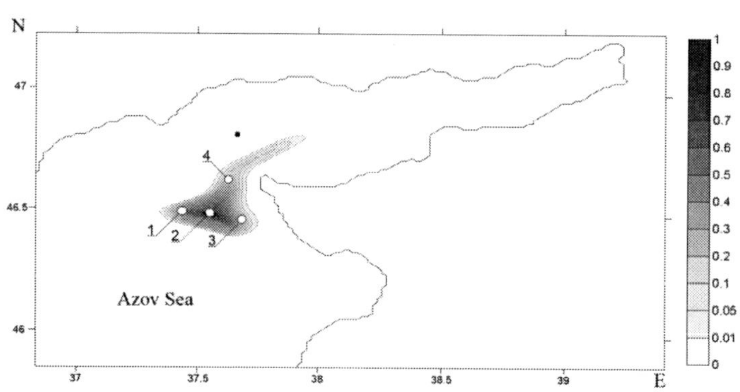

Figure 4. The location of an immediate point source, normalized concentration, and measurement point fields

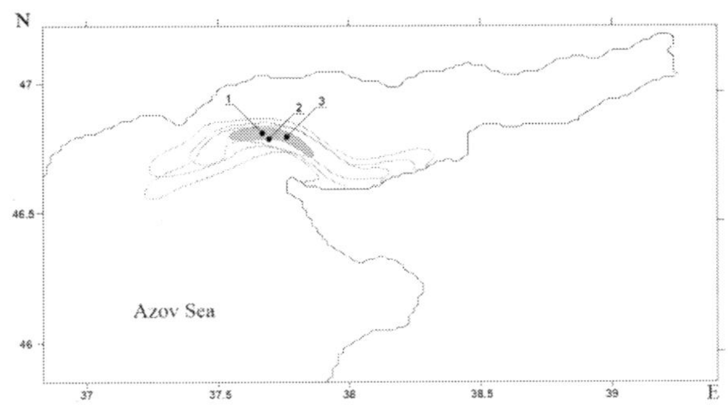

Figure 5. Adjoint problem solution, area, Ω and contamination source location

The model concentration field was calculated $Q_S = 1$, and the results of the calculations were normalized for the maximum value. Figure 4 shows the actual location of the immediate point source, marked by a black dot. Apart from the model concentration field, this figure shows 4 measurement points ($N = 4$). The integration result for four adjoint problems is shown in Figure 5. The area corresponding to these adjoint problems is marked in grey in the figure. The figure shows that the actual location of the source marked as 1 in the figure belongs to this area. Numbers 2 and 3 mark potential source coordinates.

4. Conclusion

Quantitative experiments prove that it is possible to identify an intermittent capacity point source of contamination using a three-dimensional migration model for mass impurities. Under variational identification, it is possible to produce reliable results if new measurement data are specified only in some nodes of the computational mesh. The aforesaid represents important interpolation features of the algorithm. Measurement data assimilation with random noise demonstrated the filtering action of the assimilation procedure. The calculation results prove that it is possible to identify the location of the initial contamination field and the coordinates of the immediate point source by solving a series of adjoint problems. This type of identification values the data of the impurity spread range. Quantitative experiments confirm the reliability of the algorithm based on solving adjoint problems for the mass impurity migration model. The results can be used for solving various environmental problems associated with the study of the impact of anthropogenic contamination sources in the waters of the Sea of Azov and the Black sea.

Acknowledgment
This research was carried out within the scope of state research order for Complex Multidisciplinary Studies of Oceanological Processes Influencing the Functioning and the Evolution of Coastal Ecosystems in the Black Sea and the Sea of Azov 0827-2018-0004 (Coastal studies).

References

[1] Ivanov V A and Fomin V V 2008 *Mathematical simulation of dynamic processes in offshore areas* (Sevastopol: EKOSI-gidrofizika) 363 p
[2] Fomin V V 2002 Quantitative water circulation model for the Sea of Azov *Acad. works of UHMI* **249** 246–255
[3] Marchuk G I and Penenko V V 1978 Application of optimization methods to the problem of mathematical simulation of atmospheric processes and environment *Modelling and Optimization of Complex Systems* Proc. of the IFIP-TC7 Working conf.)NewYork: Springer) pp 240–252

[4] Agoshkov V I, Parmyzin Ye I and Shutyayev V P 2013 Observational data assimilation for the circulation problem in the Black Sea and the analysis of the responsiveness of its solution *Izv. RAN. Fizika atmosfery i okeana* **49(6)** 643–54

[5] Shutyaev V P 2018 Adjoint equations in variational data assimilation problems *Russ. J. of Numerical Anal. and Mathem. Modell.* **33(2)** 137–47

[6] Kochergin V S and Kochergin S V 2015 Identification of a pollution source power in the Kazantip Bay using the variational algorithm *Phys. Oceanogr.* **2** 69–76

[7] Demyshev S G, Kochergin V S and Kochergin S V 2018 Using the Variational Approach and Adjoint Equations Method Under the Identification of the Input Parameter of the Passive Admixture Transport Model", *Physical and Mathematical Modeling of Earth and Environment Processes* Chapter 7 3rd Int. Sci. School for Young Sci. pp 51–61

[8] Marchuk G I 1982 *Mathematical simulation for environmental problems* (Moscow: Nauka) 320 p

[9] Skiba Y N and Parra-Guevarra D 2017 Application of Adjoint Approach to Oil Spill Problems *Environ. Modell. & Assessment* **22(4)** 379–95 (Springer)

Asphalt concrete on the basis of converter slags

N S Kovalev[1], E N Otarova[2], V V Gladnev[1], E A ogly Sadygov[1], P V Demidov[1] and E V Panin[1]

[1] Voronezh State Agricultural University named after Emperor Peter the Great, 1, Mitchurina str., Voronezh, 394087 Voronezh, Russia
[2] Air Force Academy named after professor N.E. Zhukovsky and Iu.A. Gagarina, 133, Marshal Nedelin str., Voronezh, Voronezh region, 394052, Russia

E-mail: NSKovalev@mail.ru, ekaterina.otarova@mail.ru, lav-78@mail.ru, elzas.sadygov@mail.ru, 79204170254@yandex.ru, ksenia.bos89@mail.ru

Abstract. The properties of converter slag, its activity and the influence of the degree of decay on the structural and mechanical properties of asphalt concrete are studied. It has been established that asphalt concrete with a degree of decay of converter slag of 1.24 % or less can be used for paving roads. It has increased heat, water and frost resistance, which will significantly increase the service life of roads from the beginning of operation to overhaul.

1. Introduction

In the process of construction, reconstruction and repair of roads, large volumes of conditioning natural mineral raw materials are used in the form of sand, gravel, gravel, mineral powder, the reserves of which are depleted over time. The main costs (about 60–70 %) in the construction of roads fall on the construction of pavement, and primarily on two structural layers: coating and base

An urgent problem is the use of local materials and industrial waste in road construction. It is possible to reduce the cost of building roads by replacing expensive imported materials with local materials. However, along with this, it must be borne in mind that the coatings and foundations of roads have sufficient durability (reliability). Only in this aspect will the use of local materials and industrial waste give a high economic effect, especially during the period of intensive road construction.

There are no deposits of strong stone materials on the territory of the Lipetsk region, and road organizations mainly use imported road-building materials. At the same time, the Novolipetsk Metallurgical Plant (NLMK) is located in the region, producing millions of tons of slag, including converter slag, annually as by-products.

The use of slag in road construction was considered in [1–4]. A specific feature of converter slag is its susceptibility to varying degrees of silicate, calcareous and ferruginous decomposition, with a large proportion of calcareous decomposition [2, 5]. This feature inhibits the use of converter slag in road construction.

We conducted studies to determine the degree of slag decomposition, which does not adversely affect the structural and mechanical properties of asphalt concrete, which will expand the range of road building materials, improve the environmental situation, reduce the area under dumps, and reduce the cost of building roads.

IOP Conf. Series: Materials Science and Engineering **905** (2020) 012040 doi:10.1088/1757-899X/905/1/012040

2. Methods and materials

The converter slag of the oxygen converter shop No. 2 of NLMK is a black-brown material with a dense, less often a porous structure, with a high density. Due to the heterogeneity of converter slag, its physical and mechanical properties vary over a rather wide range (Table 1).

The mineralogical composition of the converter slag is represented mainly by dicalcium silicate, tricalcium silicate, calcium ferrochromate, and the RO phase. The content in the slag of the first two minerals can reach 80 % [5].

Table 1. Physical properties of converter slags

Indicators	Fraction size, mm		
	5–10	10–20	20–40
	Indicator value		
Average bulk density, g/cm^3	1.68	1.72	1.69
Average density, g/cm^3	3.13–3.57	3.31–3.69	3.33– 3.45
Water absorption, % of the volume	1.40–1.90	1.00–1.46	1.10–1.90
Voidness, % of volume	45.5–51.2	46.6–52.4	45.9–51.5

One of the characteristics of slag materials is their activity. We conducted studies on the influence of the age of slag from the moment of its receipt on hydraulic activity.

To this end, the slag material was ground in a ball mill to fineness grinding mineral powder. Samples of diameter and height were prepared from the obtained slag flour with a sealing load of 10 MPa and a water-slag ratio of 0.16.

After manufacturing, the samples were kept for 7 days in air, and then placed in humid conditions. Before the test, the samples were placed for two days in water and the compressive strengths were determined. The test results of the activity of slag materials depending on the age of the slag from which the slag flour is made, and on the exposure time of the samples in wet conditions are presented in Table 2.

Table 2. Influence of slag age on its hydraulic activity

The age of the slag from which the slag flour is made, days	The compressive strength, MPa, of samples stored under wet conditions, days				
	7	28	60	120	360
20	1.23	2.00	9.05	8.05	14.50
30	1.48	1.96	8.08	9.20	15.00
60	1.70	2.30	9.60	10.11	17.00
90	1.70	2.20	9.90	10.00	17.60
120	1.58	2.20	9.00	9.30	16.50
180	1.50	2.00	8.80	9.10	17.00
270	1.44	2.00	8.70	8.80	17.50
320	1.33	1.90	9.20	9.40	18.90
360	1.28	1.60	7.50	–	–

An analysis of the results presented in table 2 indicates that the converter slag is classified as inactive slag, since the compressive strength at 30 days is less than 6 MPa. With increasing storage time of samples of slag flour in wet conditions, its strength increases, especially intensively after 30 days.

The age of the converter slag at the time of its grinding into slag flour also has a significant effect on the tensile strength. The maximum tensile strength is observed for samples at all test ages prepared from converter slag, the age of which at the time of grinding into slag flour is 60–90 days. This is explained by both the carbonization of free lime and the hydrolysis and hydration of slag minerals [1].

Silicate decomposition of slag occurs during the transition of dicalcium silicate from the β-form to the γ-form, which causes an increase in the volume of the substance and creates high internal stresses – pieces of slag crack, and some break up into flour. Glandular decomposition occurs after wetting the slag with water, when iron sulfide passes into ferrous hydroxide and pieces of slag crack under the influence of internal stresses. Lime decay occurs when the minerals of the converter slag interact with water. The decay time is long, since it occurs at the points of contact of inclusions of free lime and metal with atmospheric moisture. During the thermal treatment of slag for 6 hours, according to GOST 3344, all types of decay occur [6].

The influence of the degree of decomposition on the mechanical properties (crushability, abrasion, and frost resistance) of converter slag of a fraction of 5–10 mm was studied at different periods of slag aging under natural conditions (Table 3).

By crushing and wear, crushed stone from converter slag belongs to the 1st class of strength [6]

The analysis of the results of table 3 indicates that, despite the decay, the frost resistance of the converter slag exceeds 100 cycles.

The study of the physical and mechanical properties of asphalt concrete was carried out according to the standard method taking into account the specific features of converter slag (decomposition, chemical and hydraulic activity) [7].

The stability of the physico-mechanical properties of asphalt mixtures based on converter slag depends on the degree of decomposition. The greatest decay of converter slag is observed in the initial period from the moment of its receipt. With an increase in shelf life, the decay rate decreases.

In order to clarify the degree of decomposition at which converter slag can be used, the samples were made from slag aged for different periods in air. After each aging period, the degree of slag decomposition in the autoclave was established, the samples were molded, and the properties of asphalt concrete at the age of 2 days were determined. and after steaming in an autoclave. In addition, the physical and mechanical properties were studied after exposure to air for 28, 60, 90, and 180 days. Such a research technique will make it possible to judge the change in the properties of asphalt concrete on converter slag depending on the degree of decay.

For research, a fine-grained asphalt mixture of type B was selected, selected from the curves of dense mixtures, the composition of the mineral part of which consisted entirely of converter slag: slag fr. 5-10 mm – 40 %, slag sand fr. 0–5 mm – 50 %, mineral powder from converter slag – 10 %, bitumen grade BND 90/130 – 7 % (in excess of 100 % of the mineral part) [8].

Table 3. Mechanical properties of the converter slag of the fraction 5–10 mm and its frost resistance depending on the degree of decay

Age of slag, days	The degree of decay, %	The number of freezing cycles	Loss in mass, %	Crushability, %	Wear in the shelf drum, %
7	7,1	0	–	7.8	17.0
		15	0.35	–	–
		25	0.75	–	–
		50	1.00	–	–
		75	1.00	–	–
		100	1.25	–	–
40	3,7	0	–	14.2	17.0
		15	0.50	–	–
90	1,5	0	–	12.3	16.9
		15	0.25	–	–
150	1,4	0	–	13.1	16.1
		15	0.10	–	–
180	1,3	0	–	6.8	15.8
		15	0.10	–	–
320	1,25	0	–	7.2	15.8
		15	0.10	–	–

3. Results

For a more detailed study of the structural and mechanical properties of asphalt concrete based on converter slags with an unstable crystalline structure, the samples were formed from asphalt mixtures with different slag exposure (1, 2, 4, and 7 months) and tested at the age of 2, 28, 60, 90, 180 days from the date of manufacture (Table 4).

Analysis of the data given in table. 4 shows that samples of asphalt mixtures with a slag decomposition rate of 5.3 % have fairly high compressive strengths at temperatures of 20 and 50 °C, low water saturation, and high coefficients of water resistance and long-term water resistance. A certain increase in tensile strength over time, especially at a temperature of 50 °C, is explained by hydration processes and the interaction of bitumen functional groups with calcium and magnesium oxides of converter slag [9, 10].

The bitumen membrane on the surface of the slag material prevents the penetration of moisture, and therefore there is no visible damage to asphalt during prolonged storage in air. However, when steaming in an autoclave, asphalt concrete samples are completely destroyed. Cracks, chips, yields of slag decomposition products appear on them.

With a decrease in the degree of decay to 2.4 %, samples of asphalt concrete can withstand autoclaving, but the values of compressive strengths sharply decreased below those required by the standard. This indicates an unstable structure of slag asphalt concrete and the possibility of its destruction during operation of asphalt concrete pavements (Fig. 1).

With a degree of decomposition of converter slag of 1.27 %, asphalt concrete maintains stable properties when samples are exposed to air and after autoclaving. After steaming in an autoclave, the samples do not change shape and no cracks or chips are observed on them.

As our studies have shown, such a degree of decay occurs approximately no earlier than after 7–8 months of storage of converter slag in air with intensive watering.

Figure 1. The state of asphalt concrete after steaming with a degree of decay of converter slag 2.4 %

Thus, it follows from the foregoing that converter slag with a degree of decay of not more than 1.2–1.3 % is allowed for the production of asphalt concrete.

Insufficient corrosion resistance of asphalt leads to premature destruction of coatings. Asphalt concrete collapses intensively with prolonged exposure to water and alternate freezing-thawing [11]. Since converter slag has an unstable structure, the long-term exposure to water should accelerate these processes.

The study of long-term exposure to water was carried out with slag with different degrees of decomposition and duration of water saturation of 2, 28, 90 and 180 days (Table 5).

As a rule, with increasing storage time of samples in water, water saturation increases. This is due to the partial decompression of the samples during water saturation and the creation of pore reserves during slag decomposition, which were previously inaccessible to water migration.

With an increase in the duration of water saturation with a degree of decay of slag of 1.7 % or more, the compressive strength decreases, but remains above the requirements of the standard. However, on samples with a slag decomposition rate of 5.4 % after 180 days, cracks and chips are observed (Fig. 2a). Swelling is observed on samples made from slag with a decay rate of 1.7 % (Fig. 2 b), and these phenomena are not observed on samples from slag with a decay rate of 1.27 % (Fig. 2 c).

The compressive strengths of asphalt concrete at a temperature of 20 °C at a slag decomposition rate of 1.27 % are reduced to a water saturation duration of 28 days. and then increase. This is due to the hydraulic activity of the slag. Bitumen actively interacts with hydration products, and at the phase separation a strong adhesive contact is formed that is resistant to prolonged exposure to water.

IOP Conf. Series: Materials Science and Engineering **905** (2020) 012040 doi:10.1088/1757-899X/905/1/012040

Table 4. Influence of the degree of decomposition of the converter slag and the age of the samples on the physical and mechanical properties of asphalt concrete

Age of samples, days	Average density, g/cm3	Water saturation, % volume	Swelling, % of volume	Tensile strengths in compression, MPa, at temperature, ° C 20	50	Coefficients: water resistance / long water resistance
			slag age 1 month, decay **5.3 %**			
2	3.19	0.68	0.70	4.1	1.0	0.98/1.15
28	3.14	1.51	0.46	4.7	1.8	0.94/1.17
90	3.17	0.98	0.91	4.3	1.8	0.98/1.13
			after steaming in an autoclave			
2				samples collapsed		
			slag age 2 months, decay **2.4 %**			
2	2.85	3.81	0.72	5.2	2.1	0.98/0.86
28	2.83	3.88	0.19	6.4	1.8	0.81/0.83
60	2.83	4.18	0.58	6.6	2.2	0.82/0.98
90	2.85	4.70	0.34	7.1	2.1	0.83/0.75
			steaming in an autoclave			
2	2.71	6.05	0.00	1.7	0.6	1.11/1.05
			slag age 4 months, decay **1.7 %**			
2	2.80	5.49	0.09	4.2	2.1	0.97/0.98
28	2.87	4.06	0.07	4.3	2.2	0.93/0.91
60	2.96	3.03	0.00	5.5	1.5	0.85/1.02
90	2.81	5.87	1.76	5.0	1.1	0.77/1.08
180	2.82	4.13	0.89	4.8	1.2	0.91/0.91
			after steaming in an autoclave			
2	2.75	7.63	0.70	2.6	0.9	0.88/0.97
			slag age 7 months, decay **1.27 %**			
2	2.68	3.17	0.00	6.1	2.8	0.9/1.15
28	2.66	3.57	0.32	6.0	3,0	1.07/1.15
60	2.67	3.66	0.94	7.2	2.5	1.11/1.11
90	2.69	1.74	0.00	10.9	2.6	0.86/1.05
180	2.67	2.99	0.00	8.1	2.6	0.85/0.75
			after steaming in an autoclave			
2	2.58	7.58	0.00	5.6	1.5	1.07/1.10

The study of frost resistance was carried out with slag with a different degree of decay and the number of freeze-thaw cycles (Table 6).

As can be seen from the presented results, with an increase in the number of freeze-thaw cycles, the water saturation of asphalt concrete increases. This phenomenon is explained not only by the softening of asphalt concrete due to the transition of water into ice and an increase in its volume, but by the calcareous decay of slag grains. The absence of a pattern of changes in water saturation from the aging time of the converter slag to the moment of its use for the manufacture of asphalt concrete can be explained by the uneven distribution of lime in the slag mass.

According to the main indicators of physical and mechanical properties, asphalt concrete based on converter slag with a different degree of decay satisfies the requirements of the standard.

With a degree of decay of slag of 5.3 and 2.4 %, the compressive strength of asphalt concrete decreases, but remains within the requirements of the standard. The decrease in tensile strength after exposure to freeze-thaw cycles is associated both with decompression of asphalt concrete (destruction) and with the decay of slag.

At a slag decomposition rate of 1.7 and 1.27 %, the compressive strengths increase after exposure to freeze-thaw cycles. This is explained by the processes of hydrolysis and hydration of slag, the physicochemical interaction between the functional groups of bitumen and slag minerals with the formation of organomineral compounds (structural processes prevail over destructive ones).

Table 5. Influence of the degree of decomposition of the converter slag and the duration of water saturation on the physical and mechanical properties of asphalt concrete

Duration of water saturation, days	Average density, g/cm³	Water saturation, % volume	Swelling, % volume	Tensile strengths in compression, MPa, at temperature, °C		Water resistance factors
				20	50	
			slag age 1 month, decay **5.3 %**			
2	3.04	1.70	0.66	4.9	2.2	1.43
28	3.04	4.33	0.46	5.7	1.5	1.20
90	3.03	3.43	0.00	5.7	1.5	1.42
180	3.04	7.70	1.01	2.7	1.3	-
			slag age 2 months, decay **2.4 %**			
2	2.79	4.06	2.86	6.5	2.2	1.03
28	2.82	6.50	2.40	5.2	1.3	0.89
60	2.83	6.27	2..51	4.5	1.3	0.81
90	2.82	10.83	6.69	4.7	1.1	0.54
			slag age 4 months, decay **1.7 %**			
2	2.83	2.25	0.54	6.4	2.0	1.29
28	2.81	6.55	0.00	6,3	1.6	1.55
60	2.78	4.50	2.16	4,3	1.6	1.13
90	2.83	7.65	4.67	3,4	1.2	1.18
180	2,83	10.63	6.89	3.6	1.0	1.00
			slag age 7 months, decay **1.27 %**			
2	2.67	1.78	0.02	7.1	2.9	1.44
28	2.68	4.26	0.58	4.7	2.6	0.87
60	2.66	3.34	0.74	6.4	2.2	0.80
90	2.68	2.73	0.13	7.5	1.9	1.70
180	2.67	2.97	0.05	7.9	2.1	1.81

Table 6. Influence of decomposition of converter slag decomposition and number of freezing-thawing cycles on the physical and mechanical properties of asphalt concrete

Number of freeze-thaw cycles	Average density, g/cm3	Water saturation, % volume	Tensile strengths in compression, MPa, at temperature, °C		Water resistance factors	Frost factors
			20	50		
			slag age 1 month, decay **5.3 %**			
25	3.03	1.40	5.5	2.0	1.05	1.17
50	2.99	2.33	4.8	1.8	0.87	0.97
100	2.79	2.67	3.6	1.6	1.14	0.95
			slag age 2 months, decay **2.4 %**			
25	2.80	4.81	6.8	1.5	0.93	0.98
50	2.70	8.15	5.8	1.7	0.62	0.75
75	2.69	7.79	5.2	1.5	0.73	0.69
100	2.61	9.39	3.6	1.0	1.02	0.52
			slag age 4 months, decay **1.7 %**			
25	2.88	3.15	4.1	1.9	1.00	0.89
50	2.82	3.93	4.0	3.0	1.00	0.89
75	2.79	5.67	4.5	1.6	1.26	0.82
			slag age 7 months, decay **1.27 %**			
25	2.68	3.14	6.9	2.0	0.81	1.50
50	2.70	2.47	8.8	2.0	0.94	1.22
75	2.67	3.26	10.2	2.5	0.72	0.95
100	2.66	3.42	10.9	2.6	0.86	0.82

IOP Conf. Series: Materials Science and Engineering **905** (2020) 012040 doi:10.1088/1757-899X/905/1/012040

Figure 2. The condition of asphalt concrete after 180 days of water saturation with a degree of decomposition of converter slag of 5.4 % (a); 1.7 % (b); 1.27 % (c)

Figure 3. The condition of asphalt concrete after 100 cycles of freezing and thawing with a degree of decomposition of converter slag of 5.4 % (a); 1.7 % (b); 1.27 % (c)

However, the state of asphalt concrete samples with different degrees of decay after freezing-thawing cycles is different. With a slag decomposition rate of 5.7 %, intense cracking, chips, cracks, and roughness of the surface of asphalt concrete are observed (Fig. 3a). With a decrease in the degree of slag decomposition to 1.7 % (Fig. 3b), the intensity of crack formation and turbulence of asphalt concrete decreases significantly. Cracks appear much less frequently. At a slag decomposition rate of 1.27 %, no cracks or turbulence were observed on the samples of asphalt concrete (Fig. 3c). Therefore, when studying the frost resistance of asphalt concrete based on converter slags, attention should be paid not only to changes in physical and mechanical properties, but also to those conditions under which the integrity of asphalt concrete samples is preserved after repeated freezing and thawing cycles.

4. Conclusion

1. Converter slag with a degree of decay of less than 1.27 % can be used in road construction for the installation of asphalt concrete pavements. Asphalt concrete has a high heat, water and frost resistance.

2. The possibility of using converter slag will expand the range of local building materials, which will lead to a decrease in the cost of building roads.

3. The use of converter slag will significantly improve the environmental situation, reduce the area occupied by dumps.

References

[1] Bondarev B A, SHtefan Iu A, Goncharova M A and Shtefan G E 2005 *Slag aggregate asphalt concrete* Monograph (Lipetsk: LGTU) 183 p

[2] Volkov M I, Korolev I V and Zinov I T 1962 *Blast furnace slag in road construction* (Donetsk: Donetsk. kn. izd-vo) 70 p

[3] Dagaev B I 1973 Bitumen-mineral mixtures based on blast furnace slag with adjustable properties *Car roads* **4** 14

[4] Gezentsvei L B, Gorelyshev N V, Boguslavskii A M and Korolev I V 1985 *Road asphalt concrete* (Moscow: Transport) 350 p

[5] Aleksandrov S E, Titova L N and Privalova A I 1982 *On the stability of slag structures, Rational use of slag slag products in construction* (Voronezh: TSCHO) pp 27–9

[6] GOST 3344-*83 Crushed stone and sand slag for road construction. Technical conditions (as amended by No. 1). Enter 1985-01-01* 2007 (Moscow : FSUE Standartinform) 10 p

[7] GOST 12801-98 *Materials based on organic binders for road and airfield construction. Test methods. Enter 1999-01-01* 1999 (Moscow: Gosstroy of Russ.) 54 p

[8] GOST 9128-2013 *Mixtures of asphalt concrete, polymer-asphalt concrete, asphalt concrete, polymer-asphalt concrete for roads and airfields. Technical conditions Enter 2014-14-11* 2014 (Moscow: Standartinform) 50 p

[9] Kovalev N S and Bykova Ia A 2008 Study of the physico-chemical interaction of slag materials with bitumen *Bull. of the Volgograd State Univer. of Architect. and Civil Engineer. Ser. Construct. and architect.* **11(30)** 81–7

[10] Samodurov S I, Maslov S M and Kovalev N S 1976 On the durability of bitumen-slag coatings of roads *Univer. News. Ser. Construct. and Architect.* **8** 147–51

[11] Merentsova G S and Goriunova Iu A 2017 Factors affecting the durability of asphalt concrete and the operational reliability of asphalt concrete pavements *Polzunovsky Almanac* **2(4)** 124–7

IOP Conf. Series: Materials Science and Engineering **905** (2020) 012041 doi:10.1088/1757-899X/905/1/012041

Automation system for oil and gas service enterprises

I V Remizova[1], T S Krugovaya[1], S Yu Krotova[2] and E N Ovchinnikova[2]

[1] Department of Information Measuring Technologies and Control Systems, St. Petersburg State University of Industrial Technologies and Design, St. Petersburg, Russia

[2] Department of Informatics and Computer Technology, St. Petersburg Mining University, St. Petersburg, Russia

E-mail: remirina@bk.ru, tanysha.krugovaya@yandex.ru, Krotova_SYu@pers.spmi.ru, Ovchinnikova_E.N@pers.spmi.ru

Abstract. The paper presents the results of the development of a single integrated solution for the automation of technological processes of service enterprises of the oil and gas sector, which allows to organize the work of production and management in a single information space. Algorithms have been developed to solve individual problems that are interconnected into certain algorithmic modules. The results of the integration analysis of modern automation systems of the ERP-class are presented, in particular, the level of adaptation to the stages of the product life cycle, compliance with standards and safety requirements, the complexity of the transition to the automation system. The results of the analysis of possible risks of the project during the implementation of this algorithm are presented. The practical significance of the development of a single production management algorithm is described. Finally, the treatise evaluates the main advantages and economic effect of the introduction of an automation system for an enterprise engaged in the production of oil and gas pipes at the process control level.

1. Introduction

The concept of "oil and gas sector", along with oil refining, includes two major components: oil and gas production and oil and gas service (enterprises and organizations which activities are aimed at providing services and ensuring the necessary level of oil production and transportation).

The development trends of the Russian economy are such that the revenues generated by the oil and gas sector provide the main inflow of investment resources, the stability of the national currency and form about 45–50 % of federal budget revenues [1].

Due to the irrational planning of the production cycle and the large time required to eliminate defects, the process of manufacturing oilfield and transportation pipes of various diameters and sizes assumes large losses [2, 3]. The consequence of production losses are material losses, which are expressed in cash.

To date, the automated systems used in enterprises involved in the extraction, processing and storage of oil and gas are not suitable for the technological process of manufacturing oil pipes and organizational and managerial accounting at the enterprises of oil and gas services [4].

The development of an automated production cycle management system for oil and gas enterprises occupies a special niche in scientific and technical research. Most of the works present systems aimed at enterprises engaged in the production, processing, storage, transportation and sale of oil and gas [5–7].

However, special systems for organizing and planning production processes at the oil and gas service enterprises have not yet been created.

Thus, the development of a single integrated solution for the automation of technological processes of enterprises in the service sector of the oil and gas industry, which allows organizing the operation of production and management systems in a single information space, optimizing information flows and reducing production costs, is an urgent task. Improving financial results by increasing the manageability of the enterprise and the rational distribution of all types of resources, leads, in turn, to reduced cost of production and services.

2. Methods and materials

As a systematization of the facts for this study, we used methods of collecting information and theoretical and empirical methods: analysis, comparison, selection of factors, the establishment of relationships between them.

In particular, the basis of the analysis of modern automation ERP-systems includes investigation of integration domain, level of adaptation to the stages of the product life cycle, compliance with Russian standards and safety requirements, the complexity of the transition to an automation system.

To evaluate the economic efficiency of designing and implementing software modules of a unified algorithm for organizing and planning a production line for pipe manufacturing, the following types of calculations and analyzes were used: quantitative calculation of the economic effect, calculation of capital costs for design and implementation, calculation of savings by increasing labor productivity, and also a risk analysis of projects when introducing an algorithm for organizing and planning a technological line [8].

3. Results

The developed system for automating the activities of oil and gas service enterprises at the process control level includes an algorithm for organizing and planning a production line for the production of tubing, divided into algorithmic modules that correspond to individual tasks. The tasks of the system are interconnected in such a way that individual modules of the general algorithm function in specific production processes (Figure 1).

The form of algorithmic support of an automated process control system is based on the application of the current principle of modularity, according to which mathematical software can be decomposed into a set (or synthesized from a set) of autonomous parts of algorithmic problems.

The task "Start in the Workshop" implements the input of a new production order and allows you to load input data, either in XLS-format from any ERP-system, or manually. This indicates a high integration ability of the organization and planning algorithm.

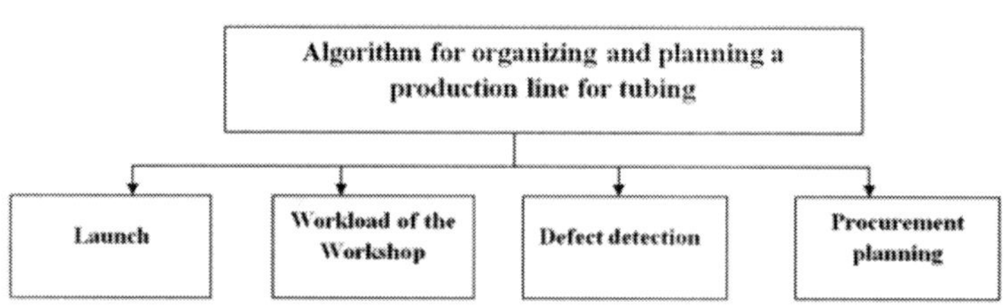

Figure 1. Algorithmic tasks of the production line of tubing

In the production process, tubing pipes go through all the operations of the production line, and each transition is accompanied by preparation of normative and reference documentation. The essence of the task is to combine many different documents (in accordance with GOST 633-80) in a single information field: the number of the order made following a particular request. Optimization and visualization of this process allows sufficiently and quickly determining the possibility of sending a new order to production, as well as to identify the overload of the production line and deviation from the deadlines.

Thus, the following tasks can be solved: to calculate the possibility of fulfilling a new production order on time; show in detail the workload of equipment included in the production line; timely prepare materials for production consumption. At the same time, the work of the supply service directly affects the economic performance of the manufacturing enterprise. Only the exact balance between the loading of production capacities and the timely execution of orders without delay can do without an overabundance of products and save on storage of finished products.

This problem can be solved by the following algorithm: the initial information is entered into the database manually, or loaded from any ERP-system in XLS-format; the order number is checked for uniqueness, if duplication of data is detected, an error message is displayed; when choosing the type of tubing, a table opens, according to GOST 633-80, where it is necessary to choose the geometric parameters of the pipe and its coupling. If, according to the technical specifications, the parameters are different from the normative documentation, you can enter them manually. Then, when sent to the workshop, the order proceeds to the first operation of the production line for the production of tubing (tubing), and the technical specifications are stored in the database.

The task “Start in the Workshop” allows entering information from the normative and reference documentation into the database, and also viewing the workload of the workshop when moving orders for operations of the tubing production line and, if necessary, periodically adjust the data.

Algorithm for performing the task "Start in the Workshop": data is entered manually or loaded from any ERP-system in XLS-format; changes are introduced at the stages of the product (order) life cycle on the production line; adjusted data is stored in the information system. When one selects the module "Workload of the Workshop", the workload of the production line is displayed and each order at the stages of the product life cycle is monitored according to the deadlines.

Figure 2 presents a block diagram of the solution of the algorithmic problems of an automation system for an oil and gas service enterprise at a process control level.

Based on the algorithms of the tasks considered in the project, software was developed for the automation system of the oil and gas sector. Problem solving is carried out using screen forms.

Screen forms for the tasks “Start in the Workshop” and “Workload of the Workshop” are presented in Figs. 3 and 4.

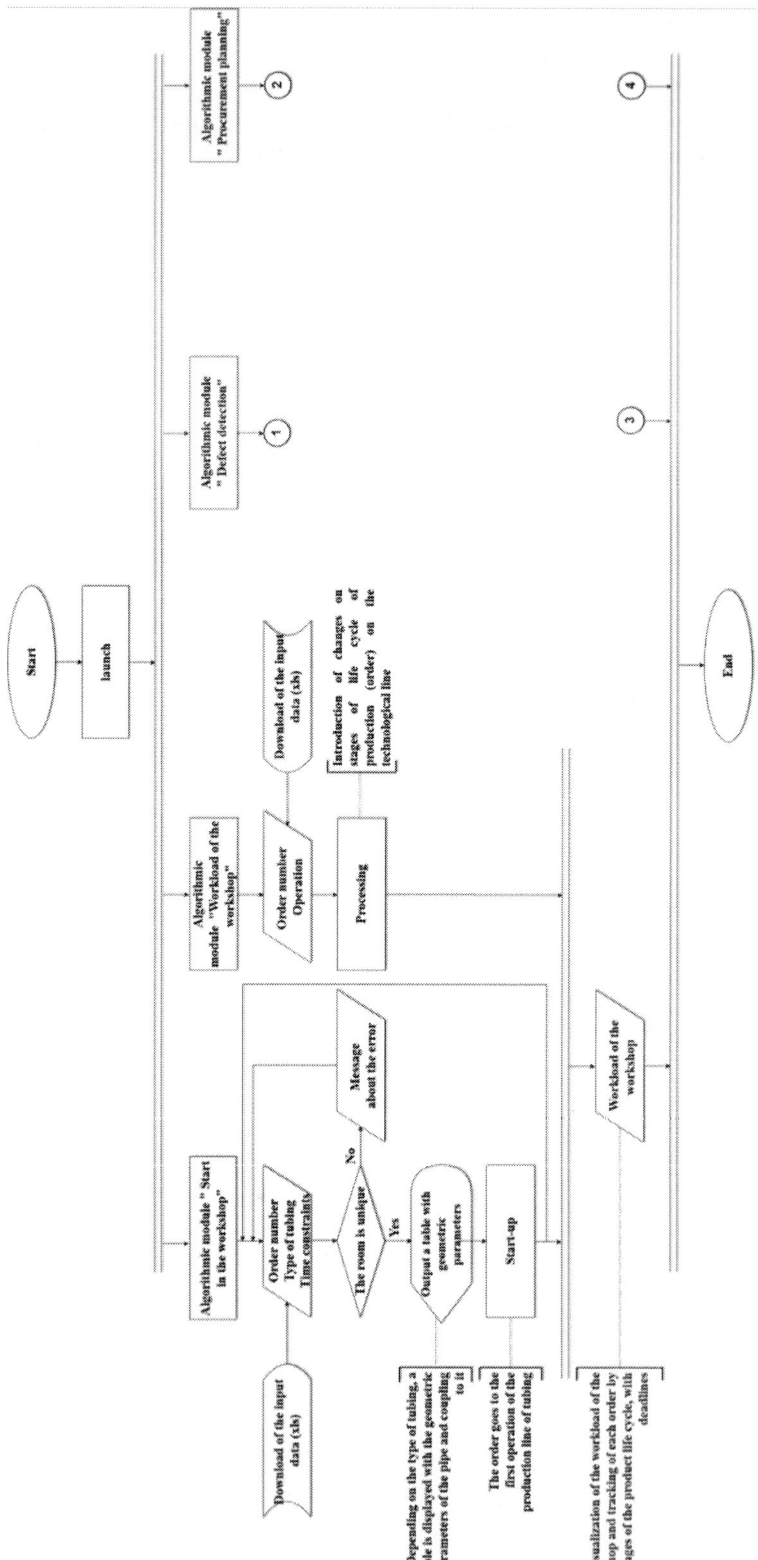

Figure 2. Flowchart for solving algorithmic problems

The task "Defect Detection" captures the operations of the technological process in which production losses occur. The database accumulates information about detected defects in tubing, and a summary table is generated for a certain period.

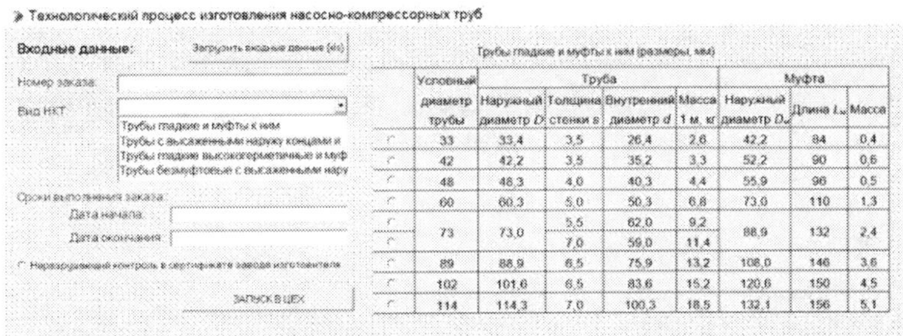

Figure 3. Screen form for the task "Launch to the Workshop".

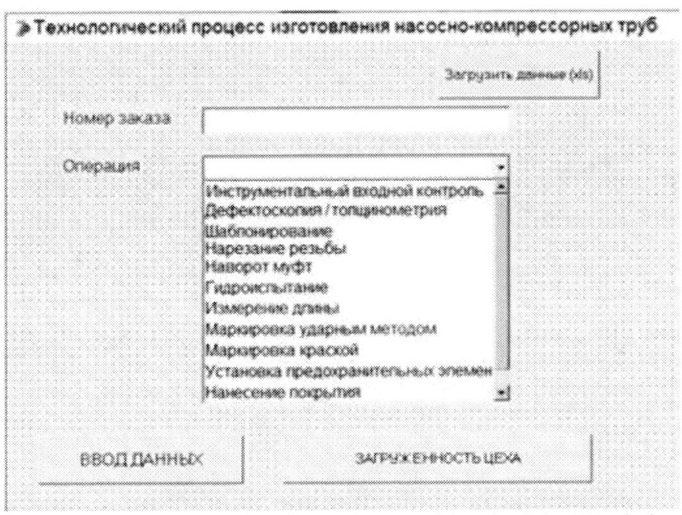

Figure 4. Screen form for the task "Workload of the Workshop"

The algorithm for performing the task "Defect Detection" is presented as a block diagram in Figure 5; the screen form of the problem is given in Figure 6.

The algorithm includes: loading data from an ERP-system in XLS-format or manual input; recording and saving data in the information system from the condition of choosing a time period; in a given period, the number of defects is calculated, the number of unfinished orders is considered; according to the results, a summary table is displayed on the number of defects on each operation of the production line and the total number of orders for the requested period.

Tabular visualization of defects in operations of the processing line displays their quantitative ratio for a certain period [3].

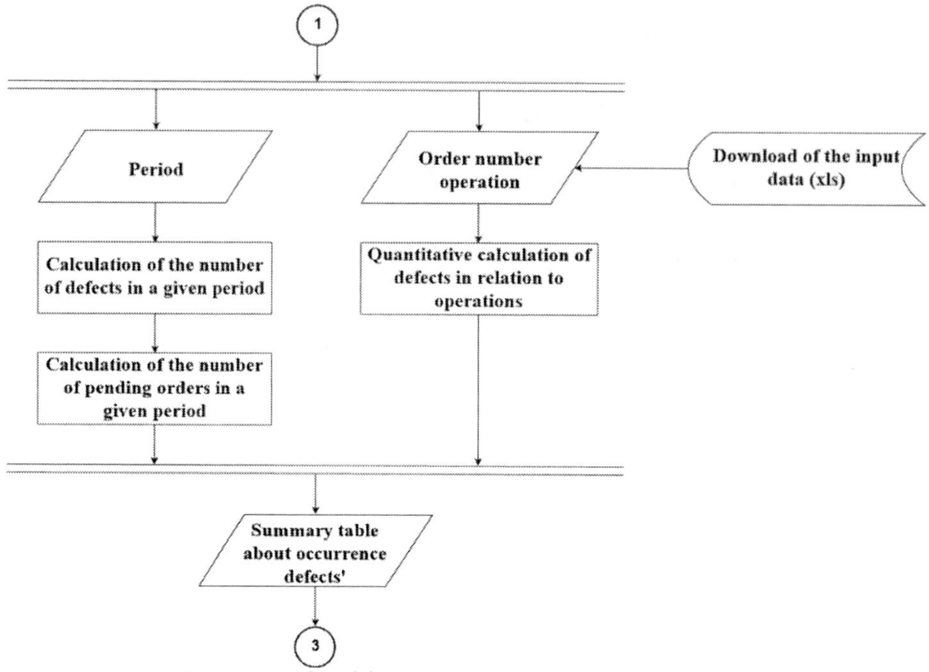

Figure 5. Algorithm for the task "Defect Detection"

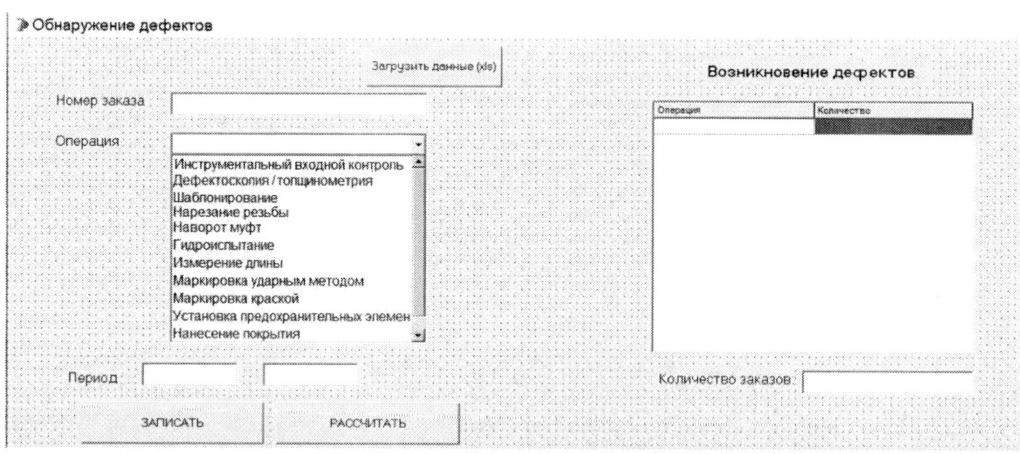

Figure 6. Screen form for the "Defect Detection" task

The task "Procurement Planning" allows moving on to rational planning of the procurement of raw materials. When placing a new order, there is a need to calculate the amount of raw materials needed to complete the technical task from the customer company.

Service companies face the problem of duality in providing production with raw materials. The nominal customer, when placing an order for pipe products, primarily relies on the required length. Products are sold in tons. From here, double conversion is carried out first by the customer from length to tons, then the manufacturer in the opposite direction from tons to length.

Differences in weight when purchasing raw materials for the manufacture of pipe blanks or selling pipe products are often a real problem for oil and gas service companies producing tubing and their customers. Excessive weight of finished products leads to financial losses of enterprises in the service sector of the oil and gas industry and unreasonable benefits for customers. Lack of weight changes the situation in the opposite direction, without eliminating the bias.

Due to the double conversion described above, production losses occur when the tonnage is transferred to length and vice versa. The solution to the situation considered was taken into account when creating the algorithm for the procurement planning task.

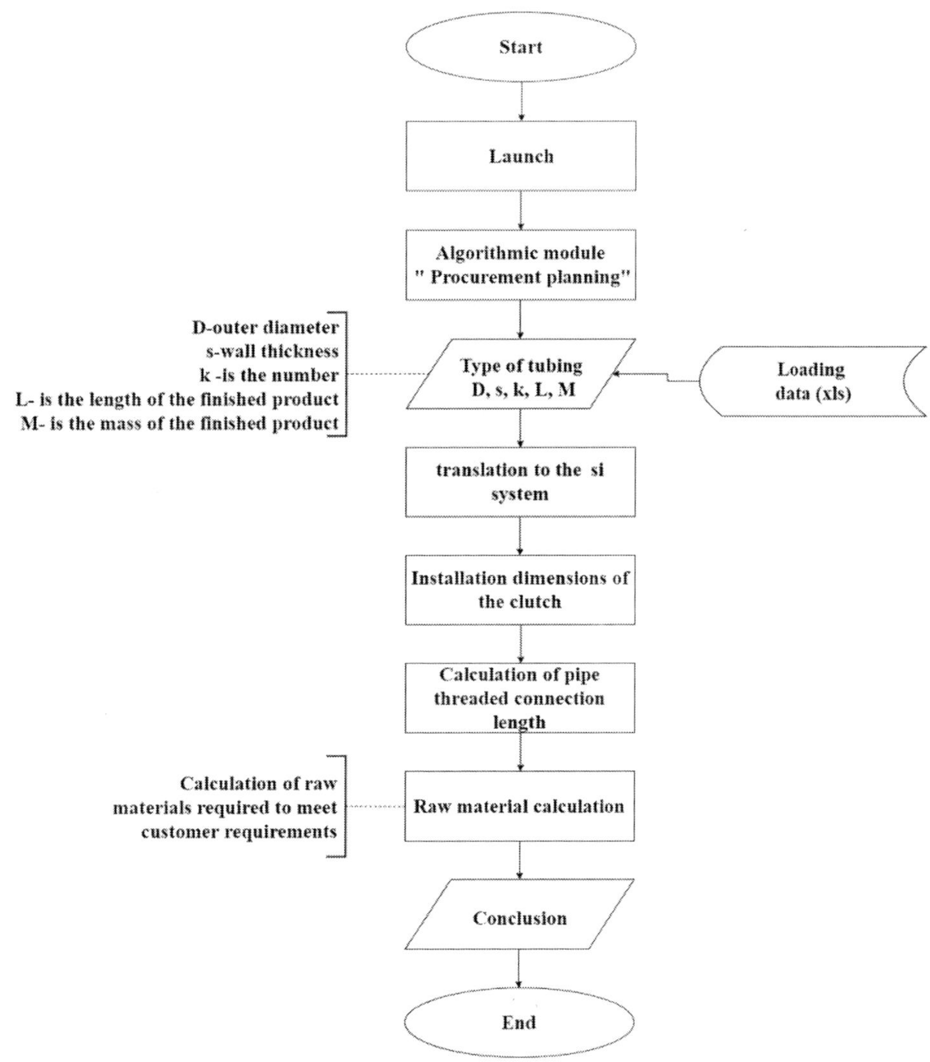

Figure 7. Algorithm for "Procurement Planning" task

The algorithm for solving the problem includes entering the source data manually, or by loading from any ERP-system in XLS format; data are referred to the International System of Units; sizes of the coupling for tubing are set; the length of the threaded pipe joint is calculated; the calculation of the raw materials necessary to meet the requirements of the customer is made. The block diagram of the algorithm for solving the "Procurement Planning" task is presented in Figure 7.

The implementation of an automation system for the activities of oil and gas service enterprises at the process control level, developed on the basis of a unified algorithm for organizing and planning a production line for tubing manufacturing, was carried out at a real service enterprise, including the production and marketing of pipe products, together with a wide range of services [9].

According to the results of calculating the economic efficiency of designing and implementing software modules for a unified algorithm for organizing and planning a production line for tubing manufacturing, the benefit is noticeable in the medium and long term [10]. Expected economic efficiency amounted to 1.6 million rubles.

4. Conclusion

An algorithm was proposed for organizing and planning a production line for the production of tubing, which underlies the developed system for automating the activities of oil and gas service enterprises at the process control level.

As a result of the implementation of the tasks of the algorithm "Start in the shop" and "Workload of the Workshop": the logical sequence of the process is determined from the site of raw materials acceptance to the receipt of finished products; stages of coordination of technological and technical communication between the main operations of the production process are displayed; it becomes possible to visualize the workload of the processing line; the possibility of fulfilling a new production order and preparation of material for production consumption in a timely manner according to the technical specifications can be calculated.

The precise balance between the loading of production capacities and the timely execution of orders without delay allows avoiding an overabundance of products and saving on storage of finished products.

The performance of the task "Defect Detection" allows systematizing the factors affecting the occurrence of production losses. Tabular visualization of defects in operations of the processing line displays their quantitative ratio for a certain period.

The developed module, which solves the problem of "Procurement Planning" and is part of the organization and planning algorithm for the production line for the production of tubing, allows increasing the accuracy of converting tonnage to length and minimizes production losses associated with irrational planning of raw material procurement.

According to the calculations of economic efficiency, confirmed at a real enterprise, the costs of introducing the organization and planning algorithm into the existing management system are minimal, and the project payback period is about one month.

Using this development leads to the adjustment of the business process itself, as tasks are completed faster. Employees can process large amounts of information during their working hours, which can be used either to reduce staff costs or to quickly develop a business with the same number of employees.

The practical significance of the scientific project lies in the fact that the developed algorithm for organizing and planning the production line is integrated under any production accounting system, since each algorithm module allows loading data in XLS-format.

The results of the project can be used at enterprises engaged in the manufacture of oilfield (tubing) pipes for the oil and gas sector.

References

[1] Tokmakov E A 2016 Contribution of the oil and gas complex to the development of the economy of the Russian Federation *Young Sci.* **17** 479–82

[2] Garcia M.V., Irisarri E., Perez F et al July 2017 An Open CPPS Automation Architecture based on IEC-61499 over OPC-UA for flexible manufacturing in Oil&Gas Industry *IFAC-PapersOnLine* **50(1)** 1231–8

[3] Fugarov D D, Gerasimenko Y Y, Nesterchuk V V et al December 2018 Methods for Revealing hidden Failures of Automation System for Technological Processes in Oil and Gas Sector *J. of Phys.: Conf. Ser.* **1118(1)** no 012055

[4] Movchan I B and Yakovleva A A 2019 Oil and gas industry: field diagnostics of pre-breakthrough condition of trunk pipelines *Int. J. of Civil Engineer. and Technol.* **10(2)** 1571–81

[5] Malyutin A G 2011 *Corporate Electronic Systems* (Omsk: OSTU) 38 p

[6] Artamonova T E, Ovsyannikova A V, Vorobyova A V and Popovich A E 2016 ERP systems. Efficiency and problems of implementing ERP-systems *Natural and Techn. Sci.* **4(94)** 173–4

[7] Makhovikov A B 2016 Principles of corporate communications management system development for mineral resources companies *J. of Mining Instit.* **217** 24–8

[8] Polyakov R K 2007 Development of an information-logical model of a risk management algorithm in an enterprise management system *Russ. Econ. Internet J.* **1** 90

[9] Krugovaya T S and Remizova I V December 2018 Automation of the activities of oil and gas service enterprises at the process control level *Prospects for the Development of Science in the Modern World* Part 1 XIII Int. Sci. and Pract. Conf. (Ufa: Ed. Dendra) pp 78–82

[10] Rozhkov A A and Karpenko M S 2015 Economic-mathematical models and algorithms for optimizing financial costs in the management of energy-saving innovations in mining enterprises *Mining Inform. and Analyt. Bull. (sci. and techn. J.)* **6** 307–13

IOP Conf. Series: Materials Science and Engineering **905** (2020) 012042 doi:10.1088/1757-899X/905/1/012042

Natural and anthropogenic landscapes of the gold-mining regions in Bashkortostan (Russia)

A N Kutliahmetov, A A Kulagin, O V Serova and I M Gatin

Department of ecology, geography and nature management, Akmulla Bashkir state pedagogical university, Ufa, Russia

E-mail: ecobspu@mail.ru, Kulagin-aa@mail.ru

Abstract. This article presents the results of laboratory and field studies of territories in the zone of influence of the gold extracting and mining enterprises of the Southern Urals. Researches were conducted on six sites of gold mining of the past years, in order to identify natural and technogenic mercury in various natural and man-made landscapes of mining Bashkortostan. A brief description of the mineralization of the study area and types of gold deposits is given. The mercury content in bedrock, in various types of ores, man-made soils (fine-grained tailings dump, tailings), soils, bed-silts, in the soil air, in water was determined. The largest technogenic anomalies are Hg dissemination flows in bed silts of watercourses and lakes. Technogenic Hg anomalies in soils are local and are formed in the areas of mining and old ores processing. The practical absence of anomalies in transit environments with intensive Hg accumulation in long-term deposition reflects the regressive residual nature of technogenic mercury anomalies.

1. Introduction

The long-term intensive exploitation of gold placer mine, gold ore deposits and complex gold-copper-zinc containing deposits in the eastern regions of Bashkortostan (Bashkir Trans-Urals) is a large-scale factor in the transformation of the nature and human environment.

Gold has been mined in the Bashkir Trans-Urals for over two hundred years [1]. Metallic mercury was used traditionally for the extraction of Au by amalgamation as in many other countries. Mercury accumulated in the tailings and dumps of runner gold-processing plants and was released from them into the environment [2, 3]. Mercury is a typical associate component of the ores of gold and sulphide deposits, which are widespread in the area [4].

2. Methods and materials

Cu-Zn-sulphide and Au are the main types of mineralization of the area. Sulphide deposits (Uchalinskoe, Novo-Uchalinskoe, Ozernoye, Zapadno-Ozernoye) are genetically related to the ore-containing Devonian volcanogenic formations. Gold deposits of the region are divided into gold-quartz, gold-quartz-sulphide, gold-sulphide and gold-polymetallic. Au in the first two types of deposits, mainly in the native form, and in the latter two – as an accompanying element in sulfides. The Uchalinsky ore region is one of the main gold-bearing regions of the Urals, gold mining in this region started at the beginning of the 19th century: numerous Au placers along the Miass, Ui, Ural and their tributaries were exploited. Then bed-rock Au deposits were exploited, the main are deposits of the Buydinsky, Mindyaksky and Ilyinsky ore fields [5]. In addition, Mn, Cr, ornamental and technical jasper, building materials (granite, limestone, clay, etc.) deposits are known in the area. The largest anthropogenic landscapes are distinguished in the city of Uchaly, Mindyak and Mansurovo villages and their surroundings. Man-made landscapes of abandoned gold placers are located in the form of narrow stripes along the valleys of the rivers Miass, Ui, Ural and their many tributaries.

Mercury is constantly present in increased concentrations in the ores of sulphide deposits and forms extended areolas around their deposits. Hg content in ores varies over a wide range: from 1 to 250 g/t and depends on the mineral composition of the ores. The lowest grades are found in sulphur-sulphide ores – 1–10 g/t, noticeably higher are in copper-sulphide ores – on average about 6–10 g/t and sharply increased in copper-zinc and zinc ores – on average 20–30 g/t. Such differentiation of Hg by type of ores is due to the different levels of the main Hg mineral concentrates in them: sphalerite and fahlores, the Hg concentration in which reaches 0.3–0.5 %, average 0.2 %. Hg own minerals are also known in sulphide ores: cinnabar and coloradoite, but their role in the overall balance of Hg is negligible. High accumulation of Hg (up to 80 g/t) in the oxidation zone of sulphide deposits is typical, where it is in metallic form (Rezenkova, Samoylova, 1975). Almost all of the Hg extracted from the subsoil with sulphide ores goes into concentrates and is removed with them to melting outside the region, and only a small amount of it (up to 5 % of the total mass) remains in the tailings and is stored in the Mining and Processing Plant's (MPP) tailing dump. Since the temperatures in the ore processing cycle at the processing plant do not exceed 100 °C, and the sublimation temperatures of Hg from its main concentrators – sphalerite and fahlores are 400–600 °C (from pyrite> 600 °C), Hg does not come into the atmosphere. This is confirmed by measurements of concentrations of Hg vapors at the site of the MPP and the territory of the city of Uchaly.

The purpose is given results of laboratory and field studies of territories in the zone of influence of the gold extracting and mining enterprises of the Southern Urals.

Samples were taken from: bedrocks, man-made soils (fine-grained part of beneficiation tailings, tailings), soils, bed silts, soil air (mercury survey), water (processed 2,742 samples). Analyses of rocks, subsoils, soils and bed silts samples for gross Hg content were performed on an AGP-01 gas-port analyser with a UV-1 thermosublimation unit. The resulting material was processed by statistical methods using Microsoft Excel.

3. Results

Thus, anthropogenic "sulphide" Hg cannot be considered as a source of mercury contamination of the environment. However, high Hg content in tailings indicates the potential environmental hazard of the tailings as a source of local secondary mercury contamination. The dumps of quarries and underground mines, which exploit sulphide deposits, in which ore-containing low-mineralized rocks with anomalous (areolas) Hg content accumulate, may be of greater importance in the input of Hg into the environment. Mercury can flow from them into landscapes as a result of air and water migration. However, so far no noticeable increase in Hg concentrations in landscape components in connection with the tailing dump, Uchaly MPP concentrator and dumps of sulphide deposits has not been established.

Mercury, in anomalous (one to two orders of magnitude lower than in sulphide deposits) contents, has been established at all types of gold ore deposits and in their areolas. The degree of Hg concentration there varies: relatively low contents (up to 0,n g/t) are characteristic of gold-quartz and gold-sulphide-quartz deposits, and higher (from n up to n100 g/t) – for gold-sulphide and, especially, gold-polymetallic [5, 8].

The fact that at the industrial site of the Bashkir Gold Mining Company, where mercury-containing Au ores are currently being processed by heap leaching, using sodium cyanide solution, the environmental mercury contamination is practically not manifested is confirmation of the decisive importance of amalgamation at the gold mining of the past years as the primary source of pollution of the environment [6]. It should be noted that the absolute majority of modern gold mining factories work with cyanation [7–10].

IOP Conf. Series: Materials Science and Engineering **905** (2020) 012042 doi:10.1088/1757-899X/905/1/012042

From the results of research by the following authors: N.Yu. Antoninova, N.V. Gonchar, A.V. Kivatskaya, V.V. Lodeyschikov, M.A. Meretukov, V.I. Sotnikov, G.V. Sedelnikova, M.I. Fazlullin, V.A. Chanturia and others – it follows that the most common technological processes for gold-bearing ores processing are based on the use of a highly active toxic substance – sodium cyanide. The use of sodium cyanide predetermines the danger of emergencies at gold ore processing enterprises, the number of which has recently been growing [11, 12].

The use of Au amalgamation in gold mining in the Uchaly district in the period of 1823–1995 led to a significant income of Hg in the environment. Considering that Hg consumption during amalgamation is about 1.5 kg per 1 kg of extracted Au, and the total volume of extracted Au exceeds 40 tons (including 29 tons before 1917 year), the environment received about 70 tons of anthropogenic Hg, which is firmly held by subsoils, soil cover and bed silts of water bodies and watercourses.

The regional Hg geochemical background (BG) in the rocks is 30 μg/kg, in soils – 40 μg/kg, and in bed silts of watercourses – 12 μg/kg. For atmospheric air the background concentrations are 2 pg /L. The content of Hg in the studied waters below the threshold of its detection is 0.01 μg/L with the used method of analysis [4].

The bottom sediments of the rivers Buydy, Zirikly and Yamyelga and the soils were studied on the dumps and in the vicinity in the village of Buyda (Figure).

The Hg content in river Yamjelga bottom sediment is near-background and typical, and only at 2 points they are anomalous (3.7 and 5 GB). This is consistent with the absence of ore occurrences and Au placers in the river basin. An almost continuous flow of Hg dissemination, approximately more than 12 km (its eastern continuation in the deposits of the Kiddysh river is not contoured), is established in the bed silts of the Buydy river. Its anthropogenic source is the heaps of the runner factory in the Buyda village. Hg concentrations are sharply increased to 0.69 mg/kg (16.5 GB) in the soils on them, and Hg concentrations in the atmospheric air at the site of the former runner factory are 61 pg/L (30 GB). The maximum values reach the concentration of Hg in the bed silts of the river directly near the dumps – 1.34 mg/kg (112 GB).

Downstream, the Hg content in bed silts generally decreases, but unevenly, depending on the slope of the river bed, increasing in areas of its flattening and, accordingly, the accumulation of silty sediments. Highly abnormal Hg content (up to 10 GB) is also noted in the bottom sediment of the Kidysh river, in its most studied eastern part.

A sharp decrease in the Hg content in the sediments upstream from the dumps is the proof of the spatial connection of the Buydin Hg dissemination flow with the erosion of the spoil dumps. The average Hg content here is 2.7 GB, and the maximum – 6 GB, although up-stream Buydy river the burial waters of the Uchalinsky mine come in along the drain channel. The latter indicates a low level of Hg content in these waters.

3 intervals with abnormal levels of Hg are established in the bed silts of the Zirikly river. Relatively long (2.5 km) and intense (up to 54 GB) flow was detected in the western part of the site outside the Buydinsky ore field. It is spatially conjugated with complex (with Hg) geochemical anomalies in rocks and soils. The second local (0.6 km) intensive (up to 12.4 GB) flow also records the continuation of the ore zone of the Uragan-Idrisovsk Au deposit. At 6 km downstream from this stream, third rather long (1.7 km) and less intensive (up to 9.4 GB) Hg flow is detected. There are no manifestations of ore mineralization and geochemical anomalies in bedrock in this site. Therefore, its formation should be attributed to the accumulation of thin Hg-rich sediments carried from the area of the Buydinsky ore field, due to the abrupt flattening of the Zirikly riverbed at its exit to the foothill plain. There are no signs of the connection of the Hg dissemination flows, detected at Zirikly river, with anthropogenic sources and its apparent conjugation with ore-bearing zones and ore-genic geochemical anomalies in rocks and soils indicates the ore-generating nature of these flows. They drastic differ from the Buydin anthropogenic flows in smaller length (up to 2.5 km), intermittent propagation and, in general, lower intensity.

4. Conclusion

As a result of the conducted research, it was established that anthropogenic Hg, associated with the gold deposits exploitation, is one of the sources of its income into the landscapes of the region and the main source of local mercury pollution of the surrounding environment. The most intensive anthropogenic source of Hg are the dumps of gold-processing plants that processed gold ores using amalgamation.

The largest anthropogenic anomalies are Hg dissemination flows in bed silts of watercourses and lakes. Hg anthropogenic anomalies in soils are local and are formed in the areas of mining and processing of gold ores using the amalgamation method.

Hg anomalies in waters are not established, and in the atmospheric air have low intensity and are fixed in the places of processing of gold ores. The practical absence of anomalies in transit environments with intensive Hg accumulation in long-term deposition environments reflects the regressive residual nature of anthropogenic mercury anomalies.

Due to the substantially absence of natural and local manifestations of environmentally hazardous anthropogenic mercury contamination (identified in Buyda, Ilyinka villages), the overall environmental situation in the region with regard to mercury contamination is rated as relatively satisfactory.

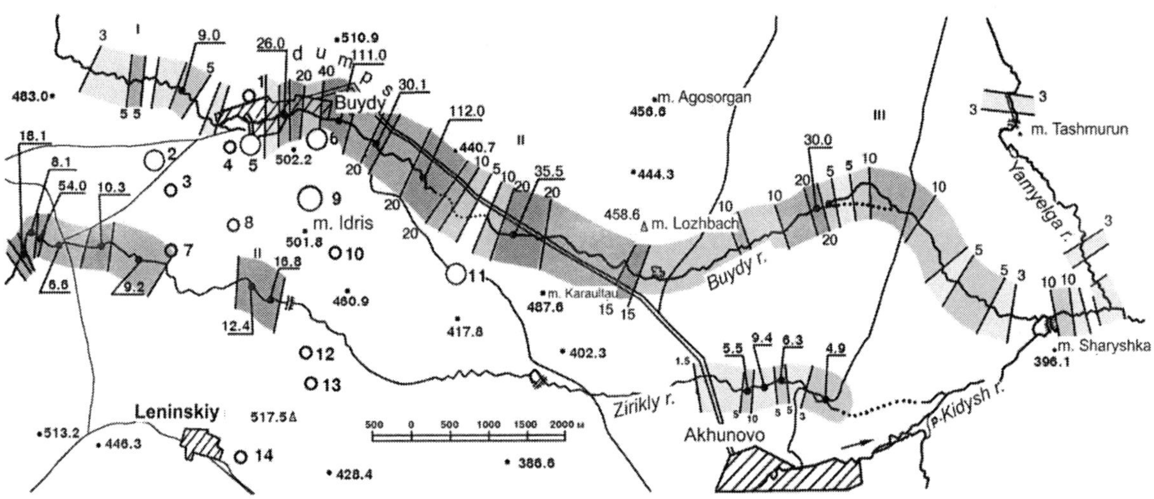

Figure 1. Buydinsky plot. Mercury dissemination flows in the bed silts of the Buydy, Zirikly, Yamyelga rivers. *Legend:* deposits and ore occurrences: 1 – Kalkan, 2 – Novo-Timofeevskoe, 3 – Vorontsovskoe, 4 – Oktyabrskoye, 5 – Staro-Timofeevskoe, 6 – Asfandyarovskoe, 7 – Ganeevskoe, 8 – Arslanovskoe, 9 – Uragan-Idris, 10 – Pik-Tau, 11 – Puchkovskoe, 12 – Vorontsovskoe, 13 – Staro-Zhurakovskoe, 14 – Leninskoye

References

[1] Makarova V N 2006 *The development of the gold mining industry of Bashkortostan during the period of two modernizations: (1880s of the 19th century – 1930s of the 20th century)* Cand. Dissertation (Ufa: BSU) pp 4–159

[2] Kutliahmetov A N, Nizamutdinova N R, Shaydulina G F and Safarova V I 2013 *Features of the impact of the gold industry on the environment* State Resource Management: Special Issue: Ecology of the Russian Federation, overview of problems, dynamics, current state and prospects (Moscow) pp 312–9

[3] Opiso E M and Aseneiro J P J 2018 Solid-state distribution of mercury in the tailings of gold artisanal mines in selected key areas of Mindanao, Philippines, and its consequences for mercury detoxification *Waste Manag. and Res.* **36(3)** 269–76

[4] Kutliahmetov A N 2002 *Mercury pollution of landscapes by mining enterprises of the Bashkir Trans-Urals* Cand. Dissertation (Ekaterinburg: BSU) pp 3–151
[5] Vishev I I 2002 *The formation and development of the gold industry in the southern Urals in the XIX century* Cand. Dissertation (Chelyabinsk: SUU) pp 5–148
[6] Gonchar N V *2003 Research and evaluation of the environmental safety of heap leaching of gold in the conditions of the Urals* Cand. Dissertation (Yekaterinburg: UGLTU) pp 3–155
[7] Kivatskaya A V 2006 *Ecological and geochemical research of gold heap leaching (on the example of OJSC "Mine Veseliy", Respublic of Altai)* Cand. Dissertation (Tomsk: TGU) pp 5–161
[8] Dreis K and Kieffer A 2016 Cyaniding of mercury-contaminated tailings: potential effects on health and environmental justice *Current environ. Reports* **3(4)** 443–9
[9] Sidelnikova G V 2001 Experience in the use of heap leaching of gold. Mineral resources of Russia *Econ. and Manag.* **1** 61–6
[10] Veiga M, Angeloci G and Nieken W 2015 Reducing mercury pollution by training Peruvian artisanal gold miners *J. of clean product.* **94** 268–77
[11] McNulty T 2001 Cyanide Substitutes *Minig magazine* **184(5)** 256–61
[12] Yang J, Jin-Xiang F, Dong-Xu Zh et al 2011 Security measures and cases of contamination of the reservoir with cyanide *Zhongguo jishui – China Water and Wastewater* **6(27)** 29–35

Environmental impact assessment of gold underground leaching technology

A N Kutliahmetov[1], A A Kulagin[1], F F Iskhakov[1], I R Rakhmatullina[1], N R Nizamutdinova[2], G F Shaidullina[2] and V I Safarova[2]

[1] Akmulla Bashkir state pedagogical university, Ufa, Russia
[2] GBU RB "UGAK", Ufa, Russia

E-mail: ecobspu@mail.ru, ugak@ufanet.ru

Abstract. The research assessed an impact of the underground gold mining (GM) on snow cover, surface and ground water of the industrial site and the adjacent area. It has been established that the water of observation wells located downstream the terrain contains specific pollutants typical for the GM technology. The snow cover contains dust, heavy metals, arsenic and some light organochlorine compounds, which are components of GM solutions. The article also gives recommendations on monitoring and assessment of the environmental impact of the PoE technology.

1. Introduction

The gold mining industry in the territory of the Republic of Bashkortostan (RB) has almost a 200-year history. During this time various technologies were used for processing of gold placers and gold-bearing ores [1, 2]. According to many professionals, the modern base of traditional technologies has reached its limits [3, 4], so the use of new geotechnological methods is timely and relevant. In particular, underground gold mining (GM) is one of the most economical innovative technologies in the gold mining industry [5–8].

At present, there are about a dozen industrial facilities at GM in the Middle Urals, which are at different stages of development. The first enterprise in Russia to apply the underground leaching method for gold mining in 1994 was ZAO Gagarka-Ai-PV [5, 19, 20].

Taking into account the economic attractiveness of the GM method, which makes it possible to conduct a complex geo-technological process at minimum cost, it is possible to assume its wide introduction into the practice of gold mining. In this regard, an objective assessment of the environmental safety of GM production is extremely important.

2. Problem Statement

The GM principle in its classical variant at favorable combination of natural factors (seismic, geo-structural, lithologic-geochemical, hydrogeochemical) is simple. Several dozens of wells are drilled in the ore body. Some (injection) wells are supplied with leaching solution, while others (pumping) are used to extract productive gold-containing solution for further processing. The GM method creates an opportunity for cost-effective mining of deposits or their individual areas, which differ from the complex mining and geological conditions of occurrence, low-power deposits, deposits with low gold content in the ore [9–12].

An important advantage of the GM technology in comparison with traditional methods of polymetallic deposits development is insignificant negative impact on the environment, as the landscape is practically not disturbed and the fact that the formation of solid waste of ore processing is prevented. It also eliminates the most expensive operations, such as ore extraction, transportation, and mechanical processing of ore prior to dressing.

Calcium or sodium hypochlorite produced by electrolysis or gaseous chlorine is used as a reagent for gold extraction [13, 14]. Due to the initial toxicity and aggressiveness of leaching reagents, special attention is paid to ensuring industrial safety, reducing the corrosion hazard for process equipment, as well as environmental protection [7, 8].

Pollution of natural environments at PoE is usually reduced to the impact on the aquifer compounds used in the technology and formed in the process of gold extraction (chlorine, active chlorine, chlorine ion, heavy metals) [8, 12, 15]. In works [5, 12, 16–18] it is noticed that pollution has local character, but represents potential danger, as in the course of underground leaching, hammocks and after hundred end at recultivation of underground aquifers.

The SP technology is characterized by a number of features: there is a large inertia of the process due to the fact that leaching takes place in the bowels; the main object of impact is the aquifer, which is difficult to manage [19]; the existing system of environmental monitoring in the zone of influence of SP technology requires improvement

3. Research Questions

On the basis of experimental data on the condition of air, water and snow samples taken in the zone of influence of gold mining companies that use the technology of underground leaching, to assess the environmental safety of the technology.

4. Purpose of the Study

The objective of the work is to assess the negative impact of the gold leaching technology on the snow cover, surface and ground waters of the industrial site and the adjacent area

5. Research Methods

Samples of natural and ground waters were taken in accordance with GOST R51592-2000, GOST 17.1.5.05-85, samples of atmospheric air and snow were taken and studied in accordance with recommendations of RD 52.04.186-89. Selection points were located at a distance of 20 m from the site boundary and at the boundary of the sanitary protection zone (300 m from the site of the underground leaching unit of BPV No. 5) along the main channels and in the direction of the prevailing wind.

Chemical-analytical studies of air, water and snow cover samples were performed in the accredited laboratory of SBURBUGAK (No. RU.0001,510312) with the use of certified methods based on classical and modern methods of analysis – titrimetry, photometry, atomic absorption spectrometry, chromatography and chromatography of mass spectrometry.

6. Findings

The impact of WP on natural waters is in the change of hydrodynamic regime, inflow to the aquifers of leaching solution containing a highly effective oxidizer – hypochlorite ion (active chlorine). As a result of chemical processes occurring during the interaction of the leaching solution with ore minerals, gold turns into a solution in the form of soluble salt [1]. Together with gold, heavy metals present in the ore are transformed into a soluble migration-active form: copper, zinc, cadmium, lead, mercury, etc. Besides, high oxidative capacity of active chlorine causes chlorination of natural organic compounds present in underground waters. These are water humus, lignin-like substances, cellulose and other natural and organic substances, with the interaction of active chlorine with which highly toxic organochlorine compounds are formed.

The technological process of obtaining gold by PV method includes the following stages: preparation of leaching solution (chlorine solution in water); injection of leaching solution into ore-bearing zone; control of solution movement in the bowels; lifting of productive solution to the surface by ellipsis; capture of residual chlorine of ellipsoidal air; clarification of the productive solution from mechanical impurities; extraction of gold, silver, other useful components, as well as contaminants from the productive solutions; capture of mechanical losses of sorption containing gold [3, 8, 9, 14, 19].

The scheme of the in-situ leaching technological process is shown in Figure 1.

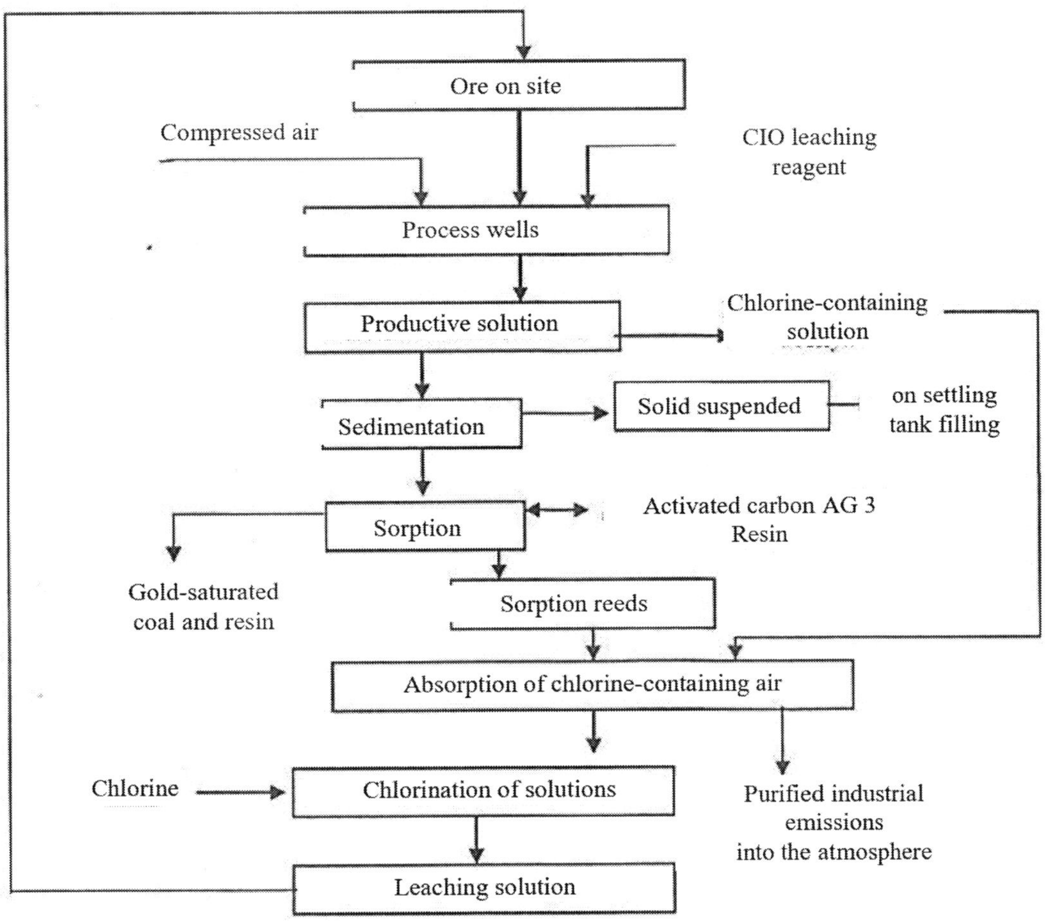

Figure 1. Scheme of the underground leaching process

Results of the survey of process and natural waters.

The results of the analysis of water samples taken in the operational area of the BPV5 (Figure 1) and in the area of its influence are shown in Table 1.

The technological productive solution of the SP enterprise is characterized by low pH (3.2), high bihromate oxidation (COD 111 mg O2/l) and mineralization (7840 mg/l), high concentrations of hypochlorite ion (118 mg/l), chlorides (4110 mg/l), macro components (calcium, magnesium, sodium, potassium, aluminum), as well as heavy metals, including highly toxic elements (mercury, cadmium, lead, etc.)).

The presence in technological solutions of a large number of volatile and low-volatile chlorinated hydrocarbons [21] creates a potential risk of their spread in the underground aquifer, outside the technological area of the SP in the event of cessation of pumping solutions.

Results of the snow cover survey. Snow cover accumulates atmospheric deposition of pollutants during the snowstorm period (about 6 months), which makes it possible to use snow as an indicator of airborne environmental pollution.

Sources of atmospheric air pollution in the area of PoE plant location are industrial emissions formed at the stage of leaching solution chlorination [21].

Investigation of snow cover samples included determination of the main indicators of water quality (pH, suspended solids and dry residue content), inorganic anions (chlorides, sulphates, hypochlorites) in melt water filtrates. The content of heavy metals and arsenic was determined separately in filtered melt water and solid snowfall. Organic compounds were determined from unfiltered snow samples.

Table 1. Content of substances in the underground waters of the industrial site of the PoE enterprise and the adjacent territory, mg/dm^3

Definable indicator	Productive mortar	Technological well		Observed wells						Water well
		higher ground	lower ground	well 1	well 2	well 4	well 5	well 6	well 7	
pH	3.2	3.2	3.7	7.8	8.9	8.0	8.1	7.8	7.9	8.2
x n k	111.0	35.0	27.0	12.5	51.0	21.5	6.0	6.0	12.0	6.0
Dry residue	7840	8410	5800	545	267	388	530	408	308	475
Hypchlorithione	118.0	208	132	<0.05	<0.05	<0.05	<0.05	<0.05	<0.05	<0.05
Chloride ion	4110	4060	3515	90.4	81.2	46.0	37.0	53.0	7.4	8.9
Sulfate ion	86.0	38.0	55.0	73.0	33.0	87	104	63	33	50
Fe	1.03	0.38	0.21	0.49	0.16	0.15	0.59	0.21	2.79	0.06
Mn	0.13	0.02	0.03	0.77	0.15	0.05	0.03	0.04	0.85	0.01
Cu	65.7	54.2	31.8	0.025	0.06	0.01	0.014	0.007	0.023	0.005
Zn	90.5	72.1	44.9	0.15	0.03	0.05	0.03	0.03	0.14	0.03
Hg	0.03530	0.0536	0.0967	0.00009	<0.00001	0.00077	0.00004	0.00001	<0.0001	<0.0001
Cd	1.416	1.053	0.612	0.0004	<0.0001	<0.0001	<0.0001	<0.0001	<0.0001	<0.0001
Pb	0.395	0.223	0.135	<0.0006	<0.0006	<0.0006	<0.0006	<0.0006	<0.0006	<0.0006
Summ of i	18.8539	17.1607	13.6547	0.0261	0.0322	0.019	0.0039	0.0016	0.00046	<0.0004

The results of the analysis of snow cover samples are presented in Tables 2 and 3.

Table 2. Content of inorganic pollutants (IP) in snow samples taken at the IP plant site and in the area of its influence, mg/dm^3

Definable ingredients	Main directions of distribution of pollutants from the industrial site										
	Industrial platform	west		north		east		south-east		The outskirts of the Semenovskoye village	MPC fish farm.
		20 m	300 m	20 m	300 m	20 m	300 m	20 m	300 m		
pH	4.2	5.0	5.0	5.2	5.6	4.5	4.2	5.5	5.5	6.8	-
Weighted substances	84	52	52	34	39	27	27	52	28	67	-
Hypochlorithione	<0.05	<0.05	<0.05	<0.05	<0.05	<0.05	<0.05	<0.05	<0.05	<0.05	-
Chloride ion	215	3.1	3.1	2.5	2.8	17.0	3.8	5.3	3.8	20.6	300
Sulfate ion	14.1	4.9	4.9	4.7	4.4	9.6	0.7	4.9	5.7	10.7	100
Dry residue	280	37	37	21.6	24	54	18.5	22.3	45.0	48.6	-
Fe	0.18* / 3.33	0.16 / 1.08	0.09 / 1.03	0.06 / 0.68	0.08 / 0.83	0.05 / 1.06	0.05 / 0.062	0.05 / 0.44	0.05 / 0.79	0.16 / 0.71	od
Mn	0.03 / 0.06	0.02 / 0.01	0.02 / 0.01	0.03 / 0.02	0.03 / 0.02	0.02 / 0.02	0.01 / 0.02	0.01 / 0.01	0.01 / 0.01	0.02 / 0.02	-
Cu	0.052 / 0.075	0.011 / 0.013	0.003 / 0.009	0.003 / 0.009	0.005 / 0.009	0.003 / 0.012	0.006 / 0.006	0.002 / 0.003	0.003 / 0.004	0.002 / 0.006	-
Zn	0.07 / 0.07	0.09 / 0.03	0.03 / 0.01	0.03 / 0.02	0.04 / 0.02	0.03 / 0.02	0.03 / 0.02	0.02 / 0.01	0.02 / 0.01	0.02 / 0.01	-
As	<0.0005 / 0.0007	0.0009 / 0.0003	0.0012 / 0.0004	0.0018 / 0.0002	0.0007 / 0.0003	0.0016 / 0.0004	0.0010 / 0.0004	0.0008 / 0.0002	0.0005 / 0.0004	0.0009 / 0.0004	-

Note: * numerator – content of dissolved forms of elements; denominator – content of forms sorbed on suspensions

Table 3. Volatile Organochlorine Compounds Content in Snow Samples Sampled at the Snow Camp Site and in the Area of Influence, mg/dm^3

Definable ingredients	Industrial platfrom	west		north		east		South-east		The outskirts of the Semenovskoye village
		20 м	300 м	20 м	300 м	20 м	300 м	20 м	300 м	
Tetrachloric carbon	0.0004	<0.0004	0.0004	0.0004	<0.0004	0.0004	<0.0004	0.0004	<0.0004	<0.0004
Chloroform	0.1000	0.050	0.0030	0.0100	0.0040	0.0120	0.0080	0.0020	0.0010	0.0010
Benzole	0.0009	0.0007	0.0010	0.0005	0.0010	0.0008	0.0010	0.0009	0.0006	0.0009
Bromdichloromethane	0.0020	<0.0004	<0.0004	<0.0004	<0.0004	<0.0004	<0.0004	<0.0004	<0.0004	<0.0004
Dibromchloromethane	0.0005	<0.0004	<0.0004	<0.0004	<0.0004	<0.0004	<0.0004	<0.0004	<0.0004	<0.0004
Toluol	0.0009	0.0005	0.0008	0.0010	0.0008	0.0010	0.0010	0.0004	0.0005	0.0007
M (p) Xylons	<0.0004	<0.0004	<0.0004	<0.0004	<0.0004	0.0005	<0.0004	<0.0004	<0.0004	<0.0004
O, Xylon.	<0.0004	<0.0004	<0.0004	<0.0004	0.0004	0.0005	0.0004	<0.0004	<0.0004	<0.0004
1,2,4-Trimethylbenzene	0.0005	<0.0004	<0.0004	0.0006	0.0004	0.0005	<0.0004	<0.0004	<0.0004	<0.0004
4-Isoyiroiltooluene	0.0016	0.0006	<0.0004	0.0010	<0.0004	<0.0004	<0.0004	<0.0004	<0.0004	0.0010
1,2- Dibrom -3-chloropropane	0.0004	<0.0004	0.0004	<0.0004	<0.0004	<0.0004	<0.0004	<0.0004	<0.0004	<0.0004

When assessing the snow cover dust content (Table 2), it was found that the content of suspended solids in the snow at a distance of 20 m from the investigated underground leaching block (BPV № 5) in the north, southeast and west directions is quite homogeneous – 34–52 mg/l; in the east direction the dust content is less – 27 mg/l. At the boundary of the sanitary protection zone – in 300 m from the industrial site of BPV № 5 the content of suspended solids in the snow, selected in the north-east, south-east and west directions, was 28–52 mg/l.

The pH of melt water varied in the range of 5.0–5.6 and was in most cases significantly lower than the pH of snow in the settlement. The lowest pH value of 4.2 was found at the industrial site, while pH 4.2–4.5 was found at a point located 20 m east of the BPV No. 5 boundary, in the site with a chlorination unit for productive solutions.

In general, the contamination of snow with inorganic anions and heavy metals is low (Table 2). The content of dry residue and individual ingredients found in snow samples within the sanitary protection zone varies in small ranges without a certain regularity.

Active chlorine in the snow cover is not detected.

To assess the degree of snow cover pollution by water-soluble forms of inorganic compounds, it is possible to use MAC values for fishery water bodies. Since snowmelt water during floods mainly enters surface water bodies, the results can be compared with MACs for fishery bodies. The content of chlorides and sulphates does not exceed MAC. The exception is the content of soluble forms of iron and manganese, which are components of dust particles, which in some points reaches 16 and 3 MACs, copper and zinc – 11 and 9 MACs, respectively.

The level of concentrations of volatile organic compounds present in snow is shown in Table 3. The constant presence of benzene and toluene in all sampled snow samples in comparable concentrations has been established, which allows them to be attributed to the constant natural components of the atmosphere. In contrast, the content of chloroform and carbon tetrachloride decreases with distance from the source of pollution – BPV-5 industrial site. Consequently, chloroform and carbon tetrachloride are marker pollutants of PV production. Apart from the above mentioned organic compounds (tetrachloride carbon, chloroform, benzene, toluene), bromodichloromethane, dibromechloromethane, 1,2,4-trimethylbenzene and 4-isopropyltholuene were found in the snow cover of the industrial site during the winter period. The most polluted is the northern direction, where tetrachloric carbon, chloroform, benzene, toluene, 1,2,4-trimethylbenzene, 4-isopropyltholuene and the eastern direction (tetrachloric carbon, chloroform, benzene, toluene, xylenes, 1,2,4-trimethylbenzene) were detected in snow with decreasing concentrations as far away from the industrial site.

The formation of aromatic hydrocarbons (benzene, toluene and their substituted ones) is probably related to the transformation of lignin in groundwater [22, 23]. The determinants of the appearance of chlorine and brominated hydrocarbons are both the presence of active chlorine and its substitution by bromine, which is present as an admixture in chlorine [24]. In the snow cover with. Semyonovskoe found chloroform, benzene, toluene, 4-isopropyltholuene. The presence of chloroform is a proof of the dispersion of organochlorine compounds from the SP plant at a distance of up to 1 km. Therefore, the SP enterprise affects the atmospheric air of the settlement.

7. Conclusion

The survey of underground water and snow cover in the area of the SP enterprise impact showed that the enterprise has an impact on ground water and atmospheric air during normal operation. The water of observation wells is polluted with specific pollutants, typical for the SP technology.

Snow cover at the industrial site of the enterprise, in sanitary protection and residential areas contains dust, heavy metals, arsenic and some organochlorine compounds, which are components of technological solutions of WP

References

[1] Kutliakhmetov A N, Safarova V I, Shaidulina G F and Nizamutdinova N R 2012 Comparative analysis of environmental safety of various methods of processing gold ores (for example, the Republic of Bashkortostan) *Ecol. of urbanized areas* **2** 49–57

[2] Salikhov D N, Kovalev S G et al 2003 *Minerals of the Republic of Bashkortostan (gold)* part 1 (Ufa: Ecology) 222 p

[3] Arens V Zh 2001 *Physicochemical geotechnology* Textbook (Moscow: Moscow State Univer.) 656 p

[4] Melentyev G B, Shelkov E M and Delitsyn L M 2008Prospects of industrial and economic revival and social and ecological rehabilitation of the Moscow lignite basin with the use of the geo-ecological and energy-technological innovations In the coll. *Technogenic resources and innovations in technoecology* (Moscow: OIVTRAKH) pp 27–40

[5] Asalkhanov V A 2005 *Development of the advanced oxychloride technology of the gold extraction from the ores with reference to the conditions of the underground leaching* Cand. Dissertation thesis (Irkutsk) 17 p

[6] Beletsky V I, Bogatkov L K, Volkov N I et al 1997 *Reference book on the uranium geotechnology* (Moscow: Energatomizdat) 672 p

[7] Melentyev G B 2009 Innovative technology and new tasks of the technological mineralogy *Ecol. of the Russ. Industry* **3** 14–28

[8] *Underground and heap leaching of uranium, gold and other metals* Vol 2 *Gold* 2005 (Moscow: Ore and metals) 328 p

[9] Dokukin Yu V and Samoilov A G 2009 About the efficiency of the gold mining by the underground leaching method *Mineral resour. of Russ.* **6** 49–53

[10] Kazakov P V and Salikhov D N 2006 *Mineral resources of the Republic of Bashkortostan (alluvial gold)* Part 2 (Ufa: Gilem) 288 p

[11] Reznik Yu N., Timoshenkov S N and Voronov E T 2008 Geotechnological method of the deep gold-bearing placers development on the basis of the chloride hydrometallurgy use *Vest. ChitGu* **4(49)** 122–8

[12] Sedov N P 2008 *Optimization of technological parameters of in-situ leaching of precious metals: the example of Dolgiy Mys* Cand. Dissertation thesis (Yekaterinburg) 22 p

[13] Zyryanov M N and Leonov S B 1997 *Chloride metallurgy of gold (in Russian)* (Moscow: JV Intermet Engineer.) 288 p
[14] Dokukin Yu V, Savenya N V and Myakotin V V *Method of extraction of precious metals from ores* Invention patent no 2137855 July 14, 1998 pp 9–14
[15] Fazlullin M I, Avdonin G I and Smirnova R N 2008 To the problem of in-situ leaching of gold *Mining Inform. and Analyt. Bull. (MIAB), seminar* **217** 207–17
[16] Glinskiy M L., Glagolev A V., Dorozhko E G., Vetrov V A. et al 2010 *Methodical recommendations on the object monitoring of the subsoil condition at the enterprises of the State Corporation "Rosatom"* (Moscow: Center for Assist. to Soc. and Environ. Initiatives of the Nuclear Industry) 192 p
[17] Rossman G I, Petrova N E and Samsonov B G 2000 Ecological evaluation of ore deposits (methodological recommendations) *Mineral. Raw Mater.* **9** 150
[18] Timoshenkov S N 2009 *Justification and development of effective technology of in-situ leaching of gold from deep alluvial deposits (on the example of South Transbaikalia)* Cand. Dissertation thesis (Chita) 23 p
[19] Zabolotskiy A I and Dokukkin Yu V 2009 First experience in the Russian Federation of the commercial gold mining by the underground leaching method from the gold-bearing weathering cords of the Gagarskiy deposit (in Russian) *Mining inform.-analyt. Bull. (GIAB)* **1** 391–402
[20] Fazlullin M I, Shatalov V V, Gurov V A et al 2002 Prospects of the underground in-situ leaching of gold in Russia *Non-ferrous metals* **10** 39–46
[21] Kutliakhmetov A N, Safarova V I, Shaidulina G F et al 2012 Formation and migration of halogen hydrocarbons in natural environments at the underground chloride leaching of noble metals *Probl. of reg. ecol.* **3** 46–53
[22] Orlov D S, Sadovnikova L K and Sukhanova N I 2005 *Soil chemistry* (Moscow: Higher school) 558 p
[23] Perminova I V 2000 *Analysis, classification and prognosis of the humic acids properties* Doct. Dissertation thesis (Moscow) 42 p
[24] Vasilieva A I 2006 *Influence of chlorination on water quality in the presence of some natural and technogenic impurities* Cand. Dissertation thesis (Ufa) 22 p

IOP Conf. Series: Materials Science and Engineering **905** (2020) 012044 doi:10.1088/1757-899X/905/1/012044

Thermal and concentration dependence of surface tension and melt density of indium-tin-lead-bismuth system

R A Kutuev

Department of Theoretical Physics, Chechen State University, Grozny, Russia

E-mail: kra-07@mail.ru

Abstract. The study of behavioral patterns and changes in the surface properties of melts as the number of components increases is of great interest. The aim of the work is to reveal theoretically predicted regularities that establish the relationship between the type of isotherms of surface tension and the density of the four-component system of indium tin-lead-bismuth from the type of corresponding side double systems. In particular, indium tin systems.

1. Introduction

This paper presents the results of the experimental study of surface tension and density of the four-component indium-tin-lead-bismuth system [1, 2]. It should be noted that the surface properties of pure fusible metals and their double alloys have been studied quite fully [3–6]. There are also experimental data on many three-component systems consisting of these metals [7–9]. Of great interest was the comparative analysis of such surface properties of fusible alloys as temperature and concentration dependence of surface tension and density. An interesting result of such work was the conclusion of the authors [10] as a result of the study of the surface properties of systems made of fusible metals, namely, the features of the concentration dependence of surface tension revealed in some double systems, in particular, in the systems of indium tin, thallium lead are manifested in three-component systems where as components are present systems In-Sn, Tl-Pb, but these features disappear with the addition of a third component to these systems. This effect is observed in the triple systems of indium tin-lead, thallium-lead-bismuth, where the minimums found on the isotherms of surface tension in the lateral double systems of thallium-lead and indium tin disappear with the addition of the third component of lead and bismuth, respectively [11, 12]. In addition, it was assumed that the surface properties of the triple systems are predicted on the basis of similar properties of the lateral double melts [1, 10, 12]. In this connection, there was an interest in studying the surface properties of melts consisting of more components. The results of the study could either confirm the revealed tendency of interrelation of surface properties of double and multi-component systems, or refute the revealed regularities.

2. Methods and materials

To investigate the melts of the indium-lead-lead-bismuth system, a combined instrument developed in our laboratory was used, in which the surface tension was measured by the method of maximum droplet pressure and an advanced hydrometer was used to measure the density of the system.

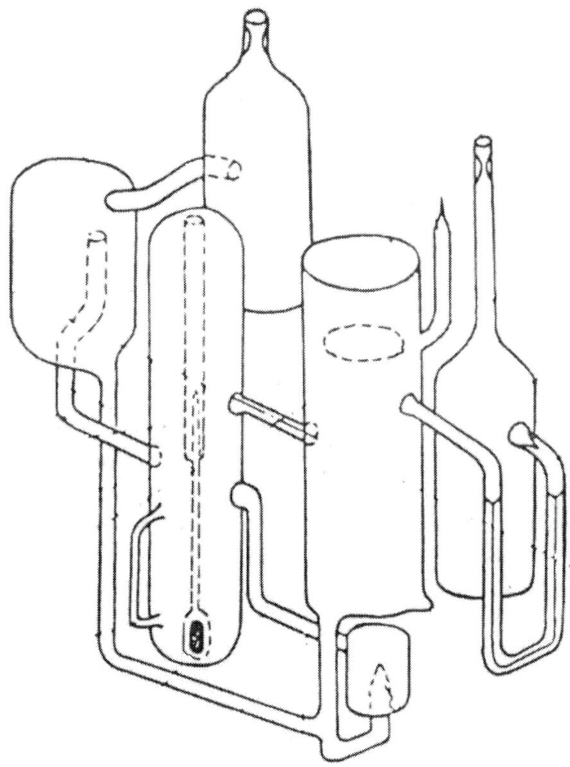

Figure 1. Combined instrument for measuring surface tension and density of multi-component melts

The device makes it possible to prepare alloys of different concentrations, to study the surface tension and density yof these alloys without opening it and, consequently, without disturbing its thermal and vacuum conditions, which has a significant impact on the reliability of the experiment and productivity, as the experimenter is free from time-consuming procedures of washing, drying and long-term thermal and vacuum treatment of the device before measuring the properties of each alloy separately.

At the same time, the consumption of expensive metals is reduced by 10 times and the time of experiments is reduced to the same extent.

The following expression is used to determine the density of the hydrometer:

$$\rho = \frac{P_0}{q(V_0 + \pi r^2 h) - \pi k R \Delta h}, \tag{1}$$

P_0 – hydrometer weight in vacuum, V_0 – total hydrometer volume, R – capillary radius, r – rod radius, h – Distance from the liquid level to the upper end of the hydrometer rod, Δh – difference in fluid levels in the hydrometric cylinder and capillary outlet.

The expression (1) is true on the assumption that the capillary outlet and the hydrometer rod are strictly cylindrical, but in practice the rods and capillaries do not always have a cylindrical form and therefore the average values of radii are used in the calculations.

The error in the calculation of the surface tension also depends on the capillary radius, which is derived from:

$$\frac{S_\sigma}{\sigma} = \sqrt{(\delta\sigma)^2 + \left(\frac{S_\rho}{\rho}\right)^2 + \left(\frac{S_r}{r}\right)^2 + \left(\frac{S_h}{h}\right)^2}. \tag{2}$$

IOP Conf. Series: Materials Science and Engineering **905** (2020) 012044 doi:10.1088/1757-899X/905/1/012044

The maximum error of PE measurement by the method of maximum drop pressure is 0.8 %, density – 0.2 %, and composition – 0.22 %.

High purity metals were used for the experiment. In-00; Sn-000; Pb-0000; Bi-0000. Temperature interval of researches from temperature liquidus to 773 K.

3. Results

The side double systems that make up this four-component system are characterized by smooth surface tension isotherms, except for the indium tin system. The dependence of the surface tension on the composition in the two-component 6 indium tin system is characterized by an isotherm with a minimum peculiarity in the region of medium compositions. On the isotherms describing concentration dependences of PH s of triple melts In-Sn-Pb [12] and In-Sn-Bi [13] the peculiarity in the form of minimum is equalized by adding the third component (bismuth or lead). Thus, the concentration dependence of PN in melts of triple systems Sn-Pb-Bi, In-Pl-Bi is transmitted by a smooth curve without extremums. Calculations have shown that the dependence of PN on the composition in the side double and triple systems, which make up the quadruple system, is described with good accuracy by the polynomial of the third degree. For this reason, we use this model to study the quadruple indium-old-lead-bismuth system.

Polynomial coefficients are calculated by means of a plan that ensures uniform concentration of experimental points on a three-dimensional symphlex in the form of a correct tetrahedron. The points of the plan, as nodes of the simplex lattice and used to determine the polynomial coefficients, are shown in Figure 2.

In order to analyze and verify the compliance of the regression equation, it is necessary to obtain the surface tension values of the four-component melts. Temperature dependences of density and PN of nine four-component melts, the compositions of which are located at the radial section originating at the top of the tetrahedron corresponding to bismuth, have been studied for this purpose. The obtained results are presented in Table 1.

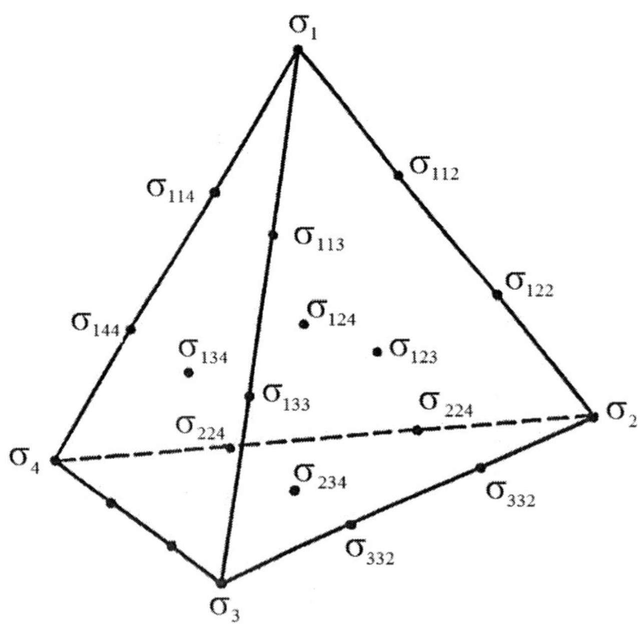

Figure 2. Scheme of the experiment for the third order model

The values of the regression equations at temperatures of 623 K and 773 K, as well as the experimental values of PN for pure metals and alloys used to determine the regression coefficients are given in Table 2.

Table 1. ρ (kg/m^3) and σ (мДж/m^2) of In-Sn-Pb-Bi melts

Composition in mol. shares				473 K		523 K	
X_{Bi}	X_{In}	X_{Sn}	X_{Pb}	ρ	σ	ρ	σ
0.00	0.33	0.33	0.34	8.877	426	8.850	475
0.06	0.31	0.31	0.32	8.905	459	8.888	458
0.14	0.28	0.28	0.30	8.937	451	8.907	450
0.22	0.25	0.25	0.28	8.991	440	8.952	439
0.29	0.22	0.22	0.27	9.017	432	8.990	430
0.34	0.21	0.21	0.24	9.062	425	9.020	423
0.40	0.19	0.19	0.22	9.075	420	9.055	419
0.43	0.18	0.18	0.21	9.126	417	9.126	417
0.50	0.16	0.16	0.18	9.288	416	9.224	415
0.54	0.15	0.15	0.16	–	–	9.292	409
0.57	0.14	0.14	0.15	–	–	9.433	407
0.63	0.12	0.12	0.13	–	–	9.494	400
0.72	0.09	0.09	0.100	–	–	–	–
0.82	0.056	0.056	0.064	–	–	–	–
0.86	0.044	0.044	0.050	–	–	–	–
0.947	0.017	0.017	0.198	–	–	–	–
1.00	0	0	0	–	–	–	–
0.00	0.33	0.33	0.34	8.816	474	8.785	472
0.06	0.31	0.31	0.32	8.857	457	8.829	455
0.14	0.28	0.28	0.30	8.884	448	8.860	447
0.22	0.25	0.25	0.28	8.933	436	8.903	435
0.29	0.22	0.22	0.27	8.951	428	8.922	427
0.34	0.21	0.21	0.24	8.996	422	8.965	420
0.40	0.19	0.19	0.22	9.020	418	8.990	415
0.43	0.18	0.18	0.21	9.083	415	–	412
0.50	0.16	0.16	0.18	9.146	411	9.066	408
0.54	0.15	0.15	0.16	9.286	409	9.177	404
0.57	0.14	0.14	0.15	9.343	402	9.287	399
0.63	0.12	0.12	0.13	9.46	398	9.400	395
0.72	0.09	0.09	0.100	9.560	393	9.500	388
0.82	0.056	0.056	0.064	9.705	386	9.663	384
0.86	0.044	0.044	0.050	9.776	383	9.737	381
0.947	0.017	0.017	0.198	9.888	379	9.811	376
1.00	0	0	0	9.965	376	9.952	375
0.00	0.33	0.33	0.34	8.722	469	8.693	468
0.06	0.31	0.31	0.32	8.760	452	8.715	449
0.14	0.28	0.28	0.30	8.798	223	8.755	441
0.22	0.25	0.25	0.28	8.848	431	8.815	429
0.29	0.22	0.22	0.27	8.860	423	8.843	421
0.34	0.21	0.21	0.24	8.905	416	8.875	414
0.40	0.19	0.19	0.22	8.920	412	8.910	410
0.43	0.18	0.18	0.21	8.922	408	8.926	407
0.50	0.16	0.16	0.18	9.003	404	8.980	402
0.54	0.15	0.15	0.16	9.075	398	9.044	396
0.57	0.14	0.14	0.15	9.205	394	9.145	391
0.63	0.12	0.12	0.13	9.283	389	9.246	387
0.72	0.09	0.09	0.10	9.408	383	9.374	381
0.82	0.056	0.056	0.064	9.557	378	9.513	376
0.86	0.044	0.044	0.050	9.620	376	9.580	373
0.947	0.017	0.017	0.198	9.730	371	9.688	368
1.00	0	0	0	9.801	367	9.758	365

Table 2. Dependence of third order polynomials on the concentration of the four-component system In-Sn-Pb-Bi

Molar fraction composition					σ_{ijk}			β_{ijk}	
X_{In}	X_{Sn}	X_{Pb}	X_{Bi}		623 K	773 K		623 K	773 K
1	0	0	0	σ_1	545	531	β_1	545	531
0	1	0	0	σ_2	535	524	β_2	535	524
0	0	1	0	σ_3	438	426	β_3	438	426
0	0	0	1	σ_4	374	365	β_4	374	365
0.66	0.33	0	0	σ_{112}	532	525	β_{12}	–41	–23
0.33	0.66	0	0	σ_{122}	530	520	γ_{12}	–9	18
0.66	0	0.33	0	σ_{113}	476	470	β_{13}	–115	–88
0.33	0	0.66	0	σ_{133}	456	448	γ_{13}	–106	–88
0.66	0	0	0.33	σ_{114}	428	418	β_{14}	–227	–221
0.33	0	0	0.66	σ_{144}	390	380	γ_{14}	–128	–117
0	0.66	0.33	0	σ_{223}	475	470	β_{23}	–104	–83
0	0.33	0.66	0	σ_{233}	452	443	γ_{23}	–63	–38
0	0.66	0	0.33	σ_{234}	418	413	β_{24}	–227	–216
0	0.33	0	0.66	σ_{244}	390	380	γ_{24}	–173	–135
0	0	0.66	0.33	σ_{334}	410	395	β_{34}	–32	–45
0	0	0.33	0.66	σ_{344}	388	376	γ_{34}	5	–9
0.33	0.33	0.33	0	σ_{123}	478	470	β_{123}	20	–59
0.33	0	0.33	0.33	σ_{134}	420	408	β_{134}	247	–828
0.33	0.33	0	0.33	σ_{124}	430	420	β_{124}	9	943
0.00	0.33	0.33	0.33	σ_{234}	416	404	β_{234}	195	106

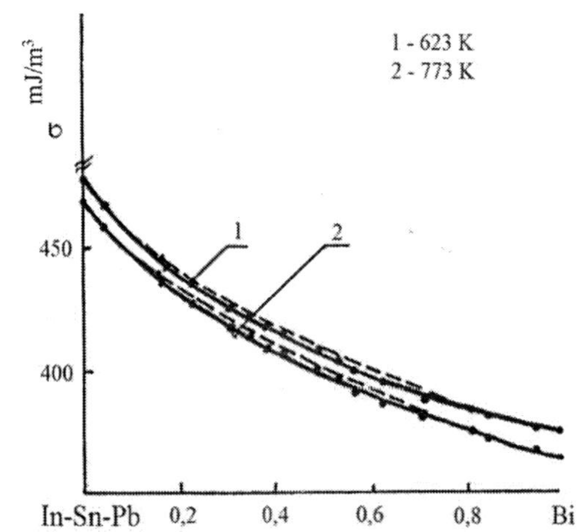

Figure 3. PH isothermal and molar volumes in the indium-tin-lead-bismuth system

As it was expected, the isotherms of surface tension in the four-component system of indium-tin-lead-bismuth are characterized by the presence of a peculiarity in the form of a minimum connected with a similar feature (minimum) on the graphs of concentration dependence s of the side two-component system of indium-tin. The existence of this feature predetermines the necessity to check the obtained equation for adequacy. In this connection, together with the data on side melts, the experimental PN isotherms for the beam section with a constant ratio of molar fractions of lead, indium and tin were used (Pb : In : Sn = 1:1:1). Comparison of experimental data on the concentration dependence of PN and calculated on the polynomials (Figure 3) showed that the experimental data and calculated values within the measurement error coincide. Also the model of the third order qualitatively describes features of influence of the fourth component on an extreme on an isothermal surface in the form of a minimum in triple systems In-Sn-Pb и In-Sn-Bi.

IOP Conf. Series: Materials Science and Engineering **905** (2020) 012044 doi:10.1088/1757-899X/905/1/012044

Thus, we managed to show that the model of the third order with sufficient accuracy reflects the real dependence of the surface tension on the composition in the studied four-component system.

Not less important is the fact that it was possible to show that the experimental values of surface tension coincide well with the PN values calculated by different theoretical equations in the assumption of the ideal solutions. The more components in the system, the better this coincidence is manifested.

Dependence of surface tension and density on temperature in the studied quadruple melt is linear with negative temperature coefficient (Figure 4, 5).

A certain interest is the concentration dependence of the temperature coefficient of surface tension on the beam section. This dependence is presented in Figure 5, from where it is seen that the isotherm $\frac{\partial \sigma}{\partial T}$ within the experimental error isotherm is described by the linear line.

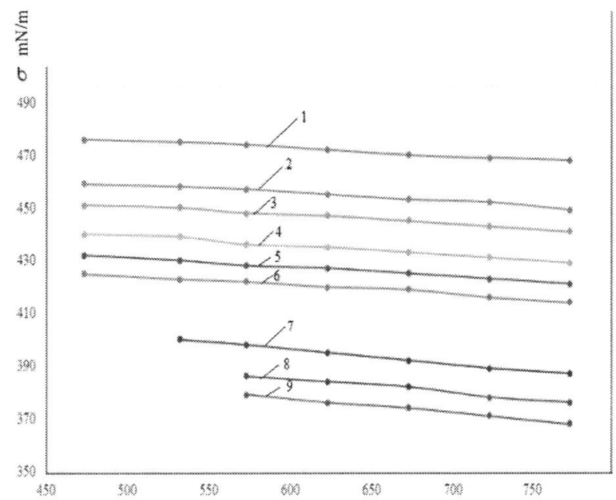

Figure 4. Temperature dependence of In-Sn-Pb-Bi melts PH at bismuth concentrations (1-0.0; 2-0.06; 3-0.14; 4-0.22; 5-0.29; 6-0.34; 7-0.54; 8-0.82; 9-1.0) mol. fractions.

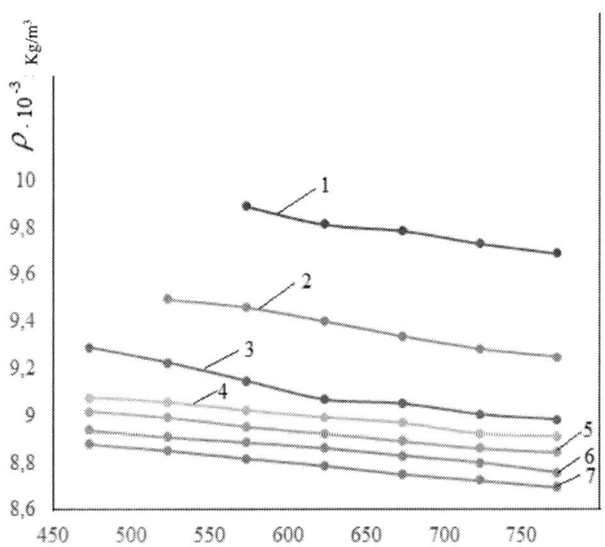

Figure 5. Temperature dependence of In-Sn-Pb-Bi melts density at bismuth concentrations (1-1.0; 2-0.63; 3-0.57; 4-0.4; 5-0.29; 6-0.06; 7-0.0) mol. fractions.

It should be noted that in contrast to the surface tension, the error of determining the temperature coefficient σ is high (up to 20 %). In our opinion, this is due to the fact that in our experiments we failed to identify the regularities of the coefficient change $\frac{\partial\sigma}{\partial T}$ with the concentration of bismuth. A common feature of the surface tension isotherms of the In-Sn-Pi and In-Sn-Bi triple side systems is the presence of a minimum in the area adjacent to the In-Sn concentration triangle. The study of the influence of the fourth component (bismuth or lead) on the behavior of this feature on the isothermal surface tension is particularly interesting.

For illustration of the "picture" of the dependence of physical and chemical properties on the composition of the multicomponent melt, the isolines of properties on different planes are used In order to illustrate the "picture" of the dependence of the physico-chemical property on the composition of the multicomponent melt, we use isolines of properties on different planes.

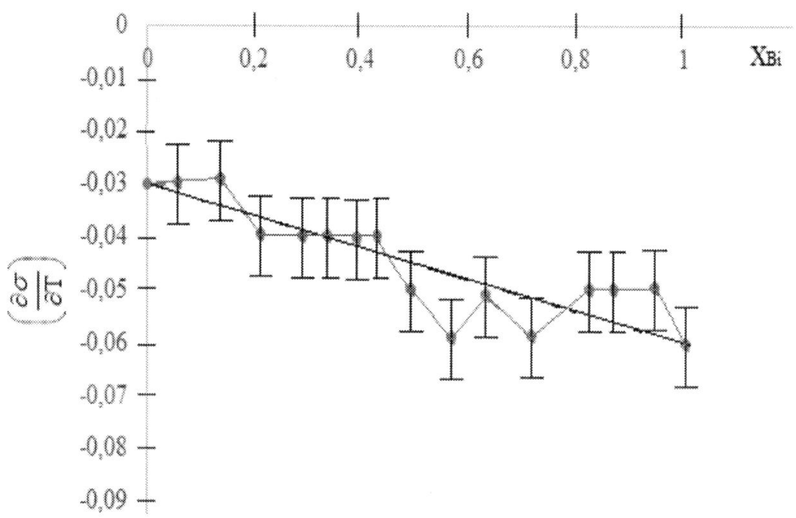

Figure 6. Concentration dependence of PN temperature coefficient in indium-tin-lead-bismuth system

Thus, Figure 7 shows the isolines of the PN sections corresponding to the planes with a constant concentration of bismuth. The diagrams clearly show how the bismuth content affects the type of isothermal surface tension of the triple indium-tin-lead system. It can also be seen from the graphs that at a low bismuth content (less than 0.05 mol. fractions) the minimum on the isothermal surface s is equalized. Further increase of bismuth content does not result in qualitative change of the isothermal surface shape.

4. Conclusion

Lead has a similar effect on the isothermal surface s of the triple indium-tin-vismuth system. At the same time, it should be noted that the maximum surface activity of lead in In-Sn-Bi alloys is somewhat less than that of bismuth in In-Sn-Pi melts. It is shown that with the addition of lead less than 0.1 molar fractions of "depression" on the isothermal surface s of the triple system In-Sn-Bi is preserved in the presence of the fourth component (lead). Isothermal surfaces of s are smoothed only at lead content exceeding 0.1 molar fractions.

Thus, we have shown that the increase in the concentration of lead and bismuth in the melt of indium tin-lead-lead-bismuth tends to equalize the peculiarities of the isothermal surface of PN. Nevertheless, with the addition of bismuth to the In-Sn-Pb system, this process is more intensive than with the addition of lead to the In-Sn-Bi system.

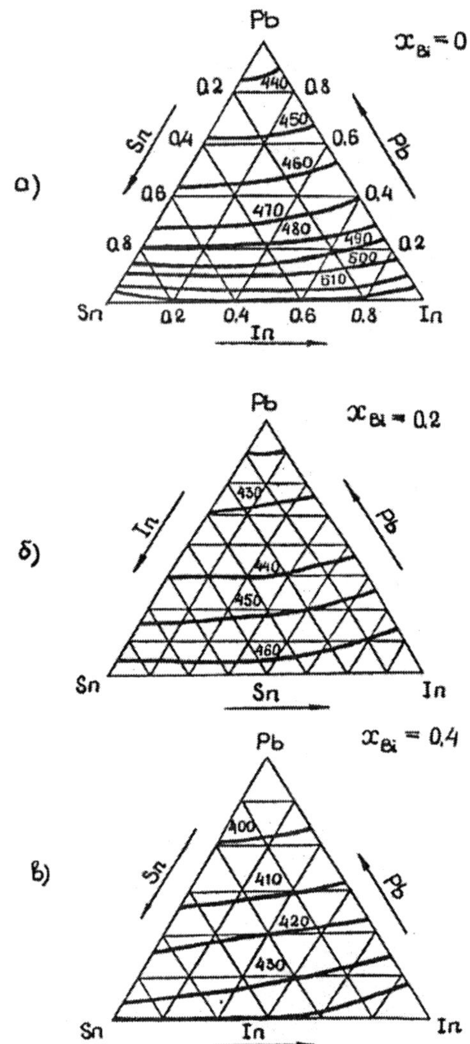

Figure 7. Surface tension lines (mJ/m^2) in the indium-tin-lead-bismuth system at 623 K and xBi=: a-0.0; b-0.2; in-0.4 mol.

References

[1] Kutuev R A, Dadashev R H and Elimkhanov D Z 2011 Surface tension of melts of multi-component systems *Izv. Kabardino-Balkarian State Univer.* **1(4)** 8–11

[2] Kutuev R A 2002 Surface tension and adsorption in indium-olovine-lead-bismuth melts *Perspective-2002* vol 2 Mater. of the All-Russ. Sci. conf. of stud., postgraduates and young sci. (Nalchik: KBGU) pp 90–4

[3] Dadashev R Kh and Kutuev R A 2019 Surface tension and density of indium-tin alloys *Engineering and Earth Sciences* Appl. and Fund. Res. Proc. of the Int. Symp. Ser. Atlantis Highlights in Mater. Sci. and Technol. pp 365–9

[4] Dadashev R H, Khokonov Kh B, Elimkhanov D Z and Bichueva Z I 2007 Concentration dependence of surface tension of double systems *News from the Russ. Acad. of Sci. Ser. of phys.* **71(2)** 264–6

[5] Dadashev R H, Kutuev R A and Ibragimov H I 2002 Surface properties and melt density of thallium-lead *Thermal and mass transfer and properties of liquid metals* vol 1 Proc. of the

Russ. interbranch conf. "Thermophysics-2002" (Obninsk: SSC RF PHEI, 29–31 October) pp 105–7, 331

[6] Alchagirov B B, Dadashev R H, Dyshekova F F and Elimkhanov J Z 2013 Temperature dependence of surface tension indium *J. of Phys. Chem.* **87(6)** 912

[7] Dadashev P X, Kutuev R A, Elimkhanov D Z and Bichuyeva Z I 2007 Surface tension in in indium-olovo-gallium melts *J. of Phys. Chem.* 81(11) 1938–40

[8] Ibragimov K, Dadashev R, Kutuev R and Shapiev S 2004 *Surface tension and adsorption in indium-tin-lead and indium-tin-gallium metals, high temperature capillarity (HTC-2004)* (Sanremo, Italy 31 March – 3 April) pp 134–61

[9] Kutuev R A 2001 Thermodynamic parameters of the surface layer of double and multicomponent melts of fusible P-metals *Physics and Mathematics* 120 p

[10] Dadashev R H, Kutuev R A and Sozaev V A 2016 *Surface properties of alloys based on lead, tin, indium, cadmium* (Moscow: Fizmatlit) 208 p

[11] Kutuev R A, Dadashev R H and Elimkhanov D Z 2012 Surface tension and density of the system of thallium-lead-bismuth *Vest. of Tambov Univer. Ser. Natural and Techn. Sci.* **17(2)** 712–3

[12] Dadashev R Kh 2008 *Thermodynamics of Surface Phenomena* (Cambridge Int. Sci. Pub. July 1) 280 p

[13] Ibragimov H I and Paskacheva B H 2006 *Thermophysical properties of metal and polymer melts* (Moscow: Intermet Engineer.) 288 p

Features of deformation and destruction of fine-grained fiber concrete

S A Maslennikov, O A Pashkova, V A Dmitrienko and D O Rud

Institute of service and business (branch) of DSTU in Shakhty, 147, Shevchenko Str., Shakhty, Rostov Region, 346500, Russia

E-mail: maslennikovsa@mail.ru, Olia51084@rambler.ru, VADmitrienko@rambler.ru, milosrud@icloud.com

Abstract. The article analyzes the conditions of buildings construction in town Shakhty after the closure of coal mining enterprises. The appearance of deformation processes after old mine workings were flooded, is usually accompanied by uneven surface deformations and cracks formation in load-bearing structures is noted in the article. The positive effect of polypropylene microfiber additives on the deformation properties of Portland cement masonry solutions was studied, as it significantly reduces the probability of cracks in building structures. The results of experimental studies of mechanical properties of compositions with optimal dosage of polypropylene microfiber are presented. An increase in flexural and compressive strength of microfiber-modified samples was noted in comparison with the control ones. The dependence of ultimate deformations on normal stresses is established, which allows predicting the permissible ultimate deformations of masonry. The module of the total solution strain was evaluated.

1. Introduction

Despite many years of experience and a huge volume of concrete used in the construction industry, the improvement and optimization of its composition has not lost relevance to date. The first studies of fiber-reinforced concrete in our country date back to the beginning of the last century, but the volume of its application is still small [1].

The question of the quality and new features of fiber-reinforced concrete stimulates numerous studies [2–7] and increase its further application in various fields of construction.

Due to the obtained results the unique operational properties of steel fiber reinforced concrete were confirmed, which allowed increasing the number of structures made of it and use it in the construction of various objects [8-12].

The most important advantages of fiber-reinforced concrete are as follows: high physical and mechanical properties; high frost resistance, moisture resistance, crack resistance, resistance to shock loadings, adhesion and abrasion resistance [13–14]. These indicators determine high performance reliability and durability of the concrete.

Such physical and mechanical characteristics of fiber-reinforced concrete meet the requirements necessary for construction in areas with high seismic activity, as evidenced by the foreign experience of extensive and widespread use as the main building material in earthquake-prone areas, for construction: tunnels, roads, bridges, industrial structures, etc. [15–16].

2. Purpose of the study

As part of restructuring the coal industry in Russia almost all enterprises of the East Donbass were closed at the end of the 90s of the last century. This led to the cessation of pumping water from the mine workings, and, accordingly, to the flooding of the developed spaces.

However, the negative consequences of mines closure provoked not only flooding. The dampening of the destroyed mountain massifs is accompanied by significant changes in the physical and mechanical properties of rocks, since a large number of clay layers are present in the stratigraphic structure of the region. This is accompanied by surface deformations and accordingly uneven deposition of the building foundations.

In the central part of the town on exploited buildings, which were guarded by pillars of coal, new cracks appear (Figure 1) and the development of previously formed ones takes place (Figure 2). Moreover, measures to strengthen structures in the undermined areas do not fully prevent crack formation due to significant soil deformations.

Figure 1. Cracks in the building of the Drama Theater.

The design scheme of the vast majority of newly built town buildings with load-bearing walls made of ceramic or silicate bricks. This circumstance contributes to the appearance of defects at newly constructed facilities (Figure 3), since additional loads on the soil will be accompanied by uneven soil deposition.

Figure 2. Opening cracks after sealing

Figure 3. Crack formation in new buildings

In this case, the use of special measures to prevent cracking is required, which is accompanied by a significant increase in the cost of construction. However, despite the construction of special structures on a number of objects, crack formation and their development are still observed.

The level of modern research in the field of regulation of the properties of Portland cement mortars and concretes makes it possible to substantially change the characteristics of mixtures and solids. Therefore, a very promising direction for reducing crack formation is to increase the deformation characteristics of masonry.

Since it is impossible to change the properties of bricks, it is possible to reduce the probability of cracking in masonry, using masonry mortar with high deformation capacity.

3. Methods and materials

Improving the performance of concrete and mortar is achieved by introducing microfiber into their formulation
[1–3]. Therefore, to assess the impact of microfiber on the characteristics of sand-cement mortar at the Department of Construction and Technosphere Safety at the Institute of Service and business (branch) of Don State Technical University in Shakhty, studies were conducted to establish the optimal dosage of polypropylene and steel microfiber in solutions.

For research, Portland cement of the Sebryakovsky plant CEM I 42.5 and fine-grained sand of Yagodinsky quarry with Mk = 1.92 were used, the content of the fraction 0.63 did not exceed 2 %. At the ratio of cement and aggregate 1:3 adopted for the control additive-free composition, the water-cement ratio corresponding to the normal consistency of the standard cone on the shaking table in accordance with the GOST 310.4-81 method was 0.44.

In the preparation of modified solutions, polypropylene fiber 12 mm long, 20 μm in diameter and steel 15 mm in length and 0.4 mm in diameter was used as an additive. The amount of mixing water was determined from the condition of equal mobility with a control composition, that is, with a cone spread of 108–112 mm. To determine the optimal amount of polypropylene microfiber, compositions were prepared with the addition of 1–3.5 % fiber from the mass of cement, and steel 3–9 %. From

IOP Conf. Series: Materials Science and Engineering **905** (2020) 012045 doi:10.1088/1757-899X/905/1/012045

each composition, 6 beam samples were made for testing under bending and compression after 28 days after mixing.

4. Results

The results of bending and compression tests of solutions with polypropylene fiber are shown in Figures 4–5, and with steel in Figures 6–7.

The analysis of the results of the performed studies shows that when testing the bending strength the introduction of 1 to 3.5 % of polypropylene fiber from the weight of cement, there is an increase in strength. At the same time, the compression test is characterized by a decrease in the effect on the strength indices with an increase in the microfiber content of more than 2 % of the mass of cement (see Figure 5). It can also be noted that the introduction of polypropylene microfiber is accompanied by an increase in bending strength by 47.4 %, and compression by 15.3 %.

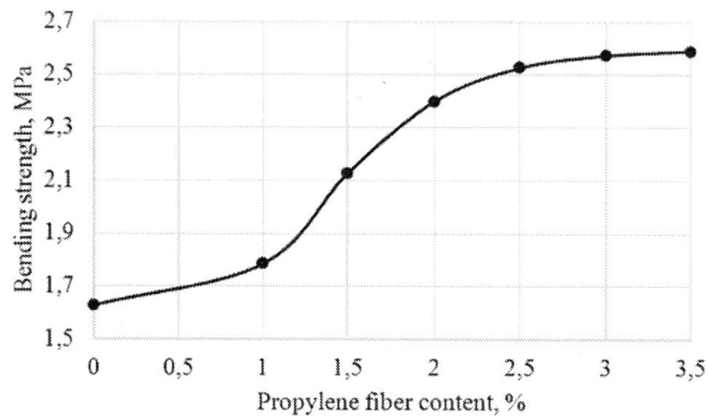

Figure 4. The graph of the dependence of the strength of the samples on bending on the microfiber content

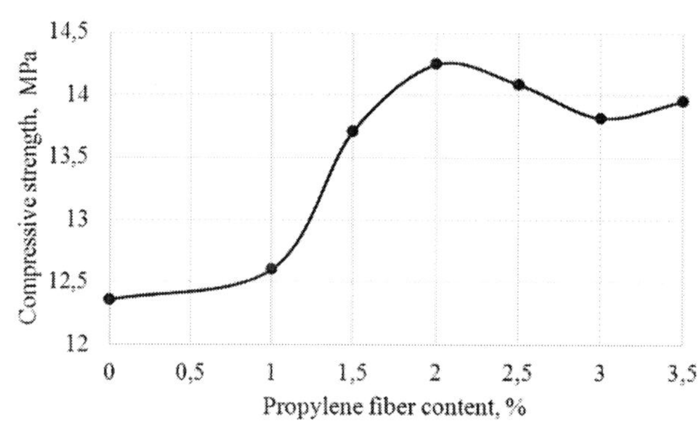

Figure 5. The graph of the compression strength of the samples on the microfiber content

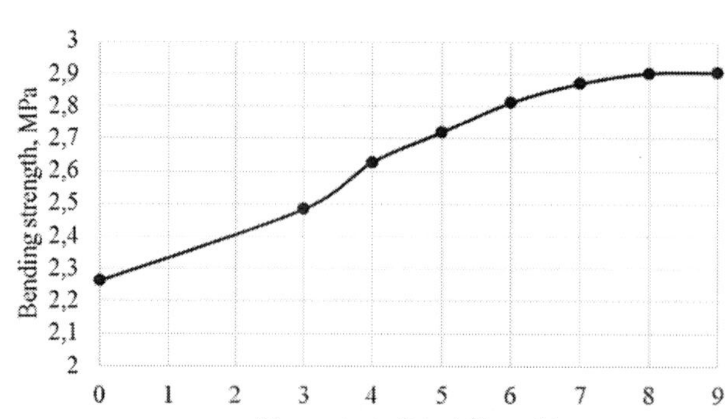

Figure 6. The graph of the dependence of bending strength of the samples on the microfiber content

The effectiveness of the addition of steel fiber is characterized by a stable increase in bending and compressive strength to an additive content of 7 % by weight of cement. However, with an increase in the amount of fiber, the strength growth sharply slows down. The most appropriate introduction of fiber is in an amount of 8 % by weight of the binder. In this case, the bending strength of the samples exceeds the control ones by 28.3 %, and the compression by 43.7 %.

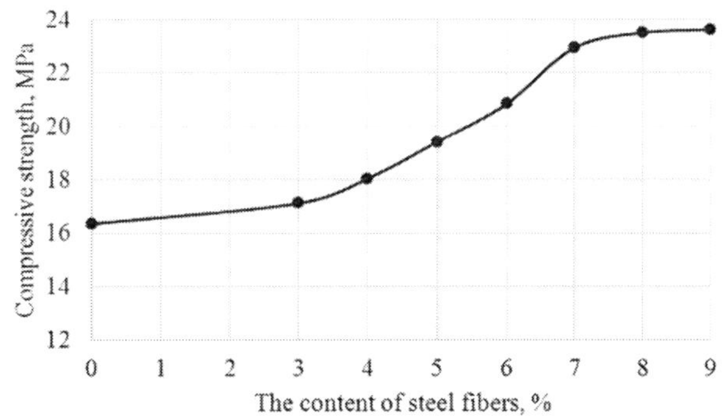

Figure 7. The graph of the dependence of the compressive strength of the samples on the microfiber content

Visual inspection of the samples shows that most of the fibers do not collapse, forming a crack during bending, but are pulled out of the sand-cement stone. This explains the decrease in the growth of bending strength of samples with steel fiber. At the same time, steel fiber better resists shear deformations on the cut inclined planes during compression.

It was also established that the strains at the ultimate stresses of solutions modified with polypropylene microfiber are several times higher than the deformations of the control and almost three times greater than for samples with steel fiber. This circumstance required a more detailed study. Therefore, in the future, when determining the bending strength and compression to control the magnitude of deformation of samples on the press E160 the indicator of clock type ICH-50 (Figure 8) is installed, which allowed fixing deformation, and applied stress at a constant rate of loading with the help of video shooting at certain points in time.

Figure 8. Study of deformation characteristics of samples

According to the results of studies, the destruction of the beams halves without additives occurs when deformation on average is 0.22 mm, and modified is 1.37–1.96 mm. The results of the study of the deformation characteristics of the halves of beams with a dosage of polypropylene microfiber of 1.5, 2 and 2.5 % of the mass of cement are presented in the 9–11 figures.

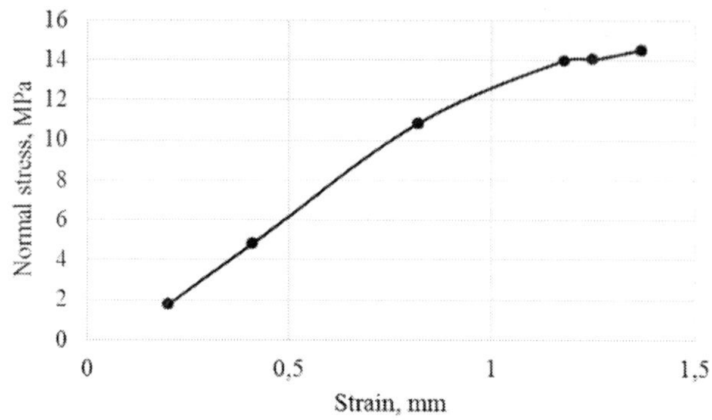

Figure 9. The dependence of the deformation of the samples on stresses with fiber content of 1.5 %

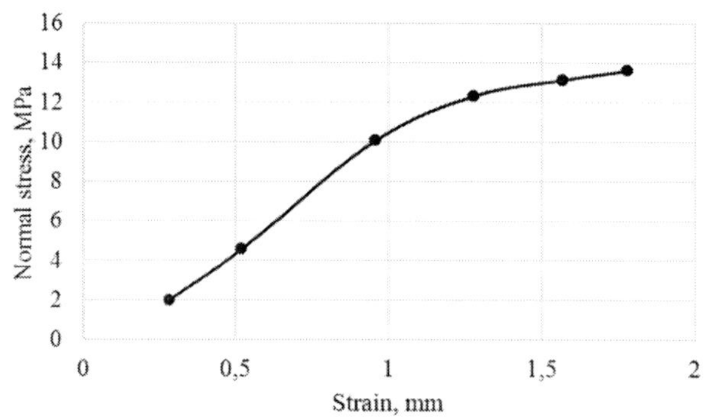

Figure 10. Sample deformation dependence on strains with 2 % of fiber content

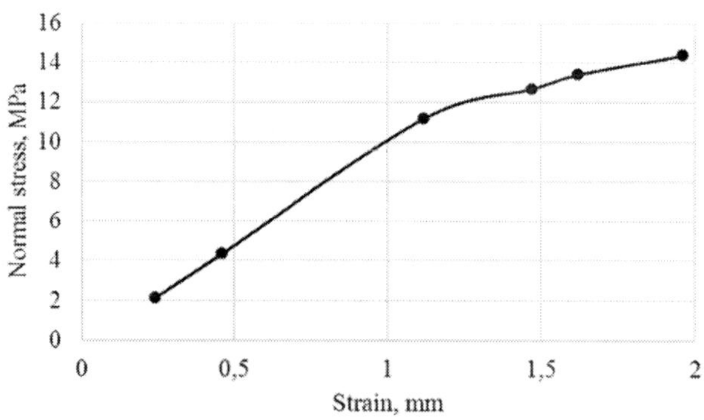

Figure 11. Dependence of deformation of samples on stresses at 2,5% fiber content

The analysis of tests of 34 samples shows that compositions keep linear dependence (elastic) of deformations on stresses approximately to 70 % from a compressive strength limit. After excluding strain values outside the linear dependence from the sample, a graph was constructed that, with a high degree of certainty, R2 = 0.9443, is approximated by a linear function (Figure 12).

The equation (1) obtained by the results of approximation will allow to determine the permissible additional deformations of masonry from uneven sedimentationof the foundation base when using a modified polypropylene microfiber solution.

$$\varepsilon = 0{,}0027\sigma + 0{,}0009, \tag{1}$$

where ε is relative deformations of samples; σ is normal compressive stresses, MPa.

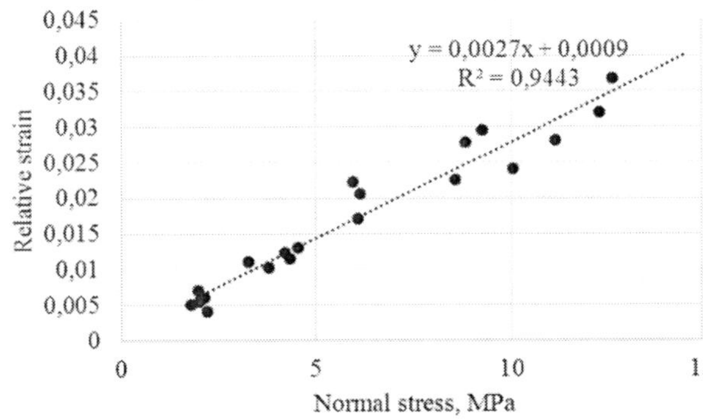

Figure 12. The graph of the dependence of the relative deformation of samples modified with polypropylene microfiber on stresses.

Figures 13–14 show the results of the destruction of the beams halves on compression.

Figure 13. The nature of the halves destruction of the control samples

The analysis shows that despite the significant deformations in compression (the samples contain dents from load plates up to 2 mm) there was no clear exfoliation on almost all 34 samples.

The nature of the destruction of concrete samples (Figure 15) modified polypropylene microfiber was investigated on samples of cubes with a rib size of 100×100×100 mm. As a result, it can be noted that the destruction of concrete without additives is accompanied by exfoliation of the destroyed zones of the side faces of the cubes with the formation of an "hourglass" (Figure 15a).

Figure 14. Samples of polypropylene microfiber modified solution after compression test

The loading of modified concrete samples is accompanied by significant vertical deformations without exfoliation of the faces of the cubes (Figure 15b). The load schedule of tests (Figure 16) is characterized by a prolonged decline in tensions after extremes. That is, the risk of avalanche-like destruction is reduced.

The tests also found that, unlike gravitational, forced mixing ensures even distribution of fibers by composition and, accordingly, a maximum increase in strength on bending and compression.

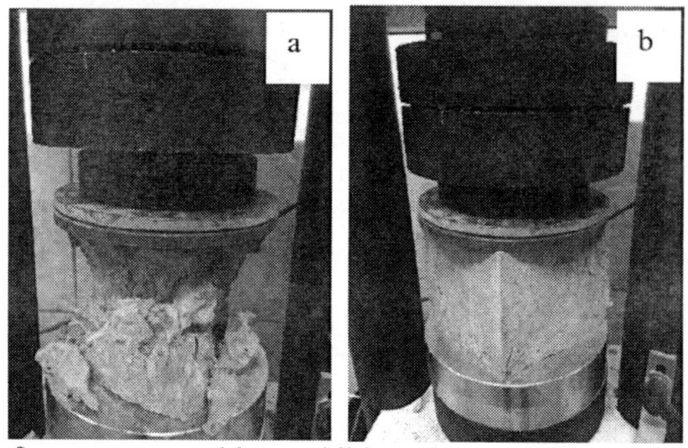

Figure 15. Destruction of concrete: a – without additives; b – modified polypropylene microfiber

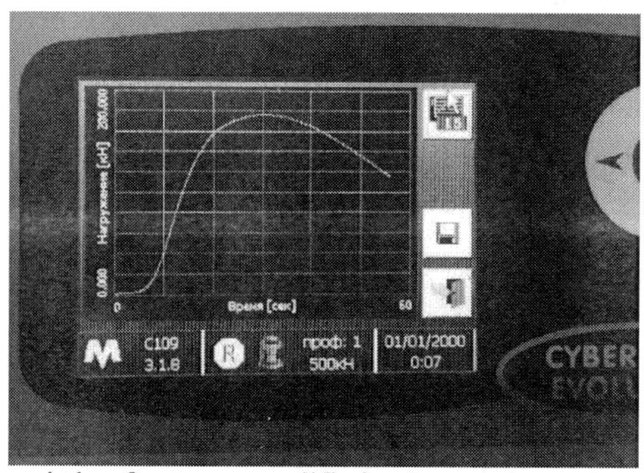

Figure 16. Loading characteristic of concrete modified polypropylene microfiber

5. Conclusion

According to the results of studies of sand-cement mortar, it was found:

- when testing the compression of the halves of control samples, the ultimate deformation (at the time of destruction) is averaged 0.22 mm;
- ultimate strains of samples of solutions modified with polypropylene microfiber were 1.37–1.96 mm;
- when a load is applied not exceeding 50 % of the limit, the deformation of the modified samples is on average 0.8 mm, the decrease in compressive strength is 11 %, which is within the range of variation of the compressive strength of samples of the same composition.

Thus, it is established that the limit deformations of samples of modified composition are 5–7 times more than the control ones, which can provide more uniform stress distribution in masonry, thereby reducing the probability of cracking. Preliminary calculations show that masonry with a height of 1 m can allow vertical deformations of up to 2.5 mm without breaking the continuity.

Since the mechanical characteristics of concrete are mainly determined by the properties of the sand-cement mortar, the use of concrete modified with polypropylene microfiber for deformable structures and the fastening of underground structures will ensure their higher reliability.

The studies were conducted as part of the implementation of the state task 7.9213.2017 / БР "Development of methods and means of construction and maintenance of underground and buried structures".

References

[1] Gorb A M and Voilokov I A *Fiber-concrete – the history of the issue. Regulatory framework, problems and solutions* Retrieved from: monolitpol.ru/files/monolitpol026.pdf
[2] Ibe E E and Shugurova A V 2017 Prospects for the use of fiber-reinforced concrete in the construction of hydraulic structures *Internet J. Sci.* **9(1)** Retrieved from: naukovedenie.ru/PDF/61TVN117.pdf
[3] Novikov N S 2016 Fire resistance and strength of fiber-reinforced concrete structures *Technosphere Safety Technol.* **3(67)** 122–7
[4] Stradanchenko S G, Pleshko M S and Armeiskov V N 2013 Development of effective fiber concrete compositions for underground construction *Engineer. Bull. of the Don* **4** Retrieved from: ivdon.ru/ru/magazine/archive/n4y2013/1994
[5] Klyuev S V, Netrebenko A V, Durachenko A V and Pikalova E K 2014 Monolithic fiber-reinforced concrete for floors of industrial buildings *Collection of sci. papers* **19(1)** 29–32
[6] Klyuev S V and Avilova E N 2013 Fine-grained fiber-reinforced concrete using polypropylene fiber for covering roads *Bull. of BSTU named after V.G. Shukhov* **1** 37–40
[7] Klyuev S V 2014 High-quality fiber-reinforced concrete for monolithic construction *Int. Res. J.* **11(2)** 29–32
[8] Klyuev A V, Klyuev S V, Netrebenko A V and Durachenko A V 2014 Fine-grained fiber reinforced concrete reinforced with polypropylene fiber *Bull. of BSTU named after V.G. Shukhov* **4** 67–72
[9] Klyuev S V 2012 The use of composite binders for the production of fiber concrete *Concrete Technol.* **1-2(66-67)** 56–7
[10] Klyuev S V 2014 The development of dispersed reinforced fine-grained concrete based on industrial sand and composite binder *Int. Sci. J.* **11(2)** 27–9
[11] Pothisiri T and Soklin C 2014 Effects of Mixing Sequence of Polypropylene Fibers on Spalling Resistance of Normal Strength Concrete Engineer. J. **18(3)** 55–64
[12] Enfedaque A, Alberti M, Galvez J C, Beltran M 2018 Constitutive relationship of polyolefin fibre-reinforced concrete: Experimental and numerical approaches to tensile and flexural behavior Fatigue & Fracture of Engineer. Mater. & Struct. **41(2)** 358–73
[13] Leone M, Centonze G, Colonna D et al 2018 Fiber reinforced concrete with low content of recycled steel fiber Shear behavior. Construct. and Build. Mater. **161** 141–55
[14] Steelfibre-reinforced self-compacting concrete with 100 % recycled mixed aggregates suitable for structural applications 2017 Construct. and Build. Mater. **156** 230–41
[15] Gafarova N E 2016 Fiber-concrete for earthquake-hazardous construction areas *Int. J.of Appl. and Fund. Res.* **9-2** 179–81
[16] Kamal M M, Safan M A, Etman Z A and Kasem B M 2014 Mechanical properties of self-compacted fiber concrete mixes HBRC J. **10(1)** 25–34

Anomalous character of temperature dependences of some $AgInSe_2$ parameters $AgInSe_2$

A Matiev[1,2], R Uspazhiev[2,3], B Khamkhoev[1], Z Torshkhoeva[1], L Blagina[4], A Gachaev[2,3], L Martazanova[1] and R Evteeva[2]

[1] Ingush State University, 7, I. B. Zyazikov ave., Magas, 386001, Russia
[2] Grozny State Oil Technical University named after M.D. Millionshchikov, korp. 1,100, prosp. Husejna Isaeva, Grozny, 364068 Russia
[3] Kh. I. Ibragimov Complex Research Institute, RAS, 21A, Staropromyslovskoe highway, Grozny, 364051, Russia
[4] Branch of the Don State Technical University, Volgodonsk, Russia

E-mail: matiyev-akhmet@yandex.ru, russlan-chr@mail.ru, khamkhoev1950@mail.ru, evlihan@mail.ru, blagina_larisa@mail.ru, gachaev_chr@mail.ru, martazanova.lida@mail.ru, roza.efteeva@mail.ru

Abstract. Abstract – Based on the experimental and X-ray studies of the absorption spectra and the temperature dependence of the crystal lattice parameters of AgInSe2 single crystals and the coefficients of thermal expansion along the corresponding crystallographic directions were calculated. It was found that the thermal expansion of AgInSe2 crystals along the a and c axes changes its sign at a temperature of 142.31 and 135, 14 K, respectively. The anomalous nature of the band gap width AgInSe2 was established.

1. Introduction

The AgInSe2 compound, which belongs to new classes of semiconductors type $A^{III}B^{III}C_2^{VI}$ and $A^{I}B^{III}C_2^{VI}$ is of scientific interest for modern optoelectronics, is currently being intensively studied [1–8, 11–16]. The AgInSe2 compound crystallizes into a chalcopyrite structure in which two types of cathinones form an ordered sublattice. A unit cell containing 8 atoms (2Cu, 2In, 4Se) is characterized by the presence of a volume center and an axis ratio close to 2 [6].

The aim of this work is to study the temperature dependence of the thermal expansion coefficients and absorption spectra of AgInSe2 single crystals. Investigations of the temperature dependence of lattice parameters up to low temperatures are of interest since they are closely related to lattice dynamics and therefore can lead to a deeper understanding of these properties in $A^{I}B^{III}C_2^{VI}$ crystals.

2. Methods and materials

The studied AgInSe2 crystals were grown by the Bridgman-Stockbarger method. The starting materials were high purity elements: Ag-ultra high purity-4-11-4; In-000; Se-ultra high purity-17-4. The oxide film and other contaminants from the silver surface were removed by etching in a 5 % HNO3 solution for 8–10 minutes, followed by washing in running distilled water. Ampoules for synthesis, made of thick-walled quartz with an inner diameter of 25 mm were first etched with a 40 % HF solution for 5 minutes, washed intensively with distilled water, and then annealed in a vacuum furnace at a temperature of 1300 K. To prevent contact of the molten substances with the quartz surface, the inner side of the ampoules was covered with a layer of graphite. After preliminary orientation on the RKOP installation in the (200) and (112) directions, plane-parallel plastics were cut off, which were subjected to grinding and polishing. The thermal expansion of AgInSe2 crystals was studied on a standard DRON-3 X-ray apparatus using a low-temperature attachment URNT-180. The rotation speed of the Dron-3 detector was 0.25 deg/min. An angular mark on the tape of the recording potentiometer was carried out every 0.1 o. To improve the accuracy, the test sample was irradiated with characteristic radiation from a copper target, as a result of which the reflexes became sharper. Cooling of the sample was carried out by a low-temperature attachment URNT-180, working by the method of blowing.

The study of temperature dependences of the band gap width of AgInSe2 crystals was carried out on the SDL-2 installation. The error in determining the band gap width values from the absorption spectra was not more than MeV. The research samples were n-type AgInSe2 single-crystal plates with carrier concentrations of ~ 10^{18} cm^{-3} and mobilities of ~ 40–70 $cm^2/V^{-1} \cdot S^{-1}$, respectively, at 300 K. The surfaces of the samples were previously subjected to grinding, polishing, and then chemical etching.

3. Results

The chalcopyrite lattice has axial symmetry, therefore, the unit cell parameters a and c change independently with temperature.

To determine the temperature dependence of parameter a, the change in the interplane distance was measured d_{200}, since $a = 2d_{200}$.

The dependence $c(T)$ was determined by the change in temperature d_{112}, where the lattice parameter c was found by the formula

$$c = \frac{2ad_{112}}{\sqrt{a^2 - 2d_{112}^2}}.$$

The measured temperature dependences of the interplane distances $d_{200}(T)$ and $d_{112}(T)$ are presented in Fig. 1 *a, b*..

The temperature dependences $a(T)$ and $c(T)$, defined from $d_{200}(T)$ and $d_{112}(T)$, are shown in Fig. 2 a,b.

Dependences $a(T)$ and $c(T)$ were approximated by relations:

$$\begin{cases} a(T) = A_0 + B_0 + \dfrac{C_0}{T} \\ c(T) = A_i + B_i \dfrac{C_i}{T} \end{cases}$$

on an electronic computing machine in the entire temperature range, as a result of which the constants A_i, B_i and C_i were determined, which amounted to 5,97 Å; 3,27·10^{-4}Å/K; 6,83 Å·K and 11,53 Å; 4,43·10^{-4}Å/K; 7,89 Å·K , respectively.

The least squares in the definition of a and c experimentally and according to the above formula are:

$a_{theor} - a_{exp}$

$$f_a = \sum (a_{\text{theor}} - a_{\text{exp}})^2 = 1{,}70 \cdot 10^{-5};$$

$$f_c = \sum (c_{\text{theor}} - c_{\text{exp}})^2 = 3{,}63 \cdot 10^{-4}.$$

The standard deviation $\sigma = \sqrt{\dfrac{f}{N-1}}$ was also $\sigma^a = 8{,}99 \cdot 10^{-4}$ and $\sigma^c = 4{,}15 \cdot 10^{-3}$. From the dependences $a(T)$ and $c(T)$ it follows that at a certain temperature value, the thermal expansion of the lattice changes sign.

It can be seen that at a certain temperature value, the thermal expansion changes its sign. From the dependencies $a(T)$ and $c(T)$ were counted the values of the coefficients of thermal expansion along the corresponding crystallographic directions by the formula

$$\alpha_a = \frac{1}{a_o} \cdot \frac{da}{dT}; \qquad \alpha_c = \frac{1}{c_o} \cdot \frac{dc}{dT}.$$

Fig. 3 represents the temperature dependences of the thermal expansion coefficients of AgInSe2 crystals. The anisotropy of thermal expansion in these crystals is weakly expressed; however, in the low temperature range (T <120 K), the ratio $\left(\dfrac{\alpha_c}{c_a}\right)$ increases with decreasing temperature and amounts to 1.91 and 1.41 at 80 and 300 K, respectively

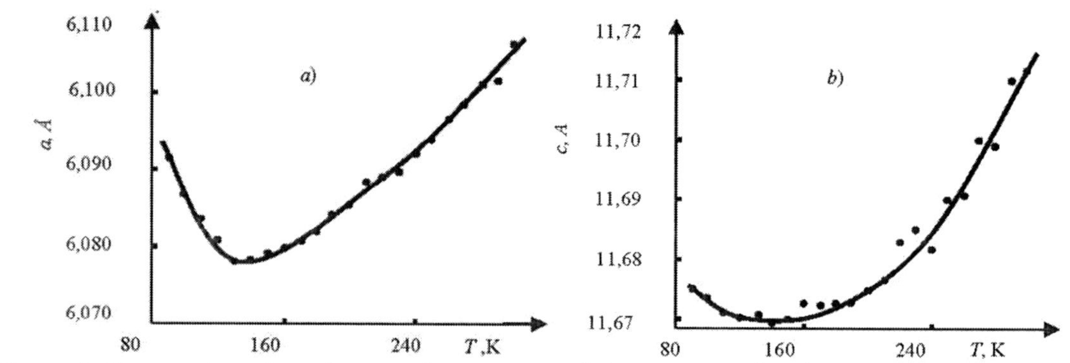

Figure 1. Temperature dependences of parameters of elementary cells of AgInSe2 single crystal

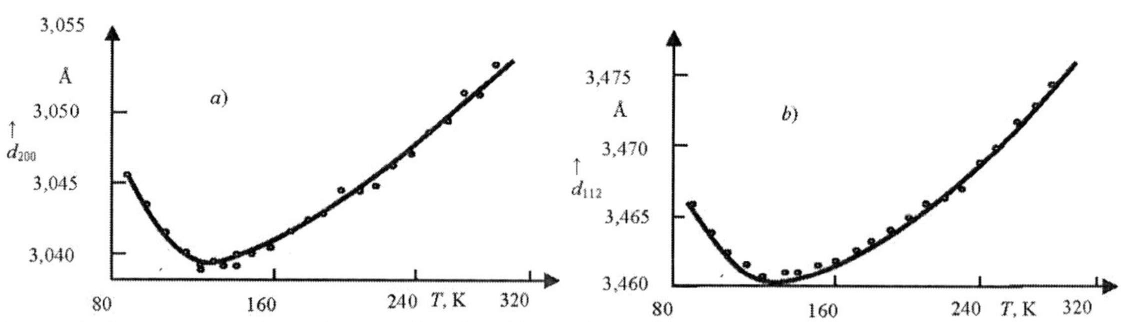

Figure 2. Temperature dependences d_{200} $d_{112}$of interplane distances

The thermal expansion of AgInSe2 crystals along the axes a and c changes its sign at a temperature of 142.31 and 135.14 K, respectively.

The values of the volume coefficient of thermal expansion calculated by the formula $\alpha_V = (2\alpha_a + \alpha_c)$ showed that α_V changes its sign at a temperature of $T_o = 139{,}44$ K.

The value of the characteristic temperature of the sign α_V change is very close in value to the Debye temperature $\theta = 138\text{K}$ for AgInSe2 crystals.

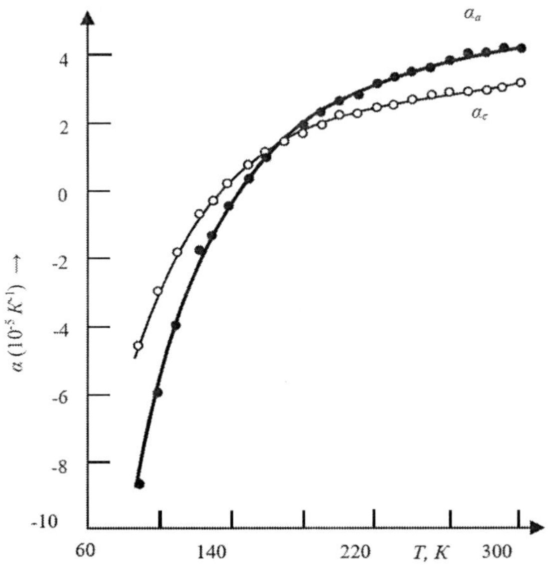

Figure 3. Temperature dependences of the coefficient of thermal expansion of AgInSe2 crystals.

Comparing the obtained results with our results on the study of the temperature dependence $E_g(T)$ on the absorption spectra [4, 6], we conclude that the volume thermal expansion in AgInSe2 to the change $E_g(T)$ seems to dominate in the entire studied temperature range. However, the ratio between the relative change in volume $\frac{\Delta V}{V}$ and $\frac{\Delta E_g}{E_g}$ in the temperature ranges $90 < T < 120$ K and $140 < T < 300$ K is very different, which suggests the existence of a competing mechanism that gives a positive change ΔE_g.

At $T > 120$K, we observed [4, 6] a clear anomaly in behavior ΔE_g in the sense that some "positive" mechanism almost completely compensates for the narrowing caused by the volume effect.

Usually, the thermal expansion of solids is described using the Grüneisen parameters.

In the case of AgInSe2, which crystallizes in a uniaxial chalcopyrite structure, it is necessary to introduce two independent Grüneisen parameters γ_a and γ_c, which are related to the thermal expansion coefficients (α_a, α_c) by the relations:

$$\gamma_a = \frac{V_m}{C_p}\left[\left(C_{11}^S + C_{12}^S\right)\alpha_a + C_{13}^S \cdot \alpha_c\right]$$

$$\gamma_c = \frac{V_m}{C_p}\left[2C_{13}^S\alpha_a + C_{33}^S \cdot \alpha_c\right],$$

where V_m – the molar volume, C_p – the specific molar heat capacity at constant pressure, C_{ij}^S – the adiabatic stiffness constant. Currently, there is no information about the experimental values C_{ij}^S and $C_p(T)$ in AgInSe in the temperature range of interest to us and this creates difficulties in conducting the quantitative analysis of temperature dependences γ_a and γ_c.

However, not excluding other explanations, it is possible to assume the existence of low-energy vibrations of the lattice modes with a negative mode of the Grüneisen parameters in AgInSe2. Unfortunately, information about the low-energy branches of the phonon spectrum is not sufficient.

The accordance between the temperature dependences of the band gap width and the lattice parameters should be noted. The anomalous nature of dependence $E_g(T)$ is also inherent in dependence $a(T)$ and $c(T)$. This experimental fact testifies to the prevailing contribution of the volume effect to the temperature dependence of the band gap width of single crystals in AgInSe2.

As it was mentioned above, according to existing concepts, the temperature dependence $E_g(T)$ of semiconductors is due to thermal expansion of the lattice (volume effect) and electron-phonon interaction. The change in the sign of the temperature coefficient $\left(\frac{\partial E_g}{\partial T}\right)_p$ indicates the competition of the two effects. It is believed that the thermal expansion causes a positive temperature coefficient E_g, and the electron – phonon interaction – its negative sign. It is natural to assume that in AgInSe2 the contribution of thermal expansion predominates below 120 K, and the electron-phonon interaction – above 120 K.

It is known that the isochoric change can be caused by a nonlinear dependence $(\delta E_g)_V$ on interatomic distances. Generalizing the obtained result to the case of an anisotropic crystal, we write

$$\langle \delta E_g \rangle = \sum_{\mu=1}^{3} \langle \delta E_g \rangle_\mu ,$$

$$\langle \delta E_g \rangle_\mu = \Xi_\mu^{(1)} \Delta\mu + \frac{1}{V} \sum_{f_\mu \lambda} \varepsilon_{f_\mu \lambda} \chi_\mu \Xi_{f_\mu \lambda}^{(2)} \tag{1}$$

where Δ_μ – the expansion and compressibility χ_μ corresponding to the main direction μ.

For the temperature derivative of (1) we have

$$\left(\frac{\partial E_g}{\partial T}\right)_p = \left(\frac{\partial E_g}{\partial T}\right)_V + \alpha_V \left(\frac{\partial E_g}{\partial T}\right)_T ;$$

$$\sum_\mu \alpha_\mu \Xi_\mu^{(1)} \equiv \alpha_V \left(\frac{\partial E_g}{\partial \Delta_V}\right)_T$$

$$\left(\frac{\partial E_g}{\partial T}\right)_V = \frac{1}{V} \sum_{f_\mu \lambda \mu} \chi_\mu C_{f_\mu \lambda} \Xi_{f_\mu \lambda}^{(2)} ; \tag{2}$$

Here α_V – the temperature coefficient of volumetric expansion, $C_{f_\mu\lambda} = \partial \varepsilon_{f_\mu \lambda} / \partial T$ – the heat capacity of the lattice oscillator (f – wave vector, λ – polarization).

Let us consider the contribution of longitudinal waves, at which

$$\langle \delta E_g \rangle = \sum_\mu \left(\langle \Delta_\mu \Xi_\mu^{(1)} \rangle + \langle \Delta_\mu \Xi_\mu^{(2)} \rangle \right); \ \sum_\mu \alpha_\mu \equiv \alpha_V$$

Using averaged parameters ε_1 and ε_2 defined by equalities,

$$\left(\frac{\delta a}{a}\right) \Xi_a^{(1)} + \left(\frac{\delta c}{c}\right) \Xi_c^{(1)} \equiv E_1 \Delta_V ; \ \sum_\mu \chi_\mu \Xi_\mu^{(2)}$$

for the high $(T > \theta)$ range $\theta = 138\ K$ – temperature range – Debye temperature [195,196] we get

$$\left(\frac{\partial E_g}{\partial T}\right)_p \approx \left(E_1 + \frac{E_2}{3\gamma_G}\right)\alpha_V\ ;\ \alpha_V \approx \gamma_G \frac{\chi}{V} 3Nk_B$$

where k_B – the Boltzmann constant, -the high-temperature value of the Grüneisen coefficient $(\gamma_G = 2)$.

Due to the lack of experimental data on E_I, we assume that the change δE_g in the range $T \leq 120\ K$ is caused only by a change in the lattice parameters a and c, i.e.

$$(\delta E_g)_{T<100} \approx E_1 \Delta_V(T);\quad \Delta_V(T) = 2\frac{\delta a}{a} + \frac{\delta c}{c},$$

from the course of the curves describing dependencies $\delta E_g, a(T)$ and $c(T)$ of the range $T \leq 100\ K$ follows $\delta E_g \approx 25\ meV$; $2\frac{\delta a}{a} \approx -0{,}0033$; $\frac{\delta c}{c} \approx -0{,}0005$;

For the parameter E_I this gives

$$E_I \approx -(0{,}025/0{,}0038) \approx -6{,}6\ eV$$

From high temperature dependences follows $\alpha_V = 1{,}15 \cdot 10^{-4}\ K^{-1}$ and $(\partial E_g/\partial T)_p = 1{,}5 \cdot 10^{-4}\ eB/K$

Substituting these values gives $\varepsilon_2 \approx 32\ eV$. More accurate estimates require experimental data on pressure coefficients.

In addition, it can be seen from (1) that the influence of transverse waves of the lattice is not excluded.

At this stage, it follows from the above analysis that the isochoric part of the change $(\delta E_g)_V$ in AgInSe2 single crystals has a positive sign, that is $\left(\frac{\partial E_g}{\partial T}\right)V > 0$, and at high temperatures $(\delta E_g)_V$, apparently compensates for the negative contribution from the thermal expansion of the lattice.

We have suggested that at low temperatures the contribution of the volume effect (thermal expansion of the lattice) to the temperature dependence E_g is predominant.

The thermal expansion of the lattice at low temperatures is closely related to the dynamics of the lattice. The temperature dependences of the change in the band gap width of AgInSe2 are presented in Fig. 4.

An analysis of the temperature dependence of the change in the AgInSe2 band gap width shows that the band gap width increases with temperature in the range of (80 ÷ 120) K (Fig. 4 and the temperature coefficient is (4.0 ÷ 6.2) · 10^{-4} eV/K.

In the temperature range (120÷300) K, the width of the band gap decreases with increasing temperature and the temperature coefficient is –(1,2÷1,9) * 10^{-4} eV/K.

The temperature dependence of the band gap width in semiconductors is caused by the thermal expansion of the lattice and the electron- phonon interaction.

Thermal expansion causes a positive temperature coefficient, and electron-phonon interaction – a negative sign [5].

In the studied temperature range, the role of each of the two mechanisms described above changes, which causes this behavior of temperature dependences of the band gap width in AgInSe2 crystals. In particular, we believe that in AgInSe2 the thermal expansion prevails below 120 K and the electron-phonon interaction – above 120 K. Unfortunately, at the moment there are no data on low-energy branches of the phonon spectrum.

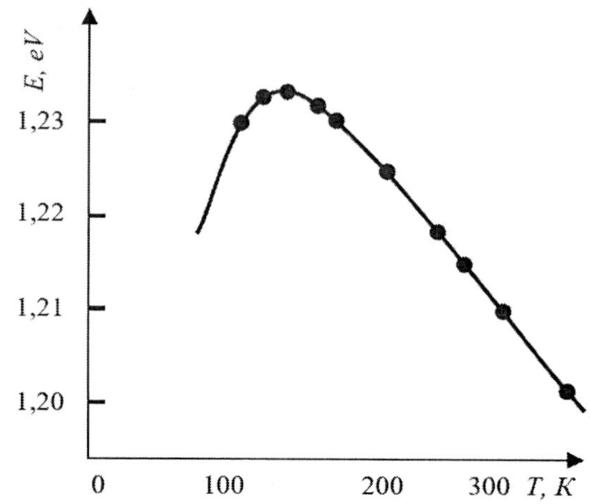

Figure 4. The temperature dependence of the band gap width in AgInSe2 single crystals

This does not allow for a detailed discussion of the role of electron-phonon interaction in the temperature dependences of the AgInSe2 width. This should be the subject of further research.

4. Conclusion

Based on the analysis of the temperature dependences of the absorption spectra and thermal expansion coefficients of AgInSe2 crystals, the following parameters are determined:

1) temperature dependences of band gap width and crystal lattice parameters;

2) It is shown that the experimentally discovered anomalous temperature dependence of the band gap width has a flat maximum at T = 120 K, which is a consequence of the anomalous behavior of the temperature dependences of the lattice parameters at this temperature;

3) having carried out a quantitative analysis on the basis of these dependencies and the corresponding theory, it was revealed that the isochoric part of the change $(\delta E_g)_V$ in this case has a positive sign, that is,

$$\left(\frac{\partial E_g}{\partial T}\right)V > 0.$$

At high temperatures $(\delta E_g)_V$ it is not decisive, but only partially compensates for the dominant negative contribution from thermal expansion.

References

[1] Siebentrit S 2004 Alternative buffers for chalcopyrite solar cells *Solar Energy* **77** 767–75

[2] Castro S L, Bailey Sh G et al 2004 Synthesis and Characterization of Colloidal CuInS2 Nanoparticles from a Molecular Single-Source Precursor *J. Phys. Chem. B* **108(33)** 12429–35

[3] Gremenok V f 2005 Investigation of photosensitive structures based on semiconductor compounds – A1B3C62 *Actual problems of solid state physics in 2 t Mater. Intl. scientific Conf. (Minsk, October 26-28)*

[4] Matiev A H and Georgobiani A N 2004 Temperature dependence of the change in the band gap CuInSe2 and AgInSe2 Optics, optoelectronics and technologies *Proceedings of the VII Intern. Conf. (Ulyanovsk, UGU, October)* p 32

[5] Matiev A H and Georgobiani A N 2005 Temperature dependence of thermal expansion AgInSe2 Opto nanoelectronics, and nanotechnologies and Microsystems *Proceedings of the VII International conference. (Ulyanovsk, Vladimir, HUGO; Wledge 27-30 June)*, p 65

[6] Matiev A X 2005 *Phase equilibria and electron-optical properties of t1b3c62-A1B3C62 systems (A-si, Ag; B-I p, VA; C-S, S2)* Doc. Dissertation (Ulsu) 399 p

[7] Bodnar I V 2008 Optical properties of solid solutions AgGax In1-x Se2 *Physics and engineering of semiconductors* **42(2)** 153-6

[8] Gremenok V F, Ivanov M S and Zalessky V B 2009 Solar cells based on semiconductor materials Alternat. *energy and Ecol.* **1(69)** 59-124

[9] Kopytov A V and kosobutsky A V 2010 Thermodynamic and elastic properties of AgInSe2 and AgInTe2 solid state *Physics* **52(7)** 1271-3

[10] Kopytov A V and Kosobutsky A V 2010 Thermodynamic and Elastic Properties of AgInSe2 and AgInTe2 *Phys. of the Solid State July* **52(7)** 1359–61

[11] Dawei D, Lingzhi O and Yueqing G 2014 Near-infraed broadly emissive AgInSe2/ZnS quantum dots biomedical optical imaging *J. of Mater. Chem. C* **2(34)** 7077–85

[12] Abdullaev M A, Magomedova D K and Alhasov A B 2014 Fabrication and properties of CuInSe2/AgInSe2/CdS double heterojunction cascade solar cells *Inorganic Mater.* **50(3)** 228–32

[13] Haloui H, Touafek K et al August 2015 The Copper Indium Selenium (CuInSe2) thin films solar cells for Hybrid Photovoltaic Thermal Collectors *Energy Proc.* **74** 1213–9

[14] Jianbo Y and Junhong J 2015 Preparation of AgInS2 nanocrystals and their application as sensitizers for TiO2 nanorod array photoelectrodes *IOP Conf. Ser. Mater. Sci. and Engineer.* **87(012114)** 1–7

[15] Alice L and Vacdonald J E 2016 Optoelectronic Properties of CuInS2 Nanocrystals and Their Origin *J. Phys. Chem. Lett.* **7(3)** 572–83

[16] Matiev A Kh, Uspazhiev R T et al 2019 Phasing Diagrams of TlInSe2 – CuInSe2 and TlInSe2 – AgInSe2 Systems *Atlantis Highlights in Material Sciences and Technology* (AHMST) vol 1 Int. Symp. Engineer. and Earth Sci.: Appl. and Fund. Res. (ISEES 2019) pp 249–52

IOP Conf. Series: Materials Science and Engineering **905** (2020) 012047 doi:10.1088/1757-899X/905/1/012047

Electrical conductivity of uniaxial-deformed monocrystals $Tl_{1-x}Ag_xInSe_2(0 \leq X \leq 0,03)$

A Matiev[1,2], R Uspazhiev[2,3], B Khamkhoev[1], V Kodin[4], N Sidorkina[4], M Nalgieva[1], R Evteeva[2] and Kh Khamsurkaev[2]

[1] Ingush State University, 7, I. B. Zyazikov ave., Magas, 386001, Russia
[2] Grozny State Oil Technical University named after M.D. Millionshchikov, korp. 1,100, prosp. Husejna Isaeva, Grozny, 364068 Russia
[3] Kh. I. Ibragimov Complex Research Institute, RAS, 21A, Staropromyslovskoe highway, Grozny, 364051, Russia
[4] Branch of the Don State Technical University, Volgodonsk, Russia

E-mail: matiyev-akhmet@yandex.ru, russlan-chr@mail.ru, khamkhoev1950@mail.ru, kodinvv2013@mail.ru, sun48@bk.ru, evlihan@mail.ru, roza.efteeva@mail.ru, hhi67@mail.ru

Abstract. We have studied the effect of uniaxial compression (stretching) deformation on electrical conductivity along the crystallographic axis [001] of monocrystals $TlInSe_2$ and derivant solid solutions. It has been found that substitution of Tl atoms with Ag atoms in crystals of solid solutions based on $TlInSe_2$ leads to a shift of the energy of indirect transitions to the long-wave zone. The obtained results have been compared to both theoretical calculations of the zone structure of $TlInSe_2$ crystals and the multi-valley model.

1. Introduction

Compounds of $A^{III}B^{III}C_2^{VI}$ class of semiconductor type are of scientific interest for modern optoelectronics and are currently being intensively studied [1–13].

$TlInSe_2$ is a typical representative of the recently discovered non-full-valent semiconductor compounds, which, like its other crystal-chemical analogues from the above class, has a specific structure of the crystal lattice, composed of two independent structural units – an octahedron with monovalent thallium ions and a tetrahedron with trivalent indium ions, surrounded by four selenium atoms with covalent-tetrahedral bond [1].

The research purpose is to study the effect of uniaxial compression (stretching) deformation on electrical conductivity along the crystallographic axis [001] of monocrystals $TlInSe_2$ and derivant solid solutions. It has been found that substitution of Tl atoms with Ag atoms in crystals of solid solutions based on $TlInSe_2$ leads to a shift of the energy of indirect transitions to the long-wave zone. The obtained results have been compared to both theoretical calculations of the zone structure of $TlInSe_2$ crystals and the multi-valley model.

2. Methods and materials

The effect of uniaxial deformation on the electrical conductivity along the crystallographic axis [001] of the crystals $TlInSe_2$, $Tl_{0,99}Ag_{0,01}$ $InSe_2$ and $Tl_{0,97}Ag_{0,03}InSe_2$ has been studied in static mode, the essence of which is as follows: constant deformation is transmitted to the crystal under study by means of a special device designed for this purpose (Fig. 1). The test sample 1 in the form of a thin ($1 \div 10$ microns in thickness) rectangular "hair" with mirror planes of natural cleavage, is glued to a flat steel beam (2), t in thickness, bent by moving the middle part to (Δh) as relating to the plane of sharp ends of the side supports (3), located at a distance l. Sputtered indium has been used as resistible contacts (up to 50 V).

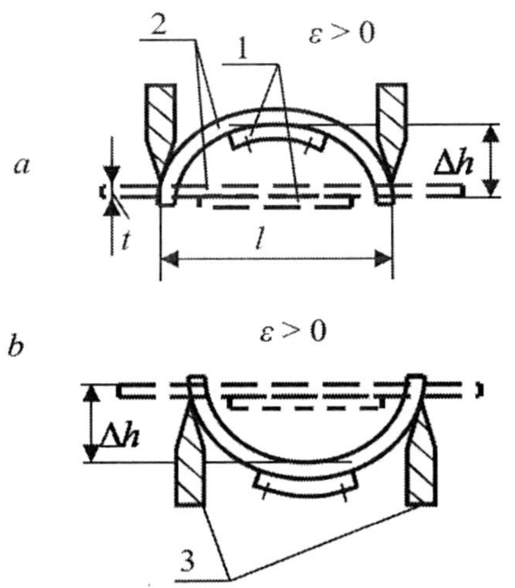

Figure 1. Diagram view of the device for measuring the relative deformation in static mode: 1 – sample; 2 – steel beam; 3 – side supports

The value of deformation has been estimated as

$$\varepsilon = \frac{4t}{2R}, \tag{1}$$

where t – thickness of the beam; R – radius of curvature.

$$R = \frac{(\Delta h)^2 + \left(\frac{l}{2}\right)^2}{2\Delta h}, \tag{2}$$

where l – length of the beam's section under pure deflection deformation;

Δh – maximum deflection of the beam.

Substituting the value from formula (1) into formula (2) we finally get

$$\varepsilon = \frac{4t}{l^2}\Delta h \tag{3}$$

Required displacement Δh (both to positive and negative deformation) has been measured with an accuracy of ± 0.001 mm by means of a time-sensitive indicator. Measurement of the sample resistance both in the absence and in the presence of deformation ($R_{\varepsilon,0}$) has been carried out in the mode of low load resistance according to the diagram presented in Fig. 2. Vertical channel of the plotter "U" has been used to register the voltage drop on the load resistance both in the absence (V_0) and in the presence (V_ε) of deformation, and the horizontal "X" – for fixing the external temperature impact. Statistics of preliminary measurements has showed a significant change in resistance both in the presence and in the absence of deformation.

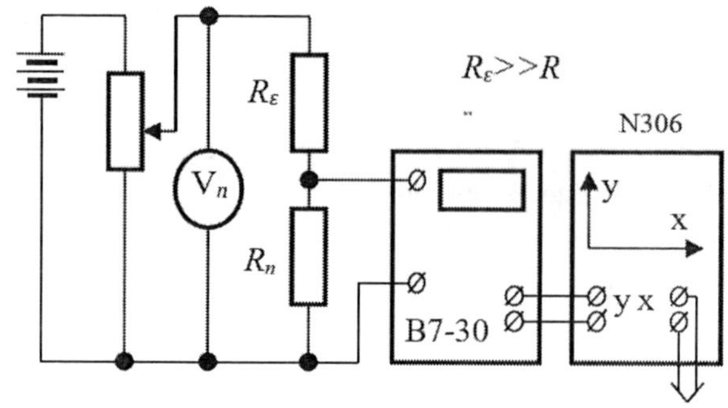

Figure 2. Basic diagram for measuring electrical conductivity in static mode

As a result of our numerous measurements on crystals carried out so far at "normal drying" and optimum thickness of suitable glue, approximately the same strain effect takes place, both at positive and at negative deformation of identical size.

3. Results

Crystals of solid solutions, as well as crystals $TlInSe_2$ show a sufficiently high sensitivity to changes in electrical conductivity under deformation (ε = $1.52 \cdot 10^{-3}$) of both stretching and compression along the crystallographic axis [001]. Analysis of temperature dependencies of electrical conductivity of crystals $TlInSe_2$, $Tl_{0,99}Ag_{0,01}InSe_2$ и $Tl_{0,97}Ag_{0,03}InSe_2$ presented in Fig. 3 *a, b, c,* both in the absence of uniaxial deformation and in its presence (stretching and compression along the crystallographic axis [001]) has allowed determining the energies of impurity levels ionization. The ionization energy in crystals *p*-$TlInSe_2$ in the absence of deformation is ΔE_α = 1.21 eV, which is quite well consistent with the value of the energy of indirect transitions $E_{ind.}$= 1.22 eV at 300 K for these crystals [1].

Substitution of monovalent thallium (Tl) atoms with silver (Ag) atoms causes shift of indirect transitions to the long-wave zone. Stretching deformation along the crystallographic axis [001] shifts the energy of indirect transitions to the short-wave zone, while compression deformation, on the contrary, shifts the energy to the long-wave zone. The obtained results are presented in Table 1. As can be seen from Table 1, the values of baric coefficients of change in the indirect energy gap both in $TlInSe_2$ crystals and in the crystals of derived solid solutions at a room temperature fluctuate within:

$G^{[001]} = (3,19 \div 3,65) \cdot 10^{-10}$ eV/Pa

It is known that the strain-resistive properties of semiconductor materials are mainly determined by elastic constants of the material in the corresponding crystal-like directions. The higher the value of the Young's modulus (E_l), the higher the sensitivity of electric conductivity to deformation.

Table 1. Displacement of the long-wave edge of indirect energy gap in crystals $Tl_{1-x}Ag_xInSe_2$, both in the presence and in the absence of uniaxial deformation

$TlInSe_2$			$Tl_{0,99}Ag_{0,01}InSe_2$			$Tl_{0,97}Ag_{0,03}InSe_2$		
P=0, Pa	P = $-2.189 \cdot 10^8$, Pa P = $2.189 \cdot 10^8$, Pa		P=0, Pa	P = $-2.189 \cdot 10^8$, Pa P = $2.189 \cdot 10^8$, Pa		P=0, Pa	P = $-2.189 \cdot 10^8$, Pa P = $2.189 \cdot 10^8$, Pa	
E_i, eV	$E_{i\varepsilon}$, eV	$G^{[001]}$, V/Pa	E_i, eV	$E_{i\varepsilon}$, eV	$G^{[001]}$, V/Pa	E_i, eV	$E_{i\varepsilon}$, eV	$G^{[001]}$, V/Pa
1.21	1.14 1.28	$3.38 \cdot 10^{-10}$ $3.19 \cdot 10^{-10}$	1.2	1.13 1.27	$3.70 \cdot 10^{-10}$ $3.70 \cdot 10^{-10}$	1.2	1.05 1.25	$3.65 \cdot 10^{-10}$ $3.60 \cdot 10^{-10}$

The value of the Young's modulus in the crystallographic direction [001] for $TlInSe_2$ crystals makes E_l=$14,4 \cdot 10^{11}$ DIN/cm^2 [1] at 300 K, which significantly exceeds the corresponding values of materials known in semiconductor tensometry.

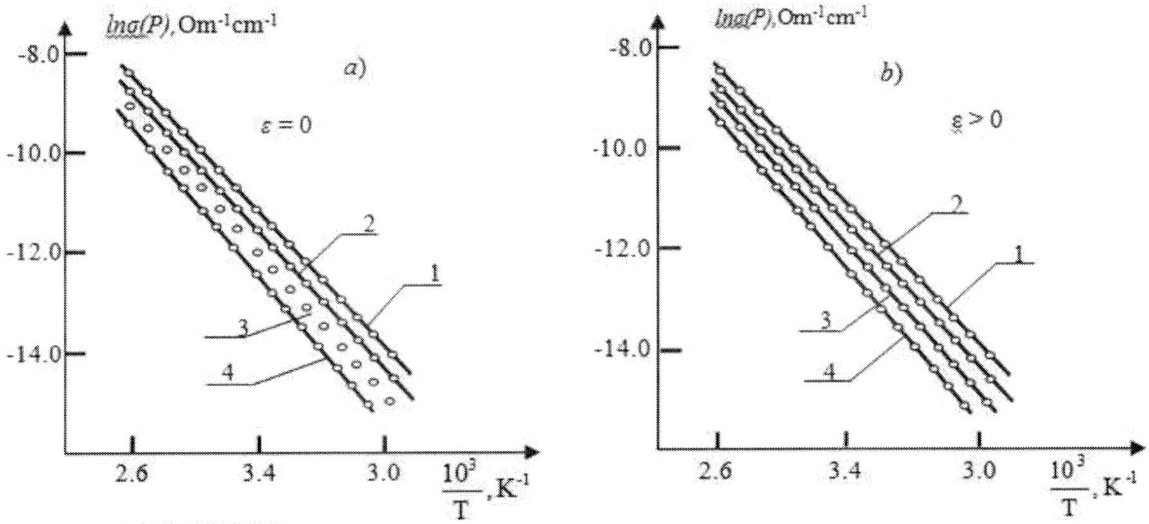

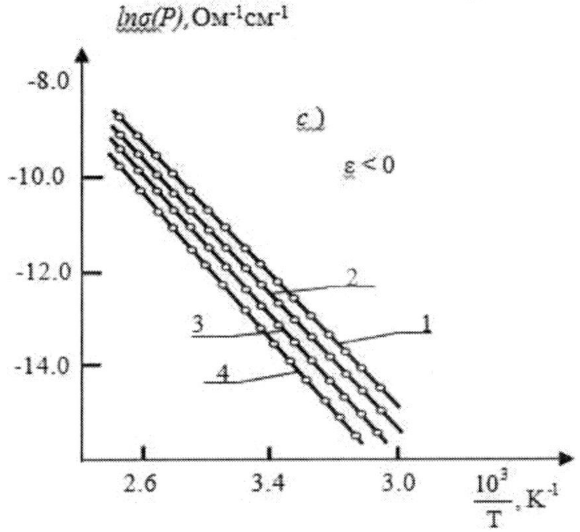

Figure 3. Temperature dependence of specific electrical conductivity of the crystals $Tl_{1-x}Ag_x InSe_2$: 1 – $TlInSe_2$; 2 – $Tl_{0,99}Ag_{0,01}Se_2$; 3 – $Tl_{0,97}Ag_{0.03}Se_2$; 4 – $Tl_{0,97}Ag_{0,03}Se_2$

Apparently, some of the main reasons for the observed relatively large piezoresistive effect in monocrystals of $TlInSe_2$ and derivative solid solutions are their complex energy structure, multi-linearity, sharp anisotropy of effective masses, mobility, etc.

As it follows from the theory of groups, in $TlInSe_2$ and TlSe crystals [1], surfaces of constant energy are three-axis ellipsoids of a general form. The observed tensor-resistive properties in crystals based on p-$TlInSe_2$ can be qualitatively explained on the basis of a 4-ellipsoid model (Fig. 4) [1] at the location of extremes at points $G(N_g$ =I ellipsoid), $T(N_T = 4\cdot\ 1/4 =$ I ellipsoid) and $N(N_N = 4\cdot 1/2 = 2$ ellipsoid) with elongated ellipsoid axes along the corresponding axes of symmetry (Fig. 3). The law of dispersion on planes for carriers in valleys has the form:

$$N - \varepsilon = \frac{h^2}{2}\left(\frac{K_x^2}{m_{//}} + \frac{K_x^2}{m_\perp}\right),$$

$$T - \varepsilon\left(\frac{K_x^2}{m_{//}} + \frac{K_x^2}{m_\perp}\right) \text{ and } G - \varepsilon = \frac{h^2}{2}\left(\frac{K_x^2}{m_{//}} + \frac{K_x^2}{m_\perp}\right) \qquad (4)$$

$m_{//}$ and $m_\perp$ are the longitudinal and transverse effective masses according to the axes of the ellipsoids. Corresponding mobilities at this:

$$\mu_{//} = \frac{q\tau}{m_{//}} \quad \mu_{\perp} = \frac{q\tau}{m_{\perp}} \tag{5}$$

If the total number of carriers in the valleys is n_0, then the number of carriers in the valleys G, T and N will be $n_0/4, n_0/4, 2 \cdot n_0/4$, respectively.

The contribution to conductivity from the respective valleys in a direction parallel to the tetragonal axis will be:

$$\sigma_g^{//} = e\frac{n_0}{4}\mu_{//};\ \sigma_T^{//} = e\frac{n_0}{4}\mu_{//};$$

$$\text{and } \sigma_N^{//} = e \cdot 2\frac{n_0}{4}\mu_{\perp}. \tag{6}$$

Thus, the total electrical conductivity along the crystallographic axis [001]:

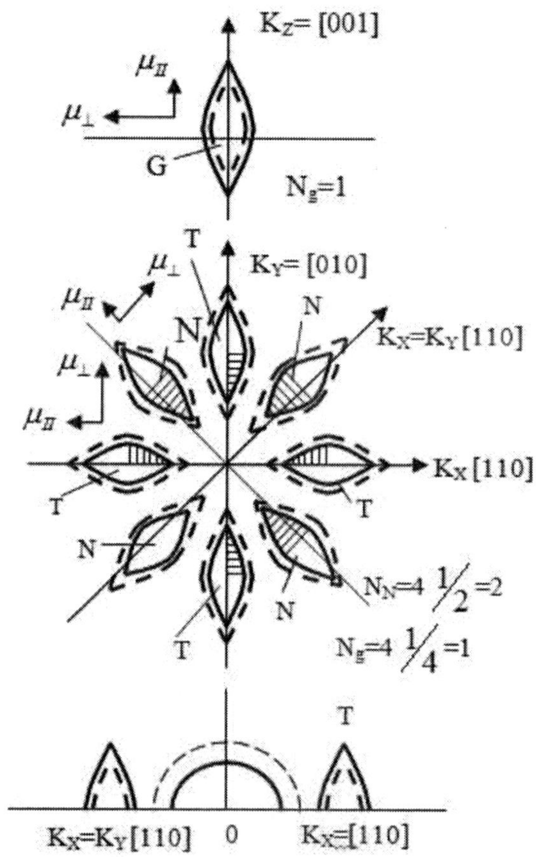

Figure 4. Four-ellipsoid model $TlInSe_2$

However, one-sided deformation of crystals along the axis [001] will change the ratio of the number of "heavy" (with mobility $\mu_{//}$) and "light" (with mobility $\mu_{\perp}$) holes, taking part in the electrical conductivity in this direction. This change, caused by the flow of carriers from valley to valley, apparently, causes tenzo- or piezoresistive effect. When stretching $TlInSe_2$ crystals and derived solid solutions along the axis [001], the relative number of heavy holes,

$$\sigma_0 = \sigma_Z = en_0\left(\frac{3\mu_{\perp} + \mu_{//}}{4}\right) \tag{7}$$

moving in this direction will increase. Therefore, the conductivity will decrease. At compression, however, the conductivity will increase. The reason for redistribution of carriers in the valleys is obviously the fact that the extremes with respect to the unilateral deformation are in different conditions. Stretching crystals of the type $TlInSe_2$ along [001] inevitably causes compression in the direction of [001] and [010]. As a result, the extremum Γ on the axis [00l] is raised, and the corresponding extremes T and N on the axes [001] and [010] are lowered (dotted lines in Fig. 4). Thus the contribution of the respective valleys to electrical conductivity in the direction [001] changes due to the flow of carriers from $T(\frac{\Delta n}{3}/4)$ and $N(\frac{2\Delta n}{3}/4)$ into the valleys $G(\frac{\Delta n}{3})$. Thus, the conductivity of crystals of both p-$TlInSe_2$ and crystals of derived solid solutions along the direction [00l] when they are stretched in the same direction:

$$\sigma_P^{//} = \frac{en_0}{4}(3\mu_\perp + \mu_{//}) + \frac{en_0}{4}(\mu_{//} - \mu_\perp) \tag{8}$$

Therefore, electrical conductivity changed due to stretching will be negative in this case:

$$\Delta\sigma_P^{//} = \frac{e\Delta n_0}{4}(\mu_{//} - \mu_\perp) < 0 \ , \tag{9}$$

since $\mu_\perp > \mu_{//}$. In case of compression:

$$\Delta\sigma_P^{//} = \frac{e\Delta n_0}{4}(\mu_{//} - \mu_\perp) > 0 \tag{10}$$

i.e. the conductivity along [00l] decreases under stretching and increases under compression, which is qualitatively consistent with the experiment results.

4. Conclusion

1. It has been found that substitution of Tl atoms with Ag atoms in crystals of solid solutions based on $TlInSe_2$ leads to a shift of the energy of indirect transitions to the long-wave zone.
2. We have determined the values of baric coefficients of changes in the width of forbidden values for uniaxial deformations of stretching and compression.
3. Stretching along the crystallographic axis [001] for the studied crystals shifts the energy of indirect transitions to the short-wave zone, while compression, on the contrary, to the long-wave zone. This is also confirmed by the data on temperature dependence of electrical conductivity and finds a qualitative explanation on the basis of the 4-ellipsoid model.

References

[1] Matiev A Kh 2005 *Phase equilibria and electron-optical properties of systems T1B3C62-A1B3C62 (A-Cu, Ag; B-In, Ga; C-S, Se2)* PhD Thesis
[2] Mustafaeva S N, Asadov M M et al 2014 Dielectric properties and charge transfer in TlInSe2) 0.1(TlGaTe2)0.9 at direct and alternating current *Phys. of Solids* **56(6)** 1055–59
[3] Hosokawa Sh, Stellhorn J R et al 2018 Lattice Distortions in TlInSe2 Thermoelectric Material Studied X-ray Absorption Fine Structure *Phys. status solidi (a)* 215
[4] Sheleg A U, Shevtsova V V et al 2013 Low-temperature X-ray studies of monocrystals TlInS2, TlGaS2 and TlGaSe2 *Surface. X-ray, synchrotron and neutron Studies* **11** 39–42
[5] Matiev A Kh, Uspazhiev R T et al 2019 Phasing Diagrams of TlInSe2 – CuInSe2 and TlInSe2 – AgInSe2 Systems *Atlantis Highlights in Materi. Sci. and Technol.* **1** 249–52
[6] Nadzhafov A I, Madatov R S et al 2015 Features of electrical conductivity of TlInSe2 at photo- and x-ray excitations *Phys. and technol. of semiconduct.* **49(9)** 1901–4
[7] Hosokawa Sh and Stellhorn J R 2017 Lattice distortions in TlInSe2 thermoelectric material studied by x-ray absorption fine structure *Phys. status solidi (a)* **1** 215
[8] Mustafaeva S N and Asadov M M 2015 Conductivity and thermo-EMF of crystals (TlInSe2)02 (TlGaTe2) 0.8 *Non-organic mater.* **51(3)** 267–71

[9] Mustafaeva S N 2010 Influence of frequency on the electrical and dielectric properties of monocrystals (TlGaS2)1-x (TlInSe2) x (x = 0.005, 0.02) *Non-organic mater.* **46(2)** 145–8
[10] Mustafaeva S N 2013 Frequency dependence of real and imaginary parts of complex dielectric permittivity and conductivity of monocrystal TlInSe2 in relaxation processes *J. of radio electr.* **7** 1–8
[11] Mustafaeva S N and Asadov M M 2013 Dielectric and thermoelectric properties of crystals based on initial compounds of TlInSe2–TlGaTe2 system *Condensed matters and interphase boundaries* **15(2)** 150–5
[12] Mustafaeva S N and Asadov M M 2014 Dielectric properties and charge transfer in TlInSe2) 0.1(TlGaTe2)0.9 at direct and alternating current *Phys. of Solids* **56(6)** 1055–9
[13] Madatov R S and Nadzhafov A I et al 2015 Features of electrical conductivity of TlInSe2 at photo-and x-ray excitations *Phys. and technol. of semiconduct.* **49(9)** 1202–5

IOP Conf. Series: Materials Science and Engineering **905** (2020) 012048 doi:10.1088/1757-899X/905/1/012048

Technology for the formation of refractory metals for micro- and nanoelectronics products

G A Mustafaev[1], A I Khasanov[2,3], N V Cherkesova[1] and A G Mustafaev[4]

[1] Kabardino-Balkarian State University, 173, Chernyshevsky St., Nalchik, Russia
[2]Chechen State University, 17аб Dudayev Boulevard, Grozny, 364060, Russia
[3] Kh. Ibragimov Complex Institute of the Russian Academy of Sciences, 21A, Staropromyslovsky ave., Grozny, Russia
[4] Dagestan State University of National Economy, building 43-a, Gadzhiyev Str., Makhachkala, Russia

E-mail: zoone@mail.ru, aslan2001@rambler.ru, natasha07_2002@mail.ru, arslan_mustafaev@mail.ru

Abstract. Metal silicides play a significant role in the preparation of ohmic contacts and Schottky barriers on silicon. The formation of ohmic contacts is carried out by applying a metal film with a thickness of ~ 100 nm to silicon, followed by annealing at temperatures of 400–600 °C, as a result of which there is a reaction between silicon and metal with the formation of silicide. With this technology, silicon diffusion leads to instability of devices. In this regard, the authors developed an improved technology for the formation of silicides of refractory metals to obtain ohmic contacts. Ion implantation allows, due to ionic mixing, forming nickel silicide on the surface of the samples when heated. It is important that, with an increase in the radiation dose, the formation of Ni2Si silicide slows down. This effect can be explained by the formation of a dielectric Ni3Si2. Similar phenomena were observed during the bombardment by oxygen and nitrogen ions.

1. Introduction

One of the methods for the production of silicides is the deposition of thin films of platinum and silicon, and the synthesis of alloys. The dissociation of SiH_4 molecules into elemental silicon and various molecular fragments leads to the deposition of silicon on the discharge electrode and its surrounding surfaces. The degree of dissociation depends on the output power of glow discharge. In the process of reactive sputtering, in which reactive gas dissociates into solid products, the stoichiometry of the films depends on the deposition rate of the target material and the substrate, re-sputtering from the target and the speed of sputtering of target material [1–2].

Thus, at low powers and high partial pressures of the reactive gas, the dissociation rate will be sufficiently high, so that the target is completely covered by the sputtering product. At high powers and low partial pressures, the sputtering rate of the target material exceeds the dissociation rate of the reactive gas, which leads to sputtering of the target material. The necessary stoichiometry of the deposited substance is obtained by selecting the atomization power and pressure of the reactive gas. Figure 1 shows the change in the Si/Pt concentration ratio in the film as a function of the SiH_4 flow rate for cathode voltages of 700 and 900 V.

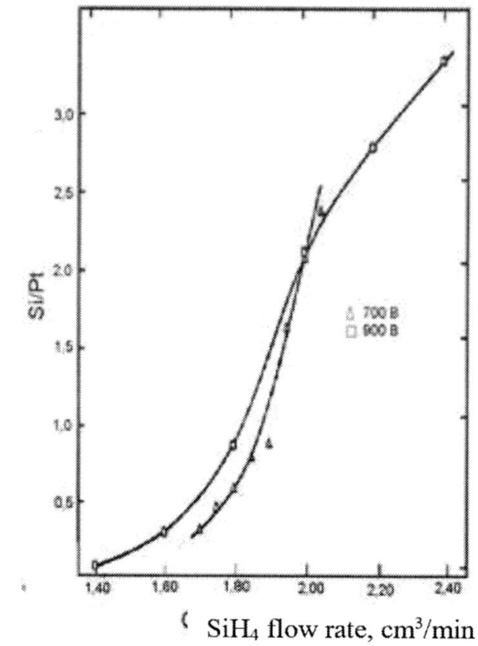

Figure 1. Ratio of Si / Pt concentrations in the film depending on the SiH4 flow rate for cathode voltages of 700 and 900 V

As it can be seen from Figure 1, at low powers, the Si/Pt ratio is very small at the beginning and rapidly increases at flow rates above 1.7. At flow rates greater than 2.2, platinum was not detected in the films. At high powers, the change in the Pt/Si ratio depending on the silane flux occurs more smoothly. This indicates that more precise control over stoichiometry can be carried out at high atomization powers.

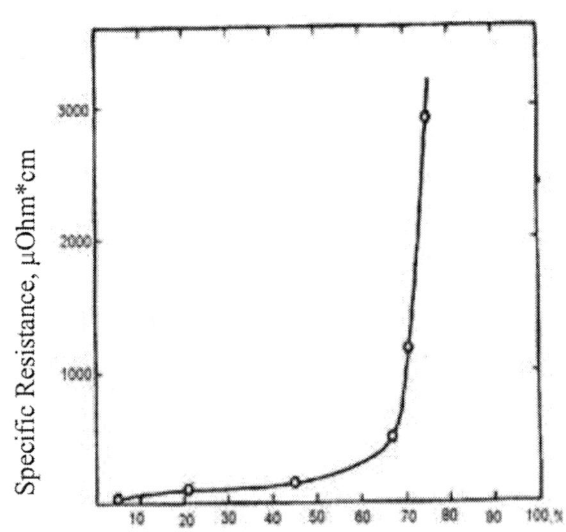

Figure 2. Dependence of resistivity on silicon concentration for films deposited at 20 ° C and 900 V at the cathode

The change in resistivity as a function of silicon concentration for films deposited at room temperature and a voltage of 900 V at the cathode is shown in Figure 2. The specific resistance of the platinum film is ~ 22 μm cm and increases linearly with increasing silicon content up to 65 %. Above 65 %, a sharp increase in resistance is observed, which indicates semiconductor conductivity. This result is consistent with x-ray diffraction and electron microscopy.

Silicides of refractory metals also attract attention as a material for metallization of IC [3, 4]. Metal disilicides Ti, Nb, Ta, Mo, W are low-resistant, their potential use is as gate electrodes in VLSI. Replacing polycrystalline Si in gate metallization with low-resistance refractory metal silicides can get around the limitations inherent in high-resistance doped polysilicon. Refractory silicide films deposited on polysilicon act as shunt layers for polysilicon gate electrodes. In order to simplify metallization schemes, these multilayer structures can be replaced by silicide films.

Refractory metal silicides are prepared on silicon substrates from metal halides, reduced by H_2 or Si to form refractory metals (for example, Mo, V or Ti); then the metal is converted into metal disilicide by diffusion of atoms at sufficiently high temperatures (in this case, it is necessary to exclude the reduction of the metal halide with silicon in order to limit the erosion of the silicon substrate).

Silicon refractory metals can also be obtained by direct diffusion of atoms from a silicon layer chemically deposited from the vapor phase into a substrate of refractory metal, such as Mo, at temperatures of 1000–1200 °C. The high temperatures necessary for the recovery of metal halides and interdiffusion of elements lead to mechanical stresses and the formation of voids at the interface between the silicide and the substrate. In addition, the above processes for the preparation of metal silicides can be used only on substrates made of silicon or refractory metal.

In the self-aligned MOS technology for the formation of refractory metal silicide films in the source and drain regions, metal films are deposited by physical methods (with subsequent interdiffusion of the metal and silicon at high temperatures). An alternative method involves the direct deposition of refractory metal silicide films based on the reaction of a silicon-containing gas (silane, chlorosilane, silicon tetrachloride) with a gaseous metal compound. In order to control the composition of films grown by chemical vapor deposition, the thermal stability of the gaseous metal compound and the silicon-containing molecules must be sufficiently high.

2. Experiment

Refractory silicides are most compatible with conventional MOS technology, and the most suitable for interconnects and contacts are the metal disilicides Mo, Ta, Ti and W. These conductive materials, chemically deposited from the vapor phase, have the necessary properties for their use in gate electrodes.

Titanium silicide films were chemically deposited on a silicon substrate from gas mixtures of TiCl4-SiH4 or TiCl4-SiH4-Ar at temperatures of 600–850 °C. The specific resistance of these films is very close to the volume value of 10 μм*cm. Chemical deposition of $TiSi_x$ films (x = 1) from the vapor phase in the plasma was carried out at 450 °C in a system with a hot wall. Polysilicon structures were grown on doped layers from an amorphous silicon layer 100 nm thick, TiSiI, 1 120 nm thick, and an amorphous silicon layer 15 nm thick. Processing in plasma helped to clean the interface between the substrate and the film. In these multilayer structures, the lower layer of amorphous silicon serves as a source of excess silicon during subsequent heat treatment, and the upper layer of amorphous silicon protects amorphous titanium silicide from oxidation. Thermal annealing of these structures at 650 °C leads to the formation of $TiSi_2$ layers with a specific resistance of 14 μм*cm. Such a process can be used to form a polycide (polysilicon silicide) gate electrode of titanium silicide deposited on polysilicon in the same cycle in which polysilicon was doped.

In order to precipitate tungsten silicide from WF_6-SiH4 mixtures in a low-pressure reactor with a cold wall, reactive gases are introduced into the reactor by separate injection and mixed near the surface of the substrate to minimize bulk reactions resulting in a powdery material. Tungsten silicide films were prepared from a gas mixture of SiH_4-WF_6-He at temperatures of 350–450 °C and a pressure of 7–40 Pa. The growth rate varied from 50 to 100 nm/min and does not depend on the substrate temperature, pressure, and SiH_4 flow rate.

IOP Conf. Series: Materials Science and Engineering **905** (2020) 012048 doi:10.1088/1757-899X/905/1/012048

The composition of the film is determined by the ratio of the flow rates of SiH_4-WF_6. They are enriched in silicon (composed of a mixture of Si + WSi_2), and become polycrystalline after annealing at a temperature of 900–1000 °C. The specific resistance is 40–50 μm*cm. The surface resistance of a Si-enriched tungsten silicide film with thickness of 250 nm chemically deposited directly into a gate oxide 65 nm thick is 1–4 Ohm/m, depending on the temperature and annealing time.

With the transition of MMSII technology of devices to the submicron range of element sizes, in order to increase the speed and functional density, the value of the series resistance of the structures increases.

In this regard, it is necessary to use self-aligned silicide contacts, which reduce the resistance of the gate, n^+ and p^+ –regions of the source and drain by 1–2 orders of magnitude. It is necessary to take into account that the formation of such contacts is not just an additional operation to the proven technology; in this process, during the formation of the silicide, a certain layer of silicon is absorbed, together with the doping impurity contained in it – the thickness of the absorbed Si layer is 1.5 times greater than the layer deposited participating in the Ti reaction. As a result, it is possible to reduce the impurity concentration near the $TiSi_2$ / Si interface and, as a result, the unacceptable resistance of the contact layer. At the same time, if the silicide layer extends to the depletion region near the edge of the oxide insulation, transition leaks may occur.

The formation of $TiSi_2$/Si contacts on n^+ and p^+ material was studied for various values of the Ti layer thickness in shallow areas of the source and drain. The effective contact resistance between the n^+ region and the silicide layer is determined by two equipotential surfaces; it is due to the finite value of the specific resistance at the $TiSi_2$/Si interface and it also includes the Si resistance under silicide.

Figure 3 shows the dependence of the contact resistance in the MOS structure on the diffusion depth of As in Si for different thicknesses of the Ti layer.

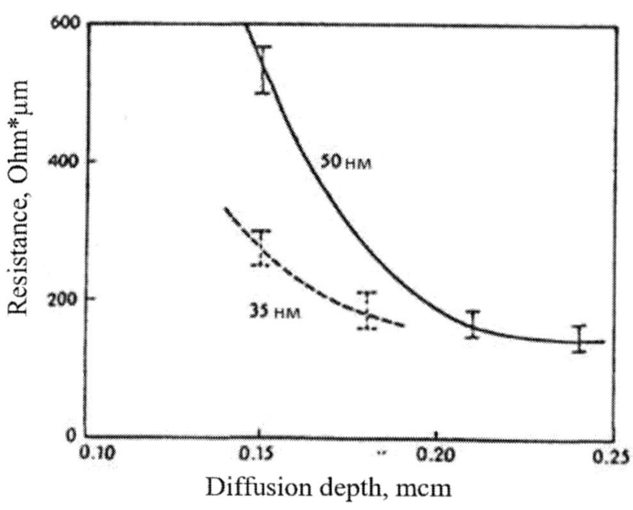

Figure 3. Dependence of contact resistance in the MOS structure on the depth of diffusion of As into silicon for different Ti layer thicknesses

The dependence of the contact resistance in the MOS structure on the diffusion depth of As in Si for different Ti layer thicknesses characterizes the contribution of the contact resistance to the series resistance of the device; both curves were determined for equal doses of As implantation ($2*10^{15}$ cm^{-2}), followed by heat treatment at 900–950 °C for 30–60 min to obtain different transition depths. With a Ti layer thickness of 50 nm, the resistance of the $TiSi_2$ layer is 4 Ohm/m, and with a transition depth of more than 0.30 μm, a low contact resistance is achieved. For a transition depth of 0.15 μm, the contact resistance is significantly higher.

The increase in the contact resistance with a decrease in the depth of the source and drain regions is reasoned by the relatively large thickness of the Si surface layer absorbed during the formation of $TiSi_2$ (i.e., a decrease in the final thickness of these regions) and can cause significant degradation of the current – voltage characteristics and dynamic conductivity at a low channel resistance (especially the submicron channel lengths), in the case of low voltage at the drain and high voltage at the gate.

At a Ti layer thickness of 50 nm, the As concentration near the $TiSi_2$ / Si interface is only $3*10^{19}$ cm^{-3}. The contact resistance can be decreased reducing the thickness of the Ti layer to 35 nm, then the As concentration near the interface will exceed 10^{20} cm^{-3}. At a diffusion depth As of 0.15 μm and a Ti layer thickness of 35 nm, a silicide layer resistance of 6 Ohm/m and a pair of contacts of no more than 300 Ohm*μm were obtained. This is quite satisfactory when compared to a minimum channel resistance of 2000 Ohm*μm in the linear region for a MOS FT with an n-channel with the length of 0.3 μm.

The resistivity of the contacts on the n^+ – layers is about 3 times lower than on the p^+ – layers; this may be due to the difference in the effective mass of the electron and the hole, since the work function for $TiSi_2$ corresponds to the same barrier height on n^- and p^-type Si layers.

Molybdenum is most acceptable in terms of adhesion to the silicon dioxide layer. In the case of thermal burning of Mo into Si, the natural oxide on the Si surface causes the process carried out at elevated temperature and impairs the reproducibility of the silicide formation, and strong stresses arise around the periphery of such contacts. In order to lower the temperature of silicide formation, interstitial ions of various elements were used, which caused Mo to be mixed with Si.

The use of inert gas ions for mixing leads to the formation of bubbles in the contact area that impair the reliability of the contacts. If mixing is performed using B, P or As, then simultaneously with the formation of the silicide, substrate doping is achieved. Mo/Sl contacts were formed for the samples created by ion implantation of phosphorus ions P^+.

The studies were performed on Si p-type plates (4.5 Ohm*cm) with an orientation of (100). Mo deposition with a thickness of 100 nm was carried out at a substrate temperature of 3000 °C using a magnetron source. Ion introduction of phosphorus P with a dose of 10^{16}–$10^{17}$$cm^{-2}$ was performed at a substrate temperature of 25, 150 and 300 °C, energy of 250 keV. X-ray diffraction analysis showed that at a dose of phosphorus ions P up to 10^{16}–10^{17} cm^{-2}, reproducibly ensures the formation of the $MoSi_2$ phase. At a dose of 1016 cm^{-2}, Mo is not completely absorbed during the formation of the silicide, and unreacted Mo remains on top of the $MoSi_2$ layer; at a dose of 10^{17} cm^{-2}, all Mo is spent on the formation of silicide.

The measured values of the contact resistance for different conditions of the ionic introduction of phosphorus ions P made by the four-probe method [5] on the samples showed that the contacts have a resistance of ~ 8 Ohms. These results confirm the possibility of reproducible formation of low-resistance contacts at doses of 10^{16}–10^{17} cm^{-2}.

After ion-doping, high-temperature annealing was performed to electrically activate impurity atoms and recrystallize the damaged Si surface layer. In the work, the samples that underwent ionic introduction of phosphorus P ions at a temperature of 25 °C were additionally annealed at 1000 °C in a H2 hydrogen atmosphere for 30 min. At doping doses of 10^{16}–10^{17} cm^{-2}, the contact resistance changed insignificantly. Annealing at 850 °C (also 30 min in a hydrogen H_2 atmosphere) for a sample with a dose of 10^{17} cm^{-2} did not cause the surface roughness of the silicide layer to appear, whereas without ionic introduction of phosphorus, the surface became rough.

Atomic mixing by the action of an ion beam of phosphorus with large doses creates the possibility of the formation of MoSi2 at a significantly lower temperature. The ion beam causes cascade mixing in the surface layer of the sample and diffusion accelerated by irradiation; in the first process, atoms are mixed during collisions, and in the second, diffusion is accelerated as a result of the increase in the density of vacancies and the concentration of interstitial atoms. Since the main effect of increasing the dose of ionic implantation P and the temperature of the sample during this treatment is the increase in grain size, accelerated diffusion can be considered as the main process, which can lead to migration of grain boundaries.

IOP Conf. Series: Materials Science and Engineering **905** (2020) 012048 doi:10.1088/1757-899X/905/1/012048

In the research, ion bombardment was also used to improve the uniformity of the properties of Ni2Si films. Nickel silicide films were formed on Si wafers, the surface of which was previously subjected to cleaning by various methods [6].

Immediately after cleaning, the plates were placed in a spraying unit. Ni films 30–50 nm thick were deposited in a vacuum of 10^{-5}–10^{-6} Pa at a rate of 2 nm/s. Then the samples were irradiated with xenon ions with energy of 300 keV at room temperature or at a temperature of 170 °C. The radiation doses were 5–10^{13}, 10^{14}, 5–10^{14} and 10^{15} cm^{-2}.

Next, the samples were placed in a vacuum oven to form nickel silicide. Ni films were annealed for 20 min at a temperature of 250 °C. An identical procedure was carried out for the case of 180 keV arsenic ion bombardments.

A study of backscattering spectra showed that immediately after xenon ion bombardment, the effect caused by it is noticeable only for a dose of 10^{15} cm^{-2}. At lower doses, the backscattering spectra corresponded to unirradiated samples. In contrast, after annealing, the effect was noticeable in the samples irradiated with small doses. After annealing the irradiated sample, a Ni2Si layer is present in it, which is more uniform than in the sample subjected only to heat treatment.

3. Conclusion

The experiment showed that the formation of a Ni_2Si layer also occurs when arsenic ions are used for implantation, and then the samples are subjected to heat treatment.

Ion implantation allows (due to ionic mixing) forming nickel silicide on the surface of the samples under heating, regardless of the type of previous surface cleaning [7, 8]. It is significant that, with the increase in the irradiation dose, the formation of Ni_2Si silicide slows down. This effect can be explained by the formation of a dielectric Ni_3Si_2. Similar phenomena were observed during the bombardment by oxygen and nitrogen ions.

References

[1] Stavitski N, van Dal M J H, Lauwers A et al May 2008 Evaluation of Transmission Line Model Structures for Silicide-to-Silicon Specific Contact Resistance Extraction *IEEE Transact. on Electr. Devices* **55(5)** 1170–6

[2] Lin C, Xiao Yi-Hsuan and Tsui Bing-Yue March 2005 High-performance poly-Si TFTs fabricated by implant-to-silicide technique *IEEE Electr. Device Letters.* **26(3)** 185–7

[3] Larrieu G, Yarekha D A, Dubois E et al Dec. 2009 Arsenic-Segregated Rare-Earth Silicide Junctions: Reduction of Schottky Barrier and Integration in Metallic n-MOSFETs on SOI *IEEE Electr. Device Letters* **30(12)** 1266–8

[4] Jin Ch-B, Kim Ch-J and Jo M-H 2006 Growth of nickel silicide nanowires by chemical vapor deposition *IEEE Nanotechnol. Mater. and Devices Conf.* (Gyeongju) pp 200–1

[5] Tokarska M June 2014 Evaluation of Measurement Uncertainty of Fabric Surface Resistance Implied by the Van der Pauw Equation *IEEE Transact. on Instrumentat. and Measurement* **63(6)** 1593–9

[6] Knoll L, Zhao Q T, Habicht S et al April 2010 Ultrathin Ni Silicides With Low Contact Resistance on Strained and Unstrained Silicon *IEEE Electr. Device Letters* **31(4)** 350–2

[7] Mustafaev G A, Khasanov A I, Mustafaev A G and Cherkesova N V 2018 Thin Layers of Oxide Coating in Very-Large-Scale Integration *Proceedings of the Int. Symp. ISEES 2018* (Atlantis Press) pp 397–9

[8] Yang Ts.-H., Luo G., Chang E.Y et al Sept. 2003 Study of nickel silicide contact on Si/Si1-xGex *IEEE Electr. Device Letters* **24(9)** 544–6

IOP Conf. Series: Materials Science and Engineering **905** (2020) 012049 doi:10.1088/1757-899X/905/1/012049

Magnetic-gravity separation in the simulation modeling paradigm

V G Nikolaev[1], V V Biryukov[1] and R M Nikitin[2]

[1] Department of Physics, Biology and Engineering Technology, MASU branch in Apatity, 29, Lesnaya str., Apatity, 184209 Russia
[2] Laboratory of new enrichment processes and apparatus, Mining Institute KSC RAS, 24, Fersmana str., Apatity, 184209 Russia

E-mail: nikviktor1@yandex.ru, birukovval@rambler.ru, remnik@yandex.ru

Abstract. An interdisciplinary approach to the simulation of magnetic gravity separation is proposed. The approach is based on revealing the laws of motion of mineral particles taking into account their magnetic properties in a computational experiment and establishing the form of an empirical multidimensional function for calculating the deposition rate of aggregated ferromagnetic particles in a multiphase multi-speed process medium. Based on the results, a comparative assessment of the convergence of the obtained deposition kinetics parameters with respect to the practice of applying the classical Stokes formula is given. The possibility of interpreting the fractional separation characteristics of the magnetic-gravity separator using the error function is shown. A method for integrating the simulation model of magnetic-gravity separation into an analytical prototype of the technological enrichment scheme is considered, for which the basic principles of constructing technological schemes for mineral processing are presented and the principles of constructing their simulation models are formulated.

1. Introduction

With the development of methods and practices of mathematical modeling, the need and the possibility of involving modern information technologies in solving problems in the field of mineral processing became obvious. These include: accumulation in specialized databases of information accumulated during many years of research on the geological and mineralogical composition of ore raw materials, physical, chemical and physico-chemical properties of the minerals included in its composition; creation of specialized knowledge bases on methods for processing mineral raw materials with known physical, chemical and physico-chemical properties; improvement of existing technological equipment and the development of new separation apparatus using specialized CAD-CAM-CAE complexes; development of models and creation of automated control systems for technological devices and technological schemes of enrichment enterprises; simulation of technological schemes for determining more efficient modes of production processes [1].

One of the methods for obtaining and using knowledge about physical and technological processes is modeling. The type of mathematical model depends both on the nature of the real object and the tasks of researching the object, the required reliability and accuracy of solving this problem. A special type of mathematical models are simulation models – computer programs that describe the structure and reproduce the behavior of a real system over time, allowing you to obtain detailed statistics on the functioning of the system depending on the input data. These include: empirical (a set of algebraic equations obtained on the basis of the dependences of internal statistics and the input and output data of the system); phenomenological (a set of algebraic or differential equations obtained from engineering, physical and (or) chemical positions and requiring calibration (material balance model)); fundamental (a set of algebraic or differential equations based on the fundamental laws of physics and chemistry and requiring minimal calibration) [2].

2. Computational Experiment

The essence of magnetic-gravity separation is the separation of ferromagnetic particles by size and magnetic susceptibility in an upward swirling liquid flow with the application of a vertical weakly inhomogeneous constant magnetic field. In the upward flow of water, a fluidized layer of particles of the initial suspension is formed, which differ in size, shape and content of the useful component magnetite, which leads to their distribution in density and magnetic susceptibility. In fluidized beds, solid particles move randomly and at the same time intensively mix, which impairs their segregation and reduces the separation selectivity. When a vertically oriented constant magnetic field is applied, ferromagnetic particles of the fluidized bed form vertically oriented chains—ferromagnetic aggregates that freeze the chaotic motion of the particles of the layer. The composition of ferromagnetic aggregates includes particles for which the interaction force of the magnetic moments of the particles exceeds the strength of their hydrodynamic resistance [3, 4].

The height of the magnetically stabilized fluidized bed of suspension particles depends on the difference in the mass flow of the feed of the initial suspension, wash water, drain and separator concentrate. Particles that are part of the ferromagnetic aggregates are discharged into the concentrator of the separator, the rest go into the drain.

Technological suspensions of enrichment plants are a mixture of finely ground mineral particles of various size fractions with water. Such a substance cannot be assigned to the class of liquids, gases, or solid deformable bodies and is defined as a heterogeneous medium. The separation of components of technological suspensions in enrichment (including magneto-gravitational) devices can be described using the mathematical apparatus of a multiphase multi-speed continuum (MMC), where the medium is considered as a superposition of interpenetrating continua, each of which belongs to its phase. Moreover, at each point in the medium, one can determine N densities, N velocities, etc., as well as the parameters characterizing the medium as a whole. Since the parameters of the phases and the whole medium vary in space and time, the substantial derivatives associated with the movement of the i-th phase and with the movement of the medium as a whole:

$$\frac{d_i}{dt} \equiv \frac{\partial}{\partial t} + \mathbf{v}_i \nabla\,; \qquad \frac{d}{dt} \equiv \frac{\partial}{\partial t} + \mathbf{v} \nabla\,.$$

With this in mind, the MMC theory is implemented with the balance relations of mass, momentum, and energy for each of the phases. The mass conservation equation for the i-th phase:

$$\frac{\partial \rho_i}{\partial t} + \rho_i v_i \nabla = \sum_{j=1}^{N} J_{ji}\,, \tag{1}$$

where J_{ji} characterizes the mass transfer intensity from the j-th to the i-th phase. The mass conservation equation for the whole medium:

$$\frac{\partial \rho}{\partial t} + \rho v \nabla = 0\,. \tag{2}$$

Momentum conservation equation for the i-th phase:

$$\rho_i \frac{d v_i}{dt} = \nabla^k \sigma_i^k + \rho_i g_i + \sum_{j=1}^{N} \left(P_{ji} - J_{ji} v_i\right), \tag{3}$$

where P_{ji} is the intensity of momentum exchange between the j-the and i-the phases, k is the upper index indicating the number of the Cartesian coordinate, σ is the surface force tensor, g is the mass force vector relating to the medium as a whole:

$$\sigma = \sum_{i=1}^{N} \sigma_i \,, \qquad \rho g = \sum_{i=1}^{N} \rho_i g_i \,.$$

The momentum conservation equation for the entire medium:

$$\rho \frac{d\mathbf{v}}{dt} = \nabla^k \sigma^k + \rho \mathbf{g} - \sum_{i=1}^{N} \nabla^k \left(\rho_i \omega_i^k \mathbf{w}\right) , \tag{4}$$

where ω_i is the radial velocity of the i-the phase at the interface and w is the velocity of relative motion of phases. The energy conservation equation for the i-th phase:

$$\rho_i \frac{d}{dt}\left(u_i + \frac{v_i^2}{2}\right) = \nabla(\mathbf{c}_i - \mathbf{q}_i) + \rho_i \mathbf{g}_i \mathbf{v}_i + $$
$$+ \sum_{j=1}^{N} \left[E_{ji} - J_{ji}\left(u_i + \frac{v_i^2}{2}\right)\right] , \tag{5}$$

where u_i is the internal energy of the i-the phase, $v_i^2/2$ is the kinetic energy of the i-th phase, c_i is the work of external forces, q_i is heat flux, E_{ji} is the energy exchange intensity between the j-th and i-the phases. The energy conservation equation for the whole environment:

$$\rho \frac{dE}{dt} + \sum_{i=1}^{N} \nabla^k \left(\rho_i \omega_i^k E_i\right) = \nabla^k \left(c_i^k - q_i^k\right) + \rho g^k v^k + \sum_{i=1}^{N} \rho_i g_i^k \omega_i^k \,. \tag{6}$$

Balance relations (1)–(6) show that the fundamental laws of conservation of mass, momentum, and energy, expressed analytically in the theory of molecular magnetic complexes, correspond to the equations of motion of the phases of a heterogeneous medium in the Euler approach. Moreover, since the phases of a heterogeneous medium, filling its entire volume, infinitely penetrate each other and retain their instantaneous volume, the necessary refinement and addition of the balance ratios is the introduction of the volume fraction of the phase. This value is a function of space and time, the algebraic sum of all volume fractions of the medium is equal to unity.

Thus, the mathematical apparatus of the MMC is ideally suited for simulating the motion of multiphase heterogeneous media in which the separation of mineral components during magnetic-gravity separation takes place and describes: representing a single technological mineral suspension in the form of a set of separate discrete fractions, each of which occupies its own volume fraction; fractions of particles with different densities, fineness and magnetic susceptibility within a single apparatus volume; change in velocities and concentrations of different fractions under forces of different physical nature, concentration, condensation and removal from the apparatus separating the products [5, 6].

The computer model that allows obtaining the values of the parameters of the magnetic-gravity separator was developed in the ANSYS Fluent software package. Figure 1 shows the geometry of the model.

The design includes a cylinder-conical housing with a drain device and a concentrate outlet pipe. A device combining the functions of supplying the initial suspension, supplying washing water and swirling it is placed in the apparatus body. Outside, the housing is surrounded by an electromagnetic solenoid, which creates a vertically directed weakly inhomogeneous magnetic field.

For the computer model, a source feeding model was developed (Table 1), which was based on a rough magnetite concentrate containing 67.84 % of total iron. In addition to magnetite particles, the initial feed included intergrowths of magnetite with weakly magnetic minerals of various density and size. In addition, slimy particles of magnetite and a non-magnetic mineral with a particle size of 0.15e-4 m were included in the initial feed. In total, 16 different particle fractions were formed. The feed rate was 0.5 kg/s or 1800 kg/h.

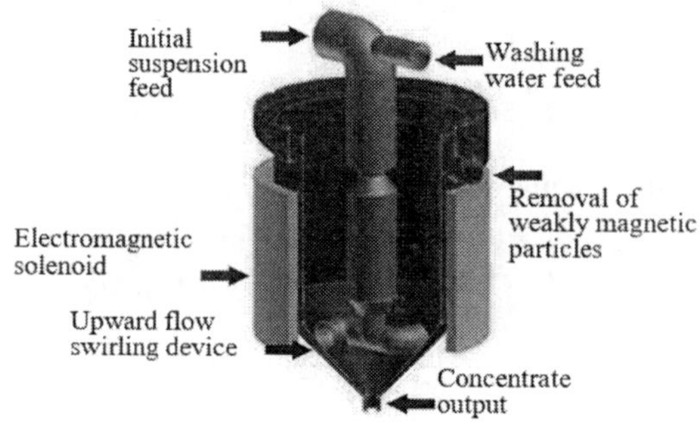

Figure 1. The geometry of the model of the magnetic gravity separator.

Computational experiments with the model made it possible to study the hydrodynamics of separation processes occurring in the apparatus. Figure 2 shows the distribution of volume concentrations of the liquid and solid phases of the suspension on a vertical section of the apparatus. It can be seen that solid particles form a magnetically stabilized layer with a pronounced boundary, which makes gravitational vibrations.

The analysis of the simulation results of the separation process of the components of the initial suspension enabled obtaining the technological characteristics of the magnetic-gravity separator. The total mass of solid particles of the fluidized bed amounted to 3623.6 kg. Table 2 shows the mass and mass content of individual fractions in the volume of the apparatus. The separator concentrate contains 70.6 of total iron.

Table 1. Initial feed parameters

Particle fraction size [m]	1.2e-4							0.85e-4			0.6e-4		0.4e-4		0.15e-4	
Magnetite mass content [%]	100	90	80	70	30	20	10	100	20	10	100	10	100	10	100	0
Fraction particles density [kg/m^3]	5100	4855	4610	4365	3385	3140	2895	5100	3140	2895	5100	2895	5100	2895	5100	2650
Fraction mass fraction in feed	0.157	0.018	0.014	0.01	0.002	0.004	0.006	0.158	0.004	0.014	0.195	0.01	0.19	0.01	0.198	0.006

Table 2. Parameters of fractions in the concentrate

Particle fraction size [m]	1.2e-4							0.85e-4			0.6e-4		0.4e-4		0.15e-4	
Magnetite mass content [%]	100	90	80	70	30	20	10	100	20	10	100	10	100	10	100	0
Fraction mass in separator magnetic fraction [kg]	663.62	298.53	164.14	50.55	0.005	0.008	0.104	947.79	0.007	0.024	796.19	0.015	442.41	0.016	259.98	0.01
Magnetic fraction mass content in separator volume	0.183	0.182	0.045	0.014	~0	~0	~0	0.262	~0	~0	0.22	~0	0.122	~0	0.072	~0

Magnetic-gravity separation is a promising separation process that allows selectively separating the components of ferromagnetic suspensions with similar magnetic and hydrodynamic properties, which ensures its use to control the quality of the concentrate, to enrich the crude magnetite concentrate and to obtain products suitable for nonblast-furnace steel production.

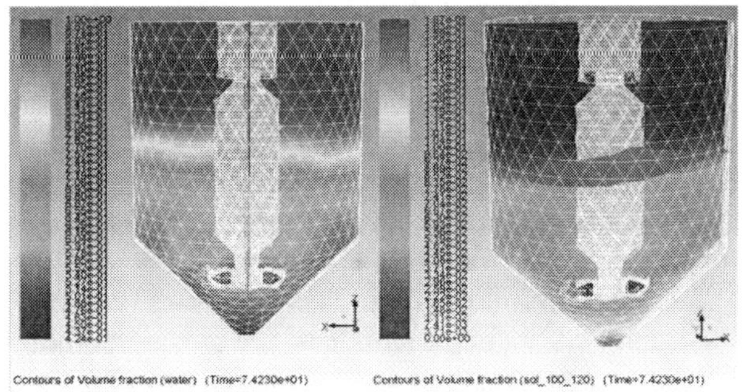

Figure 2. Distribution of volumetric fractions of water and solid particles fraction 1.2 e-4 m with a total iron content of 72.4 %.

3. Analytical aspect of transition to simulation model

Analysis of the simulation results showed that in the model of the apparatus there are three zones differing in their hydrodynamic characteristics: a downward flow zone, starting from the feed device of the initial suspension and ending in the cone of the device for swirling the upward flow of water; a zone of upward fluid flow, starting in the cone of the swirling device and extending in the wall region to the device for draining non-magnetic fractions; a turbulence zone located in a volume from a concentrate discharge device to a fluid swirling device. The turbulence zone is characterized by a sharp change in the direction of fluid movement, which contributes to a better washing of the magnetic fractions of the suspension moving in the direction of the concentrate discharge device.

In the axial zone of the downward flow, the rotational speeds are close to zero. Hydrodynamic zoning affects the distribution of concentrations of magnetic and non-magnetic fractions of the suspension in the volume of the apparatus model. In the axial region, from the initial slurry feeding device to the washing water swirling device, there is a downward flow of the initial suspension that enters the separation zone, which coincides geographically with the zone of existence of the magnetically stabilized fluidized bed. In the separation zone, non-magnetic fractions are carried upstream into the discharge of the apparatus, while fractions with high magnetic susceptibility structurize the magnetically stabilized layer and are deposited in the turbulent zone, where they undergo additional washing from non-magnetic fractions. The magnitude of the interparticle force in the dipole–dipole interaction depends on the angle of inclination of the ferromagnetic aggregate in the shear flow. In the turbulence zone, complete destruction of the ferromagnetic aggregates and the removal of thin non-magnetic fractions into the near-wall zone of the upward flow are observed.

Figure 3 shows the distribution of the downward velocity of the magnetic fraction flow.

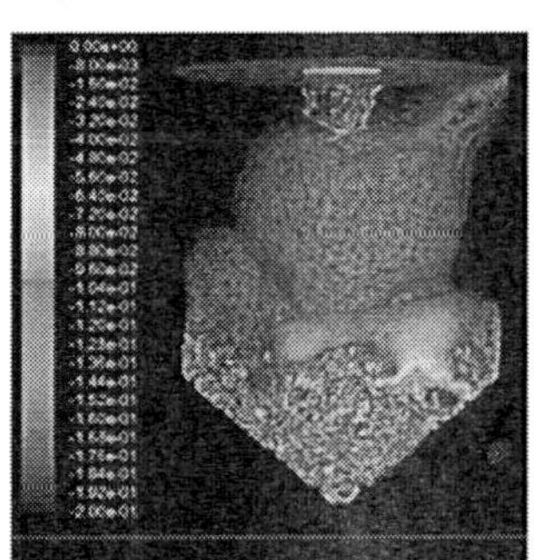

Figure 3. Distribution of the magnetic fraction velocity in the volume of the magnetic-gravity separator model

The increased viscosity of the ferromagnetic suspension in the region of existence of the magnetically stabilized layer leads to a decrease in the rotation speed of the washing liquid. Thus, a decrease in the fluid rotation rates leads to an increase in the magnetic force of the interparticle interaction with an increase in the height of the separation space. In the area of the drain device, the angle of deviation of the ferromagnetic unit from the vertical is close to zero. The strength of interparticle interaction in this area is maximum. At the same time, the upper section of the solenoid is at the level of the feed device of the initial product. In the center of the solenoid is a region of a uniform field. In the area of the drainage device, the magnetic field has a significant negative gradient, which, along with the interparticle magnetic force, creates an additional mass magnetic force that is aligned with the gravitational acceleration vector. The maximum values of interparticle and gradient magnetic forces prevent the removal of fine fractions of ferromagnetic particles, thereby increasing the separation selectivity in the apparatus.

Thus, as a result of a computational experiment, not only hydrodynamic, but also technological characteristics of the process were obtained, which made it possible to establish the form of the empirical function of the fractional separation characteristic:

$$\varepsilon(\rho) = \frac{1 + \mathrm{th}\left(\dfrac{\rho * \left(1000 + \dfrac{v}{1800 * d^2}\right.}{5.6 * 10^{-6} * H^2 - 0.133 * H + 900}\right)}{2} \quad (7)$$

where th is the function of hyperbolic tangent, ρ is the mineral density, v is the velocity of upward liquid flow, d is the particle diameter, H is the magnetic field strength.

The applicability of function (7) is limited by the absence of taking into account the deposition rate of aggregated ferromagnetic particles. The classical Stokes formula, in turn, does not allow establishing the dependence of the particle deposition rate on the intensity of an external magnetic field. Therefore, before using the mathematical editor Mathcad and then during the computational experiment, necessary clarifying analytical regularity needed to be identified. As a result, the following equation was derived:

$$V(d) = d^2 + \left(A + BC + SC^{3/2}H^2\right), \quad (8)$$

were V is deposition rate of a ferromagnetic particle in an upward fluid flow in a weakly inhomogeneous magnetic field, m/s;d is the particle diameter, m; C is the magnetite concentration in a particle, fractions of unit; H is the magnetic field strength, A/m; A, B, S are empirical coefficients depending on the magnetic susceptibility of the mineral, the degree of magnetization of ferromagnetic particles, the gradients of the magnetic field in the apparatus, the solid content in the original pulp, the geometric dimensions of the apparatus.

Comparative testing of the obtained formula for estimating the deposition rate of particles of binary ore of ferruginous quartzites was carried out for materials with pure quartz densities $\rho_1 = 2650$ kg/m^3 and a magnetite content of 10^{-5} and for materials with magnetite densities $\rho_2 = 5100$ kg/m^3 in the absence and presence of magnetic fields. As a result of testing, the data presented in the form of graphs in Figure 4 were obtained.

The deposition rate of aggregated ferromagnetic particles in a weak inhomogeneous magnetic field and an upward flow of water determines the entering of particles into the concentrate or into the separator discharge without taking into account the probabilistic characteristics of the separation process.

When the separator is working, elongated magnetite aggregates are formed. In this case, their drag coefficient changes when moving in water, the magnitude of which depends on the applied field. Magnetite aggregates are precipitated, the particles of suspension are condensed and discharged into the separator concentrate.

At a given feed rate of the wash water, the particles of the initial suspension differ in velocities relative to the apparatus body. Particles of small diameter and with a low concentration of magnetite and, therefore, having a lower deposition rate are discharged into the separator drain. Obviously, the relative particle deposition rate can be defined as

$$V_R = V(d) - V_{H_2O}.$$

At the same time, it is also obvious that there is a certain fraction of particles of magnetite suspension, which has a zero deposition rate and, therefore, is equally likely to be extracted both in the concentrate and in the discharge of the separator. To determine its size, one can use the following equation:

$$d_{50} = \sqrt{\frac{V_{H_2O}}{A + BC + SC^{3/2}H^2}}.$$

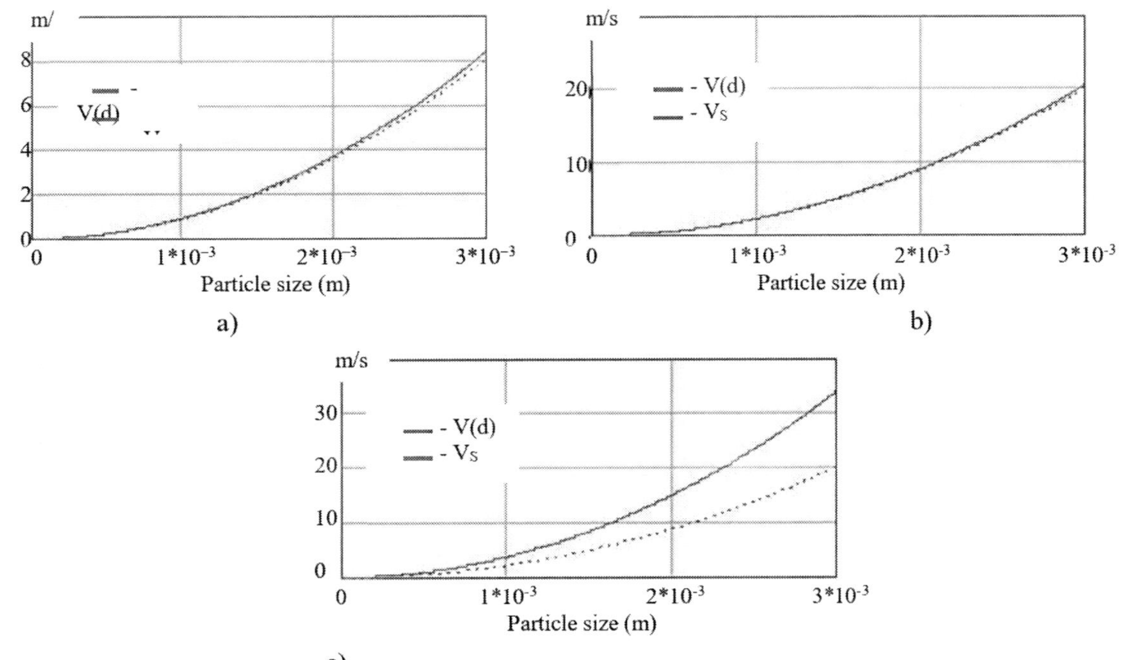

Figure 4. a) Dependence of deposition rates of particles with a density ρ_1 in water on their size, calculated by eq. (8), V (d), and according to the Stokes formula V_S; b) Dependence of the deposition rates of particles with a density ρ_2 in water on their size; c) Dependence of the deposition rates of particles with a density ρ_2 in water when applying a uniform magnetic field H = 4000 A/m on their size

This can be the basis for the principle of forming the separation characteristics of the separator, taking into account the deposition rates of aggregated ferromagnetic particles. The separation characteristic function gives the probability of extracting particles of a fraction of a certain diameter and with a certain concentration of magnetite into the separator concentrate. It is possible to determine the probability of removing the fraction of particles in the concentrate, while using the normal distribution function with a known mathematical expectation and a given variance—the function of the "probability integral"

$$E(d, c, H, V_{H_2O}) = \frac{\mathrm{erf}(Q(d - d_{50})) + 1}{2},$$

where erf(d) is the function of the probability integral; d_{50} is the diameter of the suspension particle fraction that is equally likely to be extracted into the discharge or the separator concentrate; c is magnetite concentration in a particle; H is the magnetic field strength; V_{H_2O} is the washing water feeding rate.

IOP Conf. Series: Materials Science and Engineering **905** (2020) 012049 doi:10.1088/1757-899X/905/1/012049

For example, in Figure 5, we give the separation characteristic of one of the narrow fraction of particles

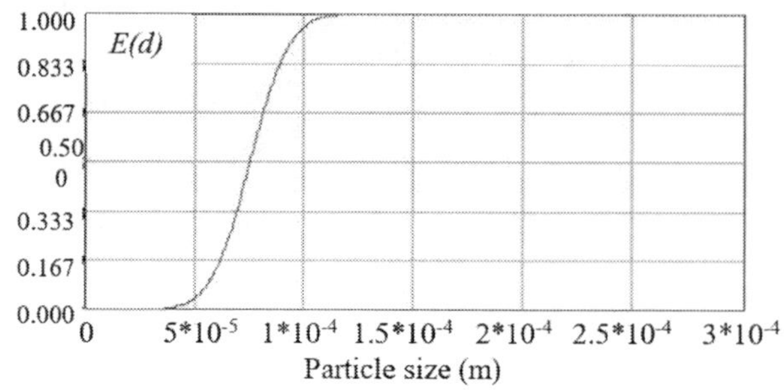

Figure 5. Separation characteristic of a magnetic-gravity separator for a fraction of suspension particles with a magnetite concentration of C = 0.5 with controlling parameters of $H = 4000$ A/m, $V_{H_2O} = 0.012$ m/s

4. Conclusion

The decisive influence on the efficiency of technological schemes for mineral enrichment is provided by their structure. Existing empirical methods for developing and monitoring the structure of technological schemes, especially in cases where a large number of types and units of technological equipment are used, do not allow reaching the expected result in terms of achieving the specified values of the criteria for optimizing separation processes. The relevance of this problem is exacerbated by the development of objectively manifest tendencies of involvement in the processing of depleted ores and the intensification of competition in the world markets of mineral resources [7, 8].

When developing models of structurally complex technological schemes for mineral processing, the principle of block structure is applied [9–11]. The allocation of blocks is made taking into account the separation of the model into stages and modes of operation. The division of the model into blocks, decomposition, can be carried out on the basis of the required degree of detail of the description of the structure, the visibility of the display of the features of functioning in it. In this regard, we can distinguish two basic principles for constructing technological schemes for mineral processing: cyclical and staged. They allow for functional (by function), component (by type of elements) and structural (by type of relations between elements) decomposition of the technological scheme, the minimum component (DMC) of which is the complex of the stage of size reduction and separation, which has, as a rule, one entrance and three outputs (Figure 6).

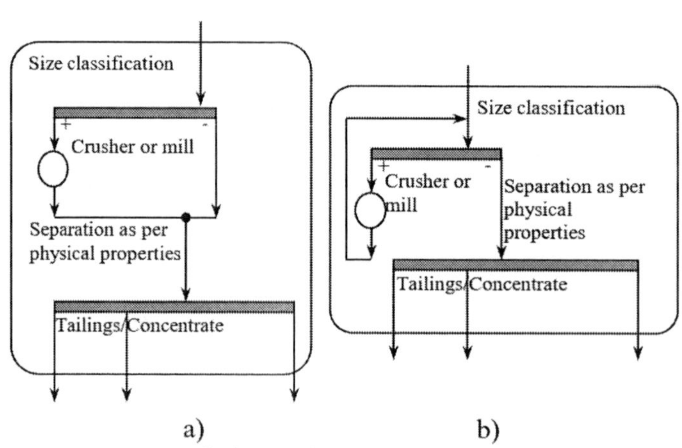

Figure 6. Mminimum component of the technological scheme decomposition. a) direct-flow component; b) with reverse flow at the stage of particle size reduction

The inputs and outputs of the DMC in the technological scheme are connected by material flows. Such a construction of a technological scheme in the form of series and parallel connected complexes allows establishing uniform input and output qualitative and quantitative technical and technological parameters based on the revealed patterns of their functioning.

The structure of any technological scheme of mineral processing is based on the possibility of using one or another separation process in it, based on the contrast of the physical properties (ξ) of solid particles. Therefore, it is the separation property, and not the content of the valuable component (β), that is responsible for the particle getting into the concentrate stream or, conversely, into the tail stream of the technological operation. Given the mass and variety of intergrowth forms of particles involved in the operation of mineral processing, direct and indirect, internal and external force factors, the separation result, expressed in numerical values of technological indicators, is probabilistic. Due to the fact that the intensity of the manifestation of the separation property varies in the aggregate of solid particles in a wide range $[\xi_{min}; \xi_{max}]$, it can be divided into smaller ranges $[\xi_i; \xi_{i+1}]$, while in the limit $[\xi; \xi + d\xi]$, referred to as fractions or, accordingly, elementary fractions. Then, it is obvious that if the ξ property, which varies in the range $[\xi_{min}; \xi_{max}]$, is inherent in the entire set of solid particles, then the range $[\xi; \xi + d\xi]$ will be inherent for every elementary fraction.

In this regard, the integration of the simulation model into an analytical prototype of the technological scheme is carried out by filling in the information structure of their direct and feedback links in a single, for each specific chain of technological ore-concentrate processing, system for monitoring the current material composition of mineral raw materials and products of its processing [12]. Therefore, as in the case with the construction of the technological scheme, the concept of constructing simulation models of mineral processing processes assumes the existence of principles that include the principle of universality of the model infrastructure, which consists in the fact that when building a digital model of the technological scheme of mineral processing, only coupled simulation models of processing processes can be used. In this context, the processing processes include preparatory, separation and auxiliary processes. Thus, simulation models should have a certain universality of the infrastructure; their coupling should be done by ensuring that the content and capacity of the information in the communication channels between each previous simulation model and the subsequent one are consistent. The correspondence between the type of mineral raw materials and the equipment used is determined by the target useful component of the mineral raw material and the method for enriching the useful product (concentrate). One should also implement the principle of unification of the modeling object, which, in turn, consists in the fact that the simulation model of the processing process is the result of a statistical interpretation of the logical and mathematical description of a series of physical and (or) computational experiments reproduced in predetermined conditions, one of which is the inverse of the forecast and refinement of the model.

References

[1] *ANSYS Acquires Dynardo* Retrieved from: https://www.ansys.com/

[2] Feoktistov A Yu 2016 Application of numerical methods of engineering analysis to solve the problems of the mining industry *High technol. Develop. and use of mineral resour.* **3** 151–4

[3] Skorokhodov V F, Khokhulya M S, Opalev A S et al 2013 Application of computational fluid dynamics methods to the study and analysis of mineral separation processes *Phys. and techn. Probl.of mining* **3** 179–87

[4] Skorokhodov V F, Khokhulya M S, Opalev A S et al 2019 Applied aspects of the application of computer simulation of the hydrodynamics of multiphase media in studies of the processes of separation of minerals during ore dressing *Phys. and techn. Probl. of mining* **2** 139–53

[5] Tu J, Yeoh G H and Liu C 2018 *Computational fluid dynamics: a practical approach* (Butterworth-Heinemann) pp 123–8

[6] Weinhold I and Parry J 2013 *The third wave of CFD" 2013 NAFEMS World Congress* (Salzburg, Austria) pp 130–4

[7] Aleksandrova T N, Romashev A O and Semenikhin D N 2015 Mineral and technological aspects and promising methods for intensifying enrichment of sulfide gold-bearing ore *Metallurgist* **59(3-4)** 330–8

[8] Potemkin V A and Albendari A M 2018 Modeling of the rheological properties of mineral suspensions using computational fluid dynamics methods *Mine survey. and subsoil use* **1(93)** 58–61

[9] Gein K and Sarson T 1992 *Structural Systems Analysis: Means and Methods* (Moscow: Eiteks) 274 p

[10] Nguyen W Ch and Petrov A V 2010 Economic model of technological processes for mineral processing *Bull. of ISTU* **5(45)** 16–21

[11] Nguyen W Ch and Petrov A V 2011 Dynamic Modeling of Mineral Processing Technological Processes *Wiener Readings* Proc. of the IV All-Russ. Conf. Part II (Irkutsk: ISTU) pp 203–10

[12] Kuzmin O V and Kuzmina V V 2016 Combinatorial polynomials of partitions in multi-stage models of mineral processing *Transport infrastructure of the Siberian region* vol 1 Proc. of the Seventh Int. sci.-pract. conf. (Irkutsk) pp 286–9

IOP Conf. Series: Materials Science and Engineering **905** (2020) 012050 doi:10.1088/1757-899X/905/1/012050

The use of wind potential in the local energy of Yakutia

I Yu Ivanova[1], D D Nogovitsyn[2], T F Tuguzova[1], V A Shakirov[1], Z M Sheina[2] and L P Sergeeva[2]

[1] L.A. Melentiev Institute of Energy System SB of RAS, 130, Lermontov str., Irkutsk, 664033 Russia

[2] V.P. Larionov Institute of Physical and Technical Problems of the North SB of RAS, 1, Oktyabrskaya str., Yakutsk, 677980, Russia

E-mail: nord@isem.sei.irk.ru, dmitry.nogovitzyn@yandex.ru, zin.scheina2016@yandex.ru, sergeevalp2015@yandex.ru

Abstract. The article is devoted to the possibility of using renewable energy sources for electricity supply to consumers in remote regions of Russia, isolated from power systems. The use of wind energy resources for power supply will reduce the consumption of diesel fuel and, thus, improve the reliability and efficiency of electricity supply to consumers. At the same time, wind power stations are considered as a complementary energy source when working together with diesel power plants, which are the main ones for power supply to decentralized consumers. The construction of wind farms in addition to existing diesel power plants in modern price conditions is cost-effective in the Arctic regions of Yakutia. The extent of the use of renewable energy sources is currently very small. Justification of the use of wind energy for each locality requires a separate specific study to assess the financial and economic efficiency with a preliminary determination of the optimal capacity of wind farms.

1. Introduction

Recently, the use of renewable energy sources (RES) for providing a more reliable power supply to remote consumers' areas of Russia isolated from the energy system is becoming increasingly relevant. The longest zone of decentralized power supply is located at the territory of the Sakha Republic (Yakutia). This is due to a number of reasons, one of which is the weak transport development of the Northern and Arctic regions of the republic where the scheme of fuel delivery is complex and multi-link (railway, waterway along rivers and through the Northern sea route, road transport) with the length over 5 thousand km. That is why the limited time of seasonal importation, dependence on fuel supplies and as a consequence its rise in price, irregular power supply to consumers which is provided with electricity from autonomous energy sources, mainly diesel generators, take place. In modern conditions the price of diesel fuel in the Northern regions is 50–60 thousand rubles/t, and the transport component in the cost of fuel exceeds 60 %. Because of the multi-link schemes of fuel delivery its loss is great – about 20 % [1]. In this regard, the cost of electricity production reaches 30–40 rubles/kWh.

The Arctic regions are the most attractive in the republic in terms of the use of wind potential for energy supply: here the average annual wind speeds at the height of the weather vane of hydro meteorological stations are 5–7 /s [2]. In the rest of the Republic this figure does not exceed 3 m/s. The use of wind energy resources for power supply will reduce the consumption of diesel fuel and, thereby, increase the reliability and efficiency of electricity supply to consumers.

IOP Conf. Series: Materials Science and Engineering **905** (2020) 012050 doi:10.1088/1757-899X/905/1/012050

At the same time, the scalability of renewable energy sources on the republic territory is currently very small and represented only by solar power plants in the amount of 13 units with a total installed capacity of 1335 kW. The only wind power plant (wind turbine) in Tiksi (250 kW) failed in 2015 as a result of strong winds.

The paper makes an analysis of preconditions for harnessing wind energy in the Republic of Sakha (Yakutia), which include the problems of energy supply to the northern settlements and available considerable wind potential. The factors that affect the efficiency of the wind potential implementation are presented. Firstly, wind speed and wind annual distribution vary depending on site location: sea coast, bay, mainland. Secondly, in the last years there has been a tendency towards a decrease in the wind speed compared to the average long-term data available in reference sources. Thirdly, the efficiency indices are greatly affected by the choice of a standard size of wind turbines because of a difference between the wind energy parameters at the height of the tower and at the height of the weather station vane. The wind speed frequency in terms of intensity and electricity output of wind turbines of different types are analyzed and the analysis results are presented.

The relationships between the wind energy potential indices and annual distribution, and the efficiency for the off-grid consumers in the Republic of Sakha (Yakutia) are presented considering energy consumption curves. The results of investigating the technical capability and cost-effectiveness of using small-capacity wind turbines in the northern regions of the Republic of Sakha (Yakutia) are demonstrated. Technical requirements for the wind turbines to be used in the northern areas, which are related to the installation and operation problems under the conditions of remoteness and low temperatures, are described.

2. Methods and materials

Using the WindMC-Analyzer program, the calculation of wind speeds repeatability was carried out by gradations and estimated the possible power generation of wind turbines for each gradation of wind speeds. The method developed by the authors [8] is used to determine the optimal capacity of renewable energy sources for a decentralized consumer.

3. Results

Low use of renewable energy sources, including wind power plants, is due not only to their high capital intensity but to the uneven and unstable presenting features of renewable natural resources energy both in time and also depending on the territorial location of an object.

One of the important characteristics that determine the energy value of wind is its average annual speed. This indicator at the height of the weather vane in the Central part of the Republic of Sakha (Yakutia) does not exceed 2–3 m/s. This is practically the only zone in the Republic where it is possible to use wind potential for energy purposes. The highest wind speeds in the Republic are observed on the coast of the Northern seas – 4–6 m/s (Fig. 1).

Despite the fact that the Arctic regions have natural and economic prerequisites for the use of wind energy resources, there is a number of limitations, also due to natural and, to a greater extent technical factors rather than economic ones.

The development of wind power in areas with cold climates has a number of specific features. It is necessary to use wind power plants of Arctic design, without hydraulic braking system, without reducer, with adjustable angle of attack of the blades, the possibility of Autonomous operation, remote control and monitoring of work. It is necessary to take into account the problems of transportation of the wind power plant and its construction at the site in conditions of poorly developed transport infrastructure. Special cold-resistant steel grades must be used in the model. Low-temperature synthetic lubricant for key bearings and generator bearings, synthetic gear oils should be used everywhere in the Arctic and cold climate, even if it is not necessary under the conditions of winter minimum temperature. Heating of the controller, wind direction sensor and anemometer is required. Protection of the blades from icing can be carried out by applying anti-icing paint. Thus, when choosing a model of a wind power plant, a detailed analysis of technical characteristics and assessment of working conditions is necessary.

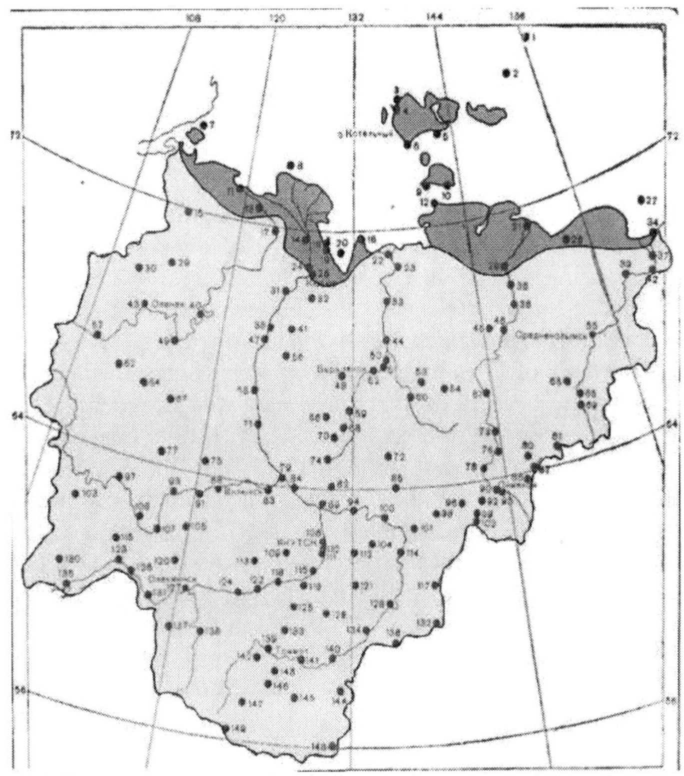

Figure 1. Distribution of the average annual wind speed at the height of the weather vane on the territory of the Republic [2]

Due to the isolation of decentralized consumers from the power system, their annual electricity consumption schedules with small loadings and prevalence in them of a municipal and household component have a pronounced effect maximum in wintertime. This imposes certain limitations on efficiency of the use of wind energy potential which has a high seasonal irregularity.

In this regard, wind power plants are considered as a complementary energy source when working together with diesel power plants which are essential for the power supply of decentralized consumers. When analyzing the feasibility of wind energy resources using for energy purposes, it is important to take into account not only price indices, but also various natural factors that affect the efficiency of wind power plants.

The annual distribution of wind speeds is essential. There can be maxima in different periods of a year. According to the results of the monitoring of average wind speeds changes during a year on the territory of the republic three groups of settlements with different graphics of annual distributions of wind potential were identified.

For the sea coasts (villages Ust-Olenyek, Ust-Yansk, Tabor, Chokurdakh, Kolymskoye*)* it is more typical to have summer maximum and the greatest values of wind speeds. At the same time in bays (Tiksi, Ambarchik) the wind potential also has high values but its distribution is different – here it is observed winter maximum. For the mainland (Sanyjyakhtakh, Eik, Okhotsky Perevoz, Olyekmisk) it is more typical to have the double autumn and spring maximum and a much lower rate of wind speeds. It should be noted that in these groups there can be locally different distribution depending on local characteristics, for example: depressions or elevations of relief, river beds, etc.

The effect of annual distribution of wind potential is that for the power supply at the same annual values, the best conditions for its use will be areas with winter maximum. This dependence is related with a great correspondence of schedules of WPP power generation and its consumption with a winter maximum of wind potential, slightly less-with spring-autumn one, and at the summer maximum, electricity generation and consumption schedules are in the opposite phase.

Another important factor for using the wind potential is the reliability of its indicators and the change in wind speeds depending on the period of measurement. The results of research conducted by A.V. Meshcherskaya, V.V. Eremin et al. on the change in wind speed in the North of Russia in the second half of the 20th century [3], which considered the testimony of 23 weather stations for 1936–2000, confirm the conclusion that in the vast majority of territories the wind speed decreased, especially in the intra-coastal areas significantly (in some regions more than 2 times).

As an example considered the comparative change of average monthly wind speeds during one year in the area of the weather station in Chokurdakh settlement (Allaikhovsky ulus) located near the coast of the East Siberian Sea. According to the long-term observations for the period of 1936–1960 during one year the prevailing wind speed was 4–5 and 6–7 m/s [2]. The average annual wind speed at the height of the weather vane of 11 m hydro meteorological station was estimated at 4.7 m/s.

The results studies on the decline of average annual rate in recent years compared to the period 1936–1960, which is reflected in the reference literature [2], fully consistent with the findings cited by A.V. Meshcherskaya, V.V. Eremin and others in [3].

With the program WindMC-Analyzer the data on average monthly wind speeds were obtained in the area of Chokurdakh weather station for the period 2001–2012. The average annual speed for this period is 3.78 m/s. The biggest difference of indicating is observed from March to October and is 1–1.6 m/s.

At the same time, according to the results of the statistic estimators of wind speed the average coefficient of variation for the period of 2001–2012 was 0.59, RMSD of wind speed is 2.22 which indicates quite favorable conditions for the use of wind energy in the area.

In work [3, 4] it is noted that the possible causes of such reductions can be: the change in the height of a wind measuring device, the equipment of a network of stations with wind vane with a heavy board, the replacement of a weather vane with a wind indicator, the transfer of meteorological sites. In addition to methodological reasons the widespread decrease in wind speed can be determined by climate change, in particular, by the change in a general circulation of the atmosphere [3].

A choice of the size of a wind turbine has a significant impact on the results of studies on justification for wind potential use, since increasing the unit capacity of the installation significantly increases the height of a wind tower. And when the wind speeds are recalculated from the height of a weather vane to the height of a wind tower, the possible generation increases by the installation of electricity. At the same time, under conditions of the Northern and Arctic regions of the republic, due to the weak transport accessibility of the territories, it is necessary to focus on wind turbines that do not require special means for installation.

To assess the impact of the repeatability of different wind speeds during a year and the height of a wind power plant tower to generate electricity in Chokurdakh settlement the units of German companies have been considered: Turbowinds T-400-34, NW 44-850 VN, NW 66-1600 NY-P. The height of Turbowinds T-400-34 wind turbine tower is 34 m, for the rest is 68 m [5–7]. Starting speed is 3–4 m/s, speed of output at rated power is 13–14 m/s.

Using the WindMC-Analyzer program, the calculation of wind speeds repeatability was carried out by gradations for the period of 2001–2012 and estimated the possible power generation of wind turbines for each gradation of wind speeds. The results of studies have shown that with the same repeatability of wind speeds with the increase of a wind tower height the greatest output of energy is shifted towards lower speeds which in general increases total electricity production for a year.

All above mentioned factors induce in estimating of use efficiency for power supply of a settlement isolated from a power system the necessity to determine the amount of electricity generation by wind farm. Thus there is a task to correlate the schedule of annual power generation at a different set of wind turbines with a schedule of consumptions.

Taking into account the high capital intensity of renewable energy sources including wind farms for the effective application of consumers isolated from power systems the definition of their optimal power has a significant importance.

The authors developed a method to determine the optimal capacity of RES for a decentralized consumer. It is described in detail in [8] and based on the ratio of the cost indicators of RES and the displaced organic fuel on the existing energy source which is more dependent on the characteristics of potential of renewable natural energy resources. The main criterion in the technique is the maximum effect with a minimum of capital expenditure, i.e. the ratio of the cost of renewable energy and the displaced diesel fuel due to its operations.

At the modern price indicators the value of optimal power to a large extent depends on the natural potential which determines the installed capacity utilization factor (ICUF) of a wind farm.

For wind conditions of the continental zone (Sanyjyakhtakh and Verkhoyansk settlements) where wind characteristics are low, and, consequently, the ICUF has a small value, the optimal capacity of a wind farm is several times higher than the maximum load of consumers'. In the coastal zone (Tajmylyr, Ust-Olenek and Chokurdakh settlements) at the ICUF more than 20 % the wind farm is close to the maximum load.

The studies conducted to assess the effectiveness of using wind farms show that the Arctic territories have the best indicators of their application [9–14]. The payback period of projects for construction of wind farms in coastal areas of the republic is 7 to 9 years.

As an example discussed discussed the comparative efficiency of wind power plants use for natural and price conditions of Chokurdakh settlement and the traditional scheme. According to the results of research the optimal capacity of a wind power plant for this settlement is 1700 kW as a part of two wind turbines on 850 kW each. The value of the wind farm optimal power is comparable to the maximum load of a consumer.

To assess the financial and economic efficiencies of wind farm applications for electricity supplies in Chokurdakh settlement it was used the production and financial model describing the dynamics of cash flows at operation of the wind farm in addition to the existing diesel power plant. The budget efficiency of the project was estimated, i.e. the costs from budget for fuel delivery and maintenance of energy sources for the selected options of a wind farm capacity in comparison with the electricity supply only from the existing DES.

Despite the significant one-time investments, the return on projects of WPP construction under conditions of Chokurdakh settlement is provided by reduction of annual costs for energy supply compared to the operation of one DES due to the fuel displacement.

The combination of graphs of the cumulative discounted costs for power supply only from DES and when working together DES and WES showed that the project of construction of wind power plant in Chokurdakh settlement with 850 kW power units is financially attractive: discounted payback period of the project is about 9 years.

4. Conclusion

The results of studies of the nature of changes in wind energy potential show that when considering the possibility to use of and justify the application of wind power plants it is necessary to take into account its distribution during a year, and use data on the magnitude of the potential for the modern period of measurements. Because ignoring this factor it would lead to overestimate the utilization factor of the installed capacity of renewable energy source in assessing the financial and economic efficiency of projects, as a result, a negative attitude towards renewable energy projects is formed due to non-receipt of the expected effect in practice. The best for use in terms of resource provision is the intra-annual distribution of wind potential which has a winter maximum.

The construction of wind farms in addition to existing DES in the current price environment is cost- effective in the Arctic regions of the Sakha Republic (Yakutia). The rationale for the use of wind energy for an each settlement requires a separate specific assessment study of financial and economic efficiency with preliminary determination of the optimal capacity of wind farms.

IOP Conf. Series: Materials Science and Engineering **905** (2020) 012050 doi:10.1088/1757-899X/905/1/012050

The value of the optimal power of wind farms is influenced by the following factors: distribution of wind potential during a year, performance characteristics of wind turbines, electricity consumption schedule, remoteness and availability of a territory and, accordingly, the cost of fuel and equipments delivery. The value of the optimal capacity of a wind farm for power supply of a consumer isolated from the power system in the Arctic zone is comparable to the maximum of its load.

For the use of unclaimed electricity from a wind farm during the day it is necessary to conduct additional studies to assess the effectiveness of its daily accumulation.

However, there are a number of technical limitations when using wind potential for energy supply under Arctic conditions. It is necessary to organize the production of Arctic wind turbines without hydraulic system braking, without a gear (direct drive), and with an adjustable angle of attack of blades, the possibility of autonomous operation, remote control and operation control [1]. In addition, it is necessary to take into account the problems of transportation of wind turbines and its installation on a construction under conditions of poorly developed transport infrastructure [15]. Models of wind turbines should provide for the possibility of installation without special means of rise. Special cold-resistant grades of steel and low temperature synthetic lubricants for key bearings and generator bearings, synthetic gear oils should be used.

It is necessary to provide heating of a wind turbine controller, the wind transducer and the anemometer. The protection of blades on a wind turbine from icing can be carried out by applying de-icing paint having a property of water repulsion. It is also possible to cover the surfaces of wind turbine modules by de-icing solution. Thus, when choosing a model of a wind turbine it is necessary to have the detailed analysis of technical characteristics and assessment of working conditions, similar presented in [15, 16]. When assessing the costs of construction and operation it is necessary to take into account all marked features and specifications.

References

[1] Voropai N I and Saneev B G (ed) 2011 *The Eastern vector of Russia's energy strategy: current state, prospection* (Novosibirsk: Acad. Publ. house "Geo") 368 p

[2] *Handbook on climate of the USSR iss. 24 Yakut ASSR Part 3 Wind* 1967 (Leningrad: Gidrometeoizdat) 271 p

[3] Shakirov V A and Artemiev A Yu 2011 Comprehensive analysis of wind power plants effectiveness in the Republic of Sakha (Yakutia) *Polzunovsky vest.* **2/2** 162–6

[4] Mescherskaya A V, Eremin V V, Baranova A A and Maistrova V V 2006 The change of wind speeds in the North of Russia in the second half of the XX century on surface and upper-air data *Meteorol. and Hydrol.* **9** 46–58

[5] *Clean energy for our future* Retrieved from: www.turbowinds.com (Date of issue: 20.12.2013).

[6] *High performance wind farm NW 44-850 VN Nomenclature standard* 2008 13 p

[7] *High performance wind farm NW 66-1600 NY-P Nomenclature standard* 2008 28 p

[8] Ivanova I Yu, Tuguzova T F and Khalgaeva N A 2014 Determination of optimal capacity of a renewable energy source for a consumer isolated from electricity power system *Proc. of the Acad. of Sci. Power* **3** 22–8

[9] Lukutin B V and Kiushkina V R 2006 *Wind farms in autonomous energy of Yakutia* (Tomsk: Publ. House of TPU) 202 p

[10] Minin V A 2010 Assessment on economic efficiency of joint work of diesel power plants and wind power plants in the North *Alternat. energy and ecol.* **7** 67–75

[11] Marchenko O V and Solomin S V 2010 A study on economic efficiency of wind power plants as part of decentralized power supply systems *Alternat. energy and ecol.* **1** 126–31

[12] Hunter R and Elliot G 1994 *Wind/diesel Systems* (Cambridge, UK: Cambridge Univer. Press) 264 p

[13] Lundsager P and Baring-Gould E I 2005 Isolated Systems with Wind Power In Ackermann T (ed) *Wind Power in Power Systems* (New York: John Wiley & Sons, Ltd) pp 299–329

[14] Minin V A and Rozhkova A A 2014 Assessment on the effectiveness of joint work of diesel power plants with wind power plants *Electr. Generat. Stat.* **6(995)** 29–32

[15] Chauhan B, Pinard J and Pinard T M 2003 *Weis Pre-feasibility analysis of wind energy for inuvialuit region in northwest territories* (Canada: Aurora res. Instit.) 66 p

[16] Tammelin B, Cavaliere M, Holttinen H et al 2000 Wind energy production in cold climate. *WECO. Meteorological publications*, vol. 41, Helsinki: Finnish Meteorological Institute, Project report, EUR 19398 European Commission, Community research 38 p

Anticorrosion properties of manganese-containing complex oxides, obtained by stoneware method

E A Nurislamova, M R Ziganshina, S N Stepin and V A Mendelson

Kazan National Research Technological University, 10, K.Marx St., Kazan, Tatarstan 420111, Russia

E-mail: a.elvira.9@gmail.com, zigmay4@mail.ru, stepin_1949@mail.ru, vallery78@mail.ru

Abstract. In recent years, in the search for alternatives to toxic anticorrosive chromium pigments, the use of complex metal oxides in this capacity has been actively developed. In this work, the anticorrosive and other physicochemical properties of complex oxides, including manganese in oxidation degrees +4 and +5, obtained by the calcination method were investigated. As a result, it has been shown that the products obtained have good prospects for being used as anti-corrosion pigments. Higher inhibitory properties of pigments containing manganese in oxidation state +5, is probably a consequence of their aqueous extract's alkaline reaction. During the experiments, it was found that the competitiveness of the synthesized pigments can be markedly enhanced by using natural ore, pyrolusite, as manganese source.

1. Introduction

The steady increase in the number of objects made of carbon steel, highly vulnerable to corrosion combined with the increased aggressive environmental impact caused by the development of industrial plants generating waste that has a damaging effect on various materials, make it obvious that this process should be counteracted. The most common way to protect steel objects from environmental corrosive effects is dyeing. The coatings applied to the surface of various substrates are usually multi-layered and the first pre-coating layer has the main protective load. Anti-corrosive effectiveness of primers is provided by including of inhibiting pigments in their structure.

Compounds containing chromium in the highest degree of oxidation have demonstrated the highest anticorrosive efficiency, among the pigments of this type that have been used until recently. The other side of chromium-containing pigments' high inhibitory ability is ecotoxicity, which is associated with their long-term circulation in the environment. In many industrialized countries, the use of pigments of this type is prohibited at the legislative level (EU Directive 2000/53/ EC allows transport vehicles to be covered with only by two grams of hexavalent chromium). However, despite active searches, it is necessary to state that a complete low-toxic alternative to chromate pigments has not yet been found.

Elemental base analysis for anticorrosive pigments capable of replacing chromates shows that manganese (d-element of the adjacent group of Mendeleev's Periodic System) is close in property to chromium. Like chromium, it can be in different degrees of oxidation and has the ability for complex formation. An important advantage of manganese-containing compounds compared to chromium-containing compounds is significantly lower toxicity. Another advantage of manganese is its high average content in the earth's crust, amounting to $8.5 \cdot 10^{-2}$ % (mass.). This fact leads to the wide use of its compounds in various fields of science and technology.

IOP Conf. Series: Materials Science and Engineering **905** (2020) 012051 doi:10.1088/1757-899X/905/1/012051

One of the areas of search for the replacement of chromate pigments in anti-corrosion coatings is the use of complex metal oxides [1]. It should be admitted that, in spite of the above mentioned, there is unfairly small amount of works related to the use of manganese-containing compounds of this type for anti-corrosion purposes. Manganese compounds are used as pigments (manganese black, umber, manganese brown, manganese blue, manganese violet, manganese green) in paints production, but their main application area is the manufacture of paints and varnishes for decorative purposes [2].

2. Experimental

2.1. Syntheses of manganates

The thermoanalytic study of the synthesis was carried out on an OD-3425-1500 derivatograph, in the range of 20–1000 °C with a heating rate of 15 deg/min.

The process of samples obtaining consisted of the following operations: load preparation, its calcination in a muffle furnace, washing with water, drying at 120 °C and grinding in a planetary mill.

2.2. Synthesis products' identification

The degree of manganese oxidation in the obtained pigments was determined by the method described in [4].

2.3. X-ray photoelectron spectroscopy method

The samples were studied by X-ray photoelectron spectroscopy on an ES 2401 apparatus using MgKα radiation (1253.6 eV). Vacuum in the spectrometer chamber 10^{-6} Pa. The relative error in determining the concentration of elements is ±3 % of the measured value. The results were processed using the CasaXPS licensed program.

2.4. Physicochemical properties the resulting products as pigments

Physicochemical properties of the processed products' as pigments.

Determination of oil absorption according to ISO 787-5-80.

The density of pigments was determined by the pycnometric method according to GOST 21119.5-75.

The mass fraction of substances soluble in water was determined by the method of hot extraction according to ISO 787-3-79.

The reaction of the pigment aqueous extract was determined according to GOST 21119.4-75.

2.5. Methods for testing pigments' anticorrosive properties

The pigments' inhibiting ability was evaluated by the values of the potential and the corrosion current of steel in an aqueous extract containing 3% sodium chloride.

The corrosion potential of steel was measured using a pH-340 potentiometer.

A three-electrode electrochemical cell was used to determine the corrosion current. It was prepared by gluing a hollow glass cylinder with 3 cm inner diameter onto the steel sample surface. The section of steel surface forming the bottom of the resulting glass was used as working electrode. 20 ml of electrolyte was poured in the resulting cell. The platinized platinum auxiliary electrode was lowered into the electrolyte solution. A silver chloride electrode that was placed in Luggin served as the reference electrode.

The electrochemical cell was connected to a PI-50-1 potentiostat with a PR-8 programmer. After the stationary potential had been established the cathode polarization of the steel sample was carried out for 15 min at minus 30 mV. Polarization curves (current versus potential) required for current calculation were recorded in the sphere of slight potential deflections from the stationary value (from minus 30 to 30 mV) using the small linear polarization method. This method guarantees minimal perturbations introduced into the metal dissolution process during electrochemical measurements. The corrosion current according to the results of polarization measurements was calculated using a computer program based on the Stern – Geary equation solving [5, 6].

3. Results and Discussion

Analysis of the composition of the above-mentioned manganese-containing synthetic pigments obtained by industrially thermal method showed that, for example, the composition of "manganese blue" ($Ba_3(MnO_4)_2 \cdot nBaSO_4 \cdot mBaO$, where n=10–16, m=1–3) includes barium/manganese composite oxide [7]. Testing has shown that this decorative pigment, intended for artistic paints production and plastics coloring, has the ability to suppress steel corrosion [8, 9]. Therefore, planning the synthetic part of the work we proceeded from analogy. Manganese (IV) oxide, barium nitrate and barium sulfate were chosen as starting components for synthesis. In the load of this composition, manganese oxide is a reducing agent, barium nitrate is an oxidizing agent, and barium sulfate, apparently, performs stabilizing function.

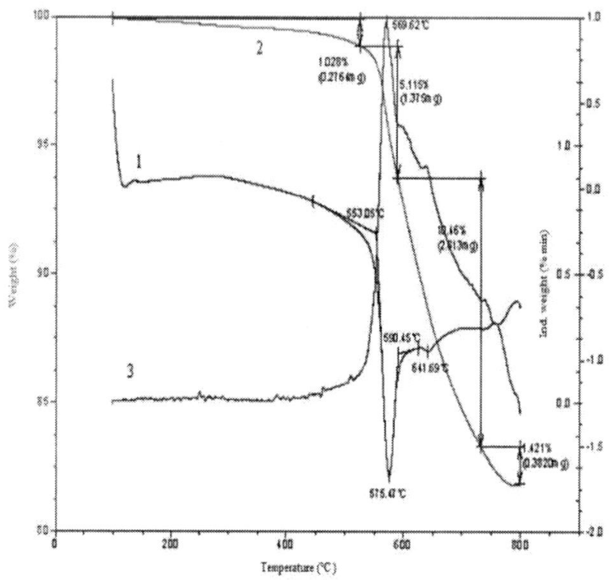

Figure 1. The results of the charge thermal analysis. 1 – differential thermal analysis (DTA) curve, 2 – thermogravimetry curve (TG), 3 – differential thermogravimetry curve (DTG).

As can be seen from the thermal study results presented in Fig. 1, DTA and DTG curves of the initial components' mixture are characterized by the presence of several effects, accompanied by a loss of mass. When the temperature rises to 100–110 °C, the removal of mainly weakly bound adsorbed and interlayer water is recorded.

Starting with a temperature of 550 °C, a reaction, leading to the formation of barium manganate sulfate (BMS) takes place. It is accompanied by significant weight loss due to decomposition of the charge nitrate component.

As a result of preliminary experiments, it was found that the anticorrosive and physicochemical characteristics of the product obtained depend on the synthesis' temperature and time conditions and the initial components' ratio.

With this in mind, a three-factor experiment was carried out in order to optimize the characteristics of the obtained product. The relationships between the components and the synthesis conditions were calculated using the Minitab 14.0 program, developed by DuPont specialists.

Temperature, time of the charge heat treatment and the charge composition were chosen as the parameters to be optimized. The response function was the corrosion current of steel in contact with aqueous extract of the synthesized product containing 3 % sodium chloride and the content of water soluble substances (CWSS), which is an important characteristic of anti-corrosion pigments. It is known that the optimal values of CWSS are in the area less than 1 % [10].

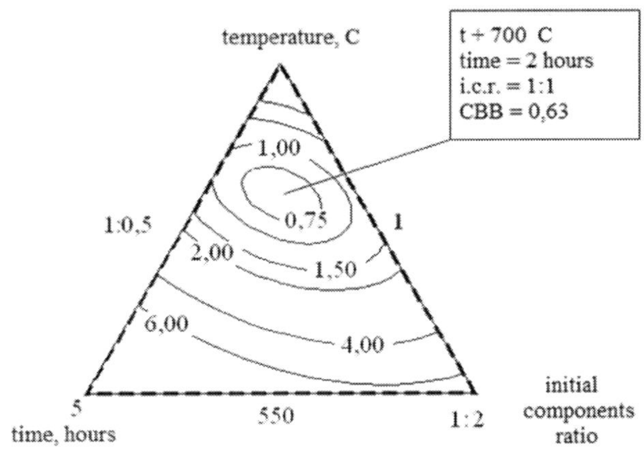

Figure 2. Contour diagram: the conditions' influence on the synthesis of SME on SVLP in the resulting products

On the basis of the results obtained, approximation was carried out with the construction of contour diagrams. For example, figure 2 shows a diagram that allowed to determine the synthesis conditions for the product with minimum CWSS of 0.63 %: time 2 h, temperature 7000 C, the ratio between the starting components: MnO2: BaSO4:Ba(NO3)2 – 2:1:1. We can take certain formal oxidation state of manganese that was +5 as confirmation of the correctness of the above gross formula of the product.

Additional assessment of different composition mixture physicochemical characteristics, calcined under different temperature and time conditions allowed to conclude the influence of these parameters on the color and stability of the obtained products. The conclusion about the instability of the latter was made in case of water extracts' coloring. Stable products were obtained only at the optimum temperature of 700 °C, their color changed as the temperature increased from blue (650–700 °C) through blue-green (700 °C) to bright green (750 °C and above). Probably stabilization is facilitated by the inclusion of sulfate component into the mixture.

In order to identify the sulfate component effect on the pigment's anti-corrosive properties, the composition of the charge was changed and barium sulfate was eliminated. As a result of the synthesis, a product was obtained with the formal oxidation state of manganese +4: barium manganite (MB) BaMnO3, characterized by SWSS 0.4 %.

Table 1 presents the main properties of the synthesized calcined pigments.

According to literature data, Mn (IV) compounds are characterized by the presence of typical effects on their thermograms [13]: the effect of water removal, the exothermic effect corresponding to crystallization, and the endothermic manganite effect — decomposition. On the differential thermal analysis' and thermogravimetry curves derived from the study of the products obtained in full accordance with [13], the effects typical of manganites were got: four endothermic and an exothermic one. In the temperature range of 20–570°C, dehydration is observed, and its most active phase goes to temperatures of 400–425 °C. This endothermic transformation extended in a wide temperature range is caused by the release of weakly bound crystallization water (dehydration) and the removal of structural water of OH-groups (dehydroxylation). In the temperature range of 570–750 °C, a complex exothermic transformation is observed with two exothermic effect maxima (640 and 710 °C). They correspond to the superposition of two processes, i.e., the crystallization of manganite and oxygen release during the decomposition of manganese dioxide impurity present in the sample. In the temperature range of 885–920 °C, manganite decomposes (permanganite effect) with a new compound formation. The next effect at the temperature of 920–965 °C probably indicates the presence of β-kurnakite formed as a result of the dissociation of MnO_2 impurity compound. With further heating, a slow mass loss is observed on the DTG curve, which may indicate the beginning of the manganite's gradual decomposition, the active phase of which begins above 1000 °C. The data obtained allow us to consider the studied substances as manganites.

IOP Conf. Series: Materials Science and Engineering **905** (2020) 012051 doi:10.1088/1757-899X/905/1/012051

This is confirmed by IR spectra analyses of the studied substances that indicates the absorption spectra have bands characteristic of manganites in the area of 940, 840, 570, and 460 cm^1 wavenumbers [14].

3.1. General physicochemical properties of synthesized manganites as anticorrosive pigments
Table 1 presents the main characteristics of calcium and barium manganites that allow us to make a conclusion about the possibility of their use as anti-corrosive pigments.

Table 1. Properties of the synthesized manganese-containing pigments

Pigment designation	Colour	pH, aqueous extract	Oil absorbtion g/100g	Density kg/m^3	Steel corrosion potential mV	Corrosion current density, $ImcA/cm^2$
MB	brown	7.5	21	5.46	125	1.7
MSB	green	9.0	10	4.33	121	0.9

*Steel corrosion current density, in contact with 3 % NaCl solution 16,5 mcA/sm^2

The data presented in table 1 allow us to conclude that both synthesized products have good prospects to be used as anti-corrosion pigments. Comparison of their properties makes it possible, in addition to stating their color difference, to conclude that they have a higher corrosion suppressive ability, which, apparently, is the consequence of its aqueous extract alkaline reaction.

In order to reduce the cost and, as a result, increase the competitiveness of the synthesized pigments, we investigated the possibility of replacing chemically pure manganese oxide (IV) with ore containing this oxide (pyrolusite) as the charge component.

Table 2. The qualities of samples synthesized using pyrolusite with different mn (IV) content

MnO_2 content, %	The content of substances solubter, %	Oil absorbtion, g/100g	Density, kg/m^3	Corrosion current density, $ImcA/cm^2$
54–65	1	13	4.58	7.30
80–83	0.65	10	4.30	1.2
94	0,78	10	4,29	1,1

Experiments have shown that the use of ore with basic substance content of at least 80 %, during the MB pigment synthesis, has a slight effect on its properties (Table 2).

Taking into account the anticorrosive efficiency and the possibility of using cheap raw materials for the production of synthesized manganese-containing pigments, it is possible to predict their high demand in case of their industrial production.

4. Conclusion

The complex oxides of manganese in the oxidation state +4 and +5 were synthesized using the method of calcination. Their physicochemical properties allow to use them as anti-corrosion pigments. The calcination product containing Mn(V) and $BaSO_4$, possibly as a stabilizer, has higher anti-corrosion properties, probably due to the alkaline reaction of the aqueous extract. The proven possibility of using natural pyrolusite as a raw material containing manganese can contribute to the significant reduction in the cost and, consequently, increased competitiveness of the synthesized pigments.

References

[1] Rozenfeld I L, Rubinshtejn F I and Zhigalova K A 1987 *Protection of metals from corrosion by paintwork* (Moscow)

[2] Yegorova S R, Ziganshina M R, Karandashov S A et al 2015 Synthesis and study of inorganic pigment properties on the basis of dead chrome catalyst *Res. J. of Pharmac., Biol. and Chem. Sci.* 6(6) 1540–8

[3] Kalendova A, Vesely D and Kalenda P 2006 A study of the Effects of Pigments and Fillers on the Properties of Anticorrosive Paints *Pigment Resin Technol.* **35(2)** 83–94

[4] Korsunskij L F, Kalinskaya T V and Stepin S N 1992 *Inorganic pigments* Reference book (St. Petersburg)

[5] Stepin S N, Vahin A V, Sorokov A V and Ziganshina M R 2000 Research method of anticorrosive properties of pigments and pigmented coatings *Paints and varnishes and their applicat.* **1** 25–27

[6] Svetlakov A P, Stepin S N and Davydova V I 1983 Electrochemical method for assessing the protective properties of phosphate coatings on steel *Paints and varnishes and their applicat.* **4** 33–5

[7] Stepin S N, Vakhin A V, Ziganshina M R and Karandashov S A 2001 The study of anticorrosive properties of pigments *Paints and varnishes and their applicat.* **10** 3–6

[8] Belen'kij E F and Riskin I V 1974 *Chemistry and pygment technology* (Leningrad: Himiya)

[9] Stepin S N, Ziganshina M R, Ahmadieva A A et al 2009 Assessment of the anticorrosive properties of manganese blue *Paints and varnishes and their applicat.* **3** 26–33

[10] Solodov V A, Ziganshina M R and Bayburina E A 2015 Anticorrosive Efficiency of Coprecipitated Manganese Compounds IV *Int. J. of Appl. Chem.* **11(5)** 601–9

[11] Stepin S N, Ziganshina M R and Peshkova M S 2005 Manganese against corrosion *Sci. in Russ.* **4** 52–3

Mineral particle disintegration model

A G Oleinik and V V Biryukov

FIC Kola Science Center RAS, 24A, Fersmana Str., Apatity, 184209, Russia

E-mail: oleynik@iimm.ru, birukov@goi.kolasc.net.ru

Abstract. The study proposes mathematical model for reducing the size of mineral particles in crushing and grinding operations, which is a form of the reflected normal Gaussian – Laplace distribution. The study considers the distribution density modification of the mineral particles depending on the magnitude of the energy impact during grinding. Comparison of the results of computational experiments with the results of laboratory studies confirms the possibility of using the proposed model of disintegration of mineral particles in the development of software modules for the operational evaluation for crushing and (or) grinding of mineral raw materials.

1. Introduction

There is a need to develop new approaches to make research and implementation of ore dressing processes, the creation of energy and resource-saving technologies and devices. This will increase the competitiveness between industries dealing with mineral processing. The processing of solid minerals are considered the most energy-intensive in the processes of size reduction (crushing and grinding) [1]. Therefore, the issues of optimizing energy costs for changing the size of mineral raw materials is still relevant in the scientific industry around the world [2, 3].

Iron ore enterprises spend 50 % of the total energy consumption for the ore grinding processes. One of the main directions of increasing the profitability of mining enterprises is to increase the efficiency of grinding processing operations in technological schemes for mineral processing.

The structure of the mineral dressing has the main influence on the effectiveness of the technological schemes. The selection of new grinding circuits is mainly based on the experience of a design engineer, knowledge of the industry, and what was previously used and worked well for ores with similar characteristics. However, the traditionally used methods for developing the structure of technological circuits of mineral dressing do not help to obtain the result expected. This issue is exacerbated by the development of objectively manifest tendencies of involvement in the processing of depleted ores and increased competition in the world mineral resource markets. Already in the late 20th century, the need for the use of mathematical modeling methods to improve approaches to the design of technological circuits of mineral dressing based on the results of laboratory studies became apparent.

Most Russian scientists made a great contribution to the formation and development of mathematical modeling of the separation of mineral component. I.T. Tikhonov developed the mathematical apparatus of diffuse mass transfer of narrow separation fractions under the action of forces of various nature. It identifies the analytical form of separation characteristics in various separation processes carried out in serial and promising enrichment devices. O.N. Tikhonov studied the energy processes of changing size [5] and the measurement of work indices for the "classical" laws of crushing [6]. L.A. Barsky particularly [7] analyzed the most important criteria for the effectiveness of the functioning of technological schemes for processing mineral raw materials. This becomes a basis for the development of cost-effective technologies for industrial production of minerals. Among foreign works, it should be noted the 1977 book A.J. Lynch [8], which presents approaches to modeling schemes for changing the size of minerals, as well as researches by A. Gupta and D. Yan. They present the mathematical apparatus and practical recommendations for the development of mineral processing equipment that provides cost-effective achieving maximum performance and product quality. The second expanded edition of the study was published in 2016 [9].

The growth of computing capabilities allowed the creation and practical use of resource-intensive computer models of multi-speed multiphase media [10–12].

Work on digital modeling of the separation processes of complex mineral complexes in relation to some designs of concentrators was developed at the Kola Scientific Center of the Russian Academy of Sciences. The work here was carried out in three main areas: the creation of an information and reference system to support research on the processes of separation of mineral components; development of simulation models of the separation processes used; development of logical and analytical models of ore processing [13, 14].

2. Modeling object and methods

The principle of block structure is applied in the development of models of structurally complex technological schemes of industrial design. The allocation of blocks during decomposition is carried out within expected stages and modes of the synthesized circuit based on the required description and visualization of the structure. In this regard, there are two basic principles of constructing the flow diagrams of mineral processing – cyclicality and staging. This allows for functional (in function), component (in the form of elements) and structural (in the form of reaction between elements) decomposition of the model of the mineral processing technological scheme, where the minimum the decomposition component (MCD) is the complex stage of the reduction in size and separation.

The inputs and outputs of the various components of the decomposition of the technological scheme are connected by material flows. Such a construction of a technological scheme establishes uniform input and output qualitative and quantitative technical and technological parameters based on the revealed patterns of their functioning.

The purpose of creating a mathematical apparatus for describing the process of disintegration of solid mineral particles is to develop a software module for the operational determination of the performance of technological equipment. This forms computer models of the minimum components of the decomposition of technological circuits in mineral processing.

There are several well-established theoretical approaches to the study and description of crushing and grinding processes. All these approaches are based on the laws of Rittinger, Bond, P.A. Rebinder and the law V.P. Kirpicheva and F. Kika [6, 15].

There are also examples of using the Fokker – Planck differential equation of diffusely convective mass transfer to describe the processes of reducing the size of mineral particles [16, 17]. Based on the Fokker-Planck partial differential equation the study describes the evolution of the probability density of the coordinates. As well as momentum of particles in processes one and two-parameter models of the flow structure in technological devices are constructed that perform predictive functions based on the use of conservation laws mass, momentum and energy. The disadvantages of such models include the complexity and cumbersomeness of numerical decision algorithms and the need for significant computing resources.

Another wide class of grinding models is based on the use of patterns of modification of the distribution function of the dispersed composition of particles in the processes of particle size reduction. Distribution functions can be constructed on the basis of either the lognormal or the Rosin-Rammler (Weibull) distribution. The disadvantages of such models include the lack of a clear algorithm for the formation of initial and final distribution functions and the difficulty of interpreting the results.

In this paper, there is possibility of constructing an analytically simple model of the disintegration of mineral particles in MCD. The researchers do not examine the actual grinding of mineral particles, but only consider their proposed approach to the mathematical description of this process. This is planned to be used to model the functioning of the MCD of the mineral processing technological scheme.

3. Disintegration model

As the object of modeling, we consider the change in the granulometric composition of the feed sample, given by the continuous distribution function of the outputs. It was gained from magnetite quartzites in the range [0; 3] mm. The initial data were obtained as a result of laboratory tests of the sample using a laser diffractometer ANALYSETTE 22 NANOTEC from FRITSCH. In this case, both integrated (cumulative) distribution graphs (curves in Figure 1) and there is discrete output density distribution curves of the studied particle size class before and after grinding the test sample obtained (Figure 1).

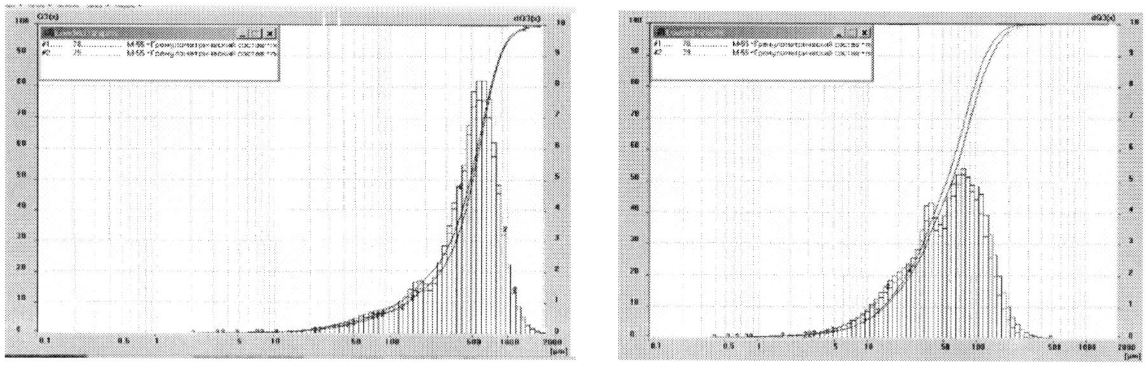

Figure 1. The distribution of the outputs of the magnetite quartzites size classes in the range [0; 2] mm, obtained in laboratory studies: a – before grinding the test sample; b – after grinding the sample

According to the Kirpichev-Kik law, energy consumption for reducing the size of mineral particles is proportional to their volume or mass. Then the specific energy (the energy spent on the destruction of a unit mass of mineral particles belonging to the same size class) is described:

$$\frac{d_1}{d_2} = exp(kE) \quad \Longrightarrow \quad E = \frac{1}{k} \cdot ln\frac{d_1}{d_2}, \tag{1}$$

d1 and d2 are the particle sizes before and after grinding, k is the coefficient of proportionality, depending on the properties of the material being ground and the particular type of equipment, E is the energy spent on grinding.

This stage of the study revealed the result of grinding particles belonging to one class of particle size, it does not depend on the result of grinding particles belonging to another class of particle size. It is proposed to use the reflected normal distribution (Gaussian – Laplace distribution) of the form for describing the distribution density of the newly formed during grinding of mineral particles:

$$f(d) = \frac{1}{\sigma\sqrt{2\pi}}\left[exp\left(-\frac{(d-d_2)^2}{2\sigma^2}\right) + exp\left(-\frac{(d+d_2)^2}{2\sigma^2}\right)\right], \tag{2}$$

here

$$\sigma = d_1\left(\exp\left(-\frac{E}{2}\right) - \exp(-E)\right); \quad d_2 = \frac{d_1}{\exp(E)}.$$

4. Results and discussion

The particle size reduction model was successively applied to all particle size classes identified in the sample feed the grinding operation with a sequential increase in energy exposure. As the energy applied for grinding is increased, the type of graph changes (Figure 2). There is a change in the type of density distribution graphs of the outputs of several model particle sizes: 2 mm; 1 mm; 0.5 mm.

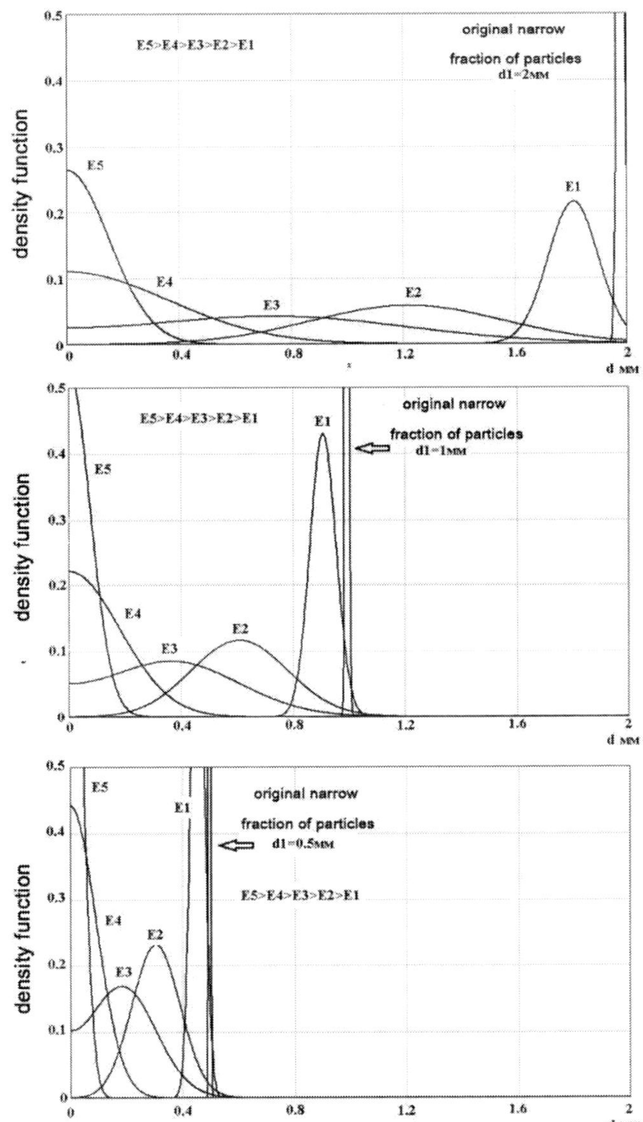

Figure 2. The calculated graphs of the density of the distribution of the output for various particle sizes (E5> E4> E3> E2> E1)

As the applied energy increases, the mathematical expectation of the distribution shifts to the left and the shape of the distribution density graph becomes asymmetric (the Gauss – Laplace distribution). Physically, this corresponds to the gradual transition of particles from "large" classes to smaller, and to the accumulation of the yield of ground material near the "0" scale of particle size.

Analysis of the dispersed composition using a laser diffractometer also showed a shift and change in the shape of the distribution functions and the distribution density of the outputs in the fineness classes of magnetite quartzites (Figure 1).

Using the proposed model of particle size reduction based on the Gaussian – Laplace distribution, a computational experiment was conducted with five model classes of particle size – 1, 0.9, 0.8, 0.7, 0.6 mm. The experiment assumed that the outputs of all model classes are the same and the proportion of each is 0.2. The total yield of all classes is 1. The grinding energy equal to 0.4 conventional units was experimented on these size classes.

Figure 3 presents the distribution density curves for various particle size classes obtained as a result of modeling the process of their disintegration. The presented summary graph of the distribution density accurately reproduces the shape and behavior of the actual grinding (Figure 1, b).

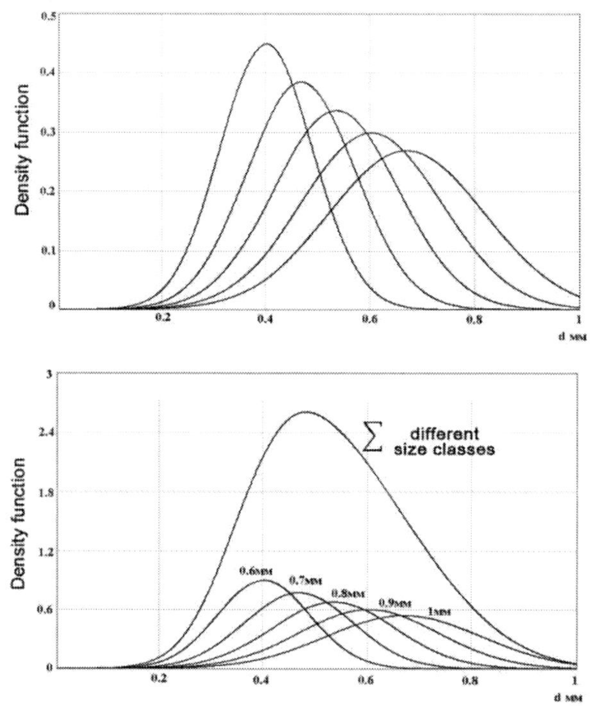

Figure 3. Calculated graphs change in the density form of the output distribution for different size classes (1; 0.9; 0.8; 0.7; 0.6 mm)

5. Conclusion

The model of disintegration of mineral particles uses the *E* coefficient of energy costs. The model will further be developed to determine the dependence of this coefficient on the physic-mechanical properties of the crushed minerals – modulus, hardness, strength. In addition, the *E* coefficient depends on the type of equipment used for disintegration.

Comparison of the obtained results of laboratory measurements suggests, the proposed model of disintegration of mineral particles can be used in the development of a software module for evaluating the performance of equipment for crushing and (or) grinding of mineral raw materials. However, the materials should be taken in the minimum components of the decomposition of mineral processing technological circuits. The proposed model can also be applied autonomously to specific processes of particle size reduction (type, energy intensity of equipment, physical and mechanical properties of solid particles) for the operational forecasting of the results of mineral raw materials.

The energy costs of the disintegration of mineral particles are considered as one of the main performance criteria. This is used in the development of information technology for the structural synthesis of multistage technological process schemes using the example of the synthesis of mineral processing schemes. The structural synthesis on elements with limited compatibility is solved as the problem of satisfying constraints (CSP – Constraint satisfaction problem). This makes to use methods and heuristics developed in the theory of satisfaction of constraints, to jointly process both quantitative and qualitative limitations of the subject area [18].

The study proposes the presented model as part of the creation of this technology within its interfacing with databases (raw materials, technologies, equipment) and the forms some restrictions for use in the procedures for deriving effective schemes of technological processes.

The assumption presented at the present stage of model development is still to be studied and verified.

The model presented in this paper based on the Gaussian – Laplace statistical distribution does not require large computational resources, but it is possible to determine the particle size distribution with sufficient accuracy finished material according to the characteristics of raw materials and equipment. The scope of the model is an operational modeling and optimization of grinding technology.

An important aspect of the model development are the processes associated with the discovery of mineral grains. When studying the grinding process, the size of the solid particles as well as other physical properties (density, magnetic susceptibility, floatability, etc.), have their influence for the results of further separation processes.

Acknowledgment
The study presented is partially supported by the RFBR grant No. 19-07-00359-a

References
[1] Pokrajcic Z and Lewis-Gray E 2010 Advanced comminution circuit design – essential for industry *The AusIMM Bull.* 38–42 Retrieved from: http://www.ceecthefuture.org/wp-content/uploads/2013/08/ 2010_aug_bulletin_038advanced_comminution_circuit_design.pdf
[2] Ballantyne G R and Powell M S October 2014 Benchmarking comminution energy consumption for the processing of copper and gold ores *Minerals Engineer.* **65** 109–14
[3] Arsentiev V A, Baranov V F and Weisberg L A 2007 Current State and Prospects for the Development of Crushing and Grinding of Mineral Raw Materials *Mining J.* **2** 10–4
[4] Pokrajcic Z 2010 *A methodology for the design of energy efficient comminution circuits* Doct. Dissertation (University of Queensland) 320 p Retrieved from: https://espace.library.uq.edu.au/view/UQ:211463
[5] Tikhonov O N 2008 Calculation of the energy of crushing and grinding, taking into account the characteristics of fineness *Ore dress.* **3** 10–4
[6] Tikhonov O N 2008 Methods of measuring work indices for the laws of Rittinger, Kik-Kirpichev and Bond *Ore dress.* **5** 10–4
[7] Barsky L A and Kozin V Z 1978 *System analysis in mineral processing* (Moscow: Nedra) 487 p
[8] Lynch A J 1977 *Mineral crushing and grinding circuits: their simulation, optimization, design, and control* (New York: Elsevier Sci.) 342 p
[9] Gupta A and Yan D S 2016 *Mineral Processing Design and Operations: An Introduction* 2rd ed 882 p
[10] Sarhan A R, Naser J and Brooks G 2016 CFD simulation on influence of suspended solid particles on bubbles' coalescence rate in flotation cell *Int. J. of Mineral Proc.* **146** 54–64
[11] Narasimha M, Sripriya R and Banerjee P K 2005 CFD modeling of hydrocyclone – prediction of cut size *Int. J. of Mineral Proc.* **75** 53–68
[12] Leahy M J, Schwarz M P and Davidson M R November 2006 An air sparging CFD model of heap bioleaching of chalcocite *Appl. Mathem. Model.* **30(11)**1428–44
[13] Skorokhodov V F, Khokhulya M S, Opalev A S et al May 2013 Computational fluid dynamics methods in research and analysis of mineral separation *J. of Mining Sci.* **49(3)** 507–13

[14] Biryukov V V, Skorokhodov V F, Nikitin R M and Oleinik A G 2017 Formation of models of technological schemes for processing rare-earth mineral raw materials using methods of system analysis *Proc. of the Kola Sci. Center of the Russ. Acad. of Sci.: Inform. Technol.* **8(3)** 124–34

[15] Pilov P I and Pryadko N S 2012 Modeling of closed grinding cycles on the basis of the Rittinger hypothesis *Zagabachenya cinnamon kopalin, VIP* **51(92)** 98–107

[16] Manita O A, Romanov M S and Shaposhnikov S V November 2015 On Uniqueness of Solutions To Nonlinear Fokker-Planek-Kolmogorov Equations *Nonlinear Analysis. Theory Methods Appl.* **128** 199–226

[17] Vasiliev P V 2012 Numerical solution of the equation of kinetics of disintegration and disclosure of polycrystalline minerals *Sci. reports of Belgorod State Univer. Ser. Econ. Comp. sci.* **7(126)** 92–100

[18] Zuenko A and Oleynik Y 2019 Constraint Programming Based on Matrix-Like Representation of Qualitative Constraints *Frontiers in Artific. Intellig. and Applicat.* **312** 264–75

IOP Conf. Series: Materials Science and Engineering **905** (2020) 012053 doi:10.1088/1757-899X/905/1/012053

Phase transformations and dielectric characteristics of FE-relaxor $Sr_{0.75}Ba_{0.25}Nb_2O_6$ ceramics and thin films

A V Pavlenko[1], G N Tolmachev[1], D V Stryukov[1], S Kh Alikhadzhiev[2], I M Aliev[2,3], K M Zhidel[4], S N Tsebaev[2] and R S Dzhambulatov[3]

[1] Federal Research Centre The Southern Scientific Centre of The Russian Academy of Sciences, Rostov-on-Don, Russia
[2] Chechen State University, Grozniy, Russia
[3] Integrated Research Institute, H.I. Ibragimova RAS, Grozniy, Russia
[4] Research Institute of Physics SFU, Rostov-on-Don, Russia

E-mail: tolik_260686@mail.ru, tolm100@mail.com, strdl@mail.ru, said366502@mail.ru, ialiew@mail.ru, karinagidele@gmail.com, ceb-salakh@yandex.ru, asldzam@mail.ru

Abstract. In the paper the results of a structure and dielectric properties study of ferroelectric-relaxor $Sr_{0.75}Ba_{0.25}Nb_2O_6$ (SBN-75) ceramics and thin films with the structure of an unfilled tetragonal tungsten bronze have been presented. The existence of orientation domains in the SBN-75 film with rotation in the interface plane only by angles of ± 18.4° relative to the axes of the MgO substrate is revealed. In the tetragonal approximation, the unit cell parameters of the film (c = 3.948 Å and a = 12.49 Å) and its strain in comparison with the cell parameters of the ceramic sample have been determined.

1. Introduction

$Sr_xBa_{1-x}Nb_2O_6$ (SBNx) solid solutions (SS) with unfilled tetragonal tungsten bronze (TBB) structure are prominent representatives of uniaxial ferroelectric-relaxors (FER) [1]. They have high electro-optical and pyroelectric coefficients, that makes them promising materials for use in nonlinear optics, holography, and piezoelectric devices [2]. The crystal structure of SBNx is a three-dimensional framework formed by connected NbO_6 octahedra with a formation of three types of channels along the c axis [1, 3]. The five-cornered channels are statistically filled with barium and strontium atoms, the four-cornered channels are filled only with strontium atoms. Three-cornered channels are unfilled according to all structural studies. In SBNx SS, the displacement of Sr, Ba, and Nb atoms relative to the oxygen planes leads to the ferroelectric polarization appearance (at polarization switching metal atoms pass through the mirror plane). The probabilistic filling the positions by barium and strontium cations (in $Sr_{0.75}Ba_{0.25}Nb_2O_6$ [4], the population of five-cornered channels by Ba and Sr is 30.9 and 57.2 %, respectively, the population of four-cornered channels by Sr is 71.5 %) cause significant fluctuations in the chemical composition in the volume of the material and as a consequence the ferroelectric-relaxor properties appears [5]. It was shown that ferroelectric-relaxor properties appear also in SBNx thin films by our studies of these structures [5, 6].

In this paper the results of a structure, microstructure, and dielectric characteristics study of $Sr_{0.75}Ba_{0.25}Nb_2O_6$ (SBN-75) ceramics and thin films presented.

2. Methods and materials

$Sr_{0.75}Ba_{0.25}Nb_2O_6$ ceramics was made from Nb_2O_5, $SrCO_3$ and $BaCO_3$ by the solid-phase reaction method followed by sintering as part of conventional ceramic technology at the Research institute of physics SFU.

RF sputtering of SBN-75 films ~ 300 nm thickness at an oxygen pressure of 0.5 T on a (001) MgO substrate was performed on a Plasma-50-SE system.

The grain structure of the ceramic was studied by a FE-SEM Zeiss SUPRA 25 scanning electron microscope at the Joint Center for Scientific and Technological Equipment of the SSC RAS.

X-ray phase analysis, structural perfection of ceramics and thin films, unit cell parameters in the normal to the substrate plane direction, and also the orientation relations between the film and the substrate were established by X-ray diffraction by a modernized DRON-4-07 diffractometer (CuKα).

Temperature dependences of the relative permittivity, $\varepsilon'/\varepsilon_0$ (ε_0=8.75·10^{-12} F/m is the dielectric constant), and the dielectric loss tangent, tanδ, of the samples obtained using an Agilent 4980A LCR meter.

3. Results

The θ-2θ X-ray diffraction patterns of the $Sr_{0.75}Ba_{0.25}Nb_2O_6$ ceramics contained only reflections typical for the TTB structure (Fig. 1). The calculated values of the ceramic unit cell parameters in the tetragonal unit cell approximation were a_c = 12.42 Å and c_c = 3.908 Å, which completely correlates with [5]. No impurity phases were found in ceramics. This is confirmed by the results of SEM analysis of the cleaved facet of studied ceramics presented in fig. 2. It can be seen that the material is quite homogeneous, impurity phase inclusions both in the form of individual grains and in the form of intragranular inclusions are not observed. The geometrical grain size of ceramics varies from 2 μm to 15 μm, their shape is mainly irregular polyhedrons with rounded borders.

The results of the $\varepsilon'/\varepsilon_0(T,f)$ and tan $\delta(T,f)$ measurements of the $Sr_{0.75}Ba_{0.25}Nb_2O_6$ ceramic are shown in Fig. 3. A temperature decrease from 800 to 300 K is accompanied by an increase in $\varepsilon'/\varepsilon_0$ (at T = 360–600 K, the value of $\varepsilon'/\varepsilon_0$ is almost independent of f) and the formation of maximum, T_{m1}, which shifts to the high-temperature region with increasing measuring electric field frequency. Taking into account [1, 3] it is caused by the phase transition in the $Sr_{0.75}Ba_{0.25}Nb_2O_6$ phase from the paraelectric (PE) phase (P4b2) to the ferroelectric (FE) phase (P4bm). At T ~ 270–280 K, an anomaly is formed on the $\varepsilon'/\varepsilon_0(T)$ curves, which is most pronounced at low frequencies. The analysis of the $(\varepsilon'/\varepsilon_0)^{-1}(T)$ dependences (Fig. 3b) shows that the Burns temperature, T_d, (the temperature of the polar nano-regions appearance in the FER, above which the relaxors behave is like classical FE in the PE phase (the Curie-Weiss law is satisfied) is ~ 520 K.

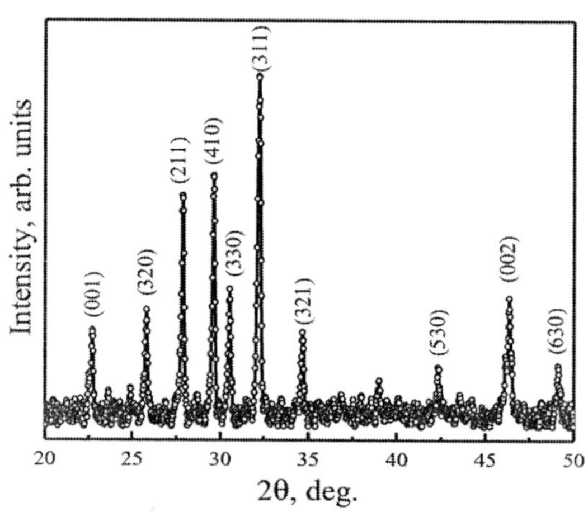

Figure 1. θ-2θ X-ray diffraction pattern of $Sr_{0.75}Ba_{0.25}Nb_2O_6$ ceramic at T = 300 K

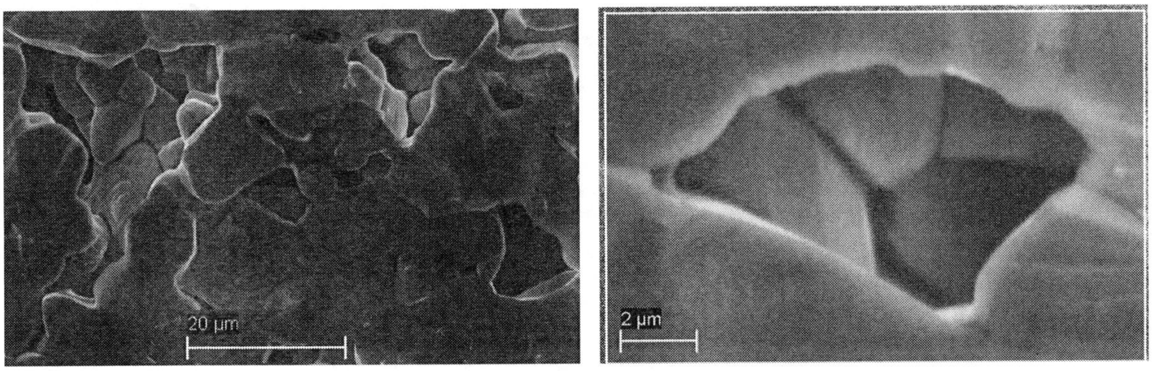

Figure 2. SEM images of $Sr_{0.75}Ba_{0.25}Nb_2O_6$cleaved facet

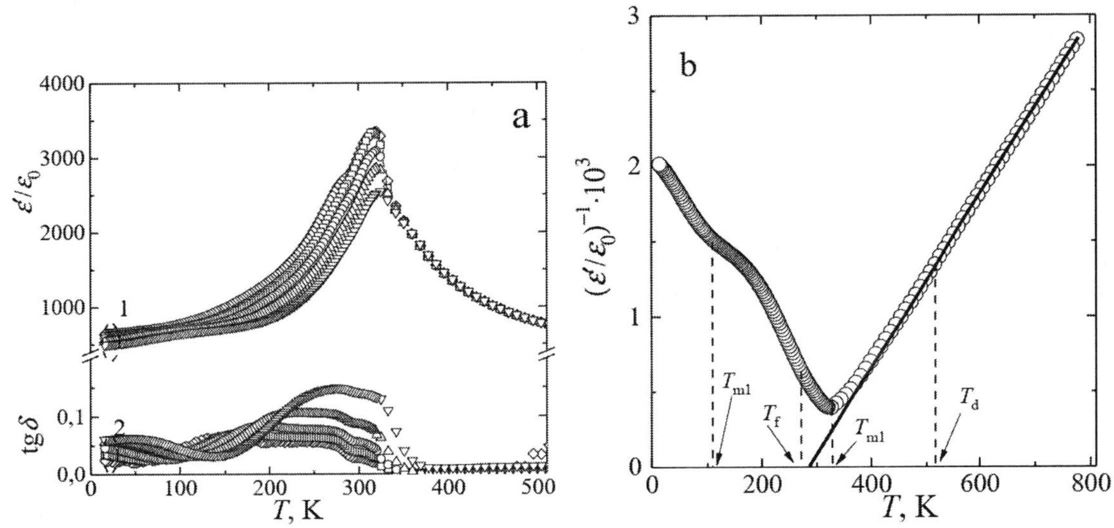

Figure 3. $\varepsilon'/\varepsilon_0(T)$ (1) and tg $\delta(T)$ (2) dependences at $T = 20$-600 K temperature rtange and $f = 10^2$–10^6 Hz frequency range, measured in the cooling mode (a) and $(\varepsilon'/\varepsilon_0)^{-1}(T)$ dependency at T = 20–800 K temperature rtange and $f = 10^6$ Hz frequency (b) of $Sr_{0.75}Ba_{0.25}Nb_2O_6$ ceramic

As can be seen from fig. 4, the dependence $T_{m1}(f)$ is described by the Vogel-Fulcher law:

$$f = f_0 \exp(E_{act}/(k\cdot(T_m - T_f))) \quad (1)$$

where f_0 is the frequency to overcome the potential barrier E_{act}, k is the Boltzmann constant, T_f is the Vogel-Fulcher temperature, interpreted as the temperature of the "static freezing" of electric dipoles or the transition to the state of a dipole glass. The calculated values of E_{act} and f_0 were 0.05 eV and 10^{12} Hz, respectively, and are comparable with those for the SBNx ceramics in [8].

The calculated value of T_f = 280 K allows to connect the anomalies in the $\varepsilon'/\varepsilon_0(T)$ and tan $\delta(T)$ dependences with the phase transformation of $Sr_{0.75}Ba_{0.25}Nb_2O_6$ from the relaxor (nanopolar) state to the macrodomain state, in which bulk FE domains coexist with near-surface nanoscale domains [4]. The values of T_f for the studied $Sr_{0.75}Ba_{0.25}Nb_2O_6$ ceramics exceed those in a single crystal [9] and are close to the values in ceramics [8].

At T <200 K, diffuse maximum (T_{m2}) are formed on $\varepsilon'/\varepsilon_0(T)$ and tan $\delta(T)$, which shift to the high-temperature region with f increasing. Taking into account the results of [10, 11] and found at T = 77 K the FE-like dielectric hysteresis loops, the diffuse low-temperature FE → FE phase transition from the tetragonal to monoclinic phase, characterized by the appearance of spontaneous polarization components in the perpendicular to the c axis direction [11], is a reason for the appearance of these hysteresis loops in $Sr_{0.75}Ba_{0.25}Nb_2O_6$.

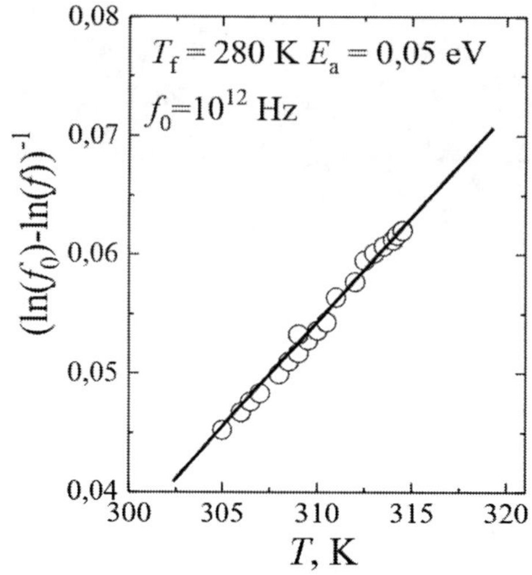

Figure 4. $(\ln(f_0)-\ln(f))^{-1}$ dependence on T_m, demonstrating the Vogel-Fulcher law in $Sr_{0.75}Ba_{0.25}Nb_2O_6$

Thus, we found that two diffuse ferroelectric phase transformations on the dielectric characteristics of $Sr_{0.75}Ba_{0.25}Nb_2O_6$ ceramics occur in the 20–800 K temperature range. The dielectric relaxation during the PE → FE transition (280–520 K) has a non-Debye character, which indicates the coexistence of the short and long polar orders in $Sr_{0.75}Ba_{0.25}Nb_2O_6$ [12] due to fluctuations in the chemical composition. These effects, in general, also occur in $Sr_{0.75}Ba_{0.25}Nb_2O_6$ polycrystalline thin films [5].

The θ-2θ X-ray diffractions patterns of $Sr_{0.75}Ba_{0.25}Nb_2O_6$/MgO heterostructure (Fig. 5) contained only reflections corresponding to $Sr_{1-x}Ba_xNb_2O_6$ solid solutions and MgO (001) substrate, which indicates the absence of impurity phases and buffer layers in the heterostructure. The φ-scans of (221) $Sr_{0.75}Ba_{0.25}Nb_2O_6$ reflection and (113) MgO reflection (Fig. 4) have proved that the barium-strontium niobate film was heteroepitaxially grown The presence of 8 reflections in the φ-scans of $Sr_{0.75}Ba_{0.25}Nb_2O_6$ indicates at two types orientation domains.

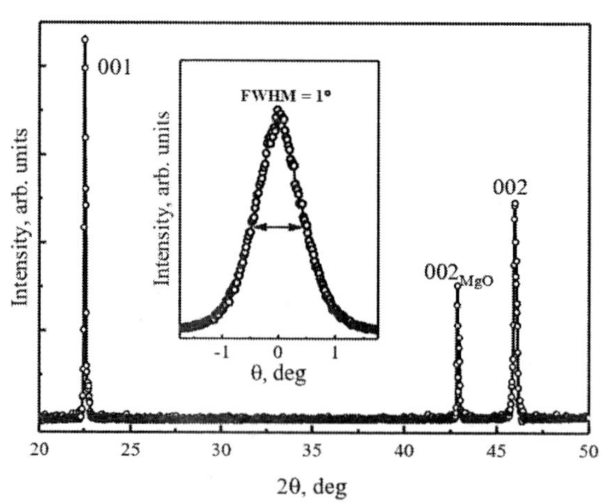

Figure 5. θ-2θ X-ray diffraction pattern of $Sr_{0.75}Ba_{0.25}Nb_2O_6$/MgO heterostructure. Rocking curve of (002) reflection in inset

A comparison of the positions for (221) reflections of the film and (113) reflections of a substrate demonstrate that the [110] crystallographic axes of two $Sr_{0.75}Ba_{0.25}Nb_2O_6$ orientation domains are rotated relative to the [110] axis of the MgO substrate at angles + 18.4° and –18.4°. The results of θ-2θ scans proved that the [001] axis of the substrate and both $Sr_{0.75}Ba_{0.25}Nb_2O_6$ orientation domains are codirectional. The formation of various orientation domains in SBNx films heteroepitaxially grown on MgO (001) substrates takes place in most works devoted to this material and it is associated with the specificity of the film and substrate crystal structures coupling [13]. By used our technology for the synthesis of $Sr_{0.75}Ba_{0.25}Nb_2O_6$ films the structure of the films is characterized by the presence of orientation domains with a rotation of only ± 18.4°.

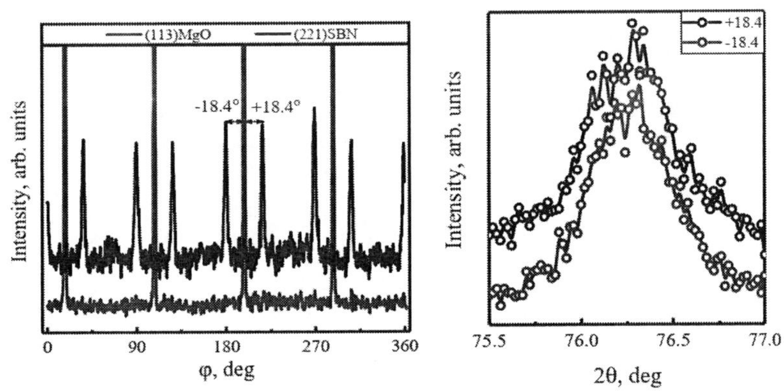

Figure 6: (a) φ-scans of (221) film reflection and (113) substrate reflection; (б) θ-2θ X-ray diffractogram of (313) reflection for two orientation domains +18.4° and –18.4°

The unit cell parameters of the Sr0.75Ba0.25Nb2O6 film calculated in the tetragonal approximation, were $a_f = 12.49 \pm 0.01$ Å, $c_f = 3.948 \pm 0.001$ Å. As can be seen from Fig. 6b both orientation domains have the same lattice parameters. The unit cell strains of the Sr0.75Ba0.25Nb2O6 film in the interface plane is $\varepsilon_{11} = (a_f\text{-}a_c)/a_c = 0.006$, and in normal to surface direction is $\varepsilon_{33} = (c_f\text{-}c_c)/c_c = 0.01$.

Thus, in the SBN-75 film magnitude of the unit cell strain is significant for the TTB structure, which can lead to a significant change in electrophysical and optical properties, to a temperature shift in the FE phase transitions. The effect of strain on SBN-75 film properties is the subject of our further research.

4. Conclusion

The results of the structure and dielectric characteristics studies of SBN-75 ceramics and thin films are presented. Based on the dielectric properties analysis of SBN-75 ceramics in the temperature range 20–800 K, the sequence and regions of phase transformations are determined. It found that the SBN-75/MgO heterostructure is epitaxial where only two types of orientation domains are existed. The crystallographic axes of orientation domains are rotated on ± 18.4 ° relative to the substrate axes. It was shown that the SBN-75 film unit cell is strongly strained relative to the bulk. The results can be used in the development and manufacture of functional materials based on $Sr_{0.75}Ba_{0.25}Nb_2O_6$ ferroelectric-relaxor thin films.

Acknowledgment

This work was carried out as part of the State Task for the SSC RAS (project no. 01201354247) and was supported by the grant MK-4100.2018.2.

References

[1] Kuz'minov Yu S 1982 *Segnetoelektricheskie kristally dlya uprav-leniya lazernym izlucheniem* (Moscow: Nauka) 400 p

[2] Ewbank M D, Neurgaonkar R R, Coy W K and Feinberg J 1987 Photorefractive properties of strontium-barium niobate *J. Appl. Phys.* **62** 374 DOI: 10.1063/1.339807

[3] Podlozhenov S., Graetsch H.A., Schneider J et al 2006 Structure of strontium barium niobate $Sr_xBa_{1-x}Nb_2O_6$ (SBN) in the composition range $0.32 \leq x \leq 0.82$ *Acta Crystallogr. Sect. B: Struct. Sci.* **62** 960–5 DOI: 10.1107/S0108768106038869

[4] Chernaya T S, Maksimov B A, Verin I V et al 1997 Kristallicheskaya struktura monokristallov $Ba_{0.39}Sr_{0.61}Nb_20_6$ *Cristallogr.* **42(3)** 421–6

[5] Santos I A, Guerra L S, Araújo E B et al 2008 Mesoscopic structure evidenced by AC dielectric nonlinearities in $Sr_{0.75}Ba_{0.25}Nb_2O_6$ relaxor ferroelectric thin films *J. of Phys.: Condensed Matter* **20(13)** 135209 DOI: 10.1088/0953-8984/20/13/135209.

[6] Pavlenko A V, Zakharchenko I N, Anokhin A S et al 2017 Structure, the lattice dynamic, and the dielectric characteristics of Sr0.5Ba0.5Nb2O6 films *Phys. of the Solid State* **59(5)** 909–13 DOI: 10.1134/S1063783417050249.

[7] Pavlenko A V and Nazarenko A V 2018 Dielectric characteristics of $Sr_{0.75}Ba_{0.25}Nb_2O_6$ ceramics *Sci. in the south of Russ.* **14(2)** 3–8 DOI: 10.23885/2500-0640-2018-14-2-3-8

[8] Zhao Y, Wang J, Zhang L et al 2016 Relaxor transition and properties of Mn-doped $Sr_xBa_{1\_x}Nb_2O_6$ ferroelectric ceramics *Ceram. Int.* **l42** 16697–702 DOI:10.1016/j.ceramint.2016.07.120

[9] Dec J, Kleemann W, Woike Th and Pankrath R 2000 Phase transitions in $Sr_{0.61}Ba_{0.39}Nb_2O_6$:$Ce^{3+}$: I. Susceptibility of clusters and domains *Eur. Phys. J. B* **14(4)** 627 DOI: 10.1007/s100510051071

[10] Buixaderas E, Savinov M, Kempa M et al 2005 Infrared and dielectric spectroscopy of the relaxor ferroelectric $Sr_{0.61}Ba_{0.39}Nb_2O_6$ *J. Phys.: Condens. Matter.* **17(4)** 653 DOI: 10.1088/0953-8984/17/4/008

[11] Pilipenko A S, Burkhanov A I and Ivleva L I 2009 Evolution of the dielectric and acoustic parameters of chromi-um-doped SBN single crystals with variations in temperature *Phys. of the Solid State* **51(3)** 577–81 DOI: 10.1134/S1063783409030238

[12] Laguta V V, Glinchuk M D and Kondakova I V 2004 Vogel-Fulcher law as a criterion for identifying a mixed ferroe-lectric-glass phase in potassium tantalate doped with lithium *Phys. of the Solid State* **46(7)** 1262–9 DOI: 10.1134/1.1778451

[13] Willmott P R, Herger R, Patterson B D and Windiks R 2005 Experimental and theoretical study of the strong dependence of the microstructural properties of $Sr_xBa_{1-x}Nb_2O_6$ thin films as a function of their composition *Phys. Rev. B.* **71** 144114 DOI: 10.1103/PhysRevB.71.144114

IOP Conf. Series: Materials Science and Engineering **905** (2020) 012054 doi:10.1088/1757-899X/905/1/012054

Proposal development for the boiler modernization at the Far North enterprise

G Z Samigullina and G R Platunova

Udmurt State University, Izhevsk, Russia

E-mail: gyzals@mail.ru, dyukina-guzel@yandex.ru

Abstract. The article presents the transfer relevance of the boiler from diesel to natural gas. There is an economic calculation of the efficiency in using a gas-equipped PKN-2 boiler with a GG-1 type burner, as well as the environmental efficiency of modernizing the boiler installation. Place: Novo Energy Services LLC enterprise, the Yamalo-Nenets Autonomous District, the Yurkharovskoye oil and gas condensate field, bush No. 2.

1. Introduction

The development of modern society is associated with the use of energy. In connection with the acceleration of scientific and technological progress fuel and energy resources are considered one of the most important issues. With the evolving structure of modern production, liquid and gaseous fuels – gas, diesel fuel, gasoline, fuel oil and various oils – play an important role. The evolution of boilers and heating systems went along with the development of human civilization. Along with the development of heating equipment, the range of materials used in the form of fuel expanded: first firewood, then coal, oil, fuel oil, diesel fuel, natural gas, liquefied gas, solar radiation, nuclear energy, etc. Along with the development of heating equipment, the struggle for fuel economy began – I wanted to burn fuel as little as possible, and get as much heat as possible. This task is becoming more and more urgent in our time, as the usual energy resources on the planet are depleted, and fuel, accordingly, is becoming more expensive [1].

The past 10–20 years the environmental component of fuel combustion became an important issue. The combustion process releases a large number of harmful elements (oxides). The most harmful of these are nitrogen oxides NOx and carbon monoxide CO. Moreover, the higher combustion temperature is the more harmful substances are released. Now there is a struggle to reduce the temperature of combustion in burner devices, and with it the reduction of emissions of harmful substances into the atmosphere [2–5].

The most acute problem of the high cost of fuel and the protection of the environment from flue gases with harmful substances is faced by the boiler plant Novo Energy Services, located in the Yamal-Nenets Autonomous District of Russia *(Nadym district, Yurkharovskoye oil and gas condensate field, bush No. 2)*.

The climate features of the territory are affected by its northern position (680 N), influx of solar radiation, as well as increased cyclonic activity and the flat nature of the underlying surface. Permafrost, the proximity of the cold Kara Sea, and the abundance of swamps, lakes and rivers influence climate formation. The continent has a significant impact, which is manifested in the pronounced winter-summer features of the transformation of air masses and the continental climate.

The climate features of the region were adopted at the nearest weather station – Tazovsky. The climate of the region is characterized by long and cold winters, strong winds and blizzards, short and cool summers with long daylight hours. There is cloudy sky and relative humidity observed throughout the year. The annual and daily amplitudes of air temperatures are large, despite the proximity of the Kara Sea and the water content of the territory.

The average annual temperature is negative (–9.3 °C). The average temperature of the coldest month (January) –26.7 °C, and the hottest (July) –13.4 °C. The absolute minimum temperature falls in January –57 °C, the absolute maximum – July +31 °C. The period with a temperature below 0 °C is on average 243 days. The intensive study of the north requires good eather conditions with a large amount of heat due to harsh climate. For these purposes small-sized boiler houses running on diesel fuel were used.

Research objective is project development for the modernization of the boiler room at the Novo Energy Services enterprise (Yamal-Nenets Autonomous Okrug of the Russian Federation, Nadymsky District, Yurkharovskoye oil and gas condensate field, bush No. 2).

2. Methods and materials

The boiler room at the Novo Energy Services LLC is equipped with a steel boiler named 'KSV-I, 9 LV "VK-3" with a burner that runs on liquid fuel. It is intended for heating and hot water supply in closed heat supply systems with a maximum cool temperature of up to 115 °C and a pressure of up to 0.6 MPa [4]. Diesel fuels represent a fraction from the boiling point of 140–200 °C to the boiling point of 330–360 °C. Chemical stability of diesel fuel considers the ability to withstand oxidative processes during storage. The presence of heteroatomic compounds, especially in combination with unsaturated hydrocarbons, contributes to their oxidative polymerization and polycondensation, thereby affecting the formation of resins and precipitates. Chemical stability is estimated by the amount of sediment formed in the fuel (mg/100 ml) according to ASTMD 2274.

Summer diesel fuel

Density: about 860 kg/m^3

Flash point: 62 °C

Pour point: –5 °C

Obtained by mixing straight-run, hydrotreated and secondary-origin hydrocarbon fractions with a boiling point of 180–360°C. An increase in the temperature of the end of boiling leads to increased coking of the nozzles and smokiness.

Winter diesel fuel

Density: about 840 kg/m^3

Flash point: 40 °C

Pour point: –35 °C

Obtained by mixing straight-run, hydrotreated and secondary-origin hydrocarbon fractions with a boiling point of 180–340 °C. Winter diesel fuel is also obtained from summer diesel fuel by the addition of a depressant additive, reducing the pour point of the fuel, and slightly changing the temperature of the maximum filterability. By the artisanal method, up to 20 % of TS-1 or KO kerosene is added to summer diesel fuel, while the operational properties are practically unchanged.

Arctic diesel fuel

Density: about 830 kg/m^3

Flash point: 35 °C

Pour point: –50 °C

Obtained by mixing straight-run, hydrotreated and secondary-origin hydrocarbon fractions with a boiling point of 180–330 °C. The boiling limits of Arctic diesel fuel approximately correspond to the boiling limits of kerosene fractions, therefore this fuel is, in fact, heavier kerosene. However, pure kerosene has a low cetane number of 35–40 and insufficient lubricating properties. To eliminate these problems, cetane-enhancing additives and mineral motor oil are added to Arctic fuels to improve lubricating properties. A more expensive way to produce Arctic diesel fuel is dewaxing summer diesel fuel [5].

Natural gases are colorless, odorless, and tasteless. The main indicators of combustible gases that are used in boiler houses are: composition, calorific value, specific gravity, combustion and ignition temperature, explosive limits and flame propagation speed. Natural gases from purely gas fields consist mainly of methane (82...98 %) and other hydrocarbons. The density of natural gas (methane) under normal conditions (0 °C and 0.1 MPa, i.e. 760 mmHg), rg = 0.73 kg/m^3 (i.e., lighter than air by about 1.8 times)

Compared to other types, gaseous fuel has the following advantages: it burns in the theoretical amount of air, which ensures high thermal efficiency and combustion temperature; during combustion does not form undesirable products of dry distillation and sulfur compounds, soot and smoke; It is relatively easily supplied via gas pipelines to remote consumption facilities and can be stored centrally; the possibility of increasing the efficiency (efficiency) of the boiler unit; facilitation of work of staff; easily ignited at any ambient temperature; it requires relatively small costs in production, which means it is a cheaper type of fuel compared to other; can be used in compressed or liquefied form for internal combustion engines; relatively high environmental friendliness, characterized by the absence of solid inclusions in the combustion products and less harmful gaseous emissions; has high anti-knock properties; during combustion it does not form condensate, which provides a significant reduction in wear of engine parts, etc.

At the same time, gaseous fuel also has certain negative properties as toxic effect, explosive mixtures formation when mixed with air, easy flow through leaks of compounds, etc. Therefore, gaseous fuel reactions require careful observance of the relevant safety rules.

Thus, there arise such issues as organization of a warehouse for storing liquid fuel stock with the accompanying toughened fire protection measures when operating a boiler plant on diesel fuel. It is necessary to provide measures against freezing, in the winter, of the liquid fuel supply pipeline to the boiler installation. Gas requires relatively low production costs, which means it is a cheaper type of fuel compared to other. Relatively eco-friendly , characterized by the absence of solid inclusions in the combustion products and fewer harmful gaseous emissions.

The study shows the most optimal method for upgrading the boiler room is to replace KSV-I, 9 LV VK-3 with a PKN-2M boiler equipped with a GG-1 burner. The need to modernize boiler houses is made by the equipment deterioration (up to 70–80 %), the presence of a large number of small boiler houses with inefficient liquid fuel boilers (actual boiler efficiency is 50–60 %, normal is 75–80 %). Almost all old boiler houses lack water treatment.

The insertion point of the gas pipeline should be made into the existing high-pressure steel gas pipeline Du 100, Ru 1.2 MPa. The PKN-2M boiler is a stationary vertical-water tube, double-drum boiler with screens included in the circulation circuit through four collectors. Screens and convective beam are made of steel pipes with 57x3 diameters. The boiler is designed to operate in steam mode.

The PKN-2M boiler is equipped with a GG-1 type burner manufactured by Perlovsky Power Equipment Plant OJSC and designed to burn gaseous fuels. GG-1 burners passed state tests. The parameters of the PKN-2M boiler and burner are presented in table 1.

For the operation of the boiler room, gas will be supplied from the Yurkharovskoye field through the main gas pipeline.

The project provides for: connection to the existing steel gas pipeline Du 100, Ru 1.2 MPa; installation of cabinet hydraulic fracturing; entering a low pressure gas pipeline into the boiler room; internal gas equipment of a boiler room, installation of automatic control and safety controls for boiler units, installation of a gas alarm according to CO and CH4 [6].

The main fuel provided is endowed natural gas according to GOST 5542-87.

The project was developed in accordance with the requirements of SNiP 42-01-2002, "Gas distribution systems" [7], "Safety rules in the gas sector" and provides for the conversion of the boiler house to natural gas.

Prior to the start of installation of gasification, a customer's technical supervision representative should be appointed from among the engineers who passed the exam for knowledge of the "Safety Rules in the Gas Sector" and the relevant chapters of SNiP.

IOP Conf. Series: Materials Science and Engineering **905** (2020) 012054 doi:10.1088/1757-899X/905/1/012054

Table 1. Boiler parameters pkn-2m and gg-1 burners

№	*Name of parameter*	*Range Unit*	*Capacity*
	BOILERS PKN-2M Art. No. 1–3.		
1	Thermal value	T/per hour	1,0
2	Vapour pressure	Kg/sm	8,0
3	Boiler inlet water temperature	°C	50
4	Saturated-steam temperature	°C	150–170
5	The efficiency of the boiler at rated load (calculated)	%	78
6	Gas-flow rate	Nm/per hour	190
7	Flue-gas temperature	°C	260
8	End-to-end dimensions		
	– length	mm	3330
	– width	mm	2160
	– height	mm	2465
9	Heating surface	m^2	57
10	Water zone	M^3	2,3
	GG-1 BURNER		
1	Amount	Pcs	1
2	Normal heat release	Cal/ per hour	1.1
3	Gas calorific rate $Q^р_н$=7960ккал/м^3	Nm /per hour	119
4	Burner inlet gas temperature	°C	0–40
5	Burner inlet water temperature	°C	10–40
6	Burner inlet gas pressure rating	kPa	0,72
7	Minimum coefficient of excess air at rated thermal power	–	1,1
8	Dimensions:		
	– length	mm	265
	– width	mm	720
	– height	mm	870

The finished project must be agreed with the gas supply organization and registered with the gas technical inspection [5].

The boiler house gas supply system includes: GRPSh with an RDNK-50 regulator; high pressure steel overhead gas pipeline (Du 50, Ru 1.2 MPa); steel internal gas pipeline of low pressure (DN 150, Ru 0.002 MPa); entering the gas pipeline into the boiler room; gas equipment of the boiler room; automatic safety of combustion, regulation and alarm operation of boilers.

The project provides for a special gas service, which is entrusted with the following tasks: continuous technical supervision of the gas industry; conducting preventive audits; repair of gas equipment and facilities on them; performing gas hazardous work in the gas industry and ensuring readiness at any time to take measures to prevent or eliminate the accident associated with the operation of gas pipelines and gas equipment; maintaining the stability of gas parameters and ensuring uninterrupted supply of gas in the quantities necessary for consumers; accounting of gas consumption and control of its rational use.

The gas service is managed by the chief power engineer of the enterprise or his deputy, and in the absence of such a position, it is run by the chief mechanic. The chief power engineer of the enterprise or his deputy is the person responsible for the gas economy of the enterprise.

Directly the gas service is headed by the head of the gas service. The gas service should be equipped with a telephone service with a gasified boiler room and the Raigaz emergency service.

Persons no younger than 18 years old who have been trained under a special program for gas boiler operators can be allowed to care for boiler units. A staffing schedule is provided for servicing the boiler house, consisting of: a senior operator; boiler room operator (2 in 1 shift); electrician.

3. Results

The result analysis of calculations of emissions dispersion of harmful substances into the atmosphere from the sources in question shows that the use of natural gas as a fuel, in comparison with solid fuel – coal, can significantly improve the sanitary and hygienic conditions in the adjacent territory, by eliminating the content of ash in the air basin, soot, dust, sulfur dioxide and reduce nitric oxide. (The reduced concentration of carbon monoxide COa = 1.0 in the exhaust gases at a load of 32 % is 0 mg/m^3, at a load of 62 % – 0 mg/m^3, the concentration of nitrogen oxides NOx a = 1.0 in the exhaust gases at a load of 32 % – 319 mg/m^3, 62 % – 365 mg/m^3. The specific content of carbon monoxide CO per 1 ton of emitted steam at a load of 32 % is 0 g/t, at a load of 62 % – 0 g/t, the specific content of nitrogen oxides NOx per 1 ton of emitted steam at a load of 32 % is 43.80 g/t, with a load of 62 % – 49.30 g/t. The hourly emission of carbon monoxide at 32 % load is 0 kg/h, at 62 % – 0 kg/h, the emission of nitrogen oxides at 32 % load ke 0.11 kg/h, 62 % load – 0.23 kg/h. Emission of carbon monoxide per 1 thousand m^3 of burnt gas with a load of 32 % – 0 kg/thousand m^3, at 62 % – 0 kg/thousand m^3, emission nitrogen oxides per 1 thousand m^3 of burnt gas with a load of 32 % – 0.431 kg/thousand m^3, with 62 % – 0.490 kg/thousand m^3).

The technological process of gas combustion is provided by an automation system, which guarantees environmental safety of operation. The automation and control system of the boiler room is equipped with the necessary number of sensors that provide complete protection of the environment from harmful emissions.

According to the "Interim Methods for the Regulation of Gas Consumption in Medium and Low Power Boilers", the control individual gas consumption rate for boilers of a given heating capacity is: for boilers PKN-2M-190 kg. t/t/h, which corresponds to an efficiency of 74.8 %. The reference individual gas flow rate for boilers of a given heating capacity is 168 kg. t/t/h, which corresponds a boiler equal to 85.0 % [8–14].

The results of the tests and the calculation of the economic side of the conversion of the boiler house to gas showed that the annual economic effect amounted to 47 million rubles in 2018 (by increasing the efficiency of installed gas boilers to 90–92 %, introducing general boiler automation, modern water treatment, and reducing costs) for transportation and storage of fuel). During the economic calculation of the return on investment through the use of highly efficient energy-saving technologies based on modern highly cost-effective equipment, it is 6.5 months. Savings for the heating season during the operation of the gas boiler will amount to 11 783 431.19 rubles. This is achieved largely by replacing diesel fuel with gas fuel. There is a decrease in emissions of pollutants into the environment, which means that the payment for emissions has decreased.

The basic principles of organizing the operation of boiler ensure reliable, economical and trouble-free operation of the equipment, therefore, it is necessary to develop a safety regulation for the maintenance of a modernized boiler plant.

The results of tests, economic calculations and an environmental assessment of the modernization of the boiler plant at the Novo Energy Services LLC enterprise revealed its high economic and environmental efficiency.

4. Conclusion

1. An effective method of upgrading the boiler installation in the conditions of the Novo Energy Installations LLC enterprise is to replace the boiler *KSV-I, 9 LV VK-3* with a burner operating on liquid fuel for a PKN-2M boiler equipped with a *GG-1 burner*.

2. Gas is an "internal" fuel cheaper than other sources and the most environmentally friendly in comparison with diesel fuel, because it is imported and the cost of its transportation is needed.

3. In the practical implementation of the proposed modernized boiler plant, there will be a reduction in emissions of toxic gases into the atmosphere.

4. The reduction in pollution charges for the environment will amount to 1,317.19 rubles (in 2018 prices).

5. To ensure reliable, economical, trouble-free operation of the boiler installation, qualified specialists and compliance with labor safety measures (including electrical and fire safety rules) are required.

References

[1] Arbatov A M 2000 Mineral'no-syr'evaja baza strany *Ekon.* **2** 13–22
[2] Samigullina G Z 2017 *Istochniki zagrjaznenija sredy obitanija* (Izhevsk: Vostochno-Evropejskij institut) 224 p
[3] Samigullina G Z and Nazarov P V *Himija okruzhajuschej sredy* (Izhevsk: Izd-vo Kamsk. Instit. Guman. i inzhener. Tehnol.) 158 p
[4] Chebatorev V P 2000 *Spravochnik rabotnikov raznoficirovannyh kotel'nyh* (Kiev: Osnova) 298 p
[5] Ahmedov R B and Cirul'nikov L M 1984 *Tehnologija szhiganija gorjuchih gazov i zhidkih topliv* (Leningrad: Nedra) 238 p
[6] *Trebovanija organov Gosgaznadzora k provedeniju naladochnyh rabot na gazoispol'zujuschem oborudovanii*, ORD-1-87. m. 1987.
[7] SNiP 42-01-2002 *Gazoraspredelitel'nye sistemy* 2003.
[8] *Metodika raschjota udel'nyh norm rasxoda gaza na vyrabotku teplovoj energii i raschjota poterj v sistemah teplosnabzhenija (kotel'nye i teplovye seti)* STO GAZPROM RD 1.19-126-2004
[9] *Pravila bezopasnosti v gazovom hozjajstve s izmenenijami no 1, Izmenenija, no 2* 1999 (Moscow: PBGH NPO OBT)
[10] *Metodika rascheta vybrosov zagrjaznja'uschih veschestv v atmosferu ot stacionarnyh dizel'nyh ustanovok* 2001 (St. Petersburg: NII Atmosfera; firma "Integral") 13 p
[11] Pugach L I 2003 *Energetika i ekologija* (Novosibirsk: NGTU) 502 p
[12] SNiP 2.35-76 *Kotel'nye ustanovki* 47 p
[13] Rosljakov P V, Egorova L E and Ionkin P A 2000 *Metody rascheta vybrosov vrednyh veschestv s dymovymi gazami kotlov* (Moscow: Izd-vo MI) 68 p
[14] PB 10-574-03 *Pravila ustrojstva i bezopasnoj ekspluatacii parovyh i vodogrejnyh kotlov* 2003

IOP Conf. Series: Materials Science and Engineering **905** (2020) 012055 doi:10.1088/1757-899X/905/1/012055

Modification as a method of directed change in the properties of ferro-piezoceramics

Kh A Sadykov[1,2], L A Shilkina[1], M R Daudov[1], I A Verbenko[1], A V Nagaenko[1] and L A Reznichenko[1]

[1] Southern Federal University, 105/42, Bolshaya Sadovaya Str., Rostov-on-Don, 344006, Russia
[2] Chechen State University, Grozny, 17a, Dudayev Boulevard, Grozny, 364060, Russia

E-mail: hizir-2010@mail.ru, lid-shilkina@yandex.ru, ilich001@yandex.ru, nagalexandr@gmail.com, reznichenko@mail.ru

Abstract. Solid-phase synthesis followed by sintering using conventional ceramic technology produced ceramics based on the binary system of solid solutions (Na, Li) NbO3. The influence of superstoichiometric modification with nickel oxide on the internal structure (crystalline, grain) and macroresponses of the analyzed media is established. Based on the analysis of piezoelectric and ferroelastic properties, the prospects of using materials in various electronic devices are shown.

1. Introduction

One of the most effective methods for changing the physical properties of ferroelectrics is the method of introducing small (no more than 10 at. %) Additives of monoxides of various elements (modification). The great potential of this method is due to the existence of a significant amount of metal monoxides and their combinations capable of dissolving in complex oxides, in particular, in ferroelectric (FE) oxides with a perovskite-type structure (OSP) over a wide concentration range, as well as oxides of elements forming when introduced into the initial objects "melt" (B2O3, SiO2, V2O5, ...), and, finally, additives of combined action that affect the properties of the initial systems due to the formation of a liquid phase (LP) and cation-exchange interaction I'm with the base (modifying glass). The undoubted advantage of the method is the ability, while preserving the complex of specific properties of the modified objects, to slightly change the values of some parameters in the right direction by slightly varying their composition.

In the case of a significant change (improvement) in the ceramic and technological characteristics of the materials being modified, as well as such changes in the dielectric and piezoelectric parameters, which can be considered as "tuning" them to the target values, the modification is represented by the parameter optimization method. Changes in these characteristics to such an extent that leads to the formation of a new set of properties, allows modification to be considered as a method of creating qualitatively new materials with adjustable parameters [1].

From a theoretical point of view, the establishment of general patterns of changes in physical properties during OSP modification provides information on the composition – structure – property relationships, which allows us to consider it as a method for studying the FE state. Thus, the modification results have a wide range of applications – from basic research to technology processes at the atomic-molecular level, in connection with this, the urgency of the FE OSP modification problem becomes clear.

According to bibliographic data, the modification of solid solutions (SS) based on the PZT system (Pb(Ti,Zr)O3) is most fully studied. At the same time, the possibilities of modifying the system itself are now practically exhausted. Multicomponent lead-containing SSs were examined quite deeply in this regard. The analysis and generalization of the vast experimental material (more than 300 combinations of modifying oxides are known for the PZT system alone) allowed us to classify all known modifiers of these systems and, on this basis, establish general patterns of change in their physical properties. This was successfully used in predicting the properties of ferro-piezoelectric ceramic materials (FPCM) and in finding the optimal ways to create them [1]. The possibilities of modifying SS based on sodium niobate (SN) have been little studied. The only exception is the NaNbO3 – KNbO3 system [2]. Recently, attention has been paid to another "niobate" system — NaNbO3 – LiNbO3, as well as multicomponent systems with its participation [3].

The advantages of the latter: increased dimensionality of the morphotropic region and, as a result, expanded choice of SS with given combinations of parameters; expanding the possibilities of SS formation with a wide variety of properties; increasing the efficiency of SS and improving their manufacturability; as well as the absence of toxic elements in their composition; – make these compositions very promising for industrial applications.

The table shows the formulas and quasi-chemical equations of cationic modification of the SS system (Na, Li) NbO3.

By cationic is meant modification by elements other than those constituting the base, by vacancy, change in the defective state of the base without a qualitative change in its chemical composition (only the concentration of the AO component or the degree of oxidation of the B component varies). Despite the conventionality of such a division (in the first case, the formation of vacancies also takes place, and in the second, the chemical composition changes), it seems legitimate, since in each of the options the leading mechanism of the formation of a new set of properties is highlighted (or, primarily, due to the influence of the crystal chemical characteristics of the modifiers, either due to the quantity and "quality" of vacancies).

The scheme presents variants of stoichiometric (isovalent vacancy-free and heterovalent with the formation of vacancies, the position of which in the crystal structure may be different) and superstoichiometric isomorphic substitution of ions in AB, AB sublattices. Moreover, with A-substitutions (I), only the case with nMA >nA * (that is, nM = 2, 3) is considered, since the valency is IIĮM = 1; for B-substitutions, the possibilities with nM<>nB are described (that is, nMB = 6, 4, 3). Isovalent modification is implemented with nMA = 1, nMB = 5. In combined modification, when ions in both cationic sublattices are replaced, a compensation variant is possible that excludes the appearance of vacancies (nMA = 2 and nMB = 4, nMA = 3 and nMB = 3), or the path associated with their appearance in A, B, AB positions.

With the introduction of modifiers in excess of stoichiometry, the variants of changes in the crystal lattice of the base can be greater than with their introduction stoichiometrically. The cation modifier can become regular nodes of the matrix crystal, which leads to the formation of vacancies in the cationic and oxygen sublattices (cation A forms vacancies in the B and O sublattices, cation B in the A and O sublattices); can be embedded in the lattice, displacing the main component, can itself be embedded in the lattice, or will be present as an impurity, distributed in the intergranular space.

In connection with the foregoing goal, the work was to establish the laws governing the formation of the internal structure (crystalline, grained) and macroresponses (dielectric, ferroelastic, piezoelectric) SS with the participation of (Na, Li) NbO3 superstoichiometrically modified with nickel oxide. The choice of this dopant is due to its crystal-chemical specificity, which provides an improvement in the entire set of electrophysical properties of the base SS [4].

IOP Conf. Series: Materials Science and Engineering **905** (2020) 012055 doi:10.1088/1757-899X/905/1/012055

2. Objects. Methods for producing and research samples

SSs based on the (Na1-xLix) NbO3 system (composition – Na0.86625Li0.12375Sr0.01Nb0.988Al0.01O3 (Basis) were chosen as objects of study. The modification was performed stoichiometrically with a concentration step of 0.5 mass %. In this case, simple NiO oxide was used as additives. Solid-phase synthesis was carried out in two stages: Tsint.1.2 = (850 – 870) °C (depending on the composition), for τ1 = τ2 = 6 hours, sintering using conventional ceramic technology at a variable temperature, Tsp., Equal to (1180, 1200) °C for 1 hour. Samples were made in different geometries to determine the effect of habit on electrophysical characteristics.

The phase composition and completeness of the synthesis were checked by X-ray diffraction on a DRON-3 diffractometer using filtered Coka radiation from Coka (Bragg-Brentano focusing). The experimental, ρexp., Density of the samples was determined by hydrostatic weighing in octane. The grain structure was studied using a JSM – 6390L scanning electron microscope (JTOL) (Japan) with a system of microanalyzers from Oxford instruments (Great Britain).

Measurements of the electrophysical parameters of SS were carried out in accordance with OST 11 0444-87. In this case, we determined the relative permittivities of polarized, ε33t / ε0, and non-polarized, ε / ε0, samples, the dielectric loss tangent of polarized samples, tgσpp, piezoelectric modules, | d31 |, d33, the electromechanical coupling coefficient of the planar oscillation mode, KP, mechanical quality factor, Qm, speed of sound,, piezosensitivity, g33, and acoustic impedance, z, (g33 = d33 / ε33т and z = ρexp., Where ρexp. Is the experimental density of the samples). The Curie temperature, TC, was established on the basis of thermofrequency measurements carried out in the temperature range (20–700) °C and frequencies of 25 Hz – 1 MHz using an Agilent E4980A precision LCR meter.

3. Experimental results, discussion

Fig. 1. shows the dependence of the parameters of the perovskite cell and the homogeneous parameter Strain SS δ versus NiO concentration. In the range $0 \leq x \leq 1.0$, two phases coexist in SS – the rhombohedral (Rh) and rhombic (R (M2) doubled monoclinic subcells) (M2), but their ratio changes with increasing x, the amount of the latter decreases and in the range 1.0 <x < 2.0 TR becomes almost single-phase – Rh.

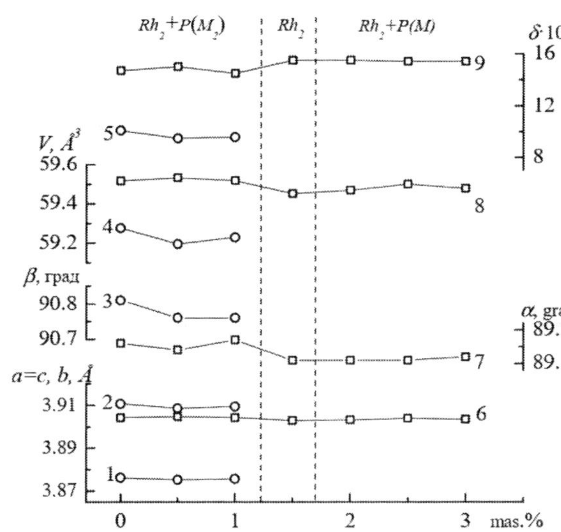

Figure 1. Dependences of the parameters of the perovskite cell of the solid solution SS based on (Na, Li) NbO3, modified NiO, on the content of the modifier: 1–b_M, 2–a_M, 3–angle β, 4–Vm, 5–δm, 6–a_{Rh}, 7–угол α, 8–V_{Rh}, 9–δ_{Rh}

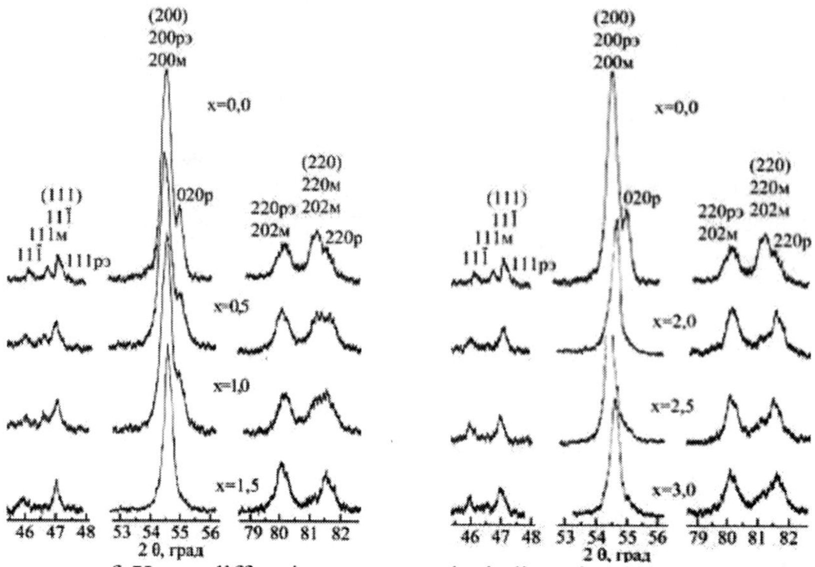

Figure 2. Fragments of X-ray diffraction patterns, including the (111) k, (200) k, and (220) k diffraction lines of the studied SSs based on (Na, Li) NbO3, modified superstoichiometrically NiO (x = wt. %)

In fig. 1. it is seen that at the point x = 1.5, the cell volume and the angle α sharply decrease: ΔV = –0.07 Å3, Δα = –0.06o. At x> 1.5, the cell parameters practically do not change with x, with the exception of the volume, which increases slightly, while the phase composition changes smoothly. In fig. 2. It is seen how, with increasing NiO concentration, the intensity of the lines of the P (M) phase gradually increases, that is, SS again becomes biphasic, but the amount of P (M) phase is much smaller than Pe.

Thus, upon superstoichiometric modification of SS based on (Na, Li) NbO3 with nickel oxide, the latter is embedded in the crystal lattice of the initial SS, substantially changing its phase composition.

In fig. 3. Fragments of the microstructure of the studied ceramics are presented, indicating fine-grained and densest packing of crystallites. The porosity of ceramics is point, intergranular, and insignificant. An increase in the NiO content leads to a refinement of the grain landscape and densification of the ceramic frame, probably due to the formation of small amounts of liquid phases.

Figures 4–6 illustrate the dependences of the dielectric, ferroelastic, and piezoelectric characteristics of SS on the concentration of the introduced dopant, and Fig. 7 – thermofrequency "behavior" ε / ε0 and tanδ.

It is clearly seen that enrichment of the base SS with nickel leads to a decrease in ε33T / ε0, tanδ, and an increase in Qm (TSP = 1190 °C), which indicates an increase in the degree of ferro rigidity of SS. The mechanism of the latter can be related to the following.

In the case of substitution of alkali metal ions for higher-charged cations (Ni2 +), anionic excess SSs are likely to form, in which excess oxygen can be located either in internodes or in positions shifted from internodes.

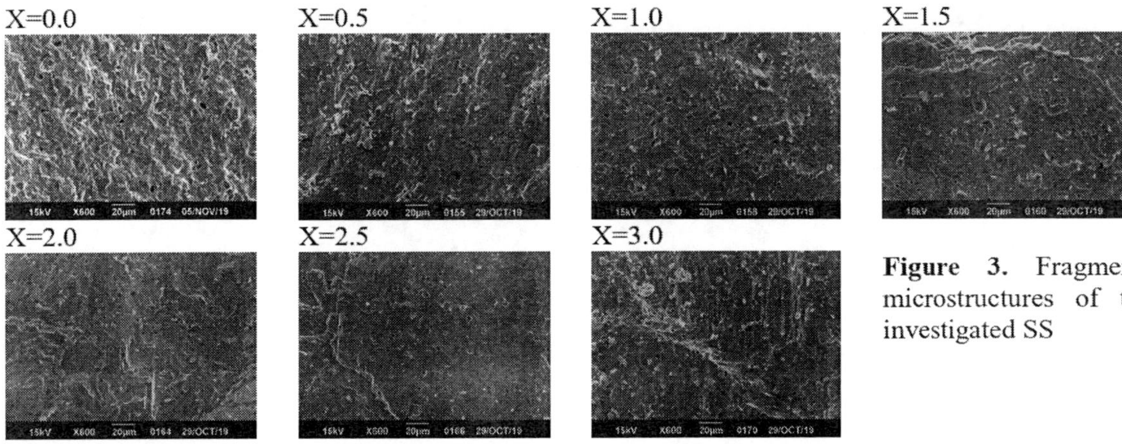

Figure 3. Fragments microstructures of the investigated SS

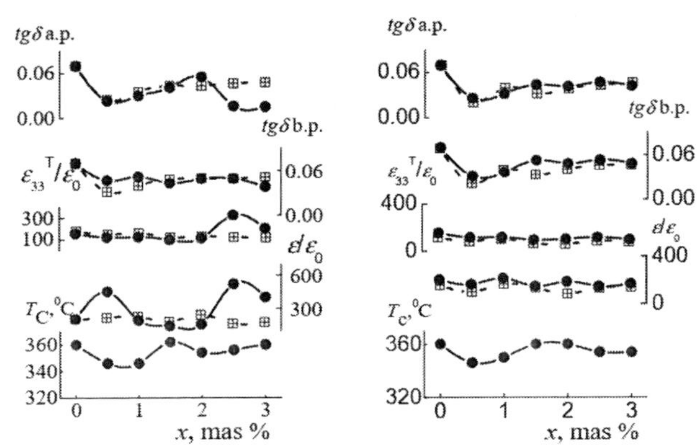

Figure 4. Dependences of the dielectric characteristics of solid solutions based on (Na, Li) NbO3 modified with NiO. (On the left Tsp = 1190 °C, on the right Tsp = 1205 °C, solid lines samples Ø 10 mm, dashed Ø 20 mm)

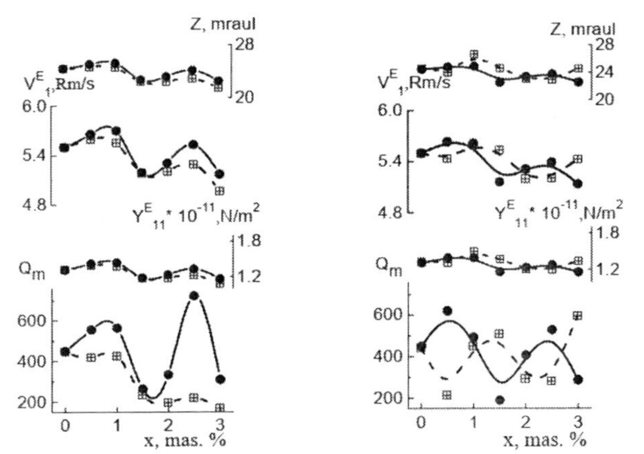

Figure 5. Dependences of the ferroelastic characteristics of solid solutions based on (Na, Li) NbO3 modified with NiO. (Left – Tsp = 1190 °C, right – Tsp = 1205 °C, solid lines – samples Ø 10 mm, dashed – Ø 20 mm)

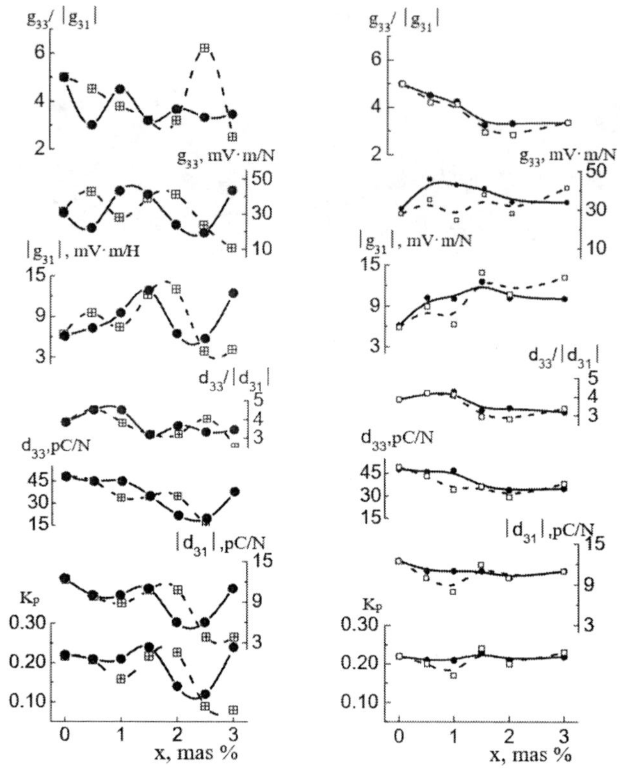

Figure 6. Dependences of the piezoelectric characteristics of solid solutions based on (Na, Li) NbO3 modified with NiO. (Left – Tsp = 1190 °C, right – Tsp = 1205 °C, solid lines – samples Ø 10 mm, dashed – Ø 20 mm)

In the latter case, such a situation forms a defective subsystem, including, among other things, Willis clusters of different types, differing in combinations of vacancies and various types of internodes.

This causes the effects of strong local ordering (short range order). Note that the excess of anions accumulates mainly on certain crystallographic planes.

O2 – ions located in internodes can initiate additional distortion of atomic groups, creating their own dipole moment. Considering the well-known tendency of niobate ceramics to self-organize defective structures [5] by ordering them and forming microdomains, we can assume the influence of the latter on the total dipole moment of the system, the degree of ferroelectric rigidity, and, consequently, the coercive force, dielectric constant, piezoelectric sensitivity, and mechanical quality factor. As shown experimentally, in the studied SS modified with NiO, the degree of ferro rigidity does increase.

The dielectric spectra of SS are characteristic of alkali metal niobates with a pronounced maximum ε / ε0 at the Curie point, the rapid increase of ε / ε0 in the paraelectrical region at temperatures higher, the higher the frequency of the electric measuring field associated with the restoration of Nb5 + → Nb4 + and the emergence of vacancies [6]. Moreover, in modified TPs, this process develops at higher temperatures [7–8].

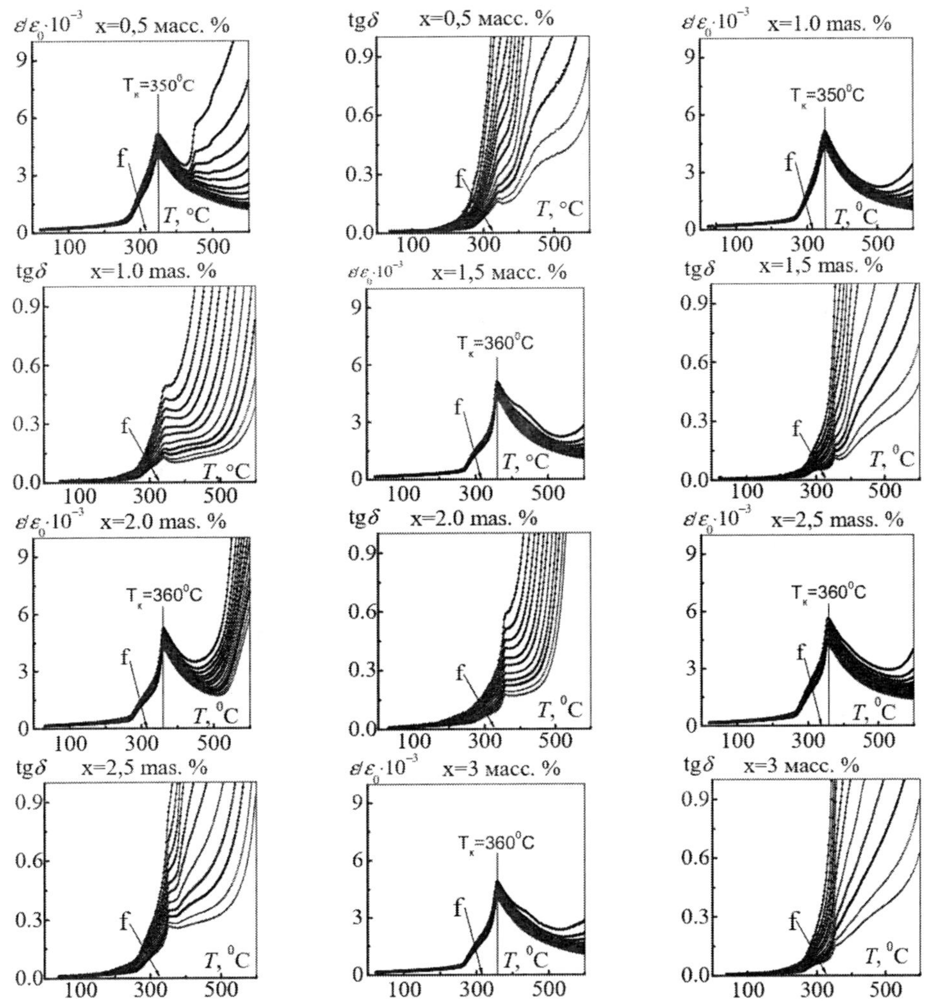

Figure 7. Thermofrequency behavior ε / ε0 tanδ of the studied SS.

4. Conclusion

The results obtained are appropriate. to consider when developing electronic devices that use the materials in question.

Acknowledgment

Research was financially supported by the Ministry of Science and Higher Education of the Russian Federation (State assignment in the field of scientific activity, Southern Federal University, 2020).

References

[1] Danziger A Ya, Razumovskaya O N, Reznichenko L A et al 2001–2002 Multicomponent systems of ferroelectric complex oxides: physics, crystal chemistry, technology In *Design Aspects of Ferroelectric Materials* vol 1, 2 (Rostov-on-Don: Publ. House of Rostov State Univer.) 800 p

[2] Higashide K, Kakimoto K and Ohsato H 2007 Temperature dependence on the piezoelectric property of (1−x)($Na_{0.5}$K0.5)NbO_3–x$LiNbO_3$ ceramics *J.of Europ. Ceramic Society* **27** 4107–10

[3] Sadykov H A, Verbenko I A, Shilkina L A et al 2013 The effect of transition 3d metal ions on the formation of the electrophysical properties of polycrystalline materials based on alkali metal niobates *News of the Russ. Acad. of Sci. The ser. is Phys.* **77(9)** 1253–5

[4] Reznichenko L A, Razumovskaya O N and Dudkina S I 1999 Anomalous behavior of dielectric constant in ferro-piezoelectric materials based on PZT with the participation of Ni (II)- and Cd (II)-containing components *Fundamental problems of piezoelectric instrumentation* vol 1 Sat. Proc. Int. sci.-pract. conf. (Piezotechnics-99) pp 109–23

[5] Wang X, Wu J, Chang X et al 2013 *Compositional dependence of phases tructure and electrical properties in* $(K_{0.50}Na_{0.50})_{0.97}Bi_{0.01}(Nb_{1-x}Zr_x)O_3$ *lead-free ceramics* **139** 8021–4

[6] Vezzoli G C 1992 Electrical properties of NbO_2 and Nb_2O_5 at elevated temperature in air and flowing argon *Physical Review B* **26(7)** 3954–7

[7] Rai R, Coondoo I, Rani R et al 2013 Impedance spectroscopy and piezoresponse force microscopy analysis of lead-free $(1 - x)K_{0.5}Na_{0.5}NbO_3 - xLiNbO_3$ ceramics *Current Appl. Phys.* **13** 430–40

[8] Yiping G, Kakimoto K and Ohsato H 2005 $(Na_{0,5}K_{0,5})NbO_3 - LiNbO_3$ lead-free piezoelectric ceramics *Mater. Letters.* **59** 241–4

IOP Conf. Series: Materials Science and Engineering **905** (2020) 012056 doi:10.1088/1757-899X/905/1/012056

Comparative tests of cements in concrete mixtures of high mobility

M S Saidumov[1], S-A U Murtazaev[2,3,1], A Kh Alaskhanov[1], M R Nahaev[1] and M R Khadzhiev[1]

[1] Grozny State Oil Technical University named after M.D. Millionshchikov, korp. 1,100, prosp. Husejna Isaeva, Grozny, 364068 Russia
[2] Integrated Research Institute named after Kh.I. Ibragimov, Russian Academy of Sciences of the Chechen Republic, 21A, Staropromyslovsky ave., Grozny, 364051, Russia
[3]Academy of Sciences of the Chechen Republic, 13, prospect named after M. Esambayev, Grozny, Russia

E-mail: saidumov_m@mail.ru, s.murtazaev@mail.ru, uspekh@ro.ru, mr-nakhaev@mail.ru, magnum5550@mail.ru

Abstract. The paper presents the results of comparative tests of imported and local cements in highly mobile concrete mixtures of various classes and brands. It is proved that on the basis of local cement it is possible to obtain concrete of any grades in terms of workability and grades in compressive strength up to B45. Various factors affecting the stability of the composition and properties of the concrete mixture are considered. Recommendations on their use in high strength concrete are proposed.

1. Introduction

In recent decades, as you know, high-rise construction is widely developed, where monolithic concrete mixes of various grades and classes are used [1, 2]. With the intensification of the construction process, the requirements for concrete mixes and concrete have noticeably increased [3]. Instead of the usual concrete mixes of grades of workability from P1 to P3, highly mobile concrete mixtures of grades P4, P5 (for draft of cone) or P4, P5, P6 (i.e. grade of mix for spread of cone) have been increasingly used recently. Undoubtedly, such high-quality, highly mobile (cast) concrete mixes require special approaches in their preparation, dosing of raw materials, mixing and compaction [5, 17].

The strength of the future composite on clean, high-quality aggregates that meet the requirements of the relevant standards, and proper compaction will also depend on the activity and normal density of the binder, W/C concrete mix, i.e. from the qualitative indicators of the latter [6, 18].

As you know, the workability of concrete mixtures is influenced by a number of factors [7–11]:

- the ratio between binder and aggregates;
- the ratio between small and large aggregates;
- water demand and quality, astringent;
- water-cement ratio of concrete mix, etc.

In addition to the factors listed above, the workability of the mixture is affected by the nature and shape of the aggregate grains, the presence of fine aggregate, mineral filler, chemical modifiers, etc. [12–16, 20–22].

At construction sites in the Chechen Republic, such as the Grozny Mall Shopping and Entertainment Center, the Akhmat Tower multifunctional high-rise complex, the Grozny TPP, etc., high-quality highly mobile concrete mixes and high-strength concrete based on them are regularly used [19, 23, 24].

To develop technological maps for the selection of optimal recipes for high-quality concrete mixtures of various grades and classes, we conducted comparative studies of local and imported binders and aggregates to increase the efficiency of use of local resources.

2. Methods and materials

The technological properties of concrete mixtures were studied, namely, water holding capacity (water separation), sedimentation (separation), uniformity (uniform distribution of aggregate grains throughout the volume) of concrete mixtures.

In the experimental and control compositions, Portland cement of different manufacturers was used – the imported PC M500 D0 Novoroscement and the local PC M500 D0 GUP Chechencement.

PC M500 D0 Novoroscement is characterized by the following indicators: normal density (ND) is 26.2 %, specific surface area of particles is 3132 cm^2/g, water separation of cement paste is not more than 16 %, setting time: start – 2 hours. 25 minutes, and the end – 3 hours. 45 minutes

The characteristics of the M500 D0 PC of the Chechencement State Unitary Enterprise are close to the M500 D0 Novoroscement PC, but there are differences in the following indicators: ND = 25.5 %, specific surface 3260 cm^2/g, water separation – more than 18 %, setting time – 2 hours. 15 minutes. (beginning) and 3 hours. 40 min (end).

The local crushed stone of the Argun deposit of ordinary quality was used as a large aggregate: the crushability mark in the cylinder was 800; bulk density – 1372 kg/3; true density – 2627 kg/m^3; the content of dusty and clay particles is 0.5–0.9 %; the content of crushed grains is more than 83 %; voidness – 41.5 %.

Fine aggregate – the local natural sand of the Chervlenskoye field, despite the fineness of its grains, is most common in the region due to the lack of higher quality sand deposits. The nature of the shape of the grains of Chervlensky sand is rough, irregular in shape, mostly monofractional: the particle size modulus is not more than 1.6-1.9, which classifies it as fine sand; the content of dusty and clay particles is 1.5–1.95 %; bulk density – 1528 kg/m3; true density – 2623 kg/m^3; voidness – 43.4 %.

The experimental and control compositions of concrete mixtures of different grades in terms of workability, obtained by calculation and theoretical methods and corrected by trial mixtures, are presented in table 1.

Experimental studies were carried out in the accredited testing laboratory of the scientific and technical center for collective use "Modern building materials and technologies" of the Grozny state oil technical university named after academician M.D. Millionschikova.

Also, in the studied compositions of concrete mixtures, a chemical additive was used – the Linamix PK superplasticizer in the amount of 1.1 % by weight of cement. The quality indicators of the Linamix PK additive are as follows: liquid (brown) with a solution density of 1.03–1.08 g/cm^3; pH value – not less than 8.0 ± 1; the recommended dosage is 0.3–2.5 % by weight of cement.

The mineral filler from concrete scrap and brick fight was not specifically used to make it easier to identify signs of water separation of concrete mixtures.

3. Results

The results of comparative tests, binders in highly mobile concrete mixtures obtained using the above raw materials are presented in tables 1 and 2.

Table 1. Compositions of high-movable concrete mixes for comparative tests of binders

Composition number	W/C	Consumption of materials per 1 m3 of concrete, kg				
		SC	S	C	A	W
1	0,41	1200	620	400*	4,4	163
2	0,44	1190	613			175
3	0,47	1180	608			188
4	0,42	1200	620	400**	4,4	166
5	0,45	1187	614			180
6	0,49	1180	610			195

Note to table 1: * – Portland cement grade PC M500 D0 GUP "Chechencement"; ** – Portland cement of the brand PC M500 D0 Novoroscement; W/C – water-cement ratio; SC – crushed stone consumption; S – sand; C – cement; A – additive; W – water.

Table 2. Results of comparative tests of binders in highly movable concrete mixtures

Composition number from the table1	Mark on the draft of the cone, cm	The density of the mixture, kg/cm^3	Water separation, % by weight	Solution separation, % by weight
1	P3	2387	0,3	2
2	P4	2383	0,4	2
3	P5	2378	0,6	4
4	P3	2392	0,1	1
5	P4	2385	0,2	1
6	P5	2380	0,3	2

Indices of uniformity and stability of concrete mixes must meet the requirements presented in GOST 7473-2010 “Concrete mixtures. Technical conditions ”[8] (table 2).

Table 3. Requirements for breadability of concrete mix [8]

The mobility or stiffness of the mixture	Separation of concrete mix, %, no more	
	water separation	solution separation
Z1 – Z5	0,2	3
P1 – P2	0,4	3
P3 – P5 и P1 – P6	0,8	4

Since the delamination of concrete mixtures characterizes the quality of the latter, in addition to the density of the concrete mix and the grade of the sediment cone, we mainly studied the indicators of solution separation and water separation, which, as a rule, evaluate the delamination of the mixture.

The solution separation of the concrete mix with coarse aggregate, characterizing its delamination under dynamic action, was determined by comparing the content of the mortar component in the lower and upper parts of the concrete mix, compacted in a measuring vessel or mold.

For this, the concrete mixture was laid out in a measuring vessel (mold) and compacted in accordance with GOST 10180-2012. The compacted concrete mixture was additionally vibrated on a laboratory vibratory platform for 10 seconds.

After additional vibration, the upper layer of the concrete mixture with a height of about half the height of the measuring vessel (mold) was selected on a previously weighed baking sheet, and the mixture remaining in the lower part of the measuring vessel (mold) was vibrated until the surface of the mixture was leveled. Then, the height of the layer of the HH mixture remaining in the lower part of the measuring vessel (shape) was measured with an error of up to 5 mm, and the height of the selected layer of the mixture of HB was calculated. The remaining mixture was laid out on a second weighed baking sheet.

Divided into two samples, the concrete mixture from the upper and lower parts of the measuring vessel (mold) was weighed and wet sifted on a sieve with holes with a diameter of 5 mm. During wet sieving, each sample of the mixture laid out on a sieve was washed with a stream of clean water until the cement-sand mortar was completely removed from the surface of the coarse aggregate grains.

The washed coarse aggregate from each sample of the concrete mixture was transferred onto a clean baking sheet, dried to constant weight at a temperature of (105 ± 5) °C and weighed.

The mass of the solution component in the samples of the upper and lower parts of the measuring vessel (form), considering the volume of the sample taken, was determined by the formulas:

$$m_{R.U} = (m_{CM.U} - m_{C.U})\frac{0,5H}{H_U} \quad (1)$$

$$m_{R.D} = (m_{CM.D} - m_{C.D})\frac{0,5H}{H_D} \quad (2)$$

where $m_{R.U.}$, $m_{R.D.}$ – the mass of the mortar component of the mixture located in the upper and lower parts of the measuring vessel (form), g;

$m_{CM.U.}$, $m_{CM.D.}$ – mass of concrete mixture selected from the upper and lower parts of the measuring vessel (form), g;

$m_{C.U.}$, $m_{C.D.}$ – mass of dried coarse aggregate contained in samples from the upper and lower parts of the measuring vessel (form), g;

H – height of a measuring vessel (form), mm;

H_U, H_D – the actual height of the upper and lower layers of the mixture.

Mortar separation of concrete mix Π_R, %, was determined by the formula:

$$\Pi_R = \frac{m_{R.U} - m_{R.D}}{m_{R.U} + m_{R.D}} \cdot 100 \quad (3)$$

The solution separation of each concrete mixture sample was determined twice and calculated with rounding to 1 % as the arithmetic mean of the results of two determinations that differ by no more than 20 % of the average value. With a larger discrepancy between the results, the test was repeated on a new sample of concrete mix.

The test results of concrete mixtures for solution separation are presented in table 2.

The water separation of the concrete mixture was determined after it was settled in a measuring vessel or mold for a certain period of time.

For this, the concrete mixture was placed in a measuring vessel (mold) and compacted on a vibrating platform for 10 seconds. The level of the concrete mixture was (10 ± 5) mm below the upper edge of the measuring vessel (mold). After that, the mold (vessel) was covered with a sheet of vapor-tight material (glass) and left alone for 2 hours.

The separated water was taken with a pipette every 15 min, collecting it in a glass with a lid and weighing it at the end of the test.

The water separation of the concrete mixture Π_U, in %, characterizing the volume of water released from the concrete mixture in 2 hours, referred to the volume of concrete mixture in a measuring vessel (form), and was calculated by the formula:

$$\Pi_U = \frac{m_U}{\rho_U \cdot V_{B.CM}} \cdot 100 \quad (4)$$

where m_U – mass of separated water, g;

ρ_U – the density of water, taken equal to 1 g/cm^3;

$V_{V.CM}$ – the volume of concentrated concrete mixture, cm^3.

Water separation was determined twice for each concrete mixture sample and was calculated with rounding to 1 %, as the arithmetic mean of the results of two determinations that differ from each other by no more than 20 % of the average value. With a larger discrepancy between the results, the test was repeated on a new sample of concrete mix.

The test results of concrete mixtures for water separation are also presented in table 2.

The analysis of tables 1 and 2 showed that with the transition to higher grades according to the cone sediment (from P3 to P5), the delamination parameters (water and solution separation) of concrete mixtures at the M500 D0 PC of the Chechencement state enterprise in comparison with concrete mixtures at the M500 PC D0 Novoroscement is much closer to the limit of permissible values

IOP Conf. Series: Materials Science and Engineering **905** (2020) 012056 doi:10.1088/1757-899X/905/1/012056

according to GOST 7374-2010 [8], according to which water separation should not be more than 0.8 % by mass, and solution separation should not exceed 4 %.

These indicators of the control compositions of concrete mixtures at the imported processing center M500 D0 Novoroscement are noticeably stable. This can be explained by the fact that this Portland cement initially has a lower water separation (no more than 16 %), when the same indicator at the M500 D0 Chechencement is more than 18 %.

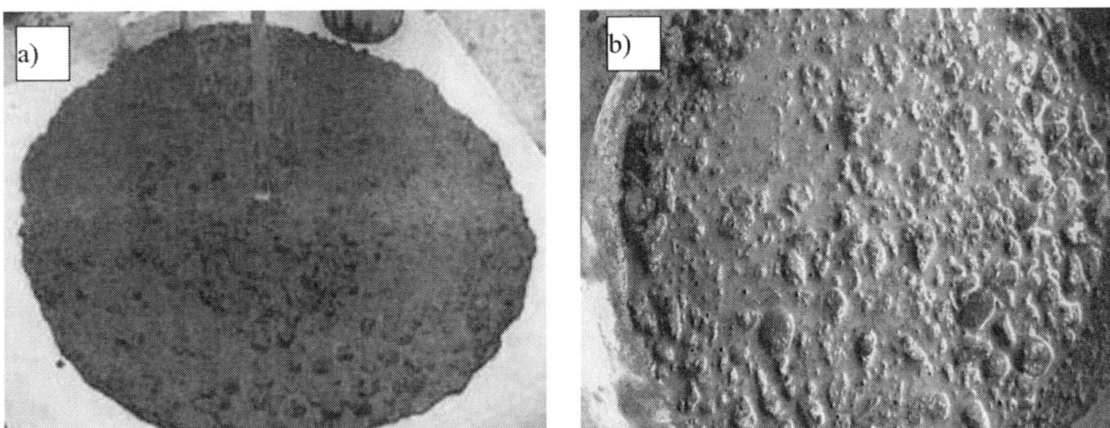

Figure 1. Comparison of concrete mixtures: a – high-quality mixture at the PC M500 D0 Novoroscement; b – with signs of water separation and stratification at the human processing center M500 D0 of the State Unitary Enterprise "Chechencement"

Naturally, the control and experimental compositions presented in Table 1 meet the requirements of GOST 7374-2010 and had a compressive strength class of B22.5. However, as additional studies of similar compositions with a higher specified strength (concrete classes B50-B60) and mobility over a cone melt of 56–62 cm (P5) and more than 62 cm (P6), in which the consumption of cement and water are much higher than in low-grade concretes, in highly mobile concrete mixtures at the M500 D0 Processing Center of the Chechencement State Unitary Enterprise, signs of water separation and delamination could be seen (Figure 1). Although there was no apparent water and solution separation.

In highly mobile concrete mixtures of low grade B25-B40 concrete at the local PC M500 D0 of the Chechencement state unitary enterprise, signs of water separation and separation are not observed. This phenomenon can be found in highly mobile concrete mixtures of higher classes of concrete (B50 and above), for which the value of the coefficient r, which characterizes the ratio of sand consumption to crushed stone consumption and determined by formula (1), decreases markedly.

$$r = \frac{\Pi}{C} \tag{5}$$

where Π – is the consumption of sand in the concrete mixture, kg; C – crushed stone consumption in concrete mix, kg.

For concrete of not high classes (B40 and lower), the index r, as a rule, is in the range of 0.48-0.6. r noticeably decreases (to 0.3-0.4) when switching to high-grade concrete (B60-B80 and higher), in concrete mixtures of which the binder consumption increases with an equal replacement of a fraction of a fine aggregate with a higher water retention capacity than grains of large aggregate (crushed stone). And the attempt to switch to high classes of concrete, while maintaining the r value by simultaneously reducing the fraction of crushed stone and sand, contributes to the violation of the high-density packing of coarse aggregate grains, the "rigid skeleton" of which forms the basis for the high strength of the future concrete composite.

4. Conclusion

Thus, on the basis of the comparative studies, it can be concluded that one of the most rational solutions to the problem of using locally produced cement in high-quality concrete mixes and concretes is its use either for middle-class concrete in compressive strength (B25-B45), or to obtain it is based on filled binders with a mineral filler of various nature with a higher water retention capacity (for example, filler from concrete scrap and brick fight), characterizing Xia high stability and rheology, which are suitable for both low and medium concrete classes, and for the high-active and high-strength concretes (B50 and above).

Acknowledgment

This work was carried out as part of research on the implementation of scientific project No. 05.607.21.0320. "Development of technology for new building composites on clinkerless alkaline binders using substandard natural and secondary raw materials" that received support from the federal target program "Research and Development in Priority Directions for the Development of the Russian Science and Technology Complex for 2014–2020". Unique Agreement Identifier RFMTFI60719X0320. This work was carried out as part of research on the implementation of scientific project No. 18-48-200001 "High-quality concrete with enhanced performance properties based on local natural and technogenic raw materials," which received support from the Russian Foundation for Basic Research (RFBR).

References

[1] Bazhenov Yu M, Demyanova B C and Kalashnikov V I 2006 *Modified high-quality concrete* A sci. publ. (Moscow: Publ. house of the Associat. of construct. Univer.) 368 p

[2] Bazhenov St Yu M, Murtazaev A Yu and Saydumov M S 2014 *Building composites based on concrete scrap and stone crushing waste* A sci. publ. (Grozny: FSUE Publ. and Print. Complex Grozny Worker) 334 p

[3] Demyanova V S and Gusev A D 2013 *Effective building materials using industrial waste* (Penza: PGUAS) 127 p

[4] Salamanova M, Khubaev M, Saidumov M and Murtazayeva T 2017 Self-Consolidating Concretes with Materials of the Chechen Republic and Neighboring Regions *Int. J. of Environ. & Sci. Ed.* 11(18) 12724–7

[5] Kaprielov S S, Sheinfeld A V and Al-Omais D 2018 Experience in the production and quality control of high-strength concrete in the construction of the high-rise complex "OKO" in Moscow City *Industr. and Civil Engineer.* **1** 18–24

[6] Lesovik B C, Murtazaev St A Yu and Saydumov M S 2012 *Building composites based on screenings for crushing concrete scrap and rocks* A sci. publ. (Grozny: Municipal Unitary Enterprise "Printing House") 192 p

[7] Bazhenov Yu M, Bataev D K-S and Mazhiev H N 2011 *Fine-grained concrete from recycled materials for the repair and restoration of damaged buildings and structures* A sci. publ. (Grozny: IE "Sultanbegova Kh S") 342 p

[8] GOST 7473-2010 *Concrete mixes. Specifications. Introduction 2012-01-01* 2011 (Moscow: Standartinform) 16 p

[9] Murtazaev St A Yu, Bataev D K-S and Ismailova Z K 2009 *Fine-grained concrete based on fillers from recycled materials* A sci. publ. (Moscow: Komtekhprint) 142 p

[10] Lermit R 2007 *Concrete Technology Problems* (Moscow: LKI Publ. House) 296 p

[11] Bazhenov Yu M 2011 *Concrete technology* (Moscow: DIA) 500 p

[12] Afonina M I and Kozyreva E V 2017 Features of the construction of sports facilities in the territories of former industrial zones *Days of Student Science* Sci. and techn. Conf. based on the results of res. work by stud. of the Instit.of Construct. and Architect. pp 390–2

[13] Shcherban E M, Stelmakh S A and Khalyushev A K 2018 Development of the composition of pozzolanic cement on volcanic tuff *Construction. Architecture. Economics* Int. Forum "Victory May 1945" pp 110–3
[14] Stelmakh S A, Shcherban E M and Serdyukov K V 2017 The influence of some characteristics of the coarse aggregate used on the properties of heavy concrete intended for the manufacture of centrifuged products and structures *Bull. of Belgorod State Technol. Univer. named after V G Shukhov* **10** 15–20
[15] Soldatov A A, Galych A V and Sariev I V 2016 Experience in the use of sodium silicate as a binder in the production of building materials", *Actual problems of construction, transport, engineering and technosphere safety* IV annual sci.-pract. Conf. of the North Caucasus Fed. Univer. pp 186–8
[16] Udodov S A 2015 Re-introduction of a plasticizer as a tool for controlling the mobility of concrete mix *Sci. works of the Kuban State Technol. Univer.* **9** 175–85
[17] Usov B A 2009 *Physical and chemical processes of building materials science in concrete and reinforced concrete technology* (Moscow: Publ. house of the Moscow State Ed. Univer.) 326 p
[18] Koryanova Yu I, Rezantsev N E and Shumilova A S 2018 Materials and structures used in the construction of high-rise buildings – from tradition to innovation *Sci. Alley* **6((4)20)** 95–9
[19] Murtazaev S A Yu and Salamanova M Sh 2018 Clinker-free binders based on finely dispersed mineral components *Ibausil conference proc.* vol 2 pp 707–14
[20] Lesovik V S, Alfimova N I and Trunov P V 2014 Reduction of energy consumption in manufacturing the fine ground cement *Res. J. of Appl. Sci.* **9(11)** 745–8
[21] Murtazaev S-A Y, Zaurbekov Sh Sh, Alaskhanov A Kh et al 2018 Impact of Technogenic Raw Materials on the Properties of High-Quality Concrete Composites *Advances in Engineer-Research Research* vol 177 Int. Symp. Engineer. and Earth Sci.: Appl. and Fund. Res. (ISEES 2018) pp 275–9
[22] Lesovik V S, Zagorodnyuk L K, Mestnikov A E et al 2015 Designing of mortar compositions on the basis of dry mixes *Int. J. of Appl. Engineer. Rese.* **10(5)** 12383–90
[23] Murtazaev S-A Y, Mintsaev M Sh, Aliev S A and Saydumov M S 2015 Strength and Strain Properties of Concrete, Comprising Filler, Produced by Screening of Waste Crushed Concrete *Canad. Center of Sci. and Ed. Received* **9(4)** 32–44
[24] Bataev D K-S, Uzaeva A A, Uzaeva S A et al 2018 The Study of Shrinking Deformations of Repair Compositions on Barkhan Sands", *Advances in Engineering Research* vol 177 Int. Symp. on Engineer. and Earth Sci. (ISEES 2018) pp 254–7

Obtaining sodium silicate solutions for the activation of mineral powders

M Sh Salamanova[1,2], Z Kh Ismailova[2], M R Nahaev[3] and R G Bisultanov[2]

[1] Department of Materials Science, Integrated Research Institute. H.I. Ibragimova of the Russian Academy of Sciences, Grozny, Russia
[2] Institute of Construction, Architecture and Design, Grozny State Oil Technical University named after Academician M.D. Millionschikova, Grozny, Russia
[3] Chechen State University, Grozny, Russia

E-mail: madina_salamanova@mail.ru, ismailova-z-h@mail.ru, mr-nakhaev@mail.ru, ramazan_bisultanov@mail.ru

Abstract. Currently, the creation of clinker-free binders that do not require high-temperature processing and a substantial natural resource, and have high technical indicators, is an urgent task. The active component of alkali metals characterizes alkaline cement. A distinctive feature of alkali cement is its ability to interact with aluminosilicate minerals during hydration, which leads to the formation of active and hardly soluble compounds. The basis for the production of concrete composites based on chemical activation binders is the domestic and foreign experience in the use of slag-alkali composites in construction. In this work, an energy-dispersive microanalysis of clinker-free compositions based on alkaline silicate of sodium solutions from silica-containing rocks is carried out. The developed formulations of clinker-free binders based on highly dispersed powders of different nature, sealed with an alkaline solution prepared by the wet method at temperatures up to 95 °C and atmospheric pressure. This technology expands the scope of alkaline cement and produces concrete with desired properties. This work was carried out as part of research on the implementation of scientific project No 05.607.21.0320. "Development of technology for new building composites on clinkers alkaline binders using substandard natural and secondary raw materials," which received support from the federal target program "Research and Development in Priority Directions for the Development of the Russian Science and Technology Complex for 2014–2020."

1. Introduction

Modern rates of development of construction require an increase in the production of concrete composites. At the same time, scientists seek to reduce their cost without compromising the reliability and durability of the designed products and structures [1–3]. However, this task is difficult. The conditions faced by developers: a deficit of the developed natural resource, a constant rise in the cost of energy, an increase in anthropogenic impact on the ecological environment. Therefore, the special attention of scientists is focused on the development of less resource- and energy-intensive technologies for producing building composites [4, 5]. Currently, the urgent task is to create clinker-free binders that do not require high-temperature processing and have high technical indicators. This category of binders can be attributed to alkaline cement. A feature of these materials is that the active component is alkali metals, capable of interacting with aluminosilicate minerals during hydration. As a result, fairly strong sparingly soluble compounds are formed that are identical to the products of hydration of a traditional binder [6–9].

2. Methods and materials

Experience in the use of such binders in construction is known [1–8]. Slag-alkali binders based on ferrous metallurgy wastes are studied in the most detail. Slag-alkali binders include blast furnace granulated slag, the chemical composition of which is represented by aluminum and silicon oxides and is similar to the composition of Portland cement clinker [9, 10]. But in many regions of our country, slag from ferrous metallurgy is an expensive and rare material. Transporting data with the material is a considerable cost. On this basis, well-known developments are widely adopted. For several years, we have been researching the production of clinker-free alkaline cement mixing [12–15]. As a result, we have obtained durable and corrosion-resistant composites. In the formulations of binders, finely dispersed mineral powders of different nature were used as a replacement for blast furnace slag of ferrous metallurgy. The following rocks represent highly dispersed mineral powders: silicified marl, limestone, volcanic tuff, sandstone, quartz sand, waste from the cement industry in the form of aspiration, and clinker dust of electrostatic precipitators. All of these powders were individually mixed in predetermined proportions with the fine aggregate of the Alagirskoye deposit (particle size modulus 2.8) and sealed with salable liquid sodium glass, with silicate modulus 2.8 and density 1.24 g/cm^3. The prepared samples solidified the first day under normal conditions at a temperature of 20 ± 2 °C. After 2 days, part of the twin samples was placed periodically in an oven at 40 °C for a couple of hours, and part of the heat-treated samples was stored in water. The formulations and test results depending on the hardening conditions are shown in Table 1.

3. Results

The test results confirm that the strength characteristics of the obtained binders depend, firstly, on the mineralogical and chemical compositions of the powders understudy, secondly, on the presence of active crystallization centers, thirdly, on the hardening conditions and specific surface area of mineral additives (Table 1). Clinker dust generated at the hot end of a clinker rotary kiln has the desired mineral composition. When mixed with an alkaline solution, clinker dust is hydrated with the formation of calcium and sodium hydro silicates. Therefore, the samples obtained are characterized by high strength values during hardening in an aqueous medium. Siliconized thermally activated marl and volcanic tuff also, when activated by a liquid-glass solution, gained good activity when storing samples in water. The aluminosilicate nature of these powders justifies the data obtained. Mineral powders from limestone, sandstone, and suction dust after activation gained strength with increasing temperature. The results are justified by the high content of alkaline earth compounds.

Thus, clinker-free hydraulic binders of alkaline activation were obtained, with a ruler of strength from 10 to 50 MPa, suitable for the production of concrete composites. But in the proposed developments, the most expensive is considered liquid commodity glass, and for the widespread use of alkaline composites. Therefore, it is necessary to find an alternative to this concoction [14, 16–18]. In connection with these circumstances, the goal of the next stage of research is to prepare an effective and less costly alkaline solution. The joint was prepared in a wet way. This method consisted of dissolving finely dispersed silica additives in a 20 % sodium hydroxide solution, followed by heating to 92–95 °C at atmospheric pressure.

The essence of the method is that the amorphous component of the studied additives during dispersion reacts with alkali to form sodium metasilicate by the reaction: $2NaOH + SiO_2 = Na_2SiO_3 + H_2O$. The alkaline solution of sodium metasilicate obtained in this way was filtered as a suspension. As a result of filtration, the solution was separated from the insoluble precipitate. The chemical composition of the proposed alkaline solutions was investigated using a QUANTA 3D 200i scanning electron microscope using an energy dispersive microanalysis system (EDAX). The studies were conducted at the Research Center for Nanotechnology and Nanomaterials, GSTU.

Table 1. Formulations and properties of linker-free cements of alkaline activation

Composition number	Highly dispersed powders	Alkaline activator, % Na_2SiO_3	MPa activity, 28 days hardening conditions		
			natural	40 °C	in water
1	Quartz powder, $S_{specific\ surface} = 810\ m^2/kg$	26.0	20.3	22.3	23.5
2	Mergel (700 °C), $S_{specific\ surface} = 1150\ m^2/kg$	29.5	39.2	42.6	46.5
3	Volcanic tuff $S_{specific\ surface} = 920\ m^2/kg$	26.5	33.1	37.6	39.4
4	Limestone flour $S_{specific\ surface} = 1060\ m^2/kg$	29.0	13.7	16.5	9.0
5	Sandstone $S_{specific\ surface} = 1020\ m^2/kg$	26.5	14.6	17.3	9.2
6	Clinker dust $S_{specific\ surface} = 240\ m^2/kg$	23.0	42.1	44.7	50.5
7	Aspiration dust $S_{specific\ surface} = 280\ m^2/kg$	24.0	15.4	15.6	10.1

As a siliceous component in the alkaline solution, the beautiful glass sands of the Shatoyskoye field (CR) and the volcanic tuff of the Kenzhenskoye field (KBR) were used. The chemical composition of mineral additives is given in Table 2.

Table 2. Chemical composition of silica additives, %

Oxide composition	Volcanic tuff	Quartz sand
MgO	2.20	6.32
Al_2O_3	13.57	14.99
SiO_2	63.67	73.83
K_2O	6.00	1.83
CaO	7.79	0.6
Fe_2O_3	2.52	0.97
TiO_2	2.85	1.32
SO_3	–	0.14
Others	1.40	–

After preparing the alkaline solution, visual inspection of the resulting suspensions was carried out for a week (Fig. 1a, 1b). Visual inspection showed that after a day, the level of sediment in the solution based on finely dispersed quartz powder is higher than in the solution with volcanic tuff. This fact can be explained by the fact that quartz is contained in an increased amount. Quartz is a reasonably solid mineral, which in the initial period, interacts more difficult with alkali. After a week, the level of sediment changed, it decreased in the same solution, it is likely that the process of dispersion of quartz in a solution of caustic soda was activated, due to the amorphous component of this additive. Energy dispersive analysis and microphotographs of the corresponding solutions were carried out both of the pure filtrate and the precipitate. The results are presented in Figures 2–5.

The results of microanalysis show that precipitates of alkaline solutions contain a more significant amount of unreacted silica. In the filtrates of an alkaline solution based on volcanic tuff, studies showed a higher SiO_2 content of up to 2.85 %, while in a solution based on quartz SiO_2 powders up to 1.99 %, and microphotographs confirm the presence of a glassy phase.

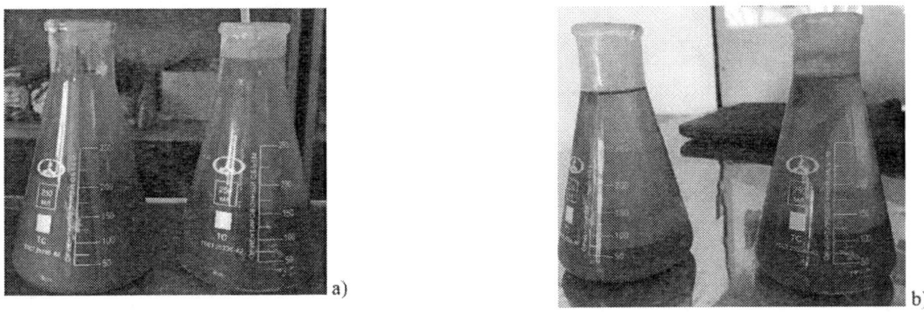

Figure 1. Pictures of alkaline solutions based on quartz sand and volcanic tuff, in a day (a) and in a week (b)

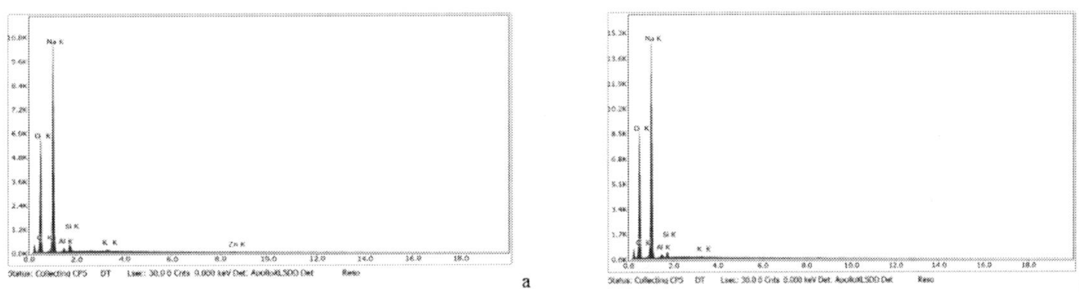

Figure 2. Energy dispersive analysis of an alkaline solution based on quartz sand, without sediment (a), with sediment (b)

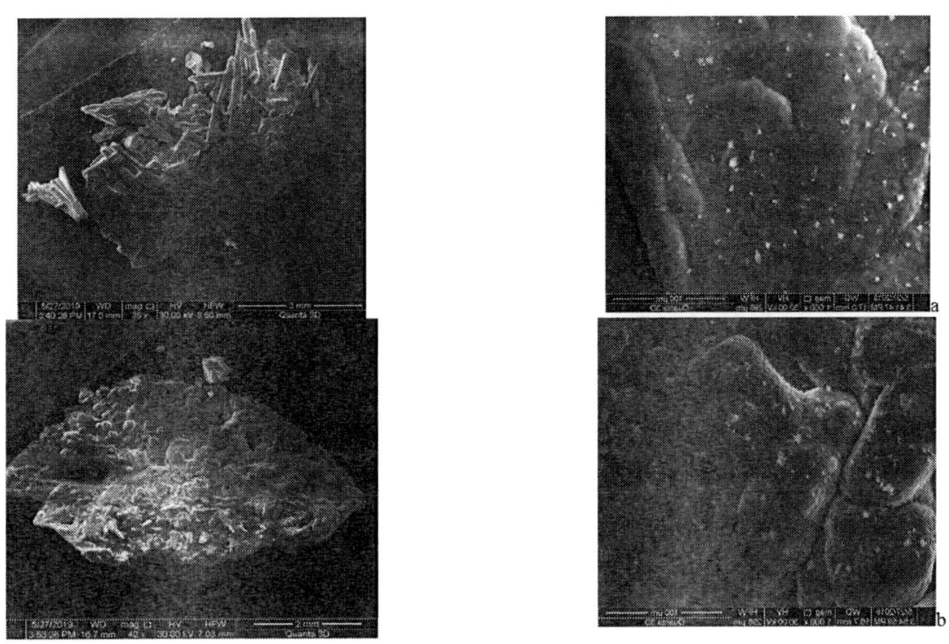

Figure 3. Micrographs of an alkaline solution based on quartz sand, without sediment (a), with sediment (b)

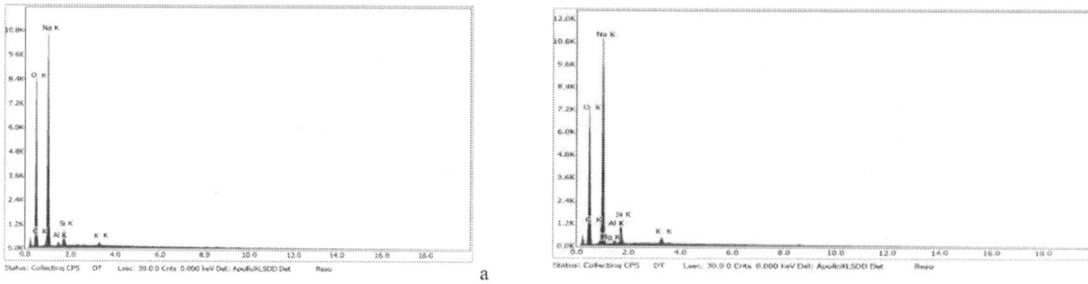

a b

Figure 4. Energy dispersive analysis of an alkaline solution based on volcanic tuff, without sediment (a), with sediment (b)

At a further stage, studies were conducted confirming the effectiveness of the developed solutions of sodium metasilicate. For this, beam samples were prepared from finely dispersed powders (Table 1), sealed with the alkaline solutions under study. After stripping, the samples were subjected to thermal action for two days in an oven (3 hours at t = 40°C). Next, one part of the samples were placed in water to establish the hydraulicity of the binders. The other part was placed in air to interact with carbon dioxide and to form a gel of silicic acid, at least on the surface layers of the samples. The research results are shown in Table 3.

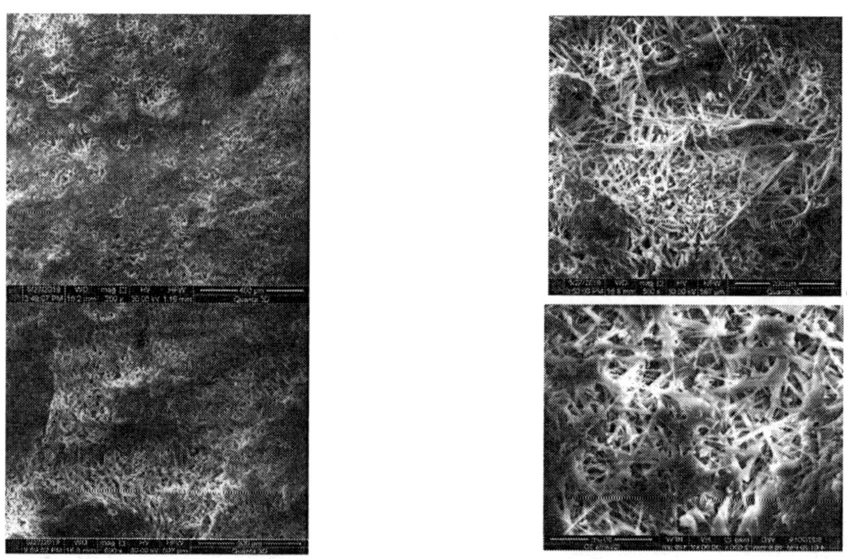

a b

Figure 5. Micrographs of an alkaline solution based on volcanic tuff, without sediment (a), with sediment (b)

4. Conclusion

The test results confirm the effectiveness of the alkaline solution based on volcanic tuff; this is especially evident in compositions No. 2 and 3. Thermally activated silicified marl and volcanic rocks of aluminosilicate nature form a durable geopolymer stone by activation with an alkaline solution. Geopolymer stone is represented by a framework aluminosilicate trapped in an alkaline medium with the formation of a three-dimensional aluminosilicate hydrogel [19, 20]. Powders of carbonate nature do not have hydraulic properties and showed low strength results.

Table 3. Formulations and properties of linker-free cements of alkaline activation

Composition number	Highly dispersed powders	Alkaline activator, % Na_2SiO_3		MPa activity, 28 days hardening conditions	
		quartz sand	volcanic tuff	natural	in water
1	Quartz powder,	26.2	–	14.7	13.5
	$S_{specific\ surface} = 810\ m^2/kg$	–	26.2	15.3	14.1
2	Mergel (700 ^0C),	29.5	–	22.9	25.4
	$S_{specific\ surface} = 1150\ m^2/kg$	–	29.6	23.4	25.8
3	Volcanic tuff	26.6	–	17.6	19.3
	$S_{specific\ surface} = 920\ m^2/kg$	–	26.6	18.2	19.7
4	Limestone flour	29.3	–	9.4	4.0
	$S_{specific\ surface} = 1060\ m^2/kg$	–	29.3	9.6	4.6
5	Sandstone	26.7	–	7.6	3.2
	$S_{specific\ surface} = 1020\ m^2/kg$	–	26.7	7.7	3.6
6	Aspiration dust	24.2	–	10.0	5.1
	$S_{specific\ surface} = 280\ m^2/kg$	–	24.2	10.4	5.0

Of course, the strength results differ from those obtained when mixed with salable liquid sodium glass. However, given the energy efficiency of this technology, it can be considered that the proposed solutions will expand the scope of the use of clinker-free alkaline binders and produce concrete with desired properties.

References

[1] Dietz A, Ramroth H, Urban T et al 2004 Exposure to cement dust, related occupational groups, and laryngeal cancer risk: Results of a population-based case-control study *Int. J. Cancer* 108

[2] Al-Neaimi Y I, Gomes J and Lloyd O L 2001 Respiratory illnesses and ventilatory function among workers at a cement Factory in a rapidly developing country *Occup. Med. (Lond)* **51** 367–73

[3] Murtazaev C-A Yu, Salamanova M Sh and Vataev U V 2014 Cement industry of the Chechen Republic *Bull. of the Acad. of Sci. of the Chechen Republic* **1(22)** 109–14

[4] Murtazaev St A Yu, Salamanova M Sh, Alashkhanov A Kh and Murtazaeva T S A 2015 Modern cement technologies using advanced dust and gas cleaning methods *All-Russ. Sci. and Pract. Conf. dedicat. to the 80th anniversary of the Honored Sci. of the Russ. Fed., Doctor of Techn. Sci. Professor Kh.I. Ibragimova* pp 99–107

[5] Nesvetaev G, Koryanova Y and Zhilnikova T 2018 On the effect of superplasticizers and mineral additives on shrinkage of hardened cement paste and concrete In the coll. MATEC Web of Conf. 27. Ser. "27th R-S-P Seminar Theoretical Foundation of Civil Engineering (27RSP) p 04018

[6] Stelmakh S A, Nazhuev M P, Shcherban E M et al 2018 Selection of the composition for centrifuged concrete, types of centrifuges and compaction modes of concrete mixtures *Physics and Mechanics of New Mater. and Their Applicat. (PHENMA 2018)* 337

[7] Shuisky A, Stelmakh S, Shcherban E and Torlina E 2017 Recipe-technological aspects of improving the properties of non-autoclaved aerated concrete *MATEC Web of Conf. Ser. Int. Conf. on Modern Trends in Manufactur. Technol. and Equipment (ICMTMTE 2017)* p 05011

[8] Soldatov A , Sariev I V, Zharov M A and Abduraimova M A 2016 Building materials based on liquid glass *Mater. of the IV annual sci. and pract. Conf. of the North Caucasus Fed. Univer.* Execut. ed N I Stoyanov pp 192–5

[9] Murtazaev St A Yu, Salamanova M Sh, Saidumov M S and Alaskhanov A Kh 2018 Study of the properties of concrete on clinker-free binders *Prospects for the development of the fuel and energy complex and the current state of oil and gas engineering education in Russia, dedicated to the 105th anniversary of M.D. Millionshchikov* pp 392–9

[10] Bazhenov Yu M, Alimov L A and Voronin V V 2008 *Technology of concrete, building products, and structures* (Moscow: Publ. House DIA) 350 p
[11] Bazhenov Yu M, Bataev D K-S and Mazhiev H N 2011 *Fine-grained concrete from recycled materials for the repair and restoration of damaged buildings and structures* A scientific publication (Grozny: IE "Sultanbegova Kh S") 344 p
[12] Shcherban E M, Stelmakh S A and Khalyushev A K 2018 Development of the composition of pozzolanic cement on volcanic tuff *Construction. Architecture. Economy* Int. Forum Victory May 1945 pp 110–3
[13] Udodov S.A 2015 Re-introduction of a plasticizer as a tool for controlling the mobility of concrete mix *Sci. works of the Kuban State Technol. Univer.* **9** 175–85
[14] Soldatov A A, Galych A V and Sariyev I V 2016 Experience in the use of sodium silicate as a binder in the production of building materials", *Actual problems of construction, transport, mechanical engineering, and technosphere safety* IV annual sci.-pract. Conf. of the North Caucasus Fed. Univer. pp 186–8
[15] Murtazaev S A Yu and Salamanova M Sh 2018 Clinker-free binders based on finely dispersed mineral components *Ibausil conf. proc.* **2** 707–14
[16] Salamanova M, Khubaev M, Saidumov M and Murtazayeva T 2017 Self-Consolidating Concretes with Materials of the Chechen Republic and Neighboring Regions *Int. J. of environ. & science ed.* **11(18)** 12719–24
[17] Lesovik V S, Alfimova N I and Trunov P V 2014 Reduction of energy consumption in manufacturing the fine ground cement *Res. J. of Appl. Sci.* **9(11)** 745–8
[18] Murtazaev S-A Y, Zaurbekov Sh Sh, Alaskhanov A Kh et al 2018 Impact of Technogenic Raw Materials on the Properties of High-Quality Concrete Composites *Engineering and Earth Sciences: Applied and Fundamental Research* vol 177 Int. Symp. (ISEES 2018) pp 275–9
[19] Murtazaev S-A Y, Mintsaev M Sh, Aliev S A and Saydumov M S 2015 Strength and Strain Properties of Concrete, Comprising Filler, Produced by Screening of Waste Crushed Concrete *Canad. Center of Sci. and Ed.* **9(4)** 32–44
[20] Bataev D K-S, Uzaeva A A, Uzaeva S A et al 2018 The Study of Shrinking Deformations of Repair Compositions on Barkhan Sands *Int. Symp. on Engineer. and Earth Sci. (ISEES 2018)* vol 177 pp 254–7

Possible solutions to problems in the cement industry

M Sh Salamanova[1,2], S A Aliev[2], S-A U Murtazaev[1,2], M S Saidumov[2] and I A Gabazov[2]

[1] Integrated Research Institute named after Kh.I. Ibragimov, Russian Academy of Sciences of the Chechen Republic, 21A, Staropromyslovsky ave., Grozny, 364051, Russia
[2] Grozny State Oil Technical University named after M.D. Millionshchikov, korp. 1,100, prosp. Husejna Isaeva, Grozny, 364068 Russia

E-mail: madina_salamanova@mail.ru, asa-fenix@mail.ru, s.murtazaev@mail.ru, saidumov_m@mail.ru, asa-fenix@mail.ru

Abstract. In this paper, we substantiate approaches to the transition to clinker-free binders and building composites with their use to replace resource-intensive cement, at least in those areas of construction where its high technical functional properties are not needed. The optimal formulations and properties of alkaline activation binders based on mineral powders are revealed and the theoretical foundations of the formation of the structure and strength of cement stone based on an alkaline activator are revealed. The results of the studies are of practical value for the construction industry, as the obtained formulations of clinker-free cements will partially replace expensive and energy-intensive Portland cement in the production of concrete and reinforced concrete structures. This work was carried out as part of research on the implementation of scientific project No. 05.607.21.0320. "The development of technology for new building composites on clinkerless alkaline binders using substandard natural and secondary raw materials" that received support from the Federal target program "Research and development in priority areas for the development of the scientific and technological complex of Russia for 2014–2020".

1. Introduction

For many years, the modern construction industry has considered concrete and reinforced concrete to be the leader in the construction market, despite the enormous costs associated with their production. And these costs, in the first place, are associated with the use of energy – and resource-intensive binder component. Therefore, the question of creating an alternative solution to this costly material is of increasing interest and is relevant.

At the moment, there are developments proposed by the school of Professor V.D. Glukhovsky, slag-alkali cements and concrete based on them, received universal vocation in the 70s of the last century. The production of these composites requires the presence of granular blast furnace slag of ferrous metallurgy in a finely divided state, for the activation of which it is necessary to use alkaline solutions such as liquid glass, soda ash, caustic soda, sodium metasilicates. There are many results confirming the effectiveness of this area, but nevertheless, we do not observe the large-scale implementation of these developments, since there is a lot of research that has not been investigated in this technology that requires further study [1–5, 24, 25].

This paper presents its view on the development of clinker-free technology. Considering the fact that in many regions of our vast country there are no reserves of ferrous metallurgy wastes, and transportation of this valuable raw material over long distances is economically disadvantageous, our efforts were aimed at finding mutually substituting, blast furnace granulated slags, materials. And this is no coincidence, the slags of ferrous metallurgy in their chemical composition are identical to the composition of clinker minerals, therefore, the processes of structure formation of cement stone will be similar for both types of binder [3, 4, 6, 9]. Due to the lack of slag, we made attempts to obtain alkaline mixing binders using finely dispersed mineral powders of various origins.

And this is possible, since the North Caucasus has huge natural potential for the development of the cement industry. In mountainous regions, reserves of cement marls, limestones, flasks, dolomites, large deposits of building and glass sands, limestone, shell limestone, sandstones, natural sand dunes in the Terskaya Lowland are concentrated [16, 21–23].

2. Methods and materials

To develop clinker-free alkaline activation cements, finely dispersed mineral powders were prepared from sedimentary and igneous rocks: silica marl and quartz sand of the Vedeno deposit, limestone of the Yarysh-Mardynsky quarry, sand dunes of the Shelkovskoye deposit, volcanic tuff from the Kabardino-Balkarian Republic.

The energy dispersive microanalysis of the powders under study, performed using a Quanta 3D 200 i scanning electron microscope, showed a significant difference in the chemical composition of the mineral additives:

- quartz sand, %: MgO = 6,32; Al_2O_3 = 14,99; SiO_2 = 73,83; K_2O = 1,83; CaO = 0,60; Fe_2O_3 = 0,97; TiO_2 = 1,32; SO_3 = 0,14.
- volcanic tuff, %: MgO = 0,20; Al_2O_3 = 13,57; SiO_2 = 73,67; K_2O = 6,00; CaO = 1,79; Fe_2O_3 = 1,52; TiO_2 = 2,85; il = 0,40.
- limestone, %: MgO = 0,72; Al_2O_3 = 1,55; SiO_2 = 5,05; K_2O = 0,60; CaO = 90,14; Fe_2O_3 = 1,40; SO3 = 0,49.
- silicified marl, %: MgO = 1,64; Al_2O_3 = 6,42; SiO_2 = 28,6; K_2O = 1,33; CaO = 16,90; Fe_2O_3 = 1,08; TiO_2 = 0,47; SO_3 = 0,29; il = 43,2.
- barchan sands, %: MgO = 2,41; Al_2O_3 = 7,81; SiO_2 = 59,54; K_2O = 1,44; CaO = 17,52; Fe_2O_3 = 2,60; Na_2O = 1,35; SO_3 = 0,21; il = 7,12.

To prepare fine powders from the studied rocks, coarse ones were preliminarily crushed in a jaw crusher, and then, like fine grains, they were subjected to fine grinding in a VM-20 laboratory vibratory ball mill. At certain intervals, samples were taken from the mill to determine the specific surface of the powders and to conduct a comparative analysis (using the PSX-12 device).

3. Results

The specific surface of the obtained powders varied from 210 to 1120 m^2/kg depending on the grinding time (Fig. 1).

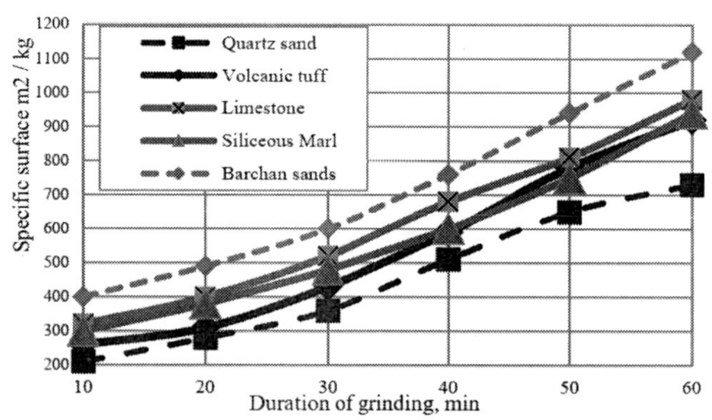

Figure 1. The dependence of the specific surface of fine powders on the duration of grinding

The main binder – in the proposed compositions is an alkaline activator, shutting them with the studied powders, we get a cement paste, which hardens with time, forming an alkaline cement stone [7–11]. It should be noted that the properties of cement stone will depend both on the nature of the mineral powder and on the type of grout [8, 12–20].

To study the effect of the specific surface area of the obtained mineral powders on the properties of the cement paste, molding mixtures were prepared from the studied components, and they were activated by various alkaline solutions: salable liquid sodium glass, caustic soda and soda ash. At the first stage, it was necessary to trace how the specific surface of mineral powders will change the normal density of alkaline cement paste. The normal density of the alkaline cement paste was determined according to the standard method, on a Vika instrument, replacing the needle with a pestle, the results are shown in table 1.

Table 1. Normal density of alkaline cement test, %

No.	Mineral powders	Alkaline solution		
		Na_2SiO_3	NaOH	Na_2CO_3
		Specific surface area Ssp = 250 m^2/kg		
1	Quartz sand	25	26	27
2	Volcanic tuff	24	24	25
3	Siliceous Marl	24	25	26
4	barchan sands	28	29	29
5	Limestone	27	28	28
		Specific surface area Ssp = 500 m^2/kg		
6	Quartz sand	26	27	28
7	Volcanic tuff	25	27	27
8	Siliceous Marl	25	26	26
9	barchan sands	29	29.5	30
10	Limestone	27	29	29
		Specific surface area Ssp = 900 m^2/kg		
11	Quartz sand	27	29	29
12	Volcanic tuff	26	28	27
13	Siliceous Marl	26	27	28
14	barchan sands	33	34	35
15	Limestone	29	31	31

The results of the study showed that the normal density of all the studied compositions of alkaline cement paste does not differ significantly depending on the alkaline solutions used, the use of caustic soda and soda ash slightly increase this indicator. A significant influence is exerted by the nature of the mineral powder and the degree of dispersion, as we observe, with an increase in the specific surface, the normal density of the alkaline cement paste increases. This indicator is especially high for highly dispersed sand dunes, the explanation for this is the presence of clay particles in the sand. I analyze the results of studies, it can be assumed that with an increase in the specific surface, the demand for powders in an alkaline solution increase. Therefore, to obtain binders of alkaline mixing, there is no need for long-term grinding of mineral additives, Ssp = 250 m^2/kg is quite enough.

At the next stage, the task was set to determine the setting time of binders for alkaline activation, on the basis of the studied powders with a specific surface Ssp = 250 m^2/kg, the preference for this degree of dispersion was given for the rational use of technological equipment and energy saving. The research results are shown in table 2.

The results obtained confirm that the setting time of the binders of alkaline mixing varies depending on the type of activator, the use of sodium metasilicate leads to a quick setting of the mixture, sodium hydroxide and soda ash lengthens this indicator for 5–7 minutes. It should also be noted that limestone mineral powders trapped with an alkaline solution are characterized by a slower start and setting time, and of course the reason for this is the petrographic composition of these rocks, it is likely that the calcite mineral prevailing in the present case reacts much later with an alkaline activator.

As we know, compositions based on liquid glass belong to the category of air binders, but the aim of these studies was to obtain durable waterproof materials that can exhibit hydraulic properties. To confirm this hypothesis, beam samples of 20x20x100 mm in size were prepared from a molding mixture consisting of the studied powders Ssp = 250 m^2/kg mixed with an alkaline solution, and coarse quartz sand with a particle size of 2.8 was used as a filler. After a day, and some samples on the second day, were redistributed. Some of the samples were placed periodically for several hours in an oven at a temperature of 40 °C. After heat treatment, part of the samples were stored in water for 28 days, and the other part of the samples hardened under normal conditions at a temperature of 18–20 °C. We studied the kinetics of a set of strength binders for alkaline activation, under various conditions of hardening. The composition of the solid components in the compositions was constant, only the consumption of the alkaline hardener changed, depending on the normal density of the mixture: mineral powder 500 kg/m^3, fine aggregate 1500 kg/m^3. The research results are shown in table 3.

Table 2. Influence of the nature of mineral powders and alkaline solution on the terms of gripping binders of alkaline activation (start of crossing / end of crossing)

No.	Mineral powders	Alkaline solution		
		Na_2SiO_3	NaOH	Na_2CO_3
		Specific surface area Ssp = 250 m^2/kg		
1	Quartz sand	0–30	0–35	0–36
		0–55	1–05	1–10
2	Volcanic tuff	0–26	0–34	0–40
		0–45	0–55	1–05
3	Siliceous Marl	0–20	0–25	0–30
		0–35	0–42	0–50
4	barchan sands	0–14	0–20	0–24
		0–23	0–35	0–40
5	Limestone	0–50	1–05	1–15
		1–22	1–34	1–45

Table 3. Kinetics of strengthening of knitting alkali closing

Alkaline activator	The limit of compressive strength, MPa per day								
	After TO			in water			normal conditions		
	28	90	180	28	90	180	28	90	180
				quartz sand					
$Na_2 \cdot SiO_3$	24.7	25.6	26.1	18.4	23.5	24.7	18.4	19.0	19.6
NaOH	20.2	21.0	22.3	16.4	17.4	18.4	16.5	17.0	17.5
Na_2CO_3	15.3	15.7	15.9	12.7	13.6	13.9	12.1	12.9	13.2
				volcanic tuff					
$Na_2 \cdot SiO_3$	44.0	46.5	48.7	42.6	44.7	45.2	39.1	40.8	42.6
NaOH	35.3	36.8	37.9	33.2	34.9	35.7	31.5	32.3	33.6
Na_2CO_3	30.2	31.8	32.6	29.8	30.1	31.6	27.5	27.4	27.9
				silicified marl					
$Na_2 \cdot SiO_3$	45.4	50.0	51.3	44.1	46.3	48.6	40.0	42.2	43.4
NaOH	36.1	37.5	39.7	34.3	37.5	39.7	33.1	35.2	36.2
Na_2CO_3	31.0	32.6	33.8	30.7	31.6	33.5	29.2	30.3	31.5
				barchan sands					
$Na_2 \cdot SiO_3$	46.6	47.9	49.5	44.1	45.4	46.7	42.5	43.7	44.8
NaOH	35.4	36.8	38.4	33.7	35.7	36.0	31.9	32.6	33.7
Na_2CO_3	34.2	34.9	35.9	31.0	31.7	32.8	30.6	31.6	32.9
				Limestone					
$Na_2 \cdot SiO_3$	9.7	10.1	10.7	6.4	6.5	6.5	4.5	4.6	5.6
NaOH	8.1	8.7	9.3	–	–	–	4.0	4.2	4.3
Na_2CO_3	7.0	7.6	7.9	–	–	–	3.4	3.9	4.4

4. Conclusion

The results showed that sodium metasilicate is the best activator of mineral powders of various origin, a solution of sodium hydroxide and soda ash is inferior to this mash. Astringent alkaline activation based on sand dunes and silicified marl showed the highest results; volcanic tuff powder is slightly inferior to them. The effectiveness of the indicated powders can be explained by their aluminosilicate nature, which, when activated by an alkaline solution, allows the formation of a strong heliopolymer stone represented by a three-dimensional aluminosilicate hydrogel. But it is important to take into account the high cost of liquid sodium glass, therefore, in the future, our studies will be aimed at creating a less energy-consuming activator, or using binary alkaline solutions [4, 6, 9].

Substandard sand sands belong to the thin class with a particle size modulus of 0.65, they are unsuitable for use in construction, as they are characterized by a high content of clay particles, but in the proposed technology this can only favorably affect the properties of composites, and besides, it can be used them in their natural form, without subjecting to fine grinding.

The obtained research results will significantly expand the scope of application of clinker-free alkaline mixing cements and become an alternative to energy and resource-consuming Portland cement, and of course, it will be possible to at least partially replace it in the construction industry.

References

[1] Glukhovsky V D (ed) 1979 *Alkaline and alkaline-alkaline earth hydraulic binders and concrete* (Kiev: Vishcha school) 232 p
[2] Glukhovsky V D and Pakhomov V A 1978 *Slag-alkali cements and concretes* (Kiev: Budivelnik) 184 p
[3] Krivenko P V and Pushkareva K K 1993 *Durability of slag-alkali concrete* (Kiev: Budivelnik) 224 p
[4] Davidovitz J 2008 *Geopolymer. Chemistry and applications* (St. Quentin: Instit. Geopolymer) 592 p
[5] Duxson P, Fernandez A and Provis J 2007 Geopolymer technology: The current state of the art *J. Mater. Sci.* **42** 2917–33
[6] Murtazaev S-A Yu, Salamanova M Sh, Saidumov M S, Ismailova Z Kh 2017 The influence of active surface centers on the reactivity of mineral additives *Modern Sci. and Innovat.* **2(18)** 168–75
[7] Murtazaev S A Yu and Salamanova M Sh 2018 Prospects for the use of thermally activated raw materials of aluminosilicate nature *Volga Sci. J.* **2(46)** 65–70
[8] Nikiforov E A, Loganina V I and Simonov E E The effect of alkaline activation on the structure and properties of diatomite *Bull. of BSTU*
[9] Salamanova M Sh and Murtazaev S-A U 2019 Alkaline activation cements: the possibility of reducing the energy intensity of obtaining building composites *Build. Mater.* **7** 31–42
[10] Nesvetaev G, Koryanova Y and Zhilnikova T 2018 On effect of superplasticizers and mineral additives on shrinkage of hardened cement paste and concrete *Theoretical Foundation of Civil Engineering (27RSP)* MATEC Web of Conf. 27. Ser. 27th R-S-P Seminar (TFoCE 2018) p 04018
[11] Stelmakh S A, Nazhuev M P, Shcherban E M et al 2018 Selection of the composition for centrifuged concrete, types of centrifuges and compaction modes of concrete mixtures *Physics and Mechanics of New Materials and Their Applications* (PHENMA 2018) p 337
[12] Shuisky A, Stelmakh S, Shcherban E and Torlina E 2017 Recipe-technological aspects of improving the properties of non-autoclaved aerated concrete *Modern Trends in Manufacturing Technologies and Equipment* MATEC Web of Conf. Ser. Int. Conf. (ICMTMTE 2017) p 05011
[13] Soldatov A A, Sariev I V, Zharov M A and Abduraimova M A 2016 Building materials based on liquid glass *Actual problems of construction, transport, engineering and technosphere safety* Mater. of the IV annual sci. and pract. Conf. of the North Caucasus Fed. Univer. pp 192–5
[14] Martschuk V and Stark T 1998 Untersuchungen zurn Frost-Tausalz-Widerstaud von Mochleistungsbetonen. *Wiss. Z. Bauhaus – Univ. Weirnar* **44(1–2)** 92–103

[15] Larbi J A and Bijen J M 1990 Effect of water-cement ratio, quantity and fineness of sand on the evolution of lime in set portland cement systems *Cem. and concr. Res.* **20(5)** 783–94

[16] Salamanova M Sh, Aliev S A and Murtazaeva R S-A 2019 The structure and properties of binders for alkaline activation using cement dust *Bull. of the Dagestan State Techn. Univer. Techn. Sci.* **2(46)** 148–58

[17] Kozhukhova N I, Chizhov R V, Zhernovsky I V and Strokova V V 2016 Structure formation of geopolymer perlite binder vs. Type of alkali activating agent *ARPN J. of Engineer. and Appl. Sci.* **11(20)** 12275–81

[18] Kozhukhova N I, Chizhov R V, Zhernovsky I V and Strokova V V 2016 Structure formation of geopolymer perlite binder vs. Type of alkali activating agent *Int. J. of Pharmacy and Technol.* **8(3)** 15338–48

[19] Larbi J A and Bijen J M 1990 The chemistry of the pole fluid of silica fumeblended cement systems *Cem. and concr. Res.* **20(4)** 506–16

[20] Kozhuhova N I, Zhernovskiy I V, Osadchaya M S et al 2014 Revisiting a selection of natural and technogenic raw materials for geopolymer binders *Int. J. of Appl. Engineer. Res.* **9(22)** 16945–55

[21] Udodov S A and Guiche M R 2015 The effect of dosage of redispersible powder on the localization of the polymer and the deformation properties of the solution *Sci. works of the Kuban State Technol. Univer.* **9** 164–74

[22] Salamanova M Sh, Murtazayev S Yu and Ismailova Z H 2018 The Use of Highly Active Additives for the Production of Clinkerless Binders *Engineering and Earth Sciences: Applied and Fundamental Research* vol 177 Proc. of the Int. Symp. (ISEES 2018) pp 355–8 ISSN Part of series: AER, ISSN: 2352-5401, ISBN 978-94-6252-637-2

[23] Salamanova M Sh and Murtazayev S Yu 2018 Clinker-free binders based on finely dispersed mineral components *Internationale Baustofftagung, Tagungsbericht.* 12–14 september 2018, Bauhaus-Universitdt Weimar. Band 1 und 2. Weimar: B. 2 pp 707–14

[24] Lesovik V S, Zagorodnyuk L K et al 2015 Designing of mortar compositions on the basis of dry mixes *Int. J. of Appl. Engineer. Res.* **10(5)** 12383–90

IOP Conf. Series: Materials Science and Engineering **905** (2020) 012059 doi:10.1088/1757-899X/905/1/012059

Structure and properties of comenic acid and complex compounds of magnesium, copper (II) and zinc

M Kh Shamsutdinova[1], L Kh Dzhabrailova[1], Yu Ya Kurbanov[2], V A Volynkin[3], N N Bukov[3], L I Ivashchenko[3], D T Lindt[3] and V T Panyushkin[3]

[1] Chechen State University, 17a, Dudayev Boulevard, Grozny, 364060, Russia
[2] North Ossetian State Medical Academy, 40, Pushkinskaya str., Vladikavkaz, 362025, Russia
[3] Kuban State University, 149, Stavropolskaya st., Krasnodar, 350040, Russia

E-mail: medina_humaidovna@mail.ru, yunus.kurbanov@lenta.ru, vva@chem.kubsu.ru, nbukov@mail.ru, chem.lindt_d.a@mail.ru, panyushkin@chem.kubsu.ru

Abstract. The paper studies the features of complex formation of comenic acid with Mg^{2+}, Cu^{2+} and Zn^{2+} ions. Besides, the paper presents the forms of binding of comenic acid with magnesium, copper (II) and zinc ions as well as the composition of complex compounds formed. Using IR and NMR (^1H, ^{13}C) spectroscopy, there was studied the structure of compounds the geometry of which was confirmed by quantum chemical calculations.

1. Introduction

Bioinorganic chemistry is a science covering both inorganic chemistry and biology, which is attracting more and more attention of researchers – theorists and practitioners – since bioinorganic chemistry studies the structure of biological clusters, which include compounds of the so-called *metals of life* and bioligands, i.e. proteins, nucleic acids, vitamins, hormones and other biologically active compounds [1]. The results of the study in the field of bioinorganic chemistry find their practical application in the following fields: medicine, agriculture, food industry, environmental protection, etc.

Pharmaceutical clusters are being created in our country where the development of biologically active substances and the production of drugs based on them will be carried out. The task of pharmaceutical clusters is to increase the share of Russian drugs in the domestic market, since now the domestic pharmaceutical industry produces only 15–20 % of vital and necessary drugs.

Comenic (CA) (5-hydroxy-4-oxo-4H-pyran-2-carboxylic) acid is $C_5H_2O_2(HO)COOH$ or $C_6H_4O_5$. There is an opinion [2] that the basis of the manifested biological activity of CA is the ability of the latter to form both simple salts and complex compounds by the *metals of life*. **The goal** of the work was to synthesize comenic acid compounds with magnesium, copper (II) and zinc, to study their physical and chemical properties and structural features both in solution and in solid state.

2. Methods and materials

Comenic acid (I) was used in the work (hereinafter, H2A) obtained by the method of [2], the acid was isolated on an Altex-112 liquid chromatograph with an Ultra-sphere ODS column and an Optilab Multiger 902B differential detector.

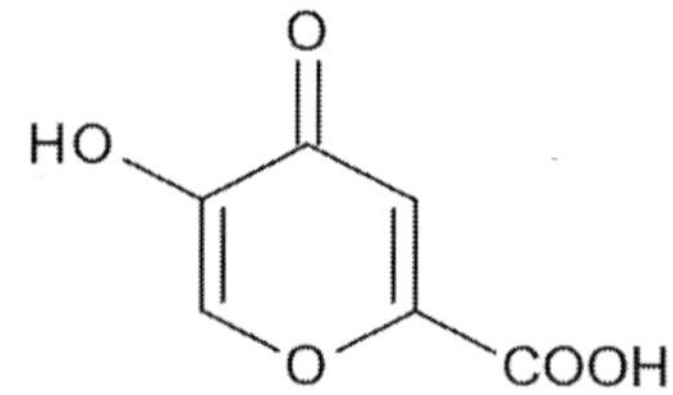

Comenic acid at a room temperature is a white crystal, which turs yellow when exposed to air. It is sparingly soluble in water and insoluble in ethanol. Judging by the data of a derivatographic study in an argon atmosphere (Fig. 1), comenic acid does not have a melting point, but sublimes with decomposition at a temperature of 295–297 °C.

Comenic acid was identified by electronic, IR and ^{1}H and ^{13}C NMR spectra. The dissociation constants (pK^{i}) obtained by potentiometric titration of comenic acid against hydrochloric acid were 1.8 and 7.3, respectively.

The salts of magnesium, copper (II) and zinc used in the work, as well as other auxiliary reagents (acids, alkalis, solvents) had qualifications of not less than AR grade. Bidistillate was used to prepare aqueous solutions for spectral and potentiometric studies.

The synthesis of complex compounds of comenic acid was carried out by mixing 0.1 M hot (80 °C) solution of an acid and a metal salt (magnesium chloride, copper (II) chloride, zinc) in an equimolar ratio. The solution was sustained for one day. After that, it had been evaporated until a fine-cristalline precipitate of the obtained compound was formed. The crystals were filtered off, washed with cold alcohol and dried out in a desiccator over calcium chloride.

The isolated compounds were analyzed for the metal content by titration and/or atomic absorption method on ICP-AEC iCAP 6500 Spectrometer, and for the carbon and hydrogen contents on Carlo Erba CHN Elemental Analyzer.

Gross formulas: $Mg(HA){\cdot}H_2O$: found %: C – 36.6; Mg – 12.8; $[Cu(H_2O)_2HA]$ Cl: found %: C – 22.11; Cu – 20.42. Calculated %: C – 22.64; Cu – 19.9; $[Zn\ (H_2O)_2HA]Cl$: found%: C – 24.68; Zn – 23.41. Calculated %: C – 25.14, Zn – 22.39.

Thermogravimetric analysis of comenic acid and its compounds was carried out on a NETZSCH STA 409 PC/PG analyzer in argon atmosphere in the temperature range from the room temperature to 400 – 1000 °C with a temperature rise rate of 5–10 K/ min.

IR spectra were recorded on a Vertex 70 IR Fourier spectrometer, NMR spectra on a JNM-ECA-400 spectrometer (400 MHz, Japan, JE02)

Quantum and chemical calculations were carried out by the semiempirical method MNDO/PM3 using HyperChem 8.0 software package.

3. Results

Being hydroxy-keto-carboxylic acid, comenic acid in an aqueous solution dissociates in two steps:

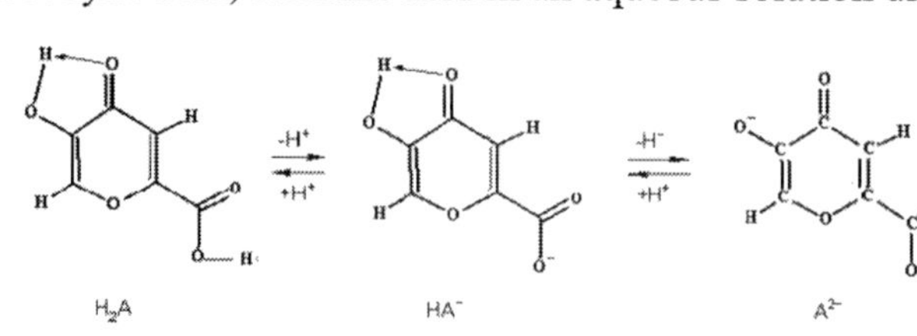

I II

where pK1 = 1.4 – 2.4 and pK2 = 7.4 – 7.8 [4-6], respectively.

Thus, the nature of epy interaction of metal ions with comenic acid can depend on the medium pH, which, in turn, is due to the dependence of the concentration of variously protonated forms of comenic acid on the pH of the medium. Although any of such form is capable of coordination with metal, their different ratio makes it possible to assert different structure of the complexes formed. It should be borne in mind that the formation of a more stable complex compound with a deprotonated form of hydroxy acid (HA^- and $A2^-$) shifts the equilibrium of dissociation reactions to the right. This, in turn, leads to the predominance of complexes with a deprotonated ligand at lower pH than it follows from the region where this form predominates in aqueous solution.

It can be assumed that completely deprotonated form of (A2−) comenic acid during complexation may not be available due to the occurrence of hydrolysis of metal ions at pH>6. However, chelation leads to a shift in equilibrium towards the deprotonated form, and in addition, some metal ions [5] are able to displace hydrogen ions.

Derivatograms of CA in an argon atmosphere (Fig. 1) showed two effects of sample mass change (TG), i.e. the removal of sorption water in the temperature range of 50–80 °C with a weight loss of the order of ≤0.3 % and pyrolysis at a temperature of 292–297 °C. Endo effects in the range of 273–274 °C, indicating the melting of CA were not found.

The pyrolysis of CA compounds is more complicated. According to thermogravimetric measurements in CA compounds with magnesium ion (Fig. 2) at temperatures up to 80–85 °C, sorption moisture evaporates with a mass loss of ≈1.3 %. Then, in the range of 153–193 °C for a monosubstituted and 225–231 °C for a bisubstituted compound, cleavage of the coordination water molecules included in the complex is observed.

The pyrolysis of a monosubstituted complex occurs in two stages in the range of 342–377 °C, and of a bisubstituted complex it occurs in one step at a temperature of 376–384 °C. In this case, the pyrolysis of bisubstituted complex is accompanied by an exo- effect, and for a monosubstituted one, the first stage at a temperature of 342 °C is endothermic, and the second one at a temperature of 377 °C is exothermic. Finally, in the range of 690–790 °C, the final carbonization of the samples occurs with a residual mass of 30.5 % for the monosubstituted and 33.1 % for disubstituted complex.

The data of IR spectra allow us to draw the following conclusions about the structure of CA and its complexes. Comenic acid in solid form is present in the form of dimers (characteristic absorption bands in the range of 3300–2600 (υOH) and 1740 (υC = O carboxyl group) cm^{-1}); 5-hydroxy group (υOH=3350 cm^{-1}) is connected by an intramolecular hydrogen bond with the keto group (υC=O=1650 cm^{-1}).

The monosubstituted complex is formed by the interaction of a metal ion with a carboxyl group (decrease in υC=O to 1620 cm^{-1}). The presence of υOH absorption bands in the region of 3620–3400 cm^{-1} suggests the linear structure of the monosubstituted complex.

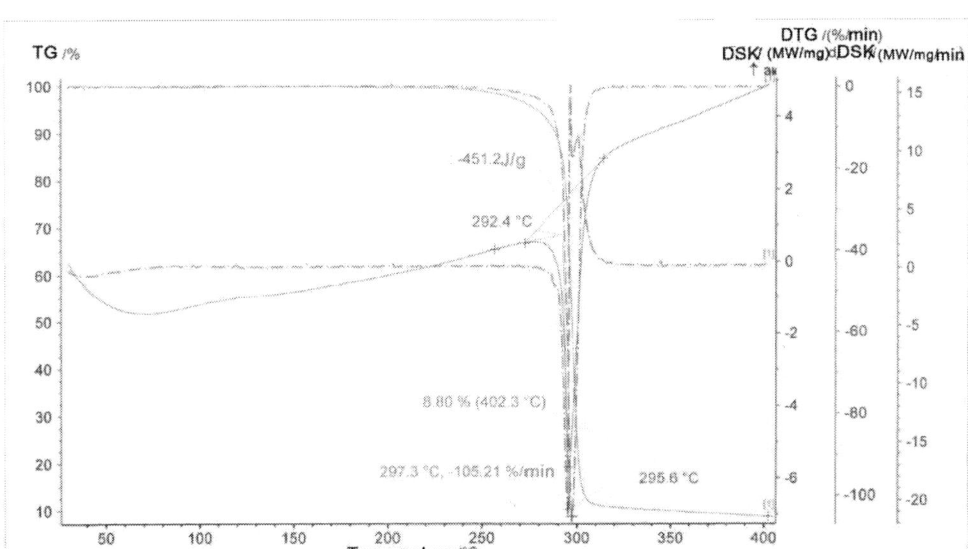

Figure 1. Thermal analysis curves of comenic acid with temperature rise rate of 10 K/min

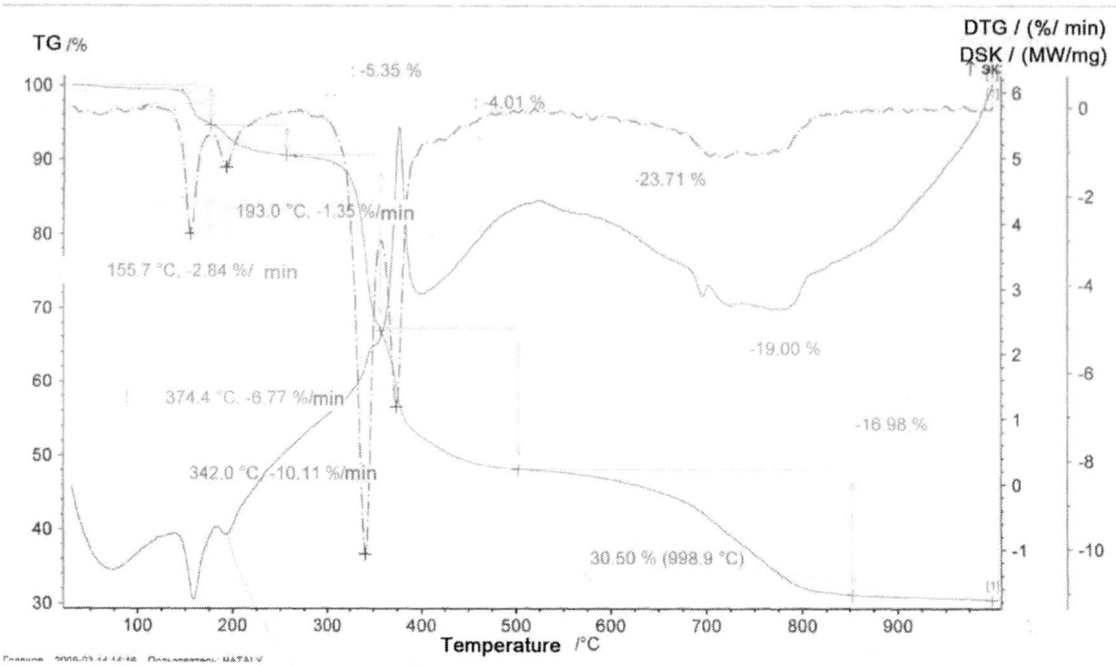

Figure 2. Thermal analysis curves of Mg (HA)·H_2O

The disubstituted complex is formed by the substitution of hydrogen by lithium ion in the 5-hydroxy group and additional coordination with the carbonyl group of the γ-pyrone. Consequently, an isle structure with additional coordination of metal ion with water molecules and oxygen atoms of the γ-pyron can be assumed. The absorption bands in the range of 480 and 820 cm^{-1} can be attributed to the absorption bands of M-O bonds. The isolated complexes of 1:1 composition according to IR spectroscopy probably have a linear structure with additional coordination at the oxygen atom of 5-hydroxy group.

In addition to protolytic equilibrium, CA is capable of forming various tautomeric forms. Therefore, CA in solution and solid form can exist in α – diketone form (**III**):

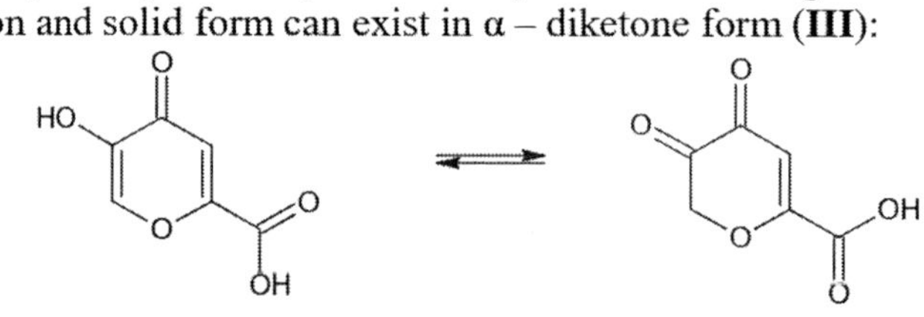

III **IV**

Our study of aqueous solutions of CA by ^{1}H and ^{13}C NMR methods does not confirm the validity of the above diagram in solution. In the ^{1}H NMR spectrum of an aqueous CA solution, there are only two signals of equal intensity from protons in positions (3) and (6). The Fig. 3 shows the dependence of chemical shifts on the pH value. The equality of the integral intensities of proton signals in positions (3) and (6) makes it possible to conclude that in an aqueous solution in the entire pH range studied, the tautomeric equilibrium is almost completely shifted toward form **III**.

^{13}C NMR data for some pH values were compared with data for a solution in DMSO-d6, in which the carboxyl group is slightly ionized. The shift of the signal of the carbon atom in position (7) to a weak field is also observed upon further deprotonation of the carboxyl group (pH 4.11). With a subsequent increase in pH, the deprotonation of hydroxo group of the pyrone ring begins (pH 6.89), which leads to stabilization of the electron system and an increase in the electron density at the carbon atom nucleus (7).

The shift of the signals of carbon atoms in position (4) and (5) corresponds to the deprotonation of hydroxyl group of the pyrone ring. In contrast to proton spectra, there is practically no change in the position of carbon signals in positions (3) and (6) with an increase in pH.

In solid state NMR spectroscopy, ^{1}H MAS spectra of organic molecules at moderate rotation frequencies (up to 20 kHz) are usually poorly resolved due to the strong dipole-dipole interaction. Satisfactory resolution was also not obtained for comenic acid in the indicated range. Therefore, several techniques for suppressing the dipole-dipole interaction were tested. The wPMLG3 technique turned out to be optimal for the objects under study. The scaling factor was 0.597.

In the CA spectrum, two proton signals are observed at positions (3) and (6) (7.2 and 8.5 ppm, respectively), a wide signal at ~ 3 ppm (coordination and adsorbed H2O) and a signal of 12.5 ppm of carboxyl group, merging with a wide signal of the OH group of the pyrone ring. The position of the proton signals in positions (3) and (6) in the powder correlates with the values obtained for the aqueous solution at pH<1 (7.27 and 8.19 ppm, respectively). Thus, in the structure in the solid state, comenic acid does not significantly differ from similar derivatives of γ-pyrons. It consists of dimers formed due to two hydrogen bonds on carbonyl oxygen and hydroxyl of the pyrone ring, which is consistent with the data of IR spectroscopy. In turn, such dimers can crosslink to form sandwich structures in which the ketone groups of one molecule are bonded to the carboxyl group of another molecule, forming structures similar to cinnamic acid.

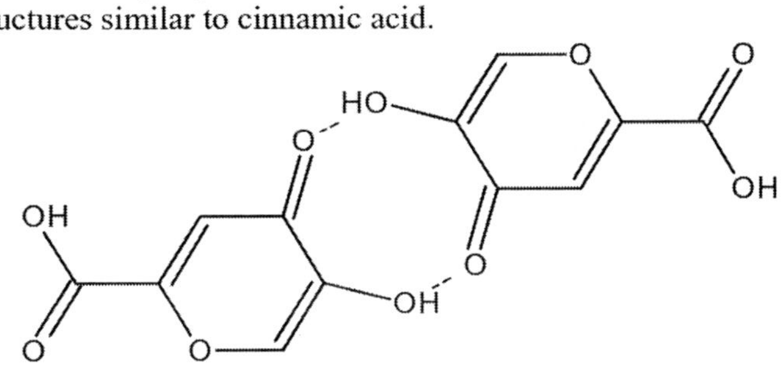

Theoretical study of the equilibrium geometry and an electronic structure of comenic acid and its compounds

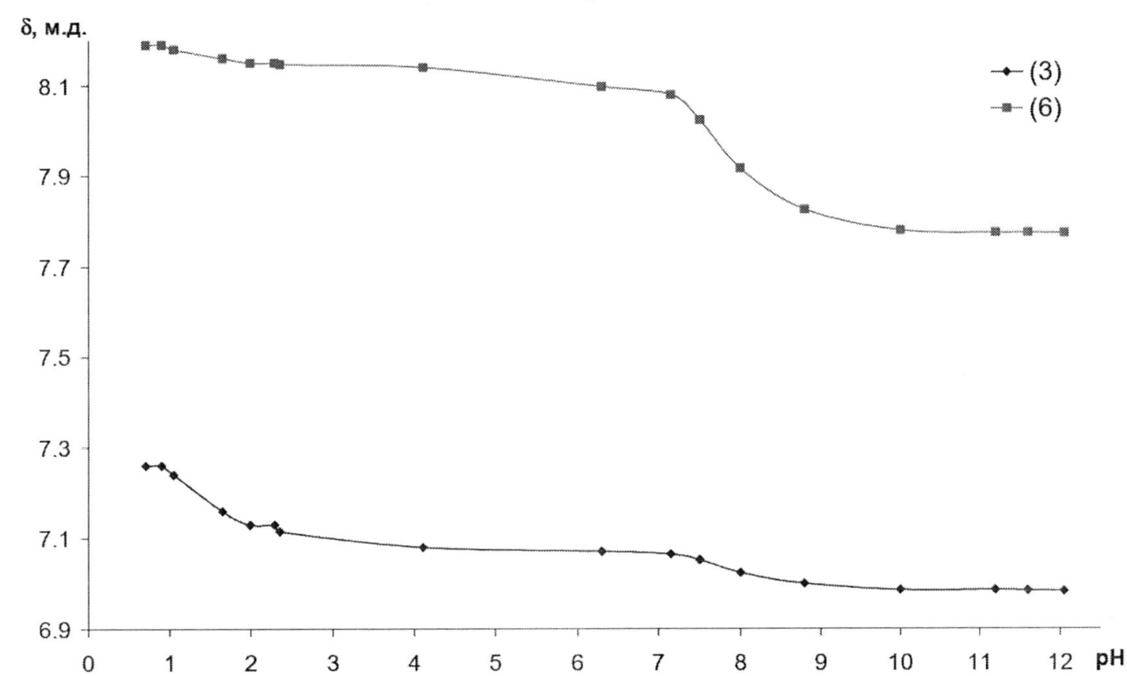

Figure 3. The dependence of chemical shifts of protons CA on the pH of the aqueous solution

To clarify the features of the geometric and electronic structure, as well as evaluate the enthalpy effects of chelation and the formation of salts and complex compounds of comenic acid, quantum chemical calculations were performed. The data obtained were compared with the results obtained previously [4]. The spatial numbering of atoms in the molecule of comenic acid compounds is shown in Fig. 4.

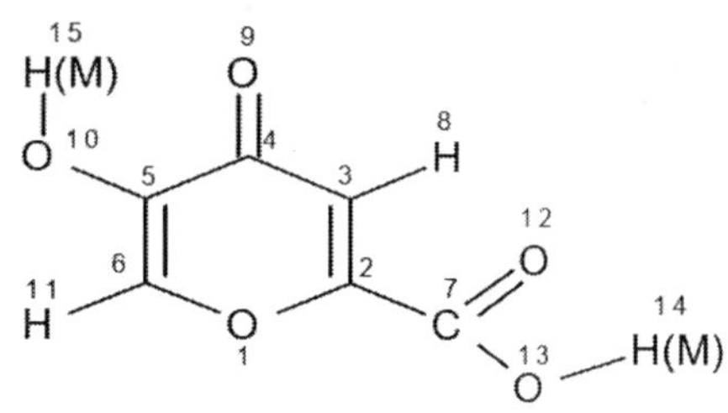

Figure 4. Geometrical structure and numbering of atoms in a comenic acid molecule

In the comenic acid compounds, the hydrogen atoms in the positions (14) and (15) are replaced by the corresponding ions of the studied metals.

According to the calculation results, the comenic acid molecule has a flat structure, which is facilitated by the system of conjugated double bonds C^2–C^3, C^4–O^9, C^5–C^6, and C^7–O^{12}. In the comenic acid molecule, the O^{10}–H^{15}–O^9 hydrogen bond (2.17 Å bond length) is confirmed, as was shown above, by the data of IR spectra. The effect of solvation is manifested in the broadening of the hydrogen bond to 2.21 Å.

In quantum chemical calculations, the molecule was optimized so that only the carboxyl group was assumed to be ionized, that is, the anion of comenic acid was considered singly charged (HA^-). This was also due to the main interest in the pH range 3–7, where this form is implemented in solution.

During ionization of the comenic acid molecule, as expected, the charges $q(O^{12})$, $q(O^{13})$ and $q(O^9)$ increase, and the charges $q(C^3)$ and $q(C^5)$ shift to the region of more negative values. In this case, the geometry of the carboxyl group changes, i.e. the conjugation of C^7–O^{12} bond with the system of double bonds disappears, which leads to an extension of C^2–C^7 bond; on the contrary, the C^7–O^{12} and C^7–O^{13} bonds come closer together. A change in the bond lengths of the pyrone ring is also observed: lengthening of O^1–C^2, C^2–C^3 and C^4–O^7 bonds, O^1–C^6, C^3–C^1 bonds are shortened. When passing from a free acid to an anion, the intramolecular hydrogen bond O^{10}–H^{15} ... O^9 is retained.

The calculation of complexation energies shows that taking into account additional corrections to the energies of solvated molecules is reflected in a 2.4–3.3 kcal/mol decrease in the absolute values of salt and complexation energies and a decrease of ~6 kcal/mol in the absolute values of complex formation energies. As one would expect, upon the transition from the gas phase to the solvated one, the absolute values of the energy effects of complex formation are greatly reduced, nevertheless, these processes remain energetically favorable.

Thus, the authors of this article are in consensus with the work [5], where the results of quantum and chemical calculations illustrate exclusively the fundamental possibility of implementing the proposed mechanism of donor-acceptor binding. Although, based on the results obtained, no conclusions can be drawn about the stability of the resulting complexes in solution.

4. Conclusion

The features of complex formation of comenic acid with Mg^{2+}, Cu^{2+}, Zn^{2+} ion were studied. The forms of binding of comenic acid and the composition of the resulting complex compounds were determined.

Using the methods of thermogravimetry, IR and NMR spectroscopy, there were studied the structures of CA and their complex compounds. It was shown that complexes of the composition 1:1 have a linear structure. The geometry of the compounds studied was confirmed by theoretical calculation by the semi-empirical method.

According to the IR spectra, the comenic acid in solid form is present in the form of dimers: υOH = 3300–2600 cм$^{-1}$, υC=O = 1740 cм$^{-1}$, υOH = 3350 cм$^{-1}$, 5-hydroxy group is linked by an intramolecular hydrogen bond to the keto group (υC=O = 1650 cм$^{-1}$). Monosubstituted complex compounds are formed due to the interaction of metal ions with the carboxyl and carbonyl groups (υC=O = 1620 cм$^{-1}$) and presumably have a linear structure. The disubstituted complexes are formed by the substitution of hydrogen ion in 5-hydroxy group and additional coordination with the carbonyl group of the γ- pyrone and has an isle structure.

According to the NMR spectra of (^{1}H, ^{13}C) in solution and in the solid state, it can be concluded that upon the transition from a solid state to a solution, the geometric structure of comenic acid does not change significantly. The main changes observed in the NMR spectrum are due to the deprotonation of corresponding groups. The ketone group of comenic acid is involved in the structure of complexes in the solid state.

References

[1] Eichhorn G (ed) 1979 *Inorganic biochemistry* vol 1 (Moscow: Mir) 712 p

[2] Pirogova A N, Bukov N N, Panyushkin V T et al 2009 Complex compounds of lithium with comenic acid *Chemistry and Chemical Technol.* **52(12)** 30–2

[3] Bukov N N, Dzhabrailova L Kh, Shamsutdinova M Kh and Panyushkin V T 2013 Synthesis and structure of complexes with comenic acid *J. of General Chem.* **85(6)** 1053–5

[4] Rogachevsky I. V and Plakheeva I B 2006 Quantum and chemical studies of equilibrium geometry and electronic structure of the genus of derivatives of γ-pyrone *J. of General Chem.* **76(11)** 1904–18

[5] Alekseev Yu T and Zhdanov Yu A 1998 Complexes of natural carbohydrates with metal cations *The success of chem.* **67(8)** 723–44

[6] Derbenev A V, Krylov B V and Shurygin A Ya Effets of neconik and comenic acids on slow solium chennels of secondary nemrons *Menbr. Cell. Boil. Zoo.* **13(3)** 379–87

[7] Panova T I 2004 The mechanism of interaction of comenic acid with the opoid receptor **1(1)** 128–32

[8] Volynkin V A, Pirogov D I and Panyushkin V T 2011 Study of comenic acid and its salts with Li^{+} and Na^{+} by NMR in solution and solid state *J. Struct. Chem.* **52(5)** 922–5

Study of changes in strength properties along section thickness of high-strength centrifuged and vibro-centrifuged concrete

S A Stelmakh, E M Scherban, A P Korobkin, K E Tkacheva, S A Osadchenko and A A Kadrov

Don State Technical University, 1, Gagarin square, Rostov-on-Don, 344000, Russia

E-mail: sergej.stelmax@mail.ru, au-geen@mail.ru, nikborock@yandex.ru, karinatkacheva@mail.ru, osa.don@yandex.ru, anton_kadrov95@bk.ru

Abstract. Improving the efficiency of precast concrete products and structures can be achieved by creating thin-walled products of effective cross-sections from high-strength concrete the formation of which requires application of centrifugal methods of compaction. The paper discusses literature data on the issue of forming the structure of centrifuged concrete by the thickness of the section wall, taking into account different densities of the components of the molded concrete mass under the action of centrifugal forces during mold rotation during concrete compaction at a constant speed. The authors conducted a series of experimental studies to establish the nature of the change in strength properties along the thickness of cross section of products from heavy concrete made by centrifugation and vibrocentrifugation. As a result of experiments, it was found out that the density of samples of centrifuged structures decreases in cross section from the outer layer to the inner one. The greatest values of ultimate strength under axial compression and tension during bending are characterized by the outer layer, the smallest are characterized by the inner layer. The middle layer in terms of strength indicators is close to the average strength of the whole product. A feature of changes in compressive and tensile strength during bending of vibro-centrifuged products and structures compared to centrifuged layers is a smaller difference between the outer and middle layers and a significant difference between the outer and inner layers. It was established that the variability of the structure of high-strength concrete is less pronounced compared to the lower-grade concrete.

1. Introduction

The development of modern construction takes the path of improving the designs and shapes of buildings, which is facilitated by the improvement of the physical and mechanical properties of building materials. The use of high-strength concrete and steels leads to the creation of lighter thin-walled reinforced concrete structures and is possible only with an advanced technology of manufacturing the elements and developed theory aimed to produce calculations for them.

The improvement of the efficiency of precast concrete products and structures can be achieved by creating thin-walled products of effective cross-sections from high-strength concrete, for the casting of which it is advisable to use centrifugal compression methods [1].

From the literature data [2–5] it is known that the concrete structure along the wall thickness of an as-formed pipe, taking into account different densities of the components of the molded concrete mass under the influence of centrifugal forces when the mold rotates during the compaction of the concrete mixture at a constant speed, becomes sharply inhomogeneous. This phenomenon is explained by the fact that centrifugation is both a method of forming pipes and a method of separating concrete mixture. In most cases, on the outer surface of the product there is concrete of regular texture, then fine-grained porous concrete, and on the inner surface there is a layer of cement stone with scarce impregnations of aggregate covered with a sludge layer that is to merge when the centrifugation is over.

The heterogeneity of the concrete structure is aggravated by the presence of pores and capillaries of various nature, formed during the hardening of as-formed pipe, since the water that previously filled both the ducts and the pores as well as capillaries of other origin evaporated. The number and size of pores and capillaries depend on the volume of water introduced into the concrete mixture, the dispersion of cement, its mineralogical composition, the quantity and the nature of mineral additives, hardening conditions, and other processing factors [6, 7].

To obtain a dense and uniform structure of centrifuged concrete, previous researchers of the centrifugation process most often attempted to select a concrete mix that would allow concrete to be created by centrifugation of the required density and strength. A fairly homogeneous concrete structure of the pipe walls can be obtained by using aggregate of different densities in one batch, which, when centrifuged, is distributed over the thickness of the product wall according to its bulk weight. However, in most cases, the pipe plants do not want to use these methods because of the difficulties in both the selection of concrete composition and the supply of aggregates or other components of the mixture. In most cases, especially in the production of non-pressure pipes, ready-mixed concrete of common composition is used.

Under the influence of normal pressure distributed unevenly along the wall thickness of the molded product from a minimum on the inside to a maximum on the outer surfaces, the liquid phase cannot simultaneously be squeezed out of the entire thickness of the concrete. When the concrete mixture is pressed with constant pressure, the aggregate grains clamp finely divided liquid fraction, interfering with squeezing out of excess mixing water.

Since the excess mixing water is squeezed out under constant pressure along radially oriented filtration channels, it is obvious that the cross-section of these channels increases as it approaches the inner surface of the compacted mixture, similar to a river bed as it approaches the mouth of the river. Moreover, the larger the initial water-cement ratio, the greater the total amount of squeezed water and, therefore, the greater the cross-sectional width of these channels. In addition, the cross-sectional dimensions of the channels also depend on the thickness of the sealing layer. This circumstance leads to an increase in the water impermeability of the pipe walls during layer-by-layer centrifugation, when the cross-section of the ducts of each layer is minimal, directly proportional to the thickness of the layer, and, in addition, the mouths of the ducts fill up when forming the next layer. At the end of centrifugation, the movement of fluid along the ducts slows down and, if the centrifugation time is too long, it stops. However, the ducts remain filled with liquid and after hardening, there are radial pores that drastically reduce the water impermeability of the products.

In addition to the amount of squeezed water and the thickness of a packed layer, the cross-sectional dimensions of the filtration ducts are significantly influenced by another factor, i.e. the speed of squeezing water through these channels. It must be said that the speed of squeezing water during the compaction process varies very significantly.

It is obvious that the maximum width of the channel in the cement paste corresponds to the maximum rate of squeezing out water. Under the traditional molding conditions and under the influence of constant pressure, the maximum amount of squeezed out water is observed at the very beginning of the centrifugation process, and then it gradually decreases. Thus, the channel of the duct, formed at the very beginning of the process of compaction of the concrete mixture, becomes redundant. The question arises whether it is impossible to squeeze out the required amount of water through the channels of a smaller cross section at a more constant speed. A smooth molding condition, which provides for a very slow growth in the sealing numbers of mold rotations, will lead to a significant decrease in the speed of squeezing out water.

Let us introduce the concept of centrifuge acceleration intensity, which characterizes a smooth (slowed down) increase in the number of mold revolutions from distribution to sealing for a certain period of time as the main factor determining the formation of the most dense and strong structure of centrifuged concrete.

In the traditional molding conditions under the constant pressure, the aggregate grains prevent free compaction of the cement paste.

Centrifugal compaction of the concrete mixture is mainly associated with the movement of coarse aggregate along the wall thickness of the molded product in the direction from the inner surface of the pipe to the outer surface, and the liquid phase of the mixture in the opposite direction. The movement of the coarse aggregate is much easier with a minimum amount of friction in the system. This can be achieved by selecting the optimal ratio of sand and gravel in the concrete mixture, as well as by selecting a water-cement ratio taking into account the moisture content of coarse and fine aggregate.

During centrifugation, the water-cement ratio during compaction is reduced. Plastic concrete mixes are better compacted; however, at the same time, with the phenomenon of separation, the amount of sludge drained from the pipe at the end of centrifugation increases. Together with the sludge, a significant part of the cement merges, which is presented in the form of suspension in the sludge. From this point of view, hard concrete mixtures are much more efficient. Nevertheless, in traditional centrifugation modes, a more or less satisfactory quality of the pipes is obtained only when using sufficiently plastic concrete mixtures with an initial water-cement ratio of more than 0.6 and a cone draft of more than 5 cm. When forming pipes in the smooth centrifuge dispersal mode, it is possible to use sufficiently slump concrete with an initial water-cement ratio of 0.45–0.5 and a cone slump of 2–4 cm.

The change in the process of separation of water from the cement-water system during centrifugation was considered. At the initial stage of compaction, with insignificant centrifugal forces, water is separated in physicomechanical bonds, which has lower binding energy than adsorption water. This period is characterized by the absence of structural bonds between particles of the solid phase. In other words, particles of a solid phase under the influence of a centrifugal force field settle in the mixture move in the liquid phase, displacing it into the inner layers. This process lasts until structural bonds arise between all cement flocs. A small amount of water remains in micropores and capillaries of the system, usually in the form of liquid-water menisci with small radii of curvature and high binding energy. In the following steps, at high centrifugation speeds, water separation decreases. During this period, the remaining capillary, as well as loose-bound adsorbed water is separated. By the end of centrifugation, the water should remain in the concrete mixture exclusively in physical and chemical bonds.

This circumstance makes it necessary to limit the duration of centrifugation, especially when only maximum pressure is applied to it. Excessively long rotation of the centrifuge at constant sealing speed of the mold will ultimately lead to dehydration of the concrete mixture and reduce the strength of the centrifuged samples. Therefore, when centrifuging, it is very important to determine the optimal parameters of molding conditions.

Since the centrifugation process is both a method of compacting the concrete mixture and a process of its separation, which, in contrast to the process of compacting the concrete mixture, negatively affects the strength and other characteristics of the concrete mixture. The concrete mixture is good if the aggregate grains move under the action of centrifugal forces to the outer layer through the cement paste, which they cannot squeeze onto the inner surface at its normal viscosity, but compact it with each other. Therefore, a well-compacted concrete mixture is characterized by a high content of cement paste with a low moisture content and, therefore, less separation of concrete mixture. On the contrary, during centrifugation of a very mobile mixture, a large separation occurs, as a result of which the compaction leaves much to be desired [8].

A decrease in the acceleration rate of the centrifuge gives another very important effect when forming tubular products designed to move liquids containing a significant amount of organic and inorganic impurities, namely non-pressure and pressure sewer and culverts. In the process of molding a denser and more uniform structure of concrete, the walls of the pipes improve the quality of their inner surface by reducing the separation of concrete mixture during smooth acceleration of centrifuge and more careful packing of the aggregate grains not only on the outer surface of the products, but also on its inner surface.

The degree of compaction during centrifugation depends on the amount of water squeezed out of the concrete mixture. At the same time, the water-holding capacity of cement, which is determined by the fineness of cement grinding, as well as the amount and type of mineral additives is of great importance. Tripoli powder is a common additive to Portland cement, but during centrifugation most of this finely dispersed additive is carried out on the inner surface of the pipe being molded and, at the end of centrifugation merges with the sludge. This fact significantly reduces the effectiveness of additives during centrifugation. It is advisable either to reduce the number of additives in cement intended for centrifuged pipes, or to use non-additive cements. However, there is no special cement production for pipe production; therefore, a significant role in saving material resources is played by the reduction of sludge during centrifugation. From this point of view, pipe molding in the conditions of smooth acceleration of centrifuge also gives an additional positive effect, since not only the absolute output of the liquid into the slurry is reduced, but also the number of fine fractions in the slurry composition due to a decrease in the separation of concrete mixture [2–5, 8].

2. Methods and materials

In the research laboratory of the Department of Technology of Binders, Concrete and Building Ceramics, DSTU, the authors conducted a series of studies to establish the nature of the change in the strength properties of the section thickness of heavy concrete products made by centrifugation and vibration centrifugation methods.

In order to streamline the collected experimental data, according to the methods described in previous works, the authors produced and studied concrete products of class B60 manufactured by centrifugation and vibrocentrifugation methods and brought to uniformity for the convenience of analytical comparison of the results [9–11]. The values of the concrete properties such as axial compression strength and tensile strength in bending were subject to comparison.

3. Results

The authors experimentally selected compositions of centrifuged and vibrocentrifuged concrete, which made it possible to obtain high-strength concrete of class B60 and higher in laboratory conditions. The cross section of centrifuged and vibrocentrifuged products was divided into three layers of equal thickness: external, medium and internal. The experimental results are presented in Tables 1–4 and graphically shown in Fig. 1–4.

Table 1. Test results of high-strength centrifuged concrete (hscc) by the layers of the products cross-section on such indicator as ultimate compresive axial strength at the age of 28 days

Concrete Design Class	Test results against cross-section layers of the samples						Section test results without layering	
	outer		middle		inner			
	Density, kg/m³	Strength, MPa	Density, kg/m³	Strength, MPa	Density, kg/m³	Strength, MPa	Density, kg/m³	Strength, MPa
B60	2531	82.2	2498	72.0	2478	72.9	2510	76.5
	2522	81.5	2508	75.3	2482	73.2	2500	72.9
	2518	82.7	2510	74.8	2488	71.7	2498	74.1
	2529	83.0	2495	73.8	2489	71.4	2512	74.5
	2525	80.9	2492	75.5	2484	71.0	2502	75.3
	2523	83.4	2500	72.6	2492	70.8	2495	75.5
Average	2525	82.8	2501	74.9	2486	72.3	2503	75.5

Table 2. Test results of high-strength centrifuged concrete (hscc) by the layers of the products cross-section on such indicator as tensile strength at the age of 28 days

Concrete Design Class	Test results against cross-section layers of the samples						Section test results without layering	
	outer		middle		inner			
	Density, kg/m³	Strength, MPa	Density, kg/m³	Strength, MPa	Density, kg/m³	Strength, MPa	Density, kg/m³	Strength, MPa
B60	2531	8.86	2498	7.64	2478	7.75	2510	8.18
	2522	8.78	2508	8.04	2482	7.78	2500	7.75
	2518	8.92	2510	7.98	2488	7.60	2498	7.89
	2529	8.96	2495	7.86	2489	7.57	2512	7.94
	2525	8.71	2492	8.06	2484	7.52	2502	8.04
	2523	9.01	2500	7.71	2492	7.50	2495	8.06
Average	2525	8.94	2501	7.99	2486	7.68	2503	8.06

Table 3. Test results of high-strength centrifuged concrete (hscc) by the layers of the products cross-section on such indicator as ultimate compresive axial strength at the age of 28 days

Concrete Design Class	Test results against cross-section layers of the samples						Section test results without layering	
	outer		middle		inner			
	Density, kg/m³	Strength, MPa	Density, kg/m³	Strength, MPa	Density, kg/m³	Strength, MPa	Density, kg/m³	Strength, MPa
B60	2527	86.3	2518	84.2	2498	76.1	2516	81.7
	2518	87.9	2514	86.6	2482	75.4	2522	81.2
	2514	85.8	2530	83.8	2499	72.7	2510	80.4
	2529	88.9	2528	86.0	2505	74.9	2500	82.6
	2525	89.5	2512	85.1	2489	75.1	2515	83.2
	2523	89,0	2519	84.3	2492	72.8	2519	79.9
Average	2523	88.8	2520	85.5	2494	75.4	2514	82.2

Table 4. Test results of high-strength centrifuged concrete (hscc) by the layers of the products cross-section on such indicator as tensile strength at the age of 28 days

Concrete Design Class	Test results against cross-section layers of the samples						Section test results without layering	
	outer		middle		inner			
	Density, kg/m³	Strength, MPa	Density, kg/m³	Strength, MPa	Density, kg/m³	Strength, MPa	Density, kg/m³	Strength, MPa
B60	2527	9.36	2518	9.10	2498	8.13	2516	8.80
	2518	9.55	2514	9.39	2482	8.05	2522	8.74
	2514	9.30	2530	9.06	2499	7.72	2510	8.65
	2529	9.67	2528	9.32	2505	7.99	2500	8.91
	2525	9.74	2512	9.21	2489	8.01	2515	8.98
	2523	9.68	2519	9.12	2492	7.74	2519	8.59
Average	2523	9.66	2520	9.26	2494	8.05	2514	8.86

4. Conclusion

As the result of the experiments, it was found out that the density of samples of centrifuged structures changes as follows: It decreases along the cross-section from the outer layer to the inner one. The same trend is observed with the strength characteristics of high-strength centrifuged concrete: the outermost layer is characterized by the highest tensile strengths under axial compression and tensile bending, and the inner layer is the smallest one. The strength indicators of the middle layer is close to the strength of the whole product (without dividing into layers). The difference in the limits of strength under axial compression and tension during bending between the outer and inner layers is about 15 %.

IOP Conf. Series: Materials Science and Engineering **905** (2020) 012060 doi:10.1088/1757-899X/905/1/012060

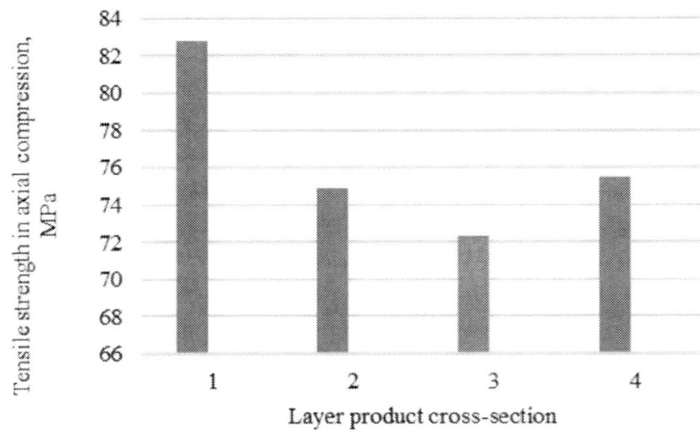

Figure 1. The dependence of the tensile strength on axial compression on the layer cross-section of the centrifuged product: 1 – outer layer; 2 – middle layer; 3 – inner layer; 4 – one piece product

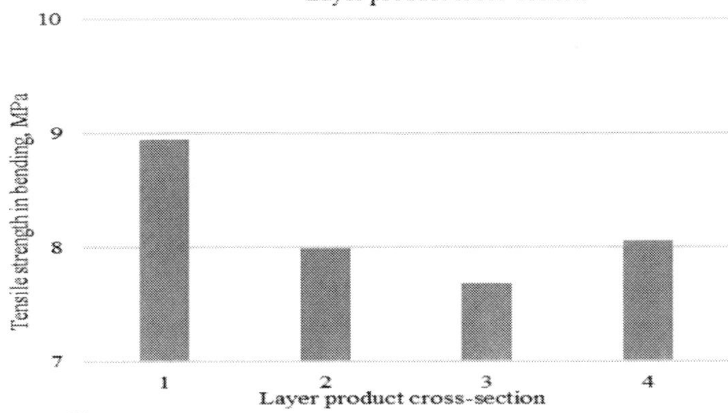

Figure 2. Dependence of tensile strength in bending on the layer of the cross-section of centrifuged product: 1 – outer layer; 2 – middle layer; 3 – inner layer; 4 – one piece product

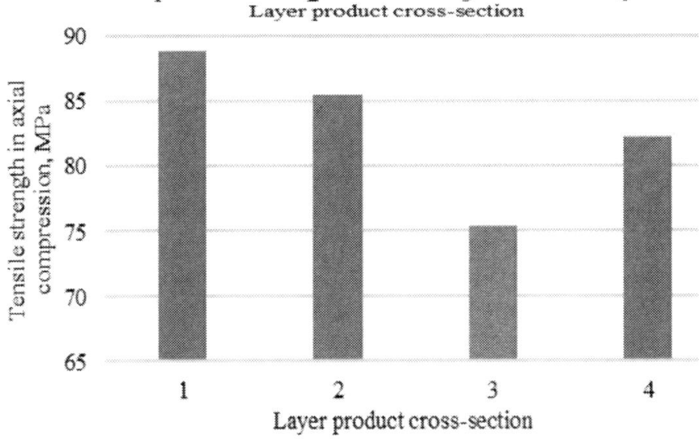

Figure 3. Dependence of tensile strength on axial compression from the layer of the cross-section of vibrocentrifuged product: 1 – outer layer; 2 – middle layer; 3 – inner layer; 4 – one piece product

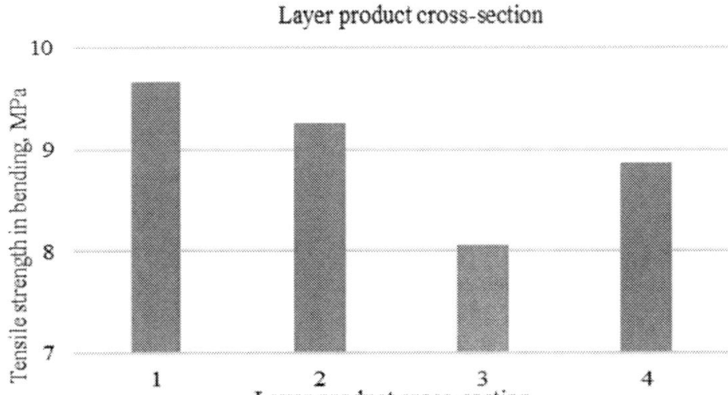

Figure 4. Dependence of tensile strength in bending on the layer of the cross-section of vibrocentrifuged product: 1 – outer layer; 2 – middle layer; 3 – inner layer; 4 – one piece product

The peculiarity of changes in compressive and tensile strength during bending of vibrocentrifuged products and structures compared to centrifuged products in layers is a small difference (3–5 %) between the outer and middle layers and a significant difference between the outer and inner layers (15–20 %).

References

[1] Pastushkov G P 1994 *Multi-storey frame buildings with load-bearing reinforced concrete centrifuged elements* (Minsk: Belarus. State Polytechnic. Acad.) p 36
[2] Akhverdov I N 1993 *Technology of reinforced concrete products and structures for special purposes* (Minsk: Sci. and Technol.) 240 p
[3] Berg P A and Yantsen T G 1991 Compaction of concrete mix by centrifugation *Concrete and reinforced concrete* **8** 23–5
[4] Soskin G M, Karpov A N et al 1991 Strength Fiber Concrete Culverts *Concrete and reinforced concrete* **1** 14–5
[5] Yundin A N, Shchutski V L, Nesvetaev G V and Fomina S V 1993 On increasing the durability of centrifuged reinforced concrete structures *Univer. News. Construct. and architect.* **10** 7–10
[6] Alaee P and Li B 2017 High-strength concrete exterior beam-column joints with high-yield strength steel reinforcements *Engineer. Struct.* **145** 305–21
[7] Ismail M K and Hassan Assem A A 2017 An experimental study on flexural behaviour of large-scale concrete beams incorporating crumb rubber and steel fibres *Engineer. Struct.* **145** 97–108
[8] Dubinina V G 2002 *Development of optimal centrifugation parameters for reinforced concrete pressureless pipes* (Nizhnii Tagil: Nizhnii Tagil Technol. Instit. of the UPI) 150 p
[9] Mailyan L R, Stelmakh S A, Kholodnyak M G et al 2018 Recommendations on accounting variatropy the calculation, design and manufacture of centrifuged structures of heavy concrete *The Eurasian Sci. J.* **4**
[10] Stelmakh S A, Shcherban'E M and Zholobova O A 2018 Prescription and technological aspects of manufacturing high-quality centrifuged products and structures from heavy concrete *IOP Conf. Ser. Mater. Sci. and Engineer.* **463(1)** 022056
[11] Suval R 1997 *The properties of centrifuged concrete and the improvement of the design of centrifuged reinforced concrete racks of power transmission line supports* (Rostov-on-Don: Rostov State Univer. of Civil Engineer.) 267 p

IOP Conf. Series: Materials Science and Engineering **905** (2020) 012061 doi:10.1088/1757-899X/905/1/012061

Force modeling of interparticle pair interaction of spheroid in composite cement

A K Khalyushev, E M Scherban, S A Stelmakh, A V Nalimova, I O Egorochkina, A K Sysoev and E A Zholobova

Don State Technical University, 1, Gagarin square, Rostov-on-Don, 344000, Russia

E-mail: khaljushev@mail.ru, au-geen@mail.ru, sergej.stelmax@mail.ru, nalimova@bk.ru, arin77@bk.ru, aksisoevrgsu@yandex.ru, info@rniiakh.ru

Abstract. The paper discusses the modeling of the forces of interparticle pair interaction of spheroidal agglomerate in composite cement. Based on the calculation of the autohesion forces that make up the spheroidal agglomerate, i.e. the Portland cement ("carrier particles") and silica fume ("guest particles"), it was found that for uncharged particles the magnitude of the electric interaction forces is negligible in comparison with molecular van der Waals and capillary forces. The bipolar charging of particles in the corona discharge field at a corona electrode potential of +18 and −18 kV sharply increases the electric component of the autohesion force, which in value approaches the value of capillary forces. In this case, the Coulomb attractive forces between oppositely charged particles begin to act before the direct contact of the particles, which contributes to the agglomeration of particles. These calculations are confirmed by the results of scanning electron microscopy (SEM) and energy dispersive spectroscopy (EDS) of composite cement samples containing Portland cement (75 %) and dry gas purification silica fume (25 %), the number of angular particles of irregular shape decreases for a mixture of these materials processed in the device for surface modification.

1. Introduction

As the fineness of grinding of composite cement increases, the water demand increases, i.e. the normal density of cement paste increases by 5–7 % when the specific surface area changes from 320 to 500 m^2/kg. To reduce water demand and, consequently, improve workability of concrete mixtures and concrete quality, optimization of particle size distribution of dispersed mineral components is needed. Particle shape, as well as the quantitative content of fine fractions, have an important effect on the water demand of the mineral components of concrete. In [1], it was shown that with an increase in the content of fly ash from TPPs, instead of Portland cement, the fluidity of the cement paste increases. This is explained from the standpoint of interparticle interactions in colloidal systems: for particles of fly ash, the value of the electrokinetic potential is $\xi = -12 \ldots -14$ mV, and for particles of Portland cement it is $\xi = +\ 2.17$ mV. With a small amount of ash in the composition of the cement paste, the particles of ash and cement flocculate due to electrostatic attraction. When the ash consumption is high (more than 50 % cement replacement), the negative integrated charge of the particles prevails in the system, and their electrostatic repulsion occurs, which increases the mobility of the paste. In addition, the adhesion of smaller particles of ash on oppositely charged particles of the clinker phases causes repulsion of the latter with the release of water immobilized in the floccules. An increase in mobility is also due to a decrease in friction between particles, since they take the form of spheroids [2, 3]. The spherical shape also minimizes the specific surface of the particles [4].

Currently, in various industries where highly dispersed powders are used, the research on the so-called dry particle coating is being intensively developed. The essence of the method consists in the fixation of finely dispersed particles with the size of 0.1–50 microns (guest particles – "guest" particles) on the surface of larger ones, i.e. 1–500 microns (host particles – "carrier particles") in the process of intensive mixing.

The percentage of "guest" particles required for the experiment is calculated based on the assumption of 100% surface coverage of the carrier particles with a monolayer of "guest" particles. It is assumed that the "carrier particles" and the "guest' particles are spherical with an appropriate size and density that do not change their shape during the coating process. Based on these assumptions, the mass fraction of "guest" particles for 100 % coverage can be determined [5] by the following formula:

$$W,\% = \frac{(Nd_{guest}^{3} \cdot \rho_{guest})}{(D_{host}^{3} \cdot \rho_{host}) + (Nd_{guest}^{3} \cdot \rho_{guest})} \cdot 100 \quad (1)$$

where W is a mass fraction of guest particles, %;

d_{guest} is the diameter of guest particles, microns;

ρ_{guest} is the density of guest particles, kg/m3;

D_{host} is the diameter of the carrier particles, microns;

ρ_{host} is the density of carrier particles, kg / m3.

With the ratio $D_{host} >> d_{guest}$ (especially, $D_{host}/d_{guest} \approx 10$):

$$N = \frac{4(D_{host} + d_{guest})^{2}}{d_{guest}^{2}} \quad (2)$$

Higher efficiency of surface modification is achieved in the installations where the mixing of the particles is accompanied by their intense collision with each other (High energy impact blending coater "Nara Hybridizer", High shear mixer "Cyclomix"). As a result of surface modification of powders, an increase in their rheological characteristics, a decrease (increase) in wettability, a change in granulometry and other properties are achieved [1, 5–7].

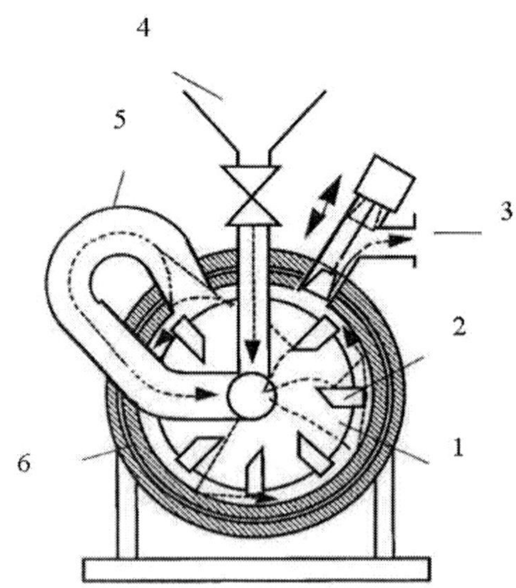

Figure 1. Drawing of the mixer – hybridizer (triboactivator): 1 – rotor; 2 – mixer; 3 – output channel; 4 – input channel; 5 – circular circulation; 6 – outer shell [12].

In this context, it is worth highlighting a promising direction that is being developed in the research of Japanese scientists, i.e. the technology for the formation of "spheroidal" cement (Spherical cement) [8–11]. The technology for producing such cement consists in processing Portland cement together with mineral additives in a special device, i.e. a mixer-hybridizer (Figure 1). In the process of high-speed mixing, an intense friction and collision of particles occur, which leads to the formation of surface charges. As a result, adhesion of small particles of gypsum, mineral additives and an intermediate phase to the surface of larger angular particles of silicate phases is ensured. Capillary forces also promote particle fixation since in the spaces between hydrophilic particles pressed against each other, water molecules are adsorbed by the condensation surface and liquid menisci are formed. During processing, the particles coarsen, reducing the total specific surface, and acquire a shape close to spherical ($k_{\Phi}\approx0.85$).

As a result of reduction in the number of "free" fine particles (diameter less than 3 microns), reduction in the surface roughness of "spheroids", and a slowdown of the early stages of cement hydration, the mobility of concrete mixtures prepared on the basis of "spheroidal" cement increases (W/C decreases by 6–8 %). When using superplasticizers in the concrete mixtures, a decrease in their adsorption per unit surface of cement was also noted (up to 20 %).

2. Methods and materials

Modeling the forces of interparticle pair interaction in a spheroidal agglomerate.

The efficiency of the formation of spheroidal agglomerates can be increased when using Portland cement compositions with highly dispersed mineral additives (silica fume, metakaolin).

Surface modification of composite cement by particles of silica fume was carried out on a special installation (Figure 2).

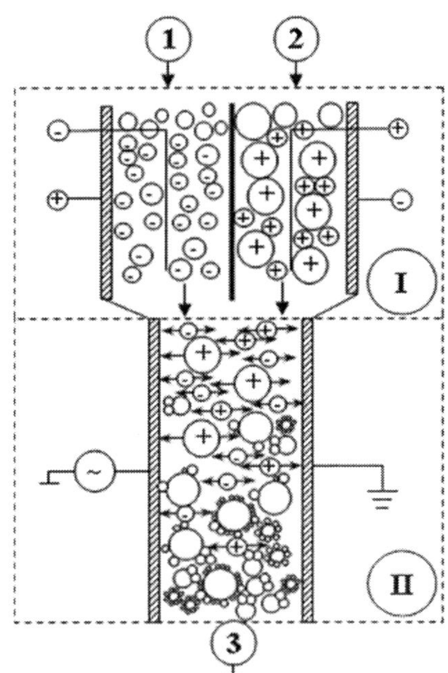

Figure 2. Installation drawing for surface modification of mineral dispersed concrete components: I – particle electrification chamber (bipolar charging) in the corona discharge field; II – agglomerator

At first, the initial dispersed mineral components can be charged in a high-voltage field of a corona discharge, while a sample of finely dispersed component ("guest" particles) is poured through an electrification chamber – (1) with a negative corona electrode, and a sample of coarse component ("carrier particles") through the chamber electrification – (2) with a positive corona electrode. The unit of electrification chamber consists of a corona electrode located in the center of the pipe, which is made of copper wire with the diameter of 0.6 mm. The current-voltage characteristics of the electrification chamber are as follows: electric field strength E = 18–20 kV/cm, current strength I = 30–50 μA.

As a result of the electric field impact created between the corona and external electrodes, the initial mineral components adsorb ions from the interelectrode space and acquire a negative (positive) charge.

After that, the bipolar charged mineral dispersed components enter the agglomerator, where under the impact of a high-voltage alternating electric field, the vibrations with different amplitudes and frequencies are carried out. Large particles oscillate with bigger amplitude and frequency than small particles, the gradient of the frequency and amplitude of oscillations causes an increase in the frequency of collisions of particles, which leads to the formation of spheroidal agglomerates.

Since the autogenesis of many particles that make up bulk material is composed of forces (molecular, electric, capillary) that manifest themselves in the individual contacts between particles, it is possible to calculate these forces by simulating the pairwise interaction of two particles that make up a spheroidal agglomerate – Portland cement ("carrier particles") and silica fume ("guest" particles). We assume that the average particle size of silica fume is 2 μm, and that of Portland cement is 20 μm, then the mass fraction of "guest" particles according to formula (1) will be 25.5 %.

Electric forces.

According to [10], the electric charge density for the particles of Portland cement and silica fume is as follows, respectively:

$$Q_1 = 9 \cdot 10^{-9}\ C/m^2 \quad (3)$$

$$Q_2 = -8 \cdot 10^{-11}\ C/m^2 \quad (4)$$

Then, some particles:

$$q = Q \cdot S = Q \cdot 4\pi r^2 \quad (5)$$

$$q_1 = 9 \cdot 10^{-9} \cdot 4 \cdot 3.14 \cdot \left(10 \cdot 10^{-6}\right)^2 = 113.04 \cdot 10^{-19} C$$

$$q_2 = -8 \cdot 10^{-11} \cdot 4 \cdot 3.14 \cdot \left(1 \cdot 10^{-6}\right)^2 = -100.48 \cdot 10^{-23} C$$

Particle electric capacity:

$$C = 4\pi\varepsilon\varepsilon_0 r \quad (6)$$

$$C_1 = 4 \cdot 3.14 \cdot 4.5 \cdot 8.85 \cdot 10^{-12} \cdot 10 \cdot 10^{-6} = 50 \cdot 10^{-16}\ \Phi;\ C_2 = 4 \cdot 3.14 \cdot 3.9 \cdot 8.85 \cdot 10^{-12} \cdot 1 \cdot 10^{-6} = 43.3 \cdot 10^{-17}\ \Phi.$$

where ε is the dielectric constant of the particle substance, ε_0 is the electric constant (F/m), r is the particle radius, m.

The potential of particle surface:

$$\varphi = q/C \quad (7)$$

$$\phi_1 = \frac{113.04 \cdot 10^{-19}}{50.02 \cdot 10^{-15}} = 2.25 \cdot 10^{-3} V; \quad \phi_2 = \frac{-100.48 \cdot 10^{-23}}{43.35 \cdot 10^{-17}} = -2.32 \cdot 10^{-6} V.$$

The electric field strength at the surface of a cement particle is calculated by the following formula:

$$E = \frac{1}{4\pi\varepsilon\varepsilon_0} \cdot \frac{q}{r^2} \quad (8)$$

where ε is the dielectric constant of the medium in which the particle is placed (air ε=1).

$$E = \frac{113.04 \cdot 10^{-19}}{4 \cdot 3.14 \cdot 1 \cdot 8.85 \cdot 10^{-12} \cdot \left(10 \cdot 10^{-6}\right)^2} = 102.7\ V/m$$

The force acting on a silica particle at the surface of a cement particle is determined by the following formula:

$$F_{el} = Eq_2 \quad (9)$$

$$F_{el} = 102.7 \cdot \left(-100.48 \cdot 10^{-23}\right) = 10.3 \cdot 10^{-20}\ N$$

Molecular forces.

Since d2 << d1, we can assume that a convex surface of radius r2 is in contact with a flat surface. The strength of the molecular interaction according to Hamaker has the following form:

$$F_m = \frac{Ar_2}{12H^2} \tag{10}$$

where A=(0.01…3.2)· 10^{-18}; J is the constant of molecular interaction, depending on the heterogeneity of solid surfaces; H=0.5· 10^{-9} m is the width of the gap between the contacting bodies

$$F_m = (0.01...3.2)\cdot 10^{-18}\cdot\frac{10^{-6}}{12\cdot 0.25\cdot 10^{-18}} = (3...1060)\cdot 10^{-3}N$$

Capillary forces.

If the shape of the formed menisci of the liquid is concave, then according to the Kelvin equation, the saturated vapor pressure above them is calculated by the following formula:

$$p = p_0 \exp\left(\frac{2\sigma v}{rRT}\right), \tag{11}$$

where p_0 is the saturated vapor pressure above a flat surface, σ is the surface tension coefficient of water, v is the molar volume of water, r is the meniscus curvature radius (negative for a concave surface), T is the absolute temperature, R = 8.31 J/(mol·K) is the universal gas constant. At a temperature of T =293 K, we have p_0=2.33 kPa, σ=73 mN/m, v=18· 10^6 m^3/mol.

Saturated vapor pressure over a concave surface is lower than over a flat surface. Thus, capillary condensation occurs at pressures lower than p_0, i.e., at lower air humidity:

$$\varphi = \frac{p}{p_0} = \exp\left(\frac{4\sigma v}{aRT}\right), \tag{12}$$

where a ~ 1 nm is the size characteristic of atomic-molecular roughness. It follows that capillary condensation occurs already at a relative humidity of ~ 10 %.

According to the Laplace equation, the fluid pressure under the concave surface is reduced by:

$$\Delta p = 2\sigma/r \tag{13}$$

which leads to the appearance of the capillary component of forces of autohesion. Substituting $r=a/2$=0.5 nm into the equation, we obtain Δp≈300 MPa. Since $d_2 << d_1$, we can assume that a convex surface of radius r_2 is in contact with a flat surface. Then the capillary force will be equal to:

$$F = 4\pi\sigma r_2, \tag{14}$$

$$F = 4\cdot 3.14\cdot 73\cdot 10^{-3}\cdot 50\cdot 10^{-9} = 4.58\cdot 10^{-8}H.$$

Thus, the calculations show that the main role in the formation of spheroidal aggregates is played by capillary and molecular van der Waals forces. The electrical component of the forces of autohesion is significantly inferior to these forces.

Let us consider the case when particles of Portland cement and silica fume are pre-bipolar charged in the corona discharge field. According to the data given in [12], fly ash particles with the size of 0.1 μm, processed at a corona electrode potential of 8 kV, acquire about ten elementary charges, and 1 μm in size acquire one hundred elementary charges. With increasing particle size and electric field strength, the particle charge increases significantly [13, 14].

For particles of Portland cement and silica fume treated accordingly with the potential of the corona electrode +18 and –18 kV, we take their charge value:

$$q_1 = 2\cdot 10^3\cdot 1.6\cdot 10^{-19} = 3.2\cdot 10^{-16}C;$$

$$q_2 = 2\cdot 10^2\cdot 1.6\cdot 10^{-19} = 0.32\cdot 10^{-16}C$$

Let us find the potential for particles in the corona discharge field:

$$\varphi_1 = \frac{3.2\cdot 10^{-16}}{0.5\cdot 10^{-14}} = 6.4\cdot 10^{-2} \quad \varphi_2 = \frac{0.32\cdot 10^{-16}}{0.43\cdot 10^{-15}} = 0.74\cdot 10^{-1}$$

Let us repeat the calculation of electric forces based on this potential value. Electric field strength at the surface of a cement particle:

$$E = \frac{3.2\cdot 10^{-16}}{4\cdot 3.14\cdot 1\cdot 8.85\cdot 10^{-12}\cdot\left(10\cdot 10^{-6}\right)^2} = 3200\ V/m$$

The force acting on a charged silica particle at the surface of a cement particle:

$$F = 32 \cdot 10^{2} \cdot 0.32 \cdot 10^{-16} = 10.2 \cdot 10^{-14} \text{ N}.$$

3. Results

Obviously, the charging of particles in the field of the corona discharge sharply increases the electric component of the forces of autohesion, which approaches the magnitude of the capillary forces in value. In this case, the Coulomb attractive forces between oppositely charged particles begin to act even before the direct contact of the particles, which contributes to the agglomeration of particles. These assumptions are confirmed by the results of scanning electron microscopy (SEM) and energy dispersive spectroscopy (EDS).

After the mixing of a mixture of materials that contain Portland cement (75 %) and dry silica fume silica (25 %), the number of angular particles of irregular shape decreases (Figure 3, c), this is true for a mixture of these materials processed in a device for surface modification (Figure 3, d). At the same time, larger aggregates up to 50-100 microns in size are found (Figure 3, a).

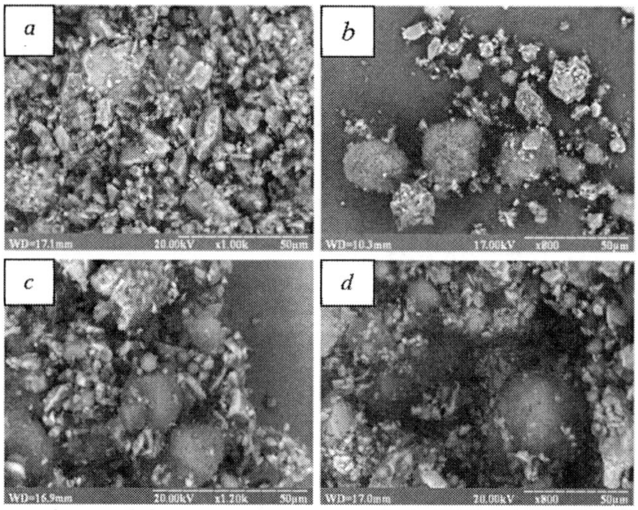

Figure 3. SEM of dispersed materials: a – Portland cement; b – silica fume; c – mechanical mixture of Portland cement and silica fume; d – a mixture of Portland cement with silica fume after surface modification.

According to the energy dispersive spectroscopy, the surface of such "spheroids" (point 1, Figure 4. b) is represented by the following oxides, %: SiO_2 (84.93); SO_3 (11.91); Al_2O_3 (1.85); MgO (0.84); K_2O (0.45), which is a characteristic of the chemical composition of silica fume. For the purpose of comparison, a particle on the surface of a "spheroid" (point 2) is represented by such a set of oxides,%: CaO (51.44); SiO_2 (37.31); Fe_2O_3 (4.21) SO_3 (2.24 %); Al_2O_3 (1.43); K_2O (3.07), which is very close to the chemical composition of Portland cement.

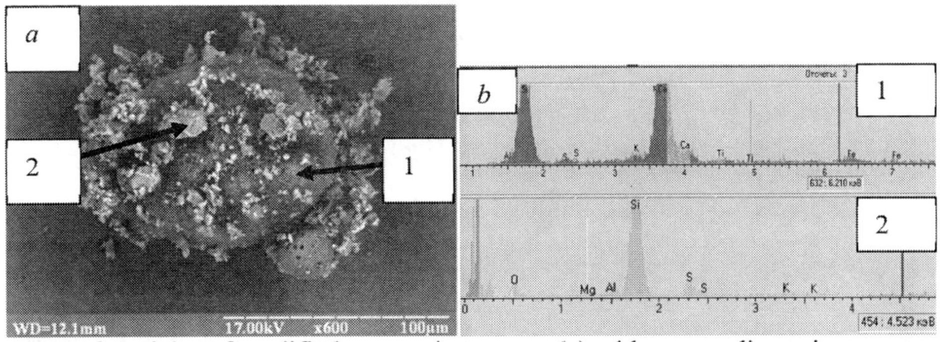

Figure 4. SEM of particles of modified composite cement (a) with energy dispersive spectroscopy (b).

The mobility of cement pastes of the composition "Portland cement – silica fume – superplasticizer" C-3 (1 %) was evaluated by the yield index G, which is determined by the formula G = F2 / 502-1, where F is the average diameter of the mixture according to Suttard. It was found that for a cement paste of the composition PC (210 g) – SF (70 g) – S-3 (2.8 g), B/B = 0.32, the index G is 1.1, while for a cement paste based on modified cement, the value of G = 1.1 is achieved with a water-binding ratio B/B=0.285 (11 % less in comparison with the control composition). This provides an increase in compressive strength of cement stone at the age of 28 days of normal hardening by 37 %. It should also be noted the high stability of spheroidal agglomerates obtained during the modification of composite cement, which, according to scanning electron microscopy, were found in the structure of a cement stone at the age of three days of normal hardening (Figure 5).

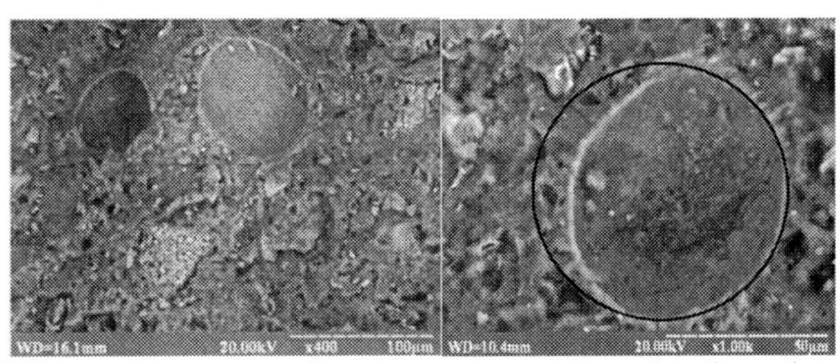

Figure 5. SEM and EDS spectroscopy of cement stone samples based on a modified composite Portland cement with the addition of ultrafine silica fume at the age of three days of hardening.

The results obtained are consistent with the data provided by Japanese scientists [8–11], who identified spheroidal agglomerates in the structure of modified cement hydrated for 28 days.

4. Conclusion

Based on the simulation of forces of interparticle pair interaction, the calculation of the autohesion forces that make up the spheroidal agglomerate, i.e. particles of Portland cement (d_1 = 20 µm) and particles of silica fume (d_2 = 2 µm) was performed. It is shown that, for uncharged particles, the magnitude of the electric interaction forces is negligible in comparison with the molecular van der Waals and capillary forces. The bipolar charging of particles in the corona discharge field at a corona electrode potential of +18 and –18 kV sharply increases the electric component of the autohesion force, which approaches the magnitude of molecular forces in value.

In this case, the Coulomb forces of attraction between oppositely charged particles begin to act before the direct contact of the particles, which contributes to the agglomeration of particles. These calculations are confirmed by the results of scanning electron microscopy (SEM) and energy dispersive spectroscopy (EDS) of samples of composite cement, which contains Portland cement (75 %) and silica fume of dry gas cleaning (25 %), the number of angular particles of irregular shape is reduced for a mixture of these materials processed in a surface modification device.

References

[1] Yang J, Sliva A, Banerjee A et al 2005 Dry particle coating for improving the flowability of cohesive powders *Powder Technol.* **158** 21–33

[2] Sabir B B, Wild S and Bai J 2001 Metakaolin and calcined clays as pozzolans for concrete: a review *Cem. Concr. Comp.* **23** 441–54

[3] Mehta P K May 2004 High-Performance, High-Volume Fly Ash Concrete for Sustainable Development *Workshop on Sustainable Development and Concrete Technol.* **1** 3–13

[4] Ferraris C F, Obla K H and Hill R 2001 The influence of mineral admixtures on the rheology of cement paste and concrete *Cem. Concr. Res.* **31** 245–55

[5] Iwasaki T, Satoh M and Ito T 2002 Determination of optimum operating conditions based on energy requirements for particle coating in a dry process *Powder Technol.* **123** 105–13

[6] Thomas G, Ouabbas Y, Grosseau P at el 2009 Modeling the mean interaction forces between powder particles. Application to silica gel – magnesium stearate mixtures *Appl. Surface Sci.* **255** 7500–7

[7] Ouabbas Y, Chamayou A, Galet L at el 2009 Surface modification of silica particles by dry coating: Characterization and powder ageing *Powder Technol.* **190** 200–9

[8] Tanaka I, Suzuki N, Ono Y and Koishi M 1999 A comparison of the fluidity of spherical cement with that of broad cement and a study of the properties of fresh concrete using spherical cement *Cem. Concr. Res.* **29** 553–60

[9] Tanaka I, Suzuki N, Ono Y and Koishi M 1998 Fluidity of spherical cement and mechanism for creating high fluidity *Cem. Concr. Res.* **28** 63–74

[10] Tanaka I, Koishi M and Shinohara K 2002 A study on the process for formation of spherical cement through an examination of the changes of powder properties and electrical charges of the cement and its constituent materials during surface modification *Cem. Concr. Res.* **32** 57–64

[11] Tanaka I, Koishi M and Shinohara K 2002 Evaluation of the wettability of spherical cement particles surface using penetration rate method *Cem. Concr. Res.* **32** 1161–8

[12] Ji J-H, Hwang J, Bae G-N and Kim Y-G 2004 Particle charging and agglomeration in DC and AC electric fields *J. of Electrostatics* **61** 57–68

[13] Stelmakh S A, Shcherban E M and Zholobov A L 2018 Prescription and technological aspects of manufacturing high-quality centrifuged products and structures from heavy concrete *IOP Conf. Series: Mater. Sci. and Engineer.* **463** 022056

[14] Stelmakh S A and Shcherban E M 2013 Influence of type of cement on the properties of foam concrete processed by a variable electric field *News of the Rostov State University of Civil Engineer.* **17** 147–8

IOP Conf. Series: Materials Science and Engineering **905** (2020) 012062 doi:10.1088/1757-899X/905/1/012062

To a question of a choice of a method of production of construction works on set of technological cards

S A Sinenko

Moscow State University of Civil Engineering, 26, Yaroslavskoye sh., Moscow, Russia

E-mail: sasin50@gmail.com

Abstract. The need to create an automated system for selecting rational organizational and technological solutions based on modern design technologies. Methods. Based on known BIM building information modeling objects are created by the Bank (information base) of typical technological cards (TC) from which the method of the study criteria based on modified genetic algorithms (GA) is the most efficient TC for this type of work. Results. The article deals with the choice of the method of production of construction works on the set of technological maps of construction and installation works. Conclusion.

1. Introduction

The efficiency of building production, its outcomes are largely determined by the degree of organization of construction processes which in turn depend on the quality of decision of tasks of designing of building and construction works and primarily, the formation of model's construction and selection based on the most rational organizational-technological decisions (OTR).

Organizational and technological design, as it is known, is the most important element of preparation of construction production and includes tasks of modeling of construction processes at construction of objects. The solution of problems of organizational and technological design of building production (without use and with use of the computer) associated with several difficulties caused by insufficient regard to the existing methods of organization and technology of construction production, the actual conditions of work and possibilities of modern computing. In this regard, there is a need to improve methods for solving problems of designing and construction of works subject to the requirements of construction practice, simulation of construction and selection of the most rational organizational-technological decisions.

As part of the automation of the choice of the method of production of construction works on the set of technological cards (TC), a model of multi-criteria optimization based on genetic algorithms was developed to select the most suitable analogue of a typical TC.

Process maps (TC), which are part of the work production project (PPR), are usually developed for complex types of work and work performed by new methods. They are the basis of scientific organization of construction processes and are developed according to current standards (SNiP, GOST, Enir, instructions).

The technological map is a set of rules, norms, instructions and schemes for the implementation of a given project or part of it. The use of TC allows to exclude various variations and interpretations of a method of action in this situation. This is an instruction that allows you to most effectively and quickly do a given type of work. The structure of the technological map: scope; organization and technology of work; requirements for quality and acceptance of work; calculation of labor costs, machine time and wages; schedule of work; material and technical resources; safety; technical and economic indicators.

The main purpose of TC is to assist builders and designers in the development of technological documentation. According to the TC, the technological sequence of construction processes is established, weekly-daily schedules and orders to produce works are made. They are used to justify the duration of construction of objects in the calendar plans and network schedules.

TC are developed by leading design and construction organizations to perform General construction and specialized works. In some cases, TC is developed for complex construction and installation processes [11–13]. It is possible to facilitate the development of TC, improve their quality and reduce the development time only based on the use of the most modern information technologies [14].

Linking the algorithm of automated selection of standard TC with the program of information modeling, it becomes possible to use the technique not only for design, but also for real production of works. For example, in the production of stone works with the help of robots-masons, whose work is carried out based on the information model of the building by selecting the necessary TC.

2. Methods and materials

The analysis showed that in the last decade, Russian scientists conducted research and proposed effective development on modeling of construction processes and the choice on their basis of organizational and technological solutions, to optimize the parameters of the flow and parallel-flow organization of construction, on the variant design of technological and organizational solutions; proposed methods to improve the regulatory framework of organizational and technological design; expanded opportunities for versatile evaluation of options. This is confirmed by the developments presented in the works of Afanasyev V.A., Antanavichyus K.A., Beletsky B.F., Budnikov M.C., Buslenko N.P., Velichkin V.Z., Voropaev V.I., Golub D.G., Isakov A.A., Mikhailov B.C., Nebritov B.N., Sadovsky V.I., Rybalsky V.I. Fokov R.I., and other scientists in the works of research, design and educational institutions of construction profile [1, 4–10].

Modern methods and techniques of construction involve the use of information modeling (BIM), which involves the creation of a common information model of construction object is the necessary basis for its creation and use and development at different stages of design and throughout construction and operation

Building information modeling (BIM) is the process of collectively creating and using building information, forming a solid basis for all decisions throughout the life cycle of the facility (from the earliest concepts to the detailed design, construction, operation and demolition). BIM tools are designed to eliminate redundancy, re-entry and loss of data, errors in their transmission and conversion.

Real-time visualization and localization require context-sensitive information (such as space, place, time) for the normal functioning of the system. In this way, a BIM model can provide spatial relationships, while sensor technology can provide location and time information. Modern localization methods (indoors) use probabilistic algorithms to estimate orientation, which often require a lot of computing power. Since the model contains the exact location of the components, the BIM model provides the following advantage: there is no need to assess the true position of the landmarks, and the complexity of the process and the machine time of the algorithm is dramatically reduced.

Information modeling technology of objects (Building Information Modeling) is a development of the computer-aided design (CAD) system generally accepted today. The main difference from the latter, in addition to three-dimensional drawing, is the presence of the model database containing detailed information about the technological, technical, architectural, engineering, construction, estimate, economic characteristics of the object. Depending on the specific requirements, the database can be supplemented with legal, operational, environmental and other information, the most important for our case is the database of technological maps for the work performed.is a development of the computer-aided design (CAD) system generally accepted today. The main difference from the latter, in addition to three-dimensional drawing, is the presence of the model database containing detailed information about the technological, technical, architectural, engineering, construction, estimate, economic characteristics of the object. Depending on the specific requirements, the database can be supplemented with legal, operational, environmental and other information, the most important for our case is the database of technological maps for the work performed [1, 7–10].

As mentioned above, for the purpose of optimal selection of TM from the information base according to the studied criteria, a technique based on modified genetic algorithms has been developed (GA). The GA methodology offers three optimization options:

- compliance with the required duration of the project,
- minimization of project cost indicators,
- minimization of mechanical strength indicators,
- minimization of indicators of labor intensity of processes.

The target function has the following form:

Minimization

$$\sum_{i=1}^{n}\sum_{j=1}^{m_i} C_{ij} * K_{ij} + I * D + P * \sum_{i=1}^{n}\sum_{j=1}^{m_i} C_{ij} * K_{ij} \quad (1)$$

When the restriction $D \leq T$, where n – is the number of works in the project; m_i – is the number of alternatives to the TC process i; Cij – direct costs – th process, when you select the alternative j; K_{ij} – is a binary variable of process i when alternative j is chosen, then Kij equals 1, otherwise equals 0 Kij; I – indirect costs for the project per unit time; D – is the execution time of the process; T – is the required execution time of the process; P – is the percentage of indirect costs.

The process should be selected one of the possible typical technological maps, the most optimal. This requires compliance with time and resource constraints, which can be represented as follows:

type relationship (end-start) S_B≥S_A+D_Aj+L_AB, (2)

link of the form (start-start) S_B≥S_A+L_AB, (3)

type relationship (ending-ending) S_B+D _ Bj≥S _ A+D _ Aj+L _ AB, (4)

type relationship (start-end) S_B+D _ Bj≥S _ A+L_ AB, (5)

where S_B – start time of process B; S_A-start time of process A; D_Aj-duration of process A when alternative j is selected; D_Bj – duration of process B when alternative j is selected; L_AB – technological break.

The model of automation based on modified genetic algorithms is developed considering the acceleration of obtaining the optimal choice of TC due to the evaluation of inefficient solutions in the process of the algorithm and their clipping. At the stage of formation of information base of TM (population) for each technological process its repeatability is checked. If there are identical individuals in the population, the final choice of the optimal variant will be carried out considering the studied criteria and the weights of their influence established by experts. In addition, each TC is checked for the possibility of application and adaptation to the specified conditions [1, 3–6].

The population is evaluated, selected, propagated, and updated until the conditions for stopping the algorithm are met. Thanks to this, the individual with the best indicators of fitness for this project will be selected.

In other words, the principle of operation of the algorithm is as follows: after selecting the most suitable process cards by name and design features, if there is a problem of further choice of TC, the algorithm gives preference to TC:

- first, with a lower cost,
- in the second place, with the shortest duration,
- in the third place, with the lowest value of mechanical strength,
- in the fourth place, with the lowest value of labor.

3. Results

Each process card is identified by a unique identification number (id) that is used to make the request. In turn, the type of work in which the card is contained is also determined by its identification number.

Since the electronic directory is focused on personal and mobile computers, and is designed for WEB resources, therefore, its compatibility is free to any platform. and further development and updating of the system will not create any problems for users, since the system is in the cloud.

To work with the database, a set of QtSql classes is connected to work with databases using the structured query language SQL and the QSqlDatabase class to represent the database connection.

To demonstrate the work of the algorithm to improve the adoption of OTR in the production of stone works, a matrix of parameters was compiled from the information base of technological cards (TC), which presents the parameters from the TC: cost, duration, mechanical strength, labor costs.

Typical process maps are compared by the weight of the impact of criteria on the project. As a result, priority is placed in the selection and the total weight of the impact of criteria on the project is greater in the selected TC, which will be selected by the algorithm for use in the production of construction works in the project for the construction of the building.

In addition, the characteristics of workspaces (type, size, and location) and the work that takes place in the activity areas change over time during construction and move in three dimensions. Without the use of 4D modeling, it is almost impossible to detect conflicts of workspaces on the construction site. The development of a model for visualization of works with 3D workspaces on the basis of the building information model and taking into account the time parameter from the construction schedule, allows you to track the performance of works and at the design stage to determine any conflicts between their workspaces [8–10].

In some cases, it is important that the designer promptly provided information base containing: requirements for the installation of scaffolding and examples of their placement; regulations of production quality control, including input, operational and acceptance control; methods of determining the need for materials, products and structures, machinery and equipment, tooling, tools, inventory and accessories; requirements for transportation, warehousing and storage of products and materials; measures for safety and labor protection, environmental and fire safety; information on other issues that are reflected in the TC.

When solving specific target tasks contained in the TC, the latter must ensure full compliance of input and output data of these tasks with the requirements of regulatory and methodological documentation.

Software modules that automate the solution of specific tasks must have computational and graphical components. There are such variants when graphic modules are implemented as some add-on over Autocad package of the last versions, and calculation modules of the program work, for example, in the environment of FoxPRO and C++ [14].

Software modules are required to provide crane selection in the parameters of the load and the height of rise; the drawing of the crane with reference to the objects of the construction site; the requirements calculation in the inventory of administrative buildings; and the development of technological schemes; computer-aided design of pits; selection of effective use of excavation and transport vehicles; calculation of dewatering of pits and trenches; selection of hoisting devices; calculation and selection of lighting equipment; calculation of loads and energy consumption for construction and installation works; calculation of the need for storage areas.

It is necessary that the output data of the calculation modules of the programs are issued according to the current standards of organizational and technological documentation in the form of calculation, graphic, formalized text materials. The documentation generated in this way could be used by the contractor during the construction of the facility without any modifications.

Currently, work is underway on the automated development of TC. The necessary condition for automation of TC development is their unification, and the degree of unification is objectively low. Now, there is no automated solution to several tasks: layout of formwork, installation of scaffolding, solutions to produce geodetic works, etc. While it is impossible to automate the coordination of joint work of construction equipment in cramped conditions. For example, several cranes in a small area may encounter boom equipment. It is extremely difficult to determine the optimal schedule of their joint work and movement [11–13].

There are, however, areas of TC, the automation of which is the prospect of the near future. This is the formation of realistic images of models of performance of specified types of work; the use of banks, databases. For example, on construction machines, tools, labor intensity base [11,12]. Based on the design and calculation data are generated tabular reports: the schedule of production of works, bill of quantities, schedules of requirements in equipment and personnel.

For registration of drawings the database of construction equipment and conditionally-graphic designations are used. Selection of load-lifting mechanisms is carried out based on the entered data: weight and sizes of cargo and load-grabbing devices; departure of cargo; height of lifting of cargo.

Cranes are selected from the base, the lifting characteristics of which meet the specified conditions of lifting the load, considering the normative reserves for lifting height and weight of the load.

The program allows you to automatically calculate the radii of the working and hazardous areas, as well as draw their designations on the drawing.

Selection of copra is carried out based on the selected size of piles. From the information database are selected copra, the technical characteristics of which allow you to immerse the piles of the selected diameter and weight. In addition, the database contains the following types of equipment: concrete mixers; concrete pumps; car lifts; bulldozers; drilling rigs; equipment for the device of bored piles; pipe-laying machines; excavators.

The new functionality and content of the program databases can be a breakthrough in the development of organizational and technical documentation. Therefore, the priority for developers is to implement the wishes received in the order of feedback.

As for users, in the ranks of experts and constructive critics, developers hope to see those who are directly associated with the development of technological maps and the introduction of new construction technologies [12].

4. Conclusion

The issues of integrated use of information resources, combining the functionality of different software tools, creating information models of various processes and objects, the development of methods and algorithms for processing and presenting information are widely discussed by the scientific community. In particular, the issues of workspace planning are considered and reflected in the works of scientists Akinci B., Fischer M., Elmahdi A., Guo S.-J., Riley D. and others.

28.07.2017 by order of the Government of the Russian Federation No. 1632-R the Program "Digital economy of the Russian Federation «was adopted. One of the directions of development of the digital economy was called the rapid transition to new standards that ensure the use of modern technologies. In addition, it is necessary to consider the possibility of applying appropriate national and international standards in the development of the Russian regulatory framework. 20.12.2017 g. at the meeting of the presidential Council for strategic development and priority projects, it was noted that at the first stage it is necessary to introduce information modeling technologies in the design process of buildings under construction within the framework of state participation, and then information modeling should be made an industry standard.

This method has been discussed at numerous conferences in NRU MSU, including the 1ST international scientific and practical conference of departments of organizational and technological profile of construction Universities and technical universities. "Experience in the implementation of postgraduate programs in the field of training "Construction". Moscow 2018; VII International scientific and practical conference held on June 5, 2019 in Penza and others [2].

5. Conclusion and conclusions on the work done

Currently, the prospects for the development of the proposed tools are seen primarily in the development of BIM technologies. Currently, there are four levels of BIM maturity, showing how advanced BIM technologies are. 1 level BIM 0 (BIM Level 0) – the level of maturity at which the exchange of information on capital construction projects at all stages of the life cycle is not carried out or is carried out on paper. 2 BIM level 1 (BIM Level 1) – the maturity level at which two-dimensional drawings and three-dimensional models of capital construction projects are created. Organized information exchange between participants in the creation of electronic information on capital construction and real estate at BIM level 1 is not made. 3 level BIM 2 (BIM Level 2) – the level of maturity at which all participants use their own information models, as well as produce a joint information exchange of data. 4 level BIM 3 (BIM Level 3) – the level of maturity at which the exchange of information on capital construction and real estate at all stages of the life cycle is carried out by all participants of the BIM project, and all work is carried out using a single information model.

In construction, in the simulation of construction production and visual the project construction organization (PIC), the project of production of works (PPR), spatial-temporal coordination of construction participants; planning, management and control (plan-fact); optimization of the construction site and logistics; improvement of geodetic marking works; control of deviations; monitoring of labor protection and industrial safety: in the modular digital production of building structures and products with Assembly on the site; rationalization of Executive documentation and its full reflection in the information model of the object. All this should lead eventually to reduction of terms of erection of buildings; reduce the cost of object creation; ensure access to reliable information about the construction progress; reduce waste; reduce the number of accidents on site and enhance the reliability and quality of the construction.

Calculation of the required parameters in the production of construction and special works, the routing, the automated selection of methods of construction and Assembly based on selection of the necessary TK through BIM technology is correct, as confirmed by the theoretical analysis and the operation of selecting the method of construction works in the aggregate routings. The experimental application of the method described above has shown its effectiveness in the development of the project of a multi-story residential building in Moscow.

References

[1] Ginzburg V M, Sapozhnikov V N, Kagan P B et al 2019 *Automation of organizational and technological design in construction* (Saratov: Univer. Ed.) 240 p

[2] Sinenko S A, Kashapov N F and Kudryavtsev I O 2019 On the possible description and mapping of the universal model of the design system *Modern science: current issues, achievements and innovations* Part 2 Sat. articles. VII Int. sci. and pract. Conf. (Penza: ICNS "Science and education") pp 184–90 5 June

[3] Sinenko S, Zhadanovsky B and Obodnikov V 2019 Construction machines requirement for the building site *Web of Conf.* **110** 01036 Retrieved from: https:// doi.org/10.1051/e3sconf/201911001036

[4] Sinenko S A and Feldman A O 2018 Efficiency Perfection of Organizational-Technological Decisionson the Basis of Information Flows in the Construction of Multi-Storey Residential Buildings *IOP Conf. Ser. Mater. Sci. and Engineer.* **463** 042010 DOI:10.1088/1757-899X/463/4/042010

[5] Skvortsov A V 2015 BIM data Models for infrastructure *CAD and GIS of highways* **1(4)** 16–23

[6] Rumyantseva E V and Manukhina L A 2015 BIM-technologies: approach to design of construction object as a whole *Modern sci.: actual probl. and ways of their solut.* **5(18)** 33–6

[7] Lushnikov A C 2015 Challenges and benefits of implementing BIM in construction companies *Bull. of civil engineers* **6(53)** 252–6

[8] Lapidus A A, Telichenko V I, Tumanov D K et al 2014 Development of methods of technology and organization of construction production for solving problems of energy efficiency *Technol. and organizat. of construct. Product.* **2** 10–6

[9] Losev K Yu, Chulkov V O and Kazaryan R R 2018 Assessment of the Functional Operator's Condition in the Environment of a Vehicle Cabin During Construction Works *IOP Conf. Ser. Mater. Sci. and Engineer.* **463** 032084

[10] Kazaryan R R 2018 System-targeted approach to the integrated use of transport in the interests of life safety *MATEC Web of Conf. (TransSiberia 2018)* **239**
[11] Oleynik P P 2017 *Organization planning and management in construction* (DIA publ. House)
[12] Oleynik P P and Brodsky V 2015 *Organizational forms of construction* Textbook 84 p
[13] Oleynik P P and Shirshikov B F 2013 *Structure of sections of organizational and technological documentation and requirements to their contents* (MGSU publ. House) 63 p
[14] Sinenko S A and Danilova E D 2018 To the question of improvement of forms of display of norms and standards on the organization and technology of construction *BST: Bull. of construct. Equipment* **5(1005)** 54–5
[15] Goroshko E I 2001 *Integrative model of free associative experiment* (Moscow: Instit. of Linguist. of the Russ. Acad. of Sci.) 320 p

Visualization of process maps for construction and installation works

S A Sinenko and S A Aliev

[1] Moscow State University of Civil Engineering, 26, Yaroslavskoye sh., Moscow, Russia
[2] Grozny State Oil Technical University named after M.D. Millionshchikov, korp. 1,100, prosp. Husejna Isaeva, Grozny, 364068 Russia

E-mail: sasin50@gmail.com, asa-fenix@mail.ru

Abstract. the article presents the principles, methods and means of visualization of organizational and technological solutions developed as a part of technological maps for the production of construction and special operations. In this paper, the authors analyze the possible ways of the development of visualization of organizational and technological solutions. The examples of 3D images of the main production schemes as a part of technological maps for the installation of beams in machine rooms in the building of hydroelectric power plants are shown. The developed display techniques can be used in the practice of the design of technological maps for the production of construction and special operations. This technology will expand the scope of modern design tools in the practice of construction production.

1. Introduction

The quality and cost of the construction project and its components are largely determined by the applied design technology. Previously, all technical documentation was created manually on drawing boards. Currently, when PCs are located directly at the workplaces of designers and technologists, any project is impossible without the use of computer-aided design (CAD) systems.

In the Russian Federation, such systems appeared as a "progressive tool to accelerate the work of designers in the early 80s of the last century. The main objective of CAD was reduced to the creation of the external surfaces of buildings and structures and their elements, strength calculations, later these systems made it possible to calculate various design schemes in order to create architectural and construction drawings, captions and dimensions, and also produce finished and completed drawings in accordance with the requirements of existing standards.

The use of CAD allows increasing the productivity of a designer and technologist by 2–3 times, increasing the efficiency of interaction between different departments, the level and quality of design work. With the help of CAD, it is possible to shorten the preparation time for construction production, free designers, expand the design and manufacturing capabilities of complex equipment, as well as create a unified design and technological base for construction and design organizations, which, of course, will positively affect the financial position of contracting organizations.

The progress of computer technology and its development over the past decades has emerged as a means of the improvement of the design process, dramatically increasing the efficiency of the intellectual activity of a designer. Finally, it becomes possible to use virtual reality technology as a medium for formation. The idea of direct design appeared as a result of the analysis of the use of virtual technology as a tool for the visualization of design decisions on the organization and technology of construction production [1]. It is based on the immersion of a civil engineer in the designed environment. An engineer being inside the projected space, determines the direction of the changes and implements these changes interactively, moving the forms in the virtual space. The implementation of the idea of "direct design" is supported by many studies in the field of virtual reality [2–12].

2. Methods and materials

As it is known, as part of the organizational and technological documentation, in particular, production work plan (PWP), a technological map is developed for the production of the most complex construction and installation works and special operations, reflecting the methods and techniques for the performance of construction work, which indicates the scope of application, the plan, means and methods of construction and installation works, work flow schemes, temporary inventory devices, machines, mechanisms and equipment temporary areas for warehousing and site joints, movement of assembly cranes and mechanized installations, devices and safety instructions.

Modern software systems "Hector-builder" (HMI, PWP, HMI CAD and others), NanoCAD, "Builder" and others allow improving the design and development of technological maps for the construction and installation work [5–10].

The main principle for the development and rationalization of process control and management documentation (PCMD) is the maximum reduction of manual labor of designers, increasing labor productivity in design and construction organizations, reducing the duration of design and improving the quality of PCMD. The improvement tool should be presented by the methods and technical means based on the use of computers, peripheral devices, reprography and organizational equipment, which allow combining the technological process and reproduction of documents into a single system [3].

The main conditions for rationalization are: the use of planar prototyping and documents – blanks, electrographic apparatus for the production of originals, microfilms of enlarged copies of documents, as well as aperture cards as a means of the automation of the copying and storage of documents [6–10].

The "Hector: Designer-Builder" program provides construction planners with the latest technologies for the development of all sections of construction management plans (CMP) and production work plans (PWP), including construction general layouts.

One of the modern directions in the development of computer graphics is the construction of photorealistic images of objects modeled on a computer [1].

This is relevant for computer aided design, motion-picture industry, advertisements, design, computer games, etc. The increase in the quality of computer graphics products is rapidly spreading in the field of practical application, and the requirements of users for realistic 3-D images are constantly growing [2].

The growth of computing power allows solving increasingly complex tasks of visualization of three-dimensional scenes (the number of objects, light sources, image resolution), the increase in complexity of the tasks themselves is ahead of it.

There are two main approaches to the visualization of 3-D models:

- The creation of pseudo-three-dimensional fragments, with their subsequent processing according to the principles of two-dimensional graphics, with little use of the simplest laws of the third dimension;
- The creation of a model of a three-dimensional world, with its further projection on screen. All modern three-dimensional games are developed in this way.

The creation of animation is a complex process, requiring attention to detail. It is frame-by-frame arrangement of objects of construction general layout, setting the trajectory of their movement and interaction with the rest of the 3D model. Depending on the power of the computer, the quality of animation and visualization can vary from photorealistic to a schematic demonstration of the capabilities of software package.

Creating an animation, it is necessary to consider what is needed to get in the result, namely, it is necessary to calculate the time of the finished animation. It is recommended to perform all the calculations in advance, before starting the animation on render, since this is a very long process. In order to save machine time and resources of their workstation, many people use video editors and edit a ready-made video file, stretching frames, thereby slowing down the movement of objects. This method has both pros and cons. The advantage is time and resources saving, and disadvantage is the resulting video.

Thus, the required number of frames was not calculated in advance, the video editor has nothing to fill in the gaps between the frames, which as a result lead to a jagged non-realistic video. A smooth and beautiful video is much preferable to both an outside viewer and a designer.

According to the above-mentioned aspects, we can conclude that the creation of any animation begins with the calculation of time and number of frames. Thus, in order to create a five-minute animation, it is necessary to take into account that the optimal number of frames per second is 30. Using simple arithmetic calculations, we get a finite number of frames equal to 9000. It takes about 30–33 seconds to calculate each frame on a modern computer, which implies that 9000 frames video is created in about 75 hours.

It is recommended to perform the creation of a visualization of the process of building a hydroelectric power station in the reverse order. That is, the building on the animation should not be built, but disassembled. This is the right approach to the visualization of the construction process. With this method, it is possible to avoid a heap of objects and, most importantly, correctly display the construction process.

In addition, it is necessary to take into account the computing capabilities of a working computer. To save machine resources during the calculation, a camera is created on the 3D model, within which the final image will be calculated. Thus, it becomes possible to avoid rendering the invisible sections of the created model and spend all the useful computer power on the calculation of the required image.

The camera is also an object of the 3D model and has its coordinates in space. Having set the coordinates of the frame-by-frame motion of the camera, we get an animation showing the existing position. It is customary to use such animations at the beginning and end of a video to show visually the difference between the starting frame and the ending.

The construction process chart is not created without the location and calculations of the used construction equipment. In animation, its presence is also necessary, since without technology it is impossible to display realistically the process of the construction of a building.

The important stage in the creation of animation is a frame-by-frame arrangement of interaction objects with a building model. It is necessary to set the interval of occurrence of each object and each part of the model at exact time. For example, the first object is a crane, which begins the installation of a crane beam. It has a trajectory of motion and the disappearance of the interacting elements, building structures, is configured on the key frame. Thus, after processing and inverting the visual range of video file, the effect of the assembly of industrial building is achieved.

This technology is one of the first steps to improve the design of technological maps for the production of construction and installation works.

One of the modern directions in the development of computer graphics is the construction of photorealistic images of objects modeled on a computer.

Although the growth of computing power allows solving increasingly complex tasks of visualization of three-dimensional scenes (the number of objects, light sources, image resolution), the increase in complexity of the tasks themselves is ahead of it.

Nowadays, the most common 3D modeling method is to lay out partially the basics of vector graphics. It is performed in the following way: any three-dimensional model (which, ultimately, builds up the virtual world consists of) is represented as a certain number of intersecting planes. The excesses of planes are cut off. As a result, two-dimensional polygons remain, placed in a three-dimensional coordinate system.

One of the main directions in the development of computer graphics is physically accurate modeling of light passage in various environments.

According to the final indicator, it was found that Autodesk 3D Max became the software environment for the realization of the virtual reality of construction general layout during the construction of an object.

Developing since the 90s of the last century, the package of three-dimensional modeling 3D max has already reached a certain level of excellence. This is confirmed by its current popularity in the thriving industry of film, television and computer games. Impressing with their believability, 3D special effects of TV screens, the unimaginably realistic virtual reality of three-dimensional computer worlds, as well as numerous high-quality architectural and design projects implemented using the 3D max package, have firmly entered our lives [2,3].

Modeling usually takes from 50% to 80% of the entire workflow. It is important to understand that the time spent on modeling, firstly, depends on the complexity of the situation during the installation of structures, on the building or premises, and not on their size. The time for texturing the model depends on the complexity of the materials and their diversity. A one-story industrial building is much easier and faster to texture than a small cottage with several types of stone and facing bricks.

Visualization involves the adjustment of lighting, surroundings and materials. At this stage, the computer model acquires a photorealistic, presentable appearance, image quality is determined. It can take from 10% to 50% percent of the entire workflow. The quality of visualization directly depends on the experience and professionalism of a designer.

The main components of the cost of visualization are the amount of work and its quality.

The process chart of construction and installation works, performed in usual format, contains all the necessary attributes.

With the so-called raising of drawing, it is necessary to know how many and what angles of an object are necessary to depict. This component affects the timing and quality of the displayed virtual reality.

The color scheme includes information on finishing materials and their colors. Here all the basic elements are included: from design to a part of a building, and material samples in the form of photographs are provided.

As a result, it is possible to obtain the images of various qualities. It is necessary to take into account the resources of a power plant. In the case, if the process of the creation of visualization, and later on animation, proceeds on the so-called render farm, then it is not recommended to degrade the output image quality. During the work on a standard computer, the quality of animation decreases, since the computer is simply not able to calculate so many objects.

The capabilities of computer graphics allow creating three-dimensional models of architectural objects of any degree of complexity and organization in a virtual environment.

3D visualization is an important tool for the attraction of investment in a construction project, since investors see not only dry calculations and technical drawings, but also a vivid image of the future building. Sometimes it is very difficult to reveal the concept of a project using only traditional tools, for example, drawings, sketches or photographs of analogues.

In order for a three-dimensional object to be displayed on a screen in real time, it must be represented as points in a three-dimensional coordinate system, where each displayed point will have the coordinates x, y and z. The points of each object, fully setting its position in space, are stored in the system memory. In order to display an object on a two-dimensional screen, it is necessary to form its image (to visualize it).

In order to build simple exterior scenes, it is possible to use templates that exist in 3D Max, starting with the 6th version.

The whole model is designed to animate the process of the construction of an industrial building; all the elements of the building are created separately and are combined into groups.

The visualization of a technological process is a way to display information about the state of technological equipment and process parameters on a computer monitor or operator panel in an automatic control system in construction, which also provides graphical methods for controlling the technological process. The display system is based on the process mnemonic diagram, a static image in a visually simple and intuitive form, showing equipment elements, possibly processed materials and products, their interaction and processing order. The static mnemonic diagram is animated, reflecting the real state of equipment and raw materials.

3. Results

The presentation of information in 3D is limited to simple and unambiguously perceived colors (red, green, yellow, white, black, gray). In this case, not small font is used in the captions.

The passion for dynamic pictures (for example, animation) will distract and tire a designer of main design solutions. The use of photorealistic images of objects will worsen perception. In order to control the technological process, the mnemonic diagram contains the elements of a graphical interface, most often typical for modern software: input-output windows, buttons, sliders. In addition, events (for example, mouse clicks) on image elements can be used.

In addition to the denominated mnemonic diagrams, the windows, located on top of the mnemonic scheme are used. These windows display the dependences of the process parameters on time, as well as text messages about the state of the system and the actions of a designer. Modern design tools, as a rule, contain built-in editors that allow implementing all visualization tasks.

In this case, it is proposed to carry out work on the creation of a technical map for the installation of beams under an overhead crane in the machine room of a hydroelectric power station in the group of Archicad 19 and Artlantis 6. In Archicad (Archicad 19), a model is created and subsequent code is made in layout, and in Artlantis (Artlantis 6) the visualization of the 3D model is performed. The model was created using standard Archicad tools:

- The top with cutouts for the turbines were performed using the "tops" tool;
- The columns were made using a combination of tools "column" and "complex profile";
- The top beams were made using the tool "beams";
- The crane was exported as "pdf" element from the finished library;
- The beams for ceiling travelling crane were created using the combination of the "beam of complex profile" tools (for the main body of the beam), "solid modeling operations" were used to create recesses in the upper part of the beams, mounting outlets were created from "cylindrical primitives" from the standard library, rails were created from beams of the special profile "corner";
- Scaffolding was created from a combination of elements "top", "columns of cylindrical section", "cylindrical beams";
- The construction bolt is created using the combination of a "cylindrical column" and "top" with a hexagonal profile;
- The bolt washer is created using "top" with a cut;
- The mounting holes for bolts are made combining a "cylindrical primitive" from the standard library of elements and "solid modeling operations";
- The landscape to indicate the location of the crane was created using 3D net tool.

After the model is created, it is exported to the Artlantis 6 program, in which the materials are set element by element and rendered, as well as images are obtained, which are put into a layout in the Archicad and saved in pdf (Fig. 1, 2)

4. Conclusion

The obtained results confirm the effectiveness of visualization of the schemes for the implementation of construction processes as a part of process charts for the construction, installation and special operations. The discussion of the obtained results and approbation proved the reliability of research directions.

Figure 1. Installation of large-span beams under the travelling crane in the building of hydroelectric power plant

Figure 2. Mounted beam

Of course, the results are still imperfect (this directly depends on the chosen observation point), but taking into account the effectiveness of this technology, we can assume that the proposed methods and means of display will expand the scope of these techniques in the practice of construction production and will allow getting realistic types of production.

Modern computer facilities and monitors as well as software tools have rich graphic capabilities. Designer full use these features. As a result, it gives more realistic images of organizational and technological solutions in process charts for construction and special operations.

Acknowledgment
This study was carried out as a part of research aimed at the improvement of the design of construction in Moscow State University of Civil Engineering.

References

[1] Sinenko S A, Ginzburg V M, Sapozhnikov V N et al 2019 *Automation of organizational and technological design in construction* (Saratov: Higher Ed.) 240 p
[2] Sinenko S A, Kuzmina T K and Oleynik P P 2015 *Customer activities in market conditions* Handbook (DIA Publ. House) 288 p
[3] Kolesnikova E B, Kuzmina T K and Sinenko S A 2015 *The solution of organizational and technological problems. Construction* Textbook (Moscow: Ed. DIA) 96 p
[4] Sinenko S, Poznakhirko T and Obodnikov V 2019 Automation of visualization process for organizational and technological design solutions *Construction Engineering and Management* 2nd Conf. for Civil Engineer. Res. Networks [2018 MATEC Web of Conferences 270, 05008 22 February]. AN 05008 Retrieved from: https://doi.org/10.1051/matecconf/201927005008
[5] Sinenko S A and Feldman A O 2018 Efficiency Perfection of Organizational-Technological Decisionson the Basis of Information Flows in the Construction of Multi-Storey Residential Buildings, *IOP Conf. Ser. Mater. Sci. and Engineer.* **463** 042010 DOI: 10.1088 / 1757-899X / 463/4/042010
[6] Skvortsov A V 2015 BIM data models for infrastructure *CAD and GIS for roads* **1(4)** 16–23
[7] Rumyantseva E V and Manukhina L A 2015 BIM-technologies: approach to designing a building object as a whole *Modern Science: Actual Problems and Solutions* **5(18)** 33–6
[8] Lushnikov A S 2015 Problems and advantages of implementing BIM technologies in construction companies *Bull. of civil engineers* **6(53)** 252–6
[9] Lapidus A A, Telichenko V I, Tumanov D K et al 2014 Development of methods of technology and organization of construction production to solve energy efficiency problems *Technol. and organizat. of construct. Product.* **2** 10–6

[10] Losev K Yu, Chulkov V O and Kazaryan R R 2018 Modeling and Assessment of a Building Intellectual Grade in the Community of Full Participant in Construction Activities *IOP Conf. Ser. Mater. Sci. and Engineer.* **463** 032085

[11] Kazaryan R R 2018 System-targeted approach to the integrated use of transport in the interests of life safety *MATEC Web of Conf.* **239**

[12] Murtazaev S-A Y, Kuladzhi T V, Taymaskhanov Kh E et al December 2018 Advances in Engineering Research *Int. Symp. on Engineer. and Earth Sci. (ISEES 2018)* vol 177 Retrieved from: Https://doi.org/10.2991/isees-18-18, 2018.48

Properties of foam concrete based on highly concentrated aluminosilicate binder suspension

M N Sivalneva, V V Strokova and V A Kobzev

Belgorod State Technological University named after V.G. Shukhov, 46, Kostyukova street, Belgorod, 308012, Russia

E-mail: 549041@mail.ru, vvstrokova@gmail.com, 9087815876@mail.ru

Abstract. Cementless binders and building materials are promising. The article presents results of production of foam concrete compositions using high-concentration aluminosilicate binder suspension based on granodiorite. Their basic physical and mechanical characteristics and microstructural features are described. The method of multifactor experiment was used, which allows to vary the parameters: water-solid ratio, amount of binder component and stabilizing additive. Thus, during steps of mathematical planning, optimal parameters were recommended, the observance of which is necessary to obtain cementless cellular composites with the best properties.

1. Introduction

In the current economic conditions, the already tested concrete materials, including heat-insulating and structural-heat-insulating non-autoclaved foam concrete, are popular.

The traditional representatives of these products are cement-based composites. Significant carbon dioxide emissions, high material and energy intensity of the cement industry, the complex manufacturing process, and the high cost of cement require the development and search for alternative types of binders for the production of non-autoclave foam concrete products using low- and non-clinker binders.

Scientists of Belgorod State Technological University n.a. V.G. Shukhov (Russia) achieved success in the development of composite cementless highly concentrated and nanostructured cementing systems, as well as their application for the production of building materials [1–14].

The team described issues of obtaining and studying a highly concentrated aluminosilicate binder suspension (HCABS) which is a cementless system based on intrusive rocks of acidic composition – granodiorites [12, 15].

This study presents the results of the development of foam concrete compositions using HCABS. It describes their structural and physico-mechanical characteristics which determine their role in the general classification of building materials.

2. Methods and materials

When designing effective construction materials, it is necessary to choose appropriate raw materials. Designing compositions with a number of standardized and specified building and technological properties determines a link in the production of both materials and structures based on them.

One of the methods to increase the research efficiency is the use of mathematical methods for planning experiments, which can reduce costs and duration of research while improving quality of the results.

The method of mathematical planning of experiments makes it possible to develop a model of the dependencies. The nature of the dependence can be unknown. But it can solve practical problems. It is enough to find its approximate expression.

The multivariate experiment method consists of the following steps:

- preliminary study of the object;
- building the mathematical model;
- interpretation of the mathematical model.

The experiment planning involves identification of the mathematical relationship between the material properties, their consumption and properties of constituent components and technological factors.

Depending on the established conditions of the problem during the experiments, all factors (input parameters – X_1, X_2, ..., X_n) vary at three levels: main (0), lower (–1) and upper (+1). In this case, the lower and upper levels differ from the main one by a certain value – the interval of variation (ΔXi).

The experiment is carried out according to the planning matrix and is determined by the number of factors and conditions.

To process the results, the method of mathematical statistics is used. It helps obtain algebraic equations that reflect the relationship between the varied factors and the results obtained during the research (output parameters).

Thus, the mathematical design of the experiment was used to develop rational foam concrete compositions. Humidity of the main binder component – HCABS (22 %), was taken into account. A stabilizing additive – polyvinyl alcohol (PVA) was used as a 5 % aqueous solution.

The optimization of properties was evaluated based on the analysis of physicomechanical characteristics of foam concrete: compressive strength and average density of the composite. The quantity of binder and additives, as well as the water-solid ratio (W/S) were used as variable parameters (Table I).

Table 1. Conditions for planning hcabs-based foam concrete composition development experiment

Factors		Variation levels			Variation interval
natural	*codified*	*–1*	*0*	*+1*	
W/S	X_1	0.35	0.40	0.45	0.05
HCABS, kg	X_2	312	390	468	78
PVA solution, kg	X_3	0.75	1.25	1.75	0.5

* the composition is calculated based on the volume per 1 m^3 of the mixture.

The study of the basic physical and mechanical characteristics of foam concrete samples was carried out according to the regulatory and technical documents of the Russian Federation: GOST 25485–89 "Cellular concretes. Specifications", GOST 10180-2012 "Concrete. Methods for determining strength of control samples", GOST 12730.1–78 "Concretes. Methods for determining density", GOST 30256–94 "Method for determining thermal conductivity with a cylindrical probe", GOST 25898–2012 "Building materials and products. Methods for determining vapor permeability", GOST 24816–2014 "Building materials. Method for determining equilibrium sorption humidity."

At the macro level, the study of structural parameters of foam concrete was carried out using a Hirox KH-7700 video microscope, at the micro level – using a high resolution scanning electron microscope Mira 3 FesSem in a high vacuum discharge mode at different magnifications.

Non-autoclave foam concrete was studied. This material was produced by mixing all components: the cementless binder, the GreenFroth protein blowing agent, water and a stabilizing additive which was used as a solution of polyvinyl alcohol. The speed of the mixing plant gradually increased to a maximum.

In the research, a highly concentrated aluminosilicate binder suspension based on the screening of crushing of holocrystalline acidic intrusive rocks – granodiorite (deposite in Pavlovsk, Russia) was used as the main binder component. The main properties of the suspension are shown in table II. The technology for producing HCABS consists in one-stage mechanical activation of a coarse fraction of granodiorite screenings (> 1.25 mm) in the presence of a modifying component, which is sodium silicate solute. It should be noted that the choice of the optimal size fraction of the raw material is determined by the content of the minimum amount of layered aluminosilicates in it, which negatively affect the mechanochemical synthesis of the suspension [15].

Table 2. The properties of HCABS and stone based on HCABS

Indicator (HCABS suspension)	Value	Indicator (stone based on HCABS)	Value
Grinding residue № 0063 [%]	less 1	Compression strength [MPa]	5.05
Viscosity [Pa·s]	17–20	Deflection strength [MPa]	2.10
Environment pH	8.0–9.0	Density [kg/m^3]	2100
Humidity [%]	20–22	Specific efficiency of natural radionuclides [Bk/kg]	60.8

3. Results and discussions

Based on the experimental planning conditions, a planning matrix was created (Table III). According to the matrix, foam concrete samples were molded. Basic physical and mechanical characteristics were determined (compressive strength and average density).

Table 3. Experiment planning matrix

No	Factors						Output parameters	
	X_1	X_2	X_3	*W/S (X_1)*	*HCABS (X_2)*	*PVA (X_3)*	*Strength, MPa (Y_1)*	*Density, kg/m^3 (Y_2)*
1	+1	+1	+1	0.45	468	1.75	1.35	541
2	+1	+1	–1	0.45	468	0.75	1.23	508
3	+1	–1	+1	0.45	312	1.75	0.85	342
4	+1	–1	–1	0.45	312	0.75	0.81	320
5	–1	+1	+1	0.35	468	1.75	1.48	527
6	–1	+1	–1	0.35	468	0.75	1.32	503
7	–1	–1	+1	0.35	312	1.75	0.92	346
8	–1	–1	–1	0.35	312	0.75	0.86	322
9	+1	0	0	0.45	390	1.25	1.07	419
10	–1	0	0	0.35	390	1.25	1.29	430
11	0	+1	0	0.40	468	1.25	1.33	513
12	0	–1	0	0.40	312	1.25	0.83	340
13	0	0	+1	0.40	390	1.75	1.03	412
14	0	0	–1	0.40	390	0.75	0.96	398
15	0	0	0	0.40	390	1.25	1.12	406

Based on the experimental data, regression equations were obtained for constructing nomograms of the dependence of compressive strength and density.

Regression equation for compressive strength:

$Y_1 = 1.103 - 0.056 \cdot X_1 + 0.244 \cdot X_2 + 0.045 \cdot X_3 + 0.094 \cdot X_1^2 - 0.006 \cdot X_2^2 - 0.091 \cdot X_3^2 - 0.013 \cdot X_1 \cdot X_2 + 0.023 \cdot X_2 \cdot X_3 - 0.008 \cdot X_1 \cdot X_3$.

Regression equation for average density:

$Y_2 = 411.039 + 0.2 \cdot X_1 + 92.2 \cdot X_2 + 11.7 \cdot X_3 + 11.109 \cdot X_1^2 + 13.109 \cdot X_2^2 - 8.391 \cdot X_3^2 + 3.125 \cdot X_1 \cdot X_2 + 1.375 \cdot X_2 \cdot X_3 + 0.875 \cdot X_1 \cdot X_3$.

Based on the regression equations, nomograms of the dependences of density and compressive strength were constructed (Figure 1).

Compliance with the given conditions allows us to produce a material with density of 320–540 kg/m^3. When analyzing the data, one should take into account the influence of all factors both individually and in combination. First of all, an increase in the average density of foam concrete is due to the amount of binder introduced and directly related to this factor. Samples with W/S = 0.45 with an increased binder content are characterized by the highest density values. A decrease in the amount of binder is accompanied by a decrease in density with the subsequent intersection of the planes W/S = 0.45 with W/S = 0.35 (Figure 1, a).

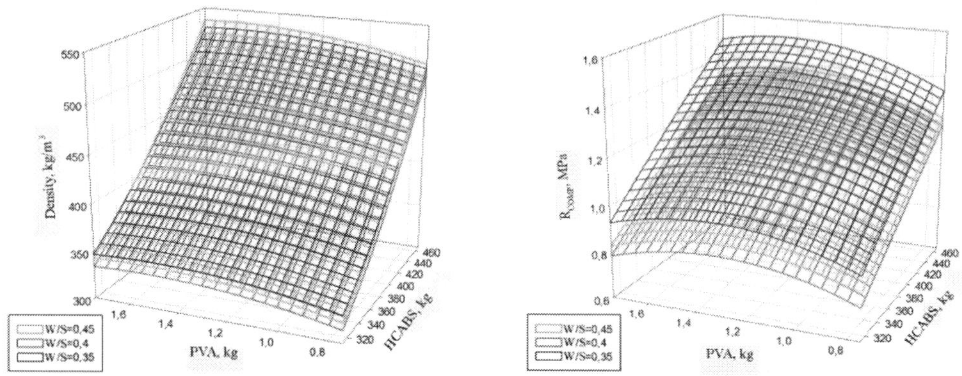

a b

Figure 1. General view of the dependence of HCABS-based foam concrete characteristics on variation factors: a – average density; b – ultimate compressive strength.

The characteristics of the compositions with different water-solid ratios have no significant differences and are equal. With a high W/S value, the system has an excess of water which leads to a deterioration in quality of the material. As a result, an irrational structure has a negative effect on the basic physical and mechanical characteristics of the composite. When forming samples with W/S = 0.45, shrinkage was revealed. It is reflected in density values. The less water in the system (up to a possible acceptable value taking into account the moisture content of the binder itself and the amount of PVA solution), the higher the quality of concrete.

The constructed dependencies are confirmed by the presented strength nomograms (Figure 1, b).

An increase in compressive strength is determined by the dependence on the water-solid ratio in the following sequence: W/S = 0.35 → W/S = 0.45 → W/S = 0.4. It is known that a large amount of water decreases the strength. In this case, the characteristics of samples with W/S = 0.45 are higher than those with W/S = 0.4. A decrease or loss of stability of the mixture occurs in the initial period of formation of coagulation structures. As a result, this leads to the subsidence and unjustified compaction of the system, and increased strength. In this case, the sample is characterized by the irrational structure of its porosity with pronounced open and torn pores; therefore, it is inferior in characteristics with W/S = 0.35.

Taking into account the regression equation, the maximum strength values were found. For D500 foam concrete, the highest value (R_{ad} = 1.52 MPa) was observed in samples with the following parameters: water-hard ratio W/S = 0.35, 468 kg of binder and 1.45 kg of PVA. For D400 foam concrete, the maximum strength (R_{ad} = 1.24 MPa) is achieved at W/S = 0.35, the amount of BABS 390 kg and PVA 1.15 kg. The minimum content of the additive (0.75 kg) does not have a significant structural effect and does not decrease and prevent destructive processes.

Thus, during the mathematical planning, the following parameters should be observed to obtain foam concrete based on HCABS with the best strength characteristics: minimum W/S (0.35), average PVA additive content, the amount of binder is determined by the required density of finished products.

Based on the analysis, the optimal compositions were selected and samples were produced. They were tested in accordance with GOST 25485–89 "Cellular concrete. Technical conditions." Foam concrete products with W/S = 0.35 and density values corresponding to D400 and D500 were developed (Table IV).

According to GOST 25485–89, the following classes of compressive strength were established: D400 – B0.75 and B0.5; D500 – B1 and B0.75. The materials satisfy and exceed the requirements of the regulatory documents. This is true for foam concrete D400. For the heat-insulating concrete, frost resistance is not standardized. Thermal conductivity indicators, in comparison with those stated in the GOST requirements decreased up to 20–30 %.

Table 4. Compositions and main characteristics of hcabs-based foam concrete

Components *			Characteristics							
HCABS, kg	*Frother, l*	*PVA solution, l*	*Density, kg/m³*	*Density Grade*	*Tensile strength under compression, MPa*	*Strength class*	*Thermal conductivity, (W/m ° C)*	*Vapor permeability, mg/(m · h · Pa)*	*Sorption humidity%* 75	95
					Experimental data					
390	6.3	23	419	D400	1.24	B1	0.08	0.23	5.6	9.3
468	7.9	28.7	524	D500	1.52	B1	0.085	0.21	5.9	10.1
					Indicators according to GOST 25485–89					
				D400		B 0.75 B 0.5	No more than 0.1	No less than 0.23	8	12
				D500		B 1 B 0.75	No more than 0.12	No less than 0.20	8	12

* calculation of the composition per 1 m3 of foam concrete

HCABS-based foam concrete is a neocomposite which requires a more detailed study.

When studying the structure of foam concrete at the macro level, the surface layer of the sample was analyzed. We investigated samples of foam concrete of a rational composition, i.e. modified with PVA; for visual comparison, we studied unmodified foam concrete.

The general structural characteristic of cellular materials consists of the description of pores and the spatial arrangement of pores, pore size distribution, maximum and average sizes, their shape, and thickness of inter-pore partitions.

Foam concrete of rational compositions is characterized by a uniform distribution of pores whose diameter is 200 μm – 2 mm (Figure 2, a). The polydisperse nature of the pore size distribution determines a denser inter-pore septum which positively affects the characteristics of the finished material as a whole.

a b

Figure 2. The macrostructure of the HCABS-based foam concrete surface: a – modified, b – unmodified

An insignificant number of large pores is observed. Its large size is due to their fusion, as indicated by the elongated shape of these cellular formations. Pores have a size of 500-800 microns, a regular shape in the form of a sphere with closed porosity. By the number of expressed peaks, their distribution should be considered bimodal.

An unmodified foam concrete sample is characterized by a different structure (Figure 2, b). The failure of this material is already evident at the molding stage, when visual signs of the initial coagulation process appear. With further hardening, significant shrinkage of the mass occurs due to the macrostructure of the sample surface.

In the unmodified foam concrete sample, a pronounced large-pore structure is observed, pore sizes are about 2 mm. Pores have an irregular elongated shape. The large distance between the pores (1 mm) indicates that the system was initially unstable, and the cellular structure was partially destroyed. This indicates the impossibility of producing foam concrete materials based on HCABS without modification.

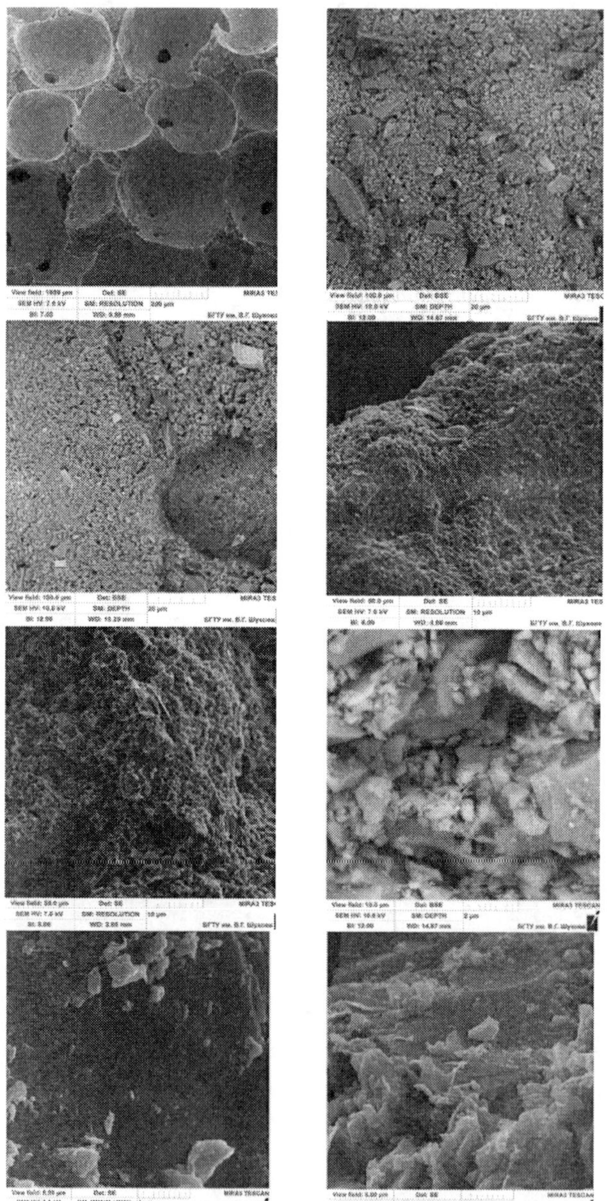

Figure 3. Photographs of the microstructure of non-autoclave foam concrete based on a cementless HCABS

Microstructural studies were carried out only with modified foam concrete compositions with a D400 density grade. The internal structure of the sample surface produced by the mechanical action was studied.

The general view of the microstructure shows the nature of the pore system expressed by cells in a large size range (Figure 3).

Porosity is determined by the presence of pores of a spherical shape with clear boundaries between them, no merging is observed which confirms the positive effect of the modifying component on the cellular structure of materials. The inter-pore septum has a different thickness of 20–100 microns. The total mass of the system is composed of a polydisperse matrix of angular particles.

With a larger-scale increase in various elements of the foam concrete structure (inter-pore septum, pore space surface), a large number of binder particles of various sizes appear on the surface.

The accumulation of nanoscale particles of regular shape attached to the surface of larger particles may indicate the epitaxial growth. This process determines the crystallization of silicic acid on quartz particles, and silicon-alumina acid – on plagioclase particles [15]. This statement is based on a previous study of the HCABS matrix system. It was found that during the hardening process of the binder suspension, the phase concentration changes. The amount of the amorphous component, consisting of colloidal solutions of silicic and alumosilicic acids, monotonously decreases, and the concentration of crystalline rock-forming components (quartz and feldspar) increase.

4. Conclusion

Thus, with application of method of mathematical planning non-autoclaved foam concrete based on a cementless highly concentrated aluminosilicate binder suspension was produced. It has the following characteristics: density – 419–524 kg/m^3; ultimate compressive strength – 1.24–1.52 MPa; thermal conductivity – 0.08–0.085 W/(m·℃); vapor permeability – 0.21–0.23 mg/(m · h · Pa); sorption humidity – 5.6–5.9 and 9.3–10.1 % (with relative air humidity of 75 and 97 %, respectively). Moreover, the products comply with the requirements of regulatory and technical documents of the Russian Federation.

The structural features of foam concrete based on HCABS were studied. The positive effect of the modifying component in the form of polyvinyl alcohol on the structure formation of materials and preservation of the porous structure was described. The accumulation of nano-particles attached to the surface of larger particles indicates the occurrence of epitaxial crystallization in the binder suspension system.

Acknowledgment

The work was carried out as part of the Presidential Scholarship SP-2116.2018.1.

References

[1] Lesovik R.V., Leshchev S.I., Ageeva M.S et al 2015 The use of zeolite-containing rottenstone powder for the composite binding production *Int. J. of Appl. Engineer. Res.* **10(24)** 44889

[2] Alfimova N I, Sheychenko M S, Karatsupa S V et al 2014 Features of application of high-mg technogenic raw materials as a component of composite binders *Res. J. of Appl. Sci.* **9(11)**779–83

[3] Glagolev E S., Suleimanova L A and Lesovik V V 2016 High reaction activity of nano-size phase of silica composite binder *Int. J. of Environ. and Sci. Ed.* **11(18)** 12383–9

[4] Lesovik V S, Glagolev E S, Popov D Y et al December 2017 Textile-reinforced concrete using composite binder based on new types of mineral raw materials", *Material Science in Mechanical Engineering* vol 327 XI Int. Conf. on Mechan. Engineer., Automat. and Control Syst. (MEACS 2017) p 032033

[5] Zagorodnjuk L H, Lesovik V S, Chernysheva N V et al 2016 The objective prerequisites for transition to composite binders *Int. J. of Pharmacy and Technol.* **8(4)** 22525–37

[6] Bazhenov Y M, Zagorodnjuk L H, Lesovik V S et al 2016 Concerning the role of mineral additives in composite binder content *Int. J. of Pharmacy and Technol.* **8(4)** 22649–61

[7] Kozhukhova N, Zhernovsky I and Voytovich E 2018 Aspects of application of industrial aluminosilicate waster in production of zero-cement construction composites *Materials Science and Engineering* vol 21 no 365 21st Int. Sci. Conf. on Advan. in Civil Engineer.: Construct. – The Format. of Living Environ. April 2018 p 032043
[8] Chernysheva N V, Lesovik V S, Drebezgova M Yu et al 2018 Composite gypsum binders with silica-containing additives *Material Science in Mechanical Engineering* vol 327 XI Int. Conf. on Mechan. Engineer., Automat. and Control Syst. (MEACS 2017) December 2017 p 032015
[9] Pivinsky Yu E 1990 Ceramic binders and ceramite concrete *Metallurgy*
[10] Pivinskii Yu E and Dyakin P V 2011 Aluminosilicate refractories based on high-alumina HCBS *Refract. and Industr. Ceram.* **52(4)** 264–71
[11] Cherevatova A V, Zhernovskii I V and Nelyubova V V 2010 Evaluation of the phase and dimensional heterogeneity of the quartz component of raw material and HCBS *Refract. and Industr. Ceram.* **51(4)** 310–1
[12] Strokova V V, Nelubova V V, Sivalneva M N et al 2016 Properties of nanostructured binding elements depending on applied resource type *Int. J. of Pharmacy & Technol.* **8(4)** 22743–8
[13] Cherevatova A V 2010 Principles of creating nanostructured binders based on HCBS *Refract. and Industr. Ceram.* **51(2)** 118–20
[14] Voitovich E V, Cherevatova A V, Zhernovsky I V et al 2014 The application of nano-structured silica based admixture in gypsum binders *Mater. Res. Society Symp. Proc.* vol 22 pp 165–70
[15] Kobzev V A, Sivalneva M N and Nelyubova V V 2018 Highly concentrated aluminosilicate binder suspension from granodiorite *Vest. of BSTU n.a. V.G. Shukhov* **1** 12–8

IOP Conf. Series: Materials Science and Engineering **905** (2020) 012065 doi:10.1088/1757-899X/905/1/012065

Diagnostic ratios of individual polycyclic aromatic hydrocarbons for their identifiaction in soils of power station

S Sushkova[1], T Minkina[1], A A Batukaev[2], T Dudnikova[1], A A Batukaev[3], M V Shavanov[2], L K Adymkhanov[2] and A Barakhov[1]

[1] Southern Federal University, 105/42, Bolshaya Sadovaya Str., Rostov-on-Don, Russia
[2] Chechen State University, 32, Sheripov Street, Grozny, Russia
[3] Chechen Agricultural Research Institute, 1, Lenin Street, Gikalo, Russia

E-mail: terra_rossa@mail.ru, tminkina@mail.ru, batukaevmalik@mail.ru, tyto98@yandex.ru, abuzar_batukaev@mail.ru, musa_vahaevich@mail.ru, adymhanov64@mail.ru, tolik.barakhov@mail.ru

Abstract. the article presents the study of the content of organic ecotoxicants – polycyclic aromatic hydrocarbons in the soils of the impact zone of the energy industry enterprise of Novocherkasskaya power plant (NPP) – the largest producer of electricity in southern Russia. The study showed a high level of total PAH content in the soils of the studied territory, compared to the background territories. The degree of contamination of soils in the zone of influence of the energy industry enterprises with Benzo(a)pyrene is classified as highly contaminated soils. The diagnostic relationships of individual PAHs showed a significant effect of the NchGRES on the soils of sites located along the line of the prevailing wind from NPP. For these sites, the pyrogen-coal origin of PAHs was established.

1. Introduction

More than 35 % of environmental emissions of various types of pollutants account for energy production [1]. According to the researchers [2], the share of the coal industry in the global energy market is 39 %, and in Russia it is about 47 % of GDP. The use of coal as a source of thermal energy is due, first of all, to savings on raw materials for energy production. Along with this, coal-fired power generating enterprises in the course of their work cause damage to all components of the biosphere in the emission zone. In the course of their work, coal-fired power plants emit various pollutants into the atmosphere in a gaseous, aerosol and solid state, which include hazardous pollutants — polycyclic aromatic hydrocarbons (PAHs) [3], most of which have carcinogenic, teratogenic, mutagenic properties [4].

PAHs are a group of macromolecular compounds the elementary unit of structure of which is benzene ring. Polyarenes are fat soluble and have a high melting point [5]. The subgroup of low molecular weight PAHs includes compounds with 2–3 condensed benzene rings, and the subgroup of high molecular weight PAHs includes compounds with 4 or more condensed benzene rings in their structure. The soil has the ability to accumulate these substances. The sorption potential of soils with respect to PAHs is primarily due to the properties of the soil itself (content and chemical composition of humus, pH, granulometric and mineralogical composition of the soil).

The list of priority pollutants of the United States includes 16 representatives of this class of compounds. In the Russian Federation, the norms are established for four substances of this class of compounds, however, the maximum permissible concentration (MPC) for soils is developed for only one representative of PAHs – benzo(a)pyrene, carcinogen and mutagen of the first hazard class (20 ng/g) [6].

The purpose of the research was to establish the composition of individual PAHs in the soils of the energy complex enterprises by their diagnostic ratio.

2. Experimental area

The object of the study was the soils of the territories adjacent to the Second Generating Company of the Wholesale Electricity Market OGK-2 Novocherkasskaya Power Station (NPP). NPP is an enterprise with a capacity of more than 19 GW. In 2017, electricity generation amounted to 63.5 billion kWh, which is about 7 % of the all-Russian electricity production. NPP takes the second place within the Russian Federation in terms of electricity generation, being second only to Surgut state district power station [7]. Power plants of 6000 kWh are classified as enterprises of the first hazard class [8].

Taking into account the above mentioned aspects, it can be concluded that NPP belongs to this type of enterprises and has a significant impact on the environmental situation, as in Novocherkassk itself, in the surroundings of which the enterprise is located (MPC of benzo(a)pyrene within the city averages 16.4 [9]), and the Rostov region and Russia as a whole. Monitoring sites of the impact zone of NPP are located along the line of the prevailing wind rose [10] (No. 4, 8, 9, 10) at a distance of up to 20 km from the emission source, and around the enterprise (sites No. 1–3, 5–7, 11) at a distance of up to 3 km (Figure 1).

Figure 1. Map of monitoring sites around Novocherkassk Power Station

Table 1. Soil properties of monitoring sites of impact zone of npp

№ Site	Direction and distance from NPP	Soil type	Physical clay, %	Slit, %	Humus, %	pH
1	1ne	CH	52	27	4.3	7.6
2	1.2e	CH	52	31	4.0	7.4
3	2.7sw	MC	67	37	4.6	7.3
4	1.6nw	CH	55	29	4.6	7.5
5	3sw	FL	7	3	3.1	7.5
6	2n	MC	55	30	4.1	7.7
7	1.5n	CH	51	27	4.1	7.6
8	5nw	MC	60	32	5.0	7.4
9	15nw	CH	52	30	4.2	7.6
10	20nw	CH	53	28	4.6	7.6
11	1.7 sw	CH	53	30	3.5	7.6

Hoplic chernozem (CH) prevails in the soil cover of the study area, but fluvisols (FL) and meadow-chernozemic (MC) soils are also found (Table 1).

3. Methods and materials

Soil sampling was carried out by the envelope method from each monitoring site to a depth of 0–20 cm in July 2018 [11]. The transportation and storage of the samples were carried out in a portable device with a cooling element. The storage of samples, according to the guidelines, was carried out no more than a week from the moment of selection until the direct extraction of pollutants from soil samples. PAHs were extracted from soil samples by saponification, where the interfering fraction of soil lipids was removed by boiling the sample with a 2 % alcohol solution of potassium hydroxide, followed by three times extraction of sample with hexane [12].

After the PAH extract in hexane was evaporated on a rotary evaporator and dissolved in 1 ml of acetonitrile, the acetonitrile solution of concentrated PAHs was transferred into vials. The concentration of PAHs in an acetonitrile solution was determined using high performance liquid chromatography on an Agilent 1260 chromatograph with fluorimetric detection no later than three days after the transfer of PAH concentrate to the vial [13]. Processing the results of mass concentrations of PAHs in the sample was performed using a computer in accordance with the calibration characteristic, taking into account concentration and losses during sample preparation.

In this study, PAHs such as naphthalene, biphenyl, anthracene, acenaphtene, acenaphthylene, phenanthrene, fluorene, which are low molecular weight and benzo(a) anthracene, fluoranthene, pyrene, chrysene, BaP, benz (k) fluorantene, benz (b) fluoranthene, dibenz (a) anthracene, benz (g, h, i) perylene, which are high molecular weight PAHs. All of the above PAHs, with the exception of biphenyl, are on the list of US priority pollutants [14].

4. Results and discussion

The total PAH content in the soils of the monitoring sites of NPP impact zone significantly exceeded the background (for Rostov region) total PAH content in the soils of the Persianovskaya Steppe specially protected natural area. The PAH content in the soils of the monitoring sites decreased with the distance from NPP in the NW direction (along the prevailing wind line) total, according to the following scheme: site No. 4 (1.6NW) – 3696.6 ± 180.0 ng / g> No. 8 (5NW) – 1420.1 ± 66.7 ng / g> No. 9 (15NW) – 1291.8 ± 58.2 ng / g> No. 10 (20NW) – 1157.1 ± 44.9 ng / g. The content of 3-ringed phenanthrene in the soils of monitoring sites located along the line of the prevailing wind prevailed quantitatively in the composition of individual low-molecular PAHs (table 2).

The content of 4 ringed PAHs such as pyrene, chrysene, and fluoranthene, as well as 5 ringed BaPs and benzo (b) fluoranthene in the soils of sites located along the line of the prevailing wind outnumbered in the composition of individual high molecular weight PAHs (table 3).

Table 2. Content of low-molecular pah in the soils of monitoring sites located on the line of the prevailing wind

№ site	PAH, ng/g						
	Naphthalene	Diphenyl	Anthracene	Acenaphthene	Acenaphthylene	Fluorene	Phenanthrene
4	144.2 ± 7.2	140.1 ±6.8	61.4 ± 3.0	22.4 ± 1.1	55.1 ± 2.8	62.6 ± 3.0	453.0 ± 25.1
8	124.6 ± 5.9	24.6 ± 1.0	20.9 ± 1.1	10.7 ± 0.5	15.4 ± 0.8	42.9 ± 2.1	130.1 ± 6.5
9	113.1 ± 5.1	16.3 ± 0.8	18.6 ± 0.8	8.6± 0.4	12.9 ± 0.6	37.6 ± 1.7	93.2 ± 4.4
10	122.4 ± 5.7	13.6 ± 0.6	15.4 ± 0.7	8.9 ± 0.4	13.7 ± 0.6	50.5 ± 2.4	82.0 ± 3.9

In the soils of sites located around NPP, the total PAH content was within 716.3 ± 30 ng / g in the AL soil of site No. 5, up to 1083.1 ± 48 ng / g in the LF soil of site No. 3. The content of 3-ring phenanthrene in the soils of monitoring sites located around NPP was quantitatively predominant in the composition of individual low-molecular PAHs (Table 4). The composition of high molecular weight PAHs was outnumbered by the content of 4 ringed PAHs: pyrene, chrysene, and fluorantene, as well as 5 ringed benzo (b) fluoranthene and 6 ringed benzo (g, h, i) perylene in the soils of monitoring sites located around NPP (table 5).

Table 3. Content of high-molecular pah in the soils of monitoring sites located on the line of the prevailing wind

№	PAH, ng/g

site	Benz(a)anthracene	Pyrene	Fluoranthene	Chrysene	BaP	Benz (b) Fluoranthene	Benz (k) Fluoranthene	Dibenz (a, h) anthracene	Benz (g, h, i) perylene
4	234.6±11.1	284.1± 4.2	562.4±27.2	418.5±21.5	347.8±18.4	515.4±26.6	182.7±9.0	127.2 ± 6.1	85.1±4.2
8	104.3±5.1	202.5±10.1	196.8±9.3	132.4±6.5	90.2±4.5	169.9±8.2	54.2±2.6	38.5±1.8	62.1±3.1
9	84.1±4.1	183.9±9.5	212.6±10.3	158.1±7.5	75.2±3.6	156.3±7.6	45.9±2.3	28.4±1.3	47.0±2.2
10	68.4±3.2	162.1±7.9	154.7±7.4	174.6±7.6	68.4±3.2	127.9±6.3	32.7±1.6	22.4±1.1	39.4±1.9

Table 4. Content of low-molecular pah in the soils of monitoring areas around npp

№	PAH, ng/g						
Site	Naphthalene	Diphenyl	Anthracene	Acenaphthene	Acenaphthylene	Fluorene	Phenanthrene
1	36.1± 1.6	19.3±0.9	13.0±1.0	11.3±0.5	32.9±1.6	16.3±0.8	213.2±10.1
2	44.0±2.2	22.5±1.0	7.5±0.3	9.5±0.4	20.2±1.0	12.8±0.6	69.6±3.4
3	42.1±2.1	23.6±1.1	8.5±0.4	18.0±0.9	20.7±1.0	13.6±0.6	233.5±11.5
5	59.3±2.9	18.7±0.9	8.5±0.4	10.2±0.5	11.7±0.5	16.9±0.8	161.1±8.0
6	66.6±3.3	44.5±2.2	10.2±0.5	16.2±0.8	26.1±1.3	16.6±0.8	265.5±26.2
7	25.5±1.1	23.1±1.1	21.3±1.0	11.2±0.5	15.5±0.7	10.3±0.5	106.4±5.0
11	56.4±2.6	23.4±1.1	3.7±0.2	8.5±0.4	10.7±0.5	14.2±0.7	153.8±7.2

Table 5. Content of high-molecular pahs in the soils of monitoring areas around npp

№ Site	PAH, ng/g								
	Benz(a)anthracene	Pyrene	Fluoranthene	Chrysene	BaP	Benz (b) Fluoranthene	Benz (k) Fluoranthene	Dibenz (a, h) anthracene	Benz (g, h, i) perylene
1	51.1±2.5	179.7±8.5	73.4±3.5	73.1±3.5	107.5±5.1	118.9±5.8	31.8±1.5	7.1±0.3	83.2±4.1
2	31.6±1.5	89.2±4.4	73.0±3.5	70.6±3.3	49.6±2.4	141.5±5.7	11.4±0.5	7.3±0.3	113.4±5.1
3	53.5±2.6	76.2±3.6	100.4±4.9	58.5±2.7	59.6±2.7	194.9±9.8	29.5±1.4	25.2±1.2	126.1±6.1
5	47.9±2.4	72.0±3.5	61.4±2.9	41.3±2.0	63.3±3.1	77.5±3.6	15.5±0.7	2.7±0.1	70.3±3.5
6	21.7±1.0	43.0±2.0	48.9±2.4	83.6±3.9	97.2±4.7	166.3±8.1	15.1±0.7	6.4±0.3	96.7±4.7
7	17.4±0.8	103.4±4.9	75.3±3.6	81.7±3.9	72.3±3.5	33.6±1.6	4.9±0.2	6.7±0.3	108.0±5.1
11	52.2±2.5	82.0±4.1	70.0±3.5	43.9±2.1	46.4±2.2	74.6±3.2	15.8±0.7	4.0±0.2	81.1±4.0

The MPC of BaP was exceeded in the soils of all monitoring sites located along the prevailing wind line from NPP, with a regular decrease with distance from the emission source in the following order: No. 4 (1.6 km NE) – 17.4> No. 8 (5 km NE) – 4.5> No. 9 (10 km NE) – 3.8> No. 10 (20 km NE) – 3.4. The content of BaP for the soils of monitoring sites located around NPP ranged from 2.3 MPC in the soils of site No. 11 to 4.9 MPC in the soils of site No. 6. According to SanPiN 2.1.7.1287-03 "Sanitary and epidemiological quality requirements for Soils "[15], excesses of MPC of BaP in the soils of all monitoring sites make it possible to classify the soils of the study area as highly contaminated soils (Table 6).

Table 6. Soil contamination categories

Substance hazard class	The 1st class	The 2nd class	The 3rd class
> 5 MPC	Very high		High
From 2 to 5 MPC	Very high	High	Average
From 1 MPC to 2 MPC		Low	

It is necessary to note that the soils of monitoring sites located along the prevailing wind line are located near settlements (the village of Grushevskaya, the village of Kushchevskaya, Novocherkassk) (Figure 1), which may have the effect of additional pollution factors on the content of PAHs in the soils of this region due to mass use of vehicles.

In order to assess the contribution of NPP directly to the PAH content in the soils of monitoring sites located along the prevailing wind line from the enterprise, we calculated the ratios of individual PAHs. This calculation method is based on theoretical knowledge of the conditions for the formation of PAHs. First of all, PAHs are formed during the burning of organogenic fuel under conditions of oxygen deficiency. Toxicants can be formed as a result of anthropogenic activities, as well as a result of natural processes.

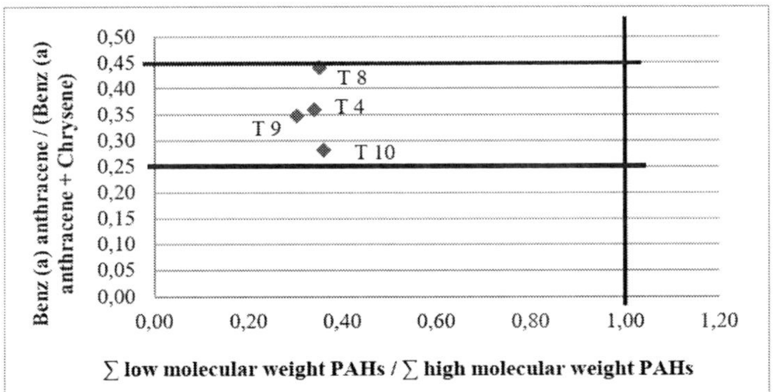

Figure 2. PAHs diagnostic ratios for soils of monitoring sites situated thrue the prevailing wind direction

However, PAHs are not formed singly they always present a mixture [16]. The ratio of some PAHs contained in the soil will depend on the material from which they are formed, for example, the chemical composition of oil or coal; kind of burnt wood, etc. [17], and from the conditions of formation: temperature, oxygen access. Diagnostic ratios determine the origin of pollution by comparing the relative concentrations of individual PAHs according to the following indicators (table 7) [18].

Table 7. Пay diagnostic ratios of individual pah. Pah source identification coefficients

Ratios	Range of values	Source
∑ Low molecular weight PAHs / ∑ High molecular weight PAHs	< 1	Pyrogenic
	> 1	Petrogenic
Benz (a) anthracene / (Benz (a) anthracene + Chrysene)	< 0.2	Petrogenic
	0,2–0.45	Coal pyrolysis
	> 0.45	Car emissions

The highlighted area in Figure 2 indicates the pyrogenic-coal origin of PAHs in the soils of monitoring sites located along the line of the prevailing wind from NPP. Since these diagnostic ratios are characteristic of soils of all monitoring sites located along the line of the prevailing wind rose from NPP, this may indicate the direct effect of the enterprise on the content of PAHs in the soils of the studied territory.

5. Conclusion

Thus, a significant excess of the content of hazardous ecotoxicants – PAHs in the soils of the impact zone of NPP was revealed, compared to the soil in the background regions. The maximum technogenic load is experienced by the soil of monitoring site No. 4 (1.6 km NW), where PAH content is exceeded by 17 times compared to PAH content in the background areas. It was established that in the soils of monitoring sites located along the line of the prevailing wind, the total PAH content decreases with distance from the emission source. In the soils of the monitoring sites located around NPP, the accumulation of PAHs depended more on the soil properties than on the distance and location of the soil of the monitoring site relative to the NPP.

The excess of MPC of BaP in soils of all monitoring sites located along the line of prevailing winds was recorded. The maximum value of MPC of BaP was found in the soils of site No. 4 (17.4 MPC). Sanitary status of soils according to BaP classifies the soils of the study area as highly contaminated soils.

The diagnostic ratios of individual PAHs to each other and the total content of low molecular weight PAHs to the total content of high molecular weight PAHs have confirmed the significant effect of NPP on the soil adjacent to the emission source. The pyrogenic-coal origin of PAHs in soils most susceptible to the influence of NPP monitoring sites is established.

Acknowledgment
The present study was done with the financial support of the Ministry of Education and Science of the Russian Federation for applied scientific research and experimental development as part of the implementation of the Federal Targeted Program "Research and Development in Priority Areas of the Development of the Scientific and Technological Complex of Russia for 2014-2020", Agreement No. 14.577.21.0292 from 04.12.2018; Project of the President of Russian Federation no. MK-2973.2019.4; RFBR no. 19-29-05265 MK and 19-34-90185.

References

[1] Lazarev G E and Shapovalov A B March 2017 The formation of dominant emissions during energy generation *Ecological and environmental problems of modern society and their solutions* XIII Int. sci. Conf. (Moscow) pp 53–66
[2] *Corrigenda to OECD Publications* Retrieved from: http://www.oecd.org/about/publishing/corrigenda.htm Electricity information 2018.
[3] Kumar V, Kothiyal N C, Saruchi P V and Sharma R Jul 2016 Sources, distribution, and health effect of carcinogenic polycyclic aromatic hydrocarbons (PAHs)–current knowledge and future directions *J. of the Chinese advan. Mater. society* **4(4)** 302–21
[4] Abdel-Shafy H I and Mansour M S December 2018 *Phytoremediation for the Elimination of Metals, Pesticides, PAHs, and Other Pollutants from Wastewater and Soil* (Singapore: Springer) pp 101–36
[5] Rovinsky F Ya, Teplitskaya T A and Alekseeva T A 1988 *Background monitoring of polycyclic aromatic hydrocarbons* p 44
[6] GN 2.1.7.2041-06 *Maximum permissible concentration (MPC) of chemicals in the soil* 2006 (Moscow)
[7] Sushkova S N, Minkina T M, Deryabkina I G et al Mach 2019 Environmental pollution of soil with PAHs in energy producing plants zone *Sci. of the Total Environ.* **655** 232–41
[8] SanPiN 2.2.1/2.1.1.1200-03 Sanitary and epidemiological rules and standards SanPiN 2.2.1/2.1.1.1200-03 *Sanitary protection zones and sanitary classification of enterprises, structures and other objects* 2003 (Moscow)
[9] *Don's Ecological Bull. On the State of the Environ. and Natural Resour. of the Rostov Reg. in 2017* 2018 10–18
[10] Sushkova S N, Deryabkina I G, Antonenko E M et al August 2018 Benzo[a]pyrene degradation and bioaccumulation in soil-plant system under artificial contamination *Sci. of the Total Environ.* **633** 1386–91
[11] GOST (State Standard) 17.4.4.02-84 *Nature Protection. Soils. Methods for Sampling and Preparation of Soil for Chemical, Bacteriological, and Helminthological Analysis* 1984 (Moscow: Izd. Standartov)
[12] Sushkova S N, Minkina T M, Mandzhieva S S et al April 2015 New alternative method of benzo[a]pyrene extractionfrom soils and its approbation in soil under technogenic pressure *J. of soils and sediments* **16(4)** 1323–9
[13] ISO 13877-2005 *Soil quality-determination of polynuclear aromatic hydrocarbons – Method using high -performance liquid chromatography* 2005
[14] Wenzl T, Simon R, Anklam E and Kleiner J Analytical methods for polycyclic aromatic hydrocarbons (PAHs) in food and the environment needed for new food legislation in the European *Union. Trend. Anal. Chem.* **25(7)** 716–25

[15] SanPiN (sanitary and epidemiological rules) 2.1.7.1287-03 *Sanitary and epidemiological requirements for soil quality* Decree of the Chief State Sanitary Doctor of the Russ. Fed. of April 2007 no 20
[16] Tobiszewski M and Namieśnik J Mach 2012 PAH diagnostic ratios for the identification of pollution emission sources *Environ. Pollut.* **162** 110–9
[17] Gennadiev A N, Pikovsky Yu I, Smirnova M A et al Mach 2016 Hydrocarbon state of soils of background taiga landscapes (south-western part of the Ustyansky plateau) *Vest. Moskovsky Univer.* **5(3)** 90–7
[18] Chunhui W A N G, Shaohua W U, Shenglu Z H O U et al February 2017 Characteristics and source identification of polycyclic aromatic hydrocarbons (PAHs) in urban soils: a review *Pedosphere* **27(1)** 17–26

Assessment of the geomorphological basis of landscapes of the Crimean Peninsula using geoinformation technologies

V A Tabunshchik[1], A A Kluchkina[2], E A Petlukova[3], I V Kalinchyk[3], M V Galkina[3], M V Penno[3] and A A Nikiforova[3]

[1] Geomatics Research Center, A.O. Kovalevsky Institute of Biology of the Southern Seas of RAS, 2, Nakhimov ave., Sevastopol, 99011, Russia
[2] T.I. Vyazemsky Karadag Scientific Station – Nature Reserve of the RAS, A.O. Kovalevsky Institute of Biology of the Southern Seas of RAS, Feodosia, Kurortnoye, Russia
[3] Taurida Academy, V.I. Vernadsky Crimean Federal University, 4, Prospekt Vernadskogo, Simferopol, 295007, Russia

E-mail: tabunshchyk@ya.ru, klyuchkinaaa@gmail.com, petlukova@mail.ru, ir_vasi@mail.ru, m_tsvetkova@mail.ua, address_mvp@mail.ru, nik.a.815@mail.ru

Abstract. The goal is, based on modern methods of geoinformatics and geomatics, to assess the landscapes of the Crimean Peninsula for tourism using a set of indicators characterizing the expressiveness of the geomorphological basis: the absolute height, steepness of the slopes, the depth of the relief and the density of the dissection of the relief. Organization and research methods: assessment of the landscapes of the Crimean Peninsula by a set of indicators characterizing the expressiveness of the geomorphological basis was carried out using the Quantum GIS (QGIS) software package by creating a series of maps reflecting the geomorphological characteristics of the relief and their subsequent overlay analysis for each of the landscapes of the Crimean Peninsula. To create maps and obtain indicator values, geoinformation, cartographic, and cartometric research methods were used. The data obtained for each landscape were further processed by statistical methods using the statistical calculation environment R and the Statistica 10 program. Results: the values of the absolute height, slope steepness, depth of the dismemberment of the relief and the density of the dissection of the relief were calculated for each landscape of the Crimean Peninsula. In addition, by implementing statistical methods, zoning of landscapes of the Crimean Peninsula was carried out using a set of indicators characterizing the expressiveness of the geomorphological basis.

1. Introduction

The Crimean Peninsula is a unique natural area that combines various combinations of geological and geomorphological conditions, climatic conditions, hydrological characteristics, soil cover, flora and fauna and, as a consequence, a variety of landscapes.

The first work on the allocation of natural landscapes on the Crimean Peninsula began at the beginning of the twentieth century. As noted by Vasily Ena [1], the first work on the allocation of natural landscapes on the Crimean Peninsula is associated with the name of Boris Dobrynin. He subdivided the Crimean Peninsula into Mountain Crimea, the steppe plain Crimea and the Kerch Peninsula and distinguished a number of smaller units.

IOP Conf. Series: Materials Science and Engineering **905** (2020) 012066 doi:10.1088/1757-899X/905/1/012066

Later, zoning issues were developed by the authors of the Council for the Study of the Productive Forces of the USSR Academy of Sciences. According to this zoning experience, the southern mountainous part of the peninsula forms the province of the Crimean-Caucasian mountainous country, and the steppe part of the peninsula forms the North Crimean province, which is part of the dry-steppe zone of the East European lowland country.

Landscape studies and zoning of certain parts of the peninsula were carried out by: Nina Dzens-Litovskaya, who identified 18 landscapes within the plain-steppe Crimea; Grigory Grishankov, indicating 5 types of localities for eastern Crimean Yail; Nina Pavlova, dividing the Steppe Crimea, including the Kerch Peninsula, into 5 landscapes and several "groups of tracts"; Zinaida Kryukova, proposing for the Crimean mountains a complex “sectoral” landscape separation scheme (12 natural districts with “landscape sectors” included in their composition). Landscape research on the Tarkhankut Peninsula was conducted by Peter Podgorodetskiy and others [1].

According to V.G. Ena [2], who proposed his scheme of zoning of the Crimean peninsula, 47 landscapes are distinguished: the landscape region of the Crimean forest middle mountains (Main Crimean ridge: landscapes 1–14); the region of the Crimean forest-sibljak sub-Mediterranean (Southern coast of Crimea: landscapes 15–27); the area of the Crimean forest-steppe foothills (Inner and Outer cuesta ridges, Southern and Northern inter-ridge depressions: landscapes 28–37); region of steppe hills (Kerch peninsula: landscapes 38–39); landscape region of the Plain-Steppe Crimea (landscapes 40–47).

In the Atlas of Natural Conditions and Natural Resources of the Ukrainian SSR, on the territory of the Crimean Peninsula, 21 categories of types and types of landscapes are distinguished, including 10 plain-steppe types of landscapes and 11 mountain forests [3].

However, the most complete is the characteristic landscape of the Crimean Peninsula made out by Grigory Grishankov [4, 5]. Natural zones, according to Grigory Grishankov, are considered as three-dimensional formations, on the one hand, and on the other, zoning as a multifactorial phenomenon. Within each landscape level stands out its own system of natural zones. “Landscape levels are planetary geomorphological formations that are relatively uniform in the nature of the relief and soil moisture, but differing in their peculiar manifestations of geographical zonality” [4]. On the territory of the Crimean Peninsula G.E. Grishankov identifies 4 landscape levels: hydromorphic, upland, low-mountain and mid-mountain [5].

2. Methods and materials

A large number of works have been devoted to the study of various landscape components [6–10]. Recently, the number of publications considering landscapes as the basis for tourism and recreational activities – the so-called landscape resources – and the number of publications where geoinformation technologies act as the leading research method, has been growing [11–22]. The study of the expressiveness of the geomorphological basis of landscapes of the Crimean Peninsula using geoinformation technologies is actively used in tourism activities to assess the aesthetic characteristics of the landscape, landscape planning of territories and water areas, making active management decisions by authorities. The basis of the study of the expressiveness of the geomorphological basis of the landscapes of the Crimean Peninsula were works [23–25].

The study includes three stages, is performed using geographic information systems (GIS) and is based on modern geographic information, cartographic, cartometric and statistical research methods. Among the old research methods, the literary and analytical method of research was actively used.

At the first stage, an analysis and selection of all available information on the research topic is performed. At the second stage, source materials are processed in geographic information systems (GIS) and statistical indicators characterizing the objects of research are obtained. At the third stage, the analysis of the obtained data is carried out, the final conclusions are constructed and conclusions are drawn.

The first stage traditionally included an analysis of all available information on the research topic. As a source of data for the analysis of the expressiveness of the geomorphological basis of landscapes of the Crimean Peninsula, a landscape map of the Crimean Peninsula was compiled by Professor Ekaterina Pozachenyuk and published in [26], which is based on a map of Grigory Grishankov. According to [26], 13 landscapes stand out on the territory of the Crimean Peninsula (Figure 1), which occupy different geographical locations and have different landscape potentials. To analyze the expressiveness of the geomorphological basis of landscapes of the Crimean Peninsula, we used the Shuttle radar topographic mission (SRTM) data [24] based on which, at the second stage of the work, maps are constructed that reflect the absolute height, steepness of the slopes, the depth of the terrain and the density of the terrain for landscapes of the Crimean Peninsula.

The second stage included the use of the Quantum GIS (QGIS) software package in the environment of which map building and calculations were performed. To build a map of landscapes of the Crimean peninsula, the paper version of the map [26] was scanned and converted to a vector format by digitizing the boundaries of landscapes using the "Layer" menu and the "Create Vector Layer" submenu.

A map of the absolute height of the territory of the Crimean peninsula was constructed by classifying the previously downloaded digital terrain model Shuttle radar topographic mission (SRTM). The construction of maps of the depth of the fragmentation of the relief and the density of the fragmentation of the relief was carried out according to the technique described in [24, 27–30]. As a result, the maps shown in Figures 3 and 4 were obtained.

The slope steepness map was constructed in the Quantum GIS (QGIS) software package using the Slope Angle tool of the Morphometric Analysis submenu of the Raster menu (Fig. 5).

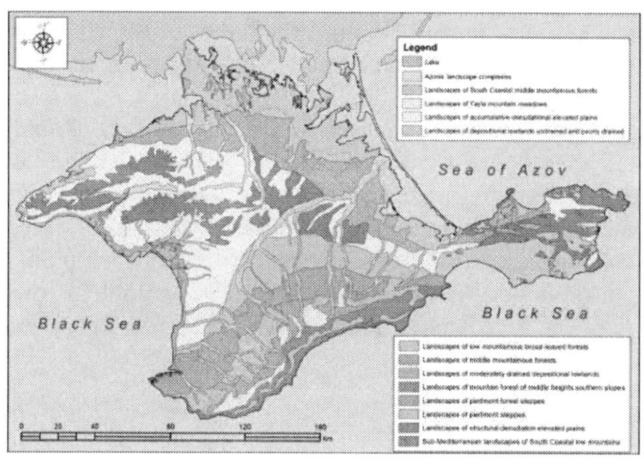

Figure 1. Landscape of the Crimean Peninsula

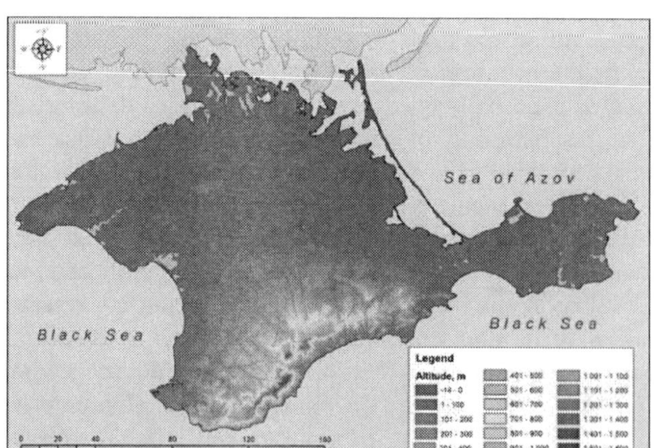

Figure 2. Absolute altitudes of the Crimean Peninsula, m

IOP Conf. Series: Materials Science and Engineering **905** (2020) 012066 doi:10.1088/1757-899X/905/1/012066

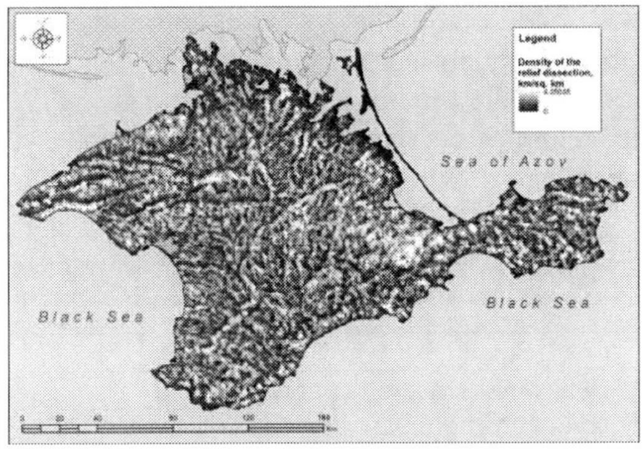

Figure 3. Density of the relief dissection of the Crimean Peninsula, km/sq. km

To obtain spatial information about the absolute height, steepness of the slopes, the depth of the topography and the density of the topography within the landscapes of the Crimean peninsula, we used the Zonal Statistics tool of the Raster menu of the Quantum GIS (QGIS) software package, which allows getting for each indicator a set of the following statistics: the minimum indicator value within the landscape, the maximum indicator value within the landscape, the amplitude of the indicator values within the landscape and the average indicator value within the landscape. Additionally, with the help of the "$.area" command, the areas of each landscape on the territory of the Crimean Peninsula were calculated.

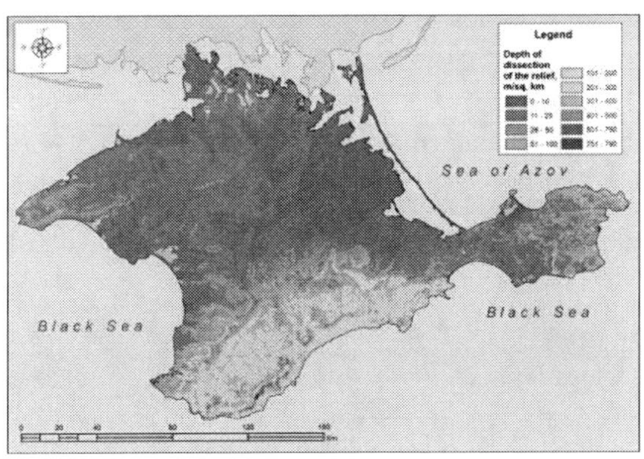

Figure 4. Depth of dissection of the relief of the Crimean Peninsula, m/sq. km

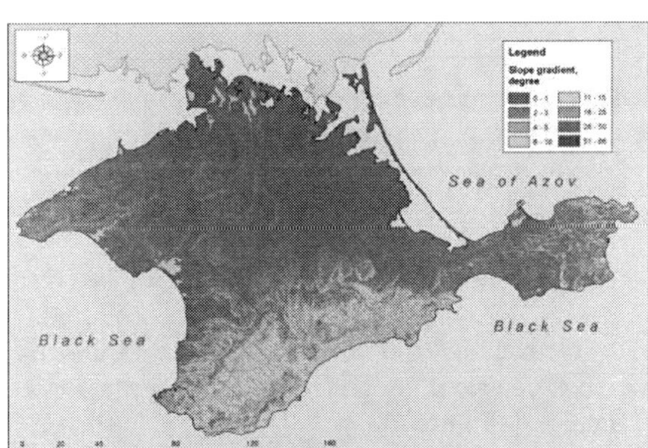

Figure 5. Slope gradient of the Crimean Peninsula, degree

IOP Conf. Series: Materials Science and Engineering **905** (2020) 012066 doi:10.1088/1757-899X/905/1/012066

3. Results

As a result of the study, it was found that the largest area within the Crimean Peninsula is occupied by the landscapes of accumulative-denudational elevated plains, while the smallest area is occupied by the landscapes of South Coastal middle mountainous forests (Table 1).

For each of the considered indicators for the landscapes of the Crimean Peninsula, maximum, minimum and average values along with amplitude values were obtained that are presented in Tables 2–4.

Table 1. Area (sq. km) and perimeter (km) of the landscape of the crimean peninsula

Landscape	Area. sq. km	Perimeter. km
Landscapes of depositional lowlands undrained and poorly drained	4829.5	2680.0
Landscapes of moderately drained depositional lowlands	2795.1	1286.3
Landscapes of accumulative-denudational elevated plains	5554.0	2892.2
Landscapes of structural-denudation elevated plains	3438.6	2066.2
Landscapes of piedmont steppes	1658.1	699.1
Landscapes of piedmont forest steppes	1800.3	835.8
Landscapes of low mountainous broad-leaved forests	919.1	554.3
Landscapes of middle mountainous forests	522.1	420.4
Landscapes of Yayla mountain meadows	427.1	256.4
Landscapes of mountain forest of middle heights southern slopes	562.0	327.2
Landscapes of South Coastal middle mountainous forests	290.7	275.4
Sub-Mediterranean landscapes of South Coastal low mountains	501.4	481.6
Azonic landscape complexes	2066.4	2882.1

Table 2. Absolute altitudes (m) of the crimean peninsula within the landscape

Landscape	Absolute altitudes. m			
	Min	Max	Range	Mean
Landscapes of depositional lowlands undrained and poorly drained	-6.0	178.0	184.0	19.7
Landscapes of moderately drained depositional lowlands	-4.0	144.0	148.0	21.5
Landscapes of accumulative-denudational elevated plains	-14.0	206.0	220.0	62.0
Landscapes of structural-denudation elevated plains	-10.0	183.0	193.0	72.1
Landscapes of piedmont steppes	12.0	386.0	374.0	178.1
Landscapes of piedmont forest steppes	-8.0	771.0	779.0	285.3
Landscapes of low mountainous broad-leaved forests	1.0	949.0	948.0	455.2
Landscapes of middle mountainous forests	251.0	1442.0	1191.0	696.6
Landscapes of Yayla mountain meadows	318.0	1531.0	1213.0	960.5
Landscapes of mountain forest of middle heights southern slopes	43.0	1500.0	1457.0	556.6
Landscapes of South Coastal middle mountainous forests	12.0	1150.0	1138.0	291.9
Sub-Mediterranean landscapes of South Coastal low mountains	-4.0	676.0	680.0	150.9
Azonic landscape complexes	-5.0	1443.0	1448.0	127.9

The resulting set of values of the considered indicators for each landscape is unique. However, given the fact that all the indicators in question have different dimensions in the classification, it is necessary to bring them to uniform series. To do this, in the Statistica 10 program, data were standardized (the "Data" menu, the "Standardize" window), as a result, comparable values were obtained.

Using the menu “Analysis – Multidimensional Exploration Analysis”, a hierarchical classification of the landscapes of the Crimean Peninsula was made according to the expressiveness of the geomorphological basis. The connection Ward method was used as a union rule, and the Euclidean distance was used as a measure of proximity. As a result, a vertical dendrogram was obtained and 3 classes of landscapes were identified (Figure 6).

The allocation of three classes of landscapes was confirmed by checking using the K-means method in Statistica 10 and using the statistical package R. So the first cluster was formed by the Sub-Mediterranean landscapes of South Coastal low mountains and Landscapes of depositional lowlands undrained and poorly drained; the second cluster was formed by the Landscapes of moderately drained depositional lowlands; Landscapes of accumulative-denudational elevated plains; Landscapes of structural-denudation elevated plains; Landscapes of piedmont steppes; Landscapes of piedmont forest steppes and Azonic landscape complexes; the third cluster was formed by the Landscapes of low mountainous broad-leaved forests; Landscapes of middle mountainous forests; Landscapes of Yayla mountain meadows; Landscapes of mountain forest of middle heights southern slopes; Landscapes of South Coastal middle mountainous forests (Table 5, Figure 7).

Table 3. Density of the relief dissection (km/sq. km) of the crimean peninsula within the landscape

Landscape	Density of the relief dissection. km/sq. km			
	Min	Max	Range	Mean
Landscapes of depositional lowlands undrained and poorly drained	0.00	2.61	2.61	0.69
Landscapes of moderately drained depositional lowlands	0.01	2.55	2.54	1.00
Landscapes of accumulative-denudational elevated plains	0.02	2.58	2.56	0.99
Landscapes of structural-denudation elevated plains	0.00	2.78	2.78	0.86
Landscapes of piedmont steppes	0.14	2.35	2.21	0.95
Landscapes of piedmont forest steppes	0.05	2.41	2.36	0.88
Landscapes of low mountainous broad-leaved forests	0.08	2.47	2.39	0.88
Landscapes of middle mountainous forests	0.10	1.94	1.83	0.82
Landscapes of Yayla mountain meadows	0.18	2.11	1.93	0.81
Landscapes of mountain forest of middle heights southern slopes	0.07	2.33	2.26	0.88
Landscapes of South Coastal middle mountainous forests	0.21	2.31	2.10	1.05
Sub-Mediterranean landscapes of South Coastal low mountains	0.01	2.45	2.43	0.72
Azonic landscape complexes	0.00	2.48	2.48	1.01

Table 4. Depth of dissection of the relief (m/sq. km) of the crimean peninsula within the landscape

Landscape	Depth of dissection of the relief. m/sq. km			
	Min	Max	Range	Mean
Landscapes of depositional lowlands undrained and poorly drained	0.0	157.0	157.0	13.3
Landscapes of moderately drained depositional lowlands	0.0	113.0	113.0	9.9
Landscapes of accumulative-denudational elevated plains	0.0	99.0	99.0	16.2
Landscapes of structural-denudation elevated plains	0.0	121.0	121.0	20.5
Landscapes of piedmont steppes	6.0	173.0	167.0	32.5
Landscapes of piedmont forest steppes	9.0	389.0	380.0	91.4
Landscapes of low mountainous broad-leaved forests	26.0	583.0	557.0	219.1
Landscapes of middle mountainous forests	24.0	510.0	486.0	161.1
Landscapes of Yayla mountain meadows	48.0	790.0	742.0	274.8
Landscapes of mountain forest of middle heights southern slopes	25.0	753.0	728.0	208.0
Landscapes of South Coastal middle mountainous forests	0.0	535.0	535.0	169.2
Sub-Mediterranean landscapes of South Coastal low mountains	0.0	50.0	50.0	12.0
Azonic landscape complexes	0.0	510	540	41.6

Obviously from Table 6 and Figure 7, the clusters under consideration represent rather independent formations, however, the inclusion of Sub-Mediterranean landscapes of South Coastal low mountains and Landscapes of depositional lowlands undrained and poorly drained into cluster 1 is very controversial and requires additional refinement, since only 4 indicators were taken into account in the analysis. Most likely this is due to the fact that the differentiation of these landscapes will manifest itself under the influence of other factors (climatic, hydrological, etc.).

IOP Conf. Series: Materials Science and Engineering **905** (2020) 012066 doi:10.1088/1757-899X/905/1/012066

Table 5. Slope gradient (degree) of the crimean peninsula within the landscape

Landscape	Slope gradient. degree			
	Min	Max	Range	Mean
Landscapes of depositional lowlands undrained and poorly drained	0.0	14.3	14.3	0.8
Landscapes of moderately drained depositional lowlands	0.0	17.0	17.0	0.5
Landscapes of accumulative-denudational elevated plains	0.0	13.0	13.0	0.9
Landscapes of structural-denudation elevated plains	0.0	15.7	15.7	1.1
Landscapes of piedmont steppes	0.0	23.4	23.4	1.8
Landscapes of piedmont forest steppes	0.0	37.0	37.0	5.3
Landscapes of low mountainous broad-leaved forests	0.6	51.8	51.2	12.3
Landscapes of middle mountainous forests	0.0	42.1	42.1	9.7
Landscapes of Yayla mountain meadows	0.2	55.4	55.2	14.8
Landscapes of mountain forest of middle heights southern slopes	0.1	58.5	58.3	11.1
Landscapes of South Coastal middle mountainous forests	0.0	61.8	61.8	9.9
Sub-Mediterranean landscapes of South Coastal low mountains	0.0	7.9	7.9	0.7
Azonic landscape complexes	0.0	42.1	42.1	2.5

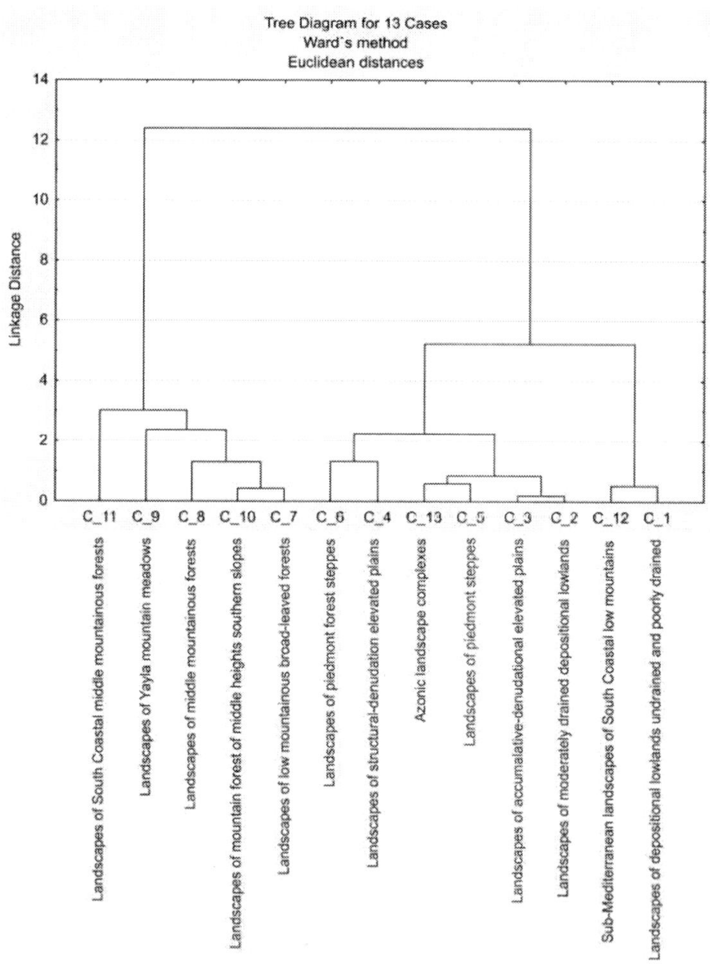

Figure 6. Clustering of the landscape of the Crimean Peninsula, m/sq. km

Table 6. Average values within the selected clusters

Criteria	Cluster		
	1	2	3
Absolute altitudes, m	85.3	124.5	592.2
Density of the relief dissection, km/sq. km	0.71	0.95	0.89
Depth of dissection of the relief, m/sq. km	12.7	35.4	2.1
Slope gradient, degree	0.75	2.1	11.6

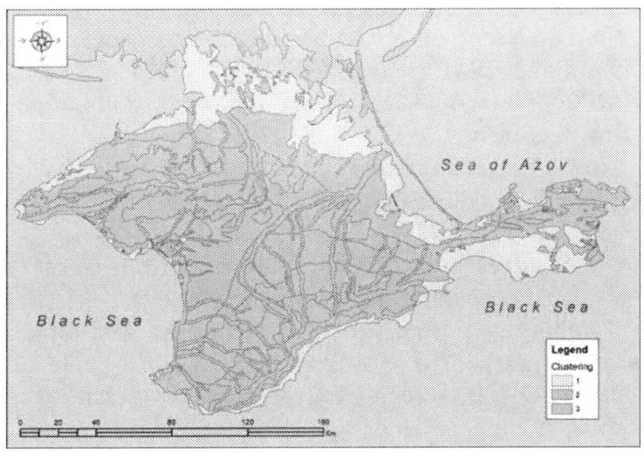

Figure 7. Clustering map of the landscape of the Crimean Peninsula

4. Conclusion

The paper shows the possibility of applying geographic information technologies in the study of the geomorphological basis of landscapes of the Crimean Peninsula. Using the methods of geoinformatics and statistics among the landscapes of the Crimean Peninsula, three relatively uniform indicators were identified that characterize the expressiveness of the geomorphological basis: the absolute height, steepness of the slopes, the depth of the dissection of the relief and the density of the dissection of the relief. The first cluster was formed by Sub-Mediterranean landscapes of South Coastal low mountains and Landscapes of depositional lowlands undrained and poorly drained; the second cluster was formed by the Landscapes of moderately drained depositional lowlands; Landscapes of accumulative-denudational elevated plains; Landscapes of structural-denudation elevated plains; Landscapes of piedmont steppes; Landscapes of piedmont forest steppes and Azonic landscape complexes; the third cluster was formed by the Landscapes of low mountainous broad-leaved forests; Landscapes of middle mountainous forests; Landscapes of Yayla mountain meadows; Landscapes of mountain forest of middle heights southern slopes; Landscapes of South Coastal middle mountainous forests.

Acknowledgment

The work presented in this paper was carried out in the framework of RF state task according to plan of scientific research of the A.O. Kovalevsky Institute of Biology of the Southern Seas of RAS («Studying the spatio-temporal organization of water and land ecosystems with the aim of developing an operational monitoring system based on remote sensing data and GIS technologies». Registration number AAAA-A19-119061190081-9).

This research was carried out with the support of the Program for the Development of the Crimean Federal V.I.Vernadsky University for 2015–2024 years during the project "Academic Mobility Network "GIS-Landscape - Technologies and techniques for the formation of geoportals of present-day landscapes of regions" in 2017, 2018 year and partially supported by the V.I. Vernadsky Crimean Federal University Development Program for 2015–2024 (VG05/18 "Landscape planning of plain

IOP Conf. Series: Materials Science and Engineering **905** (2020) 012066 doi:10.1088/1757-899X/905/1/012066

Crimea for the sustainable development of the territory" in 2018–2019 year and "Biogeographic study and conservation of relict endemic of the Mountain Crimea" in 2018-2019 year).

References

[1] Ena V 1969 Results and prospects of landscape and geographical research in Crimea In *Natural conditions and natural resources of Crimea* (Simferopol: Tavria) pp 52–5

[2] Ena V 1960 Physico-geographical zoning of the Crimean Peninsula *Bull. of Moscow Univer. Ser. 5 Geogr.* **2** 33–43

[3] Ena V 2007 *Discoverers of the Crimean land. Essays of 2500-year history of Taurida nature study* (Simferopol: Business-Inform) 520 p

[4] Grishankov G 1972 Landscape levels of continents and geographical zoning *Izv. Akad. nauk SSSR. Ser. Geograf.* **4** 3–18

[5] Pozachenyuk E 2009 *Modern Landscapes of the Crimea and Adjacent Water Areas* (Simferopol: Business-Inform) 672 p

[6] Pozachenyuk E, Lisetskii F, Kalinchuk I et al 2019 Modern landscapes of the Kerch peninsula *Int. Multidisciplinary Sci. GeoConf. Survey. Geol. and Mining Ecol. Manag. SGEM* vol 19 no 5.1 pp 477–84

[7] Silva R, Pereira J and Alves S 2019 The landscapes from Ouro Preto, Minas Gerais State: Decoding in space and time *Ornamental Horticult.* **25(1)** 9–17

[8] Prince S 2019 Dwelling and tourism: embracing the non-representational in the tourist landscape *Landscape Res.* **44(6)** 731–42 DOI: 10.1080/01426397.2018.1518520

[9] Aspinall R and Pearson D 2000 Integrated geographical assessment of environmental condition in water catchments: Linking landscape ecology, environmental modelling and GIS *J. of environ. Manag.* **59(4)** 299–319

[10] Bender O, Boehmer H, Jens D and Schumacher K 2005 Using GIS to analyse long-term cultural landscape change in Southern Germany *Landscape and Urban Plann.* 70(1-2) 111–25

[11] Kalinichenko A 2014 Spatial model of ecotourism development in South-Western Crimea *Vest. Moskovskogo univer. Ser. 5 Geograf.* **5** 62–6

[12] Los M 2017 Assessment of the Relief of the Tyumen and Tobolsk *Tourist and Recreational Framework for Tourism Development Geograf. Vest.* **4** 161–9 DOI: 10.17072/2079-7877-2017-4-161-169

[13] Sarancha M 2008 Recreational potential of the Udmurt Republic: an integrated estimation on the basis of GIS-technologies *Izv. Vyssh. Ucheb. Zaved. Severo-Kavkazskii reg. Ser. Estestv. Nauki* **4** 122–3

[14] Mu Y, Nepal S and Lai P 2019 Tourism and sacred landscape in Sagarmatha (Mt. Everest) National Park, Nepal *Tourism Geogr.* 21(3) 442–59

[15] Kreisel W and Reeh T 2011 Tourism and landscape in South Tyrol *Central Europ. J. of Geosci.* **3(4)** 410–23

[16] Drazic D, Veselinovic M, Rakonjac L et al 2014 Geographic, landscape and other natural characteristics of belgrade as the basis for development of tourism *Europ. J. of Geogr.* **5(3)** 96–122

[17] Mikulec J and Antoušková M 2011 Landscape and tourism potential in the protected landscape areas *Agricult. Econ.* **57(6)** 272–8

[18] Murphy P 1986 Tourism as an agent for landscape conservation: an assessment *Sci. of the Total Environ.* **55** 387–95

[19] Bibaeva A and Makarov A 2018 Application of Information Systems for Calculations of Indicators of Aesthetic Assessment of Landscapes *Izv. Irkutskogo gosud. univer. Ser. Nauki o Zemle* **24** 17–33 DOI: 10.26516/2073-3402.2018.24.1

[20] Khosseini S, Rubtsov V, Gabdrakhmanov N and Bulatova G 2016 Determination of Priority Zones of the Desert Tourism Development by Using Geoinformational Systems: a Case Study of Iran, Isfahan Province *Izv. Ross. Akad. nauk. Ser. Geograf.* **5** 109–18

[21] Jiang Y, Wang Y, Cheng S and Ye M 2008 The theory research on tourism landscape ecosystem *Shengtai Xuebao Acta Ecol. Sinica* **28(4)** 1786–93
[22] Pignatti S 1993 Impact of tourism on the mountain landscape of central Italy *Landscape and Urban Plann.* **24(1-4)** 49–53
[23] Kurlovich D 2013 Morphometric GIS analysis of the relief of Belarus *Zemlya Belarusi* **4** 42–8
[24] Tabunshchik V and Petlukova E 2019 Density of the Relief Dissection on the Territory of the Crimean Peninsula *Izv. Vyssh. Ucheb. Zaved. Severo-Kavkazskii reg. Ser. Estestv. Nauki.* **1** 95–100
[25] Kurlovich D, Fedorako S and Ylanchik E 2012 Assessment of suitability of landscapes of Volozhin district for tourist activity and development of ecotourism routes *Hieahrafija: probl. vykladannia* **2** 3–14
[26] Pozachenyuk E 2003 *Landscape, Atlas. The Autonomous Republic of Crimea* (Simferopol: TNU) pp 38–9
[27] Florinsky I 2016 *Digital Terrain Analysis in Soil Science and Geology* (Amsterdam: Elsevier/Academic Press) 486 p
[28] Piriyev R 1983 *Metody morfometricheskogo analiza rel'yefa (na primere territorii Azerbaydzhana)* (Baku: Elm) 119 p
[29] Pogorelov A and Dumit J 2009 *Rel'yef basseyna r. Kubani (morfologicheskiy analiz)* (Moscow: GEOS) 220 p
[30] Orekhova G and Novykh L 2017 Natural diversity of the springs of the upper reaches of the Seversky Donets and Vorskla river basins *Nauchn. vedomosti Belgorodskogo gosudarst. univer. Yestestv. Nauki* **40(18(267))** 131–9

IOP Conf. Series: Materials Science and Engineering **905** (2020) 012067 doi:10.1088/1757-899X/905/1/012067

Suggestions for improving the organization of underwater search and rescue operations in hard-to-reach areas

K S Talirovsky[1], A L Shidlovsky[1], A N Lapin[2], M S Khasikhanov[3], M D Malaev[3], A A Dzhankhotov[3], Kh I Khamsurkaev[3] and Kh K Butaeva[3]

St. Petersburg University of State Fire Service of EMERCOM of Russia, 149, Moskovskiy prospect, St. Petersburg, 196105, Russia
[2] Arctic Rescue Training and Scientific Center "Vytegra", Vologda region, Ustye village, Russia
[3] Grozny State Oil Technical University named after academician M.D. Millionschikov, Grozny, Russia

E-mail: talirovskiy@igps.ru, ppspsf@igps.ru, lapa-spas@mail.ru, ipkggntu@mail.ru, musa.malaev@mail.ru, alikhan1201@mail.ru, hhi67@mail.ru, xadizhat.butaeva@mail.ru

Abstract. The inability of vehicles in some cases to deliver lifeguards and equipment to hard-to-reach places has contributed to the development of the idea of using horses in such situations. It would not be difficult for horses to overcome places considered difficult to access, and this work would be done in a much shorter time. In this case, divers would not have to engage in heavy physical labor to carry weights. The developed methods of horse training and rescue training were tested, which confirmed the research of specialists

1. Introduction

As practice shows, very often when conducting diving search and rescue operations, violations of the labor protection requirements for divers are admitted, especially in relation to the recreational regime of diving personnel during diving descents.

In accordance with the Intersectoral rules on labor protection during diving operations (POT R M 030 2007), before diving, the diver must have a good rest [1–4]:

1. Medical support for divers during diving

2. In the pre-dive period, the person providing medical support for the diving descent must:

make sure that the working and resting regime of divers before descent meets the established requirements;

make sure that the diver, who worked at the depth, rests according to the regime provided for by the Rules.

3. After diving, divers should also be provided with rest.

However, in the daily activities of the search and rescue units of the EMERCOM of Russia, these requirements of the Rules are extremely rarely observed. Very often, the Ministry of Emergency Situations automobile transport cannot deliver a diving group directly to the site of search and rescue operations due to impassability and the diving team has to move on foot, carrying heavy diving equipment manually over long distances.

As a result of such trips, divers have to go under water in a state of severe overwork, which significantly affects the safety of diving (Tables 1 and 2).

Table 1. Duration of full rest of divers before descent, depending on the planned depths, duration and severity of ground work

Descend depth [m]	Full rest time [min]	
	Medium-severity ground work (work under ice)	Heavy ground work
up to 12	30–45	45–60
13–40	60	90

Notes: 1. Full rest of divers is considered to be their release from all types of work. During this period, only a medical examination, preparation and operational verification of equipment are allowed. 2. In addition, 1 hour before the start of a full rest, divers should be freed from heavy physical work.

Table 2. Duration of full rest of divers after descent, depending on the depth and severity of ground work

Descend depth [m]	Full rest time [min]	
	Medium-severity ground work (work under ice)	Heavy ground work
up to 12	30–45	45–60
13–40	120	180

Note: After full rest, divers should be freed from heavy physical work by the end of the working day.

After descents and work, the divers also do not rest, but are engaged in the transfer of heavy diving equipment to the transport left in the distance, which can lead to various diving diseases [5–8]. In many cases, even a cross-country vehicle does not help. There is a growing need to search for alternative vehicles that can overcome any obstacles that arise along the route of the search-and-rescue group to the place of work. As a means of optimizing the conduct of search and rescue operations (SAR), one can use equestrian and cynological services organized in recent years in rescue centers of the Russian Ministry of Emergencies.

2. Methods and materials

2.1. Possible tactics for using a mobile equestrian diving group

Option 1. The use of horses without additional technical means of delivery

If the place of the upcoming SAR is located near the SAR unit, the rescuers can get to the place of work on horses without using an automobile. A complete diving group mounts on horses, loads necessary equipment and diving station, and moves out towards the SAR site.

After completing the work, the diving group also returns to the unit ahorse not exhausted by heavy physical activities. But this option is possible only if the SAR site is no more than 25 km away from the unit, because according to regulatory documents, a horse can make only up to 50 km per day.

Option 2. The use of horses with pre-delivery vehicles

If the SAR site is located far from the SAR unit, then first of all horses, divers and diving equipment must be delivered as close as possible to the site using existing roads.

Then, from the closest and most convenient point, the diving group, in full or in part (depending on the number of horses), mounts on the horses, loads the necessary equipment and diving station, and sets off to the SAR site.

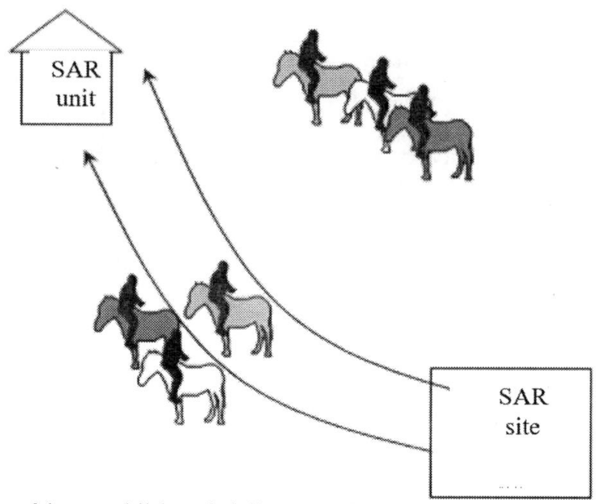

Figure 1. Use of horses without additional delivery vehicles

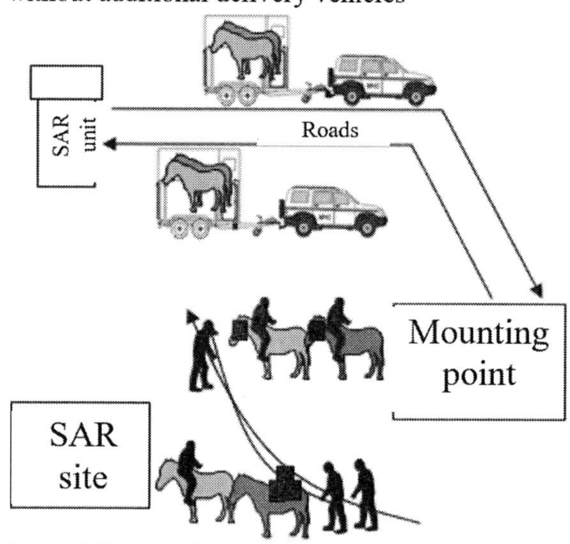

Figure 2. Use of horses with pre-delivery vehicles

In case of insufficient number of horses, the priority for riding is the diver, which will dive first (a working diver). The second horse should be used to transport diving equipment in order to exclude the heavy physical work of divers before descents. The rest of the rescuers (providing and insuring divers) continue walking light on foot without any heavy load in their hands.

2.2. Structure and composition of the mobile equestrian diving group

The structure and composition of the mobile equestrian diving group depend on the number of horses available and on the materials and equipment of the search and rescue group. Depending on these factors, the following options for organizing a mobile group are possible:

Option 1. 3 horses + 3 divers.

Figure 3. Mobile group organization: 3 horses + 3 divers.

Each lifeguard-diver moves ahorse. Diving equipment is put in special pack bags also mounted on horseback.

Option 2. More than 3 horses + 3 divers + supporting rescuers.

Figure 4. Organization of a mobile group: more than 3 horses + 3 divers + supporting rescuers

If it is necessary to attract more people to the SAR, additional horses are involved for their transportation, if available.

If there are not enough horses, supporting rescuers can move on foot, because unlike divers, they do not need to observe a resting time rules.

Figure 5. Mobile group organization: 3 horses + 3 divers + supporting rescuers

Option 3. 2 horses + 3 divers.

Figure 6. Mobile group organization: 2 horses + 3 divers

If only 2 horses are available, then a working diver must be riding, who will be the first to descend into the water. The second highest priority for riding is the standby diver. As for the supporting diver, in any case he moves on foot, because he will have enough time to rest. If he will need to descend, this will happen only after a considerable time.

In the event that a large amount of hardware and heavy equipment is required for the SAR, then two divers (supporting and standby) should go on foot, while one horse is used only for transportation of equipment.

Figure 7. Organization of a mobile group with a large amount of hardware and heavy equipment

It is option 3 that will be used more often than others when the SAR site is very far the SAR unit. The reason is the limited capacity of the horse trailer. The trailer of B-class vehicles can only accommodate two horses.

IOP Conf. Series: Materials Science and Engineering **905** (2020) 012067 doi:10.1088/1757-899X/905/1/012067

Figure 8. Organization of a mobile group when using cars with trailers

However, when using two cars with trailers, it is possible to transport four horses. In this case, various options are possible for the composition of the mobile equestrian diving group. But this is only possible in those SAR teams where there is a sufficient number of trailers and cars.

Materials and equipment of mobile equestrian diving team

For more effective work of the group and its normal functioning in SAR units, in addition to horses, there should be the following supplies:

Packing (baggage) bags for transportation of diving equipment and other hardware necessary for work. In this situation, combined luggage saddles can be used.

Currently, there are many different modifications of packing bags and saddles (Figure 9). All of them are suitable for transporting heavy diving equipment.

Figure 9. Packing (baggage) bags for transportation of diving equipment and other hardware

In accordance with regulatory documents, a horse can carry up to 170 kg. One set of packs for one horse is enough to transport a diving station.

If a lifeguard is riding a horse, then the weight of equipment in the bags should be minimal so as not to overload the horse, especially when riding off-road and in dirt.

If the horse goes riderless and is used only for transporting hardware, then full load is allowed. However, the quality of the road should be taken into account.

The diving station consisting of three sets of diving equipment weighs about 140 kg, which is quite manageable for one horse, provided that the cargo is compact in the packing bags (Figure 10).

Figure 10. Horse transporting diving station

Special horse trailers.

Currently, there are many different trailers for transporting horses (Figure 11). As a rule, they are designed for towing by a car and can accommodate two horses.

Figure 11. Horse trailers

Thanks to such light trailers, horses can be transported over considerable distances to a SAR site to provide the work of a diving group.

3. Results

The first experiment to create a mobile equestrian diving group was carried out in the Vytegra Training and Rescue Center of the ENERCOM of Russia in August 2013.

Practice has shown that for the full and well-coordinated work of lifeguard divers with horses, it is necessary to carry out preliminary educational work on the preparation of divers and horses.

When practicing various exercises with divers and horses, the following problems were identified:

- horses are frightened of strangers and work only with familiar employees of the Center. The reason is the lack of preparedness of horses;
- horses are frightened by rescuers dressed in diving or other rescue equipment. The reason is the lack of preparedness of horses that are not accustomed to an unusual type of equipment;
- horses are afraid of water and do not want to swim. The reason is the lack of preparedness of horses that must be trained in water;
- horses allow to ride them only within the stable and arena; in other places it is very difficult to ride on them. The reason is the lack of preparedness of horses;
- the horses are scared of the noise of working equipment. The reason is the lack of preparedness of horses;
- rescuers do not know how to ride. The reason is the lack of preparedness of the rescuers;
- mounting on a horse in full diving equipment is difficult. The reason is the insufficient preparedness of rescue divers, however, after training this is not difficult.

In this regard, it is possible to determine the range of tasks that need to be addressed during the preparation of the mobile equestrian-diving group.

The main exercises for the preparation of a mobile equestrian-diving group:

1. Training horses to work with strangers (Figure 6). It is necessary to train horses' contact with different rescuers as often as possible.
2. Getting the horses used to various equipment, especially to people dressed in such unusual equipment. Horses should not be afraid of lifeguards in diving equipment.
3. Training horses for water. Training horses swimming with lifeguards. The horse must calmly overcome water, ground and relief obstacles.
4. Training rescuers to work with horses. Training rescuers to interact with the horse.

Such preparation is a whole range of activities:

a) horse care;

Before starting work with a horse, rescuers must learn to care for it. They get acquainted with animals, master the methods of handling horses and caring for them. At the end of the training, the rescuer must be able to quickly clean and prepare the horse for a saddle.

b) horse riding equipment;

Rescue divers familiarize themselves with horse equipment and the rules for its use. At the end of the training, the rescuer must be able to quickly and correctly saddle the horse. Including fastening the packing bags on it.

c) mounting on a horse and dismounting, the basics of proper riding;

Initial training is carried out on a wooden model of a horse. Divers learn to sit properly in the saddle and dismount from the horse. Also, the correct sitting in the saddle and the basics of managing the horse are being worked out.

The peculiarity is that the diver performs these actions fully equipped.

d) the basics of riding, principles and techniques of horse control;

Riding in the open riding arena, a specially prepared, enclosed rectangular area with soft ground, is used to teach how to ride in the saddle and how to correctly control a horse. After successful training in the arena, horseback riding is also possible in the field, where riders can learn to overcome slopes, climbs, force shallow rivers and other simple natural obstacles.

If in the search and rescue group there are inexperienced riders, then the movement at the rear is possible when the experienced rider rides first and leads the column. The remaining rescuers, inexperienced riders, follow him in the column. Horses usually follow the front horse. In this way, inexperienced lifeguards can be transported on horseback.

e) training of rescuers to work with horses in diving equipment.

Mounting on a horse in full diving equipment is very difficult without help. But for training, a lightweight and compact diving balloon with a capacity of 6-7 liters is used. Such a mounting is advisable if the diver intends to ride on a horse into the water and dive directly from the horse's back.

The following exercises serve to train such actions.

f) working out entry into the water on horseback;

d) working out the diver's entry into the water in equipment from the horse's back.

In the course of practicing all of the above exercises, such elements as practicing mounting on a horse in various places, and not just in the riding arena, training horses to work in noisy conditions near working equipment, are worked out in parallel.

During the experiment in the Vytegra Training and Rescue Center of EMERCOM of Russia, all exercises and training for divers were conducted by Nadezhda Nikolaeva, master of physical education, trainer of horses and riders from St. Petersburg. She is one of the leading experts in this field in the Russian Federation.

She also identified the principles for selecting horses that are more suitable for these purposes:

The choice of horses suitable for search and rescue is determined by the following factors:

- work in hard-to-reach areas with changing terrain;
- climatic conditions;
- strangers, sounds;
- exercise stress;
- emotional stress.

The horse used for search and rescue must be friendly, oriented to the person, get along well with other animals. The predominant type of temperament is sanguine. Native breeds, such as the Mezen breed, were recognized as the most suitable breeds.

4. Conclusion

As a result of the research with the aim of using horses for diving search and rescue operations, the following main conclusions were drawn:

- the ability of the equestrian diving group to reach inaccessible SAR sites, due to the higher cross-country ability of horses compared vehicles;
- the ability to exclude violations of the Intersectoral rules on labor protection when conducting diving operations regarding the working regime and rest time rules for divers;
- the ability to move hardware and equipment over rough terrain;
- the ability to transport horses in special vehicles with divers over long distances in the presence of roads;
- increased efficiency of diving operations, the ability of a diver to enter water directly from the horse.

References

[1] *Metodicheskie rekomendacii (vremennye) po primeneniyu konnoj sluzhby MChS Rossii (konnyh spa-satel'nyh podrazdelenij)* 2014 (Moscow) 11 p

[2] Shidlovskij A L, Kulakov I Yu and Komarov M I 2015 *Vodolaznaya podgotovka* (St. Petersburg: SP-b univer. GPS MChS Ross.)

[3] Borovikov P A 2005 *Vodolaznoe delo Rossii* (Moscow: Mysl') 240 p

[4] *Mezhotraslevye pravila po ohrane truda pri provedenii vodolaznyh rabot* Zaregistrirovany v Minyuste RF 23 iyulya 2007 no 9888

[5] Smolin V V, Sokolov G M and Pavlov B N 2003 *Vodolaznye spuski do 60 metrov i ih medicinskoe obes-pechenie* (Moscow: Slovo) 696 p

[6] Slesarev O M and Rybnikov A V 1996 *Vodolaznoe delo* Spravochnik (Moscow: SPREK) 320 p

[7] *Vodnyj kodeks Rossijskoj Federacii* 16 November 1995 no 167-FZ (v redakcii ot 30.06.2003)

Preliminary assessment of the earth's surface movement in the impact zone of a subway tunnel construction

G V Alekseev, E A Medvedev and A A Tereshin

Institute of Scientific and Technological Construction Support, MGSU, 26, Yaroslavskoye Shosse, Moscow, 129337,Russia

E-mail: alexeev_german@bk.ru, apb_2004@bk.ru, 280100@mail.ru

Abstract. Numerical modeling of the displacement of the earth's surface during the construction of deep tunnels is considered on the example of one of the Moscow metro lines. Based on the results of numerical simulation of the driving of one and two tunnels using several calculation schemes, the forecast parameters of the displacement trough, such as subsidence, slopes, curvature, horizontal deformations of the earth's surface, are determined. According to the criteria of deformations of the earth's surface, a tunnel influence zone is constructed. In the influence zone of tunnels, it is necessary to create observation stations for surveying (geodetic) monitoring of subsidence of buildings and the earth's surface, including using automated monitoring systems. According to monitoring data, the parameters of the calculation models are adjusted to increase the reliability of the simulation results. A technique is proposed for a preliminary assessment of the displacement of the earth's surface in the influence zone of the construction of subways tunnels.

1. Introduction

Deep underground tunnels are constructed by an enclosed (mining) method without opening the land surface, at a depth of not less than 15 m. Let us consider a preliminary assessment of the development of the process of shifting the soil mass and the earth's surface during the construction of deep underground tunnels using an example of one of the Moscow metro lines.

To assess the displacement of the soil mass and the earth's surface after driving one and two subway tunnels, numerical modeling by the finite element method in the Plaxis 2d program was performed. As the experience of using Plaxis shows, numerical modeling is very sensitive to the quality of the source data: minor changes in the physicomechanical properties of soils, tunnel attachment parameters, or groundwater level can quite strongly, sometimes multi-fold change the calculation results.

For example, in [1], the influence of soil type on the vertical displacements (subsidence) of the earth's surface was estimated when modeling in Plaxis for the U-8-Nord underground line under construction in Munich (Germany). The tunneling was simulated in various types of soils, the results of modeling by the criterion of the parameters of the tear plate are compared with full-scale surveying (geodetic) observations. Based on the calculated and actual data on monitoring the movement of the earth's surface, the soil parameters optimal for the calculation model were determined.

In [2], a geomechanical analysis of subsidence of the earth's surface during the construction of a deep underground station in St. Petersburg (Russia) was performed. To assess the reliability of the calculation model and the results of numerical modeling, we also used the method of comparing the calculated trough of displacement and the data of geodetic monitoring during the construction of pylon stations in similar engineering and geological conditions.

In [3], a study was made of the influence of the depth of the tunnel on the stress-strain state of the massif and the displacement of the earth's surface when driving a double-track haul subway tunnel in St. Petersburg. It is shown that for a soil mass composed of four layers of soil with different physical and mechanical properties, an increase in the depth of the tunnel from 20 m to 40 m causes an increase in subsidence of the earth's surface by about 1.2 times. In addition, the zone of influence of the tunnel is determined from the simulation results.

In [4], the monitoring results of the earth's surface movement during the construction of a number of metro stations in St. Petersburg were analyzed. According to field observations of the subsidence of the earth's surface, the slopes and curvature in the trough of displacement were additionally calculated. Typical strain distribution curves were obtained in the displacement trough for mining and geological conditions for the construction of escalator tunnels in St. Petersburg. The data obtained were used to predict displacements and deformations of the earth's surface during the construction of tunnels in similar conditions.

In [5], when modeling changes in the stress-strain state (SSS) of a soil massif, it was shown that the initial data on the physicomechanical properties of soils, in particular, the soil deformation modulus, significantly influence the modeling results.

Thus, one of the necessary conditions for obtaining reliable results of a preliminary assessment of the process of displacement of the earth's surface and soil mass during modeling in Plaxis are:

- the adequate set of source data, including the physicomechanical properties of the soil mass,
- the use in assessing not only the criterion of the earth's surface subsidence, but also others, such as slopes, curvature, horizontal strains of compression-tension of the earth's surface,
- comparison of simulation results with field monitoring data (geodetic, geotechnical) and adjustment, if necessary, of the initial data of the calculation models.

2. Research Methods

To draw up calculation schemes for numerical modeling, a conventional engineering-geological section has been used, on which three types of soils, called *"Sand"*, *"Clay"*, *"Limestone"*, are represented.

Table 1 shows the physicomechanical properties of these types of soils adopted in the model, which are defined as average according to [6], as well as according to the data given in the publication [7]:

Table 1. Physico-mechanical properties of the soil calculation model

Parameter / Designation / Unit			*"Sand"*	*"Clay"*	*"Limestone"*
Specific gravity	γ	t/m^3	1.7	2.0	2.6
Deformation modulus	E	MPa	22	35	150
Poisson's ratio	ν		0.3	0.35	0.25
Adhesion	c	kPa	2	30	3000
Angle of internal friction	φ	°	35	20	33

For numerical simulation, three design schemes have been compiled, differing in different depths of the tunnels:

- Scheme 1: the laying depth is 15 m, the tunnels lie in *"Sand"*;
- Scheme 2: laying depth is 25 m, the axis of the tunnel passes through the contact *"Sand"*–*"Clay"*;
- Scheme 3: laying depth is 35 m, the tunnels lie in *"Limestone"*,
- distance between the axes of the tunnels in the plan: 13 m;
- diameter of the tunnel: 6.5 m;

- groundwater table: 10 m.

The dimensions of the calculation model are taken as 100 m along the horizontal axis X and 50 m along the vertical axis Y. The boundary conditions in the model are set as follows: horizontal movements are forbidden on the lateral boundaries, horizontal and vertical movements are forbidden on the lower boundary.

Figure 1 shows the calculated simulation scheme with a tunnel laying depth of 15 m:

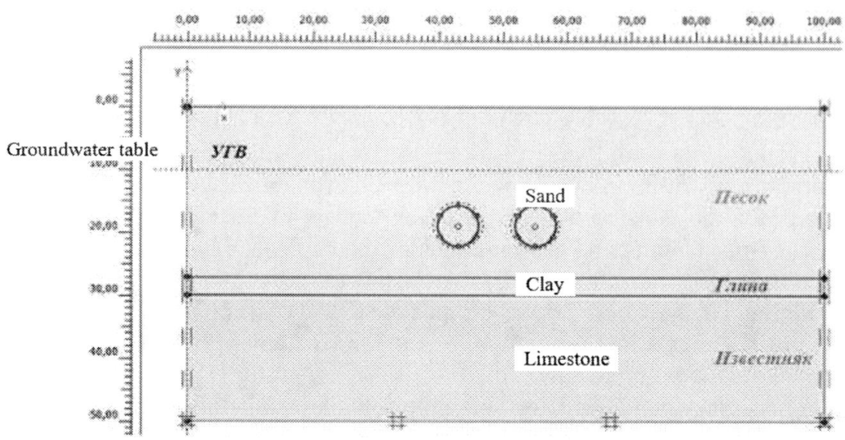

Figure 1. Settling scheme No. 1 (tunneling depth 15 m. Vertical axis (X) is the depth from the surface, m; horizontal axis (Y) are the distances in the plane perpendicular to the axis of the tunnels, m; "groundwater table" is the groundwater level; "Sand", "Clay", "Limestone" are conventional soil types accepted in the calculation model.

Modeling includes three stages:

Stage 1 – calculation of the initial (initial) stress-strain state (VAT) of the soil mass;

Stage 2 – change in the initial VAT of the soil mass after boring the 1st tunnel;

Stage 3 – changing the VAT of the soil mass after boring the 2nd tunnel.

The comparison of the simulation results of tunneling in various design schemes is presented in Table 2.

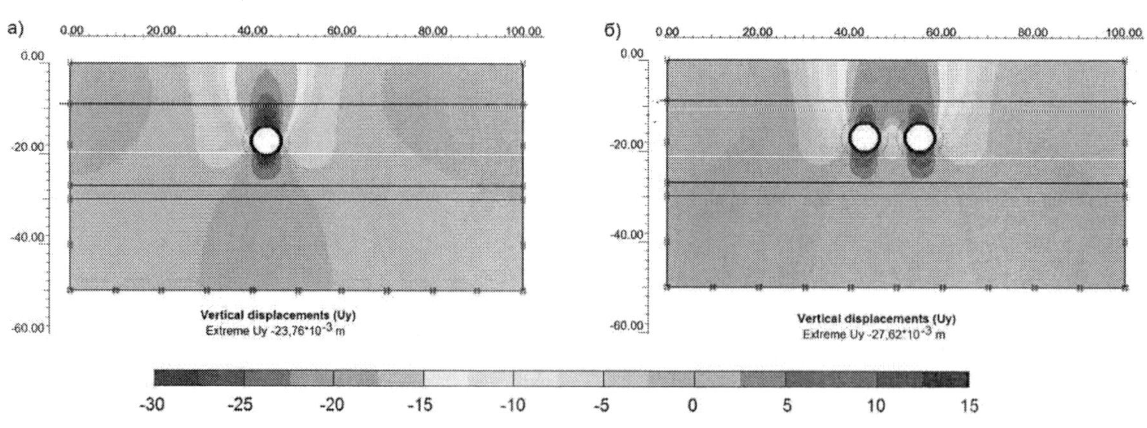

Figure 2. Estimated vertical displacements of the soil massif (mm) after boring the first (a) and second (b) tunnels

Table 2. Comparison of the results of modeling tunneling in various design schemes according to the criteria of vertical and horizontal displacements of the soil mass

Calculation scheme	Estimated vertical and horizontal displacements of the massif of soil [mm]			
	after boring 1st tunnel	after boring 2nd tunnel	displacement increment [mm]	displacement increment [times]
Scheme No. 1 (tunnels lie in "Sand", depth is 15 m)				
vertical displacements				
on surface	14	21	7	1.5
in whole massif	24	28	4	1.2
horizontal displacements				
on surface	4	7	3	1.8
in whole massif	8	8	-	-
Scheme No. 2 (axis of tunnel passes through contact of "Sand" and "Clay", depth is 25 m)				
vertical displacements				
on surface	12	25	13	2.1
in whole massif	41	50	9	1.2
horizontal displacements				
on surface	4	9	5	2.3
in whole massif	14	19	5	1.4
Scheme No. 3 (tunnels lie in "Limestone", depth is 35 m)				
vertical displacements				
on surface	4	9	5	2.3
in whole massif	18	22	4	1.2
horizontal displacements				
on surface	2	3	1	1.5
in whole massif	7	8	1	1.1

Based on the simulation results, we can conclude that the tunneling of the second tunnel causes an increase in vertical displacements of the earth's surface by an average of 1.5–2.3 times. Figure 2 shows the distribution of the estimated vertical displacements of the earth's surface after the first and second tunnels are drilled using the calculation scheme No. 1 as an example.

3. Brief analysis of the results of numerical simulation

The danger to buildings and structures on the earth's surface is not vertical movements (subsidence) as such, but their unevenness in the trough of displacement, which leads to the appearance of such types of deformations of the earth's surface as slopes, curvature, horizontal tensile/compression deformations.

In accordance with [8],

1. the boundary of the zone of influence of the underground structure or the boundary of the trough of displacement is the contour of the zone on the earth's surface, determined by the deformations:

- inclinations $i=0.5 \times 10^{-3}$,
- horizontal tensile strains $\varepsilon=0.5 \times 10^{-3}$, with interval lengths of 15–20 m.

2. the boundary of the zone of the dangerous influence of the underground structure is the contour of the zone on the earth's surface, determined by deformations:

- inclinations $i=4 \times 10^{-3}$,
- curvature $k=0.2 \times 10^{-3}$
- horizontal tensile strains $\varepsilon=2 \times 10^{-3}$, with interval lengths of 15–20 m.

Let us analyze the calculated zone of the earth surface displacement trough for each of the calculation schemes according to the simulation results.

According to the results of numerical modeling, the values of the deformations, inclinations and curvature of the earth's surface in the trough of displacement from the penetration of two tunnels with an interval of 2.5 m along the X axis are calculated, and the calculated zone of influence of the tunnels is determined. Figure 3 shows the calculated values of subsidence (a), slopes (b), curvature (c) and horizontal deformations (d) of the earth's surface according to the simulation results (the influence zone of the tunnels is shown by the horizontal dashed line, the dashed-dotted vertical line shows the position of the axes of the left and right tunnels.

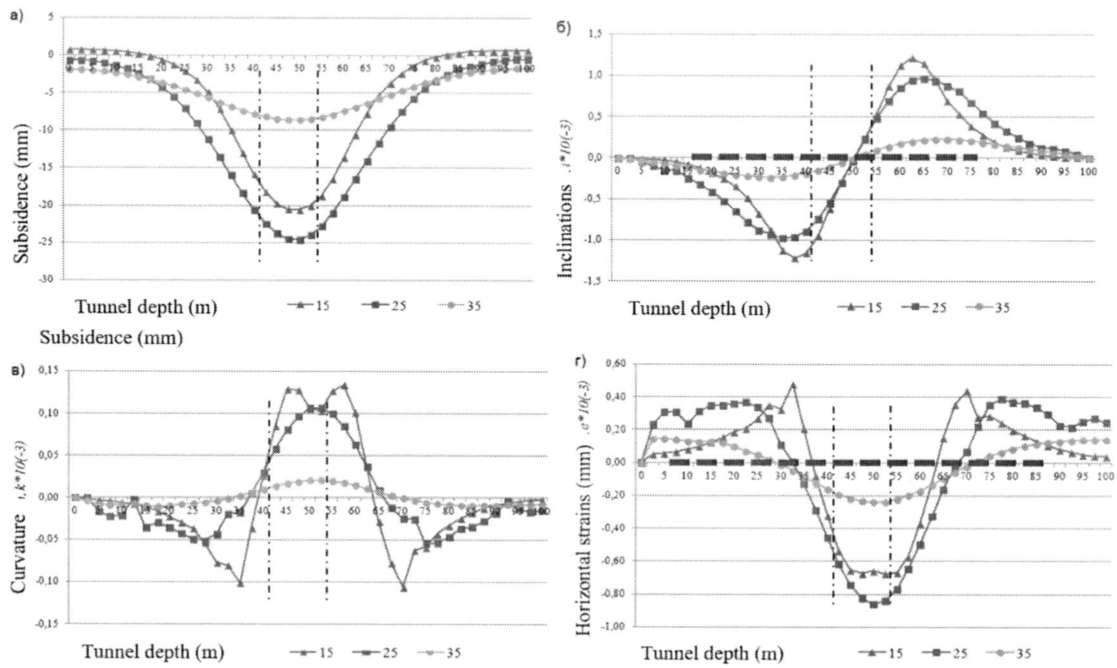

Figure 3. Calculated values of the subsidence (a), slopes (b), curvature (c) and horizontal strains (d) of the earth's surface in the displacement trough from the penetration of two tunnels according to the simulation results

In [9], according to the results of modeling a large number of geomechanical situations with various technological parameters and properties of the soil mass, regression expressions were compiled to calculate the maximum subsidence of the earth's surface η_{max} along the center line of the trough of displacement above the axis of the tunnel:

$$\eta_{max} = \frac{2 \cdot u}{1 + 0{,}42\left(\frac{H}{D}\right)^{1{,}2}} \cdot (c \cdot E)^{-0{,}08}, \quad m$$

where u is the construction gap in the arch of the tunnel lining, H is the depth of the tunnel, D is the diameter of the tunnel, c and E are the adhesion and soil deformation modulus.

In [8], the maximum subsidence of the earth's surface is recommended to be determined by the formula:

$$\eta_{max} = q_0 \cdot u\sqrt{n_1 \cdot n_2}$$

where q_0 is the coefficient that takes into account the nature of the attenuation of displacements from the excavation to the earth's surface, fluctuates for the conditions of the Moscow region in the range from 0.7 to 0.9 (the stronger the rocks, the less q_0); u is the subsidence value of the roof; n_1 and n_2 are underworking coefficients, determined from the expressions:

$$n_1 = 0,9\sqrt{D_1/H}$$

$$n_2 = 0,9\sqrt{D_2/H}$$

where $D1$ and $D2$ are the transverse and longitudinal dimensions of the underground site; H is the depth of the site. For values $n1$ and $n2> 1$, $n1$ and $n2 = 1$ are accepted. Since in tunnels the longitudinal dimension $D2$ is always very large, the coefficient $n2 = 1$.

Figure 4 shows the curves of the maximum possible settlement of the earth's surface depending on the depth of the tunnels for the parameters of the calculation model adopted in this paper, constructed according to the methods of [8] and [9]:

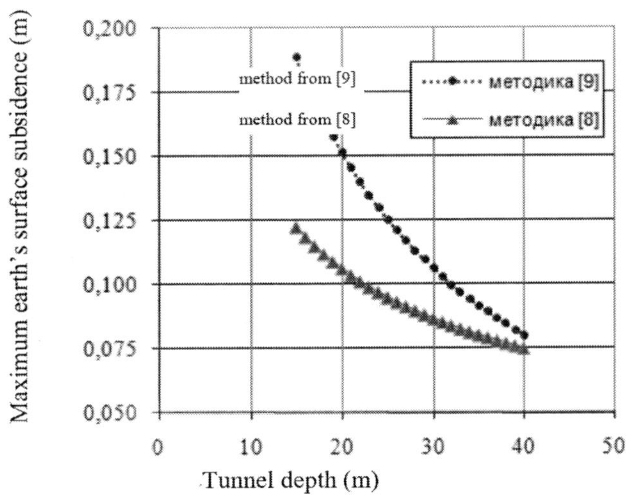

Figure 4. Calculated curves of the maximum possible subsidence of the earth's surface depending on the depth of the tunnel by the methods [8] and [9]

According to the simulation results, the maximum subsidence of the earth's surface during the tunneling of two tunnels was 25 mm with a tunneling depth of 25 m, which does not exceed the maximum possible calculated values by the methods [8] and [9].

According to surveying observations of precipitation of buildings and structures in the zone of influence of the construction of metro tunnels similar in parameters, the maximum values of subsidence of deformation marks on buildings are in the range of 5–25 mm, which is in good agreement with the modeling results in Plaxis.

According to the simulation results, according to the criterion of slopes and horizontal deformations, the maximum size of the displacement trough (zones of influence of tunnels) on the earth's surface is determined to be about 80 m.

According to the calculated value of the zone of influence of tunnels, the boundaries of the zone of influence and the boundaries of the zone of dangerous influence, if any, are plotted on the general plan of the surface. For buildings and structures on the earth's surface falling within the zone of influence of tunnels, it is necessary to conduct surveying (geodesic) monitoring to monitor precipitation and banks. According to the monitoring data, the correspondence of the simulation results to the actually observable is determined, and adjustments are made to the calculated modeling schemes.

As a rule, based on the results of direct measurements of the displacements of deformation grades on buildings and structures, it is possible to evaluate the displacement of the soil mass only at a certain stage, when most of the deformations of the soil mass are already implemented. For more operational control of the movement processes, along with direct surveying (geodetic) measurements, automated monitoring systems can be used.

Indeed, in [10], on the example of the construction of one of the metro lines in St. Petersburg, the use of an automated monitoring system is considered, which allows controlling the movement of the soil mass throughout the depth from the lining of the tunnel to the earth's surface at the initial stages, directly in the process tunnel penetration. In [11], an automated geotechnical monitoring system was described during the construction of the 1st stage of the IZMIR LRTS (Light Rapid Transit System) in Izmir (Turkey). An integrated system of geotechnical monitoring includes monitoring the movement of the earth's surface and the deformations of the structures of the tunnel itself.

In [12], emergency situations were considered during the construction of subways. An example of an emergency analysis caused by the removal of flooded sands into a tunnel and resulting in the deformation of a precast concrete waterproof lining of a tunnel shows the need for continuous automated monitoring of tunnel deformations.

4. Conclusion

In the event that the amount of data on the engineering and geological conditions of the construction site, physical and mechanical properties of soils, etc. is insufficient to build the correct calculation schemes for numerical modeling, or the available data need to be clarified when comparing the simulation results with field data, a preliminary assessment of the movement of the earth's surface in the zone of influence of the construction of subway tunnels can be performed according to the following method:

1) Engineering-geological sections are selected along the tunnel route, along which numerical modeling of changes in the stress-strain state of the soil mass after tunneling will be performed;

2) If there is a sufficient amount of data on the physicomechanical properties of soils from these sections, they are accepted as the source for numerical modeling;

3) In the absence of such data, or in case of insufficient volume, the physicomechanical properties of soils are accepted in the range "from ... to ..." according to the normative documentation for this type of soil (sands, clays, limestones, etc.), as well according to data for similar tunnel construction conditions;

4) A numerical simulation is performed, according to the results of which the calculated values of vertical and horizontal displacements of the soil mass, horizontal deformations of the earth's surface, slopes, curvature are determined;

5) Based on the simulation results, zones of influence and zones of hazardous effects of tunnel construction are determined;

6) The zones of influence are applied to the general plan of the surface, the buildings and structures falling into the zone of influence are determined;

7) For buildings and structures falling within the zone of influence of the construction of tunnels, special observation stations are created for surveying (geodetic) monitoring of deformations, including automated monitoring, including observations of subsidence, banks and horizontal deformations of buildings and structures;

8) The results of direct monitoring are correlated with the calculated data, and if necessary, the calculated models are adjusted to fully comply with the results of field observations.

References

[1] Strokova L A 2008 Modeling of surface subsidence during tunnel tunneling using the shield method *Bull. of the Tomsk Polytech. Univer.* **312(1)** 45–50

[2] Demenkov P A 2013 Geomechanical evaluation of the phased construction of a deep underground metro pylon station *Notes of the Mining Instit.* **204** 220–2

[3] Protosenya A G and Suprun I K 2012 Forecast of the stress-strain state of the double-track haul tunnel of the Frunze radius of the St. Petersburg metro *Notes of the Mining Instit.* **197** 148–50

[4] Novozhenin S U and Vystrchil M G 2016 New Method of Surface Settlement Prediction for Saint-Petersburg Metro Escalator Tunnels Excavated by EPB TBM *Proc. Engineer.* **150** 2266–71
[5] Zertsalov M G, Kazachenko S A and Konyukhov D S 2014 Study of the impact of excavation on the surrounding area *Vest. MGSU* **6** 77–86
[6] SP 22.13330.2011 *Foundations of buildings and structures* 2011 Ministry of Regional Development of Russia
[7] Osipov V I and Medvedev O P (ed) 1997 *Moscow. Geology and the city* (Moscow: Moscow Textbooks and Cartolithogr.)
[8] *Instructions for observing the movements of the earth's surface and objects located on it during the construction of underground structures in Moscow* (RD 07-166-97) 1997 (Moscow: Gosgortekhnadzor; STC "Industrial Safety")
[9] Skvortsov A A 2012 *Development of a method for predicting the processes of deformation in undermined massifs during the construction of collector tunnels* (Moscow: Moscow State Mining Univer.)
[10] Lebedev M 2016 Automated Systems as a Part of Geotechnical Monitoring in Construction and Operation of Transport Tunnels *Proc. Engineer.* **165** 448–54
[11] Unlutepe A and Ozener H 2008 Geotechnical measurements at Izmir LRT system tunnels *Tunnell. and Underground Space Technol.* **23(6)** 734–41
[12] Shvetsov V A, Merkin V E, Piskunov A A et al 2014 Abnormal situations in the construction of metro facilities. Causes and liquidation of consequences *Bull. of Eurasian Sci.* **5(24)** 1–18

IOP Conf. Series: Materials Science and Engineering **905** (2020) 012069 doi:10.1088/1757-899X/905/1/012069

Neural network application for predictive modeling

K E Tokarev[1,2], Yu A Orlova[2], A F Rogachev[1], A Yu Rudenko[1] and M P Protsyuk[1]

[1] Volgograd State Agricultural University, 26, University ave,Volgograd, 400002, Russia
[2] Volgograd State Technical University, 28, Lenin ave., Volgograd, 400005, Russia

E-mail: tke.vgsha@mail.ru, yulia.orlova@gmail.com, rafr@mail.ru, rudenkovgsxa@mail.ru, maria_protsuk@mail.ru

Abstract. The article analyzes the approaches to the construction of predictive models based on the apparatus of artificial neural networks, in particular, the method of back propagation of errors by iterative adjustment of weight coefficients.

1. Introduction

Artificial neural networks (ANN) are representatives of the cybernetic direction in science, based on the belief that natural systems in the form in which we see them today, are a community of winners in the centuries-old competition for survival.

Human, as the creator of artificial worlds, only need to understand the key mechanisms of their sustainable functioning and self-development in order to use this understanding in their own practice.

Artificial neural networks are computer systems that simulate the work of biological neurons with the possibility of parallel processing of information, the ability to learn and summarize the accumulated knowledge.

The evolution of ANN has led to the fact that currently some of them to some extent really simulate biological neural networks, while some have forgotten about the nature of its creation. Artificial neural networks were originally conceived as some abstractions, mathematical models that mimic the activity of the human brain. Distinguish the following properties harmoniously connecting with each other biological and artificial NS: the ability to learn; the ability to find solutions based on noisy, distorted and even contradictory data; fault tolerance of hardware-implemented neural networks).

In this regard, the development of methods for constructing neural network models of productivity, substantiation of their architecture and technological parameters is an important theoretical and applied task, the solution of which allows to increase the efficiency of planning and management of agricultural production by making more informed management decisions based on the use of tools for multivariate analysis of alternatives.

2. Methodology

Artificial neural networks are biological structures consisting of interconnected simple elements – formal neurons. The core of the concepts used is the idea that neurons can be modeled with fairly simple automata, and all the complexity of the brain, the flexibility of its functioning and other important qualities are determined by the connections between neurons. Each link is represented as a very simple element that serves to transmit the signal. It is assumed that the system of connections is rich enough in its capabilities and redundant enough to compensate for the poverty of the choice of elements, their unreliability, the possible destruction of the links[1, 2, 7].

The most well-deserved and probably the most important element of neurosystems is an adaptive adder. The adaptive adder computes the scalar product of the input vector x by the parameter vector. We call it adaptive because of the presence of a vector of adjustable parameters. For many problems it is useful to have a linear inhomogeneous function of the output signals. Its calculation can also be represented by an adaptive adder having n+1 input and receiving a constant single signal at the 0-th input (Figure 1).

The nonlinear signal converter is shown in Figure 2. He receives a scalar input x and transforms it into φ(x). The most commonly used sigmoid function:

An interesting property has a derivative of the sigmoid-it exists on the entire real axis and φ'=φ*(1-φ).

$$\varphi(x) = \frac{1}{1+e^{-x}}$$

An interesting property has a derivative of the sigmoid-it exists on the entire real axis and φ'=φ*(1-φ).

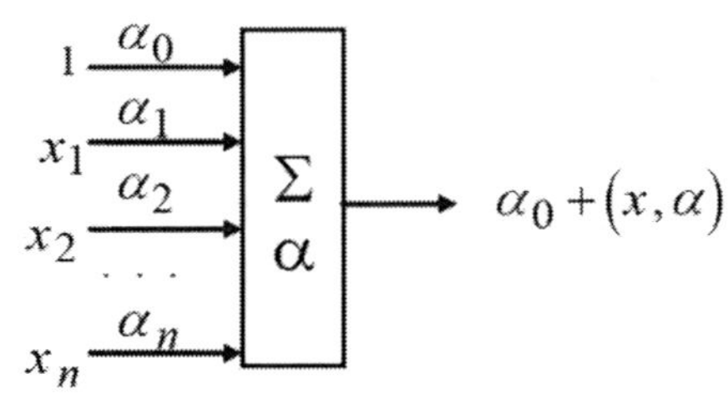

Figure 1. Inhomogeneous adaptive adder

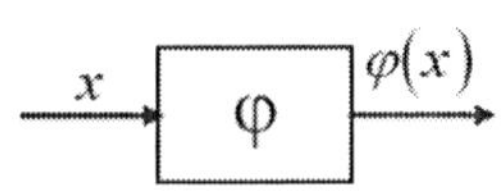

Figure 2. Nonlinear signal converter

The branch point is used to send one signal to several addresses (Figure 3). It receives the scalar input x and transmits it to all its outputs.

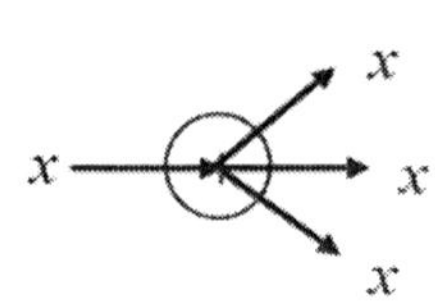

Figure 3. Branch point

The branch point is used to send one signal to several addresses (Figure 3). It receives the scalar input x and transmits it to all its outputs. The standard formal neuron is composed of an input adder, a nonlinear transducer, and a branch point at the output (Figure 4).

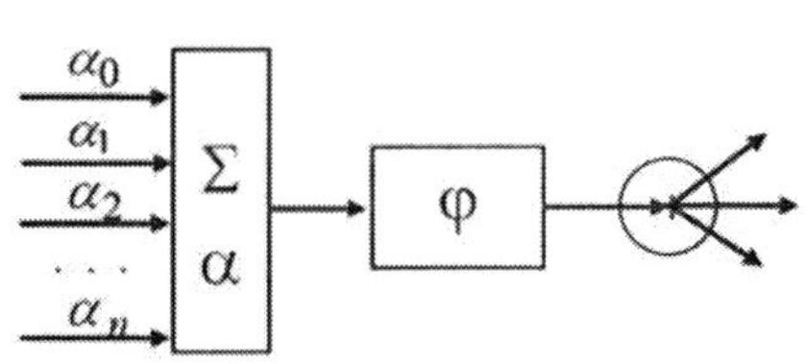

Figure 4. Formal neuron

Among NS, the most common and most studied class is layered networks (as it is otherwise called, direct distribution networks).

In such networks neurons are located in several layers (Figure 5). The neurons of the first layer receive input signals, convert them and transmit them through the branching points to the neurons of the second layer. Next, the second layer is triggered, etc.to the k-th layer, which outputs signals to the interpreter and the user. If the opposite is not specified, then each output signal of the i-th layer is fed to the input of all neurons i+1. The number of neurons in each layer can be any and has nothing to do with the number of neurons in other layers. Standard input method: all neurons in the first layer receive each input signal. The first layer is called input, the last-output, the rest-hidden.

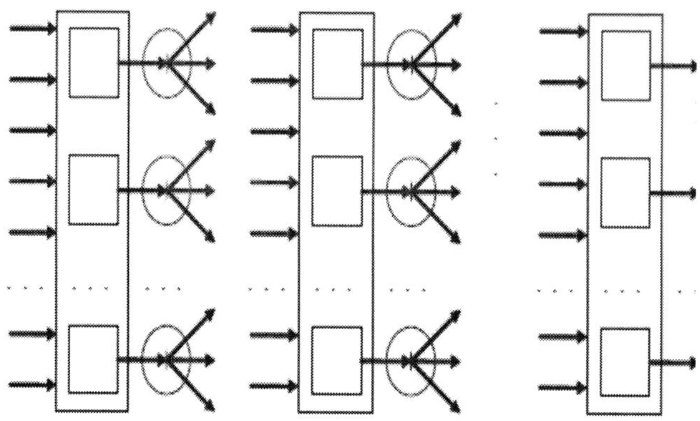

Figure 5. Artificial neural networks with layers

Note that in the literature in different ways can be called the layered network. Most often, the first (input) layer is not taken into account when naming NS of direct distribution, because its neurons are simply network inputs, but they are not fully functional neurons. Thus, a two-layer network consists of input, hidden and output layers. It can also be called a direct distribution network with one hidden layer.

The activation function of neurons (characteristic, transfer function) – a nonlinear converter that converts the output signal of the adder (Figure 4) – can be the same for all neurons in the network.

In this case, the network is called homogeneous (homogeneous). Drawing up a network of neurons of the standard form is not mandatory.

Learning NS is the process of finding the synaptic weights of all neurons that are optimal for some functional quality, called learning error. It describes an integrated measure of the closeness of the outputs of the network y(M)(k) and the instructions of the teacher:y*(k), k=1,...,K – K training examples:

$$J(w) = \sum_{k=1}^{K} Q(\varepsilon(w,k))$$

where Q(ε (w,k)) is called the instantaneous quality criterion.

Here W=(w(M)t,w(M-1)t,...,w(1))t is the vector of weights of all networks, ε(w,k)=y(M) (k) – y*(k). Often taken Q(ε(w,k))=ε(w,k)trε (w,k), R – a positive definite matrix.

D.E. Rumelhart [3, 4] developed an algorithm for iterative adjustment of the weight coefficients of the NS, called the method of back propagation of the error. This algorithm is not without its drawbacks, but its role is very large for training NS direct distribution.

Let the network consist of M layers with nμ neurons in the μ-th layer (μ=1,...,M). A pair (μ, i) will denote the I-th neuron of the μ-th layer. Inhomogeneous adaptive summation of this neuron:

$$net^{(\mu,i)} = w_0^{(\mu,i)} + \sum_{j=1}^{N_{\mu-1}} w_j^{(\mu,i)} y_j^{(\mu-1,i)} = w^{T(\mu,i)} u^{(\mu,i)},$$

w(μ,i)=(w0(μ,i),w1(μ,i),...,wn(μ,i)), μ=1,...,M, be the vector of weight coefficients (n=Nμ-1 – number of neurons in the previous layer), y(μ-1,i)=(1,y1(μ-1,i),...,yn(μ-1,i)), μ=1,...,M, – advanced input vector of the neuron μ-th layer (output neuron of the previous layer), here y(0,i)=x(i) is inputs of network. The output of the neuron is written as: y(μ,i)=F(net,I), I=1,...,Nμ; μ=1,...,m.

Let us specify a set of K training pairs {x(k),y*(k)}, k=1,...,K. the Training (finding the weight coefficients) as follows.

1. Given η>0 (usually taken as 0.2) is the learning constant (step length), and Emax>0 is the maximum allowable learning error (optimization accuracy).
2. The weights of the network are initialized by small random numbers (if equal weights are set, and different weights are required for the proper functioning of the network, the network will not be able to learn). E:=0 – accumulated error, k:=1 – training pair number.
3. The input vector x(k); x:=x(k), u – extended input vector is supplied to the network input, the vector y:=y(M) – output of this network (the last layer) is calculated at a given input.
4. The error ε=y-y*=s(M) of the network operation is calculated for this example.
5. The weights of the last layer are corrected: w(M):=w(M)- ηεF'(net(M))u (M).
6. μ:=M-1 is the number of the last hidden layer.
7. i:=1 – the number of the first neuron in the layer.
8. w(μ,i):=w(μ,i)-NS(μ,i)F'(net(μ,i) u(μ,i)T, where s(μ,i)=s(μ+1) Td(μ+1,i), and d(μ+1,i)=F'(net(μ+1)) w1 (μ+1) – vector.
9. If i<Nμ, then i:=i+1 and return to step 8.
10. If μ>1, μ:=μ-1 and return to step 7.
11. The generalized error of the network functioning is calculated: E:=E+0.5 * ε^2.
12. If k<K, then k:=k+1 and return to step 4.
13. If E<Emax, then the training is over, otherwise E:=0, k:=1 and return to step 4.

This algorithm is called the standard error back propagation algorithm. It should be noted that adjusting the weights is a minimization of the error function. The standard error back propagation method is actually a gradient method for finding the optimum of the error functional. The weights change according to the rule: w(t)=w(t-1)-nwq(ε(w (t-1),k)). Here wQ(ε(w (t-1),k)) is the gradient of the instantaneous functional – the descent direction, and η is the step length in the descent direction. It is worth noting the very slow convergence of this method.

Currently, more efficient methods of adjusting the network weights are used. For this purpose, subject to such well-known optimization techniques to search for the direction of decreasing functions as conjugate direction methods and the well-known quasi-BFGS and DFP methods [5,6].

The use of ANN in forecasting is quite transparent. They can be used in several ways, the most famous of which is similar to the methods of statistics, called the method of Windows. The Windows method assumes the use of two windows Wi and Wo with fixed sizes n and m respectively. These Windows are able to move with some step in the time sequence of historical data, starting with the first element, and are designed to access the data of the time series, and the first Wi window, receiving such data, transmits them to the input of the neural network, and the second Wo–to the output. The resulting Wi ->Wo pair at each step is used as a training sample element (a recognizable image, or observation).

One-step forecasting is used for short – term forecasts, usually absolute sequence values. The forecast is only one step ahead, but the actual value is used instead of the predicted value to make the forecast for the next step. Multi-step forecasting is used to make a long-term forecast and is designed to determine the main trend and the main points of trend change for a certain period of time in the future. In this case, the predictive system uses the obtained (output) data for the time moments k+1, k+2, etc. as input data for forecasting at time points k+2, k+3 and etc.

3. Results

As an experiment the problem of approximation of a number of dynamics of power generation of electric energy, i.e. construction of model on a final set of points on concrete examples about generation of the power released by the companies entering into Uniform power system of Russia taking into account two influencing factors was considered: losses in networks at transfer and average daily temperature. In this case, the following tasks were set: search for influencing factors on the dynamics of power generation; search for the optimal representation of statistical data on the dynamics of generation and the results of the approximation of the behavior of the series under study. The initial data for solving the problem are shown graphically in Figure 6, 7.

To solve the problem of constructing a model of the dynamics of the power generation process, taking into account the selected influencing factors using the neural network apparatus, the RBF network structure was chosen (Figure 8).

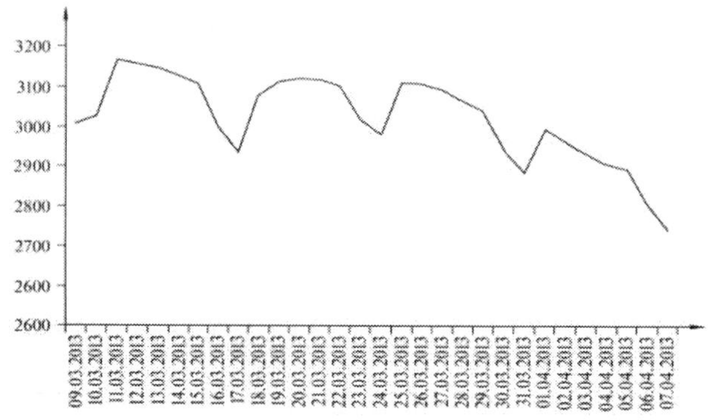

Figure 6. Power generation, GWh

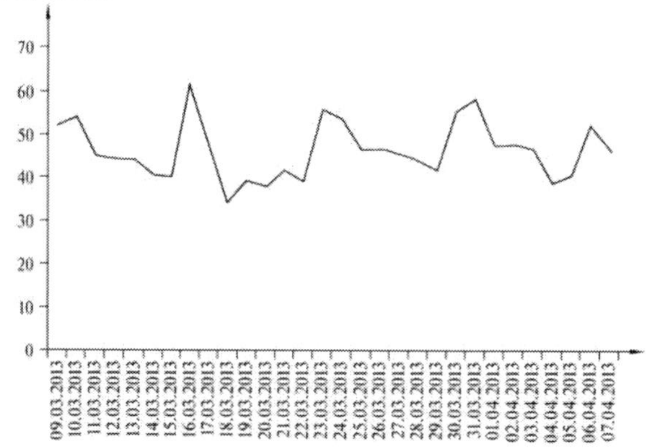

Figure 7. Network losses, GWh

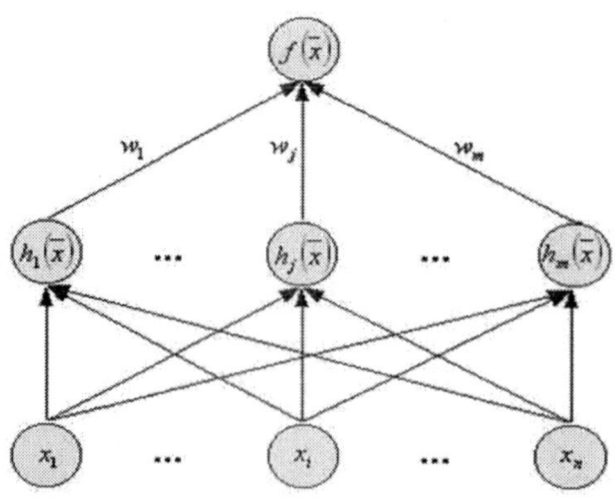

Figure 8. The structure of RBF network

The RBF network has an intermediate layer of radial elements, each of which reproduces the Gaussian response surface. Since these functions are nonlinear, it was enough to take one intermediate layer, determining only the desired number of radial elements. In order to get out of the network, it was necessary to combine the outputs of hidden radial elements. It is enough to create a linear combination of radial elements [8, 9].

The great advantage of ANN is their ability to work in multidimensional spaces, including spaces of mixed type, in which some variables are continuous and some are discrete. But the implementation of this ability implies the presence of the developer of the relevant data, as well as the ability to properly structure them in a form that is easy to remember. In the presence of powerful computer technology and excellent optimization programs, this remains the most difficult stage in the development of neuromodels, poorly amenable to formal description due to the specificity of the application problem to be solved in each case.

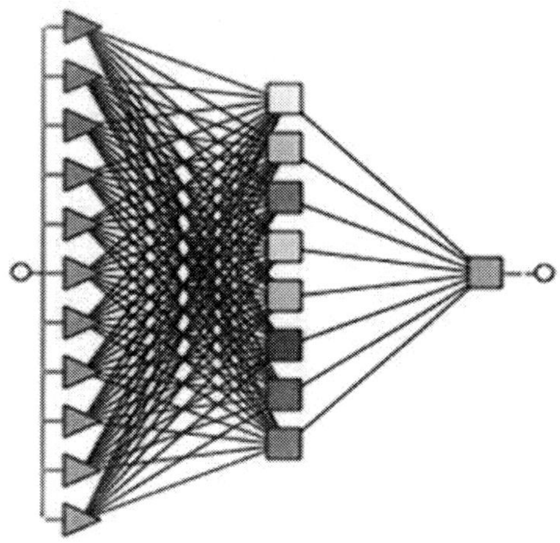

Figure 9. Structure of RBF network components for forecasting economic indicators in the energy sector

Thus, the RBF network has an output layer consisting of elements with linear activation functions (Figure 9, 10).

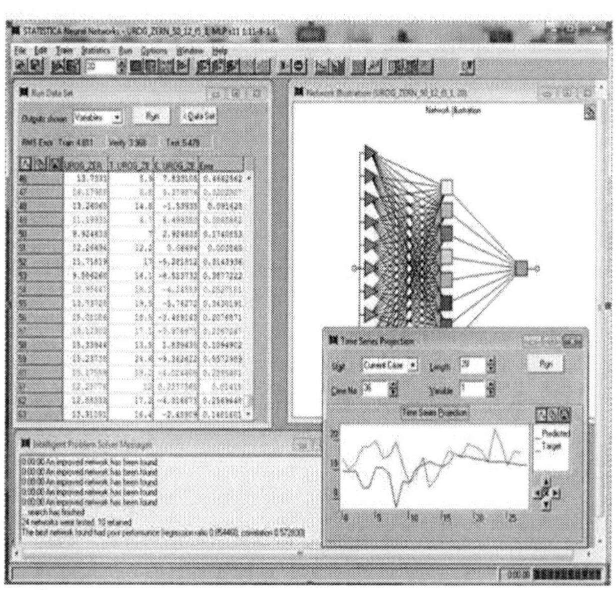

Figure 10. The neural interface system

In this sense, modeling with the help of ANN continues to be something of an art. Not meeting the expectations that they can be used to create models with the usual level of versatility and transparency, neuromodels, however, are much better prepared for use for the management of specific objects [10, 11].

Comparing the advantages and disadvantages of ins with the features of other modeling technologies, one can unwittingly come to the conclusion that their purpose is not in the total displacement of the classical models of the functional type, but in solving those problems in which traditional technologies have shown weakness. ANN provide such an opportunity, and with such a powerful tool in hand, it is unwise to ignore its capabilities.

4. Conclusions

Thus, the experiments show that artificial neural networks can be successfully used to predict the time series of power generation, released by the companies of the energy complex, as well as other economic indicators of energy. Moreover, in some cases, for example, when solving problems with unknown regularities, they give more accurate results than traditional forecasting models, and are more resistant to "noise" and adapted to environmental changes.

In addition, the range of computational problems of energy, in which the use of neural network modeling is quite extensive:

- short-term forecasting (forecasting of demand for electric energy, forecasting of demand for thermal energy, forecasting of demand for natural gas, forecasting the marginal price of the system in the energy market, forecasting the volume of electricity generation by wind farms, forecasting the temperature of the outside air;
- diagnostics and fault localization of station and network equipment;
- assessment of power system stability;
- modeling of the intensity of environmental pollution by power units.

Another disadvantage of neural models is the significant time and other resources required to build a satisfactory model. This problem is not very important if you are investigating a small number of time sequences. However, a typical predictive system in production management can include from several hundred to several thousand time sequences.

However, despite these shortcomings, the model has a number of advantages. There is a convenient way to modify the model as new observations appear. The model works well with time sequences in which the observation interval is small, i.e. a relatively long time sequence can be obtained. For this reason, the model can be used in areas where we are interested in hourly, daily or weekly observations. These models are also used in situations where you need to analyze a small number of time sequences. In addition, neural network forecasting is particularly effective when using additional information, i.e. when taking into account various factors affecting the subject of study.

Artificial neural networks are a universal tool for solving multidimensional problems of modeling and control in power engineering. The relatively slow pace of development of transients in energy systems, combined with the dramatically increased computing power of personal computers allows neural networks to make their way in new and new applications. Their application opens up real opportunities for the introduction of new, more sophisticated approaches to the methodology of modeling and management, promising to rethink the practice of exit the operation of energy facilities and to make energy consumers more quality and cost-effective.

In order for ANN to begin to bring real benefit to the domestic energy sector, it is necessary to bring them closer to the energy sector itself, namely, to give into the hands of those who work at energy enterprises, determines the situation on the ground. This requirement is in some contradiction with the way of allocation of responsibility between scientific, design, Supervisory and operational units of the power complex, in accordance with which the scientific service of energy is carried out. To be more precise-carried out in soviet times.

Today the system of scientific service of energy, left without proper financial support of the state, almost lost its former influence on scientific and technical progress in the industry. Therefore, energy companies have no other choice but to Finance scientific research on their own in order to strengthen their own competitiveness by improving the practice of technological, organizational and economic management.

ANN, in this respect, have excellent chances to attract attention from the new owners of energy companies, representing a market-like tool for the study of "particulars", rather than protruding "communities".

And the last circumstance which can promote wide introduction of neural network models in power is closer coordination of practice of technological management of objects with economic indicators and criteria, and also active use in the course of planning of work of the enterprise of prognostic models. So, the wholesale market of the electric power already works today. There is no doubt that local thermal energy markets will also work. Short-term focus shift forecasting the risks and costs of the energy enterprise will make ANN a more popular tool, create incentives for their active development by the management personnel of the energy sector.

Acknowledgment
The reported study was funded by RFBR and Volgograd region according to the research project № 19-416 -343006.

References

[1] Alekseev A A, Rozaliev V L, Orlova Y A and Zaboleeva-Zotova A V 2016 Context-sensitive image analysis for coloring nature images *Advan. in Intellig. Syst. and Comput.* **451** 133–41

[2] Gagarin A G and Rogachev A F 2012 Prognozirovanie urozhajnosti na osnove analiza kross-regional'nyh dannyh Proc. of Lower Volga Agrouniver. Complex: Sci. and Higher Vocat. Ed. **4** 339–46

[3] Lisetti C L and Rumelhart D E 1997 Recognizing facial expressions with a neural network *Proc. of the nineteenth annual conf. of the cognit. Sci. society* p 1100

[4] Widrow B, Rumelhart D E and Lehr M A 1994 Neural networks – applications in industry, business and science *Communicat. of the ACM* **37(3)** 93–105

[5] Ovchinnikov A S and Bocharnikov V S 2012 New technical solutions improve the efficiency of resource-saving irrigation methods *Proc. of Lower Volga Agrouniver. Complex: Sci. and Higher Vocat. Ed.* **1(25)** 119–24

[6] Timoshenko M A, Rogachev A F, Medvedeva L N and Tokarev K E 2019 Analysis and support of decisionmaking with the use of latest information and communication technologies for the development of social and economic area with regard to capabilities of different categories of population *Advan. in Intellig. Syst. and Comput.* **726** 989–1004

[7] Nikitin N A, Rozaliev V L, Orlova Y A and Alekseev A A 2018 Automated sound generation based on image colour spectrum with using the recurrent neural network *CEUR Workshop Proc.* **2212** 399–408

[8] Kayvan T and Alireza S 2010 Hopfield Neural Networks as Pseudo Random Number Generators. in North America *Fuzzy Information Processing Society (NAFIPS)* pp 1–6

[9] Wen S, Zhigang Z, Tingwen H and Yide Z 2014 Exponential adaptive lag synchronization of memristive neural networks via fuzzy method and applications in pseudorandom number generators *IEEE Transact. on Fuzzy Syst.* **22(6)** 1704–13

[10] Kashem M A., Jlasmon G B., Mohamed A and Moghavvemi M 1998 Artificial neural network approach to network reconfiguration for loss minimization in distribution networks *Electr. Power & Energy Syst.* 20(4) 247–58

[11] Anikin I and Khaled A 2016 Pseudo-random number generator based on fuzzy logic In *Control and Communications* (IEEE (SIBCON) pp 1–4

IOP Conf. Series: Materials Science and Engineering **905** (2020) 012070 doi:10.1088/1757-899X/905/1/012070

Alternative refrigerants with low global warming potential for refrigeration and air-conditioning industries

O B Tsvetkov, Yu A Laptev, A V Sharkov, V V Mitropov and A V Fedorov

ITMO University, St. Petersburg, Russia, bldg. A, 49, Kronverksky Pr., St. Petersburg, 197101, Russia

E-mail: tsvetkov@itmo.ru, laptev_yua@mail.ru, avsharkov@itmo.ru, v_mit@mail.ru, afedorov@itmo.ru

Abstract. Historically the use of natural and ecologically safe refrigerants was a strategy to eliminate environmental problems and avoid uncertainties with synthetic replacement fluids. Since ammonia is toxic, carbon dioxide provide high pressure, and the hydrocarbons are flammable, the general conclusion is often drawn that natural fluids gave safety problems. This paper will describe the possibilities of application as working fluids in refrigeration, heat pumping and organic Rankine cycles of the hydrofluoroolefins (HFOs) as alternative refrigerants with low global warming potential.

1. Introduction

A large number of compounds have been proposed to substitute the high global warming potential (GWP) hydrofluorocarbons (HFCs) in refrigeration, air-conditioning and heat pumping industry. In addition to natural fluids studies have shown that there are halogenated olefins characterized by the presence of C=C double bond in the molecule that are potentially suitable for a number of relavant applications.

The use of ecologically safe working fluids is a robust strategy to eliminate environmental problems and uncertainties with synthetic replacement working fluids. Some information on the typical characteristics of hydrofluoroolefins (HFOs) as new working fluids will be presented, an outline of the early history will also be given.

2. Possibilities and tradeoffs for low-GWP refrigerants

Fluorinated greenhouse gases including CFCs, HCFCs and HFCs have a significant impact on the climate change. It was recently estimated that they accounted for 12 % of all radiactive forcing of long-lived greenhouse gases, caused by increasing greenhouse-gas levels since the beginning of the industrial revolution [1–2]. Due to the phase-out of CFCs under the Montreal Protocol atmospheric concentrations of these gases are declining, while those of HCFCs and HFCs used as replacement fluids rising rapidly. HCFCs are dealt with by the various Montreal Protocol amendments and were been prohibited in new systems in the EU since 2000. HCFCs will also be banned in other countries: before 2020 in other developed countries and before 2030 in developing countries [3].

The original Montreal Protocol signed in 1987 and ratified to date by more than 190 countries asked for a reduction in the consumption and production of CFCs and halons. As a result CFCs and halons have been phased out in the developed countries by January 1996 and HCFCs will be gradually phased out over the period 1996–2020. The Montreal protocol itself does not directly address the choice for certain replacement chemicals, its only aim is to phase out ozone depleting chemicals.

A «perfect» refrigerant should be environment friendly with zero ODP (Ozone Depletion Potential) and low GWP (Global Warming Potential): it should have a medium pressure range, with evaporating pressure above atmospheric to avoid air intake into the system; it should be safe (non-toxic, non-flammable). HFC-134a was developed to be the substitute to the CFC-12. In a range of physical properties relatively close to HCFC-22 the commercially available HFCs are HFC-125, HFC-32 and HFC-143a. Each of them has prohibitive disadvantages: HFC-32 and HFC-143a are flammable, HFC-125 is not flammable, but has a very bad cycle efficiency. Some blends were developed to go below the limit of flammability. In practice HFC-125 + HFC-32 (HFC-410A) and HFC-125 + HFC-143a (HFC-507) both give nonflammable quasi azeotropes at 50/50 % concentration. The near azeotrope of HFC-32and HFC-125 is significantly different from the others in that it operate at very high pressure, approximately 50 % higher than HCFC-22. If lower pressures are desired some HFC-134a must be added. The most common blends involving HFC-134a are: HFC-404A practically identically HFC-507 with 4 % of HFC-134a added; HFC-407C is practically some HFC-410A with 52 % HFC-134a added [4].

Global warming is a big challenge for world society today [6–7]. Emissions of HFCs today contribute around 1 % to global greenhouse gas emissions, but growing by 8–9 % annually because of rapidly increasing demand for refrigeration and air conditioning in emerging economies. As anticipated for some time, an amendment to the Montreal Protocol adopted in October 2016 requires a significant reduction of the weighted value of global warming potential (GWP) of fluids used in air-conditioning and refrigeration equipment. Consequently hydrofluorocarbon (HFC) refrigerants having a high GWP will be eliminated or their use will be significantly reduced. As a result, efforts are made in order to search alternative refrigerant for high GWP refrigerants replacement. In the search for new refrigeration a number of other criteria must also be ment. These include stability, short atmospheric lifetime. Also important are thermodynamic properties, low flammability and toxicity, cost and compatibility with the materials of construction. The «optimum» refrigerant depends on the constraints of the day and as in the case today with the concern of the contribution of HFCs to anthropogenic global warming.

Taking into account various criteria particularly low toxicity, low flammability, low global warming potential (GWP) and high energy efficient, the halogenated olefins characterized by the presence of a C=C double bond in the molecule that are potentially suitable for a number of relevant applications [8–9]. The hydrofluoroolefins are the synthetic refrigerants having zero ODP, low GWP and extremely short atmospheric lifetime. The fluorinated isomers also can be identified as for example HFO-1234ye, HFO-1234yf. When a carbon-carbon double bond exist is stereoisomers may exist. For example R1234ye has two stereoisomers trans-1.2.3.3-tetrafluoropropene (R1234ye(E)) and zis-1.2.3.3-tetrafluoropropene (R1234ye(Z)) [6].

The hydrofluoroolefins (HFOs) have been proposed as novel third generation of refrigerants, particularly to replace conventional refrigerants, for example R134a in the automobile air conditioning system which has GWP of 1430. Both refrigerants R1234ye(E) and R1234ye(Z)have a GWP value of 2.3 and 2.9 respectively. Table 1 provides estimated fundamentals parameters and thermodynamic properties for some of these isomers. In the case for the fluorinated propene and ethane isomers and stereoisomers the required inputs are the acentric factor (ω), the critical state properties T_c, p_c, the normal boiling temperature T_b, molecular mass and GWP-index [9, 10].

Table 1. Thermodynamic properties of HFOs-fluids

Refrigerant	M, [kg/kmol]	T_c, [K]	p_c, [kPa]	T_b, [K]	GWP	ω
R1234yf	114.04	367.85	3382.2	243.70	<1	0.2760
R1234ze(E)	114.04	382.51	3634.9	254.18	6	0.3120
R1234ze(Z)	114.04	423.27	3533.0	282.90	1.4	0.3271
R1233zd(E)	130.50	439.50	3623.7	291.41	7	0.3025

The fluorinate propene isomers with bases 1252 and 1261, i. e. isomers containing 4H/2F and 5H/1F respectively are most likely highly flammable. As the number of hydrogen atoms in the molecule relative to the total number of hydrogen (H) and fluorine (F) atoms increases, the flammability of the molecule will increase, which would indicate increasing flammability from the R1225 isomers to the R1243 isomers.

The use of hydrofluoroolefins (HFOs) as a pure-component refrigerant causes concern due to their mildly flammability. The mixing of two refrigerants may offer a good alternative to pure substances. Such mixtures can meet the refrigerants of high cycle efficiency, b environmentally friendly and not dangerous for human beings. The most promising mixtures are composed HFO-1234yf or HFO-1234ze(E) with either an HFC or a natural refrigerant for example with CO_2. Such mixtures make to reduce the flammability of HFOs.

Toxicity for the fluorinated isomers is playing an important role in the development efforts being expended of these isomers. Several of the R1225 isomers are no longer being developed because of toxicity concerns. During the past years various thermophysical properties of the HFOs such as vapor pressures, critical properties, specific heat capacities, densities and transport properties were investigated [10]. The R 1234yf is an appropriate near «drop-in» replacement for R134a in existing applications and R1234ze(Z) is appropriate as a near «drop-in» replacement for R114 in high-temperature heat pumping applications.

As alternatives for HFCs has been proposed some compositions of HFC-HFO (table 2).

Table 2. Blends of HFC–HFO

Refrigerant	Composition	GWP
R450A	R1234ze/R134a	605
R456A	R32/R1234ze/R134a	687
R513A	R1234yf/R134a	631
R513B	R1234yf/R34a	596
R448A	R32/R125/R1234yf/R134a/1234ze	1386
R444A	R32/R152a/R1234ze	92
R459B	R32/R1234yf/R1234ze	144

3. Conclusion

Hydrofluoroolefins (HFOs) have been proposed as new generation alternative refrigerants with low global warming potentials to replace the hydrofluorocarbons (HFCs) – greenhouse gases mainly used as refrigerants.

References

[1] ANSI/ASHRAE Standard 34–2015 *Designation and safety classification of refrigerants NE* (Atlanta, GA) 30329–2305
[2] *Regulation (EU) no 517/2014 of the European Parliament and of the Council on Fluorinated Greenhouse Gases and Repealing Regulation (EC)* EU April 16 2014 no 842/2006
[3] *Montreal Protocol on substances that deplete the ozone layer* 1987 (Nairobi, Kenya, United Nations Environment Program)
[4] ISO, ISO/R1662-1971 *Refrigerating Plants Safety Rules*
[5] COP 22 2017 The role of refrigeration in the combat against climate change *Int. J. Refrigerat.* **74** 1–2
[6] Brown J S 22–29 August 2009 HFOs-new, low global warming potentials refrigerants *ASHRAE-Journal* **51(8)**

[7] International milestone agreement on the phase-down of HFC production and consumption in Kigali, Editorial 2017 *Int. J. Refrigerat.* **73** V–VI
[8] Domanski P A, Brown J S, Heo J et al 2014 A thermodynamic analysis of refrigerants: performance limits of the vapor compression *Int. J. Refrigerat.* **38** 71–9
[9] Brown J S 2008 Methodology for estimating thermodynamic parameters and performance of alternative refrigerants *ASHRAE Transact.* **114(1)** 230–8
[10] Bobbo S, Nicola G Di, Zilio C et al 2018 Low GWP halocarbon refrigerants: A review of thermophysical properties *Int. J. Refrigerat.* **90** 181–201

IOP Conf. Series: Materials Science and Engineering **905** (2020) 012071 doi:10.1088/1757-899X/905/1/012071

Magnetic and magnetoelastic properties of rare earth intermetallides based on $TbFe_2$

Z S Umkhaeva[1], A S Ilushin[2], S A Nikitin[2], T A Aleroeva[3], N Yu Pankratov[2] and I S Tereshina[2]

[1] Kh. Ibragimov Complex Institute, RAS, Grozny, Russia
[2] Moscow State University named after M.V. Lomonosov, Moscow, Russia
[3] Chechen State University, Grozny, Russia

E-mail: zargan.umhaeva@yandex.ru, sols146i@phys.msu.ru, nikitin@phys.msu.ru, tamila_hinata.ru@mail.ru, pankratov@phys.msu.ru, irina_tereshina@mail.ru

Abstract. The paper presents the results of the study of magnetic and magnetostrictive properties of rare-earth intermetallides based on $TbFe_2$ compound: $Tb_{1-x}Zr_xFe_2$, $Tb_{1-x}Sm_xFe_2$ and $(Tb_{1-x}Y_x)_{0.8}Sm_{0.2}Fe_2$. These alloys have a cubic crystal structure of the Laves phase C15 throughout the substitution region. For the system of alloys $Tb_{1-x}Zr_xFe_2$ the single-phase area is limited to the replacement parameter $0 \leq x \leq 0.2$ and $0.8 \leq x \leq 1.0$. The compounds of the $Tb_{1-x}Sm_xFe_2$ system were studied via X-ray dilatometry. The magnetostriction of alloys within the $Tb_{1-x}Zr_xFe_2$ and $(Tb_{1-x}Y_x)_{0.8}Sm_{0.2}Fe_2$ systems was studied in the temperature range of 80–320 K in magnetic fields of up to 17 kOe via a strain-gauge method. It was found that each system has the phenomenon of sign inversion of magnetostriction constants: in the system $Tb_{1-x}Zr_xFe_2$ in the area of replacement parameter $x = 0.8$, in $Tb_{1-x}Sm_xFe_2$ – in the area $x = 0.45$, and in the system of alloys $(Tb_{1-x}Y_x)_{0.8}Sm_{0.2}Fe_2$ at $x = 0.6$. It was demonstrated that magnetic compensation and spin reorientation phenomena occur in these systems. The obtained results are discussed in the model of alloys with competing exchange interactions.

1. Introduction

In recent years, great attention has been paid to the study of magnetic properties of intermetallic compounds of rare-earth metals with transition metals [1–3]. First of all, these are the compounds of RM_2 stoichiometry, where R – rare earth metal (REM) and M – 3d transition metal of iron group (Fe, Co, Ni or Mn) with Laves phase structure [4, 5, 6–10]. They are of particular interest because they coexist with two different types of electrons that are magnetic moment carriers: localized 4f-electrons of REM ions and band 3d electrons [11]. This is why, with a relatively simple crystal structure, they have unique magnetic properties, such as giant magnetostriction [12], significant magnetocaloric effect [13] and sufficiently high Curie temperatures [11]. It is known that the compounds of this type can have 2 types of structures: cubic structure of $MgCu_2$ type (structure C15) and hexagonal structure of $MgZn_2$ type (structure C14) [2].

Intermetallic REM compounds with 3D transition metals are widely known in science and technology due to their unique magnetic properties. These compounds are characterized by high magnetostrictive properties. They include alloy $TbFe_2$ well-known due to its "giant" magnetostriction. Thus, the study of the structure and magnetic properties of the Laves phases is sufficiently relevant for both the condensed matter physics and their practical use.

To date the most promising area is the creation of alloy systems consisting of many elements, including rare earths with different sign of magnetocrystine anisotropy for complete or partial compensation of magnetic anisotropy constants. As shown, such compensation can be achieved in alloys of mixed compositions of quasi-binary or more complex systems [1, 14–17].

REM intermetallides can be considered as magnetics with two magnetic sublattices: REM and 3d-transition metal. Accordingly, the magnetic properties of RFe_2 compounds are defined by three types of exchange interactions: R-R, Fe-Fe and R-Fe. In order to find optimal compositions with high FES and magnetostriction values, it is necessary to examine compounds with different substitutions of 3d-, 3p-, and 4f elements. Compound $TbFe_2$ has sufficiently high Curie temperature (~ 711 K) as well as induced and spontaneous "giant" magnetostriction [18].

The [purpose was to study the magnetic and magnetostrictive properties of rare earth alloys based on – $(TbR^IR^{II})Fe_2$ in which the atoms of the heavy rare earth element Tb are replaced by those of the light rare earth element Sm and/or non-magnetic atoms Y or Zr.

The choice of alloy systems is caused by the fact that in the case of $Tb_{1-x}Zr_xFe_2$ system, the substitution of terbium atoms in the rare-earth sublattice is non-magnetic, as well as yttrium, zirconium atoms. In alloys of $Tb_{1-x}Sm_xFe_2$ and $(Tb_{1-x}Y_x)_{0.8}Sm_{0.2}Fe_2$ systems, compounds $TbFe_2$ and $SmFe_2$ having high magnetostriction constants of the opposite sign are used as the basis [5, 18–20]. In this system, there is competition between the exchange interactions Tb-Fe and Sm-Fe depending on the concentration of yttrium introduced into the rare earth sublattice. Therefore, the study of alloys of this system will make it possible to study not only the intergranular exchange interaction R-Fe at the dilution of the rare-earth sublattice with non-magnetic yttrium atoms, but also the exchange interaction in the rare-earth sublattice itself.

2. Methods and materials

Alloys of quasi-binary systems $Tb_{1-x}Zr_xFe_2$, $Tb_{1-x}Sm_xFe_2$ and quasi-ternary system $(Tb_{1-x}Y_x)_{0.8}Sm_{0.2}Fe_2$ were synthesized and studied. The first two systems were prepared in the entire concentration region of substitution x = 0 ÷ 1 with a step 0.1, the alloys of the third system were synthesized in the same concentration region but with a step of 0.2.

Intermetallic compounds of the systems of alloys $Tb_{1-x}Zr_xFe_2$, $Tb_{1-x}Sm_xFe_2$ and $(Tb_{1-x}Y_x)_{0.8}Sm_{0.2}Fe_2$ were prepared via an arc melting method in the atmosphere of spectroscopically pure helium $Tb_{1-x}Zr_xFe_2$ and $Tb_{1-x}Sm_xFe_2$) and in the atmosphere of argon at normal pressure – $(Tb_{1-x}Y_x)_{0.8}Sm_{0.2}Fe_2$. The X-ray diffraction analysis showed that in $Tb_{1-x}Zr_xFe_2$ system alloys it is possible to obtain uniform solid solutions only in the limited region of zirconium concentration $0 \leq x \leq 0.2$ and $0.8 \leq x \leq 1.0$ [21]. The alloys of $Tb_{1-x}Sm_xFe_2$ system were single-phase for all values of the substitution parameter [22]. Quasi-binary alloys $Tb_{0.8}Sm_{0.2}Fe_2$, $Y_{0.8}Sm_{0.2}Fe_2$ are single-phase [4]. More complex structures $(Tb_{0.8}Y_{0.2})_{0.8}Sm_{0.2}Fe_2$, $(Tb_{0.6}Y_{0.4})_{0.8}Sm_{0.2}Fe_2$, $(Tb_{0.4}Y_{0.6})_{0.8}Sm_{0.2}Fe_2$, $(Tb_{0.2}Y_{0.8})_{0.8}Sm_{0.2}Fe_2$ were characterized by the phenomenon of phase separation. These samples contained a second phase RFe_3, which amount varied from 5 to 10 %.

Magnetostriction measurements were carried out in the area of low and room temperatures in magnetic fields of up to 17 kOe via the strain-gauge method. Magnetostriction error did not exceed 6%. The compounds of $Tb_{1-x}Sm_xFe_2$ system were studied via the X-ray dilatometry. Field and temperature magnetization dependencies were measured using standard equipment in the temperature range from 4.2 to 300 K. Besides, temperature dependencies on susceptibility were obtained in the temperature range from 300 to 800 K.

3. Results

3.1. Magnetostriction of alloys $Tb_{1-x}Zr_xFe_2$

Magnetization of $(Tb,Zr)Fe_2$ system compounds was studied in a wide range of temperatures in magnetic fields of up to 15 kOe. It was shown that for compounds with high concentration of terbium ions ($TbFe_2$, $Tb_{0.9}Zr_{0.1}Fe_2$ and $Tb_{0.8}Zr_{0.2}Fe_2$) the magnetic field of up to 12 kOe is not sufficient to achieve saturation magnetization, which can be explained by the presence of a large magnetic anisotropy typical for Tb atoms. In compounds with low terbium content $Tb_{0.2}Zr_{0.8}Fe_2$, $Tb_{0.1}Zr_{0.9}Fe_2$ and $ZrFe_2$ the magnetization saturation is observed in the fields above 6 kOe.

In order to determine the saturation magnetization, the magnetization dependencies μ(1/H) on the inverse value of the magnetic field were built. The saturation magnetization (μ_S) for structures with high concentration of terbium ions is determined by the extrapolation of dependence μ (1/H) to the maximum field (1/H → 0).

The method of thermodynamic coefficients was used to define the Curie temperatures, which allowed determining the Curie temperatures according to magnetic data. The results for saturation magnetization and Curie temperature of the compounds $(Tb,Zr)Fe_2$ are summarized in Table 1. The Curie temperature of compounds with increasing Zr concentration decreases linearly from 711 K for $TbFe_2$ to 628 K in $ZrFe_2$. This is caused by the decrease of the energy of exchange interactions due to the decrease of interacting pairs of magnetoactive Tb-Fe atoms.

The results of magnetic measurements of $(Tb,Zr)Fe_2$ system compounds together with the literature data [23] make it possible to conclude that there are large exchange fields acting on rare-earth ions from iron sublattice in cubic RFe_2 compounds. Exchange R-Fe interaction causes magnetic ordering of rare-earth sublattice and serves one of the reasons for giant magnetostriction deformations observed in $(Tb,Zr)Fe_2$ alloys.

The alloys of quasi-binary system $Tb_{1-x}Zr_xFe_2$ are highly strictional materials, similar to $TbFe_2$ compound. In these compounds, the giant magnetostriction typical for terbium atoms is observed not only at low but also at room temperatures. The longitudinal magnetostriction of $TbFe_2$ compound is positive and in field of 17 kOe reaches $\lambda_{\|} \approx 0.9 \cdot 10^{-3}$ at room temperature. Giant magnetostriction is observed in all compounds of $Tb_{1-x}Zr_xFe_2$ system in the field of small concentration of zirconium ($0 \leq x \leq 0.2$). This magnetostriction is caused by the contribution of Tb sublattice and emerges from single-ion magnetoelastic contributions of its ions.

Figure 1 shows field dependencies of longitudinal and cross magnetostriction of $Tb_{0.1}Zr_{0.9}Fe_2$ compound. At room temperature, the saturation of magnetostrictive curves is observed in the field above 12 kOe. For compounds with 10 % substitution of Tb ions with zirconium the longitudinal magnetostriction at T=80 K reaches $\lambda_{\|} \approx 0.9 \cdot 10^{-3}$. With 20 % substitution of terbium atoms with zirconium the longitudinal magnetostriction makes $\lambda_{\|} \approx 0.5 \cdot 10^{-3}$.

Table 1. Magnetic characteristics: Curie temperature (T_C) and saturation magnetization (μ_S) at a temperature of 4.2 K of $Tb_{1-x}Zr_xFe_2$ system compounds

Compounds	T_C, K	μ_S, μ_B/f.u.
$TbFe_2$	711	5.8
$Tb_{0.9}Zr_{0.1}Fe_2$	699	4.6
$Tb_{0.8}Zr_{021}Fe_2$	690	3.6
$Tb_{0.2}Zr_{0.8}Fe_2$	640	2.4
$Tb_{0.1}Zr_{0.9}Fe_2$	636	2.9
$ZrFe_2$	628	3.4

Figure 2 shows the concentration dependence of longitudinal and cross magnetostriction in the compounds of $Tb_{1-x}Zr_xFe_2$ system. As the zirconium concentration increases, the iron content in the system increases, resulting in the prevalence of the magnetostriction of iron sublattice, which is opposite to the magnetostriction of terbium sublattice. In the field of concentration x> 0.8 the longitudinal magnetostriction is negative and equals approximately $\lambda_{\|} \approx -0.04 \cdot 10^{-3}$.

The insertion to Figure 2 shows the concentration dependence of the magnetostriction constant $\lambda_a = \lambda_{\|} - \lambda_{\perp}$ at a temperature of 80 K. The figure shows that the sign of magnetostriction constant changes near the concentration x = 0.8.

$TbFe_2$ and $ZrFe_2$ compounds are isostructural to the C15 Laves phase. The compound $TbFe_2$ – ferrimagnetic with Curie temperature of 711 K, $ZrFe_2$ – ferrimagnetic with T_C = 628 K. The vectors of spontaneous magnetic moment in both compounds are oriented along crystallographic direction [111], so the increase in the concentration of non-magnetic zirconium atoms leads only to dilution of rare-earth sublattice [23]. As a result, at a certain concentration the magnetism of the 3d sublattice begins to prevail.

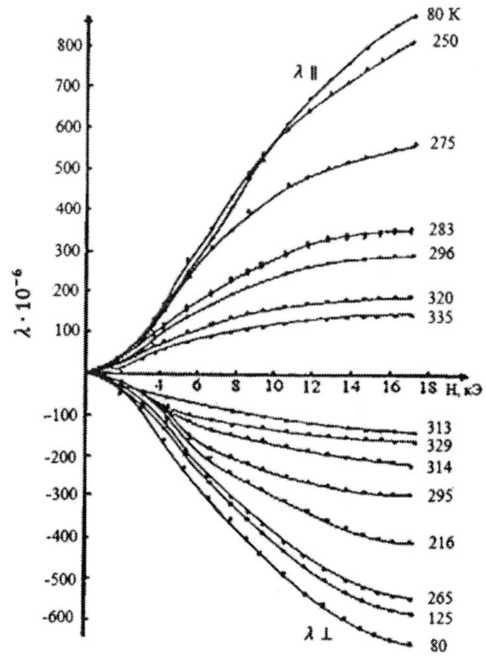

Figure 1. Isotherms of $Tb_{0.1}Zr_{0.9}Fe_2$ alloy magnetostriction.

In the areas of single-phase condition of samples $0 \leq x \leq 0.2$ and $0.8 \leq x \leq 1$ longitudinal and cross magnetostriction decreases linearly with the growth of zirconium content (Fig. 2), which illustrates a single-ion mechanism of magnetostriction. Another interesting phenomenon here is the inversion of the sign of magnetostriction constants, which combines this system with $Tb_{1-x}Sm_xFe_2$ and $(Tb_{1-x}Y_x)_{0.8}Sm_{0.2}Fe_2$ systems.

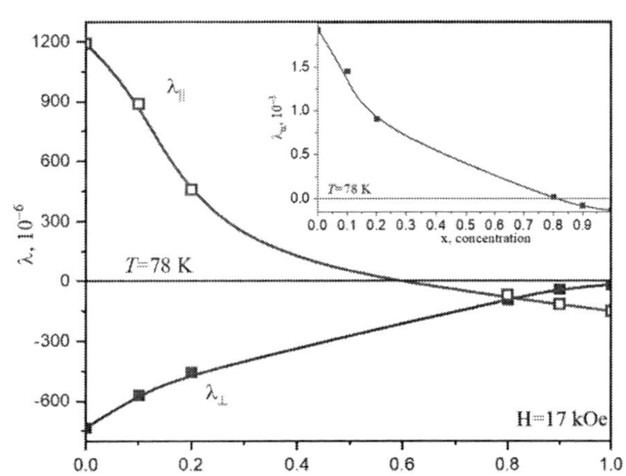

Figure 2. Concentration dependence of longitudinal $\lambda_{\parallel}$ and cross $\lambda\perp$ magnetostriction of alloys of $Tb_{1-x}Zr_xFe_2$ system

3.2. Magnetic properties of $Tb_{1-x}Sm_xFe_2$ system

Compounds $TbFe_2$ and $SmFe_2$ are quite interesting since they possess not only record values of magnetostriction constants of the opposite sign, but also different signs of intersublattice exchange interactions. The Tb-Fe exchange interaction is negative and leads to ferrimagnetic ordering of magnetic moments of terbium and iron, while the Sm-Fe interaction is positive, so the order of orientation of atomic magnetic moments of samarium and iron is ferrimagnetic [22]. In $(Tb,Sm)Fe_2$ system as Tb atoms are replaced by Sm atoms, in addition to the inversion of spontaneous magnetostriction sign the competition between Tb-Fe and Sm-Fe exchange interactions results in the change of direction of the light magnetization axis from the direction [111] typical for $TbFe_2$ towards direction [110] typical for $SmFe_2$ compound [5, 18–20]. In work [22], we showed that in the field of temperatures from 5 to 300 K in the concentration area x <0.45 in $Tb_{1-x}Sm_xFe_2$ system there are two spin-reorientated phase transitions from the state with the direction of light magnetization axis [111] to the state with the direction [110], which are followed by the formation of angular magnetic phases ‹hkl› (Fig. 3) [22].

The occurrence of angular phases results in complex distortions of the crystal structure and the decrease of crystal symmetry.

We revealed that in terbium concentration area $x = 0,45$–0.5 there is structural phase transition from the rhombohedral phase in purely cubic, i.e. in $Tb_{1-x}Sm_xFe_2$ system of alloys there was a distortive structural phase transition. It is in the region of these concentrations that a compound, in which magnetocrystalline anisotropy is almost completely compensated, can be obtained and giant magnetostriction can be expected.

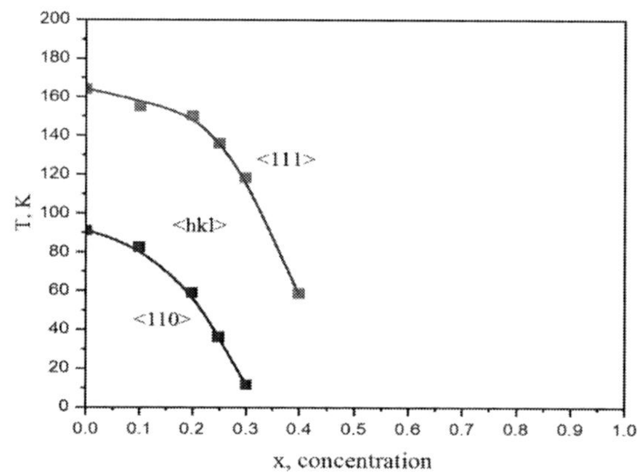

Figure 3. Magnetic spin-reorientation phase diagram of the $Tb_{1-x}Sm_xFe_2$ system.

3.3. Magnetic properties of multi-component alloys $(Tb_{1-x}Y_x)_{0.8}Sm_{0.2}Fe_2$

Magnetization study of $(Tb_{1-x}Y_x)_{0.8}Sm_{0.2}Fe_2$ system compounds was carried out in small magnetic fields of 2 kOe in the temperature range from 2 to 300 K.

Figure 4 shows the concentration dependence of magnetization in the field of 2 kE for temperatures from 2 to 300 K. We see the magnetic compensation of rare-earth and iron sublattices in the range of the parameter x = 0.6 in $(Tb_{1-x}Y_x)_{0.8}Sm_{0.2}Fe_2$ system.

For compound $(Tb_{0.4}Y_{0.6})_{0.8}Sm_{0.2}Fe_2$ the magnetization is almost reduced to zero compared to the other connections, i.e. the sum of magnetic moments of sublattices (directed antiparallel to each other) compensate each other. The composition and temperature at which the compensation effect occurs are referred to as the compensating composition and the compensation temperature, respectively [24, 25]. The most important factors for the transformation of this effect are the concentration of atoms and temperature. This phenomenon has great practical importance: by varying the concentrations of atoms substitution near the compensating composition, the magnitude of spontaneous magnetization can be controlled. Temperature studies of $(Tb_{0.4}Y_{0.6})_{0,8}Sm_{0.2}Fe_2$ (x = 0.6) magnetization show that at T_{comp} = 136 K there is a complete compensation of magnetization for this composition.

The Curie temperature values of alloys of this system are defined on the basis of the results of magnetic susceptibility study and are shown in Table 2. By diluting the rare-earth sublattice of $Tb_{0.8}Sm_{0.2}Fe_2$ compound with nonmagnetic yttrium ions, a monotonic decrease in Curie temperature occurs with an increase of Y concentration caused by the decrease in R-Fe exchange interaction integral. The Curie temperature of the ferrimagnetic compound $Y_{0.8}Sm_{0.2}F_{e2}$ almost coincides with the T_C of YFe_2 compound, but is significantly lower than the T_C of $SmFe_2$ compound. This fact makes it possible to conclude that the strong dilution of the rare-earth $SmFe_2$ compound with yttrium ions suppresses the Sm-Fe exchange interaction. For light rare-earth ion Sm, the sign of the exchange integral of Sm-Fe interaction is positive and the direction of the magnetic moment coincides with the direction of the magnetic moment of the iron sublattice. For heavy rare-earth ion Tb, the Tb-Fe exchange interaction integral is negative and hence the magnetic moment of Tb sublattice is directed anticollinearly to the magnetic moment of the iron sublattice.

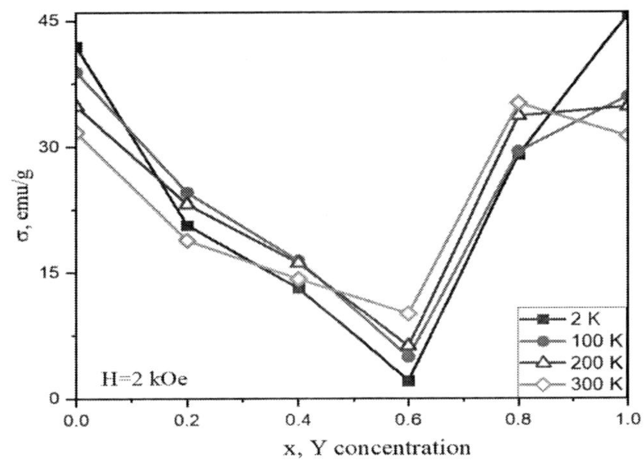

Figure 4. Dependence of magnetization of alloys $(Tb_{1-x}Y_x)_{0.8}Sm_{0.2}Fe_2$ on replacement parameter *x* in the temperature range from 2 to 300 K.

It is shown that in order to connect this system the saturation magnetization is achieved in the fields above 10 kOe. The amount of saturation magnetization was determined by the curves μ(1/H) and is shown in Table 2.

Table 2. Magnetic characteristics: Curie temperature (T_C) and saturation magnetization (μ_S) at a temperature of 4.2 K of $(Tb_{1-x}Y_x)_{0.8}Sm_{0.2}Fe_2$ system compounds.

Compounds	T_C, K	μ_S, μ_B/f.u
$Tb_{0.8}Sm_{0.2}Fe_2$	673	3.6
$(Tb_{0.8}Y_{0.2})_{0.8}Sm_{0.2}Fe_2$	658	2.2
$(Tb_{0.6}Y_{0.4})_{0.8}Sm_{0.2}Fe_2$	640	1.5
$(Tb_{0.4}Y_{0.6})_{0.8}Sm_{0.2}Fe_2$	609	0.5
$(Tb_{0.2}Y_{0.8})_{0.8}Sm_{0.2}Fe_2$	589	2.4
$Y_{0.8}Sm_{0.2}Fe_2$	541	2.0

Figure 5 shows the longitudinal and cross magnetostriction of $(Tb_{1-x}Y_x)_{0.8}Sm_{0.2}Fe_2$ system alloys depending on the replacement parameter x. It is found that the longitudinal magnetostriction of all alloys of $(Tb_{1-x}Y_x)_{0.8}Sm_{0.2}Fe_2$ system with the replacement parameter $x \leq 0.8$ in the external magnetic field of H = 12 kOe is positive, while the cross magnetostriction is negative. In alloys with $x \geq 0.9$, on the contrary, the longitudinal magnetostriction is negative, and the cross – positive. Thus, in this system, the sign inversion of magnetostriction constants occurs in the value range of x = 0.8.

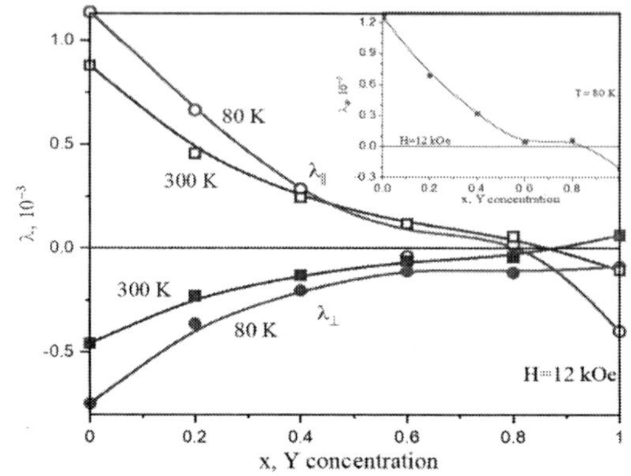

Figure 5. Longitudinal $\lambda_{\parallel}$ and cross $\lambda_{\perp}$ magnetostriction of alloys $(Tb_{1-x}Y_x)_{0.8}Sm_{0.2}Fe_2$ depending on the replacement parameter x.

4. Discussion

The behavior of magnetization of alloys $(Tb,Zr)Fe_2$ can be considered in the model of two-sublattice ferrimagnet with magnetic sublattices Tb and Fe. In this case, the exchange interactions Fe-Fe, Tb-Fe and Tb-Tb shall be taken into account. The exchange interaction integral Tb-Fe is negative.

In compounds RFe_2 the most potent is the Fe-Fe exchange interaction, it is this type of interaction that determines high Curie temperatures in these compounds. Our results for $ZrFe_2$ magnetization allow defining the magnetic moment of iron ions $\mu_{Fe} = \mu_S(ZrFe_2)/2 = 1.7\ \mu_B/Fe$. The magnetic moment of Tb ion makes $\mu_{Tb} = 9\ \mu_B$ [18]. The magnetization of Tb sublattice as a function of the concentration of Tb ions in $Tb_{1-x}Zr_xFe_2$ compounds can be calculated as $(1-x)\mu_{Tb}$. Accordingly, within the collinear ferromagnet model, it can be shown that the magnitude of the total magnetic moment is linearly dependent on the concentration of the replacement element (Zr):

$$\mu_{cal}=2\mu_{Fe} - (1-x)\mu_{Tb}. \tag{1}$$

Similar reasoning can be applied to $Tb_{1-x}Sm_xFe_2$ compound system. However, the magnetization behavior of a given alloy shall be considered in a model of a three-sublattice collinear ferrimagnetic in which the magnetic moment of Tb ions is directed anticollinearly to the moments of Sm and Fe ions. In this case, it is necessary to consider a larger spectrum of exchange interactions: inner-sublattice Fe-Fe, Tb-Tb, Sm-Sm and inter-sublattice Tb-Fe, Sm-Fe, Tb-Sm. The integral of Tb-Fe exchange interaction will be negative, and for the interaction Sm-Fe – positive. Respectively, the contribution from Sm sublattice to magnetization of $Tb_{1-x}Sm_xFe_2$ compounds makes $x\cdot\mu_{Sm}$, where $\mu_{Sm} = 0.7\ \mu_B/Sm$).

Within the collinear ferrimagnetic model with three magnetic sub-sublattices (Tb, Sm, Fe) the magnitude of the magnetic moment is linearly dependent on Sm concentration:

$$\mu_{cal}=2\mu_{Fe} + x\mu_{Sm} - (1-x)\mu_{Tb}. \tag{2}$$

Applying the above reasoning, it is possible to show the linear dependence of the total magnetic moment in the system of compounds $(Tb_{1-x}Y_x)_{0.8}Sm_{0.2}Fe_2$ on Y concentration:

$$\mu_{cal} = 2\mu_{Fe} + 0.2\ x\ \mu_{Sm} - 0.8\ (1-x)\mu_{Tb}. \tag{3}$$

Figure 6 shows the experimental saturation magnetization at T = 4.2 K for alloys of systems $(Tb,Zr)Fe_2$ (unpainted symbols) and $(Tb,Y)_{0.8}Sm_{0.2}Fe_2$ (painted symbols). The lines show the calculations made according to formulas (1–3).

Since formulas (1–3) produce both positive and negative values (for substitution parameter x in the range from 0 to 1), the figure shows the μ_{cal} modulus. As rare-earth Tb is replaced by non-magnetic Zr, Y or light REM ions, the samarium decreases the total magnetic moment μ_{cal} (x) for three systems studied in this work. Our experimental values are near the calculated lines. Linear dependencies $\mu_{cal}(x)$ reach zero at x near 0.6 for all systems, confirming our conclusion that $(Tb_{0.4}Y_{0.6})_{0.8}Sm_{0.2}Fe$ compound is compensatory. Similar behavior of magnetization is also found in the system of multicomponent alloys $(Tb_{0.2}Pr_{08})_xY_{1-x}Fe_2$ – introduction of nonmagnetic Y leads to the compensation composition at yttrium concentration of x = 0.6 [4]. It shall be noted that the change in the magnetostriction sign is observed for compounds with x = 0.8, as shown in the inserts to Figures 2 and 5.

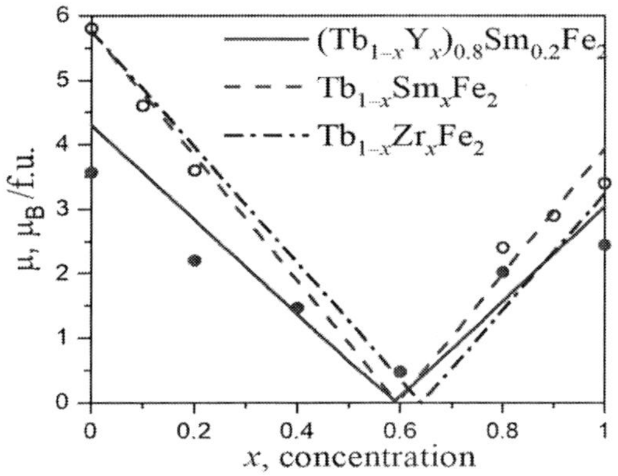

Figure 6. The concentration dependence of saturation magnetization of $Tb_{1-x}Zr_xFe_2$ system compounds (unpainted symbols), $(Tb_{1-x}Y_x)_{0.8}Sm_{0.2}Fe_2$. Solid lines indicate the design values of the magnetic moment of the specified compounds

5. Conclusion

Alloys $(TbR^IR^{II})Fe_2$ are shown to be sufficiently convenient for model studies of many properties of rare-earth alloys: magnetic, magnetoelastic, spectroscopic and optical. A distinctive feature of these compounds is that $TbFe_2$ exhibits positive spontaneous and field induced magnetostriction, whereas $SmFe_2$ and $ZrFe_2$ exhibit reverse sign magnetostriction. This results in a number of interesting effects in the systems as the concentration of the non-magnetic diluent (yttrium or zirconium) changes. In each of the studied systems, there is an inversion of the sign of magnetostriction constants in the area x = 0.8 and a phenomenon of magnetic compensation near x = 0.6. In the alloys of $Tb_{1-x}Sm_xFe_2$ system, two spin-reorientation phase transitions were established to form regions of angular magnetic structures. In $(Tb_{1-x}Y_x)_{0.8}Sm_{0.2}Fe_2$ alloy system the magnetic compensation by concentration and temperature is observed.

In our opinion, the reason for all these features in $Tb_{1-x}Zr_xFe_2$, $Tb_{1-x}Sm_xFe_2$ and $(Tb_{1-x}Y_x)_{0.8}Sm_{0.2}Fe_2$ is that the combination in the same rare-earth sublattices of light and heavy element ions (Tb and Sm), and then dilution by its non-magnetic atoms lead to the change of the sign and value of exchange interactions Tb-Fe and Sm-Fe, at the same time the exchange interactions also change in magnetic sublattices Tb-Tb, Tb-Sm, Fe-Fe. Competition between the above types of interactions resulted in the composition with complete magnetic compensation for magnetization over a wide range of temperatures. Exchange interactions, their type, sign and magnitude ultimately determine the entire spectrum of magnetic and magnetostrictive properties of this type of compounds. The study of compounds with Laves phase structure presents a great applied importance. This is caused by the fact that due to the variety of properties these intermetallides (and their related ones) find their application in metallurgy, laser engineering, semiconductor electronics, and are used as neutron absorbers in nuclear engineering. Besides, alloys with such properties can be used in robotics.

References

[1] Wang N J, Liu Y, Zhang H W et al 2016 Fabrication, magnetostriction properties and applications of Tb-Dy-Fe alloys: a review *China Foundry* **13(2)** 75–84

[2] Ren W J and Zhang Z D 2013 Progress in bulk MgCu2-type rare-earth iron magnetostrictive compounds *Chin. Phys. B* **22(7)** 077507

[3] Wang Y, Ma T Y, Wu C et al 2017 Correlation between magnetostriction and magnetic structure in pseudobinary compounds Tb(Co1-xFex)2 *AIP Advances* **7(7)** 075311

[4] Zhang G B, Liu Y D, Kan C X et al 2019 Structure and magnetostriction in (Tb0.2Pr0.8)xY1-x Fe1.93 Laves compounds synthesised by high-pressure annealing *Mater. Res. Bull.* **112** 174–7

[5] Politova G A, Karpenkov A Yu, Kaminskaya T P et al 2019 High-magnetostriction Laves-phase alloy of the samarium-iron system: the structure and phase transformations *St. Petersburg Polyt. State Univer. J. Phys. and Mathem.* **12(1)** 28–38

[6] Cwik J, Koshkid'ko Y S, Oliveira N A et al 2017 Magnetocaloric effect in laves-phase rare-earth compounds with the second-order magnetic phase transition: Estimation of the high-field properties *Acta Mater.* **133** 230–9

[7] Chzhan V B, Tereshina I S, Karpenkov A Y and Tereshina-Chitrova E A 2018 Persistent values of magnetocaloric effect in the multicomponent Laves phase compounds with varied composition *Acta Mater.* **154** 303–10

[8] Tereshina I, Cwik J, Tereshina E et al 2014 Multifunctional Phenomena in Rare-Earth Intermetallic Compounds With a Laves Phase Structure: Giant Magnetostriction and Magnetocaloric Effect *IEEE Trans. Mag.* **50(11)** 2504604

[9] Politova G A, Pankratov N Yu, Vanina P Yu et al 2019 Magnetocaloric effect and magnetostrictive deformation in Tb-Dy-Gd-Co-Al with Laves phase structure *J. Magn. Magn. Mater.* **470** 50–4

[10] Zhou Ch, Bao H, Matsushita Y et al 2019 Thermal Expansion and Magnetostriction of Laves-Phase Alloys: Fingerprints of Ferrimagnetic Phase Transitions *Materials* **12(11)** 1755

[11] Tureka K, Ruszb J and Divis M 2005 Electronic structure and volume magnitostriction of rare-earth metals and compounds *J. Magn. Magn. Mater.* **290** 357–63

[12] Stashkova L A, Gaviko V S, Mushnikov N V and Terent'ev P B 2013 Hydrogen ordering in rare-earth intermetallic (Er, Tb)Fe2 compounds with giant spontaneous magnetostriction *Phys. Metals and Metal.* **114** 1068–74

[13] Gschneidner Jr K A., Pecharsky V K and Tsokol A O 2005 Recent Developments in Magnetocaloric Materials *Rep. Progr. Phys.* **68**, 1479–539

[14] Engdahl G 2000 *Handbook of giant magnetostrictive materials* (Academic Press) 386 p

[15] Belov K P 1987 *Magnetostriction Phenomena and their Technical Application* (Moscow: Nauka) 159 p

[16] Belov K P, Katayev G I, Levitin R Z, Nikitin S A, and Sokolov V I 1983 The giant magnetostraction *Uspekhi Fizicheskikh Nauk* **140**, 271–313

[17] Stein F, Palm M and Sauthoff G 2004 Structure and stability of Laves phases Part I. *Critical assessment of factors con-trolling Laves phase stability. Intermetallics* **12** 713–20

[18] Tang Y J, Luo H L, Gao N F et al 1995 Magnetic properties and magnetostriction in $TbFe_2$ compound with the addition of manganese or gallium *Appl. Phys. Lett* **66** 388–90

[19] Samata H, Kasai T R, Taniguchi T and Nagata Y 2003 Al substitution effects to the magnetic properties of the compounds in the Sm–Fe system *J. of Solid State Chem.* **171** 339–44

[20] Liu X N, Lin K, Gao Q L et al 2018 Structure and Phase Transformation in the Giant Magnetostriction Laves-Phase $SmFe_2$ *Inorganic Chem.* **57(2)** 689–94

[21] Ilyushin A S, Aleroeva T A, Tereshina I S, Umkhaeva Z S et al 2019 Structural, magnetic, and Thermal Properties of the Compound $Tb_{0,8}Sm_{0,2}Fe_2$ with a Phase structure *Phys.of the Solid State* **61(12)** 2503–8

[22] Ilushin A S, Solodov Ye V and Umkhaeva Z S 2013 Structural and magnetic transformations in pseudobinary systems alloys $(Sm_{1-x}Tb_x)Fe_2$ compounds *Perspekt. Mater.* **11** 42–7

[23] Bisliyev A M, Umkhayeva Z S, Nikitin S A and Ilyushin A S 1989 Structure state of alloy $Tb_{1-x}Zr_xFe_2$ quasibinary system *Phys. Metals and Metal.* **67** 118–21

[24] Zhang Hong-guo, Wang Weng-hong, Liu En-ke et al 2013 Compensation effect and magnetostriction in $CoCr_{2-x}Fe_2O_4$ *Phys. Status Solidi B*, **250(7)** 1287–92

[25] Kumar A and Yusuf S M 2015 The phenomenon of negative magnetization and its implications *Physics Reports* **556** 1–34

Unconventional raw materials in silicate materials designing based on technogenic metasomatism knowledge

A Volodchenko, V Lesovik, A Volodchenko and I Cherepanova

Belgorod State Technological University named after V.G. Shoukhov, 46, Kostyukova street, Belgorod, 308012, Russia

E-mail: alex-0904@mail.ru, naukavs@mail.ru, volodchenko@intbel.ru, Inst703@intbel.ru

Abstract. A large number of scientific schools is working over improving existing building materials and designing some new composites for various functional purposes, where the composites properties play a very important role. However, priority is given to the initial properties of building composites but one of the main indicators by which these properties could be evaluated after prolonged use and under the influence of natural and man-made processes is durability. The study on building materials under changing operating conditions, based on unconventional natural and man-made materials is becoming very interesting and relevant for today. The theoretical basis for increasing the durability of building materials may be the theoretical knowledge of technogenic metasomatism in building materials science. It has been established that the use of unconventional natural and technogenic raw materials, in particular unconventional clay rocks for production of hydrothermal hardening silicate materials will help to achieve the necessary structurization level and maximum physic-mechanical properties of highly efficient wall products under various operating conditions.

1. Introduction

Currently, one of the main tasks of building materials science is to reduce the energy intensity of building composites production through the use of non-traditional, inter alia, technogenic raw materials [1–7]. The theoretical basis for solving this urgent problem is the transdisciplinary science of geonics (geomimetics).

Many scientific schools of all world countries are working on creating new building composites, developing technologies for their production and studying composites' properties, etc. However, priority is given to the initial properties of building composites but one of the main indicators by which these properties could be evaluated after prolonged use and under the influence of natural and man-made processes is durability. So, during operation, the same building composites fall into specific conditions, various processes are proceeding in them. Some buildings and structures during exploitation do not withstand external influences and for these reasons they are destroyed. Therefore a set of measures is needed aimed at more detailed studying the building materials under certain exploitation conditions, taking into account the environmental changes with increasing ecological stresses. The theoretical basis for increasing the building material durability can be the theoretical knowledge of technogenic metasomatism in building materials science. This is a stage in the evolution of building materials, characterized by the adaptation of the composite to some changing conditions during the exploitation of buildings and constructions.

The building industry evolves not only in the direction of production but also focuses on chemical and physicochemical processes. For example, the use of binders is being improved in the course of studying the processes which occur during the hardening and subsequent operation of materials produced on their base [9–11]. Using knowledge about chemical processes in developing building materials will dramatically improve the properties of composites and produce new-generation materials. Understanding these processes is extremely important when solving the problem of increasing the efficiency of any production technology and the quality of building materials.

The trend relating to deterioration of the quality and properties of materials is caused due to undesirable processes, not only external ones – exposure to CO_2, aggressive environment, but also internal ones – corrosion, recrystallization, aging, etc. For example, over the past 200 years, the carbon dioxide content in the Earth's atmosphere has been steadily increasing. In the pre-industrial era, it did not exceed 280 volume parts per million (ppm), and today it reaches 365 ppm and has a tendency to increase. Based on some statistics, in 2012, world emissions of CO_2 amounted to 34 billion tons and exceeded the 1990 figures by 50%. According to the data announced at the International Economic Forum (IWR) in the city of Münster, if the current trend continues, the volume of global carbon dioxide emissions will increase by 20 % by 2020 and amount to more than 40 billion tons. For comparison: in 1990, global carbon dioxide emissions barely reached 22.7 billion tons. The concentration of carbon dioxide in the atmosphere has reached a high level over the past 420 thousand years. All this must be taken into account when designing and using building composites.

The whole set of complex physical and chemical transformations, that is, technogenic metasomatism, when designing the material, will allow the composite, in response to external influences, to self-heal the defects that form during operation, restore their original characteristics and lead to hardening of the material with an increase in load under technogenic stress conditions.

One of the most economical and common among wall materials are silicate materials and products from them. The raw materials for their producing are environmentally friendly materials: mineral binder, siliceous component and water. Portland cement and ground lime are used as a binder. It is possible to replace lime and Portland cement with ground slag (metallurgical, fuel, etc.). Being dependent on this type of raw material, enterprises lose the opportunity to develop in a rapidly developing competitive industrial world. Also, the process of synthesis of calcium hydro-silicates in composites based on quartz sand and lime proceeds at high temperatures and pressures [12-13]. In the process of hydrothermal processing of silicate materials, organic impurities that accidentally fall into the raw material blend can burn out and volatilize. Therefore, such products do not emit harmful components during exploitation under various internal and external impacts.

The development of technology aimed at silicate brick manufacturing is based on the achievements of many researchers who carry out experimental studies on development of theories giving the knowledge of structurization of silicate materials, as well as on modernization of the technology targeted at increasing some properties of the silicate brick. This allows creating a theoretical basis for further improving the technology of autoclave materials. As a result, new types of raw materials are used in the technology of silicate brick production.

An analysis of the data within the research studying the raw material base of autoclave hardening materials allowed theoretically substantiating and experimentally confirming the possibility of using instead of sand the clay rocks of an incomplete phase going in mineral formation processes [14–15]. Such clay rocks are widespread, and also in large quantities are simultaneously recovered in the extraction of ore minerals. The specificity of these rocks is the presence of thermodynamically unstable compounds, such as imperfect hydromica structure, mixed-layer minerals, finely dispersed weakly rounded quartz, as well as a small amount of montmorillonite and kaolinite. This raw material, which has the properties of natural nano-sized particles, allows to control the synthesis of neoplasms to obtain materials with the desired properties.

Clay rocks have a very diverse mineral composition and properties. In recent decades, when the modern research methods are used, the structures of clay minerals and their properties have been studied in detail. It was found that the elementary layers and the spaces between them in the clay system are nano-scale and have a highly developed active surface [16–17].

The use of a specially selected alumino-silicate binder based on non-traditional clay raw materials in producing the silicate wall materials of hydrothermal hardening will help to achieve the necessary level of structurization and maximum level of some physical and mechanical properties of products not only at high pressures and temperatures, but also at temperatures up to 100 °C.

2. Methods and materials

For research, the Aeolian-eluvial-deluvial clay rock of the Quaternary age was used. These rocks are widespread in the region of the Kursk magnetic anomaly.

Visually, the mineral is unconsolidated (loose) rock of brown colour. It has a clay-silty structure. The size is dominated by silt and pelitic particles. By chemical composition, the rocks are classified as acidic with a high content of free silica (39.96 wt. %). In addition, a fraction of less than 0.005 mm also contains a significant amount of free quartz (33.60 wt. %).

The clay fraction of the rock is represented by Ca^{2+} montmorillonite with d_{001}= 14.81–15.87 Å (fig. 1). These layers of the mineral structure are easily replaced by polar organic molecules; therefore, when saturated with ethylene glycol, the main interplanar distance increases. During calcinations at 600 °C for two hours, the reflections characteristic of d_{001} decrease to 9.99 Å, due to the fact that interlayer water is easily removed at this temperature.

A series of reflexes of 9.99, 4.99 and 3.32 Å, the values of which do not change either after saturation with ethylene glycol or after calcination, indicate the presence of hydromica in the rock. The interplanar distances that are multiples of 7.14 Å, which disappear upon calcination within 2 hours, allows identifying kaolinite.

Quicklime clod lime with an activity of 87.34 % was used as an astringent component. As siliceous aggregate, quartz sand with a particle size modulus of 1.54 was used. The content of SiO_2 was 92.4 wt. %

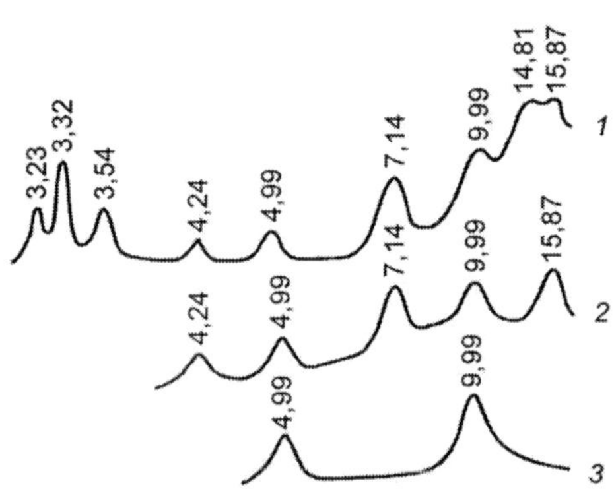

Figure 1. X-ray diffraction patterns of clay rocks of a fraction of less than 0.005 mm: 1 — original sample; 2 – saturated with ethylene glycol; 3 – calcined for 2 hours at 600 °C.

The composition of the raw blends was determined based on the conditions of hydrothermal treatment:

- for hydrothermal treatment at elevated pressures and temperatures, raw blends were prepared with a content of 8 % active CaO, the content of aeolian-eluvial-deluvial clay rock in the raw material blend was changed from 5 to 70 wt. %. As a control, a calcareous sand blend was used. The molding moisture content of the raw material blend depended on the clay content and amounted to 7–10 %. The rock was added to the raw material blend in the form of a calcareous-sandy-clay binder, obtained by joint grinding of the rock and lime to a specific surface of 500 m^2/kg;
- for hydrothermal treatment under atmospheric pressure at a temperature of 95–0659 C, a mixture was used to produce the samples, including a pre-prepared aluminosilicate binder obtained by

co-grinding clay and lime, and as an aggregate, the original aeolian-eluvial-deluvial clay rock. The moisture content of the raw blend was 8–12 %.

All samples were obtained by semi-dry molding at a specific pressing pressure of 20 MPa. The phase composition of neoplasms was studied by x-ray and thermo-graphic analysis. The physical and mechanical properties of the samples were determined in line with the requirements of the regulatory documents.

MicroSizer 201 was used to determine the particle size distribution of materials. In order to determine the material composition of the clay rocks used, as well as the emerging composition of the neoplasms, X-ray phase analysis (model diffractometer – ARL X'TRA) and differential thermal analysis (Derivatograph Q – 1500 D) were used in this research. For scanning electron microscopy (SEM), a MIRA 3 LM scanning electron microscope was used.

3. Results

For autoclaved samples, the compressive strength, average density and water absorption were determined (fig. 2). The experimental data showed that the clay rock has a positive effect on increasing the strength of silicate materials. The optimal content of loam, corresponding to the maximum strength of the samples, is 20–30 wt. %. The strength of the samples increases from 20 to 32 MPa (1.6 times). When the rock content is in the amount of 5 wt. %, there is a slight decrease in strength. The average density increases from 1780 to 1980 kg/m^3 with a rock content of 30 wt. %. The minimum water absorption (9.3%) corresponds to a rock content of 20 wt. %

There were carried out some microscopic studies of calcareous and sandy samples with a clay rock content of 30 wt. %. It has been established that on the surface of quartz grains of calcareous-sandy images and at the contacts between them, there are some substanes, indistinct and smudgy by nature (difficult ot identify). Cementing compounds are formed by the reaction of lime with fine and coarse aggregate quartz. Grains of quartz in samples with clay content are evenly distributed in the total amorphous mass.

The surface of many small particles of quartz is subject to corrosion and is surrounded at the edges by a gel-like film of neoplasms.

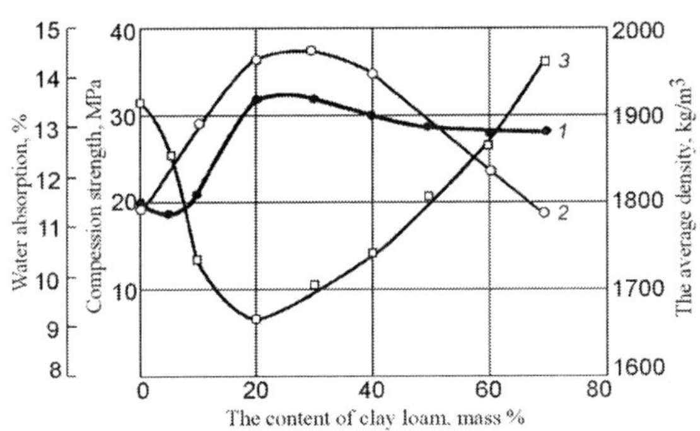

Figure 2. Physico-mechanical properties of silicate materials of autoclave hardening depending on clay conyent (active CaO content 8 wt); 1 – compressive strength; 2 – verage density; 3 – water absorption

From the data presented, it can be concluded that in the lime-sand blend in the presence of clay minerals, the formation of a cementitious compound occurs mainly due to the interaction of calcium hydroxide with clay minerals and partially finely divided quartz. Coarse quartz with lime practically does not react.

The cementitious compound of calcareous (control) samples is represented by low-basic calcium hydro-silicates of the CSH (B) group, detected by the exothermic effect at 835–1659 °C in the thermo-gram and reflections of 3.04, 2.80, 1.82 Å in the X-ray diffraction pattern (fig. 3).

Calcium hydrosilicates CSH (B) are also formed in samples based on a calcareous binder. The shift of the exothermic effect to higher temperatures (870–880 °C) is probably associated with an increase in the basicity of calcium hydrosilicates.

With a content of 5 wt. % of loam in the raw blend shows the hydrogranates formation, the amount of which increases with increasing rock content that can be judged by the increase in the intensity of the endothermic effect at 340 °C and reflections of 5.00, 2.75, 2.00 Å, increases (fig. 3).

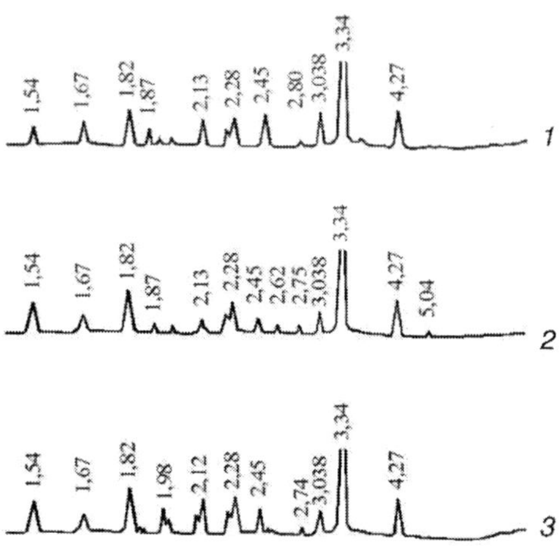

Figure 3. X-ray diffraction patterns of samples: 1 – calcareous; loam content, wt. %; 2 to 5; 3-30

The data allow us to conclude that the maximum permissible content of sand-clay rock in the raw material blend determines the amount of clay component of the rock and the content of lime.

The increase in the samples strengthing occurs as a result of the formation of a stronger microstructure of the cementitious substance due to an increase in the packing density of the material, as well as the synthesis of hydrogranates. Isometric crystals and hydrogranate plates have a small specific surface and are microfiller, which is cemented by a submicrocrystalline gel phase from low-basic calcium hydrosilicates. An increase in average density and the formation of denser packaging associated with this leads to a decrease in water absorption. This is confirmed by the fact that the samples with the maximum average density correspond to the minimum water absorption (fig. 2).

To obtain hydrothermal silicate materials under atmospheric pressure, aeolian-eluvial-deluvial clay rock of the Kursk magnetic anomaly region was used as in previous experiments.

To conduct an experiment, a raw blend was prepared with a percentage of CaO of 15 wt. % (table 1). The samples were obtained by pressing, where the pressing pressure was 20 MPa. The moisture content of the raw blend depended on the composition, and was in the range of 8–12 %. After 2 days, the samples were tested to identift a compressive strength. Part of the samples of each composition withstood 1 year in a tap water. After that, the samples were tested for tensile strength in compression in a water-saturated state. The results are shown in table 2 and 3.

Table 1. Compositions of raw blends based on a calcareous binder

Composite No.	Lime content in raw materials blend, wt. %	Lime ration: clay in a binder	Specific surface of a binder, m^2/kg
1	15	1 : 1	1000
2	15	1 : 1.5	950
3	15	1 : 2	700

The physicomechanical parameters of silicate samples indicate that the studied raw materials actively interact with lime under steaming conditions at a temperature of 90–95 °C.

This time as well, the physicochemical processes are going that lead to the synthesis of a complex binder, forming a strong frame. According to the differential thermal and x-ray phase analysis, the neoplasms are represented mainly by weakly crystallized calcium hydrosilicate compounds — CSH (B) and C_2SH_2.

Table 2. Physico-mechanical properties of silicate materials based on an aluminosilicate binder

Physoco-mechanical properties	Composite No.		
	1	2	3
Compressive ultimate strength, MPa	22.58	21.8	21
Softening coefficient	0.8	0.85	0.9
Average density, kg/m^3	1815	1825	1800
Water absorption, %	11.6	9.8	9

The test results of the samples cured for 1 year in water (table 3) showed a significant increase in strength compared to the samples that were not subjected to long-term storage in water.

Table 3. Compositions of raw blends based on a calcareous binder

Physico-mechanical properties	Composite No.		
	1	2	3
Compressive ultimate strength, MPa	34.7	32.5	32
Softening coefficient	0.73	0.8	0.8
Average density, kg/m^3	1830	1820	1810
Water absorption, %	10.2	9.5	8.8

This is due to the fact that the rock-forming minerals, in particular, its nanodispersed component, provide the synthesis of a cementing compound with hydraulic properties. The cementitious compounds are likely to possess these properties both due to the synthesis of highly basic compounds, which undergo further hydration, and due to recrystallization of gel-like calcium hydrosilicates that leads to densification of the neoplasm structure.

In the microstructure of the original sample there is an accumulation of globules up to 0.5 μm in size, which are connected by a network of neoplasms, which are weakly crystallized low-basic calcium hydrosilicates (fig. 4).

Neoplasms also cover the surface of the aggregate. A crystallization structure was formed here, providing high strength and water resistance.

The chemical composition of the globules of the original sample was studied by energy dispersive x-ray spectroscopy. According to the emission spectra, the elemental composition of the globules was calculated. The main elements that make up the globules are wt. %: O – 33.57–44.42; Ca – 28.2-40; Si – 12.44-24.63; Al – 7.37-10.92. It follows that the globules are formed mainly due to the interaction of lime with clay minerals and the X-ray amorphous phase and as a result, they are the amorphous aluminates and calcium hydrosilicates. It can be concluded that globules are intermediate compounds formed during the synthesis of neoplasms from the finely dispersed part of the rock-forming minerals, and, above all, the clay component and lime.

The microstructure of the sample after a year of storage in water is significantly different from the original sample, which was not subjected to long-term storage in water (fig. 5).

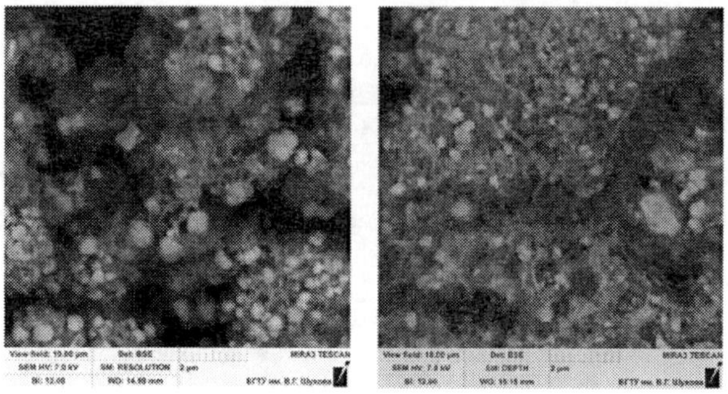

Figure 4. The microstructure of the composite obtained by hydrothermal treatment under atmospheric pressure

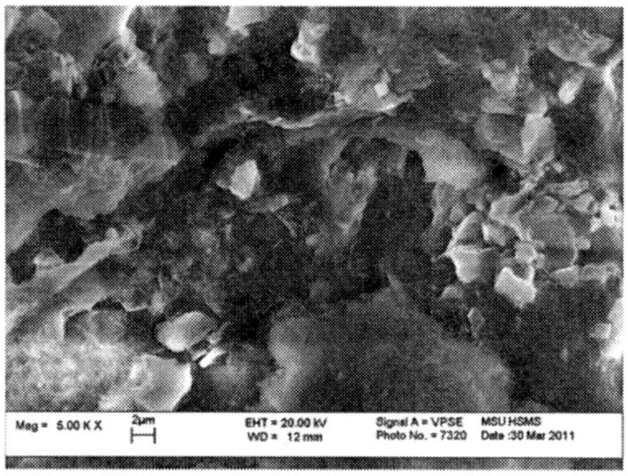

Figure 5. The microstructure of the composite obtained by hydrothermal treatment under atmospheric pressure in one year

The globules in the structure of the cementitious compound are practically absent. At the same time, the number of weakly crystallized calcium hydrosilicates increases, which, forming a continuous grid, almost completely fill the pores, cover the surface of the filler and fasten its grains together. Consequently, the globules exhibit thermodynamic instability and the formation of weakly crystallized calcium hydrosilicates based on them continues after hydrothermal treatment of products. In the water environment, the process of calcium hydrosilicates formation continues over time, which leads to the formation of a more durable microstructure of the cementitious compound. In addition, weakly crystallized calcium hydrosilicates are an unstable phase, capable of undergoing recrystallization in time, and especially in the water environment, which also leads to a change in the structure of the cementitious compound. Probably, these processes provide the hydraulic properties of the silicate materials.

4. Conclusion

Thus, using aluminosilicate raw materials which are represented by clay rocks, non-traditional for the construction industry; it is possible to obtain high-hollow silicate wall composites with improved properties, both in autoclave synthesis and in heat and moisture treatment. The use of pure quicklime or a specially prepared binder, as a binder component, in order to obtain composites with higher properties depends on the specific rocks used and their composition.

The studies have established the influence of operating conditions on the structurization mechanism of the composites based on unconventional clay raw materials. We identified that initially, when clay minerals and finely dispersed quartzas interact with lime, under conditions of heat and heat treatment, intermediate compounds are formed, in a shape of the globules, which are amorphous hydroaluminates and calcium hydrosilicates. Further, based on them, a spatial network of neoplasms is formed, consisting of weakly crystallized calcium hydrosilicates, as well as aluminum-substituted calcium hydrosilicates and hydrogranates. After prolonged use in humid conditions, these neoplasms recrystallize, thus compacting the structure at the nano, micro, and macro levels.

Acknowledgment
The reported study was funded by RFBR according to the research project № 18-03-00352.

References

[1] Suleymanova L A, Lesovik V S, Kara K A et al 2014 Energy-efficient concretes for green construction *Res. J. of appl. Sci.* **9(12)** 1087–90
[2] Dmitrieva T V, Strokova V V and Bezrodnykh A A 2018 Influence of the genetic features of soils on the properties of soil-concretes on their basis *Construct. Mater. and Products* **1(1)** 69–77
[3] Kuprina A A, Lesovik V S, Zagorodnyk L H and Elistratkin M Y 2014 Anisotropy of materials properties of natural and man-triggered origin *Res. J. of Appl. Sci.* **9(11)** 816–9
[4] Murtazaev S-A Y, Zaurbekov A Kh, Saydumov M S et al 2018 Impact of technogenic raw materials on the properties of high-quality concrete composites *Advances in Engineering Research* **177** 275–9
[5] Pucharenko J and Morozov V 2013 Structural model and strength predicting of fiber-reinforced concrete *World appl. Sci. J.* **23** 111–6
[6] Elistratkin M Yu and Kozhukhova M I 2018 Analysis of the factors of increasing the strength of the non-autoclave aerated concrete *Construct. Mater. and Products* **1(1)** 59–68
[7] Pukharenko Yu V, Letenko D G, Nikitin V A and Morozov V I 2017 Obtaining the nanomodifier for cement composites based on the "dealtom" carbon nanotubes *Mater. Phys. and Mechan.* **31** 59–62
[8] Lesovik V S and Volodchenko A A 2015 The problem of technogenic metasomatose in materials science *Bull. of BSTU named after V G Shukhov* **4** 38–41
[9] Murtazaev S-A Y, Salamanova M, Zaurbekov S et al 2016 Composite Binders with the Use of Fine Raw Materials of Volcanic Origin *Int. J. of Environ. & Sci. Ed.* **1(18)** 12711–6
[10] Strokova V V, Sumin A V, Nelubova V V and Shapovalov N A 2015 Modified binder with application of nanostructured mineral components *Bull.of BSTU named after V G Shukhov* **3** 36–9
[11] Pukharenko Yu V, Letenko D G, Nikitin V A and Morozov V I 2017 Obtaining the nanomodifier for cement composites based on the "dealtom" carbon nanotubes *Mater. Phys. and Mechan.* **31** 59–62
[12] Klimesch Danielle and Ray Abhi 2002 Evaluation of phases in a hydrothermally treated CaO-SiO2-H2O system *J. of Thermal Anal. and Calorimetry* **70(3)** 995–1003
[13] Bernstein S and Fehr T K 2012 The formation of 1.13 nm tobermorite under hydrothermal conditions: 1. The influence of quartz grain size within the system CaO–SiO2–D2O *Progress in Crystal Growth and Characterizat. of Mater.* **58(2-3)** 84–91
[14] Volodchenko A A, Lesovik V S, Volodchenko A N and Zagorodnjuk L H 2015 Improving the efficiency of wall materials for "green" building through the use of aluminosilicate raw materials *Int. J. of Appl. Engineer. Res.* **10(24)** 45142–9
[15] Volodchenko A A, Lesovik V S, Volodchenko A N et al 2016 Composite performance improvement based on non-conventional natural and technogenic raw materials *Int. J. of Pharmacy and Technol.* **8(3)** 18856–67
[16] Olad Ali 2011 Polymer/Clay Nanocomposites *Advances in Diverse Industrial Applications of Nanocomposites, InTech* 113–38
[17] Kiliaris P and Papaspyrides C D 2010 Polymer layered silicate (clay) nanocomposites: An overview of flame retardancy *Progress in Polymer Science* **35** 902–8

IOP Conf. Series: Materials Science and Engineering **905** (2020) 012073 doi:10.1088/1757-899X/905/1/012073

Mathematical modeling of the transmission pipeline under the influence of earthquake seismic waves

D M Zaripov and S F Khizbullina

Ufa State Petroleum Technological University, 14, Pervomayskaya str., Ufa, 450000, Russia

E-mail: damir.zaripov@gmail.com, svetahf@mail.ru

Abstract. The stress-strain state of an underground transmission pipeline under the action of earthquake seismic waves is considered. The main objective of this study is the construction of a mathematical model with the analytical solution for lower estimation of the influence of earthquake seismic waves on the transmission pipeline. The spectra of the design seismic actions of the earthquake are determined on the basis of parameters of real seismograms of past earthquakes. The generalized functions of seismic impulses are constructed. The transmission pipeline is modeled by a thin bar. Equations of pipeline motion using the Laplace transformation in time are solved, stresses are determined. The numerical example is considered. The results of calculations on the theories of Kirchhoff and Timoshenko are compared. Graphs of stress changes depending on the earthquake's strength in points are given. The limiting value of the earthquake magnitude resulting in large stresses and deformations is determined. As a result, the transmission pipeline is destroyed.

1. Introduction

Past data shows that earthquake damage to underground transmission pipeline can lead to extended outage and unpredictably high socio-economic losses in society, not to mention the environmental consequences that will be manifested in the future. In a relatively recent study [1], an extensive review of the literature on this issue was carried out. The analysis of existing models for calculating the stress-strain state of the transmission pipeline under the influence of earthquake seismic waves. This work makes a disappointing conclusion that modern seismic standards and guidance documents do not reflect the important results of the latest research. There are also no general principles with detailed step-by-step methodology for designing aseismic underground pipeline systems. On the contrary, there is basically a set of empirical recommendations that are sometimes incomplete or outdated.

The nature of deformation and destruction of constructions under the influence of earthquake seismic waves indicates that underground structures are less damaged than above-ground structures. There are several reasons for this. First, underground objects are constrained by the surrounding soil, which significantly reduces the oscillations amplitude. Second, seismic wave energy is dissipated into the surrounding soil. Third, the most dangerous surface waves have no greater effect on underground objects than on above-ground structures. However, in the book of Poliakov S.V. [2], it is noted that some underground structures have experienced significant deformation and damage during major earthquakes of the 20th century, including the San Francisco USA in 1906, Tokyo Japan in 1923, Chile in 1943, etc. The book [3] is devoted to the review of experimental studies of soil movement in surface and deep layers.

It should be noted that in soils with significantly different mechanical characteristics, an increase in the intensity of the influence of earthquake seismic waves. They cause deformation or destruction of long underground structures, which leads to their failure or reduce performance. It is accepted to allocate two factors of influence of ground conditions on the seismic stability of structures. The first factor is due to the dynamic characteristics of soils in the surface layers. The second is determined by the decrease of dispersed soil bearing capacity arising from oscillations. As a result, this leads to a significant change in the stress-strain state of the underground structure. At certain combinations of characteristics, in particular, the frequencies of natural oscillations of surface layers with the frequencies of forced oscillations of bedrock, resonance phenomenon in surface layers may occur [2].

In the work [4], the regularities of propagation of the incident quasicompressional and transverse seismic waves in the intermediate monoclinic layer between strongly aeolotropic soil layers were analyzed. It is established that the critical angles for incident quasicompressional and transverse waves are different. It is shown that the angles of incidence and the width of the soil layer have a significant effect on the resultant displacement, velocity and reflectance in the considered soil layer.

Work [5] explores the influence of seismic waves on an underground structure – power station. It is established that the deformation of the underground structure is obviously affected by the angles of incidence of the transverse SV-wave and the longitudinal P-wave. The deformation reaches a maximum at an angle of incidence of 30° for SV-and P-waves. Weak parts of the underground structure of the power station under seismic load are places with a with a large free surface, such as arches, roofs and floors. It is shown that the state of the contact surface between the surrounding rock and the underground structure plays a key role in the strength of the underground structure. The surrounding soil has a damping effect on the underground structure. It is also a favorable factor for the underground structural competence under seismic load.

The purpose of this work is the lower estimate of the influence of earthquake seismic waves on the transmission pipeline. In the opinion of the authors of this work, it is extremely relevant to determine the limit earthquake magnitude, at which the destruction of the pipeline may occur, leading to environmental disasters.

A seismic wave, or a seismic impulse, as can be seen from the seismograms, is a short-term oscillation with a duration of 3-4 dominant oscillation periods. In the first and terminal phases, oscillations with a high frequency and low amplitude are registered. The wave coming from great depths from the seismic center to the earth surface, is refracted, reflected from layers with curved interface boundaries with different wave propagation velocities, and wave interference occurs. Therefore, as the distance from the seismic center increases, the wave pattern on the earth's surface changes and becomes more complex. The impulse shape also changes depending on the soil properties at the observation point, attenuation constant and oscillation frequency. The most intense oscillations are recorded in the second – main phase. Oscillations occur with high amplitude and relatively low frequency (fig. 1). In work [6], 200 seismograms recorded during earthquakes in different regions of the Russian Federation and neighboring countries are presented. To perform mathematical modeling for the seismogram shown in figure 1, an approximation of the seismogram was constructed (fig. 2).

Oscillations with high amplitude and low frequency carry the maximum fraction of seismic impulse energy and affect structures more strongly than the oscillations of the first and last phases [7]. The main phase of seismic impulses is usually approximated by expressions [8]:

$$v_1(t) = \dot{V} t e^{-\eta t} \sin \Omega\, t, \quad v_2(t) = V e^{-(\eta t)^2} \sin \Omega\, t, \tag{1}$$

$v_1(t)$ and $v_2(t)$ – soil particles displacement; $\dot{V}$, V – velocity and displacement amplitudes; η – attenuation constant; Ω – oscillation frequency; t – time. The choice of expressions (1) is based on the superficial similarity with the main phase of seismograms (fig. 2). The velocities and accelerations of soil particles can be found by differentiating expressions (1):

$$\ddot{v}_1(t) = -\dot{V}\Omega e^{-\eta t}\left\{\left[\left(1-\frac{\eta^2}{\Omega^2}\right)\Omega t + 2\frac{\eta}{\Omega}\right]\sin \Omega\, t - 2(1-\eta t)\cos \Omega\, t\right\},$$

IOP Conf. Series: Materials Science and Engineering **905** (2020) 012073 doi:10.1088/1757-899X/905/1/012073

$$\ddot{v}_2(t) = -V\Omega^2 e^{-(\eta t)^2}\left\{\left(1+2\frac{\eta^2}{\Omega^2}-4\frac{\eta^2}{\Omega^2}(\eta t)^2\right)\sin\Omega\, t + 4\frac{\eta}{\Omega}\eta t\cos\Omega\, t\right\}.$$

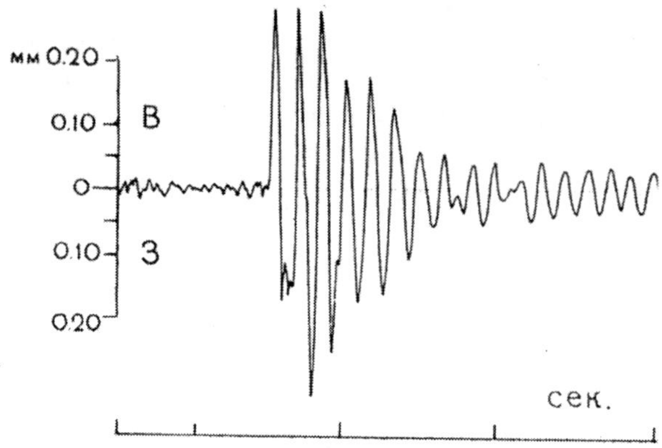

Figure 1. Seismogram [6]

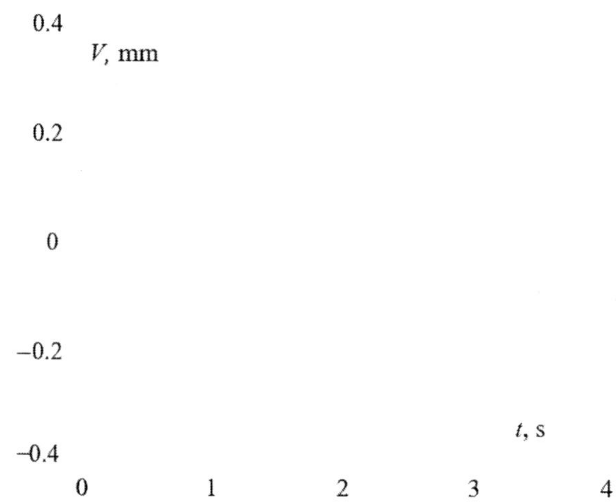

Figure 2. Seismogram approximation

However, engineering seismology manuals refer to earthquake manifestations as successive shocks and aftershocks that follow one after another after a certain period of time [9]. The theory of seismic stability, based on the assumption that the soil motions are performed according to harmonic law, ignores the phenomenon of "shocks" and "aftershocks". Eyewitness accounts show that even with a relatively weak earthquake, there are facts of displacement or overturning of individual solid objects [9].

Below, the harmonic functions (1) are replaced by generalized functions [10] and the action of the seismic impulse on the transmission pipeline is considered as a sequence of shocks.

2. Generalized seismic wave functions

We place the origin of the rectangular coordinate system *x, z* on the pipeline axis at the point before the seismic impulse, directing the x-axis along its axis, the z-axis perpendicular to the earth surface. We compatible the main phase initiation of seismic waves with the origin of the coordinate axes. At time:

$$t_* = \left(\frac{1}{4}+n\right)T_1 = \frac{2\pi}{\Omega\sqrt{1-(\eta/\Omega)^2}}\left(\frac{1}{4}+n\right) \approx \frac{2\pi}{\Omega}\left(\frac{1}{4}+n\right), \quad n = 0, 1, 2...$$

The function $\sin\Omega t$ takes the largest positive values, but $\cos\Omega t$ equals zero. Therefore, in expressions (1) and their derivatives, the terms containing $\cos\Omega t$ are assumed to be zero. We also neglect the value of η/Ω, as small in comparison with the unit. As a result we get:

$$\ddot{v}_1(t) = -\dot{V}\Omega^2 te^{-\eta t}\sin\Omega t, \quad \ddot{v}_2(t) = -V\Omega^2 e^{-(\eta t)^2}\sin\Omega t. \qquad (2)$$

Then, in expressions (1) and (2), the function $\sin\Omega t$ is replaced by its values at times t_*, and time t is replaced by time t_*. Thus, the following generalized functions of the seismic impulse are obtained:

$$v_1(t) = \frac{2\pi\dot{V}}{\Omega}\sum_{n=0}^{j-1}\left(\frac{1}{4}+n\right)exp\left[-\frac{2\pi\eta}{\Omega}\left(\frac{1}{4}+n\right)\right],$$

$$v_2(t) = V\sum_{n=0}^{j-1} exp\left[-\frac{2\pi\eta}{\Omega}\left(\frac{1}{4}+n\right)\right]^2,$$

$$\ddot{v}_1(t) = -2\pi\dot{V}\Omega\sum_{n=0}^{j-1}\left(\frac{1}{4}+n\right)exp\left[-\frac{2\pi\eta}{\Omega}\left(\frac{1}{4}+n\right)\right],$$

$$\ddot{v}_2(t) = -V\Omega^2\sum_{n=0}^{j-1} exp\left[-\frac{2\pi\eta}{\Omega}\left(\frac{1}{4}+n\right)\right]^2, \qquad (3)$$

where $n = j-1$, j – number of shocks during seismic impulse action. Impulsive forces are displaced along the pipeline with the compressional velocity a_* and act at intervals $T_1 \approx T = 2\pi/\Omega$.

3. Pipeline movement equations according to Timoshenko model and their solution

The transmission pipeline is modeled by a thin bar. Using dimensionless quantities and notation in the form:

$$\xi = \frac{x}{r},\ \tau = \frac{c_1 t}{r},\ w = \frac{W}{r},$$
$$c_1^2 = \frac{E}{\rho},\ c_2^2 = \frac{k'G}{\rho},\ r^2 = \frac{J}{F},\ \gamma = \gamma_*\left(1+\frac{m_2}{\rho F}\right),\ \gamma_* = \frac{c_1^2}{c_2^2}, \qquad (4)$$

Bar motion equations in displacements, taking into account the shear deformation and rotary inertia, are written as follows:

$$\frac{\partial^2 w}{\partial\xi^2} - \frac{\partial\theta}{\partial\xi} - \gamma\frac{\partial^2 w}{\partial\tau^2} = -\frac{rp(\xi,\tau)}{\rho F c_2^2},$$
$$\frac{\partial w}{\partial\xi} - \theta + \gamma_*\left(\frac{\partial^2\theta}{\partial\xi^2} - \frac{\partial^2\theta}{\partial\tau^2}\right) = 0. \qquad (5)$$

In expressions (4) and (5): r – radius of inertia; E, G, ρ – elastic, shear modulus and density of pipe material; $k' = 1.1$ – form of section factor; F, J – cross-sectional area and axial moment of inertia; w, θ – deflection and angle of rotation due to the bending moment; τ – time; $m_2 = \rho_2\left[D(h-0.39D) + h^2\,\mathrm{tg}(0.7\varphi)\right]$ – soil mass above the pipe; h – distance from the top of the pipeline burial to its pitch line. Pressure on the pipeline by seismic wave has the form:

$$p(\xi,\tau) = p_n(\xi_0)\delta(\xi - \tau/\beta) = \beta p_n(\xi_0)\delta(\tau - \beta\xi),$$

where δ – Dirac delta function, $\xi_0 = x_0/r$, $\beta = c_1/a_*$.

Equations (5) are solved using the Laplace transformation in time:

$$\frac{d^2\overline{w}}{d\xi^2} - \frac{d\overline{\theta}}{d\xi} - \gamma s^2 = k\gamma_* e^{-\beta s\xi},$$
$$\frac{d\overline{w}}{dx} + \gamma_*\frac{d^2\overline{\theta}}{d\xi^2} + \left(\gamma_* s^2 + 1\right)\overline{\theta} = 0, \qquad (6)$$

where $k=-r\beta p_n(\xi_0)/EF$; $\overline{w}$, $\overline{\theta}$ – images of deflection w and angle of rotation θ. From the first equation (6) is excluded θ:

$$\frac{d^4\overline{w}}{d\xi^4}-(\gamma+1)s^2\frac{d^2\overline{w}}{d\xi^2}+\gamma s^2\left(s^2+\frac{1}{\gamma_*}\right)\overline{w}=kf_2e^{-\beta s\xi},$$

$$\overline{\theta}=\frac{\gamma_*}{\left(\gamma_* s^2+1\right)}\left[\frac{d^3\overline{w}}{d\xi^3}-\left(\gamma s^2-\frac{1}{\gamma_*}\right)\frac{d\overline{w}}{d\xi}+k\beta\gamma_* se^{-\beta s\xi}\right], \quad (7)$$

$$f_2=\left(\beta^2-1\right)\gamma_* s^2-1.$$

The system solution in images (7) is:

$$\overline{w}=A_1e^{-\lambda_1\xi}+A_2e^{-\lambda_2\xi}+\frac{kf_2e^{-\beta s\xi}}{b_1s^2\left(s^2+a_1^2\right)},$$

$$\overline{\theta}=\left(\lambda_1^2-\gamma s^2\right)\frac{A_1}{-\lambda_1}e^{-\lambda_1\xi}+\left(\lambda_2^2-\gamma s^2\right)\frac{A_2}{-\lambda_2}e^{-\lambda_2\xi}+\frac{k\beta e^{-\beta s\xi}}{b_1s\left(s^2+a_1^2\right)},$$

where $A_{1,2}$ – integration constants; $b_1=(\beta^2-\gamma)(\beta^2-1)$, $a_1^2=\gamma/(\gamma_* b_1)$; $\lambda_{1,2}$ – two (of four) roots of the characteristic equation:

$$\lambda^4-(\gamma+1)s^2\lambda^2+\gamma s^2\left(s^2+\frac{1}{\gamma_*}\right)=0,$$

satisfying the condition of attenuation $\overline{w}$ and $\overline{\theta}$ to infinity.

Bar cross-section $\xi=0$ is a sliding restraint, the angles of rotation and shear of the cross-section are equal to zero:

$$\xi=0,\ \overline{\theta}=0,\ \frac{d\overline{w}}{d\xi}-\overline{\theta}=0. \quad (8)$$

The integration constants are determined by boundary conditions (8). The image of the dimensionless bending moment is found by the formula:

$$\overline{m}(s,\tau)=\frac{Mr}{EJ}=\frac{d\overline{\theta}}{d\xi},$$

where M – dimensional bending moment.

The originals of deflection and bending moment are determined using the inversion formula:

$$\frac{1}{2\pi i}\int_{c-i\infty}^{c+i\infty}F(s)e^{\tau s}ds=\begin{cases}f(\tau), & \tau>0,\\ 0, & \tau<0.\end{cases}$$

The integrands have simple poles and branching points. Calculations were made by formula:

$$I=\sum\text{res}(s)-\sum\int_{\gamma_i},$$

where $\text{res}(s)$ – residue, γ_i – integration paths along the banks of the cut and arcs of a circle of infinitesimal radius. In the case of the action of a unit mobile force for a bending moment the expression is obtained:

$$m(s,\tau)=\sum_{i=1}^{5}\text{res}(s_i),\quad 0\le\xi\le\xi_3=\tau/\beta. \quad (9)$$

The expanded expression (9) can be written in the form:

$$m(s,\tau)=\frac{r\beta p(\xi_0)}{EF}\sum_{i=1}^{3}R_i,$$

$$R_1=\frac{\beta}{4\sqrt{2}a_1b_1}\left[æ_1\left(\sin\varepsilon_1-\sin\varepsilon_2\right)+2æ_2e^{\frac{-a_1æ_1\xi}{\sqrt{2}}}\cos a_1\tau_2\right],$$

$$R_2 = \frac{\beta\left(2\beta^2-\gamma-1\right)}{4\sqrt{2}(\gamma-1)b_1\sqrt{a^2+a_1^2}}\left[æ_1\left(\sin\varepsilon_1-\sin\varepsilon_2\right)-2æ_2 e^{\frac{-a_1æ_1\xi}{\sqrt{2}}}\cos a_1\tau_2\right]$$

$$R_3 = \frac{\beta^2}{a_1 b_1}\sin\left[a_1\left(\tau_2-\beta\xi\right)\right],$$

$$\varepsilon_{1,2} = a_1\left(\tau_2 \pm \frac{æ_1\xi}{\sqrt{2}}\right),\quad æ_{1,2} = \left[(\gamma-1)\sqrt{1+\frac{a^2}{a_1^2}}\pm(\gamma+1)\right]^{\frac{1}{2}},$$

$$a^2 = \frac{4\gamma}{\gamma_*(\gamma-1)^2},\quad \tau_2 = \tau-\tau_*,\quad \tau_* = \frac{\pi c_1}{2\Omega r}.$$

In the case of the action of several forces following each other, using the principle of composition of forces, the result is the following expressions for bending moments:

$$m_I(\xi,\tau) = \frac{2\pi r\beta\dot{V}\Delta}{\Omega EF}\sum_{n=0}^{j-1}\left(\frac{1}{4}+n\right)exp\left[-\frac{2\pi\eta}{\Omega}\left(\frac{1}{4}+n\right)\right]\sum_{i=1}^{3}R_i,$$

$$m_I(\xi,\tau) = \frac{r\beta V\Delta}{EF}\sum_{n=0}^{j-1} exp\left[-\frac{2\pi\eta}{\Omega}\left(\frac{1}{4}+n\right)\right]^2\sum_{i=1}^{3}R_i. \quad (10)$$

4. Solution of the equation of the bar bending technical theory

Using dimensionless quantities (4), we write the equation of motion according to Kirchhoff's theory:

$$\frac{\partial^4 w}{\partial\xi^4}+\gamma\frac{\partial^2 w}{\partial\tau^2} = \frac{rp(\xi,\tau)}{EF},$$

where $\gamma = 1+(m_2/\rho_2 F)$. After the Laplace transformation in time we get:

$$\frac{\partial^4\overline{w}}{\partial\xi^4}+4\lambda^4\overline{w} = ke^{-\beta s\xi}.$$

The dimensionless bending moment is found by the formula:

$$\overline{m} = \frac{d^2\overline{w}}{d\xi^2}. \quad (11)$$

The inverse transformation (11) leads the expression of the bending moment to the form:

$$m(\xi,\tau) = \frac{2\beta rp_n(\xi_0)}{\sqrt{\gamma}EF}\sin\left[\frac{\sqrt{\gamma}}{\beta^2}\left(\tau_2-\beta\xi\right)\right]. \quad (12)$$

Table 1. Calculation of maximum bending moment values

Line №	№ of seismogram for [1]	Parameters							
		$\dot{V}\cdot 10^{-2}$, *m/s*	$V\cdot 10^{-3}$, mm	*J, point*	Ω, s^{-1}	η/Ω	j	m_I	m_{II}
1	156(III)	9	12	9	67	0.04	14	0.021	0.062
2	163(IV)	8	19	9.6	39	0.038	11	0.059	0.091
3	119(IV)	8.5	21	9.75	38	0.039	12	0.067	0.099
4	95(I)	8.3	20	9.7	30	0.05	10	0.061	0.079
5	type II, p. 86		0.9	5	21	0.043	3	0.00034	0.0027
6	type II, p. 86		11	8.75	27	0.056	3	0.0101	0.045
7	100(9)		0.31	1-4	44	0.034	7	0.00087	0.0046
8	8(3)		0.51	1-4	4.1	0.018	11	0.0056	0.0069
9	20(4)		3.8	7.25	4	0.038	6	0.0162	0.043
10	141(5)		3.75	7.2	32.5	0.034	7	0.035	0.054
11	31(1)		3.2	7	31.4	0	1	0.0014	

In the case of the action of the seismic impulse shown in Figure 1, the maximum moment is determined by formulas:

$$m_{II} = \frac{4\pi r\beta \dot{V}\Delta}{\sqrt{\gamma}\Omega EF}\sum_{n=0}^{j-1}\left(\frac{1}{4}+n\right)\exp\left[-\frac{2\pi\eta}{\Omega}\left(\frac{1}{4}+n\right)\right],$$

$$m_{II} = \frac{2r\beta V\Delta}{\sqrt{\gamma}EF}\sum_{n=0}^{j-1}\exp\left[-\frac{2\pi\eta}{\Omega}\left(\frac{1}{4}+n\right)\right]^2. \qquad (13)$$

It should be noted that the solution on Kirchgoff model is much easier than on Timoshenko model.

5. Model results

We will determine the stresses in the transmission pipeline during earthquakes, the seismograms of which are given in [6]. Table 1 contains the number of the seismogram record, the peak amplitudes of the velocity and displacement, the oscillation frequency of the main phase and the earthquake intensity on the scale [11], depending on the displacement value of the seismometer pendulu. In the work [11] it is also noted that seismic oscillation of the soil have a number of features. In particular, particles on the soil surface make spatial oscillation, and the components of the displacement vector along the coordinate axes are comparable with each other.

The amplitude ratios of the different maximum impulse phases depend on: η: the smaller η, the greater the amplitude ratio of the second maximum to the amplitude of the first maximum. By plotting graphs $v_i(t)$ $(i=1,2)$ using the formulas (3) for different values of the ratio η/Ω and visual comparison of these graphs with seismograms, Ratio value η/Ω is determined. The parameters of the seismogram in fig. 1 are given in line 7 of table 1.

Initial data: the pipeline with a diameter $D=1$ m and a wall thickness of $h_0=0.01$ m is laid in sandy soil at the depth of $h=1.5$ m. Soil consistency $\rho_2=1.53$ t/m^3, accept $a_*=660$ m/s, $b=310$ m/s, angle $\varphi=30^\circ$, coefficient $K_2=2.4\cdot10^3$ MPa/m. Elastic modulus, shear modulus and density of pipe material are equal $E=2\cdot10^5$ MPa, $G=0.8\cdot10^5$ MPa, $\rho=8$ t/m^3, $c_1=5\cdot10^3$ m/s, $c_2=3.3\cdot10^3$ m/s, $k'=1.1$. Bending moment calculations were performed according to formulas (10), (13) and they are given in Table 1, stresses are determined according to formula $\sigma=\sqrt{2Em}$.

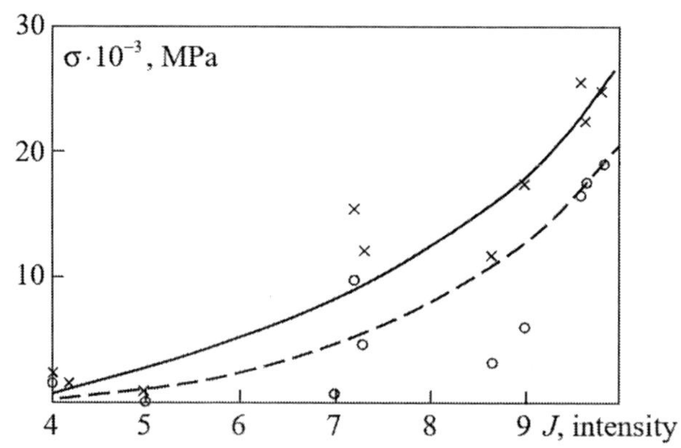

Figure 3. Models: Timoshenko $(\circ-\sigma_I)$, Kirchhoff $(\times-\sigma_I)$.

Figure 3 shows the stresses in the pipeline depending on the earthquake intensity. The solid and dotted lines show the maximum allowable stress values for the above initial data for the Kirchhoff and Timoshenko models respectively. As can be seen, the technical theory of beam bending (Kirchhoff model) gives overestimated stress values compared to the Timoshenko model. It should be noted that the stress level in the pipeline along with the earthquake magnitude is significantly affected by the number of shocks and aftershocks. With an increase in the number of shocks, stresses in the pipeline increase, see, for example, lines 5, 6, and 11 and lines 9, 10, and 11 of table 1.

From the obtained results of calculations, it follows that for this pipeline, an earthquake of 4 points is the limit.

6. Conclusion

Mathematical calculations of the influence of seismic waves of an earthquake on the transmission pipeline were carried out using two analytical models. It is shown that the technical theory of beam bending gives overestimated stress values compared to the Timoshenko model. From the obtained results of calculations, it follows that for this pipeline, an earthquake of 4 points is the limit. Of course, the obtained research results are approximate, but having a fairly simple analytical model, a lower estimate can be made. It should be noted that depending on various terrain conditions, the presence of voids near the pipeline, as well as the water cut of the surrounding soil, there may be an increase in the effect of earthquake seismic waves on the transmission pipeline, due to the occurrence of resonance phenomenon as previously noted [2]. In addition, these simple mathematical models can be complicated by introducing various methods and ways of protecting structures from the effects of seismic waves of an earthquake. This will certainly lead to nonlinearities in the differential equations of the transmission pipeline movement, which can be solved only numerically or with the use of complex software systems such as ANSYS.

References

[1] Psyrras N and Sextos A 2018 Safety of buried steel natural gas pipelines under earthquake-induced ground shaking *Soil Dynam. and Earthquake Engineer.* **106** 254–77

[2] Poliakov S V 1978 *Consequences of strong earthquakes* (Moscow: Stroyizdat)

[3] Taucer F and Apostolska R 2015 *Experimental Research in Earthquake Engineering* (New Jork: Springer)

[4] Kumariand P and Sharma V K 2019 Dynamics of seismic waves in highly anisotropic triclinic media with intermediate monoclinic layer *Appl. Mathem. Modell.* **71** 375–93

[5] Wang X, Chen Ju and Xiao M 2019 Seismic responses of an underground powerhouse structure subjected to oblique incidence SV and P waves *Soil Dynam. and Earthquake Engineer.* **119** 130–43

[6] Rustanovich D N 1974 *Earth surface oscillations in epicentral zones of strong earthquakes* (Moscow: Science)

[7] Kirillov F A 1962 On the study of the seismic effect of explosions at the Institute of Physics of the Earth of the USSR Academy of Sciences *Proc. of the Instit. of Phys. of the Earth, USSR Acad. of Sci.* **21(188)** 123–38

[8] Berzon I S, Epinateva A M, Pariiskaia G N and Starodubtseva S P 1962 *Dynamic characteristics of seismic waves in real environments* (Moscow: Publ. house of the USSR Acad. of Sci.)

[9] Nazarov A G 1959 *Seismic force engineering analysis method* (Yerevan: Publ. house of Acad. of Sci. of the Armenian SSR)

[10] Gelfand I M and Shilov G E 1959 Generalized functions and actions on them iss 1 *Generalized functions* (Moscow: Science)

[11] Medvedev S V 1962 *Engineering seismology* (Moscow: Gosstroyizdat)

IOP Conf. Series: Materials Science and Engineering **905** (2020) 012074 doi:10.1088/1757-899X/905/1/012074

External impacts on structure and relaxation properties of thermotropic liquid crystal copolyester

G S Zhukova[1], Z R Radjabov[2], A K Kurbanmagomedov[2] and G V Boykova[2]

[1] Financial University under the Government of the Russian Federation, 49, Leningradsky ave., Moscow, 125993, Russia
[2] Moscow Polytechnic University, 38 , Bolshaya Semyonovskaya str., Moscow, 107023Russia

E-mail: galsevzhukova@mail.ru, pmdekanat@mail.ru, akzanzibar@aol.com, boykovagv@mail.ru

Abstract. The methods of thermally stimulated depolarization (TSD) and X-ray diffraction analysis were used to study the impact of electric and temperature fields on the relaxation properties, structure, and molecular mobility of a liquid crystal (LC) copolymer based on polyethylene terephthalate (PET) and n-hydroxybenzoic acid (n – HBA). It was shown that polarization in the LC state besides the main relaxation transition leads to the formation of additional high-temperature maximum on current thermograms of thermally stimulated depolarization (TSD). The data of X-ray diffraction analysis indicate a change in the orientation of mesogenic units after polarization in an electric field. The results were explained by the combined action of electric and temperature fields on the conformational structure of the liquid crystal copolyester, namely, a change in the degree of crystallinity of the studied copolyester under the influence of external fields.

1. Introduction

The interest of many researchers in thermotropic LC copolyesters with mesogenic structural units in the main chain is primarily due to their ability to form filaments from these units in thermoplastic matrix flowing into the melt, which leads to self-reinforcement of such materials after cooling down the melts. The ability of LCs to change properties depending on the external influences (temperature, pressure, electric and magnetic fields, radiation, etc.), a special structure, a combination of ordering in the arrangement of mesogenic structural units of macromolecules with high mobility make it possible to widely use liquid crystal materials in various industrial fields.

Therefore, recently, an intensive study of the influence of the electric fields on the structure and properties of thermotropic LC polymers is in progress. The results of research are summarized in a number of monographs by domestic and foreign authors [1–10]. However, a special attention of researchers was focused on the study of electrical and optical properties of liquid crystals in alternating electric and magnetic fields. It was explained by the fact that in constant electric fields, the processes of charge accumulation and injection of charge carriers, electric transport, and polarization of the liquid crystal phase make it almost impossible to obtain electrical and optical properties with optimal characteristics [11, 12]. The study of the listed properties of liquid crystals was apparently restrained, for the reasons related with certain experimental difficulties, especially when it comes to direct current [6, 13, 14].

On the other hand, in many electrical and optical effects, the above phenomena play a decisive role and can provide quite useful information about the structure and properties of liquid crystal polymers.

When analyzing the above-mentioned works, the following tendency is observed: firstly, the processes associated with the dielectric effect of Fredericks, electrohydrodynamic instability, and ionic conductivity are mainly considered; secondly, the questions related to the relaxation of structural units after the removal of a constant external electric field are not studied.

Given the foregoing, the aim of the work is to study the impact of constant electric and temperature fields on the relaxation and phase properties of thermotropic LC copolymers and to establish the relationship between them.

2. Methods and materials

A liquid crystal copolymer obtained by polycondensation of n-HBA and PET with a molar concentration of 65 mol% n- HBA and 35 mol% PET is one of the thermotropic LC complex copolyesters with nematic mesophase in the temperature range of T ~ 190–320 °C [15–18].

The introduction of ethylene glycol residue with ether oxygen, which is a part of PET, into the HBA monomer provides possibility of rearranging the hard mesogenic fragments of copolymer macromolecules, this makes it possible to form an LC mesophase. As indicated above in [8, 17] by TSD method, it was shown that this copolymer passes into the nematic LC phase at a temperature of T * = 190 °C.

Therefore, depending on the temperature range of studies, this copolymer can be considered in two different phase states: it is a partially crystalline copolymer below 190 °C [19]; accordingly, it is in a liquid crystal state above this temperature. Therefore, it seems interesting to study the impact of an external electric field on the copolymer properties of below and above the phase transition temperature.

The reaction of the copolymer to the action of electric and thermal fields was studied by TSD and X-ray diffraction analysis.

The copolymer samples with a thickness of 250 μm were first heated to the polarization temperature and kept in a constant electric field of 0.6 kV/mm for 20 min, then, in the pores of the liquid nitrogen they were cooled to room temperature at a rate of 4^0 / min without removing the field.

The polarization temperature varied in the range of 100–250^{0}C. The TSD currents of polarized samples were measured using an U5-9 electrometric amplifier with the recorded measurement results on H306 two-coordinate recorder. The linear heating and linear cooling at a speed of 2 deg/min was provided by the BTP-78 thermal programmer. The constant electric field was set by the source B5-50. The measurements were carried out in a standard cell with clamping electrodes, where a sample with ground electrodes made of aluminum foil was placed.

X-ray diffraction patterns at large diffraction angles were recorded on an IRIS-3.0 setup (CU Kα, Ni-filter, flat cartridge) at 20 °C.

3. Results

Depending on the polarization temperature, several peaks are observed in the thermograms of the TSD currents due to different levels of molecular organization. The maximum at 80°C manifests itself regardless of the polarization temperature, a peak at a temperature of 65–70 °C occurs only at polarization temperatures T_p<120 °C, and corresponds to the glass transition temperature of amorphous PET.

If the samples are polarized in the LC state when T_p>T*, where T* is the phase transition temperature, then on the thermograms of the TSD currents there is one more maximum of the currents at a temperature of 120 °C and a shoulder in the temperature range of 190–220 °C (Figure 1).

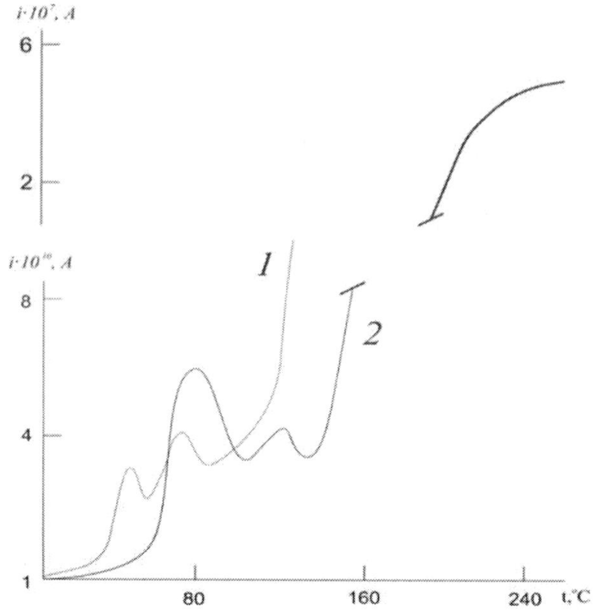

Figure 1. The temperature dependence of TSD currents of copolymer at various polarization temperatures: (curves 1 and 2, polarization temperature T_p = 170 and 250 °C, respectively)

This result is unexpected, since, according to the existing theory of relaxation spectrometry, maxima in thermograms with certain values of T_{max} should appear at any polarization temperature $T_p > T_{max}$. The appearance of this maximum after the impact of constant electric field on the copolymer in the LC state indicates that the influence of the field on it is determined by its phase state, which is determined by the temperature.

The main relaxation process at $T\alpha_{max}$= 80 °C is apparently responsible for the dipole-segmental mobility in the crystalline region of PET fragments oriented during stretching. A similar process was observed by the authors of [13], where a special study of the impact of crystallinity on this maximum was carried out. However, the work does not mention the presence of high-temperature relaxation at a temperature of 120 ^{0}C, which manifests itself for crystalline PET.

Any further increase in the polarization temperature leads to a decrease in the intensity of the main relaxation process. This indicates a limitation of the mobility of the kinetic units responsible for this process, which is apparently related both to an increase in the degree of crystallinity of PET with increasing temperature and to the orientational action of the electric field. The increase in the half-width of the peak of the TSD currents from 10 to 20 °C associated with this factor indicates redistribution of the spectrum of relaxation times.

However, upon transition to a liquid crystalline state, the intensity of the described process increases sharply, and the peak responsible for the relaxation α-process becomes similar to monorelaxation process. The temperature position of the peak of α-relaxation does not change. If the copolyester is polarized at temperatures above the temperature of the phase transition temperature T*, then the maximum of the α-relaxation process will be shifted towards high temperatures.

The results can be explained if to assume that the copolymer below the phase transition temperature is a polymer system with a heterophasic structure with three phases: a partially crystalline phase of PET that has not reacted with n-HBA; pure crystalline phase of n-HBA and mesomorphic copolyester obtained by copolycondensation of PET and n-HBA.

The following arguments speak in favor of this phase separation. It is known [1–3, 8, 15] that the copolymer macromolecule consists of hard fragments of n- HBA, terephthalic acid and flexible spaces of ethylene glycol and essential oxygen bound to them. Since the copolymer under study has a molar concentration of components of 65 mol% n- HBA and 35 mol% PET, even with perfect and complete copolymerization, local regions of its components cannot but remain. Consequently, the mechanisms of relaxation peaks in the thermograms of TSD currents become clear: the relaxation process in the temperature range 65–70⁰C is apparently responsible for the dipole and segmental mobility in amorphous PET.

This temperature region is typical of amorphous PET and corresponds to the glass transition temperature of the latter. At a temperature of T_g = 80 ⁰C, the α_1-relaxation process is responsible for similar mobility in crystalline PET. However, it is possible for the process to describe molecular mobility in the copolyester, the structure of which has sufficient thermodynamic affinity with the structure of PET. It can be seen from the TSD current curves that it is a fraction of monorelaxation process typical of pure polymers. It involves several types of kinetic units with a wide spectrum of distribution of relaxation times. The apparent activation energy calculated for this process is U = 0.6 eV and is typical of polyesters. This confirms that the ideas presented are correct.

Thus, it was shown that in the copolymer there are local regions of crystalline PET that did not react with n-HBA. The conclusion made is consistent with the results of [13], where crystallization of amorphous PET during polarization was noted; however, in our opinion, the attention is not not pay to the fact that during thermopolarization of a sample, the changes occurring in it are associated not only with the polarization temperature, but also with the orienting action of the electric field. It is obvious that the TSD peak current at 80⁰C is responsible for the segmental mobility in these areas. Or it points to the same mobility in the hard mesogenic unit of copolymer.

To clarify the influence of thermal field on the structure and physical properties of the liquid crystal copolyester, the sample was subjected to heat treatment at different temperatures corresponding to different phase states. The results obtained in this case do not indicate noticeable changes in the structural and phase properties. An analysis of the current curves of the TSD shows that in the glassy state of PET can exist both in amorphous and crystalline phase states. In this case, the transitions from one state to another are determined by the temperature.

In particular, at a temperature of T~50 ⁰C PET transitions to an elastic state, while the mobility of some parts of macromolecules (segments) increases, but there is still no mutual movement of macromolecules. Therefore, a notable crystallization process has not yet been observed. Active crystallization begins at T~100 ⁰C and ends at T~150 ⁰C with the formation of a clear crystalline structure consisting of folded lamellas.

Consequently, the crystallization rate of amorphous regions increases significantly with the simultaneous electric and thermal fields. In our opinion, the thermal field defrosts the mobility of the kinetic units of copolymer responsible for crystallization, while the electric field directs them, thereby contributing to the ordering of the structure. This conclusion is confirmed by the absence of a low-temperature maximum responsible for thawing the molecular mobility of amorphous PET on the thermogram of the TSD currents of samples of copolymer polarized in the LC state and indicates a modification of the structure of copolyester under the influence of external influences.

Based on the foregoing, it can be argued that crystallization in the thermotropic LC copolymer can be achieved by the simultaneous action of thermal and electric fields.

It is known [8, 11] that the interaction energy of the dipole moment of a macromolecule with an external electric field is tens of thousands of times less than the energy of thermal motion of molecules. Therefore, the electric field practically does not affect the degree of ordering of copolymer macromolecules in true crystalline states.

The reaction of copolymer to an electric field in a liquid crystalline state similarly to the low molecular weight of LCs [3], in our opinion, will be collective in nature, therefore an external electric field applied to the LC layer, even of low intensity, will lead to the redirection of all mesogenic units, since the director's redirection is accompanied by a spatial redistribution of a charge.

Considering the foregoing, the appearance of the second maximum during the polarization of copolymer in the LC state and its absence when the copolymer is polarized at temperatures can be explained in the following way:

$$120\ ^0C < T_p < T^*.$$

The copolymer under consideration in the LC state, due to the molding conditions (extrudates were placed parallel to the substrate plane) has a planar orientation. When exposed to an electric field on the copolymer in the LC state, the mesogenic units with non-zero dipole moments are oriented perpendicular to the electrode plane, that is, they are polarized. In this case, a double electric layer appears at the insulator-electrode interface, the sign of which, due to injection of a homo-charge is opposite to the sign of the electrode. The surface density of a homo-discharge depends on the polarization temperature, the higher the T^p, the more dipoles are oriented along the field, and the higher the surface charge density. On the thermograms of TSD currents this circumstance is reflected as an increase in the intensity of the second maximum with an increase in the polarization temperature (Figure 1). At the temperatures T_p lower than the phase transition temperatures, this orientation is excluded. In addition, the injection of a homo-charge does not take place. In this regard, even at the temperatures T_p>120 °C, but lower than the phase transition temperature, the process under discussion cannot be implemented.

The analysis of the obtained X-ray confirms everything mentioned above. Two types of X-ray diffraction patterns were obtained for each sample (initial and polarized at 170 and 250 ^{0}C): the first pattern indicated that the primary beam was directed perpendicular to the plane of the sample in the direction of the electric field vector (shooting in the plane), and the second pattern indicated that the shooting was carried out in the direction perpendicular to the electric field vector (shooting at the end).

The pattern of the intensity distribution substantially depends on the region of the sample chosen for the study, which indicates morphological heterogeneity, as well as the anisotropy that arose during the formation of this sample. It can be argued that the predominant orientation of mesogenic units or larger macromolecule formations found in the initial sample, mechanically incorporated, is retained and even enhanced at a polarization temperature above 170 ^{0}C (Figure 2a, 2a').

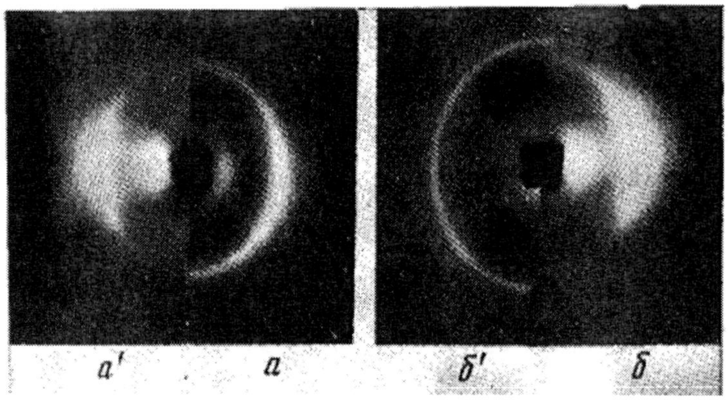

Figure 2. X-ray diffraction patterns of copolymer polarized at 170°C (a, a') and 250°C (b, b'). a and b is shooting "at the end", a'i b' is shooting "at the plane"

An increase in grain-orientation during thermal treatment below phase transition temperatures was also observed by the authors of [13].

Apparently, at the temperatures considered, recrystallization processes begin, which in this case are due to the heterophasic structure of the copolymer, since it contains crystallites from n-HBA blocks distributed in an anisotropic liquid crystal matrix. The authors of [6, 8] indicate a similar fact.

The combined actions of the temperature and the electric fields lead to the orientation effects indicated above and can be explained on the basis of the Fredericks effect, which is a characteristic of nematic LCs placed in a constant electric field. This conclusion is confirmed if we analyze the X-ray patterns obtained for the samples polarized in the mesophase at T_p = 250 °C. When shooting at the end, the X-ray diffraction pattern almost coincides with that for the initial sample and polarized at 170 ^{0}C; however, when shooting at the plane, the orientation ordering is not observed (Figure 2a'). In this case, the pattern of the intensity distribution when shooting at the end, with the primary beam perpendicular to the vector of the intensity of polarizing field, substantially depends on the selected region of the sample. Based on this, it can be argued that as a result of the Fredericks effect we observe the regions of different director orientations.

It is important to note that when the electric field is removed and the sample is heated as a result of a decrease in polarization, the copolyester texture is more ordered. This is evidenced by the results obtained from a repeated study of samples that were in the LC phase in an electric field at the temperatures below the phase transition temperature. As a result of such actions, the relaxation properties were stabilized. At high polarization stresses, periodic ordering deformation was observed in nematic liquid crystals [17–19].

4. Conclusion

1. The relaxation mechanisms of the thermotropic LC copolyester based on n-HBA and PET do not fit into the classical theory of relaxation spectrometry due to the heterophasic structure of the copolymer; as a result of the Fredericks effect, the regions with different director orientations are observed.

2. Apparently, a flow induced by constant electric field arises in copolymer in a nematic state. At the same time, mesogenic units oriented at different angles in the XOY plane acquire the same orientation, that is, the structure of the nematic is arranged in better way. This leads to an increase in the elastic coefficients of the liquid crystal copolyester, as a result of which the intensity of the α-relaxation process in the thermograms of the TSD currents decreases.

3. The combined action of a constant electric and thermal fields leads to an increase in the degree of ordering of the structure of the liquid crystal copolymer and changes its structure at the morphological level. The discovered fact of the influence of external fields on the structure of copolymer makes it possible to control it and obtain materials with various electrical and physical properties.

References

[1] Blinov L M 2013 *Liquid crystals: Structure and Properties* (Moscow: Librokom) 448 p
[2] Pikin S A 1982 *Structural Transformations in Liquid Crystals* (Moscow: Nauka) 336 p
[3] Talrose R V and Plate N A 1988 Structural Transformations of Thermotropic Liquid Crystal Polymers in Electric and Magnetic Fields In *Liquid Crystals* (Moscow: Khimiya) pp 245–96
[4] De Gennes P 1977 *The Physics of Liquid Crystals* (Moscow: Mir) 400 p
[5] Chandresecar S 1980 *Liquid Crystals* (Moscow: Mir) 343 p
[6] Kapustin A P 1978 *The Experimental Studies of Liquid Crystals* (Moscow: Nauka) 368 p
[7] Belyaev V V 2002 *Viscosity of Nematic Liquid Crystals* (Moscow: Fizmatlit) 224 p
[8] Sluckin T J, Dunmur D A and Stegemeyer H 2004 *Crystals that Flow* (London: New York: Taylor & Francis) 760 p
[9] Yang D K and Wu S T 2006 *Fundamentals of Liquid Crystal Devices* (John Wiley&Sons, Ltd) 378 p
[10] Yevdokimov Yu M, Salyanov V I, Skuridin S G et al 2011 *The CD Spectra of Double-Stranded DNA Liquid-Crystalline Dispersions* (New York: Nova Sci. Publ. Inc) 103 p
[11] Gritsenko N I, Kucheev S L and Litvin P M 2004 Long-Term Change of Nematic Alignment at Defect of SiO2 in LCoS *Proc. of Int. Conf. on Advanced Display Technologies* pp 56–60 (Minsk)
[12] Kanduc M, Dobnikar J and Podgornik R 2009 Counterion-Mediated Electrostatic Interactions between Helical Molecules *Soft. Matter.* **5** 868–77
[13] Lusheikin G A 1986 Relaxation Transitions and Molecular Mobility in Polymers In Yu N Venevtsev and V I Muromtsev (ed) *Systems of Special Temperature Points of Solids* (Moscow: Nauka) 272 p
[14] Radzhabov Z R., Bubman S Z, Magomedov G M et al 1993 Relaxation Properties of Composites Based Liquid Crystalline Copolyethers *Mechan. of Compos. Mater.* **3** 425–7

[15] Synthetic high polymer materials 1986 *Express Inform. VINITI* **36** 38–42
[16] Okolnikova G E, Grishin G E, Kurbanmagomedov A K and Bronnikov D A 2019 Experimental Study of Modified High-Strength Coarse-Grained Concrete *Syst. Technol.* **31** 25–30
[17] Vshivkov S A and Rusinova E V 2007 Phase and Structural Transformations of Liquid Crystal Polymer Systems in Magnetic and Mechanical Fields *Izv. of HEIs. Chem. and Chem. Technol.* **3** 10–3
[18] Denisova O A 2013 One of the Methods of Experimental Studies of Liquid Crystals *Electr. Inform. Syst.* **9(2)** 107–13
[19] Delev V A and Krekhov A P 2017 Structural Transformations in the Nematic Liquid Crystals with Hybrid Orientation *J. of Experim. and Theoret. Phys.* **152((6)12)** 1414–30

Morphology of geophysical fields of Terek-Caspian trough and placement patterns of oil and gas structures

E A Abubakarova[1,2]

[1] Department of problems of the fuel and energy complex, Integrated Research Institute named after H.I. Ibragimov of the Russian Academy of Sciences, Grozny, Russia
[2] Department of Applied Geophysics and Geoinformatics, Grozny State Oil Technical University named after academician M.D. Millionshchikov, Grozny, Russia

E-mail: eliza_ggni@mail.ru

Abstract. The paper considers the characteristic morphological features of geophysical fields caused by fault tectonics of the Tersko-Caspian trough. The results of the interpretation of the gravitational and magnetic fields of the region using the Cascade 3D spectral-correlation analysis of data are presented. Computer technology made it possible to analyze three-dimensional digital geophysical information using the methods of a probabilistic-statistical approach. Linear positive anomalies of geophysical fields are identified that extend in the fault zones. A comparative analysis of the maps of transformants of the gravitational and magnetic fields and maps of oil and gas structures. Based on the results of comparative analysis, a dependence was established that characterizes the confinement of the known oil and gas deposits in the region to the zones of positive linear anomalies of geophysical fields of a sub-latitudinal strike.

1. Introduction

In geological terms, the Tersko-Caspian trough has a complex structure; the foundation is characterized by a fault-block structure with the chaotic immersion of blocks. Fault tectonics of the study area is essential in the oil and gas and geological zonality of the region, which determines the migration paths of hydrocarbons [1, 3, 6, 10]. Known oil and gas-bearing structures are mainly located in the sub-latitudinal strike of the study area, which is shown in the diagram of the oil fields and oil and gas prospective areas of the Tersko-Caspian trough (Fig. 1).

A regional description of the geophysical fields of the studied territory of the North Caucasus is given in many works [5, 7]. Materials of detailed gravimetric surveys and their geological interpretations were processed on the territory of the study area. The analysis of gravimetric maps on a scale of 1: 200000 within the described territory made it possible to identify many features of the gravity field. The region's gravitational field (Fig. 2) is characterized predominantly by the sub-latitudinal strike of the iso-anomalies. The entire studied area is characterized by large anomalous zones of sub-latitudinal and northwestern orientations. In the north of the described territory, a zone of smooth decrease in the gravitational field is distinguished. In the northern zone, the gravitational field is gradually expanding from west to east.

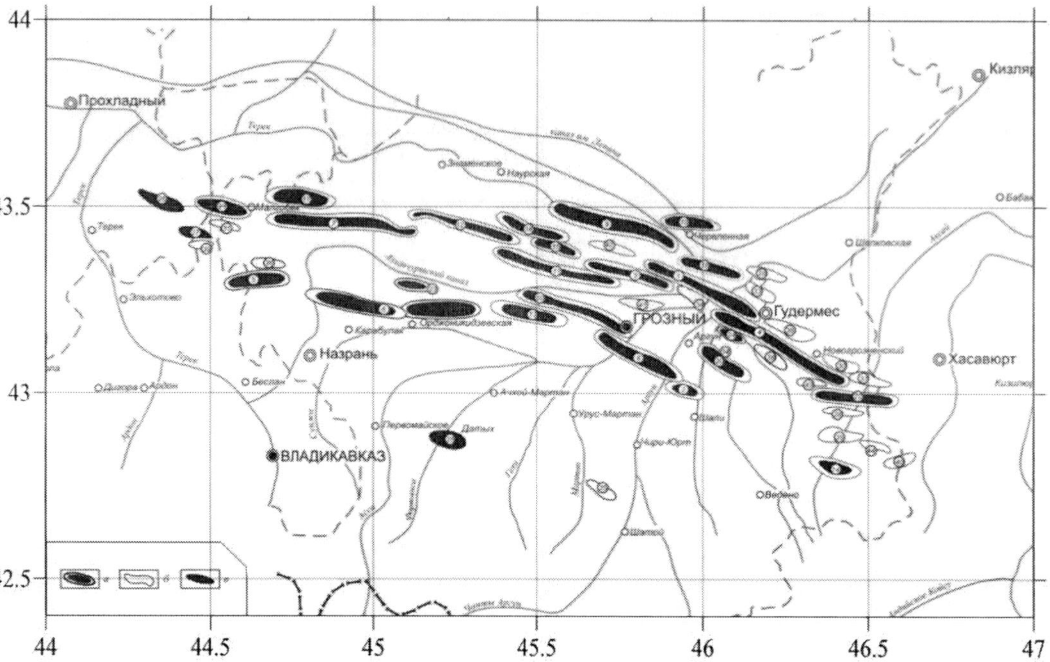

Figure 1. The layout of oil fields and promising areas (Designation: a – deposits under development; b – deposits located in exploration; c – perspective areas).

Oil fields:

1 – Bragunskoye; 2 – Goit-Kortovskoye; 3 – Goryacheistochnenskoye; 4 – Gudermesskoye (with Petropavlovskim); 5 – Zamankul'skoye; 6 – Karabulak-Achalukskoye (with Sernovodskim); 7 – Malgobek-Gorskoye; 8 – Oktyabr'skoye; 9 – Pravoberezhnoye; 10 – Starogroznenskoye; 11 – Khayan-Kortovskoye; 12 – El'darovskoye; 13 – Andreyevskoye; 14 – Benoyskoye; 15 – Mineral'noye; 16 – Severo-Mineral'noye; 17 – Severo-Malgobekskoye; 18 – Severo-Bragunskoye; 19 – Chervlennoye; 20 – Khankal'skoye; 21 – Akhlovskoye; 22 – Mesketinskoye; 23 – Alkhanchurtskaya; 24 – Severo-Dzhalkinskaya; 25 – Datykhskoye; 26 – Arak-Dalatarekskoye; 27 – Kharbizhinskoye; 28 – Lesnoye;

Exploration area:

29 – Belorechenskaya; 30 – Koshkel'dinskaya; 31 – Kurpskaya; 32 – Groznenskaya; 33 – Terskaya; 34 – Il'inskoye; 35 – Nozhay-Yurtovskaya; 36 – Magistral'naya; 37 – Suvorovskaya; 38 – Rodnikovaya; 39 – Yuzhno-Kharbizhinskaya; 40 – Zandakskaya; 41 – Groznenskaya; 42 – Priterechnaya; 43 – Yuzhno-Gudermesskaya; 44 – Severo-Sayasanovskaya.

In the central part, near the Nadterechny and Chervlensky lows, the described zone is the narrowest. In the eastern part, the Pereterennaya zone axis of minima is shifted to the south in a rock-like fashion. The large and complex Khasavyurt minimum has a sub-latitudinal orientation and is complicated in the east by the Kizilyurt local minimum. In the extreme western part, the Mozdok minimum of the gravity field has vague contours. It passes into the zone of a gradual increase in the gravitational field in the direction of the Mineralovodsky ledge. The described zone of minima corresponds to the submerged sections of the Paleozoic basement of the deflection.

Further south, the Terek zone of linear minima stands out, which corresponds to the Tersk anticline zone. Within the Tersky zone, the Arak-Dalatarek, Malgobek-Voznesensky, Eldarovsky, Aliyurt-Orlinny, Khayan-Kortovsky, Bragunsky and Guderme maximums are distinguished. The Arak-Dalatarek maximum, complicated by three local maximums, stretches from the southwest to the southeast from Prokhladny to Urusheva. It is due to the eponymous anticlinal uplift. The maxima located east of Malgobek-Voznesensky, Eldarovsky, and Aliyurt-Orliny represent a narrow strip complicated by local maxima of various intensities, bounded from the north and south by highly gradient zones. The described maxima correspond to the largest anticlinal within the Terek anticlinal zone of the Malgobek-Voznesenskaya-Aliyurt-Gorsky anticline, which stretches for 75 km and has a width of 4 km.

To the east, there is a Kalausky complex-built minimum corresponding to the depression of the same name. The maximum allocated to the east of the Khayan-Kortovsky maximum also represents a narrow strip 40 km long, limited from the north and south by highly gradient zones. In the extreme east, the anomaly is complicated by a sizeable local maximum in the area of Yastrebinaya. With the maximum of Mount Yastrebinaya in the east, the Bragunsky linear maximum articulates articulately. The anomaly is clearly expressed and corresponds to the anticlinal uplift of the same name. The Bragunian anomaly in the southeast passes into the Guderme maximum, which has a southeast direction. The Gudermes anomaly corresponds to the anticlinal uplift of the same name, within which the Tersk anticlinal zone merges with the Myatli ledge of the Montenegrin monoclinic zone. To the south, the Alkhanchurt linear minimum zone is noted, separating the Tersky and Sunzhensky maximum zones, and corresponding to the synclinal zone of the same name.

The Sunzha linear maximum zone includes the Zamankul, Sunzhen, and Starogroznensky maxima. It corresponds to the anticlinal zone of the same name, within which the Zamankul, Karabulak – Achalukskoye, Sernovodskoye, Oktyabrskoe, and Starogroznenskoye uplifts are distinguished. The Zamankul maximum with a length of 50 km and a width of 7–9 km has a sub-latitudinal strike in the central part and southwest in the west. In the far west (Zmeyskaya station), the maximum again changes its orientation to sub-latitudinal and smoothly passes into the Argudansky maximum. The Zamankul anomaly is complex and limited by highly gradient zones from the north and south.

In the western part, the Zamankul maximum is divided into Harbizhinsky and Malokabardinsky branches. Sophisticated Sunzhensky maximum stretches from Achaluki to Zakanyurt for more than 50 km, is complicated by local anomalies and is limited from the north and south by highly gradient zones. The articulation of the Zamankul and Sunzhen maxima is rocky. The Starogroznensky maximum, stretching east to the city of Grozny in the form of a narrow strip, is characterized by a limited gradient zone from the north and a less gradient zone from the south.

Analysis of the density characteristics of the section indicates the presence of several density boundaries. Still, the most contrasting are the roof of the Upper Cretaceous deposits and the pre-Jurassic base.

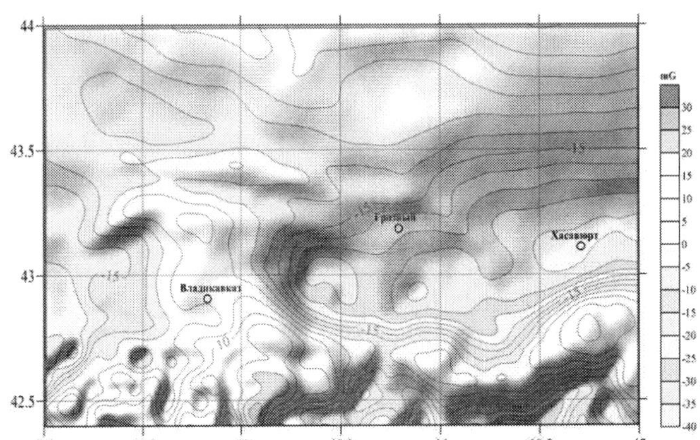

Figure 2. Anomalous gravitational field of the Tersko-Caspian trough Anomalous magnetic field ΔTa of the Tersko-Caspian trough.

The block structure of the basement and the faulty, folded structures of the sedimentary cover (mainly of the Upper Cretaceous deposits) determine the nature of the anomalous gravitational field of the studied area. The zones of development of saline sediments of the Upper Jurassic also have a significant effect on the character of the gravitational field of the studied region. When analyzing the dominant nature of the indicated gravitational boundaries, it is necessary to take into account that other factors (intrusive and effusive magmatism, inhomogeneities of the upper part of the section, and lateral density changes) may dominate in some areas.

A characteristic feature of the gravitational field of the region is the presence of highly gradient zones due to fault-block tectonics. The anomalous gravitational field of the studied region has a complex morphology. It is characterized by the presence of zones of maxima and minima of various energies corresponding to the corresponding tectonic elements.

The TKP magnetic field is characterized by the presence of a broad Grozny regional maximum, stretching in the Caucasus direction (Fig. 3). The depth of magnetoactive masses varies in the range from 10 to 18 km. The regional magnetic maximum is determined by the intrusive complex of rocks associated with a significant grave fault. The area of the negative magnetic field located in the north corresponds to the critical occurrence of the Precambrian basement.

On the meridian of Vladikavkaz, the Grozny maximum is articulated with a transverse magnetic maximum, apparently of the same nature. The region southwest is characterized by a large number of small anomalies of both signs, most due to magmatism in the fault zones.

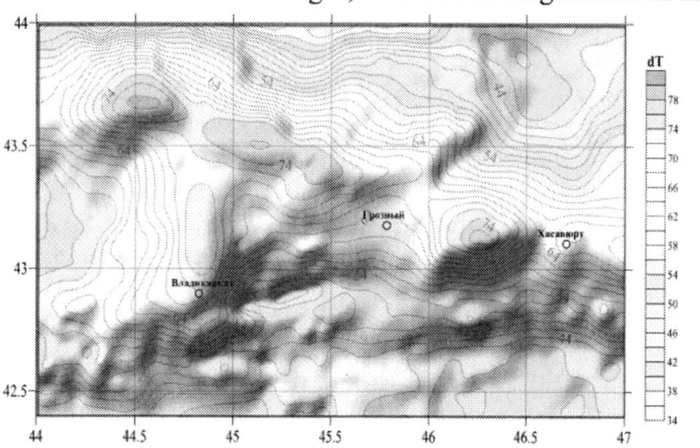

Figure 3. Anomalous magnetic field ΔTa of the Tersko-Caspian deflection.

2. Methods and materials

The study aims to establish an interdependent relationship between the anomalies of geophysical fields and the oil and gas potential of the studied region. The solution to the problem posed is based on one of the methods for interpreting digital geophysical information of the Cascade 3D software for spectral correlation analysis of data – tracing the axes of anomalies.

3. Results

A comparative analysis of the transformant of the gravitational field obtained using the above method and the location scheme of oil and gas structures was carried out (Fig. 4). According to the results of the comparative analysis, we can observe the relationship, i.e., confinement of most hydrocarbon deposits to linear maxima of gravity. Comparing the axes of the anomalies in the local component of the gravitational field and the deposits of oil fields, they are confined to the local maxima of gravity [2].

Compared with the transform of the gravitational field, the result of the interpretation of the magnetic field is displayed less informatively. In this connection, the following processing algorithm was used – filtering the results of tracing by the initial magnetic field. The software module that implements this stage of processing made it possible to isolate the estimates of the shape of weak anomalies, the amplitude of which is commensurate with the level of interference. The analysis of the compared transform obtained from the results of the final interpretation and the location scheme of the oil fields of the studied region is carried out. As a result of the analysis, it was found that Oil and gas-bearing structures can be traced in the sub-latitudinal tracing zone of positive anomalies of different magnetic field energies (Fig. 5) [4, 8, 9].

4. Conclusion

The analysis of the transformants of the gravitational and magnetic fields obtained by the interpretation of digital geophysical information using modern computer technology. This analysis shows that basically, the maxima of the gravitational and magnetic fields coincide.

A comprehensive interpretation of the regional digital geophysical materials of the studied area was made. The similar scheme of oil and gas structures of the Tersko-Caspian trough with transformants of gravitational and magnetic fields is analyzed. This study allowed establishing a relationship characterizing the confinement of known oil and gas deposits in the region to the zones of positive linear anomalies of geophysical fields of the sub-latitudinal strike.

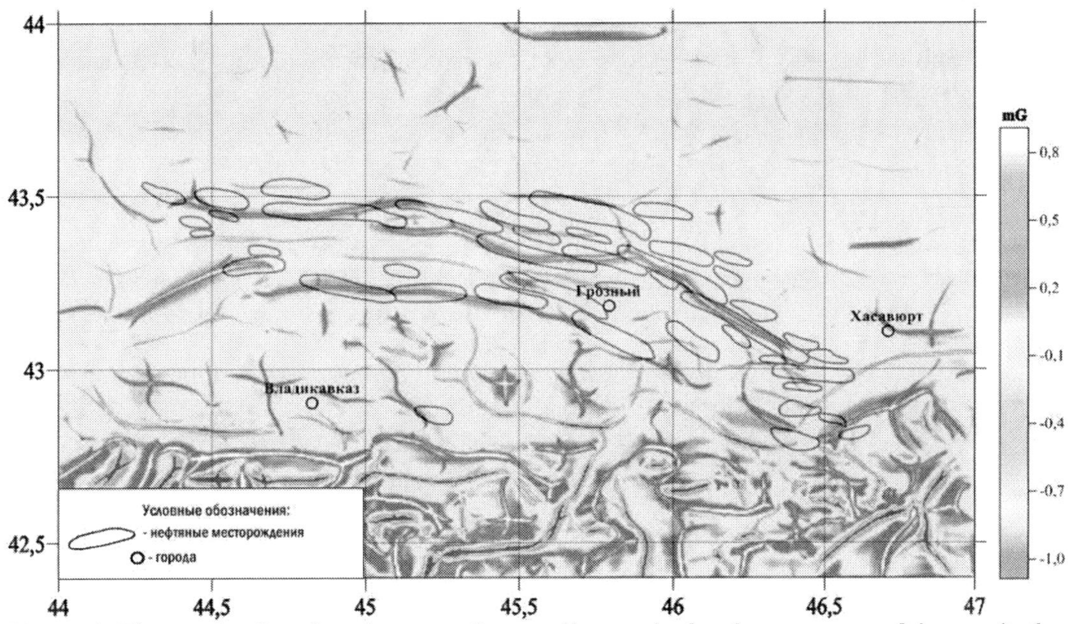

Figure 4. The result of tracing the axes of anomalies on the local component of the gravitational field and the placement of oil and gas bearing structures

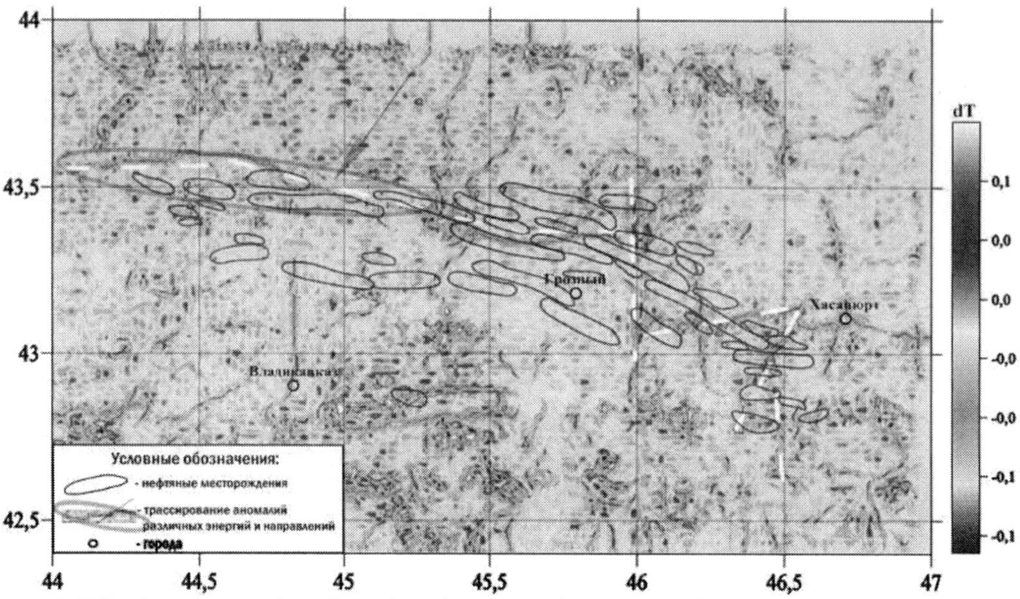

Figure 5. Filtering the results of tracing the axes of anomalies in the initial magnetic field and the placement of oil and gas structures

References

[1] Daukaev A A, Bachaeva T Kh, Abubakarova E A et al 2018 Deep Structure, and Oil and Gas Content of Submerged Zones of Western Part of Terek-Caspian Trough, Series: Advances in engineering research *Engineering and Earth Sciences: Applied and Fundamental Research* Proc. of the Int. Symp. (ISEES 2018) Retrieved from: https://doi.org/10.2991/isees-18.2018.85

[2] Volodarsky R F and Landa T I 1970 *Geological interpretation of gravitational and magnetic fields using a computer* (Moscow: Nedra) 200 p

[3] Gavrilov V P 2008 Possible mechanisms for the natural replenishment of reserves in oil and gas fields *The geol. of Oil and gas* **1** 56–64

[4] Demura G V, Zinovkin S V and Petrov A V 2013 New opportunities and prospects for magnetic exploration – computer technologies for volume modeling and filtering the results of high-frequency fruitful surveys *Subsoil use of the XXI century* **4(41)** 20–6

[5] Kerimov I A 1990 Prediction of structural features of the deeply submerged horizons of the Tersko-Caspian trough according to the data of gravity and seismic surveys *Univer. Proc. Oil and gas* **11** 24–30

[6] Kerimov I A, Daukaev A A and Bachaeva T Kh 2011 Natural oil yields in Chechnya and Ingushetia *Bull. of the Acad. of Sci. of the Chechen Republic* **1(14)** 74–9

[7] Kerimov I A 2011 *The F-approximation method for solving problems of gravimetry and magnetometry* (Moscow: Fizmatlit) 264 p

[8] Petrov A V and Trusov A A 2000 *Computer technology for statistical and spectral-correlation analysis of three-dimensional geoinformation, Geophysics* no 4 (Moscow: EAGO) pp 29–33

[9] Strakhov V N, Kerimov I A and Stepanova I E 2009 *Development of the theory and computer technology for constructing linear analytical approximations of gravitational and magnetic fields* (Moscow: IFZ RAS) 254 p

[10] Sudarikov Yu A, Serkerov S A, Kholina I et al 1976 *Use of geological and geophysical data to study regional tectonics of Oil and gas-bearing areas* (Moscow: Subsoil) 165 p

IOP Conf. Series: Materials Science and Engineering **905** (2020) 012076 doi:10.1088/1757-899X/905/1/012076

Surface Properties of Melts Based on Tin

Yu N Kasumov[1], R A Kutuev[2], A R Manukyants[1] and V A Sozaev[1]

[1] North Caucasus Mining and Metallurgical Institute (State Technological University), 44 Nikolayev street, Vladikavkaz, 362021, Russia
[2] Chechen State University, 17a Dudayev Boulevard, Grozny, 364060,Russia,

E-mail: kasumov60@mail.ru, kra-07@mail.ru, artmanukyants@mail.ru, sozaeff@mail.ru

Abstract. This paper presents the results of the surface properties of the tin-based melts. It has been shown that the best wetting of Al-4 at. % Li grade aluminum, silicon, and piezoceramics are achieved by the eutectics and terectics of tin-aluminum, tin-zinc, and aluminum alloys, respectively. Preliminary photon annealing of the substrates for 3–4 sec improves the wetting of silicon and ceramics. Solder wetting is also improved by the addition of copper and nickel micro powders, alkaline, and alkaline earth elements.

1. Introduction

Reliable data on the surface properties of tin-based alloys are needed to design lead-free solders for aluminum brazing. Knowledge of the surface properties of alloys, such as Sn-Bi and Sn-Sb, is also of interest in connection with the fact that these alloys are promising as anode materials for lithium-ion batteries [1].

Solders based on the Sn-Zn system are widely used for soldering laptop nodes [2]. Similar studies, using Sn-Ag based solders, were carried out in [3-6]. However, surface properties (surface tension, wetting angles, spreading kinetics), as well as patterns of formation of intermetallic compounds using such solders, have not been studied enough.

Particular attention in the study of tin-based surface properties is given to eutectic or near-eutectic solders of the Sn-Ag system: Sn-Ag-Cu, Sn-Ag-Bi, Sn-Ag-Sb, Sn-Cu, Sn-Ag-Cu-Ni; and also Sn-Bi systems: Sn-Bi-Cu. The use of small additives, for example, Cu and Bi, allows lowering the surface tension and reduce the rate of formation of Cu6Sn5 intermetallides during brazing (recommended solder is Sn-3.5Ag-0.5Cu) [7], as well as Ni3Sn4 intermetallides during brazing of Ni and Ni / Au metallization system [4]. Additives Bi, Sb, Ni improve mechanical properties, fluidity, regulate the temperature of solder and KTP of the soldered joint. Solders based on Sn-3.5 mass% Ag eutectic (melting point Tm ~ 221 °C) with rare-earth metals additives provide high shear strength when soldering semiconductors: Si (> 11 MPa), GaAs (> 8 MPa), GaN (> 14 MPa), diamond (> 14.5 MPa), SiC (> 7.8 MPa), dielectrics: Si3N4 (> 12 MPa), Ta2O5 (> 5.6 MPa), Al2O3 (> 11.5 MPa), SiO2 (> 7 MPa) [8]. These semiconductors are used in many optical and electronic devices.

High adhesion during brazing of sapphire is achieved by Sn-0.5 wt.% Al melt [9]. Melts based on the Sn – Al system well-wet moisten aluminum nitride [10].

2. Research results

The surface tension of tin-based alloys has been studied in many works, and until 1980, these data were generalized in a reference [11]. A study of the surface tension of Sn-based alloys in recent years has been carried out in the following works: alloys based on the Sn-Ag system [12–14], alloys of the Sn-Sb system [15], alloys: Sn-Ni [16], Sn-Si [17], Sn-Al [18.19], Sn-Tl [20] and other alloys.

Theoretical estimates of the surface tension of tin-based binary and ternary alloys are usually carried out based on the Butler-Zhukhovitsky equation [21, 22]. The analysis of theoretical and experimental works on tin-based surface properties shows that they do not take into account the presence of micro and nanoparticles in alloys. At the same time, it is precisely micro- and nanostructured melts that may turn out to be the best soldiers. These solders can be widely used for soldering aluminum and its alloys.

Aluminum and its alloys are increasingly used in electronic technology. In particular, aluminum is used as bases to which ceramic substrates of integrated circuits are soldered [23, 24].

In known methods of soldering integrated circuits, fluxes are usually used, the pairs of which disable active IP elements. In this regard, it is of interest to develop solders for flux-free tinning and to solder integrated circuits with aluminum substrates. These solders should have a sufficiently high melting point, such that heating the integrated circuits during its operation does not lead to the melting of the solder. At the same time, there are restrictions from above in temperature - no higher than 300–350 °C. Otherwise, when soldering above this temperature, diffusion zones of active elements of integrated circuits can be violated.

It is also desirable that the solder be corrosion-resistant, resistant to thermal cycles, wet aluminum, and metalized ceramics well and provide a sufficiently high strength of the solder joint.

So far, not a single solder has simultaneously satisfied all these conditions. Known solders used for flux-free brazing and tinning of aluminum [25–30] contain components that either worsen the mechanical properties of the solder and aluminum product, either containing expensive metals (silver, gold) or containing volatile metals with toxic properties (lead, cadmium). These solders also have insufficient oxidation resistance.

Of great interest for flux-free brazing of aluminum are solders containing tin and aluminum. A solder (wt. %) Was proposed in [31]: Al 0–17.2, Cu 0–4.6, Zn 12.0–41.6, (Fe, Pb, Cd, Na, Si, Ca, Mg) <2.1, Sn – the rest. The solder is interesting in that it covers the composition and practices of the Sn-Zn-Al system. Using a solid-liquid peritectic state in [32], we proposed modes and methods of tinning and brazing of aluminum.

The use of the solid-liquid state of alloys [33] may turn out to be promising in the development of solders and methods for brazing aluminum and its alloys. The high strength of the solder joint can be ensured by the mechanism of creating a dispersed medium at the solder-aluminum interface.

In the case of eutectics and peritectic, the particles of the solid phase of small sizes contained in them, interacting with the surface of aluminum, are deposited on it. As a result, a dispersed medium is formed, which, due to the side effects of dissolution [34] and capillary forces, enhances the dissolution of the oxide film and accelerates mass transfer between the liquid and solid phases.

In [43], we attempted to study the polytherms of the contact angles of tin and eutectic Sn-Al melt of aluminum alloys Al + 4 at.% Li (Fig. 1), which are widely used in the aircraft industry. At a temperature of 810–820 K, a threshold for wetting the substrates with a pure tin melt is observed (Fig. 1). For curves 1 and 2, the experiment is repeated twice. When a drop melts, intense interaction with the substrate (reactive wetting) occurs with the formation of craters. After these temperatures, tin completely spreads over the substrate, $\theta \rightarrow 0$.

When Sn-Al is wetted with an eutectic melt, the wetting angles decrease (curve 3 in Fig. 1), and the wetting threshold is observed at a lower temperature: T = 765 K. Then again $\theta \rightarrow 0$.

Such a liquid-solid dispersed medium can be created artificially. To this end, powders of refractory metals with a size of 20–100 μm were introduced into the solder of the Sn-Al system, which would be wetted by melts of the Sn-Al system. For example, Ni was used. We proposed one such solder in the study [35].

Solders compared with the known ones provide higher strength properties of soldered joints but have low fluidity. For solving this problem, reliable data on the surface properties of solders in the region of low Al concentrations are needed. The surface tension of the Sn – Al system was first studied in [18]. However, in the region of low Al concentrations in Sn (covering the eutectic of the system), the surface tension has not been studied sufficiently. Due to the dimensional effects of physicochemical properties, the addition of micro- and nanopowders to Sn-based melts leads to the intensification of mass transfer at the interface between solids and melts, enhancement of wetting, solubility, and other processes.

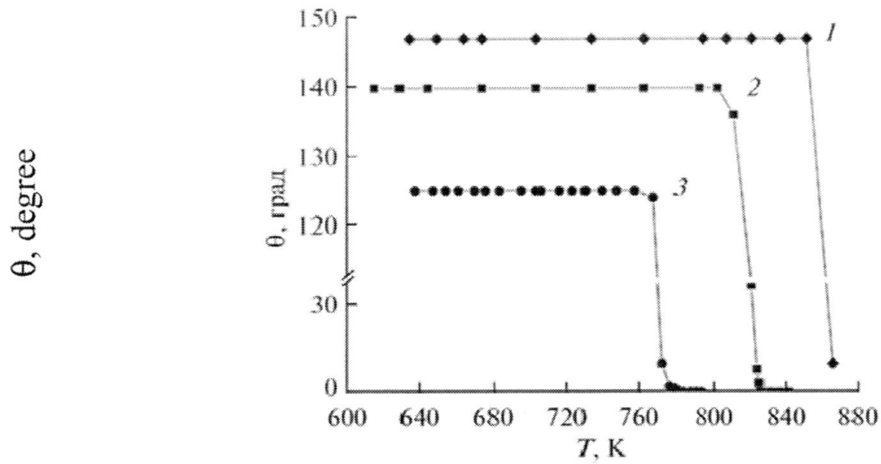

Figure 1. Temperature dependence of tin wetting angles of pure aluminum (1) and aluminum alloy Al + 4 at. % Li (2); 3 – wetting of an Al + 4 at. % Li melt (Sn – Al) eut.

An urgent problem is the development of solders and metallization systems of the piezoceramics of the TsTS brand [19].

The most significant application in metallization of ceramics was found in silvering, based on the burning of a coating 0.02–0.03 mm thick into piezoceramics. However, such a coating is not sufficiently resistant to thermocycling loads, and at high burning temperatures, the functional properties of ceramics are lost.

Therefore, this coating is underway for other metallization systems and solders for soldering PZZ piezoceramics with a lower temperature. In this regard, solders based on the Sn – Ag – Cu – Me (Me: Co, NI, Zn, Zr, Ti) systems [36] may be of interest. But more research is needed, and especially of capillary properties (polytherm, surface tension isotherms, wetting angles).

Currently, to improve the wetting by solders of substrates, attempts are made to change the surface morphology of the substrates [37].

It was shown in [37] that pretreatment of films by photon annealing (ultraviolet radiation) in an argon atmosphere improves wetting. When tin – strontium melts spread on the surface of aluminum films on silicon, preliminarily treated with photon annealing for 4 s, wetting thresholds are detected at temperatures> 850 K. Tin – barium melts and pure tin only partially wet aluminum films on silicon [45].

Data on the polytherms of the wetting angle helps to calculate the adhesion energy, Wa, of the melt to silicon, according to the Dupre formula [1–6, 37–42]. For example, at T = 550 K, estimates show that the adhesion energy of pure tin to silicon is Wa = 70.3 MJ/m^2, which is consistent with theoretical calculations [45–47].

Fig. 2 presents the results of experiments on the wetting of aluminum films on silicon. The results show that pure tin does not wet the substrate in any of the studied cases. With increasing temperature, the contact angles decrease. In experiments with an unannealed aluminum film, the most considerable wetting angles were observed. Perhaps this value is associated with a massive disorder of the surface structure, a higher roughness. The smallest contact angles were found on substrates on which aluminum films were subjected to photon annealing for three seconds.

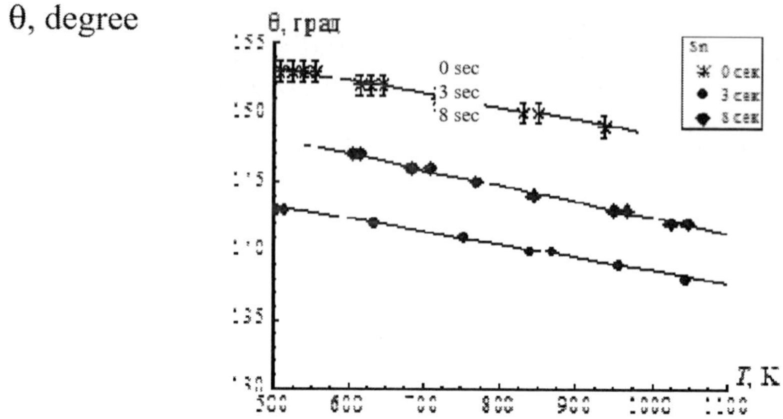

Figure 2. Polytherms of the contact angle of aluminum films on silicon with pure tin, depending on the duration of photon annealing

The connection of the contact angle on a rough surface θeff with the contact angle on a smooth surface gives the Wenzel – Deryagin equation. If the roughness coefficient k> 1, | cos (θeff) |> | cos (θ) |. Thus, the surface roughness improves wetting: θeff <θ for θ <π / 2 and worsens it for θ> π / 2. In our case, θ> π / 2. Therefore, a decrease in surface roughness under the action of photon annealing should improve wetting, which is observed in experiments [48].

The results can be explained by the fact that, at an annealing time of 4 s, the structure is ordered on the surface of the samples, and the aluminum film becomes smoother [49]. And at high temperatures, oxides are destroyed on aluminum films, which is confirmed by a sharp drop in contact angles. In the case of a melt, Sn-0.106 at. Sr this occurs at temperatures above 850 K, and in the case of a melt Sn-1.928 at. % Sr at T>865 K.

In Fig. Figure 3 shows the polytherms of the contact angles of the surfaces of the substrates with tin-strontium melts. Slight wetting of aluminum films on silicon by Sn melts – 0.106 at. Sr and Sn – 1.928 at. was found at annealing times τ = 0, 2, 6, and 8 τ = 2 and 6 s, respectively. With a photon annealing time of aluminum films on silicon of 4 s, the wetting angles are the smallest; moreover, at temperatures T> 850 K in the case of Sn – 0.106 at. and T> 856 K in the case of Sn – 1.928 at. % Sr, the wetting angle drops sharply (to 24 ° and 20 °, respectively (shown by arrows in Fig. 3 A and 3 B) [50].

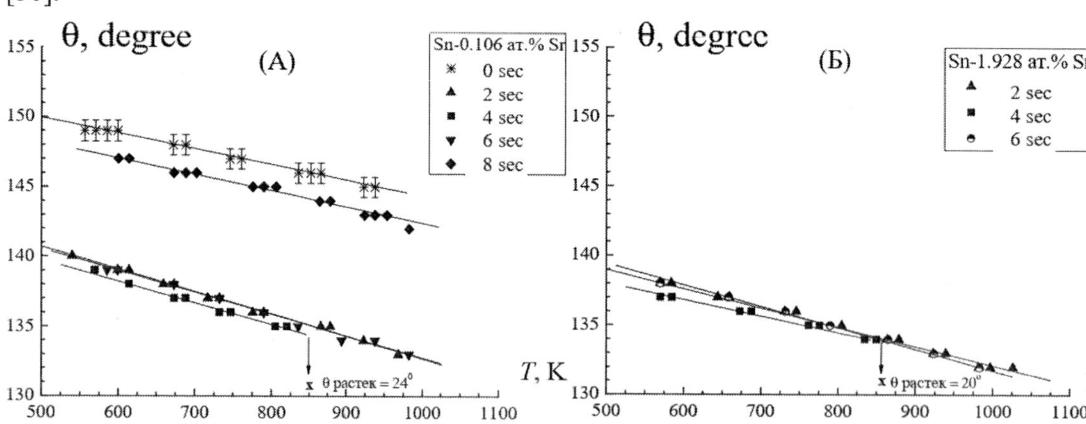

Figure 3. Polytherms of the contact angles of aluminum films on silicon with tin-strontium melts, depending on the duration of photon annealing of the substrates

Thus, it was found that photon annealing of aluminum films on silicon leads to a decrease in the wetting angle by pure tin and Sn – Sr melts. The maximum decrease in contact angles is achieved at annealing times corresponding to the minimum surface resistance of the films, i.e., 3...4 c.

It is interesting to note that when a droplet spreads over a substrate with an aluminum film on silicon, dendritic structures arise [50], which apparently indicates the formation of a complex eutectic in the tin-strontium-aluminum-silicon system.

Micro and nanopowders can be used in the manufacture of porous tapes, which, by impregnating with tin-based melts, can create new lead-free composite solders [44].

Thin ultrafine metal plates impregnated with solders are used as spacers for dispersion (composite) welding and brazing of metals with metals, semiconductors, and ceramics [51–59].

The data on polytherms of the contact angles of metallic melts of porous metals in the literature is insufficient.

Known results of studies of wetting angles are associated with the design of composite solders and, as a rule, contained a lead-containing low-melting component.

It should also be noted that in the literature, there are no data on the wetting of porous nickel and copper with melts containing additives of alkaline and alkaline earth elements. The presence of such additives opens the way to the creation of new composite cathodes.

We studied the polytherms of the contact angles of porous nickel (porosity of 30 %) and copper by Sn-Ba and In-Na melts.

It has been established that Sn-0.152 wets porous nickel at. % Ba, and Sn-0.396 at. % Ba at temperatures above 475 °C, while Sn-0.061 wets copper at. % Ba and Sn-0.396 at. % Ba melts already at 275 ° C.

Additional thresholds for wetting nickel and copper with In-0.3 at. % Na and In-0.5 at. % Na are detected at 250 and 225 °C, respectively. Research morphology of samples after crystallization. The results showed the presence of various structures of small sizes, including intermetallic formations.

In all cases, at the first moment, the melts do not wet the substrates, which is explained, first of all, by the high oxidizability of In-Na, Sn-Ba melts, the oxide films of which are destroyed with increasing temperature. But, possibly, the gas presence in the pores and the high degree of roughness of the substrates led to the partial wetting of the Cassi-Baxter [38]. With increasing temperature and destruction of oxide films, the melts come into closer contact with a rough surface, and the Wenzel mechanism is realized [38]. As the porous substrate is saturated with melts, the heterophase substrate is wetted, and the contact angle takes the form [40]: $\cos\theta_\Pi = (1-\Pi)\cos\theta + \Pi$, where Π is the porosity, θ is the contact angle according to the Dupre formula [40]. However, this formula is valid in the absence of chemical interaction of the melts with a capillary-porous substrate.

There is no theoretical description of the porous wetting and heterophase systems during the formation of intermetallic compounds.

3. Conclusion

The surface properties of tin-based melts are analyzed to construct solders for brazing aluminum, silicon, and piezoceramics.

It is shown that the smallest contact angles are achieved using tin-aluminum eutectics and tin-zinc-aluminum peritectic with the addition of copper and nickel micro powders. A decrease in wetting angles is also achieved by preliminary photon annealing of the substrates.

References

[1] Trifonova A, Wachtler M, Witnter M and Besenhard J O 2002 Sn-Sb and Sn-Bi Alloys as Anode Materials for Lithium-Ion Batteries *Ionics* **2**

[2] Suganuma K and Kim K-S 2007 Sn-Zn Low Temperature Solder *J. Mater. Sci: Mater Electro* **18** 121–7

[3] Chang Ch-A, Chen S-W, Chiu Ch-N and Huang Yu-Ch 2005 Primary Solidification Phases of the Sn-Rich Sn-Ag-Cu-Ni Quaternary System *J. of Electr. Mater.* **34(8)** 1135–42

[4] Iver E 2007 Anderson Development of Sn-Ag-Cu and Sn-Ag-Cu-X Alloys for Pb-free Electronic Solder Applications *J. Mater Sci: Mater Electron* **18** 55–76
[5] Li Bo, Shi Yaowu, Lei Yongping et al 2005 Effect of Rare Earth Element Addition on the Microstructure of Sn-Ag-Cu Solder Joint *J. of Electr. Mater.* **34(3)** 217–24
[6] Song Jenn-Ming, Huang Chi-Feng, and Chuangl Hsin-Yi 2006 Microstructural Characteristics and Vibration Fracture Properties of Sn-Ag-Cu-TM (TM 5 Co, Ni, and Zn) alloys *J. of Electr. Mater.* **35(12)** 2154–63
[7] Sharif A and Chan Y C 2005 Comparative Study of Interfacial Reactions of Sn-Ag-Cu and Sn-Ag Solders on Cu Pads during Reflow Soldering *J. of Electr. Mater.* **34(1)** 46–52
[8] Kang S K, Choi W K, Yim M J and Shih D Y 2002 Studies of the Mechanical and Electrical Properties of Lead-Free Solder Joints *J. of Electr. Mater. Spec. iss. Paper* **31(11)** 1292–303
[9] Li J G, Chatain D, Coudurier L and Eustathopoulos N 1988 Wettability of Sapphire by Sn-Al Alloys *J. of Aterials Sci. Letters* **7** 961–3
[10] Naidich Yu V and Taranets N Yu 1998 Wettability of Aluminium Nitride By Tin-Aluminium Melts *J. of Mater. Sci.* **33** 3993–7
[11] Nizhenko V I and Floka L I 1981 *The surface tension of liquid metals and alloys* (Moscow: Metallurgy) 208 p
[12] Mozer Z, Gasior W and Pstrus J 2001 Surface Tension of Liquid Ag-Sn Alloys: Experiment Versus Modeling *J. Phase Equilib.* **22(3)** 254–8
[13] Mozer Z, Gasior W and Debski A 2005 Database on Pb-free Soldering Materials, Surface Tension and Density, Experiment vs modeling *Data Sci. J.* **4** 195–208
[14] Liu X J, Inohana Y, Ohnuma I et al 2002 Experimental Determination and Thermodynamic Calculation of The Phase Equilibria and Surface Tension of Ag-Sn-In System *J. Electron. Mater.* **31** 1139–5
[15] Gasior W, Moser Z and Pstrus J 2003 Surface Tension, Density, and Molar Volume of Liquid Sb-Sn Alloys: Experiment Versus Modeling *J. of Phase Equilibria* **24(6)** 504–10
[16] Nizhenko V I and Smirnov Yu I 1994 Temperature dependence of the density and surface tension of Ni-Sn double melts *Metals* **1** 29–32
[17] Nizhenko V I and Smirnov Yu I 1994 Temperature dependence of the density and surface tension of Si-Sn double melts *ZhTF* **68(4)** 752–4
[18] Nizhenko V I and Smirnov D I 1996 The temperature dependence of the density and surface tension of the melts of the Al-Sn system *Rasplavy* **1** 3–8
[19] Akhkubekov A A, Ponezhev M Kh and Sozaev V A 1999 The study of soldered joints of ceramic boards with aluminum bases *Fundamental problems of piezoelectric instrumentation* vol. 1 Tr. Int. sci. and pract. Conf. Piezotekhnika-99 (Rostov-on-Don) pp 20–5
[20] Arsamikov U V, Dadashev R Kh, Ibragimov Kh I and Yushaev S S 1990 Surface tension and density of tin-thallium melts *Adhesion of melts and solder. of mater.* **23** 10–3
[21] Ohnuma I, Ishida K, Mozer Z et al 2006 Pb-free solders: Part II *J. of Phase Equilibria and Difusion* **27(3)** 245–54
[22] Dadashev R Kh 2007 *Thermodynamics of surface phenomena* (Moskow: Fizmatlit) 278 p
[23] Metelkin I I, Skalskaya V P, Chepyuk L I and Cheredina L F 1985 Study of junctions of ceramic boards with metal bases *Electr. equipment. Ser. Microwave Electron.* **6** 367
[24] Tsykin A V, Otmakhova N G and Kalinnikova V B 1984 Investigation of the thermal cyclic resistance of soldered joints of a polycor core with an aluminum alloy base *Electr. equipment. Ser. Microwave Electron.* **7** 367
[25] Smirnov G N 1981 *Progressive methods of brazing aluminum* (Moscow: Metallurgy) 240 p
[26] Nikitinsky A M 1983 *Soldering aluminum and its alloys* (Moscow: Engineering) 190 p
[27] Timofeev G I and Grekov S G *Solder for brazing aluminum and its alloys* A.S. USSR. N1299029. cl. B23 K 35/26 from 27. 05. 85.
[28] Bremat S A *Alloy and its use for brazing aluminum* Application 2492410, France. Claim 10.16.80. N8022203, publ. 23. 04. 82. MKI C22 18/00, V23K1 / 19

[29] Nikitinsky A M, Grekov S G, Chernov M A et al *Solder for brazing aluminum and its alloys* A.S. USSR, N1077158, class. B23 K 35/26 from 19.04. 82.

[30] US patent N4106930, CL B23 K35 / 12, publ. 15.08.78

[31] British patent N907734, CL B23 K35 / 26, 1962

[32] Akhkubekov A.A., Melnikov S.V., Sozaev V.A et al The method of brazing aluminum and its alloys A.S. USSR, No. 1792023, Cl. In 23L1 / 00.

[33] Bochvar A A and Novikov I I 1952 On the solid-liquid state of alloys of different compositions during their crystallization *Izv. AN SSSR. OT* **2** FROM 217–24

[34] Morokhov I D, Chizhik S P, Pushkov V A et al 1977 *Dispersion methods of joining materials* (Moscow: Central Res. Instit. "Electronics") 142 p

[35] Sozaev V A, Shidov H T and Shukhostanov A K Solder for tinning and brazing of aluminum and its alloys USSR patent, N1274907, B23 K 35/26. publ. 11/07/92. Bull. no 41

[36] Liu X J, Kinaka M, Takaku Y et al 2005 Experimental Investigation and Thermodynamic Calculation of Phase Equilibria in the Sn-Au-Ni System *J. of Electr. Mater.* **34(5)** 670–9

[37] Dadashev R Kh, Kutuev R A and Sozaev V A 2016 *Surface properties of alloys based on lead, tin, indium, cadmium* (Moscow: Fizmatlit) 208 p

[38] Dmitriev A S 2015 *Introduction to Nanotechnology* (Moscow: BINOM. the lab. Signs) 790 p

[39] Zimon A D 1974 *Adhesion and wetting* (Moscow: Chemistry) 416 p

[40] Missol V 1978 *Surface energy of phase separation* (Moscow: Metallurgy) 176 p

[41] Ivashchenko Yu N and Evdoshchuk L V 1974 On the calculation of the contact angle by the size of a lying drop In *Adhesion of Melts* (Kiev: Naukova Dumka) 84 p

[42] Lesev V N and Sozaev V A 2011 *Study of the statics and dynamics of small drops* (Germany, Lambert Acad. Publ.) 128 p

[43] Dalakova N V, Elekoeva K M, Kashezhev A Z et al 2014 Polytherms of the contact angles of aluminum and aluminum of a lithium alloy by tin-based melts *Surface* **4** 60–3

[44] Kashezhev A Z, Kutuev R A, Manukyants A R et al 2014 Polytherms of wetting angle of tin-barium and indium sodium porous nickel *Physico-chemical aspects of the study of clusters, nanostructures, and nanomaterials* pss 6 Interuniver. Coll. of sci. papers (Tver: TVSU) pp 149–54

[45] Dalakova N V, Elekoeva K M, Kashezhev A Z et al 2011 Temperature dependence of the contact angles of silicon with tin-barium melts *Bull. of the Russ. Acad. of Sci. The ser. is phys.* **75(8)** 1150–2

[46] Varchenya S A and Upit G P 1981 *Strength and energy characteristics of Adhesion of metal condensates to silicon and quartz* The preprint (Salaspils: Instit. of Phys. of the Acad. of Sci. of the Latvian SSR) 36 p

[47] Kozyrev E N, Obukhova O E and Sozaev V A 2004 *Adhesion in a metal-semiconductor system* (Vladikavkaz: Publ. House of the SKGMI) 32 p

[48] Ponezhev M Kh, Kashezhev A Z, Manukyants A R et al *Solder for tinning an aluminum film on silicon* RF patent for the invention no 2477206. Publ. in bull. Fig. no 7, dated 03/10/2013.

[49] Ponezhev M Kh, Kashezhev A Z, Manukyants A R et al *The method of smoothing the Surface of aluminum films on a silicon substrate* RF patent for the invention no 2477204. Publ. in bull. Fig. no 7, dated 03/10/2013.

[50] Elekoeva K M, Kashezhev A Z, Manukyants A R et al 2014 Polytherms of the contact angles of tin-strontium melt of aluminum films on silicon before and after photon annealing *Bull. of the Russ. Acad. of Sci. The ser. is Phys.* **78(4)** 512–4

[51] Morokhov I D, Chizhik S P, Pushkov V A et al 1977 *Dispersion methods of joining materials* vol 3 (Moscow: Central Res. Instit. "Electronics") 142 p

[52] Chizhik S P, Grigoryeva L K and Kuklin R N 1984 To the study of the kinetics of physical and chemical processes during soldering through porous media *Metals* **3** 181–5

[53] Naidich O V *A method for soldering optical quartz glass parts with copper parts* A.s. 219770 USSR // B.I. 1968 no 19

[54] Gusev O V, Chistyakov O D and Yakovlev G A 1980 Flux-free composite brazing of silicon with metals *Electr. Industry* **5** 20–5
[55] Yakovlev G A 1978 Investigation of the physicomechanical properties of lead – nickel sponge solders and their soldered joints *Electr. Engineer. Ser. Microwave Electron.* **8** 84–9
[56] Yakovlev G A and Yakovlev V G 1979 The use of composite solders Pb – Ni for the assembly of powerful semiconductor devices *Electr. Engineer. Ser. Semiconductor devices* **127(1)** 80–8
[57] Gusev O V, Yakovlev G A and Chistyakov Yu D 1982 Physico-chemical aspects, technology, and prospects for the use of composite soldering of semiconductors and metals with lead-based solders *Adhesion of melts and solder. of mater.* **9** 64–77
[58] Akhkubekov A A, Kanchukoev V Z, Ponezhev M Kh et al *Method of compositional soldering* Patent no 2236331. B23K 1/00. Publ. 09/20/2004. B.I. no 26
[59] Elekoeva K M, Kasumov Yu N, Kashezhev A Z et al 2007 Capillary properties of lead-free solders for electronics *Bull. of KBSU. Ser. of phys. Sci.* **11** 31–7

Determination of multistoried buildings stiffness properties in the inverse dynamics solution

V I Sobolev, I V Sobolev, T Ya Druzhinina

[1] Irkutsk State Transport University, 15 Chernyshevsky Street, Irkutsk, 664074, Russia
[2] SP Sobolev I.V., 4, 8 microdistrict Berezovyj, Irkusk region, 664528, Russia

E-mail: vladsobol@yandex.ru

Abstract. The article is devoted to the development of methods and evaluation of the accuracy of determining the horizontal stiffness of multi-storey buildings basing on the results of instrumental measurements of the self-induced vibrations parameters carried out with the help of high-precision instruments. The proposed studies are extremely relevant in estimating the unsoundness degree of the buildings that have passed a certain period of operation or buildings that have been subjected to severe exposure. The use of dynamic methods of load-bearing structure condition analysis has undeniable advantages, since it excludes the need for a detailed examination, often associated with the necessity to open the enclosing structures and evict the residents of the building. The advantages of using this approach are particularly evident in the survey of a large array of residential buildings that are in heterogeneous operating conditions and require rapid results. The practical application of the methods was tested in the survey of the high-rise apartment buildings of 1-335 series in the city of Irkutsk. However, the need to use dynamic models of small dimension in the evaluation of stiffness parameters makes it inevitable and urgent to solve the problems of assessing the accuracy of such approximations.

1. Introduction

One of the simplest versions of the inverse dynamics problem is to determine the stiffness of a multi-storey building by the actual values of natural oscillations measured using high-precision instruments. Among the various problems of inverse dynamics [1-3] the determination of the stiffness properties of the building is certainly stable in computational terms and does not belong to the category of ill-posed problems [3-7].

However, to evaluate the stiffness parameters based on the results of determining the base frequency, one has to use some approximation of the design scheme, which introduces errors in assessing the building's inherent parameters [4, 8].

It should be noted that the determination of the limpness value characterizes the quantitative assessment of the accumulation of defects of the building exposed to a certain period of operation [4], which affects the functional properties of the bearing structures [8].

Such estimation methods basing on the dynamic tests have obvious advantages, as they exclude the necessity of the enclosing structures demolishing for the internal bearing building designs inspection [9].

It should be noted that the mandatory periodic determination of the natural vibration frequencies of the fundamental tone of the building for seismic regions are identified by the conditions of state standards [9].

The solution of the inverse dynamics problem is not the only reason conditioning the need to assess the accuracy of the fudamental dynamic parameters determination; it is necessary to give examples of the necessity for such an assessment when choosing a discrete design model and determining the level of discretization, which, in most cases, is too high [7, 10, 11].

The overwhelming predominance of overestimation of the dynamics problems dimension in computational practice does not exclude the possibility of forming the dynamic models of "small" dimension, that do not reflect the necessary dynamic properties of the calculated object with a sufficient degree of approximation [12].

Among other problems of estimating the error of determining the intrinsic response parameters of multi-storey buildings, the problem of specifying the error when using simple design models with lumped inertia is relevant [6, 12]. Some common methods for determining solutions to such problems are mainly based on the use of similarity transformations [13, 14-16].

The use of such estimates does not take into account the specifics of the construction elements of the calculated structure, and the solution of such a problem is, in our opinion, extremely important.

2. Theoretical Basis and Method

Let us consider the error values obtained by approximating various construction elements with single-mass dynamic models.

Using a dynamic deflection method [19-20], the design scheme of such a structural element can be represented as a vertical member with a distributed mass of intensity ρ, length l and shear stiffness GF, where G is the shear modulus, F is the cross-sectional area (Fig. 1).

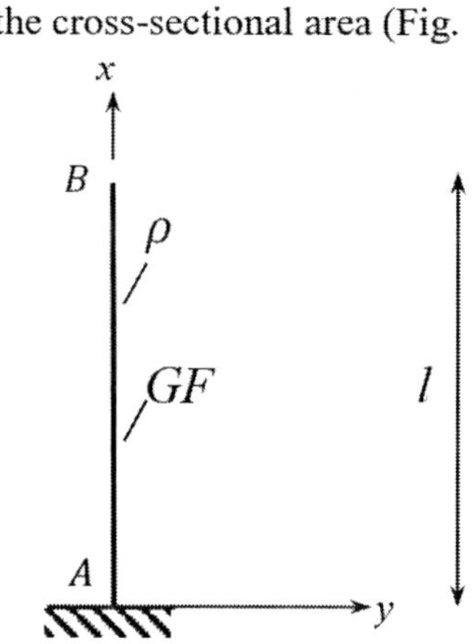

Figure 1. Design scheme of the element with shear deformations, with distributed mass ρ.

The equation of self-induced vibrations of such an element is a partial differential equation and has the form:

$$\frac{\rho\partial^2 v}{\partial t^2}+\frac{\partial^2 v}{\partial x^2}\cdot GF=0, \tag{1}$$

Where v - the horizontal displacement of the element points during the self-induced vibrations, t – the time parameter [21-24]. Let us assume that the self-induced vibrations of such a system are carried out with a certain frequency ω.

Therefore, using the method of separating variables [15, 25-29], it can be written:

$$v(t,x)=y(x)\cdot\sin(\omega t), \tag{2}$$

When substituting (2) into (1), there is an ordinary differential equation:

$-\rho\cdot\omega^2\cdot y(x)\cdot\sin(\omega t)+GF''(x)\cdot\sin(\omega t)=0$

After the simplification of which, the following equation can be obtained

$$y''^{(x)} - \frac{\rho\omega^2}{GF} \cdot y(x) = 0, \tag{3}$$

The characteristic equation for (3) is:

$$\mu^2 - \frac{\rho\omega^2}{GF} = 0$$

The roots of this equation are:

$$\mu_1 = \omega\sqrt{\frac{\rho}{GF}} \qquad \mu_2 = -\omega\sqrt{\frac{\rho}{GF}}.$$

Or $\mu_1 = \omega \cdot \mu;$ $\mu_2 = -\omega \cdot \mu,$

where $\mu = \omega\sqrt{\frac{\rho}{GF}}$.

The general solution of equation (3) is a linear combination of two linearly independent partial solutions forming a basic vector-function:

$h(x) = [sin(\omega\mu x), cos\,(\omega\mu x)]$.

In this case, the solution is given by the following:

$$y(x) = C_1 \cdot sin(\omega\mu x) + C_2 \cdot cos(\omega\mu x),$$

where C_1, C_2 - the coefficients of the linear combination determined by the boundary conditions of the equation (3) [16]. If $x = 0$, there is an equality $C_2 \cdot cos(\omega\mu x) = 0$, from which $C_2 = 0$.

Thus, taking into account the locking of joints at the point x=0, the solution is

$y(x) = C_1 \cdot sin(\omega\mu x)$.

Since the function of the oscillatory form during the self-induced vibrations is determined within the accuracy of a factor, let us take C_1 equal to 1, then

$y(x) = sin(\omega\mu x)$.

When the point *B* moves to a certain amount Δ, the equality is:

$sin(\omega\mu l) = \Delta$.

In particular, for $\Delta = 1$, the equality of the form is obtained:

$$C_1 \cdot sin(\omega\mu l) = sin\left(\omega\sqrt{\frac{\rho}{GF}} \cdot l\right) = 1.$$

If the point *B*, moved by the value $\Delta = 1$, imposes a lateral brace (Fig. 2), then there is a reaction H_B, equal to:

$$EI \cdot \frac{\partial y(x)}{\partial x}\Big|_{x=l} = cos(\omega\mu l) \cdot EI \cdot \omega\mu.$$

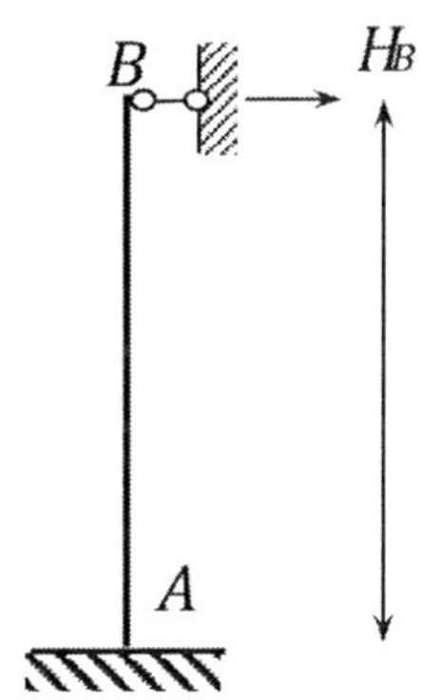

Figure 2. Calculation scheme for determining the reaction H_B in the linear connection of the upper node of the element AB.

3.

Under the influence of a single force in the connection line the amplitude of harmonic displacement *y(l)* is determined by the expression:

$$y(l) = \frac{1}{\cos(\omega\mu l)\cdot EI\cdot\omega\mu} = \frac{\sqrt{GF}}{\cos(\omega\mu l)\cdot EI\cdot\omega\sqrt{\rho}} = \frac{1}{\cos(\omega\mu l)\cdot\omega\sqrt{EI\rho}}, \quad (4)$$

In case of the magnitude of the resonance, the amplitude is achieved at the denominator of expression (4) equal to zero. Herewith,

$cos(\omega\mu l) = 0.$

The lowest frequency of self-induced vibrations is determined in this case from the equality:

$$\gamma_1\mu l = \frac{\pi}{2}$$

Thus, the frequency γ_1 of the lower tone oscillations of such a system is determined by the expression:

$$\gamma_1 = \frac{\pi}{2\mu l} = \frac{\pi}{2l}\cdot\sqrt{\frac{GF}{\rho}}.$$

Let us consider the approximation of the lowest frequency of the self-induced vibrations of wall structures with the masses localized in the levels of flooring.

Let us discuss the approximation of the simplest dynamic system represented by a two-storey building. The natural-vibration frequencies of such a system can be determined quite simply. There is also the opportunity to simply evaluate the approximation of the natural vibration frequency of the basic tone in a one-dimensional system.

The design scheme of a two-storey structure with the masses localized in the level of flooring is shown in Fig. 3.

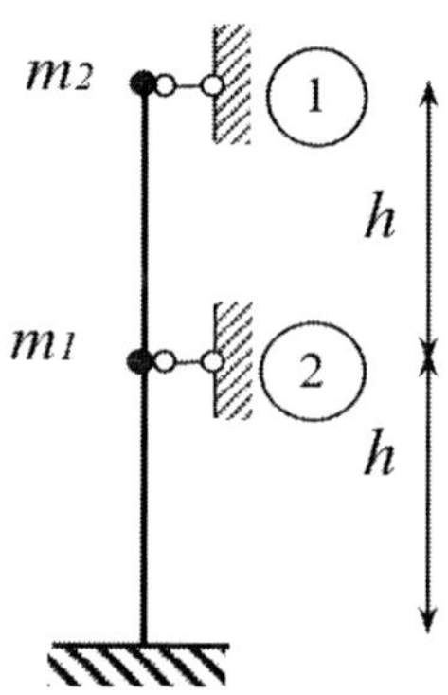

Figure 3. Dynamic model of a two-storey building with wall bearing structures – 1, 2 - link numbers.

To construct a system of dynamic equilibrium equations, the method of displacements is used, for which linear relations in the horizontal direction are imposed on the displacements of the discrete mass points (Fig. 3).

Localizing the masses of the building at the points, let us denote the values of the masses as m_1 and m_2.

The matrix R of single reactions of such a system has the form:

$$R = \begin{bmatrix} \frac{2GF}{h} & -\frac{GF}{h} \\ -\frac{GF}{h} & \frac{GF}{h} \end{bmatrix}$$

The matrix D of dynamic reactions of the model has the form:

$$D = \begin{bmatrix} \frac{2GF}{h \cdot m_1} & -\frac{2GF}{h \cdot m_1} \\ -\frac{2GF}{h \cdot m_2} & \frac{2GF}{h \cdot m_2} \end{bmatrix} \quad (5)$$

If $m_1 = m_2 = m$, the natural-vibration frequencies can be determined from the assumption

$$|D - \lambda \cdot E| = 0, \quad (6)$$

where λ – eigenvalues, E – a unit matrix.
In expanded form, the equation (6) can be written as follows:

$$\left(\frac{2GF}{h \cdot m} - \lambda\right)\left(\frac{GF}{h \cdot m} - \lambda\right) - \left(\frac{GF}{h \cdot m}\right)^2 = 0 \quad (7)$$

The roots of the equation (7) are defined as:

$$\lambda_{12} = 3GF/(h \cdot m) \mp \sqrt{9\left(\frac{GF}{h \cdot m}\right)^2 - \left(\frac{GF}{h \cdot m}\right)^2}$$

$$\lambda_{12} = \frac{GF}{h \cdot m}(3 \mp 2\sqrt{2})$$

Thus, the minimum eigenvalue is:

$$\lambda_1 = \frac{GF}{h \cdot m}(3 - 2\sqrt{2}),$$

The minimum natural-vibration frequency γ_1 is defined as:

$$\gamma_1 = \sqrt{\lambda_1} = \sqrt{\frac{GF}{h \cdot m}} \cdot \sqrt{(3 - 2\sqrt{2})} \quad (8)$$

Let us consider the error in determining the minimum natural-vibration frequency when replacing the two-mass model by one-dimensional model with a mass lumping of 2m at one point of it (Fig. 4).

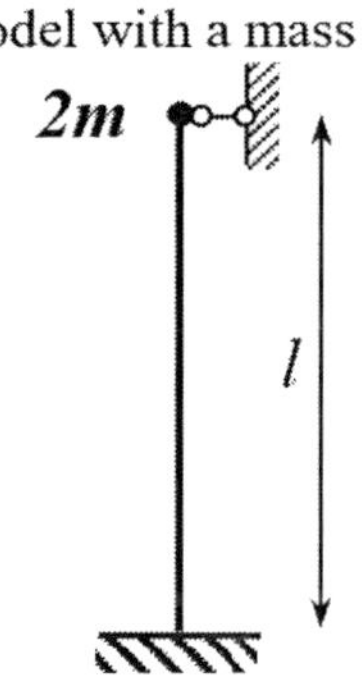

Figure 4. Simplifying of a two-dimensional dynamic model with a one-dimensional console model.

The frequency ω_1 of the natural oscillations of such a one-dimensional system has the form:

$$\omega_1 = \sqrt{\frac{GF}{3 \cdot h \cdot m}}.$$

As is seen from above, ω_1 is much larger than γ_1.

This circumstance is explained by the fact that the actual center of mass in the oscillations of the system in the first oscillatory form must take into account the unevenness of the mass movements of the system. At the same time, the most distant mass from the bearing point has greater displacements than the mass located below.

Let us form the center of concentration taking into consideration the first oscillatory form. The components of the vector A_1 of self-induced vibrations are determined from the solution of a system of equations:

$$(D-\lambda_1 \cdot E)A=0, \tag{9}$$

where $A = (a_1 \ a_2)^T$ - the eigenvector of a two-dimensional dynamical system corresponding to the eigenvalue γ_1.

As $Det(D-\lambda_1 \cdot E)=0$, in the system of equations there are two linearly dependent equations.

To define A, let us assume that $a_1 = 1$, then:

$$1\cdot\left(\frac{2GF}{h\cdot m}-\frac{GF}{h\cdot m}\cdot(3-2\sqrt{2})\right)-\frac{GF}{h}a_2=0.$$

From here it follows that $a_2 = 2\sqrt{2}-1$.

Thus, the vector A is written as:

$$A_2=(1,\ \ 2\sqrt{2}-1)^T$$

4. Conclusion

The advantages of using this approach are particularly evident in the survey of a large array of residential buildings that are in heterogeneous operating conditions and require rapid results. The practical application of the methods was tested in the survey of the high-rise apartment buildings of 1-335 series in the city of Irkutsk. However, the need to use dynamic models of small dimension in the evaluation of stiffness parameters makes it inevitable and urgent to solve the problems of assessing the accuracy of such approximations.

References

[1] Adams R D, Cawley P, Stone B JA 1978 *Vibration technique for non-destructively assessing the integrity of structures* 560

[2] Cawley P, Adams R D 1979 *The location of defects in structures from measurements of natural frequencies*

[3] Berman A 1984 *System identification of structural dynamic models – theoretical and practical bounds* 487

[4] Sobolev V I, Pinus B I 2019 Determination of the residual stiffness parameters of defective buildings on the basis of laser mappings of vibrations and solving the inverse problem of dynamics *ESSUTM Bulletin* **1** (**72**) 55-67

[5] Amelkin V V 1987 *Differential equations in applications* (Moskow: Nauka) p 160

[6] Konyuhov V Yu, Gladkih A M, Semenov V V 2019 Measures to the improvement of efficiency of a repair enterprise *Journal of Physics: Conference Series* **1353** 012046

[7] Kondratiev V V, Karlina A I, Guseva E A, Konstantinova M V, Kleshnin A A 2018 Processing and Application of Ultra disperse Wastes of Silicon Production in Construction *IOP Conference Series: Materials Science and Engineering* Vol. 463 **3** 032068

[8] Pinus B I, Morgaev D E 2003 Estimation of residual resource of 1-335 KC series buildings seismic stability in Irkutsk. Abstracts of the V Russian National conference on antiseismic construction and seismic zoning with international participation. Research Center of structures seismic stability, (Moskow: SUE TSNIISK by V.A. Kucherenko) 81

[9] State Standard (GOST 31037-2011) *«Buildings and structures. Rules of inspection and technical condition monitoring»*, 67

[10] Sobolev V I 2000 Calculation of high-rise buildings, various structural systems for horizontal seismic effects taking into account spatial deformation *Mathematical modeling in mechanics of continuous media on the basis of methods of boundary and finite elements: Proceedings of*

the XVIII International conference, SPb: SPbGU NIIH **1** 2-17
[11] Argyris J H, Boni B, Hinderlang V 1982 Finite element analysis of two- and three dimensional elastoplastic frames - the natural approach *Comp. Meth. Appl. Mech* Vol. 35 **2** 221-248
[12] Eisenberg Ya M 1997 *Development of concepts and norms of antiseismic design* (TSNIISK, SSC "Construction", Moscow) 73
[13] Davies E B, Gladwell G M L, Leydold J S, Peter F 2001 Discrete nodal domain theorems. *LinearAlgebraAppl* **336** 51-60 (2001)
[14] Konyuhov V Yu, Konstantinova M V, Gladkih A M 2019 Determination of restored units spectrum of equipment and development of the assembly unit repair method at industrial enterprises *Journal of Physics: Conference Series* **1353** 012047
[15] Konyuhov V Yu, Gorban A V, Gladkih A M 2019 Investment in the improvement of maintenance service efficiency of processing equipment of an industrial enterprise *Journal of Physics: Conference Series* **1353** 012102
[16] Ikramov Kh D 1991 *Nonsymmetric problem of eigenvalues. Numerical methods* (Moskow: Nauka) 240
[17] Koloushek V 1965 *Dynamics of building structures* (Moskow: Publishing House of Construction Engineering Literature) 632
[18] Farlow C 1985 *Partial differential equations for scientists and engineers: transl. from English* (Moskow: Mir) 384
[19] Kondratiev V V, Karlina A I, Guseva E A, Konstantinova M V, Gorovoy V O 2018 Structure of Enriched Ultradisperse Wastes of Silicon Production and Concretes Modified by them *IOP Conference Series: Materials Science and Engineering* Vol. 463 **4** 042064
[20] Konstantinova M V, Balanovskiy A E, Gozbenko V E, Kargapoltsev S K, Karlina A I, Shtayger M G, Guseva E A, Kuznetsov B O 2019 Application of plasma surface quenching to reduce rail side wear *IOP Conference Series: Materials Science and Engineering* **560** 012146
[21] Guseva E A, Kargapoltsev S K, Balanovskiy A E, Karlina A I, Shtayger M G, Gozbenko V E, Konstantinova M V, Sivtsov A V 2019 Comparative evaluation of corrosion resistance of wheel and rail steels in various media *IOP Conference Series: Materials Science and Engineering* **560** 012181
[22] Grechneva M V, Balanovskiy A E, Gozbenko V E, Kargapoltsev S K, Karlina A I, Shtayger M G, Karlina Yu I, Govorkov A S 2019 Quality and reliability improvement of "tube-tube plate" welded joints during welding by pulse pressure *IOP Conference Series: Materials Science and Engineering* **560** 012142
[23] Shtayger M G, Balanovskiy A E, Kargapoltsev S K, Gozbenko V E, Karlina A I, Karlina Yu I, Govorkov A S, Kuznetsov B O 2019 Investigation of macro and micro structures of compounds of high-strength rails implemented by contact butt welding using burning-off *IOP Conference Series: Materials Science and Engineering* **560** 012190
[24] Balanovskiy A E, Shtayger M G, Karlina A I, Kargapoltsev S K, Gozbenko V E, Karlina Yu I, Govorkov A S, Kuznetsov B O 2019 Surface hardening of structural steel by cathode spot of welding arc *IOP Conference Series: Materials Science and Engineering* **560** 012138
[25] Balanovsky A E, Shtayger M G, Kondrat'Ev V V, Van Huy V, Karlina A I 2018 Plasma-arc surface modification of metals in a liquid medium *IOP Conference Series: Materials Science and Engineering* **411** 012013
[26] Kargapoltsev S K, Khomenko A P, Gozbenko V E, Minaev N V, Karlina A I 2017 Development of new lubricants for reducing the wear of the elements of the path and running parts of rolling stock *International Journal of Applied Engineering Research* Vol. 12 **22** pp 12362-12368
[27] Voropai N, Ukolova E, Gerasimov D, Suslov K, Lombardi P, Komarnicki P 2019 Simulation approach to integrated energy systems study based on energy hub concept *IEEE Milan PowerTech, PowerTech* 8810666
[28] Voropai N, Ukolova E, Gerasimov D, Suslov K, Lombardi P, Komarnicki P 2019 A Study on Cost-Effectiveness of Energy Supply Based on the Energy Hub Concept *IEEE PES Innovative*

Smart Grid Technologies Europe, ISGT-Europe 8905736
[29] Krylov A N 1950 *On some differential equations of mathematical physics, having an application in technical matters* (Moskow: Gostehizdat)

IOP Conf. Series: Materials Science and Engineering **905** (2020) 012078 doi:10.1088/1757-899X/905/1/012078

The prospect of creating of the plant for the production of liquid fuel from coal in territory of Eastern Siberia

R S Zott[1], **V Yu Konyuhov**[2], **A M Gladkih**[2]

[1] LLC "Tarasovsky coal", 1 Dzerzhinsky Street, Irkutsk, 664003, Russia
[2] Irkutsk National Research Technical University, 83 Lermontov Street, Irkutsk, 664074, Russia

E-mail: konyuhovvy@ex.istu.edu

Abstract. In this work the shortage level of oil and gas is considered. Receiving various products from coal, exceeding prime cost of mineral raw materials in tens of times. The prospect of creation of the plant of receiving synthetic fuel from coal in territories of Eastern Siberia.

1. Introduction

Today, about half of all energy produced in the world is generated from oil, including almost all energy for autonomous mobile consumers. But for almost ten years, oil prices have been growing rapidly, and its reserves are growing more slowly, and production is becoming more expensive. Oil is a very valuable natural raw material, but according to the estimates of the World Energy Council experts, the known reserves of oil for mankind can last only about 56 years, gas for 60 years or more, and coal for at least 270 years. Consequently, coal is the most profitable mined and subsequently processed raw material. Coal is the most abundant fossil raw material, the Fischer-Tropsch coal process can find much wider application in the near future, giving not only gasoline, but also gas oil and new generation fuels [1-6].

The Fischer – Tropsch process is a well-developed technology that is already being applied on a large scale, although its proliferation is hindered by high capital costs, high operating and repair costs, and relatively low crude oil prices. In particular, the use of natural gas as a feedstock becomes appropriate when "stranded gas" is used, i.e. sources of natural gas located far from the main cities, which are inexpedient to operate with conventional gas pipelines and LNG technology.

2. Purpose of the Study

In terms of coal production, Russia ranks fifth in the world (after China, the USA, India and Australia), 3/4 of coal mined is used for energy and heat production, 1/4 - in metallurgy and chemical industry. A small part is exported, mainly to Japan and the Republic of Korea [2, 7-11].

Open coal mining in Russia is 2/3 of the total. This method of extraction is considered the most productive and cheapest. However, this does not take into account the strong disturbances of nature associated with it - the creation of deep quarries and extensive dumps of overburden rocks. Coal mining in the mines is more expensive and has a high accident rate, which is largely determined by the deterioration of mining equipment. The role of a coal basin in the territorial division of labor depends on the quality of coal, the size of reserves, technical and economic indicators of production, the degree of preparedness of reserves for industrial operation, the size of production, and features of the transport and geographical location. According to the totality of these conditions, inter-district coal bases – the

Kuznetsk and Kansk-Achinsk basins, which account for 70% of coal production in Russia, as well as the Pechora, Donetsk, Irkutsk-Cheremkhov and South Yakut basins are sharply distinguished [2-3, 12-16].

Today, the Russian coal industry accounts for 354 million tons of mining, 132 million tons of export coal, 165 thousand jobs and 500 thousand in related industries, about 54 billion rubles of taxes and social contributions annually [1, 17-24].

More than 400 different products are obtained from coal, the cost of which is 20-25 times higher than the cost of coal itself. When coking coal, that is, with strong heating without access to air, receive the following main products:

1) Coke oven gas containing methane, hydrogen, ammonia, ethylene, benzene and toluene.
2) Coal tar containing benzene, toluene, xylenes, other homologues of benzene, naphthalene, phenol.
3) Suprasmall water containing ammonia and phenol.
4) Coke, which is almost pure carbon.

And then by-products are synthesized, such as: explosives, various solvents, plastics, dyes and medicinal substances that exceed the cost of the main products, thereby the introduction of innovative technologies in the field of processing will allow to extract the largest amount of useful components from the coal mass, ensuring maximum value added [5, 25-26].

3. Results and Discussion

Even in the USSR, enterprises were built for the production of synthetic fuel from coal. There were some difficulties: bulky equipment was very expensive, and complex technology led to the fact that due to multi-stage, frequent heating and cooling, only 35-40% of the energy contained in coal was transferred to the final liquid products. The listed reasons rather determined the termination of the production of synthetic fuels in the 40s, the reconstruction of enterprises for their production into oil refineries and petrochemical plants. The efficiency of coal-chemical production was lower than that of petrochemicals. Especially in conditions when the volume of oil production in the country is constantly growing. New large oil fields between the Volga and the Urals were explored and put into operation. At the Angarsk plant since the mid-1950s. installation of the first refineries began. Oil for them was brought by rail in tanks. In August 1960, the first industrial gasoline from oil was obtained [3, 27].

But at the moment there is a problem of reducing oil reserves and, consequently, a search is underway for alternative methods for producing products derived from oil. Projects that closed in the 40s for the production of liquid fuel from coal, become promising and should be taken into consideration, and then improved, in view of modern technologies, with the aim of: improving the quality of synthetic fuel, lowering energy costs for production, etc [1, 28].

To convert the organic mass of coal into an oil-like substance, three chemical tasks must be solved:

1. remove oxygen from it, and with it such harmful impurities for fuel as nitrogen and sulfur, in the form of water, ammonia and hydrogen sulfide, respectively, having spent a lot of hydrogen for this, which is already so low in coal;
2. add to the organic mass of hydrogen to the ratio of hydrogen and carbon in oil;
3. disaggregate the macromolecules of the organic mass of coal to the molecular weight of the oil components.

Of all these tasks, the third is easiest -for a long time, thermal and thermocatalytic processes have been used in the processing of combustible minerals, in these processes the vibrational movements of atoms in molecules increase under the influence of heat, and the least strong chemical bonds are broken and large molecules turn into smaller ones [2].

The development of the technology for the production of liquid fuel from coal requires significant investment, while one should not expect a quick return on investment. Plants for the production of liquid fuels from coal can be profitable only temporarily, in the event of a rise in price of oil. Graphs of changes in oil prices in recent years show that periods of falling oil prices are invariably replaced by periods of rising prices. Fuel made from coal is competitive at an oil price above $ 40 per barrel. In recent years, periods during which the production of liquid fuels from coal have been cost-effectively prevailing. If this dynamics of oil prices continues, then the production of coal into liquid fuel will be generally profitable, although the income from this type of activity will not be stable. America has successfully used periods of rising oil prices, increasing shale gas production, and curtailing production in the event of a fall in oil prices. Testing the methodology for quickly deploying or stopping the production of liquid fuel from coal will increase the attractiveness of investments in such a project.

The payback period for the production of liquid fuel from coal is from 7-8 years and above. It is clear that such projects are a long-term investment of world-class players. This slows down the development of commercial production of synthetic liquid fuels from coal. At the same time, the attractiveness of the formula "gasoline from coal" does not leave alone many manufacturers. Therefore, requests for the development of technology to specialists in the deep processing of coal are constantly received. The economic feasibility of the production of liquid fuel from coal is determined by the sufficient reserves of coal in our country and commercial efficiency comparable to the distillation of oil.

Currently, the widespread use of coal processing technology in liquid fuels is also hindered by environmental problems caused by this production. In particular, the problem of carbon dioxide emissions. Carbon dioxide emissions from the production of synthetic liquid fuels from coal are twice as high as from the production from petroleum. An important problem is also the high water consumption in the production process. For every ton of liquid fuel produced, 5 to 7 tons of water is consumed. Solving these problems will also help to increase the competitiveness of liquid fuel and coal plants.

In China, the world's largest single coal-liquid fuel processing plant was officially commissioned in the Ningxia Hui Autonomous Region. The project was implemented, as the country is rich in coal resources, but is experiencing a shortage of oil and gas, which Russia expects in 50 years. At the current stage, the country imports more than 60 percent oil.

The total investment in the project amounted to about 55 billion yuan / 7.9 billion US dollars. The plant is capable of turning more than 20 million tons of coal into 4 million tons of oil products annually, including 2.7 million tons of diesel fuel, 980 thousand tons of naphtha and 340 thousand tons of liquefied gas, said the deputy general director of the coal mining company at Shenhua Corporation in Ningxia Yao Ming. Among the by-products of production are 200 thousand tons of sulfur, 75 thousand tons of various alcohols and 145 thousand tons of ammonium sulfate, he added [4].

Another example of the commercial application of liquid fuel technology from coal. In 2015, the INFRA Group, which developed and patented a new generation of liquid synthetic fuel production technology based on the Fischer — Tropsch synthesis process from natural or associated gas (GTL), biomass and coal (XTL), commissioned a catalyst factory. Production of 15 tons per year produces a patented Fischer — Tropsch synthesis catalyst developed by the company's specialists. The factory's task is to manufacture the catalyst for the INFRA GTL plants, as well as to develop processes for the production of new catalyst modifications on an industrial scale. In 2016, INFRA designed and built a GTL (gas-to-liquid) modular transportable refinery for the processing of natural and associated gas into M100 synthetic oil in Wharton (Texas, USA). The company plans include commercial operation of the plant and sale of synthetic oil. At the request of the oil and gas company, the INFRA group began designing the GTL plant, which is planned to be located in the Nenets Autonomous Okrug. A plant with a capacity of 20 thousand petroleum products per year will produce winter diesel fuel and high-octane gasoline from natural gas from the Vasilkovsky gas condensate field. The implementation of the construction plan for the GPP using the latest INFL GTL technology from INFRA will ensure the market of the Nenets Autonomous Okrug with high-quality commercial fuel - diesel and gasoline - and significantly reduce the cost of purchasing an expensive northern delivery. The feasibility study for the construction was carried out in 2017, the design was completed in 2019 [4].

Projects created for the socio-economic development of Eastern Siberia are either postponed or completely closed, not invested. While, on the example of the Kemerovo region, projects receive approval and are launched. Yes, of course, the Kuznetsk coal basin is considered one of the largest coal basins, but investing in the development of enterprises in Eastern Siberia, namely the Krasnoyarsk

Territory, Irkutsk Region, where the largest coal deposits are located, they will bring additional considerable profit.

An example of a project implemented in the Kemerovo region is the development of mini-plants for the production of liquid fuel from coal. Project participants are:

- NP "South Ural Technical Society" (Miass, Chelyabinsk Region) is responsible for the development of technical documentation, maintaining contractual relations, organizing the manufacture and delivery of components.

- LLC "Kvant" (Novokuznetsk, Kemerovo region) - development of technology, manufacturing of an electric discharge installation, assembly and factory testing of the kit, commissioning.

- NPO SPb EC LLC (Saint-Petersburg) - is a technology partner, is responsible for the promotion, replication of the development and its active implementation at industrial enterprises in the country and abroad.

The main difference between the project and existing methods of coal processing is precisely in a significant reduction in unit costs, equipment costs due to the exclusion of processes occurring at high temperatures and pressures, and the exclusion of catalytic reactions. Technology is much cheaper and simpler. There is no need to build entire factories with a giant production cycle, to bear huge costs.

Kansk-Achinsky coal basin, the second largest coal producer after the Kuzbass. It is located several hundred kilometers east of Kuzbass on the territory of the Krasnoyarsk Territory and partially in the Kemerovo and Irkutsk Regions. This Central Siberian basin has the most significant reserves of brown energy coal mined by open pit mining. Coal mining in the basin in 2006 exceeded 40 million tons per year, the largest coal mining enterprise is the largest in Russia coal mine "Borodinsky" - 20 million tons / year. The large sections also include "Berezovsky", "Nazarovsky", "Pereyaslovsky", "Kansky" sections. Having created and launched a project for processing coal into liquid products, such as phenols, cresols, xylenols, the demand for which is very high (this is the raw material for the production of high-tech plastics), good profit will be realized.

In the Irkutsk region in 1945, the plant No. 16 was built to produce motor fuel from Cheremkhov coal. The Irkutsk-Cheremkhovsky basin stretches along the Upper Sayan from Nizhneudinsk to Lake Baikal. It is divided into the Baikal and Prisayansk branches. The production volume is 3.4%, the production method is opening and the advantage of such a production method is that it is relatively cheap and safe for workers. The deposit is far from large consumers, delivery is difficult, therefore, local coal is mainly used in Irkutsk enterprises. The stock is about 7.5 billion tons of coal. Therefore, a promising project will be the resumption of the plant and technologies for the production of liquid fuel from coal in the Irkutsk region. The total cost of the project will be approximately 5 billion rubles [5].

4. Conclusion

Thus, modern Russia is able to solve many problems and circumvent the difficulties that existed 60-80 years ago, using new production technology, powerful devices and erudite specialists. Oil reserves are diminished every year, as are gas reserves, so coal chemistry has a great prospect and interest in new coal processing projects. Especially, such projects should be considered in Eastern Siberia, the further development of which will ensure the main increase in coal production in the industry, and further, the creation of coal processing plants with the prospects of introducing new technologies for producing liquid products.

There are examples of successful implementation of projects to create and commercialize the production of liquid fuel from coal. In China, America, and Germany, much attention is paid to the development of this type of production. Kemerovo region, rich in coal reserves, is actively developing in this area. Eastern Siberia has large reserves of coal and every reason to attract investment and expand the production of liquid fuel from coal.

References

[1] Voskresenskaya O 2015 The triumph of coal is yet to come *Russian Newspaper - Federal Issue* **6191**

[2] Glukhikh O 2010 Synthetic fuel is the future of Russia *Newspaper Vesti* Issue 4/04/2010

[3] *The history of the Angarsk. NUK "Expedition InterBAYKAL"* 2008-2018

[4] *CTL - liquid fuel from coal in China* [Electronic resource] https://slanceviy-glas.livejournal.com/394326.html (Date access: 03/25/2018)

[5] Konyuhov V Yu, Gladkih A M, Galyautdinov I I, Severina Y D 2019 Economic aspects of green technologies *IOP Conference Series: Earth and Environmental Science* **350**(1) 012036

[6] Konyuhov V Yu, Gladkih A M, Galyautdinov I I, Kiseleva T Yu 2019 Ecological architecture: The green roofs *IOP Conference Series: Earth and Environmental Science* **350**(1) 012035

[7] Gladkih, A M, Konyuhov V Yu, Galyautdinov I I, Shchadova E I 2019 Green building as a tool of energy saving *IOP Conference Series: Earth and Environmental Science* **350**(1) 012032

[8] Taimarov M A, Shakirov I M 2013 Improvement of the complex for producing coal oil *Bulletin of Kazan Technological University* **23** pp 183-185

[9] Balanovsky A E, Gozbenko V E, Kargapol'Tsev S K, Karlina A I, Karlina Y I 2020 Evaluation of influence of technological parameters on width of strengthened layer in plasma surface hardening of structural steels *IOP Conference Series: Materials Science and Engineering* **759(1)** 012002

[10] Gozbenko V E, Balanovsky A E, Kargapol'Tsev S K, Karlina Y I, Karlina A I 2020 Researches of metal texture after friction stir welding *IOP Conference Series: Materials Science and Engineering* **759(1)** 012009

[11] Leonovich D S, Zhuravlev D A, Karlina Yu I, Govorkov A S, Karlina A I 2020 Automated assessment of the low-rigid composite parts influence on the product assemblability in the GePARD system *IOP Conference Series: Materials Science and Engineering* **760(1)** 012038

[12] Kargapol'Tsev S K, Balanovsky A E, Gozbenko V E, Karlina Y I, Karlina A I 2020 Possibility of obtaining complex form details using additive surface technology *IOP Conference Series: Materials Science and Engineering* **759(1)** 012011

[13] Karlina A I, Gozbenko V E, Kargapol'Tsev S K, Karlina Y I 2019 Methods for controlling the vibration state of technical facilities *Journal of Physics: Conference Series* **1384(1)** 012019

[14] Gozbenko V E, Kargapol'Tsev S K, Karlina Yu I, Karlina A I 2019 Approximation of amplitude-frequency characteristics using equidistants *Journal of Physics: Conference Series* **1384(1)** 012014

[15] Kargapol'Tsev S K, Gozbenko V E, Kuznetsov B O, Karlina Yu I, Karlina A I 2019 The effect of the periodic driving force on a system with two degrees of freedom *Journal of Physics: Conference Series* **1333(5)** 052009

[16] Gozbenko V E, Kargapol'Tsev S K, Kuznetsov B O, Karlina A I, Karlina Yu I 2019 Determination of the principal coordinates in solving the problem of the vertical dynamics of the vehicle using the method of mathematical modeling *Journal of Physics: Conference Series* **1333(5)** 052007

[17] Sheshukov O Y, Mikheenkov M A, Egiazaryan D K 2020 Features of zinc extraction from sulfide forms *Materials Science Forum* **989 MSF** pp 228-234

[18] Sheshukov O Y, Nekrasov I V, Egiazaryan D K 2020 Features of tri-calcium silicate formation suppression by sulphur and possible ways of its preventing Materials Science Forum **989 MSF** pp 221-227

[19] Shestov V V, Antipov V V, Ryabov D K 2017 Corrosion Resistance and Mechanical Properties of Layered Structural Material Based on Aluminum Alloy and Fiberglass Thin Sheets *Metallurgist* **60(11-12)** pp 1191-1196

[20] Ryabov D K, Antipov K V, Fesenko T V, Dynin N V 2017 Influence of cobalt alloying on structure and long-term strength of Al–Cu–Mg–Ag alloy system *Inorganic Materials: Applied Research* **8(1)** pp 159-165

[21] Gnedenkov S V, Sinebryukhov S L, Egorkin V S, Vyalyi I E, Mashtalyar D V, Nadaraia K V, Ryabov D K, Buznik V M 2017 Formation and properties of composite coatings on aluminum alloys *Russian Journal of Inorganic Chemistry* **62(1)**

[22] Ryabov D K, Shestov V V, Buznik V M, Antipov V V 2016 Design concept of multifunctional layered composite materials for operation in arctic regions *Inorganic Materials: Applied Research* **7(4)** pp 547-553

[23] Gergert A P, Molostova I I, Ryabov D K 2013 Development of 1960 auminium alloy for gas centrifuges *Tsvetnye Metally* **(9)** pp 33-36

[24] Kolobnev N I, Ber L B, Khokhlatova L B, Ryabov D K 2012 Structure, properties and application of alloys of the Al - Mg - Si - (Cu) system *Metal Science and Heat Treatment* **53(9-10)** pp 440-444

[25] Solodusha S, Suslov K, Gerasimov D 2016 Applicability of Volterra integral polynomials in the control systems of electric power facilities *Proceedings of 2016 International Conference "Stability and Oscillations of Nonlinear Control Systems" (Pyatnitskiy's Conference)* 7541227

[26] Voropai N, Ukolova E, Gerasimov D, Suslov K, Lombardi P, Komarnicki P 2019 Simulation approach to integrated energy systems study based on energy hub concept *IEEE Milan PowerTech, PowerTech* 8810666

[27] Ilyushin P, Suslov K 2019 Operation of automatic transfer switches in the networks with distributed generation *IEEE Milan PowerTech, PowerTech* 8810450

[28] Suslov K, Solonina N, Gerasimov D 2018 Assessment of an impact of power supply participants on power quality *Proceedings of International Conference on Harmonics and Quality of Power, ICHQP* pp 1-5

On reasons for the deterioration of the bottomhole formation zone and recommendations for maintaining reservoir properties

G S Dubinskiy[1], V E Andreev[1], F Y Kanzafarov[2] and V S Mukhametshin[3]

[1] Laboratory of oil and gas research, SASI "Institute of strategic research of the Republic of Bashkortostan", Ufa State Petroleum Technological University, Ufa, Republic of Bashkortostan, Russia
[2] Laboratory of field chemistry, JSC «NizhnevartovskNIPIneft», Nizhnevartovsk, Russia
[3] Department of Oil and Gas Field Exploration and Development, Ufa State Petroleum Technological University, Branch of the University in the City of Oktyabrsky, Oktyabrsky, Republic of Bashkortostan, Russia

E-mail: intnm-gsd@yandex.ru, intnm@yandex.ru, kanzafarovfy@nvnipi.ru, vsh@of.ugntu.ru

Abstract. The literature on the deterioration of reservoir properties of the bottomhole zone during various operations in the wells and when acting on the reservoir is reviewed. It was revealed that bullhead well control operations and impacts on the bottom-hole formation zone have a very large negative impact if the current geological and physical conditions of the object are not taken into account. Laboratory and field tests of several hydrocarbon-based emulsion formulations were performed. Experiments and tests have shown the effectiveness of the proposed formulations and allow recommend them for use in well repair.

1. Introduction

The *Samotlorskoe* field, like most oil and gas fields, is developed using the method of flooding oil deposits. At present, the water cut of products has exceeded 90 %, but the achievement of the planned oil recovery can no longer be achieved only by applying the methods of increasing oil recovery. The provision of rational development is possible with a working stock of wells, the bottom-hole zone of the formation of which does not create obstacles to the influx of hydrocarbons.

Field experience in the development of gas and oil fields shows that during the primary and secondary opening of productive deposits, development, bullhead well control and flushing of wells, in general, during the operation of wells, the filtration-capacitive properties of BHZ (bottom hole zone) as a result of leaching of filtrates and solid phase from various technological solutions based on water, which are often used at present. For reservoir resuscitation, a number of additional impacts on the formation have to be performed, and this leads to a significant increase in the cost of repair work, lengthening of the repair time, and the natural reservoir properties of the producing formation are not always restored.

The appearance of residual oil in the BHZ and in remote reservoir zones is determined by the geological and physical conditions of the deposits, the type of wettability of the pore surface, the composition, physicochemical characteristics of the fluids saturating the reservoir, the evolution of the reservoir-fluid system during reservoir development and operation of individual wells [1–4].

Well operation depends on a number of factors: geological heterogeneity of section, quality of opening and development of productive formations, technological factors associated with the operation of wells in field development, etc. [1, 3–6]. The factors causing the deterioration of filtration-capacitive properties of BHZ can be geological, and technical and technological. They can be conditionally divided into groups:

- geological and physical characteristics of reservoir (reservoir type, porosity, permeability, composition and properties of formation fluids, etc.);
- technical and technological parameters of drilling (drilling method, drilling fluid, the duration of contact of the exposed formation with process fluids, etc.);
- technical and technological parameters of the secondary opening and development of reservoir (type of perforator, perforation fluid, method of development, etc.);
- technical and technological parameters of well operation (bottomhole pressure, temperature in the bottomhole formation zone – BHZ, etc.).

A significant reason for the decrease in BHZ permeability during well development is the penetration of solid clay particles of drilling fluid into the formation, and this phenomenon is enhanced in highly permeable reservoirs [1, 4, 5]. With increasing well downtime, i.e. contact of the formation with foreign liquids, these negative consequences become even greater. For example, for the wells of the *Langepasskoe* group of fields, the radius of contaminated zone is 0.08–0.8 m [7], for the *Shaimskoe* group of fields, the filtrate penetration into the BHZ is more than 0.1 m [8], and for the *Megionskoe* group of fields up to 1.5 m, and in some cases the filtrate penetration zone can reach 3 m or more [5]. Already at an early stage of oil field development, the state of BHZ is significantly worsened, and the decrease in permeability compared with the natural one reaches 50 % or more [5].

BHZ mudding drastically reduces drainage coverage. It is known that the coverage of a productive formation by drainage (selection in production wells) depends on the coverage of the productive formation by injection in injection wells [6–9]. So, for oil fields in Bashkiria, Tatarstan, Western Siberia, etc., the coefficient of formation coverage by water flooding along the thickness of the layer in injection wells is 0.2–0.6, and water is absorbed mainly by the most permeable layers.

The presence in the formation of clay interlayers of different thicknesses or clay components in the composition of the mineral (maybe 1–3 to 25 %) skeleton strongly affects the injectivity coefficient of water injection wells. If water is pumped into the oil reservoir that is chemically different from highly saline or fresh water, then it interacts with clay inclusions in the rock, destroying them, and the rock swells. Such a process causes clogging of the filtration channels, a decrease in the permeability of BHZ and the coefficient of coverage by waterflooding of the reservoir in thickness.

At West Siberian oil fields, it was revealed that the average duration of the development process after repair of wells plugged in front of it with an aqueous solution of NaCl reaches 4–5 days, the operating mode established before the repair is reached only after 40–60 days. Thus, there is a shortage of oil associated with the development and withdrawal of wells to a stable operating mode equal to 400–600 tons for one well repair [10–12]. For example, until 1988, in the wells of the *Urengoiskoe* and *Iamburgskoe* oil and gas condensate fields, work on opening, killing and preservation was carried out using clay and aqueous solutions of calcium chloride. After such plugging and further repair of the well, the rate of gas condensate wells in *Valanzhiiskie* strata fell by 60–63 % of the initial (or design), for the *Senomanskie* gas wells, the average drop was 20 %. Since 1988, the use of invert-micellar fluids has begun. As a result, the duration of well development decreased, but with a natural decrease in reservoir pressure in the period 1994–1996, this time reached 40–50 days. And for wells where water-based jamming solutions were used, the development time was 120–130 days. Recently, a decrease in the duration of well development to 12–14 days has been noted, which is due to the use of hydrocarbon-based bullhead well control solutions.

2. Methods and materials

Analytical studies were carried out to clarify factors of deterioration of formation properties around wells and established characteristic reasons for the decrease in reservoir characteristics. Laboratory studies were conducted to develop recommendations for preserving reservoir properties. In the studies, core samples with different permeability and mineral composition were used; compositions containing surfactants, fillers were filtered. To verify the results of laboratory experiments, field trials of developed formulations of process fluids were conducted.

3. Results

The reviewed literature on the mechanisms of mudding of the bottom-hole formation zone showed that most of the damage to the formation in the bottom-hole zone occurs when the formation is opened during drilling and well development, repair work in wells (bullhead well control, flushing, etc.). They significantly reduce filtration-capacitive properties of formation of bottom-hole treatment zone, which do not take into account geological and physical features, as well as the change in thermodynamic conditions in formation.

One of the ways to preserve reservoir properties at BHZ is to modify bullhead well control fluids. As modifying agents can be used, for example, IVV-1, GF-1, Seanec-TU.

Filtration experiments were carried out to pump surfactant compositions through physical models of the *Kalchinskogo*, *Poluniakhskogo*, *Severo-Demianskogo* and *Severo-Kachkarskogo* deposits. The studies show that the coefficient of recovery of permeability of core samples after pumping through bullhead well controls, which contains KCl (or K_2CO_3) and GF-1, reaches 1.02–1.03, i.e. there is an improvement in reservoir properties of the rock, due to a decrease in surface tension at the phase boundary 'fluid of bullhead well control' there is an increase in core permeability.

Alternative formulations for fluids of bullhead well controls and water-based wash solutions are hydrocarbon-based compositions, invert and hydrophobic-emulsion solutions, the external phase of which is represented by a non-polar medium.

A hydrocarbon-based solution was developed containing oil, low molecular weight water-soluble alcohols C_1-C_3, aluminum naphthenate (thickener), barite / bentonite clay (weighting agent). The aluminum salt of naphthenic acids with the general formula $Al(OH)_2(OCOR)$ (here R is naphthyl), dissolving in oil and / or hydrocarbons forms high molecular weight associated complexes that thicken the oil or hydrocarbon solution.

For bullhead well controls of gas and gas condensate wells, hydrocarbon-based invert emulsion formulations have been developed. Such compositions can be used as a fluid of bullhead well controls and / or blocking composition to protect the collector from clogging and damage. An emulsifier (e.g., an emulsifier), a structure-forming agent, a fibrous filler (FF) with diphilic properties and a finely dispersed filler (DF) of a carbonate nature were introduced into the composition of these systems; the dispersed phase is an aqueous solution of calcium or potassium chloride. For filtration experiments, samples of polymictic sandstone with various reservoir properties were used. Physical modeling of the emulsion system (ES) of gas and gas condensate reservoirs was performed.

The analysis of experimental results showed that the permeability of cores is not completely restored when using compositions without fillers. In low-permeability core samples (less than $50 \cdot 10^{-3}$ μm²), permeability is restored more than with average values ((50–100) $\cdot 10^{-3}$ μm²) and at high permeability (1–2 μm²) (table 1). In the experiments, a base emulsion without fillers was used, having the composition: diesel fuel – 40 %, saturated NaCl solution – 60 %, emulsion – 4 %, FFAT-1001 – 1 %.

Table 1. Results of technological liquid studies without application of fillers

Air permeability, 10^{-3} microns²	Recovery factor, K_B
247	0.380
731	0.125
902	0.157
1796	0.205

When using ES with DF in all cases, a crust formed on the end surface of core samples. The thickness and stability of the crust from DF depend on the structural features of the pore space, i.e. from permeability. On samples with high permeability, the formation of a crust is thicker, with greater stability. With a small sample permeability, the DF crust will be thinner and less durable.

When the depression reached 5.0 MPa, the process of "removing" the introduced emulsion system from the pore space easily began, and the filtration began almost immediately when the depression reached 5.0 MPa. This result was obtained in all tests performed. When an inflow is induced in gas wells, the depression on the formation can reach 10–15 MPa, in oil wells, a depression of 5.0 MPa can also be achieved practically.

The results of filtration experiments are shown in table 2.

Table 2. Results of researches of blocking liquid with asbestos and chalk as fillers

№	Type of solution	DF, %	FF, %	Air permeability of the sample, · 10^{-3} μm^2	Permeability recovery coefficient, kV
1	Base emulsion	–	–	–	–
2	№ 1+	1.0	–	194	0.71
3	№ 1+	3.0	–	216	0.66
4	№ 1+	–	1.5	504	0.69
5	№ 1+	–	1.5	278	0.62
6	№ 1+	–	3.0	177	0.90
7	№ 1+	–	3.0	300	0.60
8	№ 1+	–	5.0	243	0.86
9	№ 1+	–	5.0	401	0.87
10	№ 1+	–	5.0	550	0.92
11	№ 1+	–	5.0	998	0.95
12	№ 1+	–	5.0	1233	0.70
13	№ 1+	–	8	1673	0.93
14	№ 1+	–	10	1480	1.0
15	№ 1+	–	10	2195	1.0
16	№ 1+	1.0	3.0	300	0.98
16	№ 1+	5.0	3.0	297	1.0
17	№ 1+	5.0	5.0	401	1.0
18	№ 1+	5.0	6.0	1480	1.0
19	№ 1+	5.0	8.0	1673	1.0
20	№ 1+	5.0	10.0	2195	1.0
21	№ 1+	6.0	11.0	2258	1.0

Laboratory studies have shown the high efficiency of using ES with fillers to preserve reservoir properties in BHZ. Field trials showed an acceleration of the output of wells into operation by 3–5 days in comparison with the start-up of wells after repairs, where other process fluids were used. Therefore, it can be recommended for the use of ES with fillers during repair work in production wells.

The investigated emulsion systems can be used as washing liquid, killing liquid, and some compositions can be used as blocking pastes for installing a "viscous" packer in the interval of the reservoir to protect it from damage.

4. Conclusion

Separate laboratory studies and field trials have shown that technological fluids for well repair can protect the bottom-hole formation zone from damage and ensure permeability recovery after repair. The call of inflow and well output to the operating mode is significantly accelerated.

It is proposed to use hydrocarbon-based process fluids for killing wells and for flushing, as well as to modify these fluids with the addition of water repellents and special fillers, which protect the formation from clogging with filtrates and mechanical impurities and ensure recovery of reservoir permeability after repair work in the well.

References

[1] Mukhametshin V V, Andreev, V E, Dubinsky et al 2016 The Usage of Principles of System Geological-Technological Forecasting in the Justification of the Recovery Methods *SOCAR Proc.* **3** 46–51. DOI: 10.5510/OGP20160300288

[2] Odeh A S, Stephen Melzer L, Kuuskraa V A and Jonathan G K 2006 The Origin and Resource Potential of Residual Oil Zones. SPE-102964-MS *SPE Annual Techn. Conf. and Exhibit.* (San Antonio, Texas, USA 24–7 September)

[3] Oluwaseun Ayodele Taiwo, Abbas Mamudu, Cyrusba Dagogo-Jack, Dala Joshua and Olalekan Olafuyi 2007 "Grain Size Effects on Residual Oil Saturation," SPE-184296-MS, Petroleum Engineering Handbook, vol. V: Reservoir Engineering and Petrophysics *SPE Nigeria Annual Int. Conf. and Exhibit.* (Lagos, Nigeria 2–4 August)

[4] Tarek Ahmad 2019 *Reservoir Engineering Handbook* 5th ed. (Elsevier Inc.) 1524 p

[5] Valeev A S, Kotenev Yu A, Mukhametshin, V Sh and Sultanov Sh Kh 2019 Substantiation of the recovery of residual oil from low-productive and heterogeneous formations in Western Siberia by improving the waterflood system using gas and water-gas impacts *IOP Conf. Ser. Mater. Sci. and Engineer.* **560(1(012204)),** 1–6. DOI: 10.1088/1757-899X/560/1/012204

[6] Baker R O, Yarranton H W and Jensen J L 2015 *Practical Reservoir Engineering and Characterization* (Elsevier Inc.) 534 p

[7] Dubinsky G S, Sultanov V G and Konesev G V 1999 Influence of applied technological fluids on the quality of opening a productive reservoir *Coll. of abstracts of sci. and pract. Conf. reports Probl. of developm. of oil fields in Bashkortostan* (Ufa: Bashnipineft) pp 77–8

[8] Sidorovsky V A and Fain Yu B 1969 Testing methods of impact on the reservoir in Western Siberia *Probl. of oil and gas in Tyumen: sci. J. tech. sat.* 51–6

[9] Kuleshova L S, Kadyrov R R., Mukhametshin V V and Akhmetov R T 2019 Auxiliary equipment for downhole fittings of injection wells and water supply lines used to improve their performance in winter *IOP Conf. Ser. Mater. Sci. and Engineer.* **560(1(012071)),** 1–6. DOI: 10.1088/1757-899X/560/1/012071

[10] Bulatov A I and Savenok O V 2010 *Complete of oil and gas bore holes: theory and practice* (Krasnodar: Publ. Enlightenment-Yug) 539 p

[11] Akhmetov R T and Mukhametshin V V 2018 Range of application of the Brooks-Corey model for approximation of capillary curves in reservoirs of Western Siberia *Advan. in Engineer. Res. (AER)* **157** 5–8. DOI: 10.2991/aime-18.2018.2

[12] Rogachev M K and Mukhametshin V V 2018 Control and regulation of the hydrochloric acid treatment of the bottomhole zone based on field-geological data *J. of Mining Inst.* **231** 275–80. DOI: 10.25515/PMI.2018.3.275

IOP Conf. Series: Materials Science and Engineering **905** (2020) 012080 doi:10.1088/1757-899X/905/1/012080

Application of RAPID technology for well drilling

E L Guseynova

Department of Mechanics and Machine Building Technology, Ufa State Petroleum Technological University, Branch of the University in the City of Oktyabrsky, Oktyabrsky, Republic of Bashkortostan, Russia

E-mail guseinova_elena@hotmail.com

Abstract. The article is devoted to multilateral drilling, it provides a brief history, relevance of application, development prospects, advantages. The history of the emergence and development of the unified international classification of multilateral wells the TAML is also described in detail. The article summarizes the characteristics of the six difficulty levels of the TAML Classification System. The use of advanced RAPID brand joints is described and analyzed in detail. Various designs of joints of the Rapid brand are considered. The authors also propose the construction of a multilateral well using a connector of the Rapid X type.

1. Introduction

Russia's share in world oil production from 2000 to 2015 increased from 8.9 to 12.4 %. Today, Russia, along with Saudi Arabia and the United States, determines the dynamics of prices in the oil market. Russia is the main supplier of petroleum products and oil to the European market, while there is an increase in oil supplies to the countries of the Asia-Pacific region. In 2016, Russia set a new production record, there were 547 million tons of oil and gas condensate produced exceeding the level of production in 2015 by 1.3 %. The highest level of oil production is observed at Rosneft [1, 2].

At the same time, the use of technology for drilling multilateral wells is promising. The technology of drilling multilateral wells is not new, but also it cannot be considered sufficiently mature and developed in the oil industry. Nowadays there are a lot of wells built using this technology in the world. The technology of multi-hole drilling is used almost everywhere, and its further development has many perspectives [3].

Despite the fact that the technology for constructing multilateral wells was first patented in the USA in 1929, this method of drilling wells continues to evolve to date. For the first time, the successful application of multilateral drilling technology was carried out in Bashkiria in 1953.

The advantages of this drilling method are the following: increased coefficient of coverage of the reservoir, reduced overall costs for drilling and completion, increasing productivity, providing a more effective oil flow, ensuring increased oil recovery. The volume of oil production in the world is significantly increasing due to the construction of multilateral wells. Since sometimes there is a lack of technical ability to drill new wells from the surface, or the implementation of a new project requires too high economic costs [4].

A multilateral well drilled in a good location can be much more efficient than several wells constructed using traditional technologies. Since in this case the efficiency increases, construction costs are reduced, and the volume of oil coming from the reservoir increases. Proper application of multilateral drilling technology causes more efficient field development [5].

Drilling multilateral wells can be effective for developing fields with depleted reserves or when operating formations with low reservoir pressure.

2. Methods and materials

The beginning of the development of the terminology and classification of multilateral drilling was laid in 1997. In March 1997, Eric Diggin (Shell spokesman) was invited to convene an international forum called Technology Advancement – Multi-Laterals (TAML) held by a non-profit organization. During the forum experts from leading world oil companies (BP, Statoil, Norsk Hydro, Esso UK, Mobil, Exxon, Phillips, Texaco, Total, Maersk, Chevron, Shell International E&P, Shell Oil and Shell UK Expro) shared their experience in drilling horizontally- branched wells, and also developed a unified approach and determined the further direction of technology development for drilling multilateral wells.

In 1998, the unified international classification of multilateral wells (The TAML Classification System), as part of a Joint Industry Project (JIP), was developed and published in terms of complexity and functionality.

A meeting of TAML members was organized in Calgary in November 2002, the goals of the organization, which was transformed into a non-profit based on membership, were adjusted, and created an opportunity for new members to join. The classification has been changed, taking into account the latest developments in the field of this technology.

The term Multi-Lateral Technology is widely used in English-language publications. It extends to the entire set of applicable technologies for drilling various types of multilateral wells. As for Russian papers, there are different terms used to describe different types of wells and sidetracks.

Nowadays, according to the accepted classification of Technology Advancement - Multi-Laterals (TAML), there are 6 difficulty levels for drilling multilateral wells. Difficulty levels increase in accordance with increasing level number.

The first level of difficulty is characterized by drilling the main trunk with lateral branches without casing fastening. In this case, the properties of the rocks are of great importance, since the joint strength and isolation of the layers depend on them.

At the second level of complexity, the main trunk is cased and cemented, and the lateral trunk is either equipped with a shank or has an open bottom.

The third level of complexity is characterized by a cased and cemented main trunk, while the lateral trunk is cased without cementing; fixing at the branch point is acceptable.

The main and side trunks are cased and cemented, and a shank in the fourth difficulty level is installed in the side trunk.

At the fifth level of complexity, casing and cementing of the main and side shafts are also carried out, the fastening of technological equipment is performed using a packer. The joints of the main trunk with the lateral sections are one of the most important elements for drilling multilateral wells. The fifth level involves a hermetically sealed joint, either cemented or not.

And the most difficult sixth level is distinguished by the presence of a downhole branching on the main trunk, and the equipment is fixed for the possibility of separate production. In this case, the joint is sealed [6].

Currently, a complete TAML classification (TAML Classification System) has been developed, which includes the second part and is called *Functionality classification*. This classification provides: descriptions of wells and intersection points of shafts (joints); Unified coding of multilateral wells.

The technology of drilling multilateral wells has evolved gradually. Wells drilled since 1953 belong to TAML grades 1 and 2. The widespread use of multilateral wells has contributed to the development and complexity of the technology. Shell in 1993 in the Canadian province of Alberta drilled the first well, which corresponded to the third level of complexity. The first multilateral well of the fourth difficulty level was drilled there in 1994. In 1995, BP drilled the first multilateral well of the fifth complexity in the Gulf of Mexico, USA.

There are multilateral wells are being drilled by foreign companies Schlumberger, Weatherford, Halliburton, Baker Hughes in Russia.

3. Results

Great importance is attached to the mechanical and hydraulic tightness of the joints of the main trunk with side bends. Reliability and durability are ensured by the use of advanced RAPID articulations. There are several different varieties of such joints (Figure 1).

Modern technologies allow us to build more complex and productive wells. Using a Rapid joint, we can drill an additional wellbore in the opposite direction from the main one and turn it into another multilateral well, and then casing it. Thus, modern connectors build two wells, having an extensive system of sidetracks, from one well. That allows you to include in the work all the layers and deposits, isolated from each other, both simultaneously and separately. It will also save a huge amount of material resources and reduce the impact on the environment. Various designs of Rapid articulations have been developed and are currently in use.

The Rapid Tie Back brand system used for drilling side branches provides factory-made cutouts in the casing, which are internally closed by a drillable pipe. This technology is characterized by a short period of time required for its installation. The system has up to four lateral shafts located at an angle of ninety degrees to the main trunk. This system can be installed above the reservoir. In this case, the rate of climb of the angle of inclination of the well and the amount of lateral deviation are reduced. Installing the Rapid Tie Back system inside the reservoir allows you to drill inclined wells at large angles using equipment with small radii. The advantages of the technology is that it provides a quick exit from the casing string, with no metal chips, lower risk of subjecting the casing string to destruction. This technology can be used when drilling deposits with heavy oil, as well as in low permeability formations.

The Rapid Access multi-trunk well drilling system provides for cutting oriented side windows in the casing when mounting joints with sidetracks. In this case, there is a nozzle with a special groove used (a casing dividing sleeve), the use of which allows you to drill a well without a preliminary orientation of the cut windows. The use of several bushings allows you to collect several lateral joints, while the reservoir is opened by several wells, which allows you to get a greater effect when developing fields. The separation sleeve is an important element which causes a long-term field development planning and reservoir management. It also allows repeated access through an articulation into a horizontal well by installing an oriented deflecting wedge. This open-hole joint technology can be used for drilling shale and stable compacted rocks.

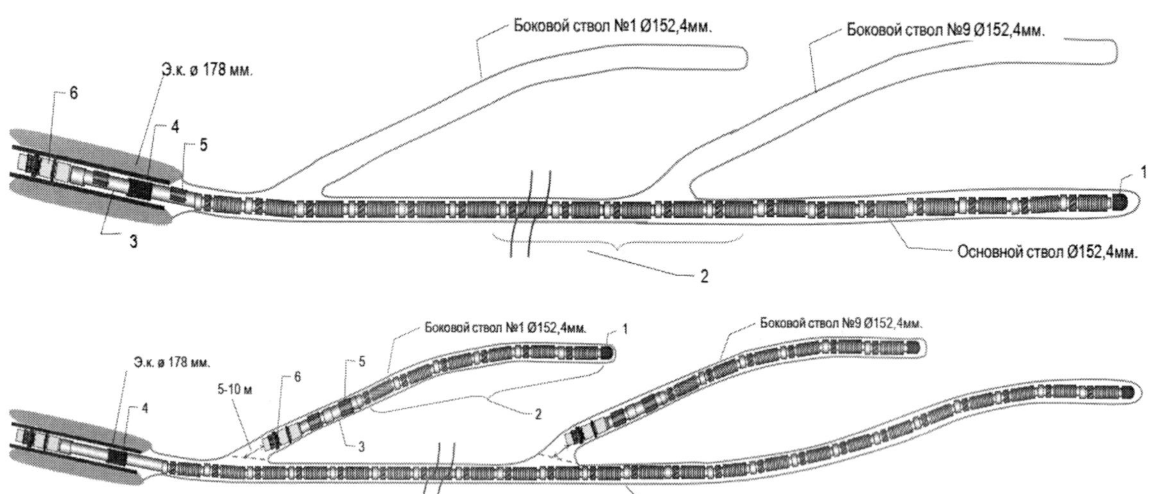

Figure 1. Typical designs of multilateral wells, the sidetracks of which can be cased and not cased: 1 – shoe BK 114.3; 2 – filter FS 114.45×30×6000-У-E-OTTM; 3 – casing OTTM 114.3×8.6"E"; 4 – swelling packer (elastomer); 5 – centralizers TZTZH-114/146-200; 6 – head of the PCH shank (distance from the main trunk 5–10 m)

Rapid Connect and Rapid Exclude articulations developed on the basis of Rapid Access represent a structural connection between the casing and side shanks. It allows selective entry into the main well and branches. Lateral branches are cased, but only the main well is cemented. Joint designs provide mechanical integrity in the conditions of loose, unstable, fragile and weakly cemented rocks.

The multi-well Rapid X connection, corresponding to the fifth level of complexity according to the TAML classification, allows for the modernization of multi-hole horizontal wells. The articulation is distinguished by the presence of a continuous fixing rail, which provides extremely high joint strength. Selective descent is carried out in both trunks on the cable, coiled tubing and drill pipes. It is possible to install several TAML 5 levels in one trunk on top of each other; selective access will still work. Using the joint allows you to maintain tightness with internal and external pressure drops up to 17000 kPa. There is also the possibility of combining the Rapid X joint with inflow control valves, downhole pressure and temperature sensors. The ability to monitor inflow control remains. The joint is fully retrieved. There is no need to orient the casing during the descent. Rapid X can be used in wells that require control of sand removal, water and gas breakthroughs, and there is a need to increase the drainage area of the reservoir. The joint is applicable in compaction wells in developed fields with a limited number of slots both new or old, oil or gas, production or injection. Also, formations with non-uniform pressure and flow regimes can be drilled with similar technology, layered and fractured, divided into blocks. The advantage of the technology is the reduction of risks during drilling, saving time and costs due to the simplicity of operations.

The success of drilling horizontal sections of multilateral wells is influenced by the shape of the well profile, the length of BHA hard part, the intensity of the set of curvature in the interval of cuts.

Figure 2 shows the design of the multilateral well proposed by the authors, the Rapid X type connector is used in the construction of the well. The well has two sidetracks. The side trunks are cased. One to nine additional trunks depart from each side trunk.

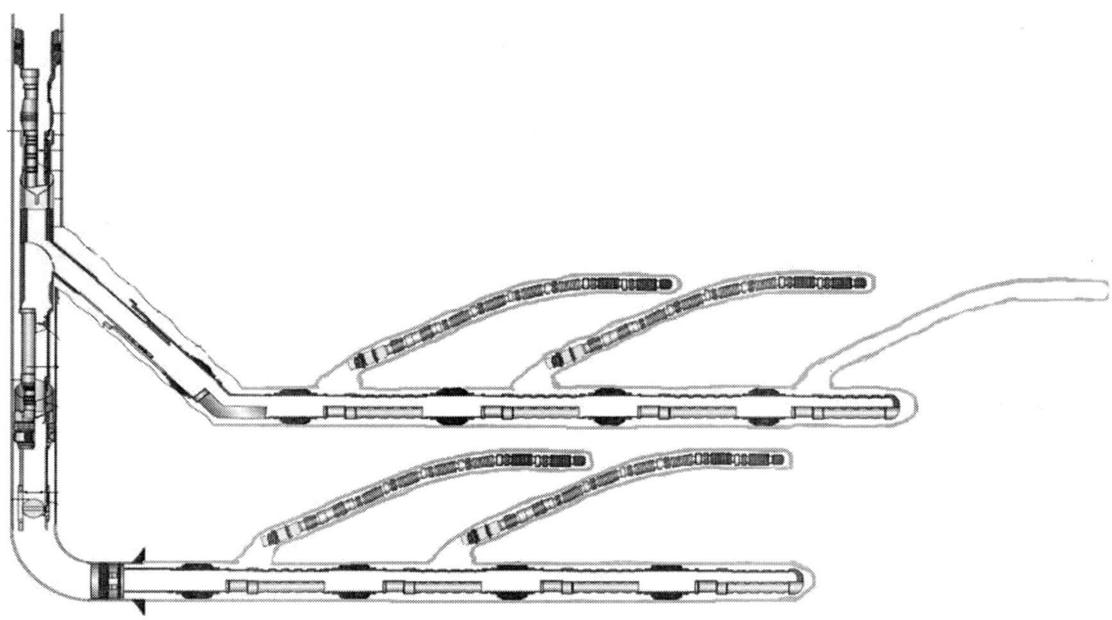

Figure 2. Proposed Well Design

4. Conclusion

Multilateral drilling as an effective technology increases the productivity of long-developed oil fields and develops new oil-fields with maximum efficiency.

The use of technology for multilateral well construction has following advantages: economic costs are reduced, the extracted volume of oil is increased, the use of several available well increases the reservoir development area. Further, there is a more accurate assessment of oil reserves made. Drilling multilateral well develops small-volume and peripheral satellite deposits which can be used in difficult geological and technical conditions. The use of multi-hole drilling technology reduces negative impact on the environment, since there will be less wells and therefore the disposal of drilling mud and cuttings will be carried out in smaller sizes. The use of multi-hole drilling technology is also considered effective when drilling on offshore platforms with a limited number of wells [7–10].

References

[1] Zaripova L M and Gabdrakhimov M S 2019 Layout for drilling seismic wells by vibration method *IOP Conf. Ser. Earth and Environ. Sci.* **350(1)**. DOI: 10.1088/1755-1315/350/1/012048

[2] Gilmanova A M 2019 Metrological support of geophysical equipment for acoustic logging *IOP Conf. Ser. Earth and Environ. Sci.* **378(1)**. DOI: 10.1088/1755-1315/378/1/012080

[3] Davydova O V 2019 Energy saving when measuring the flow rate of Russian oil wells *IOP Conf. Ser. Earth and Environ. Sci.* **378(1)**

[4] Shangareeev R R 2015 Relevance of assessment science research concerning occupational hazards of drilling operations *Ecol., Environ. and Conservat. Paper* **21** 175–80

[5] Bykov L I, Shammazov A M, Gumerov A G, Tokarev A P and Safiullina A R 2019 Reduction of vibrations of mechanical and power plant pipelines using vibration insulators with design power characteristics based on guides of a special shape *IOP Conf. Ser. Earth and Environ. Sci.* **378(1)**. DOI: 10.1088/1755-1315/378/1/012037

[6] Yangirov F N, Yakhin A R, Mustafin T S and Dikhtyar T D 2018 Substantiation of lubricant selection for drilling technology, Bulletin of the Tomsk Polytechnic University *Geo Assets Engineer.* **329(1)** 51–8

[7] Fraiya Kh., Omer E., Pulik T et al 2003 New Approaches to the Construction of Multiple Horizontal Wells *Oil and Gas Rev.* 44–67

[8] Akhmetov R T, Mukhametshin V V and Kuleshova L S 2019 Grouping of objects using a limited number of parameters characterizing geological and physical properties of layers *Atlantis Highlights in Material Sciences and Technology (Int. Symp., dedicated to the 85th anniversary of H.I. Ibragimov, Engineering and Earth Sciences: Applied and Fundamental Research (ISEES 2019))* vol 1 pp 9–13. DOI: 10.2991/isees-19.2019.3

[9] Ugbogu I N A and Nagib M B 2013 A new well control model to optimize drilling performance *Kuwait Oil and Gas Show and Conf. (KOGS 2013)* (Mishref, Kuwait: Society of Petroleum Engineers, 7 October 2013) pp 206–19

[10] Elabsy I, AlTurkey S, Alhazzaa S, Dutta A, Belhouchet M and Abba A 2018 Integration of Penta-combo LWD measurements for petrophysical and geomechanical feasibility Abu Dhabi *Int. Petroleum Exhibition and Conf. (ADIPEC 2018)* (Abu Dhabi, United Arab Emirates: Society of Petroleum Engineers, 12–15 November 2018)

Prospects of applying formation water and heavy brines derived therefrom in oil production and national economy

R R Kadyrov[1], V V Mukhametshin[2], I F Galiullina[1], L S Kuleshova[1] and A R Safiullina[1]

[1] Department of Oil and Gas Field Exploration and Development, Ufa State Petroleum Technological University, Branch of the University in the City of Oktyabrsky, Oktyabrsky, Republic of Bashkortostan, Russia
[2] Department of Oil and Gas &Oil Field Development and Operation, Ufa State Petroleum Technological University, Ufa, Republic of Bashkortostan, Russia

E-mail: ramzis.k@mail.ru, vv@of.ugntu.ru, irida2017@mail.ru, markl212@mail.ru, safiullina.a.r@yandex.ru

Abstract. In recent years, the amount of minerals recovered from the earth interior exceeds their production over the entire past human history. The reserves of some rare elements have been significantly reduced and the demand for them has increased dramatically through the use of new techniques and technologies. The task of developing new types of mineral raw materials from formation waters has become ever more relevant. In the process of formation water preparation during the production of table salt the fluids and heavy brines are obtained, which are suitable for their further use in oil production. The fluid obtained through deferrization of formation water has reduced content of magnesium ions. The use of this fluid makes it possible to increase the mechanical strength of the cement stone, the tightness of the "column-cement stone" contact zone, and to reduce magnesia corrosion.

1. Introduction

The strength enhancement of cement stone is achieved by combined use of calcium and sodium chlorides of formation water of an oil field in tempering fluid with simultaneous reduction of magnesium ion concentration by 10 and more times (up to 0.15 %) [1]. The magnesium chloride content in the formation water reaches 1.76 %. The most typical values of basic salts in formation waters of the Devonian horizon are given in Table 1.

Table 1. Typical values of basic salts in formation waters of the Devonian horizon

n/n	Salt content, %			
	$CaCI_2$	$MgCI_2$	NaCI	Total
1	5.48	1.57	16.42	23.70
2	5.29	1.76	16.41	23.69
3	5.40	1.45	16.80	23.88
4	5.32	1.57	16.75	23.87

IOP Conf. Series: Materials Science and Engineering **905** (2020) 012081 doi:10.1088/1757-899X/905/1/012081

2. Methods and materials

Magnesium ions are substituted with calcium ions in crystal lattice nodes of the cement stone. Magnesium ions have smaller radius ($0.78 \cdot 10^{-10}$ m) compared to calcium ions ($1.06 \cdot 10^{-10}$ m), and the substitution of calcium ions with magnesium ions weakens the crystal lattice. Besides, if non-magnesium-purified water is used as the tempering fluid, magnesium chloride reacts with calcium hydroxide thus forming slightly soluble magnesium hydroxide. The solubility of $Mg(OH)_2$ is almost one hundred times less than that of $Ca(OH)_2$: 0.018 g/L and 1.6 g/L, respectively, so the reaction proceeds towards the formation of magnesium hydroxide. The binding of hydroxyl groups into magnesium hydroxide is accompanied by pH reduction of the pore fluid (up to pH=10), which creates favorable conditions for gradual dissolution and hydrolysis of hydrate neoplasms in cement stone, which in turn leads to the development of magnesia corrosion [2–4].

3. Method description and its evaluation

The density of the cement stone contact to rock on one side, and on the other side – to a casing string determines the tightness of the annular space, reliability of separation of gas, oil and water-containing formations. In this respect, the main criteria of suitability of consolidated cement slurry is impermeability, good adhesion of stone to pipes and rocks [1]. The effect of sodium chloride additives on the adhesion strength of cement stone with metal and clay rock is shown in Table 2.

Table 2. Effect of sodium chloride additives on adhesion strength of cement stone with metal and clay rock

n/n	NaCI concentration in tempering fluid, %	Adhesion strength, MPa	
		Separation from metal	Separation from rock
1	0	0.3	0.05
2	6	1.0	0.10
3	10	1.0	0.20

Osmotic phenomena that cause fluids to flow through the cement stone cause cracks in the stone and thus contacting rocks. As a result of osmosis the formation fluid water gets into the well. Reverse flows directed towards the formation contaminate productive horizons and can cause reduction of oil recovery. Significant reduction of cement stone impermeability is achieved by the addition of 5–10 % sodium chloride to the fluid [5].

Thus, the increase in the strength of the cement stone is achieved by the combined use of a predetermined ratio of calcium and sodium salts of the formation water in the tempering fluid while reducing the magnesium ion concentration by 10 times (from 1.7 to 0.15 %). Besides, the use of the specified amounts of formation water salts of the oil field in the tempering fluid provides an additional positive result: increased impermeability, adhesion to casing and well wall, reduced magnesia corrosion of cement stone [5].

Formation water of the Devonian horizon was used to prepare the tempering fluid with the specified ratio of sodium, calcium and magnesium salts. The construction lime (GOST 9179-77) was used to remove $MgCl_2$ from the formation water of the oil field in the form of $Mg(OH)_2$. Since the solubility of $Mg(OH)_2$ (0.018 g/L) is significantly lower than $Ca(OH)_2$ (1.6 g/L), the reaction proceeds towards the formation of $Mg(OH)_2$:

$$MgCl_2 + Ca(OH)_2 = CaCl_2 + Mg(OH)_2.$$

When calculated 14 g/l of lime is added to the formation water, the amount of magnesium chloride is reduced to only 0.78 %, which does not provide an acceptable reduction of magnesia corrosion. As indicated in [5], all calcium hydroxide will be consumed at 0.259 % magnesium sulfate concentration. Therefore, lime is introduced in excess for more complete settlement of magnesium chloride. The introduction of 20 g/l of construction lime into formation water reduced magnesium chloride up to 0.30 %. By further diluting the formation water to prepare the tempering fluid 2 times or more, the magnesium chloride content is reduced to 0.15 % and less. This concentration is taken as the upper limit of the magnesium chloride content of the tempering fluid.

In order to determine the effect of salts ($CaCl_2$, $MgCl_2$ and NaCI in different combinations) on the main characteristics of the cement slurry and cement stone for the preparation of cement slurry with constant water-cement ratio of 0.5, the initial formation water with the density of 1.189g/cm^3, fresh tap water and calcium chloride solutions were used as the tempering fluid. The density of purified and diluted formation water prior to calcium chloride content of 5, 4, 3 and 2 % was 1.147, 1.102, 1.072 and 1.051g/cm^3, respectively.

The results of the laboratory tests are shown in Table 3. The table shows that the concentration of low salt solution does not significantly increase the mechanical strength of the cement stone. The use of formation water, diluted to 2 % sodium chloride and 1 % calcium chloride as tempering fluid if compared to fresh water results in only a slight increase in strength properties after 2 days of solidification. The bending strength was 26 MPa for fresh water and 30 MPa for salted water, and the compression strength was 6.1 and 8.2 MPa, respectively. The strength properties of cement stone increase considerably with the content of 3 % calcium chloride and 6 % sodium chloride in fluid. Bending and compression strengths were 4.8 and 11.8 MPa, but the flow of the cement slurry was reduced. When 5 % calcium chloride and 10 % sodium chloride was contained in the tempering fluid, the flow rate made 14 cm. The cement slurry with such flow rate retains mobility only for 30 minutes after tempering, then it thickens. Using standard tempering techniques, it is almost impossible to pump and pour cement slurry into the well during this time. Thus, the use of refined Devonian formation water will allow increasing the strength properties of cement stone, accelerating the hardening of cement slurry and eliminating the purchase of imported calcium chloride by using local raw materials [6].

During repair and insulation works at oil wells the well killing operation is a standard technological procedure, which is necessary to prevent the discharge of process liquids. This process includes the balance of the formation pressure with the pressure of the hydrostatic fluid column at a given density. Such solutions are prepared on the basis of imported calcium chloride, zinc bromide, etc. [7–9].

The volumes of calcium chloride-based killing fluid with a density of 1300 kg/m^3 used in various oil companies are presented below in tons (data derived from requests from company process departments):

- PJSC Tatneft – 10000;
- OAO Samaraneft – 15000;
- OAO Udmurtneft – 1000;
- OAO Bashneft – 2500;
- OAO Orenburgneft – 3500.

In order to prepare such solutions, it is necessary to purchase technical calcium chloride dissolved in fresh water. This, firstly, is too expensive, secondly, the iron impurities contained in the technical calcium chloride lead to blockage of productive formations and reduction of oil recovery. Therefore, we propose to use associated heavy brines obtained during formation water evaporation as the killing fluid [10–12].

In order to evaluate the possibility of using heavy brine as killing fluid, the model tests were carried out on tubular formation models. The purpose of the studies was to identify the effect of heavy brines on the reservoir properties. For this purpose, the formation model of one-meter-long and 27 cm diameter was packed with quartz sand, after which the initial formation water was passed through it. Heavy brine was then pushed through the model and left alone for 24 hours, after which the initial formation water was passed through. At the same time for models with permeability of 0.2; 0.7; 1.0; 1.5; 2.0 μm^2 the filtration attenuation was not observed at pressure gradients corresponding to formation pressure. Similar studies were carried out with oil-saturated models, which also did not show a decrease in filtration.

Thus, the model tests show that there is no deterioration of reservoir properties of productive formations when they are in contact with heavy brine and the possibility of using such brine as killing fluid.

Table 3. Laboratory test results

Tempering fluid		Salt content, %			Cement slurry characteristics			Setting time		Bending/compression strength, kg/cm^2 Sample storage period in formation water				
n/n	Type of tempering fluid	$CaCl_2$	$MgCl_2$	NaCl	Spreadability, ...	Density, g/cm^3	Water return, cm^3/30 min	beginning	end	2 days	7 days	28 days	3 months	6 months
1	Fresh water	–	–	–	21.5	1.81	189.6	8-00	9-40	26/61.3	28/64	39/68.4	70/75.4	87/95
2	Initial formation water	2	0.5	5.5	21	1.85	139.0	6-30	7-40	36/110	38/115	44/120	72/124	90/129
		4	1.0	11.2	16	1.89	120.1	4-15	4-50	39/116	44/116	45/119	60/120	80/121
3	Magnesium chloride-free formation water	1	0.04	2	21	1.83	142.4	7-30	8-10	30/82	33/96	49/100	84/116	109/120
		2	0.08	4	20.5	1.85	149.2	6-00	6-40	38/116	58/126	70/123	92/129	114/134
		3	0.12	6	19	1.87	140.0	5-45	6-05	48/118	64/120	80/125	103/138	119/146
		4	0.15	8	18	1.89	137.6	5-30	6-00	48/120	69/123	89/126	108/130	117/138
		5	0.20	10	14.5	1.90	130.0	2-30	3-10	47/119	63/121	84/120	100/129	115/131
4	Calcium chloride brine	2	0	0	18	1.86	158.0	6-03	7-35	37/112	43.5/115	58.4/119	66/119.5	97/123
		4	0	0	18	1.86	145.7	4-55	5-30	40/116	47/118	60/121	65/123	75/124

The works of Zaripova L.P., Shakirov Sh.K. show the huge role of food preservatives in feeding farm animals. They lay the foundations for the use of formation waters and their processing products for the preparation of food preservatives. The scope of food preservatives includes the improvement of silo quality, which now, for example in Tatarstan, makes about 4 million tons per year and the preservation of fodder grain – up to 300000 tons from 1000000 tons of fodder grain required in the republic, which are used as additives in animal diet to fill the deficiency of table salt, as well as salts of calcium, magnesium, iodine, iron, selenium.

Major problems in the use of preservatives:

- relative cost of traditionally industrialized green feed preservatives;
- need to bring them from other countries and regions of the Russian Federation;
- narrow spectrum of action compared to available and widely used preservatives, limited to inhibition of fermentation and not providing for compensation of deficient nutritional additives in the form of macro- and microelements.

The expediency of forage grain preservation is based on the following:

- possibility of preserving grain with high humidity, which is particularly important in high yields and unfavorable weather conditions;
- possibility of storing preserved fodder grain in concreted trenches and burrows;
- lower degree of exposure of preserved fodder grains to rodent eating.

The need for feed supplements is dictated by the following reasons:

- additives of table salt, iodine salts and other macro- and microelements should be added to the nutrition of all animals and birds;
- there is a deficiency of calcium salts and microelements in the diet of pigs and birds.

In order to solve these problems in the Republic of Tatarstan, it is proposed to produce and introduce into feed production a new cheap food preservatives based on highly concentrated multicomponent water salt brine formed through processing of formation waters of the Devonian horizons.

The analysis of the composition of formation waters of the Devonian horizon produced along with oil in Leninogorskneft oil-and-gas production department shows that there are large amounts of salts extremely important for livestock and poultry production – sodium chloride and calcium chloride. In particular, the formation water of wells 9187 and 23556 of D_2 horizon has the following composition (wt %): calcium chloride – 5.48, sodium chloride – 16.42, magnesium chloride – 1.60, potassium chloride – 76.27.

Thus, 92.29 % of salts are of great value and are actually imported into Tatarstan.

This formation water can be used directly as extracted from the subsoil. However, to achieve high efficiency while compensating for salt deficiency and to avoid overdose, it is necessary to separate salts for each diet of each species of animals and birds.

The TatNIPIneft Institute of PJSC Tatneft developed a design plant for complex processing of formation waters, which will be located in the village of Karabash of the Bugulminsky district of the Republic of Tatarstan (Leninogorskneft oil-and-gas production department). The project implies the processing of these waters in order to obtain food table salt (30,000 tons per year, which fully meets the needs of Tatarstan for table salt) and concentrated salt solution (concentration of salts – 43 %, of which more than two thirds of calcium chloride), which can be used in the oil production industry as heavy killing fluid with a density of 1300 kg/m^3. The planned volume of production of the killing fluid is 40,000 tons per year.

The brine composition after sodium chloride isolation and dilution with fresh water to a density of 1300 kg/m^3 has the following composition:

- calcium chloride – 25 %;
- sodium chloride – 2.5 %;
- magnesium chloride – 7.2 %;
- potassium chloride – 1.0 %;
- water – 64.3 %.

The analysis of properties and composition of the resulting heavy killing fluid (acid properties pH = 2.4 and high concentration of calcium salts, presence of sodium chloride, iron salts – 797 mg/l, iodine – 23.5 mg/l and other biogenic elements) shows that it can be used as an effective food preservative. Some adjustment of the composition of this solution will make it possible to use it through preserved silo mass, preserved forage or by direct addition to food to fully compensate for the deficiency of a number of vital elements in animal diets. The planned volume of production of this saline solution can fully cover the need of the Republic of Tatarstan for silage preservatives.

The need for livestock and poultry farming of Tatarstan in feed salt is about 40,000 tons per year, the amount of food preservatives with a composition close to the above is also about 40,000 tons per year. Thus, if the first salt plant covers the needs of Tatarstan in food table salt and the needs of oilmen in the killing fluid, for the needs of agriculture it will be necessary to build a second plant with the productivity of not less than the first.

The proposed food preservatives replace the known preservatives with very high cost and do not have the ability to enrich feed with elements, the acute deficiency of which is relevant for the animals in the territory of the republic. On the basis of concentrated salt solutions, by adding some deficient components, more efficient compositions of food preservatives can be prepared to compensate for the deficiency of nitrogen, phosphorus and other elements not contained in the brine under consideration.

Due to the shortage of feed salt at the first stage, the use of both formation water and heavy fluid presents the greatest practical interest. Since in the first case the ratio of calcium chloride to sodium chloride is 1:3 and in the second – 10:1, the different quantitative ratio can satisfy any requirements of a wide variety of animal and poultry diets.

4. Conclusion

The use of the proposed cheap food preservatives not so much saves by not importing expensive preservatives, but rather, in the face of a shortage of working capital for agricultural enterprises, creates a prerequisite for the widespread use of food preservatives. This will fundamentally improve the quality of feed, its balance in biogenic elements and, ultimately, the productivity of animals. On the basis of the above, in order to produce food preservatives in solid form it is recommended to convert the excess heavy fluid with a density of 1400 kg/m^3 obtained by additional evaporation into a fluid with a density of 1500–1600 kg/m^3, which after cooling crystallizes into a solid. Thus, using another evaporation step, it is possible to obtain food preservatives in granular solid form. This will make it possible to eliminate additional transportation costs when carrying unnecessary water in food preservatives to agricultural objects. The studies show that the formation water and its processing products are promising reagents that will be used in oil production and agriculture.

References

[1] Zhou Z, Cai X, Du X, Ma D and Zang H 2019 Strength and filtration stability of cement grouts in porous media *Tunnell. and Undergr. Space Technol.* **89** 1–9

[2] Teptereva G A, Konesev G V, Ismakov R A, Kantor E A and Dikhtyar T D 2017 Obtaining drill reagents by modification of neutral-sulphite alkali phosphonic compounds *Bull. of the Tomsk Polytech. Univer., Geo Assets Engineer.* **328(9)** 94–101

[3] Kuleshova L S, Kadyrov R R, Mukhametshin V V and Safiullina A R 2019 Design changes of injection and supply wellhead fittings operating in winter conditions *IOP Conf. Ser. Mater. Sci. and Engineer.* **560(1)** 1–5. DOI: 10.1088/1757-899X/560/1/012072.

[4] Andreev V E, Chizhov A P, Kotenev Yu A et al 2019 Prediction of Gas Exposure in the Conditions of the Oil Fields of the Volga-Urals *Atlantis Highlights in Material Sciences and Technology (Int. Symp., dedicated to the 85th anniversary of H.I. Ibragimov, Engineering and Earth Sciences: Applied and Fundamental Research (ISEES 2019))* vol 1 pp 666–9. DOI: 10.2991/isees-19.2019.132

[5] Agzamov F A, Makhmutov A N and Tokunova E F 2019 Study of corrosion stability of a cement stone in magnesia aggressive environment *Georesursy* **21(3)** 73–8

[6] Mukhametshin V Sh, Zeigman Yu V and Andreev A V 2017 Rapid assessment of deposit production capacity for determination of nanotechnologies application efficiency and necessity to stimulate their development *Nanotechnol. in Construct.* **9(3)** 20–34. DOI: 10.15828/2075-8545-2017-9-3-20-34

[7] Galiullina I F and Kadyrov R R 2018 Technical and economic background for siting production of well-killing liquid at oil fields *IOP Conf. Ser. Earth and Environ. Sci.* **194(8)** 1–6

[8] Umar A A, Saaid I B M., A A Sulaimon and Pilus R B M A review of petroleum emulsions and recent progress on water-in-crude oil emulsions stabilized by natural surfactants and solids (Malaysia: Univer. Teknologi PETRONAS)

[9] Puls R and Sanders L D Water-Energy Nexus and Environmental Aspects of Oil and Gas Production In *Competition for Water Resources: Experiences and Management Approaches in the US and Europe* pp 160–77

[10] Abou-Sayed A S, Zaki K S, Wang G, Sarfare M D and Harris M 2007 Formation water management strategy and water injection best practices: Design, performance, and monitoring *SPE Product. and Operat.* **22(1)** 59–68

[11] Titley M 2014 Wheatstone development – Challenges and solutions in formation water treatment and disposal *Proc. of the Annual Offshore Technol. Conf. vol. 2* pp 1013–16

[12] Abou-Sayed A S, Zaki K S, Wang G, Sarfare M D and Harris M H 2005 Formation water management strategy water-injection best practices-design, performance, monitoring *Int. Petrol. Technol. Conf. Proc.* pp 869–78

Cyclic impact on productive formations of high-viscosity oil deposits

Yu A Kotenev[1], V Sh Mukhametshin[2], Sh Kh Sultanov[1] and A Yu Kotenev[2]

[1] Department of Geology and Exploration of Oil and Gas Field, Ufa State Petroleum Technological University, SASI "Institute of strategic research of the Republic of Bashkortostan", Ufa, Republic of Bashkortostan, Russia
[2] Department of Oil and Gas Field Exploration and Development, Ufa State Petroleum Technological University, Branch of the University in the City of Oktyabrsky, Oktyabrsky, Republic of Bashkortostan, Russia

E-mail: geokot@inbox.ru, vsh@of.ugntu.ru, ssultanov@mail.ru, kotenyov@mail.ru

Abstract. It is shown that the priority condition for reliable forecasting of the cyclic impact on productive formations of high-viscosity oil is the joint research and study of rheological and filtration properties of high-viscosity oils, including thixotropic properties, as well as mathematical calculations to choose the right mode of the working fluid injection. For the Stepnoozerskoe field conditions, the optimal injection period is 10 days.

1. Introduction

The main task of the oil industry of the country is to create a scientific and methodological basis regarding geological and technological justification to increase the development efficiency of hard-to-recover deposits [1–11].

The peculiarity of the present and forthcoming development stage of oil and gas industry is the development of high-viscosity, heavy oil, super-heavy oil deposits in order to increase the resources of hydrocarbon raw materials and create technologies for their production. The prospects for the development of high-viscosity oil deposits can be determined by a methodological approach, including the detailed study of geological structure; the study of the experience of similar fields; numerical, laboratory and experimental adaptation of working fluids to specific geological-technological conditions of the deposit.

Domestic reserves of high-viscosity and heavy oil account for ~ 13 % of the total oil resources in Russia. According to the Institute of Oil Chemistry of the Siberian branch of the Russian Academy of Sciences, the total reserves of the Volga-Ural and West Siberian oil and gas deposits account for more than 71 % of the all-Russian reserves of heavy oil. Volga-Ural types of high-viscosity oil, compared to West Siberian, contain more sulphur, paraffin, resin, vanadium, but less dissolved gas. More than 450 deposits and natural bitumen and high-viscosity deposits are found on the territory of the Republic of Tatarstan. They are mainly located on the west and south-east slopes of the South Tatar Vault. Relative to the balance of initial reserves in Tatarstan, the share of high-viscosity oil reserves (more than 30 mPa·s) accounts for about 43 % [1]. The main share of high-viscosity oil reserves is found in the deposits of lower and middle Carboniferous period. It was found that as the depth productive formations decreases, the weighting of degassed oil, increase of sulphur content in them, reduction of gas factor and content of light hydrocarbons are observed.

2. Methods and materials

The development of high-viscosity oil fields has its peculiarities, and in conditions of complex geological structure the choice of oil production technology requires the detailed study. The Stepnoozerskoe oil field located in the Republic of Tatarstan is characterized by the specific structure of lower Carboniferous deposits, extensive development of erosion cuts of both area and channel types. Two important points shall therefore be borne in mind. First, the development of many deposits of the western slope of the South Tatar Vault revealed hydrodynamic interaction of oil deposits of the Tournaisian and the Visean stages. Since the erosion cuts filled with loose sand or slightly cemented sandstone are nothing more than natural oil storage caverns having a considerable size and therefore draining a large volume of carbonate reservoirs, it is obvious that the flow of oil from carbonates should be the more significant the better the reservoir properties of the terrigenous formation and the larger the area of the cavern. Second, there is a considerable multiple formation effect in the terrigenous thickness of the Stepnoozerskoe field. At the same time, natural modes of the Bobrikovian formations development are different: if in erosion cuts the formations can be operated quite effectively in a rigid natural mode, the lying mantle-like formation $Bв_1$ is not developed effectively without the hydrodynamic modes. It shall also be noted that high viscosity of oil may reach up to 620 mPa·s. Any solution in the development of such deposits needs detailed preliminary study and justification. The problems related to the definition of technological effects from the application of enhanced oil recovery methods and experimental investigations of rheological and filtration characteristics of oils are particularly highlighted. At present, a feasible method of field development is the effective maintenance of formation pressure. Systematic research and development of scientific bases of non-stationary impact technology were laid down in the works of I.N. Sharbatova, M.L. Surguchev. The most common technology implies that the water injection is periodically stopped at the site while maintaining the average injection rate for a long period, i.e. during the first half-period the injection wells are shut down, and then during the second half-period the water is injected at increased flowrates. Together with K.M. Fedorov (TumSU), the cyclic impact was justified, calculated and modeled.

3. Results

Effective application of cyclic action is possible in case of heterogeneity of formations along the vertical section and the presence of sufficient vertical relationship in the formation. In order to analyze the efficiency of cyclic impact, it is necessary to clarify the parameters of formations in section for each analyzed horizon. Geological-statistical sections of horizons are given in Table 1.

The preliminary analysis of this data shows that the development object of the lower Carboniferous of the Stepnoozerskoe deposit meets the conditions of successful cyclic flooding.

The parameters of cyclic action were calculated using two-dimensional numerical model of two-phase filtration in laminar- heterogeneous formation. It was assumed that the formation is developed by a row of wells, with the distance between injection and production rows equal to 300 m. The task was solved in a two-dimensional version along the vertical section without taking into account capillary and gravitational forces. The vertical permeability of the formation, which determines the hydrodynamic bond of interlayers in the oil-saturated horizon, was 0.01 times of the horizontal component. The cyclic impact period consisted of equal half-periods corresponding to the shutdown mode of only injection wells. During the half-cycle of water injection, the injection pressure was increased by 30 %, which ensures the execution of the injection plan. The initial saturation of the formation with water and oil was determined from the problem of flooding until the moment of reaching the current water cut of the product within the studied formation (70 %). The calculations were made for several dozen exposure cycles until 97 % water cut was achieved.

Figure 1 shows the example of oil production rate calculations during normal flooding and cyclic impact.

Table 1. Geological-statistical sections of oil-saturated horizons

Formation	Weighted average net oil thickness, m	Porosity, unit fraction	Hydrocarbon saturation, unit fraction	Permeability, μm^2
Kashirskian-3	1.57	0.18	0.71	0.151
Kashirskian-1	1.9	0.22	0.60	0.151
Vereiskian-5	0.91	0.131	0.73	0.089
Vereiskian-4	1.13	0.145	0.754	0.157
Vereiskian-3	2.10	0.151	0.761	0.351
Vereiskian-2	1.47	0.147	0.732	0.274
Vereiskian-1	0.80	0.143	0.745	0.253
Baskirian-2	2.31	0.14	0,75	0.235
Baskirian-1	5.18	0.14	0.75	0.345
Bobrian-10	1.04	0.206	0.866	0.320
Bobrian-01	4.71	0.239	0.938	1.185
Bobrian-02	3.0	0.242	0.95	1.878
Bobrian-03+04	4.7	0,271	0.948	5.144
Tournaisian	3.73	0.154	0.702	0.053

The accumulated oil production during flooding and cyclic impact is equal to the area under the corresponding curves (Figure 1). The difference of accumulated volumes of oil at cyclic impact and flooding constitutes additional oil production from impact. This difference for the entire development period is shown for different impact periods in Figure 2 (area of the design section –1000 m^2). As the period increases, the additional production increases to a maximum of 10 days and then drops to negative values (Figure 2). Thus, the optimal cycle period and additional oil production for the analyzed polygon object are determined. The maximum value of additional oil production (10000 t/Ha) corresponds to 3.5 % of the oil recovery factor increase of the impact area.

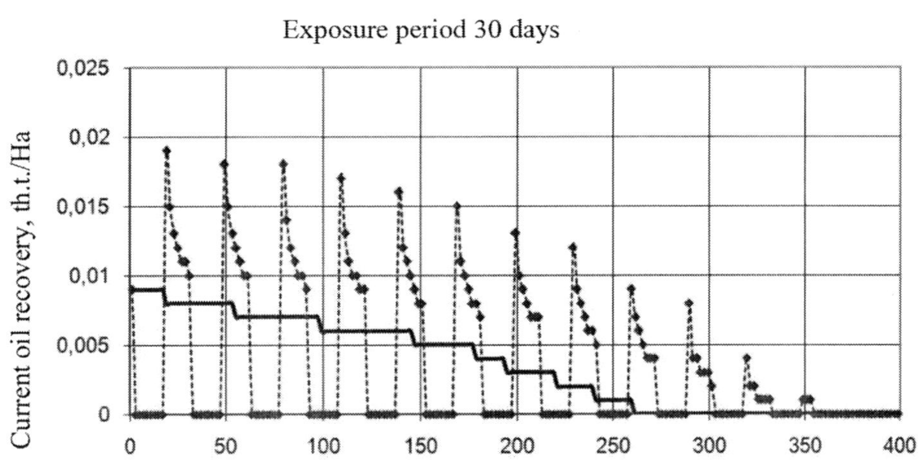

Figure 1. Example of oil production rate calculation during flooding (solid curve) and cyclic impact (dashed curve) on the second object

The performed numerical calculations require additional studies of rheological and filtration properties of the formation oil. In order to ensure the cyclic effect, it is important to study the thixotropic properties of oil. Together with M.K. Rogachev (Mining University), the experimental studies covered rheological and filtration properties of oil samples taken from 14 wells, which opened productive horizons of lower and middle Carboniferous, including 10 wells of the Stepnoozerskoe field. The research was limited to obtaining experimental relations between bulk oil flows through a capillary or rock sample and pressure drops at their ends. The experiments were conducted under "fixed volume

IOP Conf. Series: Materials Science and Engineering **905** (2020) 012082 doi:10.1088/1757-899X/905/1/012082

flow rates – changing pressure drops". The pressure drop was measured under steady-state oil flow conditions.

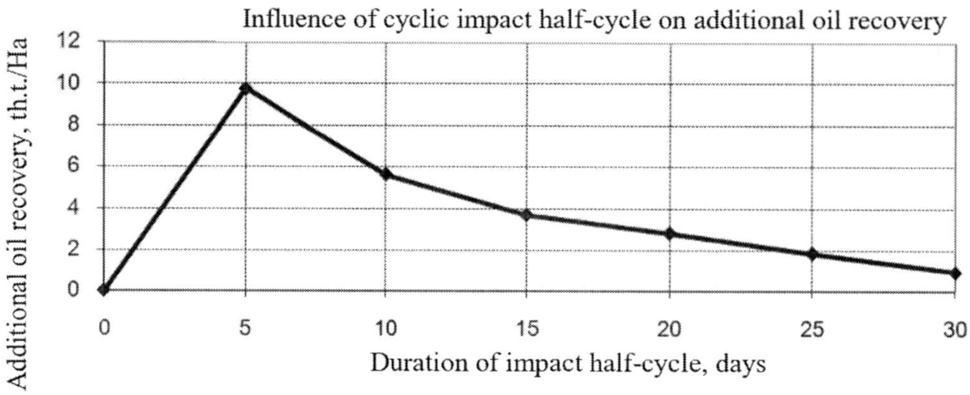

Figure 2. Additional oil production at cyclic impact limited to a full period of exposure.

With the help of rheological lines and filtration curves, the main rheological and filtration parameters of degassed and formation oil were determined: yield point (YP), structure destruction yield point (SDYP), dynamic pressure shear gradient (DPSG), structure destruction yield point gradient (SDYPG), oil viscosity anomaly and mobility anomaly indices (VAI and MAI).

The results of these studies for one of the wells are presented in Tables 2 and 3, as well as in Figure 3.

Table 2. Results of experiments on rheological properties of formation oil of the Stepnoozerskoe field

Shear rate $\gamma \cdot 10^2$, s^{-1}	Shear stress τ, Pa		Effective viscosity μ, mPa·s	
	Forward stroke	Return stroke	Forward stroke	Return stroke
1.12	0.018	0.010	1610	893
1.46	0.021	0.012	1440	822
2.94	0.028	0.017	952	578
5.11	0.031	0.020	607	391
11.2	0.042	0.042	375	375
14.6	0.054	0.054	370	370

Table 3. Results of experiments on filtration properties of formation oil of the Stepnoozerskoe field (in rock sample c k = 0.003 μm^2)

Filtration velocity $v \cdot 10^{10}$, m/s	Pressure gradient grad p, MPa/m		Effective motion $(\kappa/\mu) \cdot 10^5$, μm^2/mPa·s	
	Forward stroke	Return stroke	Forward stroke	Return stroke
1.47	0.062	–	0.24	–
1.92	0.066	–	0.29	–
3.91	0.122	–	0.32	–
6.78	0.212	–	0.32	–
14.7	0.459	–	0.32	–

With the help of obtained results for each well the rheological and filtration parameters of studied reservoir oils are determined as shown in Table 4.

Experimental studies of rheological and filtration parameters of the Stepnoozerskoe oilfield showed the following distinctive features:

- high density and viscosity of degassed oils, which is explained by increased content of asphalt-resin substances in their composition;
- high viscosity of formation oils at low gas factor;
- high nitrogen content in associated gas at low formation oil gas factor.

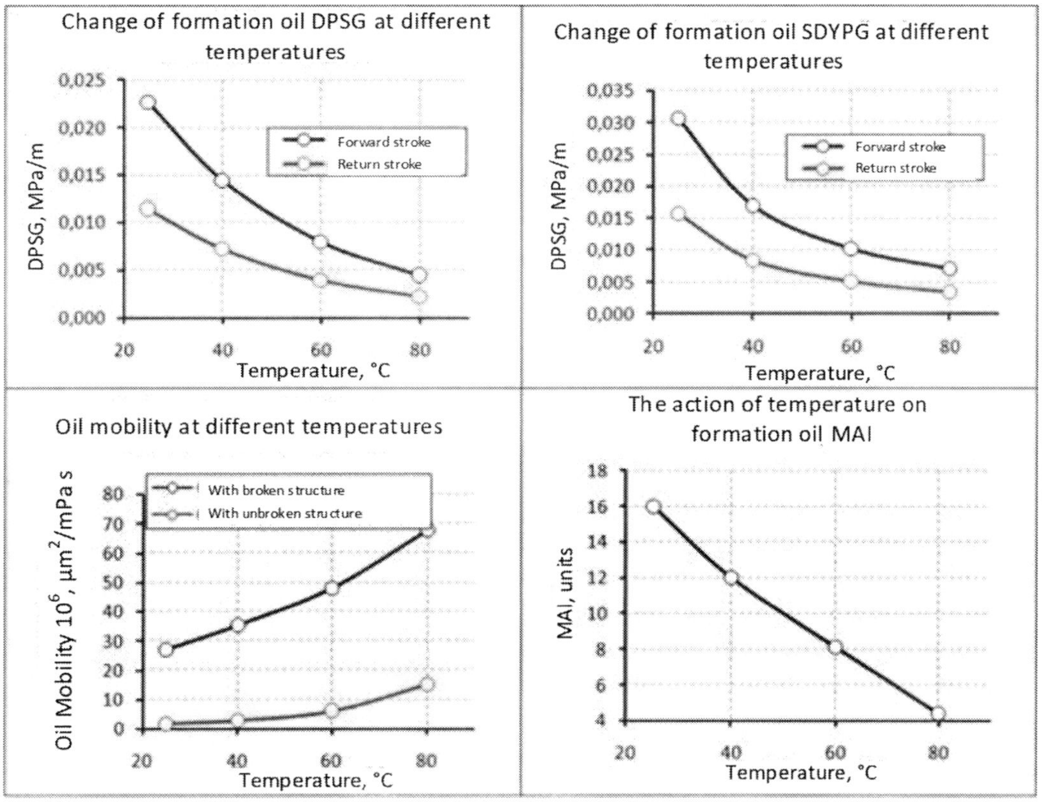

Figure 3. Rheological lines and filtration curves of formation oil of the Stepnoozerskoe field

The samples of studied oils at a temperature equal to the formation temperature have viscosity and mobility anomalies, and they can be classified as non-Newtonian oils. This behavior of oils is caused by the presence of a spatial structure of asphaltenic particles in them.

Filtration of studied oils through natural rock samples takes place at high values of boundary pressure gradients, low and variable values of oil mobility, which will certainly have a negative impact on the process of its extraction from the formation and should be taken into account when planning the development of these oil deposits. Besides, all studied oil has thixotropic properties, which are expressed in the increase of the boundary shear stresses and pressure gradients when the oil is at a standstill due to strengthening of its spatial structure. These properties should be taken into account when designing the cyclic impact methods on the productive formation, as well as when planning the operation of wells in periodic modes.

4. Conclusion

Thus, the priority condition for reliable forecasting of the cyclic impact on productive formations of high-viscosity oil is the joint research and study of rheological and filtration properties of high-viscosity oils, including thixotropic properties, as well as mathematical calculations to choose the right mode of the working fluid injection. For the Stepnoozerskoe field conditions, the optimal injection period is 10 days.

Table 4. Rheological and filtration parameters of degassed oil of the Stepnoozerskoe field

Indicator	**Well oil**									
	2462	**2172**	**2220**	**637**	**2233**	**2129**	**687**	**2245**	**2252**	**132**
Rheological parameters										
YP at forward stroke, Pa	0.026	0.015	0.023	0.014	0.019	0.019	0.019	0.025	0.026	0.018
YP at return stroke, Pa	0.014	0.008	0.012	0.007	0.010	0.010	0.010	0.013	0.013	0.010
SDYP at forward stroke, Pa	0.036	0.021	0.031	0.019	0.026	0.026	0.026	0.034	0.035	0.025
SDYP at return stroke, Pa	0.019	0.011	0.017	0.010	0.014	0.014	0.014	0.018	0.019	0.014
Viscosity of oil with broken structure, mPa·s	230	375	200	220	364	516	220	575	370	274
Viscosity of oil with unbroken structure, mPa·s	982	1250	890	893	1430	1700	982	1960	1610	1160
VAI	4.3	3.3	4.4	4.1	3.9	3.3	4.5	3.4	4.4	4.2
Filtration parameters										
Permeability of rock sample, μm^2	0.010	0.010	0.010	0.010	0.003	0.010	0.003	0.010	0.003	0.010
DPSG at forward stroke, MPa/m	0.025	0.014	0.021	0.013	0.036	0.018	0.037	0.023	0.049	0.017
DPSG at return stroke, MPa/m	0.012	–	0.011	0.006	–	–	–	–	–	–
SDYPG at forward stroke, MPa/m	0.033	0.019	0.029	0.017	0.049	0.024	0.050	0.032	0.067	0.024
SDYPG at return stroke, MPa/m	0.017	–	0.015	0.009	–	–	–	–	–	–
Mobility of oil with broken structure, μm^2/mPa·s	$2.17 \cdot 10^{-5}$	$1.33 \cdot 10^{-5}$	$2.51 \cdot 10^{-5}$	$2.27 \cdot 10^{-5}$	$0.33 \cdot 10^{-5}$	$0.98 \cdot 10^{-5}$	$0.54 \cdot 10^{-5}$	$0.87 \cdot 10^{-5}$	$0.32 \cdot 10^{-5}$	$1.88 \cdot 10^{-5}$
Mobility of oil with unbroken structure, μm^2/mPa·s	$0.67 \cdot 10^{-5}$	$0.86 \cdot 10^{-5}$	$0.74 \cdot 10^{-5}$	$1.05 \cdot 10^{-5}$	$0.30 \cdot 10^{-5}$	$0.70 \cdot 10^{-5}$	$0.33 \cdot 10^{-5}$	$0.52 \cdot 10^{-5}$	$0.24 \cdot 10^{-5}$	$0.77 \cdot 10^{-5}$
MAI	3.2	1.5	3.4	2.2	1.1	1.4	1.6	1.7	1.3	2.4

References

[1] Rogachev M K and Mukhametshin V V 2018 Control and regulation of the hydrochloric acid treatment of the bottomhole zone based on field-geological data *J. of Mining Inst.* **231** 275–80. DOI: 10.25515/PMI.2018.3.275

[2] Andreev V E, Chizhov A P, Kotenev Yu A et al 2019 Prediction of Gas Exposure in the Conditions of the Oil Fields of the Volga-Urals *Atlantis Highlights in Material Sciences and Technology (Int. Symp., dedicated to the 85th anniversary of H.I. Ibragimov, Engineering and Earth Sciences: Applied and Fundamental Research (ISEES 2019))* vol 1 pp 666–9. DOI: 10.2991/isees-19.2019.132

[3] Hartman R L, Lecerf B, Frenier W W, Ziauddin M E and Fogler H S 2006 Acid-Sensitive Aluminosilicates: Dissolution Kinetics and Fluid Selection for Matrix-Stimulation Treatments *SPE Product. & Operat.* **21(2)** 11. DOI: 10.2118/82267-PA

[4] Akhmetov R T and Mukhametshin V V 2018 Range of application of the Brooks-Corey model for approximation of capillary curves in reservoirs of Western Siberia *Advances in Engineering Research (AER) Int. Conf. Actual Issues of Mechanical Engineering (AIME 2018)* vol 157 pp 5–8. DOI: 10.2991/aime-18.2018.2

[5] Zakaria A S and Nasr-El-Din H A 2016 A Novel Polymer-Assisted Emulsified-Acid System Improves the Efficiency of Carbonate Matrix Acidizing *SPE J.* **21(3)** 14. DOI: 10.2118/173711-PA

[6] Mukhametshin V V and Andreev V E 2018 Increasing the efficiency of assessing the performance of techniques aimed at expanding the use of resource potential of oilfields with hard-to-recover reserves *Bull. of the Tomsk Polytech. Univer. Geo Assets Engineer.* **329(8)** 30–6

[7] Lestz R S, Clarke J N, Plattner D and Byrd A C 2002 Perforating for Stimulation: An Engineered Solution *SPE Drill. & Complet.* **17(1)** 8. DOI: 10.2118/76812-PA

[8] Mukhametshin V V 2018 Rationale for trends in increasing oil reserves depletion in Western Siberia cretaceous deposits based on targets identification *Bull. of the Tomsk Polytech. Univer. Geo Assets Engineer.* **329(5)** 117–24

[9] Zeigman Yu V, Mukhametshin V Sh, Sergeev V V and Kinzyabaev F S 2017 Experimental study of viscosity properties of emulsion system with SiO_2 nanoparticles *Nanotehnol. in Construct.* **9(2)** 16–38. DOI: 10.15828/2075-8545-2017-9-2-16-38

[10] Kuleshova L S, Kadyrov R R, Mukhametshin V V and Akhmetov R T 2019 Auxiliary equipment for downhole fittings of injection wells and water supply lines used to improve their performance in winter *IOP Conf. Ser. Mater. Sci. and Engineer.* **560(1)** 1–6. DOI: 10.1088/1757-899X/560/1/012071

[11] Kuleshova L S., Kadyrov R R., Mukhametshin V V and Safiullina A R 2019 Design changes of injection and supply wellhead fittings operating in winter conditions *IOP Conf. Ser. Mater. Sci. and Engineer.* **560(1)** 1–6. DOI: 10.1088/1757-899X/560/1/012072

IOP Conf. Series: Materials Science and Engineering **905** (2020) 012083 doi:10.1088/1757-899X/905/1/012083

Justification for increasing the performance of hydrochloric acid treatment in wells of fields with carbonate reservoir

V Sh Mukhametshin

Department of Oil and Gas Field Exploration and Development, Ufa State Petroleum Technological University, Branch of the University in the City of Oktyabrsky, Oktyabrsky, Republic of Bashkortostan, Russia

E-mail: vsh@of.ugntu.ru

Abstract. The geological and technological parameters, which have a prevailing influence on the success of the treatment, were established on the basis of analysis and synthesis of experience of hydrochloric acid treatments in the conditions of Tournaisian oil deposits of the Volga-Ural oil and gas-bearing province. The algorithms, which allow selecting wells for hydrochloric acid treatment, determining the parameters of treatment taking into account peculiarities of geological structure of formations, time factor, peculiarities of technology of wells and deposits operation, were proposed. The conditions for the most successful implementation of this type of treatment on the bottomhole zone to intensify oil production and reduce the water cut in order to increase the level of profitability of deposits in carbonate reservoirs with hard-to-recover reserves were established.

1. Introduction

Based on the peculiarities of formation structure and technological peculiarities of wells and deposits it is important to select those values of technological parameters during hydrochloric acid treatment on the bottomhole zone that will allow obtaining the maximum profit [1–6].

In order to solve this problem, a geological and field analysis of the treatment on the success and efficiency of hydrochloric acid treatment (HAT) of various geological and technological parameters was carried out in carbonate reservoirs of the Tournaisian formation of the North-West part of Bashkortostan.

The analyzed Tournaisian objects were drilled along both a uniform and non-uniform triangular array with distances ranging from 400 to 600 m between the wells. The central, most productive areas of deposits, where the well density (WD) ranged from 16 to 36 ha/well were the densest well array drilled. Less productive peripheral areas were drilled with single wells in an extremely uneven manner. The density of the well array in the outer oil-bearing circuit varied from 17 to 240 ha/well. The operation mode is elastic water drive, however, in the areas where water injection is not organized as a result of weak hydrodynamic connection of the oil-saturated volume of the formation with the bottom waters, the dissolved gas mode is developed. At the same time, the formation pressure in some areas of deposits decreased to 1–3 MPa.

Fluid is mainly produced from the very beginning of operation by means of rod pumps, the capacity of which is determined by production capabilities of wells. Many wells ensure commingled production of oil from two horizons, thus making it difficult to efficiently control the development.

Water production of extracted products is different. For example, from the very beginning and for quite a long time (up to a drop in formation pressure of 2–5 MPa) most wells produce 10–15 % of water, indicating that there is no connection to the bottom water, although for some wells the water production is rapidly increasing, indicating the presence of lithological, permeable "windows", through which water cut cones are formed. Well watering is largely determined by the distance between the bottom holes of the perforation and the surface of the water-oil contact.

As shown by field studies, under the conditions of the accepted density of producing wells array there is no interference between wells during development of deposits in natural modes, which is a consequence of highly developed geological inhomogeneity, low reservoir properties, lens structure of permeable layers.

The rate of oil and fluid extraction, as well as oil recovery in the absence of flooding, are determined by productivity and density of the producing wells array. Depending on these parameters the final oil recovery will be from 1 to 34 % by the end of development.

The analysis of contour flooding carried out in separate areas of deposits shows that along with the increase in the production rate of surrounding production wells there is also lack of oil production growth, i.e. the displacement process takes place only within separate local zones. At the same time, with the increase of distances between production and injection wells (more than 500 m) the efficiency of flooding is significantly reduced.

In order to improve the applied development systems, further drilling of deposits is implied along a uniform triangular and quadrangular array of wells with distances between wells from 300 to 600 m and organization of area (according to five- and seven-point scheme) and focal contour waterflooding by both drilling of new injection wells and transferring the watered production wells on separate areas with favorable geological characteristic. Flooding is not implied in areas with deteriorated reservoir properties due to low efficiency at accepted density of well array.

The following were considered as independent variables: net oil thickness ($H_Э$), mean thickness of oil-saturated interlayers ($H_П$) and their quantity (n), porosity ratio in a well ($M_Г$), share of reservoir rocks in the total formation thickness ($K_П$), time from the beginning of well operation to the moment of HAT (t), maximum well flow rate before HAT (Q_{Hmax}), oil flow rate (Q_{H1}), water cut (f_1), accumulated oil production ($Q_{нак}$) of a well at the moment of HAT, volume (V_K) and maximum pressure ($P_{зак}$) of acid injection into the formation. The following efficiency parameters were used as response functions: absolute ($Э_1$) and relative ($Э_2$) oil yield gain; absolute ($Э_3$) and relative ($Э_4$) water cut reduction; total oil production growth during the effect ($Э_5$); relative increase of well flow index ($Э_6$).

Geological and technological parameters, which have the greatest influence on HAT efficiency, were identified using the Wald's sequential procedure. Information value was determined by Kulbak's criterion.

The Wald's method of sequential diagnostic procedure is based on the Bayes formula, which is written as follows:

$$D_A < \frac{P(X_1^1/A)}{P(X_1^1/B)} \cdot \frac{P(X_2^2/A)}{P(X_2^2/B)} \cdot \ldots \cdot \frac{P(X_j^i/A)}{P(X_j^i/B)} < D_B,$$

where D_A, D_B – threshold values determined on the basis of a priori level of errors of assignment of an object of one class to another, or on the graph of distribution of sums of diagnostic coefficients;

$(P(X_j^i/A))/(P(X_j^i/B))$ – diagnostic coefficient (recognition coefficient) – ratio of frequency detected at i interval of j characteristic of class A to the frequency of the same range and characteristic, but in class B.

Diagnostic coefficient $D(X_j^i)$ for i interval of j characteristic change is calculated by the following formula:

$$D(X_j^i)=10 \cdot \lg[P(X_j^i/A)/P(X_j^i/B)],$$

and the information value $J(X_j^i)$ – by Kulbak's formula:

$$J(X_j^i)=D(X_j^i)\cdot 0{,}5[P(X_j^i/A)-P(X_j^i/B)].$$

Informative were the features, the information value of which, according to Kulbak's formula, is more than 0.5.

Such a large number of efficiency criteria and the use of different amounts of information is caused by the need to solve the task at different stages of development and in conditions of limited information about deposits (due to insufficient volume of field studies, due to organizational and financial reasons), as well as by changes in the tactics and strategy of the enterprise in market conditions. This will make it possible to respond flexibly to changes in internal and external operating conditions [7-12].

2. Results

The analysis of well distribution depending on the change of informative geological and technological parameters, where the incremental oil rate served the efficiency criterion, showed that the success significantly increases for wells with worse reservoir properties and greater geological heterogeneity. At $H_П$<4.2 m, n>3, $M_Г$<13.3 %, $K_П$<0.66 unit fractions, the success is more than 50 %. Low reservoir properties and increased geological inhomogeneity cause weak formation development in the well drainage zone and cause the reserve of oil production on one side and presence of significant filtration resistances in the bottomhole zone of the formation on the other side.

The fact that it is desirable to treatment the wells, which service life is less than 160 months and with accumulated oil production less than 24 thousand tons, indicates the need for oil production reserve. At the same time, the probability increases across wells with higher $Q_{H\,max}$. This fact is caused by the clogging of the bottomhole zone due to the precipitation of resins, asphaltenes and paraffins as a result of intensive oil production, and a significant reduction in formation pressure in the drainage zone.

An important point in HAT is the determination of the volumes of acid injected into the formation. These volumes depend on geological features of the formation at the site of its drilling by a well and technological features of wells and deposits. The use of significant dependencies resulted in inequalities in the determination of acid injection volumes and injection pressures:

$$V_K > 0{,}71\ H_Э;$$
$$V_K > 1{,}49\ H_П;$$
$$V_K > 0{,}059\ Q_{H1};$$
$$V_K > 0{,}175\ f_1;$$
$$V_K > 4{,}4\cdot 10^{-4}\ Q_{нак};$$
$$P_{max} < 0{,}98\ V_K.$$

Inequalities make it possible to diagnose the required volume of acid according to the maximum value of the parameter.

However, the results are probabilistic. In order to be able to answer the question of success unambiguously, the values of total diagnostic coefficients (TDC) were calculated and intervals, in which the effect is unambiguously positive, were established. This interval varies from 8.9 to 66.8.

By changing the process parameters of the treatment the use of TDC distributions allows transferring wells from zones with the negative effect and zones of uncertainty to the zone with the positive effect.

The distribution of wells depending on the change in the values of informative geological and technological parameters, where the efficiency criterion was the reduction of the water cut, showed that at $H_Э$>11.5 m, $H_П$>4.3 m, n<3, $K_П$<0.73 unit fractions, the success of the treatment is more than 50 %. It is desirable that the process parameters be as follows: t<118 months, $Q_{нак}$<17.8 th. t, $Q_{H\,max}$<349 t/mon., Q_{H1}>77 t/mon., V_K<7.5 m^3, $P_{зак}$<7.6 MPa.

The obtained inequalities also make it possible to diagnose the success of treatments at a qualitative level both within the analyzed fields and close to them according to geological and field characteristics by sampling of wells according to the corresponding informative features, as well as to approximately estimate the treatment parameters. However, these results, as in the case where the efficiency criterion was the increase in oil yield, only allow assessing the probability of effective treatment and cannot give a clear answer. Thus, the TDC values were calculated. It was established that if TDC changes from 14.4 to 40.9, the success of HAT is definitely positive.

The comparison of results of the study for both versions showed that a significant number of wells in the first version were in the group for which a positive effect was obtained, and in the second version – in the group for which no effect was obtained, and vice versa, i.e. often with an increase in oil production there is also an increase in water cut.

The comparison of intervals of change of significant parameters, in which the success of the treatment is more than 50 %, in both variants indicated the presence of common intervals. The values of these intervals are shown in Table 1.

It is the general intervals that explain the fact that in some wells after exposure, along with the increase in oil production, there was a decrease in the water cut of the produced product.

Table 1. Intervals of change of significant geological and technological parameters, in which the success of HAT is more than 50 %

Interval by variants		Common interval for variants 1 and 2
1	2	
$H_Э>7.7$	$H_Э>11.5$	$H_Э>11.5$
$M_Г<13.3$	$M_Г<13.0$	$M_Г<13.0$
$K_П<0.66$	$K_П<0.79$	$K_П<0.66$
$t<160$	$t<118$	$t<118$
$Q_{H\,max}>180$	$Q_{H\,max}<349$	$280<Q_{H\,max}<349$
$Q_{HI}<125$	$Q_{HI}>77$	$77<Q_{HI}<125$
$Q_{нак}<24.6$	$Q_{нак}<17.8$	$Q_{нак}<17.8$
$7{,}5<P_{max}<9.0$	$P_{max}>7.6$	$7.6<P_{max}<9.0$

Besides, the simultaneous increase in oil production and well water cut is caused by the absence of common intervals for individual parameters (Table 2).

Table 2. Intervals of change of geological-technological parameters, in which the success of HAT is more than 50 %

Intervals by variants	
1	**2**
$H_П<4.6$	$H_П>4.2$
$n>3$	$n<3$
$V_K>7.8$	$V_K<7.5$
$V_K/H_Э>0.71$	$V_K/H_Э<0.65$
$V_K>0.71\,H_Э$	$V_K<0.65\,H_Э$

The absence of a common interval of change in volumes and specific volumes of injected acid shall be particularly noted. The results show that in pursuit of additional oil production it is possible to obtain a significant increase of the water cut.

3. Conclusion

The studies made it possible to conclude the following:

- geological and technological parameters, which have a prevailing influence on the process of treatment on the carbonate reservoir of the bottomhole formation zone, are identified;
- a set of algorithms is proposed, allowing selecting wells for bottomhole formation zone treatment, technological parameters of operations in different geological conditions;
- the limits of the most successful hydrochloric acid exposure are established to increase the profitability of development of carbonate oil deposits with hard-to-recover reserves.

IOP Conf. Series: Materials Science and Engineering **905** (2020) 012083 doi:10.1088/1757-899X/905/1/012083

References

[1] Xie X, Weiss W W, Tong Z J and Morrow N R 2005 Improved Oil Recovery from Carbonate Reservoirs by Chemical Stimulation *SPE J.* **10(3)**. DOI: 10.2118/89424-PA

[2] Mukhametshin V V, Andreev V E, Dubinsky G S, Sultanov Sh Kh and Akhmetov R T 2016 The Usage of Principles of System Geological-Technological Forecasting in the Justification of the Recovery Methods *SOCAR Proc.* **3** 46–51. DOI: 10.5510/OGP20160300288

[3] Andreev V E, Chizhov A P, Kotenev Yu A et al 2019 Prediction of Gas Exposure in the Conditions of the Oil Fields of the Volga-Urals *Atlantis Highlights in Material Sciences and Technology (Int. Symp., dedicated to the 85th anniversary of H.I. Ibragimov, Engineering and Earth Sciences: Applied and Fundamental Research (ISEES 2019))* vol 1 pp 666–9. DOI: 10.2991/isees-19.2019.132

[4] Akhmetov R T, Mukhametshin V V and Kuleshova L S 2019 Grouping of objects using a limited number of parameters characterizing geological and physical properties of layers *Atlantis Highlights in Material Sciences and Technology (Int. Symp., dedicated to the 85th anniversary of H.I. Ibragimov, Engineering and Earth Sciences: Applied and Fundamental Research (ISEES 2019))* vol 1 pp 9–13. DOI: 10.2991/isees-19.2019.3

[5] Sergeev V V, Belenkova N G, Zeigman Yu V and Mukhametshin V Sh 2017 Physical properties of emulsion systems with SiO_2 nanoparticles *Nanotechnol. in Construct.* **9(6)** 37–64. DOI: 10.15828/2075-8545-2017-9-6-37-64

[6] Gonzalez I J F, Gammiero A and Llamedo M A 2012 Design of a Neural Network Model for Predicting Well Performance after Water Shutoff Treatments Using Polymer Gels *SPE Latin Amer. and Caribbean Petrol. Engineer. Conf.* (Mexico: Mexico City, 16–18 April 2012) p 9. DOI: 10.2118/153908-MS

[7] Mukhametshin V V 2018 Efficiency estimation of nanotechnologies applied in constructed wells to accelerate field development *Nanotechnol. in Construct.* **10(1)** 113–31. DOI: 10.15828/2075-8545-2018-10-1-113-131

[8] Zeigman Yu V, Mukhametshin V Sh, Sergeev V V and Kinzyabaev F S 2017 Experimental study of viscosity properties of emulsion system with SiO_2 nanoparticles *Nanotehnol. in Construct.* **9(2)** 16–38. DOI: 10.15828/2075-8545-2017-9-2-16-38

[9] Olarte J D, Haldar S, Said R et al 2011 New Approach of Water Shut off Techniques in Open Holes – and World First Applications of Using Fiber Optic Services With Tension-Compression Sub *SPE/DGS Saudi Arabia Section Technical Symposium and Exhibition* (Al-Khobar, Saudi Arabia, 15–18 May 2011) p 10. DOI: 10.2118/149116-MS

[10] Rogachev M K and Mukhametshin V V 2018 Control and regulation of the hydrochloric acid treatment of the bottomhole zone based on field-geological data *J. of Mining Inst.* **231** 275–80. DOI: 10.25515/PMI.2018.3.275

[11] Gdanski R D 2000 Kinetics of the Primary Reaction of HF on Alumino-Silicates *SPE Product. & Facilit.* **15(4)** 279–87. DOI: 10.2118/66564-PA

[12] Mukhametshin V V and Kuleshova L S 2019 Justification of Low-Productive Oil Deposits Flooding Systems in the Conditions of Limited Information Amount *SOCAR Proc.* **2** 16–22. DOI: 10.5510/OGP20190200384

On the use of geophysical research data aiming to increase the efficiency impact on the bottomhole zone of wells

V V Mukhametshin[1] and L S Kuleshova[2]

[1] Department of Oil and Gas &Oil Field, Development and Operation, Ufa State Petroleum Technological University, Ufa, Russia
[2] Department of Oil and Gas Field, Exploration and Development, Ufa State Petroleum Technological University, Branch of the University in the City of Oktyabrsky, Oktyabrsky, Russia

E-mail: vv@of.ugntu.ru, markl212@mail.ru

Abstract. Under the conditions of terrigenous Jurassic deposits in Western Siberia, we conducted a study on the influence of deposits' geological structure features, reflected by the results of geophysical surveys of wells on the effectiveness of the impact on the bottomhole zone using the hydrochloric acid composition Aldinol-20. It was established that in predicting the effectiveness of the impact on the bottomhole well zone, it is necessary to use the data of geophysical studies. At the stage of putting the field into development, it is enough to limit this information in the absence of it. The article proposes a technique that allows us to solve the tasks at different stages of development. The technique is based on the use of geological and statistical models and it is built based on using various performance criteria for various amounts of information. It is to be conducted in the conditions of a limited amount of information about deposits due to insufficient volumes of field research of organizational and financial nature, as well as in case of changing the tactics and strategies of enterprises in market conditions.

1. Introduction

In field practice, when enough information is accumulated to generalize, the results of its generalization are often used to increase the efficiency of decisions. One of the areas of such activity is the generalization of the experience of impacts on the bottomhole formation zone (PZP) [1–3]. However, the weak point of such studies is the lack of the possibility of their use when changing conditions and scenarios for the development of situations in the future both on the analyzed object itself and on analogous objects. An important point in this case is the availability of the opportunity to use the experience gained under the conditions:

- various stages of development;
- lack of complete necessary information;
- the availability of heterogeneous information;
- the availability of a limited amount of information;
- changes in the market situation, etc.

As the experience of geological-field generalization of the results of the impact on the bottomhole zone of wells [4–6] shows, the effectiveness of the impact is determined by three large groups of factors:

1 – geological, physical and physic-chemical properties of the formations and the fluids saturating them;

2 – technological indicators of the operation of wells and deposits;

3 – technological parameters of the impact.

The degree and nature of their influence on performance indicators is different in the conditions of various objects and their groups [7–9], which indicates the need for a differentiated approach in the analysis and use of the results.

In order to eliminate the above-mentioned weaknesses, the article describes a generalization of the experience of conducting activities to intensify oil production in low-permeable terrigenous strata SE11 of Western Siberia using an acidic composition based on the Aldinol-20 composition. The main objective of this generalization was to establish the degree of influence of the initial geological and geophysical parameters, as well as the filtration-capacitive and thickness properties of the layers determined by them, on various impact indicators, taking into account the peculiarities of conducting hydrodynamic and flow metric studies of wells. The choice of this group of parameters is due to accessibility, representativeness and common standards that allow them to be used regardless of place and time.

2. Methods and materials

The acid composition of Aldinol-20 consists of a mixture of polyhydric alcohols, cations and nonionic surface-active substances, a corrosion inhibitor, hydrochloric acid and the modifying additive Aldinol-MK. The addition of the Aldinol-MK modifier to acid compositions increases the effectiveness of the latter due to their deeper penetration into the formation, destruction of the hydrocarbon film on the surface of mechanical impurities, and additionally inhibits the corrosiveness of acids, which allows transporting and storing the acid composition without compromising on quality.

Distinctive features of the composition Aldinol-20 are:

- low (less than 0.1 mN / m) interfacial tension at the oil-acid composition boundary, which ensures the destruction of persistent water-oil emulsions formed in the bottomhole zone of production wells and the effective displacement of residual oil in the bottomhole zone of the well;
- slower than an aqueous acid solution, the reaction rate with the carbonate components of the oil reservoir rock, which altogether leads to an increase in both the permeability of the reservoir and the penetration zone (processing radius);
- the ability to reduce the strength of the adsorption layers at the water-oil interface, formed by surface-active substances that make up oil (asphaltenes, resins, naphthenic acids) and mechanical impurities.

When such a composition is pumped into the bottomhole zone of the formation, processes of effective dispersion and removal of components of asphalt-resin-paraffin deposits (ARPD) from inorganic sediments that clog the formation (particles of clay, iron sulfide, hydroxides, calcium, magnesium, aluminum carbonates, etc.) occur simultaneously. The latter, having lost the acid-resistant "protective shell" of the above-mentioned organic components of ARPD, begin to actively react (dissolve) in a water-acid medium. Moreover, during the chemical reaction of some of them, gaseous components are released, which, in turn, loosen sediments and float the volume of surface-active substances-acidic dispersion medium, individual components of these sediments, which significantly speeds up the process of cleaning the bottom-hole formation zone.

An important distinguishing feature of Aldinol-20 from known surface-active substances-acid compositions is that the acid composition not only cleans the bottomhole zone of the formation well from pollution, but also retains, after its neutralization, high oil-displacing properties (due to the residual surface-active substances in the system) and hydrophobizing properties (due to adsorption of cationic surface-active substances) [10].

It is recommended to use this method in wells that produce a low oil flow during development, open low-permeability formations, and have a reduced production rate compared to surrounding wells, with reduced filtration characteristics in the near-hole zone of the formation. The formation had a decrease in production rate during operation with a constant or growing formation pressure with a high degree of heterogeneity of reservoir properties of the reservoir and small working thicknesses; as well as with high water cut of the product.

3. Results

More than 120 wells studied the effect on impact indicators, reflected using the absolute (Y_1) and relative (Y_2) increase in oil production, absolute (Y_3) and relative (Y_4) decrease in water cut, the total increase in production during the effect (Y_5) and the integrated efficiency parameter (Y_6) [11], geological and physical properties of the formation (according to geophysical studies), technological parameters of the operation of wells and deposits, and technological parameters of the impact.

The geological and physical properties of the formation in the well were reflected using the following parameters: total (H_{total}, m), effective (H_e, m), effective oil-saturated ($H_{e\text{-}oil}$, m) thickness of the formation, sandiness (K_n), stratification (K_p), open porosity (m), oil saturation (K_n); permeability coefficients (K_{perm}, 10^{-3} μm^2), self-polarization (α_{cn}); readings in the reservoir according to BC (M_{BC}, Om·m), PS amplitude (A_{PS}, Om·m), on GZ1 (A0.4M0.1N) ($A_{0.4}$, Om·m), on GZ2 (A1.0M0.1N) ($A_{1.0}$, Om·m), on GZ3 (A2.0 M0.5N) ($A_{2.0}$, Om·m), on GZ4 (A4.0M0.5N) ($A_{4.0}$, Om·m), on PZ (A0.5M6.0N) ($A_{0.5}$, Om·m), gamma ray logging (γ, mcr/h), large tubing probe (β'_{tub}), small tubing probe (β''_{tub}), induction probe (β_{ip}, sim); specific electrical resistance of the probe by IK (ρ_{IK}, Om·m), on BC (ρ_{BC}, Om·m), on a set of electric logging probes (A_{compl}, Om·m), penetration zones for a complex of electric logging probes ($A^{\text{зп}}_{\text{КОМП}}$, Om·m), the ratio of the diameter of the zone of penetration of the filtrate of the washing fluid to the diameter of the well (d_{ZP} / d_C).

The parameter values were used both as a whole along the section (X_i) and along the perforated part of it ($\overline{X}_i$).

The technological parameters of the operation of wells and deposits characterized the parameters: time (t, month), maximum oil production (Q_{Hmax}, t/month) from the beginning of the operation of the well until the impact; flow rate (Q_{H1}, t/mon), water cut (f_1, %), cumulative oil production (Q_{cum}, т) at the time of exposure; initial oil production (Q_{start}, t/mon), frequency of exposure (N).

Among the parameters reflecting the exposure technology, the following were used: consumption of Aldinol-20 (ALD, t), 22 % hydrochloric acid (HCl, t); the volume of the injected solution (V, m^3), as well as various specific indicators.

Using systematic regression analysis, we built models for the dependence of all performance indicators on the considered geological and technological parameters in three versions:

1 – using the whole complex of parameters;

2 – using parameters reflecting the geological and physical properties of the formation and the technology of exposure;

3 – using parameters reflecting the technological features of the work of wells and deposits, as well as the impact technology.

Models that reflect the influence of parameters on the absolute increase in oil production have the following form:

- option 1:

$$Y_1 = -1333{,}083 - 90{,}481K_p + 22{,}523K_p - 2{,}255A_{PC} + 14{,}912\overline{\overline{H}}_{e\text{-}oil} + 42{,}403\overline{\overline{K}}_P + 1650{,}692\overline{\overline{m}} + \\ + 157{,}254\overline{\overline{\alpha}}_{sp} + 61{,}853\overline{\overline{\rho}}_{ik} + 5{,}145\overline{A}_{0.4} - 68{,}513\overline{\overline{d}}_{zp}/\overline{\overline{d}}_c + 9{,}664\overline{\overline{\beta}}'_{tub} - 0{,}001Q_{cum} + 0{,}46t + \\ + 0{,}057Q_{max} + 1{,}735\ f_1 - 0{,}466\ Q_{H1} - 125{,}543ALD + 37{,}437V + 46{,}205N + 79{,}679V/\overline{\overline{H}}_{e\text{-}oil} + \\ + 1009{,}308ALD/V - 31{,}301ALD/\overline{\overline{H}}_{e\text{-}oil} \quad (R = 0{,}793); \tag{1}$$

- option 2:

$$Y_2 = -745{,}687 - 159{,}44K_p + 16{,}781K_p - 1{,}668A_{PC} + + 21{,}271\overline{\overline{H}}_{e\text{-}oil} - 56{,}541\overline{\overline{K}}_p + 1636{,}759\overline{\overline{m}} - \\ - 418{,}022\overline{\overline{\alpha}}_{sp} + 39{,}659\overline{\overline{\rho}}_{ik} - 16{,}85\overline{A}_{0.4} - 7{,}752\overline{d_{\text{зп}}/d_C} + 6{,}135\overline{\overline{\beta}}'_{tub} - 195{,}85ALD + \\ + 72{,}25V + 76{,}62N + 55{,}79V/\overline{\overline{H}}_{e\text{-}oil} + +1362{,}594ALD/V - 29{,}818ALD/\overline{\overline{H}}_{e\text{-}oil} \quad (R = 0{,}640); \tag{2}$$

- option 3:

$$Y_3 = -385{,}698 - 0{,}185t + 0{,}019Q_{max} + 0{,}086\ f_1 - 0{,}425Q_{H1} - 142{,}704ALD + 66{,}176V + \\ + 52{,}021N - 29{,}26V/H_{e\text{-}oil} + 943{,}908ALD/V + 63{,}45ALD/H_{e\text{-}oil} \quad (R = 0{,}518). \tag{3}$$

A similar series of geological and statistical models was obtained for other performance indicators. The values of the multiple correlation coefficients (R) presented in the table allow us to use them in diagnosing and selecting wells for impact at a quantitative and qualitative level. This level is explained by the fact that the errors of the obtained models vary from 20 to 70% and make up on average 32%, although in the calculations for 7–10 wells the forecast is quite satisfactory.

Table 1. Values of multiple correlation coefficients of obtained models

Option	Performance indicator					
	Y_1	Y_2	Y_3	Y_4	Y_5	Y_6
1	0.793	0.824	0.753	0.793	0.818	0.727
2	0.640	0.742	0.705	0.640	0.739	0.681
3	0.518	0.533	0.547	0.518	0.602	0.534

An analysis of the results shows that naturally the smallest error in the prediction has models built using the whole complex of geological and technological parameters considered. In the absence of parameters reflecting the technological parameters of the work of wells and deposits (option 2), which is typical for the stage of putting the fields into development, the error increases and the accuracy decreases. However, at this stage, the use of geological and geophysical data allows us to solve the problems of predicting the effectiveness and selection of wells for impact with a sufficient degree of accuracy.

In the absence of parameters characterizing the geological and physical properties of the formation in the well (option 3), a significant decrease in the predictive abilities of the models is observed, which indicates the need to use geophysical research data for solving such problems. Although in some situations, this group of models can also be used in practice for quantitative and qualitative assessment.

4. Conclusion

1. It was established that in predicting the effectiveness of the impact on the bottomhole zone of the wells, it is necessary to use the data of geophysical studies. At the stage of putting the field into development, it is enough to limit this information in the absence of it.

2. The article proposes a technique that is based on the use of geological and statistical models. It is built on the basis of using various performance criteria for various amounts of information, which allows us to solve the problems at different stages of development. Moreover, it is to be conducted in the conditions of a limited amount of information on deposits due to insufficient volumes of field research of organizational and financial nature reasons; as well as in case of changing the tactics and strategies of enterprises in market conditions.

References

[1] Economides J M and Nolte K I 2000 *Reservoir stimulation* (West Sussex, England: John Wiley and Sons) 856 p

[2] Kuleshova L S, Kadyrov R R, Mukhametshin V V and Safiullina A R 2019 Design changes of injection and supply wellhead fittings operating in winter conditions *IOP Conf. Ser. Mater. Sci. and Engineer.* **560(1**(012072)) 1–5. DOI: 10.1088/1757-899X/560/1/012072

[3] Valeev A S, Kotenev Yu A, Mukhametshin V Sh and Sultanov Sh Kh 2019 Substantiation of the recovery of residual oil from low-productive and heterogeneous formations in Western Siberia by improving the waterflood system using gas and water-gas impacts *IOP Conf. Ser. Mater. Sci. and Engineer.* **560(1**(012204)) 1–6. DOI: 10.1088/1757-899X/560/1/012204

[4] Webb K J, Black C J J and Tjetland G A 2005 Laboratory Study Investigating Methods for Improving Oil Recovery in Carbonates *Int. Petrol. Technol. Conf.* (Doha, Qatar, 21–23 November 2005) pp 1–6. DOI: 10.2523/IPTC-10506-MS

[5] Kuleshova L S, Kadyrov R R, Mukhametshin V V and Akhmetov R T 2019 Auxiliary equipment for downhole fittings of injection wells and water supply lines used to improve their

performance in winter *IOP Conf. Ser. Mater. Sci. and Engineer.* **560(1**(012071)) 1–6. DOI: 10.1088/1757-899X/560/1/012071

[6] Nasr-El-Din H A, Van Domelen M S, Sierra L and Welton T D 2007 Optimization of Surfactant-based Fluids for Acid Diversion *Europ. Format. Damage Conf.* (Scheveningen, The Netherlands, 30 May – 1 June 2007) pp 1–11. DOI: 10.2118/107687-MS

[7] Mukhametshin V Sh, Zeigman Yu V and Andreev A V 2017 Rapid assessment of deposit production capacity for determination of nanotechnologies application efficiency and necessity to stimulate their development *Nanotechnol. in Construct.* **9(3)** 20–34. DOI: 10.15828/2075-8545-2017-9-3-20-34

[8] Ibragimov N G, Ismagilov F Z, Musabirov M Kh and Abusalimov E M 2014 Analysis of well stimulation pilot projects in Tatneft OAO *Oil Industry* **7** 40–3

[9] Yakupov R F, Mukhametshin V Sh, Khakimzyanov I N and Trofimov V E 2019 Optimization of reserve production from water oil zones of D3ps horizon of Shkapovsky oil field by means of horizontal wells *Georesursy* **21(3)** 55–61. DOI: 10.18599/grs.2019.3.55-61

[10] Zemtsov Yu V 2014 *Development and improvement of repair and insulation works in Western Siberia* (St. Peterburg: Nedra Publ.) 320 p

[11] Stenkin A V, Kotenev Yu A, Mukhametshin V Sh and Sultanov Sh Kh 2019 Use of low-mineralized water for displacing oil from clay productive field formations *IOP Conf. Ser. Mater. Sci. and Engineer.* **560(1**(012202)) 1–5. DOI: 10.1088/1757-899X/560/1/012202

IOP Conf. Series: Materials Science and Engineering **905** (2020) 012085 doi:10.1088/1757-899X/905/1/012085

Accounting of time factor during bottomhole zone treatment

V V Mukhametshin[1] and L Se Kuleshova

[1] Department of Oil and Gas Field Exploration and Development, Ufa State Petroleum Technological University, Ufa, Republic of Bashkortostan, Russia
[2] Department of Oil and Gas Field Exploration and Development, Ufa State Petroleum Technological University, Branch of the University in the City of Oktyabrsky, Oktyabrsky, Republic of Bashkortostan, Russia

E-mail: vv@of.ugntu.ru, markl212@mail.ru

Abstract. It was found that when treating the bottomhole zone with acid compositions preventing emulsification in terrigenous reservoirs, the time factor has a significant influence on the treatment efficiency, much more than the features of the geological structure of an objects. The algorithm, which allows reducing the negative influence of the time factor on the efficiency of oil production intensification by application of technological impact indices, is proposed. Geological-statistical models were obtained, which allow planning the treatment efficiency, selecting wells and technological treatment indicators taking into account the time factor in real time.

1. Introduction

The analysis of numerous studies on the bottomhole zone treatment (BHZ) using hydrochloric acid solutions shows that, for a particular method, its efficiency is determined by three large groups of parameters that reflect the following:

- geological-physical and physical-chemical properties of formations and saturating fluids;
- technological characteristics of wells and deposits performance;
- treatment technology [1–4]

The second group of parameters has a significant influence on the efficiency, including the values of time from the moment of well commissioning to the moment of treatment, water cut of well production, cumulative oil production and its ratio to extracted reserves, current formation pressure and its relative decrease to the initial one, well flow rate at the moment of treatment. This group characterizes the time factor.

Over time, the treatment efficiency is significantly reduced, which is caused by the increase of well water cut, reduction of formation pressure and production of oil reserves, although in some cases there is no clear dependence of operation efficiency on the duration of well operation. This seems to be bound to other factors [5–8].

In a certain way, the water cut of extracted products affects the efficiency of salt-acid treatment (SAT). The efficiency of both traditional and foam-acid formation treatments is significantly reduced at water cut of more than 50 %. Under the conditions of carbonate deposits of Tatarstan, traditional salt-acid treatment produce the greatest effect in wells, the water cut of which does not exceed 20 % until the bottomhole treatment, and when creating bottomhole caverns, the greatest effect is obtained in wells, where the water cut does not exceed 40 %. In the deposits of Belarus, high SAT efficiency is observed at the water cut of up to 18 %, at its further growth the relative number of successful operations is significantly reduced, and at the water cut of more than 80 % the success is only 28 %.

The treatment efficiency decreases as the formation is developed. Thus, when creating caverns, it is desirable that the specific total production of oil in wells does not exceed 200 t/m, although in some cases, especially during the initial period of development, with the increase of cumulative production, there is a slight increase in the effect.

At the reduction of formation pressure during reservoir development under depletion mode the effective application of salt-acid treatment is significantly deteriorated. This is caused by the negative effect of reaction products on oil inflow, for the removal of which from the bottomhole zone pores the available formation energy is not sufficient. There is also a decrease in the effect on wells operating the formation zones, where the pressure decreased below the bubble point pressure.

As practice shows, there is no consensus on the effect of well flow rate at the time of treatment on SAT efficiency [9, 10].

However, although the time factor has a significant impact on the bubble point pressure of bottomhole treatment:

- the measures are often not planned in real time;
- the role and influence of the time factor on the efficiency of treatment methods, wells and technologies is not always sufficiently taken into account when carrying out measures to intensify oil flow rate and reduce the water cut;
- the extent and nature of the influence of the time factor on different performance indicators when using the analogy method is rarely taken into account;
- the different influence of the time factor on the efficiency forecast in different geological-physical conditions of deposits is poorly taken into account.

2. Methods and materials

To a large extent it concerns deposits of the Vasyugan suite in terrigenous reservoirs of the North Vartovsky monocline with hard-to-recover reserves of oil where the bottomhole treatment with the use of hydrochloric acid preventing emulsification (KSPEO) and the modifying additives were intensively used.

For the purpose of increasing the efficiency of treatment planning with the use of this technology in real time taking into account the time factor, the field experience of this type of oil production intensification with creation of models and their analysis was summarized.

The absolute ($Э_1$) and relative ($Э_2$) gains of an output of oil, absolute ($Э_3$) and relative ($Э_4$) of decrease in water content of products, the general gain of production during effect ($Э_5$), the complex parameter of efficiency ($Э_6$) were used as response functions.

The parameters characterizing the following were used as independent variables:

- geological-physical properties of formations in a well: general ($H_{общ}$, m), effective ($H_э$, m), effective oil-saturated ($H_{эн}$, m) formation thickness, sandiness ($K_п$), compartmentalization (K_p), open porosity (m), oil saturation ($K_н$); permeability coefficients ($K_{прон}$, 10^{-3} μm^2), induced potential logging ($\alpha_{сп}$); readings in the formation according to lateral logging ($M_{БК}$, Ohm·m), SP amplitudes ($A_{ПС}$, Ohm·m), on GP1 (A0.4M0.1N) ($A_{0.4}$, Ohm·m), GP2 (A1.0M0.1N) ($A_{1.0}$, Ohm·m), GP3 (A2.0 M0.5N) ($A_{2.0}$, Ohm·m), GP4 (A4.0M0.5N) ($A_{4.0}$, Ohm·m), PZ (A0.5M6.0N) ($A_{0.5}$, Ohm·m), gamma logging (γ, μR/hr), big probe TNPH ($\beta'_{ННКТ}$), the small probe TNPH ($\beta''_{ННКТ}$), induction probe ($\beta_{ИЗ}$, sym); specific electrical resistance of the probe by IL ($\rho_{ИК}$, Om• m), LL ($\rho_{БК}$, Ohm·m), electric logging probe complex ($A_{комп}$, Ohm·m), penetration zones by electric logging probe complex ($A^{зп}_{комп}$, Ohm·m), relation of diameter of a zone of penetration of a washing fluid filtrate to well diameter ($d_{зп}$ / $d_С$). The values of parameters were used both on a section (X_i) in general, and on its perforated part ($\overline{X}_i$);
- time factor including process parameters of wells and deposits: time (t, month), maximum oil flowrate (Q_{Hmax}, t/month) since the beginning of well operation until treatment; flowrate (Q_{H1}, t/month), water cut (f_1, %), cumulative oil production ($Q_{нак}$, t) at the time of treatment; initial oil flowrate ($Q_{нач}$, t/month);

- treatment technology: consumption of 22 % hydrochloric acid (HCl, t); consumption of modifiers and hydrofluoboric acid [MK-V(K), t], volume of pumped solution (V, m^3), treatment frequency (*N*).

Using multivariate regression analysis, the models of response function dependencies on considered parameters are obtained in three options:

- 1 – when using the whole set of parameters;
- 2 – without using parameters characterizing the time factor;
- 3 – when using parameters characterizing the time factor and treatment technology.

3. Results

The analysis of models presented in Tables 1–3, the values of multiple correlation coefficients (*R*) and data on the percentage of contribution of parameters included in the models to general variability of efficiency parameters (Table 4) shows that the influence of parameters characterizing the time factor is more significant in comparison with parameters characterizing the features of geological structure of formations. If using the whole set of parameters (option 1), and the average percentage of their contribution is 52, then in the absence of parameters characterizing geology, this percentage decreases to 41, and in the absence of parameters characterizing the time factor – to 31. It should be added that earlier similar studies using acid compositions injection in the same geological and field conditions with the addition of Aldinol-20 [21] showed a different scenario. The influence of parameters characterizing geological-physical properties of formations is more significant than the influence of parameters characterizing the time factor. However, the impact of the time factor remains very significant, as with the use of KSPEO (Table 4).

Table 1. KSPEO efficiency forecast models (option 1)

$Э_1 =$	$282.032 - 112.468K_{\text{п}} + 8.373K_{\text{р}} + 4.633A_{\text{ПС}} - 6.746\overline{H}_{\text{эн}} + 14.533\overline{K}_{\text{п}} - 349.155\overline{m} - 68.856\overline{\alpha}_{\text{сп}} + 12.293\overline{\rho}_{\text{ИК}} - - 4.138\overline{A}_{0.4} + 149.33\overline{d}_{\text{зп}}/\overline{d}_{c} + 12.87\overline{\beta}'_{\text{ННКТ}} + + 0.007Q_{\text{нак}} - 0.485t + 0.042Q_{max} - 1.602f_1 - 0.581Q_{H1} - - 12.94HCl - 110.184V + 52.81N - 39.924V/\overline{H}_{\text{эн}} + 1157.887\text{МК-В(К)} - 234.295HCl/V + + 282.976HCl/\overline{H}_{\text{эн}} - 5607.338\text{МК-В(К)}/V - 1110.68\text{МК-В(К)}/\overline{H}_{\text{эн}}$ (R = 0.714)
$Э_2 =$	$42.565 - 9.616K_{\text{п}} - 1.997K_{\text{р}} - 0.349A_{\text{ПС}} - 0.443\overline{H}_{\text{эн}} - 3.803\overline{K}_{\text{п}} + 79.773\overline{m} + 1.191\overline{\alpha}_{\text{сп}} + 0.759\overline{\rho}_{\text{ИК}} - 1.688\overline{A}_{0.4} - 17.395\overline{d}_{\text{зп}}/\overline{d}_{c} - 3.275\overline{\beta}'_{\text{ННКТ}} - 0.001Q_{\text{нак}} + 0.018t + 0.003Q_{max} - 0.467f_1 - 0.028Q_{H1} + 5.656HCl + 8.59V + 4.438N - 1.575V/\overline{H}_{\text{эн}} - 110.976\text{МК-В(К)} + 16.592HCl/V - 15.173HCl/\overline{H}_{\text{эн}} + 676.27\text{МК-В(К)}/V + 37.919\text{МК-В(К)}/\overline{H}_{\text{эн}}$ (R = 0.666)
$Э_3 =$	$390.774 - 40.495K_{\text{п}} + 1.473K_{\text{р}} + 1.701A_{\text{ПС}} - 1.188\overline{H}_{\text{эн}} + 0.758\overline{K}_{\text{п}} - 606.928\overline{m} - 74.817\overline{\alpha}_{\text{сп}} + 1.265\overline{\rho}_{\text{ИК}} - - 2.413\overline{A}_{0.4} + 10.463\overline{d}_{\text{зп}}/\overline{d}_{c} + 1.938\overline{\beta}'_{\text{ННКТ}} + 0.002Q_{\text{нак}} - 0.257t + 0.02Q_{max} + 0.296f_1 - 0.136Q_{H1} - - 12.94HCl - 56.913V + 32.37N - 2.697V/\overline{H}_{\text{эн}} + 624.804\text{МК-В(К)} - 64.984HCl/V + 144.686HCl/\overline{H}_{\text{эн}} - - 2883.951\text{МК-В(К)}/V - 703.162\text{МК-В(К)}/\overline{H}_{\text{эн}}$ (R = 0.732)
$Э_4 =$	$1692.192 - 674.811K_{\text{п}} + 50.238K_{\text{р}} + 27.796A_{\text{ПС}} - 40.475\overline{H}_{\text{эн}} + 87.199\overline{K}_{\text{п}} - 2094.93\overline{m} - 413.138\overline{\alpha}_{\text{сп}} + + 73.759\overline{\rho}_{\text{ИК}} - 24.828\overline{A}_{0.4} + 895.981\overline{d}_{\text{зп}}/\overline{d}_{c} + 77.222\overline{\beta}'_{\text{ННКТ}} + 0.039Q_{\text{нак}} - 2.908t + 0.254Q_{max} - 9.614f_1 - - 3.487Q_{H1} - 77.638HCl - 661.106V + 316.859N - 239.546V/\overline{H}_{\text{эн}} + 6947.321\text{МК-В(К)} - 1405.772HCl/V + + 1697.856HCl/\overline{H}_{\text{эн}} - 33644.026\text{МК-В(К)}/V - 6664.082\text{МК-В(К)}/\overline{H}_{\text{эн}}$ (R = 0.714)
$Э_5 =$	$336.536 - 156.553K_{\text{п}} - 9.289K_{\text{р}} - 0.364A_{\text{ПС}} - 4.792\overline{H}_{\text{эн}} - 28.292\overline{K}_{\text{п}} + 1322.526\overline{m} - 122.53\overline{\alpha}_{\text{сп}} + + 12.068\overline{\rho}_{\text{ИК}} - 6.794\overline{A}_{0.4} + 9.822\overline{d}_{\text{зп}}/\overline{d}_{c} - 14.039\overline{\beta}'_{\text{ННКТ}} - - 0.002Q_{\text{нак}} + 0.003t + 0.056Q_{max} - 3.659f_1 - - 0.13Q_{H1} + 91.917HCl - 50.348V + 47.276N + 110.134V/\overline{H}_{\text{эн}} + 83.745\text{МК-В(К)} - 319.119HCl/V - - 159.007HCl/\overline{H}_{\text{эн}} + 628.325\text{МК-В(К)}/V - 563.851\text{МК-В(К)}/\overline{H}_{\text{эн}}$ (R = 0.750)
$Э_6 =$	$48.697 + 6.787K_{\text{п}} + 0.49K_{\text{р}} + 0.034A_{\text{ПС}} + 0.718\overline{H}_{\text{эн}} - 1.436\overline{K}_{\text{п}} - 61.706\overline{m} - 6.067\overline{\alpha}_{\text{сп}} - 0.442\overline{\rho}_{\text{ИК}} + + 0.096\overline{A}_{0.4} + 0.096\overline{d}_{\text{зп}}/\overline{d}_{c} + 0.897\overline{\beta}'_{\text{ННКТ}} - 0.034t - 0.004Q_{max} + 0.166f_1 - 0.002Q_{H1} - 1.24HCl - 5.9V - - 0.407N + 0.807V/\overline{H}_{\text{эн}} + 61.902\text{МК-В(К)} - 19.433HCl/V + 13.491HCl/\overline{H}_{\text{эн}} - 326.073\text{МК-В(К)}/V - - 62.943\text{МК-В(К)}/\overline{H}_{\text{эн}}$ (R = 0.760)

Relatively high values of multiple correlation coefficients allow proposing the obtained models to forecast the impact efficiency, wells and technologies in conditions of identification series deposits both at the stage of deposits introduction into development (option 3) and in case of the absence of representative geological-geophysical material (option 2). It should also be noted that the role of the time factor in evaluating different performance indicators is also different.

Table 2. KSPEO efficiency forecast models (option 2)

$Э_1 =$	$135.258 - 50.362K_{\text{п}} + 0.66K_{\text{р}} + 2.464A_{\text{ПС}} + 19.236\overline{H}_{\text{эн}} - 100.214\overline{K}_{\text{п}} + 335.695\overline{m} - 215.091\overline{\alpha}_{\text{сп}} + 15.047\overline{\rho}_{\text{ИК}} - 4.811\overline{A}_{0.4} + 192.54\overline{d}_{\text{зп}}/\overline{d}_{\text{с}} + 14.398\overline{\beta}'_{\text{ННКТ}} - - 1.807HCl - 106.205V + 44.039N - 36.046V/\overline{H}_{\text{эн}} + 1003.897\text{МК-В(К)} - 322.976HCl/V + 253.506HCl/\overline{H}_{\text{эн}} - 5687.843\text{МК-В(К)}/V - - 582.092\text{МК-В(К)}/\overline{H}_{\text{эн}}$ (R = 0.514)
$Э_2 =$	$48.323 - 16.565K_{\text{п}} - 2.018K_{\text{р}} - 0.329A_{\text{ПС}} - 1.076\overline{H}_{\text{эн}} + 5.459\overline{K}_{\text{п}} + 60.612\overline{m} - 7.414\overline{\alpha}_{\text{сп}} - 0.658\overline{\rho}_{\text{ИК}} - - 1.218\overline{A}_{0.4} - 40.661\overline{d}_{\text{зп}}/\overline{d}_{\text{с}} - 3.334\overline{\beta}'_{\text{ННКТ}} + 0.362HCl + 10.887V + 3.388N + 14.978V/\overline{H}_{\text{эн}} - 105.189\text{МК-В(К)} + + 83.722HCl/V - 40.325HCl/\overline{H}_{\text{эн}} + 712.229\text{МК-В(К)}/V - 26.012\text{МК-В(К)}/\overline{H}_{\text{эн}}$ (R = 0.561)
$Э_3 =$	$302.52 - 5.592K_{\text{п}} + 1.017K_{\text{р}} + 1.058A_{\text{ПС}} + 6.836\overline{H}_{\text{эн}} - 34.452\overline{K}_{\text{п}} - 338.942\overline{m} - 86.505\overline{\alpha}_{\text{сп}} + 3.743\overline{\rho}_{\text{ИК}} - - 2.906\overline{A}_{0.4} + 55.792\overline{d}_{\text{зп}}/\overline{d}_{\text{с}} + 3.138\overline{\beta}'_{\text{ННКТ}} - 2.506HCl - 62.854V + 31.821N - 12.409V/\overline{H}_{\text{эн}} + + 605.718\text{МК-В(К)} - 184.298HCl/V + 166.358HCl/\overline{H}_{\text{эн}} - 3039.927\text{МК-В(К)}/V - 547.966\text{МК-В(К)}/\overline{H}_{\text{эн}}$ (R = 0.575)
$Э_4 =$	$811.549 - 302.174K_{\text{п}} + 3.96K_{\text{р}} + 14.785A_{\text{ПС}} + 115.417\overline{H}_{\text{эн}} - 601.283\overline{K}_{\text{п}} + 2014.171\overline{m} - 1290.545\overline{\alpha}_{\text{сп}} + + 90.284\overline{\rho}_{\text{ИК}} - 28.866\overline{A}_{0.4} + 1155.238\overline{d}_{\text{зп}}/\overline{d}_{\text{с}} + 86.387\overline{\beta}'_{\text{ННКТ}} - 10.84HCl - 637.232V + 264.236N - - 216.276V/\overline{H}_{\text{эн}} + 6023.38\text{МК-В(К)} - 1937.855HCl/V + 1521.034HCl/\overline{H}_{\text{эн}} - 34127.056\text{МК-В(К)}/V - - 3492.552\text{МК-В(К)}/\overline{H}_{\text{эн}}$ (R = 0.514)
$Э_5 =$	$431.848 - 218.009K_{\text{п}} - 6.738K_{\text{р}} - 1.585A_{\text{ПС}} - 0.586\overline{H}_{\text{эн}} + 18.248\overline{K}_{\text{п}} + 1328.007\overline{m} - 112.014\overline{\alpha}_{\text{сп}} + + 5.237\overline{\rho}_{\text{ИК}} - 2.757\overline{A}_{0.4} - 183.653\overline{d}_{\text{зп}}/\overline{d}_{\text{с}} - 20.767\overline{\beta}'_{\text{ННКТ}} + 55.895HCl - 38.007V + 38.697N + + 258.05V/\overline{H}_{\text{эн}} + 138.803\text{МК-В(К)} + 164.145HCl/V - 354.194HCl/\overline{H}_{\text{эн}} + 830.042\text{МК-В(К)}/V - - 1202.711\text{МК-В(К)}/\overline{H}_{\text{эн}}$ (R = 0.599)
$Э_6 =$	$42.132 + 9.641K_{\text{п}} + 0.322K_{\text{р}} + 0.05A_{\text{ПС}} + 1.313\overline{H}_{\text{эн}} - 4.904\overline{K}_{\text{п}} - 62.345\overline{m} - 5.215\overline{\alpha}_{\text{сп}} - 0.013\overline{\rho}_{\text{ИК}} - - 0.207\overline{A}_{0.4} + 11.367\overline{d}_{\text{зп}}/\overline{d}_{\text{с}} + 0.884\overline{\beta}'_{\text{ННКТ}} + 1.266HCl - - 7.065V - 0.037N - 6.087V/\overline{H}_{\text{эн}} + 59.143\text{МК-В(К)} - - 50.476HCl/V + 23.444HCl/\overline{H}_{\text{эн}} - 350.406\text{МК-В(К)}/V - 23.137\text{МК-В(К)}/\overline{H}_{\text{эн}}$ (R = 0.576)

Table 3. KSPEO efficiency forecast models (option 3)

$Э_1 =$	$288.664 + 0.007Q_{\text{нак}} - 0.638t - 0.009Q_{max} + 0.138f_1 - 0.479Q_{\text{Н1}} - 61.907HCl - 36.46V + 31.968N - - 24.935V/\overline{H}_{\text{эн}} + 633.324\text{МК-В(К)} + 296.039HCl/V + 118.364HCl/\overline{H}_{\text{эн}} - 3357.343\text{МК-В(К)}/V - - 478.708\text{МК-В(К)}/\overline{H}_{\text{эн}}$ (R = 0.617)
$Э_2 =$	$- 21.575 + 0.032t + 0.002Q_{max} - 0.585f_1 - 0.029Q_{\text{Н1}} + 10.427HCl + 4.844V + 4.859N + 0.096V/\overline{H}_{\text{эн}} - - 104.856\text{МК-В(К)} - 23.273HCl/V - 11.342HCl/\overline{H}_{\text{эн}} + 612.429\text{МК-В(К)}/V + 61.553\text{МК-В(К)}/\overline{H}_{\text{эн}}$ (R = 0.564)
$Э_3 =$	$214.275 + 0.003Q_{\text{нак}} - 0.26t - 0.002Q_{max} + 0.583f_1 - 0.128Q_{\text{Н1}} - 26.445HCl - 38.389V + 27.467N - - 14.081V/\overline{H}_{\text{эн}} + 492.946\text{МК-В(К)} + 46.767HCl/V + 123.829HCl/\overline{H}_{\text{эн}} - 2460.441\text{МК-В(К)}/V - - 453.08\text{МК-В(К)}/\overline{H}_{\text{эн}}$ (R = 0.656)
$Э_4 =$	$1731.983 + 0.043Q_{\text{нак}} - 3.826t - 0.056Q_{max} + 0.826f_1 - 2.875Q_{\text{Н1}} - 371.442HCl - 218.761V + 191.806N - 149.611V/\overline{H}_{\text{эн}} + 3799.946\text{МК-В(К)} + 1776.233HCl/V + 710.182HCl/\overline{H}_{\text{эн}} - 20144.057\text{МК-В(К)}/V - - 2872.25\text{МК-В(К)}/\overline{H}_{\text{эн}}$ (R = 0.617)
$Э_5 =$	$96.889 - 0.001Q_{\text{нак}} + 0.103t + 0.038Q_{max} - 3.857f_1 - 0.123Q_{\text{Н1}} + 97.016HCl - 38.202V + 47.066N + + 168.269V/\overline{H}_{\text{эн}} - 121.119\text{МК-В(К)} - 259.326HCl/V - 264.205HCl/\overline{H}_{\text{эн}} + 1630.056\text{МК-В(К)}/V - - 373.33\text{МК-В(К)}/\overline{H}_{\text{эн}}$ (R = 0.675)
$Э_6 =$	$48.514 - 0.034t - 0.004Q_{max} + 0.167f_1 - 0.005Q_{\text{Н1}} - 2.046HCl - 5.559V - 0.291N - 2.938V/\overline{H}_{\text{эн}} + + 66.472\text{МК-В(К)} - 18.688HCl/V + 17.944HCl/\overline{H}_{\text{эн}} - - 351.622\text{МК-В(К)}/V - - 64.134\text{МК-В(К)}/\overline{H}_{\text{эн}}$ (R = 0.708)

Table 4. Percentage of contribution of parameters included in models to overall variability of efficiency parameters

Option	Efficiency parameter					
	$Э_1$	$Э_2$	$Э_3$	$Э_4$	$Э_5$	$Э_6$
1	51	44	54	51	56	57
2	26	31	33	26	36	33
3	38	32	43	38	46	50

4. Conclusion

Thus, the following conclusions can be made from the study of the bottomhole tratment using KSPEO in terrigenous reservoirs of the Vasyugan suite:

- the time factor has a significant influence on the treatment efficiency, much more than the features of the geological structure of an objects;
- in the conditions of different objects and treatment technologies the time factor has different influence, which should be taken into account when using the cumulative experience in other geological conditions;
- when assessing the treatment performance using different performance indicators, the degree and nature of the impact of the time factor is not constant and this should be taken into account in planning;
- the algorithm, which allows reducing the negative influence of the time factor on the efficiency of oil production intensification by application of technological impact indices, is proposed;
- geological-statistical models were obtained, which allow planning the treatment efficiency, selecting wells and technological treatment indicators taking into account the time factor in real time.

5. References

[1] Akhmetov R T, Mukhametshin V V and Kuleshova L S 2019 Grouping of objects using a limited number of parameters characterizing geological and physical properties of layers *Atlantis Highlights in Material Sciences and Technology (Int. Symp., dedicated to the 85th anniversary of H.I. Ibragimov, Engineering and Earth Sciences: Applied and Fundamental Research (ISEES 2019))* vol 1 pp 9–13. DOI: 10.2991/isees-19.2019.3.

[2] Behrmann L, Huber K, McDonald B et al 1996 Quo Vadis, Extreme Overbalance? *Oilfield Rev.* 18–33

[3] Sergeev V V, Belenkova N G, Zeigman Yu V and Mukhametshin V Sh 2017 Physical properties of emulsion systems with SiO_2 nanoparticles *Nanotechnol. in Construct.* **9(6)** 37–64. DOI: 10.15828/2075-8545-2017-9-6-37-64

[4] Zeigman Yu V, Mukhametshin V Sh, Sergeev V V and Kinzyabaev F S 2017 Experimental study of viscosity properties of emulsion system with SiO_2 nanoparticles *Nanotehnol. in Construct.* **9(2)** 16–38. DOI: 10.15828/2075-8545-2017-9-2-16-38

[5] Lestz R S, Clarke J N, Plattner D and Byrd A C 2002 Perforating for Stimulation: An Engineered Solution *SPE Drill. & Complet.* **17(1)** 8. DOI: 10.2118/76812-PA

[6] Mukhametshin V Sh, Zeigman Yu V and Andreev A V 2017 Rapid assessment of deposit production capacity for determination of nanotechnologies application efficiency and necessity to stimulate their development *Nanotechnol. in Construct.* **9(3)** 20–34. DOI: 10.15828/2075-8545-2017-9-3-20-34

[7] Kuleshova L S and Mukhametshin V V 2019 Estimation of the wells hydrodynamic drag level based on wells geophysical survey data *Atlantis Highlights in Material Sciences and Technology (Int. Symp., dedicated to the 85th anniversary of H.I. Ibragimov, Engineering and Earth Sciences: Applied and Fundamental Research (ISEES 2019))* vol 1 pp 727–9. DOI: 10.2991/isees-19.2019.145

[8] Hutton A C 1987 Petrographic classification of oil shales *Int. J. of Coal Geol.* **8(3)** 203–31. DOI: 10.1016/0166-5162(87)90032-2

IOP Conf. Series: Materials Science and Engineering **905** (2020) 012085 doi:10.1088/1757-899X/905/1/012085

[9] Andreev V E, Chizhov A P, Kotenev Yu A et al 2019 Prediction of Gas Exposure in the Conditions of the Oil Fields of the Volga-Urals *Atlantis Highlights in Material Sciences and Technology (Int. Symp., dedicated to the 85th anniversary of H.I. Ibragimov, Engineering and Earth Sciences: Applied and Fundamental Research (ISEES 2019))* vol 1 pp 666–9. DOI: 10.2991/isees-19.2019.132

[10] Khayredinov N Sh, Popov A M and Mukhametshin V Sh 1992 Increasing the flooding efficiency of poor-producing oil deposits in carbonate collectors *Oil industry* **9** 18–20

IOP Conf. Series: Materials Science and Engineering **905** (2020) 012086 doi:10.1088/1757-899X/905/1/012086

Development of oil deposits in tectonic conditions

F S Salimov[1], Yu A Kotenev[2], V Sh Mukhametshin[3], Sh Kh Sultanov[2] and A R Safiullina[3]

[1] Department of Enhanced Oil Recovery Technology, TPE Povhneftegaz, LLC LUKOIL–Western Siberia, Kogalym, Tyumen region, Khanty-Mansi Autonomous Area – Yugra, Russia
[2] Department of Geology and Exploration of Oil and Gas Field, Ufa State Petroleum Technological University, SASI Institute of strategic research of the Republic of Bashkortostan, Ufa, Republic of Bashkortostan, Russia
[3] Department of Oil and Gas Field Exploration and Development, Ufa State Petroleum Technological University, Branch of the University in the City of Oktyabrsky, Oktyabrsky, Republic of Bashkortostan, Russia

E-mail: Farid.Salimov@lukoil.comgeokot@inbox.ru, vsh@of.ugntu.ru, ssultanov@mail.ru, safiullina.a.r@yandex.ru

Abstract. The paper presents the detailed analysis of structural and tectonic structure of the foundation surface and Jurassic deposits of Western Siberia, which allows identifying complex, small-sized geological objects promising for the detection of small oil deposits, clarifying the location of exploration wells, searching for deposits in subvertical destruction zones in deposits of the middle and lower Jurassic, foundation rocks in territories with already discovered overlying oil deposits. The zones of excessive fissuring, connected, among other things, with the zones of horizontal shear, which made it possible to change the expected contours of existing deposits and to reveal perspective excessive fissuring zones.

1. Introduction

One of the most important tasks that the Russian oil industry is facing these days is the creation of a scientific and methodological basis for improving the management efficiency of hard-to-recover oil deposits development [1–10]. Such objects include the Jurassic deposits complicated by tectonic deformations.

The presence of fault block tectonics complicating the structure of Jurassic deposits was determined and proved in many fields of Western Siberia. Some tectonic deformations originate in foundation rocks and penetrates the sedimentary mantle to a different depth, the others have limited propagation within separate suites, and the third are almost not mapped by 3D seismic survey in view of low amplitude or due to formation of separate blocks at horizontal shifts.

In our opinion, one of the tools that can solve this issue is seismodynamic analysis. In Kogalym region it was first used at Povkhovsky field in 2013, using 16 dynamic attributes in the PARADIGM software complex. The zones of excessive fissuring associated with tectonic deformations were determined (Figure 1).

Using the seismodynamic analysis of the wave field the identification and determination of direct connection of extended zones and sections of excessive fissuring, low-amplitude vertical movements, having different area and shape, with disjunctive disorders of foundation rocks, which, in turn, are confined to the boundaries of structural elements of elevated troughs (Figure 2), as well as with horizontal and vertical low-amplitude failure of rocks of lower and middle Jurassic introduced additional important factors in the strategy and principles of exploration drilling, operational drilling, selection of the main types of well interventions, such as sidetracking and hydraulic fracturing, optimization of the development system.

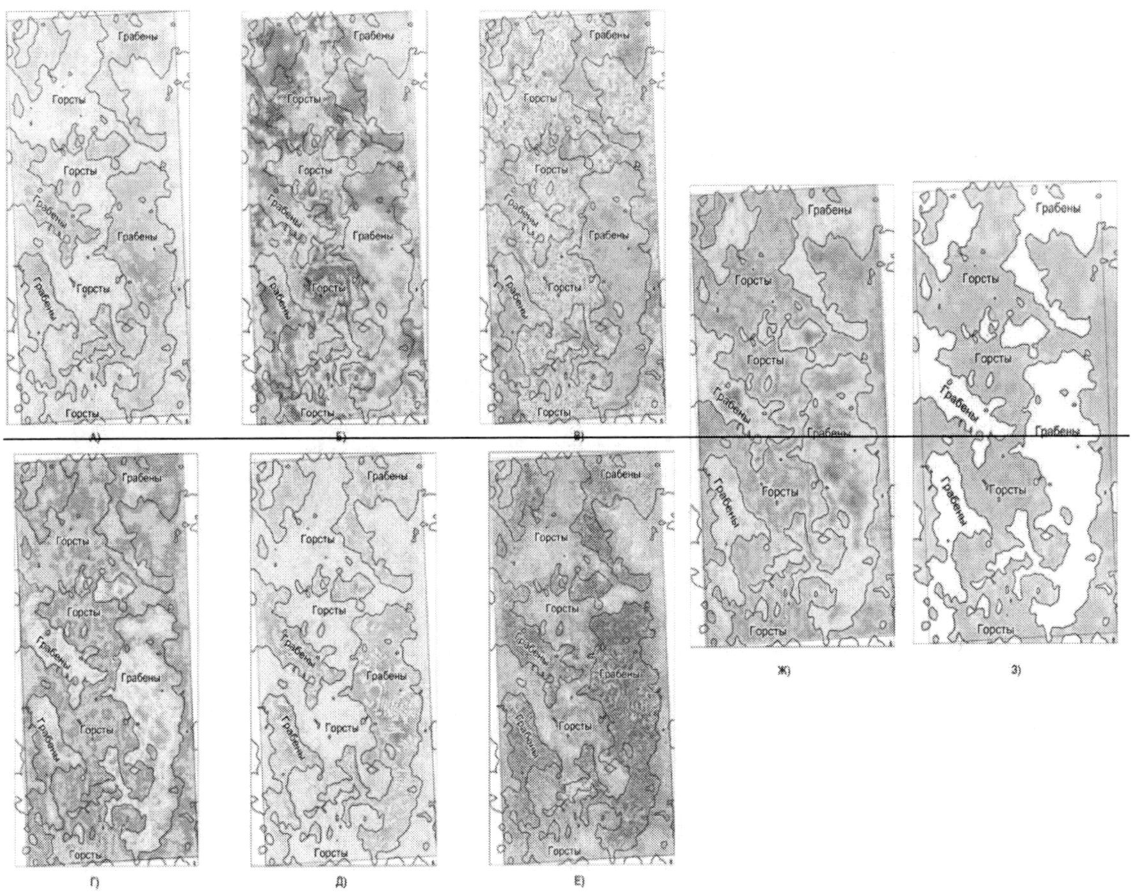

Figure 1. Comparison of results of dynamic analysis of various attributes to identify the tectonic zones in the pre-Jurassic complex: A – value of upper extremum of seismic signal amplitude; B – instantaneous phase of seismic wave signal, instantaneous phase value of the seismic wave signal at the top of the analysis window; C – instantaneous frequency band of the seismic wave, the average of the absolute values of the instantaneous frequency band of the seismic wave; D – envelope of the seismic wave signal, the value of the envelope of the seismic wave signal in the upper part of the seismic horizon; E – amplitude weighted by the instantaneous phase of the seismic wave, the value of the amplitude weighted by the instantaneous phase of the seismic wave in the upper part of the seismic horizon; F – amplitude weighted by the instantaneous phase of the seismic wave, the average value of the amplitude weighted by the instantaneous phase of the seismic wave; G – amplitude weighted by the instantaneous frequency of the seismic wave, upper extremum of the amplitude spectrum weighted by the instantaneous frequency of the seismic wave; H – zones of active tectonics (their boundaries) revealed through the trend analysis of the II order dynamic attribute values

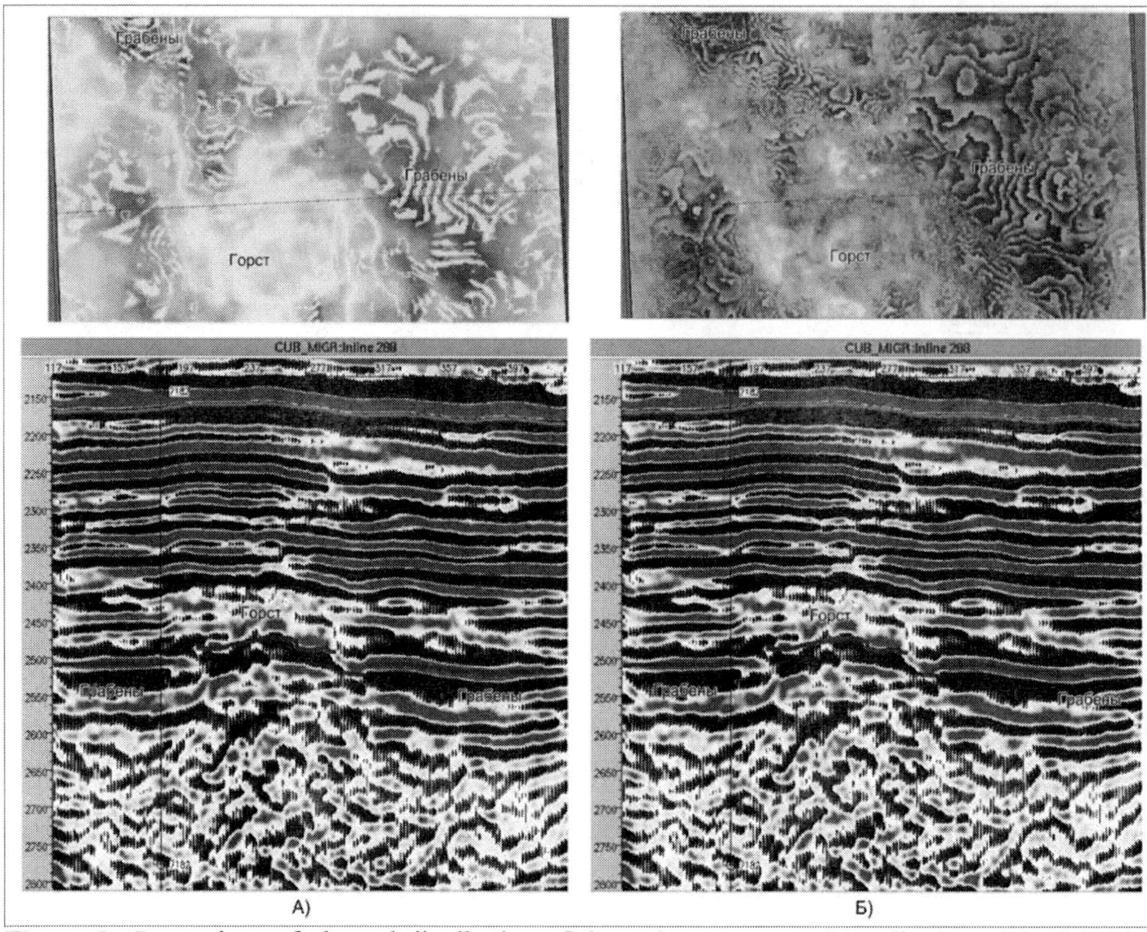

Figure 2. Comparison of planned distribution of dynamic parameters "Amplitude weighted by the instantaneous phase of the seismic wave in the upper part of the seismic horizon" (A) and "Average amplitude weighted by the instantaneous phase of the seismic wave" (B) with time section along the InLine 288

2. Methods and materials

The results of dynamic analysis should be considered in conjunction with seismic survey data with the results of geological and field analysis, geophysical well survey data, downhole logging data and other special surveys, if available.

On the basis of the performed analysis of the West Povkhovsky section, it was developed and recommended to drill perspective edge regions of a developed deposit located in the zones of excessive fissuring and subvertical destruction, sidetracking with horizontal and inclined direction result.

3. Results

The southern part of the deposit, the area of well No. 82P, is partially covered by 3D seismic survey (Figure 3), where the dynamic analysis of the "instantaneous acceleration" attribute recommends sidetracking of 4 wells, 3 of which are horizontal and 2 new production wells.

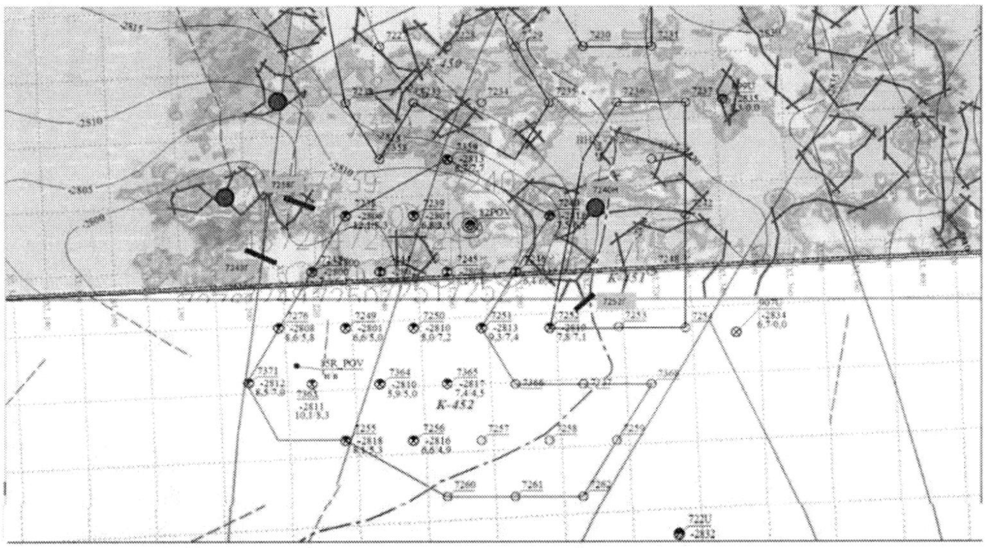

Legends:

- exploration well (well number) absolute depth mark of reservoir top of SE-1 (in meters) net thickness/net oil thickness (in meters)
- production well (well number) absolute depth mark of reservoir top of SE-1 (in meters) net thickness/net oil thickness (in meters)
- design wells
- a well that produced waterless oil influx through SE-1 formation testing
- a well that produced water with oil through SE-1 formation testing
- a well that penetrated SE-1 formation, oil-saturated according to borehole survey
- a well that penetrated SE-1 formation, oil with water according to borehole survey
- a well that penetrated SE-1 formation, water-saturated according to borehole survey
- a well with mud SE-1 formation, according to borehole survey

Disjunctive breaks, excessive fissuring boundaries detected according to trend analysis of the II order of dynamic parameters

Disjunctive breaks detected by OOO KogalymNIPIneft, 2013

Disjunctive breaks detected according to 3D seismic interpretation

Water-oil contact detected according to dynamic analysis

Elevations of SE-1 formation top

Planned location of horizontal second borehole

A well planned for drilling

Figure 3. Mapping of planned measures on the distribution of the maximum absolute value of instantaneous acceleration of the seismic wave signal in the interval of SE-1 formation in the area of well No. 82P of the West Povkhovsky section

The wells for sidetracking were selected with critical current water cut, which made 99 % or more, except for well No. 7238, which is considered for the future in case of water increase (Figure 4).

The main criterion is the location of wells within the zones of tectonic deformations, or not at a long distance, not more than 200 m.

The geological structure of the productive part of the formation on the site is very similar, the sedimentation conditions are identical, the conclusion is made from the analysis of electric logging curves. The capacity of SE-1 reservoir is 9–10 m, porosity –15–16 %, permeability according to geophysical well survey – 16–40 mD, oil saturation – more than 0.6 unit fractions, interpreted as oil in all wells. In the northern part and the eastern part, the oil saturation of edge wells is 0.47–0.5 unit fractions (Figures 5, 6). Under these parameters we have a significant difference in fluid and oil production, which directly depends on the affiliation with excessive fissuring zones.

IOP Conf. Series: Materials Science and Engineering **905** (2020) 012086 doi:10.1088/1757-899X/905/1/012086

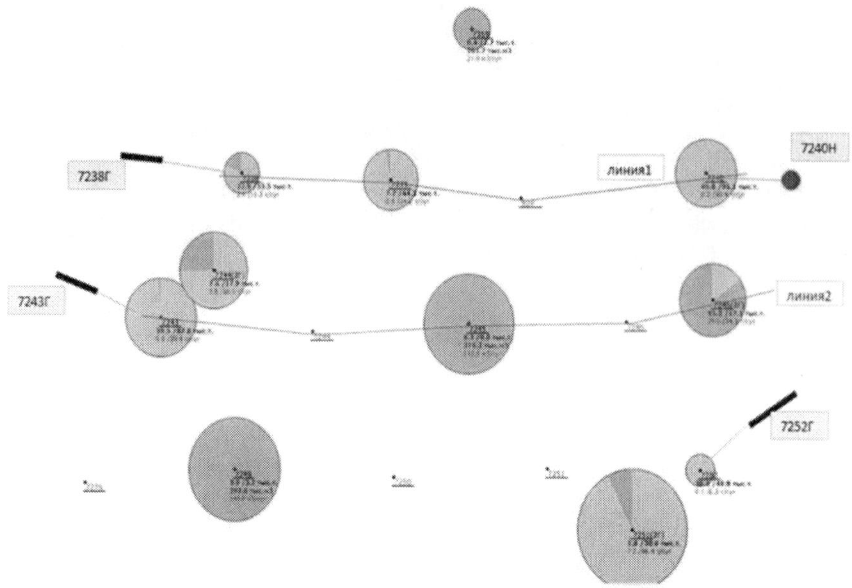

Legends:

- Development idle well
- Injection idle well
- Operating development well
- Operating injection well
- Planned location of horizontal section of the second borehole
- A well planned for drilling
- A well planned for drilling dependent on the previous
- A latral well planned for drilling

Figure 4. Extract of SE-1 formation development map, well area No. 82P

For example, the edge well No. 7240 located on the eastern side of the site in the zone of tectonic disturbances, was put to operation after drilling with fluid flow rate of 60 m^3/day, oil flow rate – 57 t/day without hydraulic fracturing. The neighboring wells located in relatively calm zones were put to operation with lower fluid flow rates – 2-3 times, which indicates the worst reservoir properties due to lower permeability directly related to the intensity of rock fracturing. Well No. 7246 – fluid flow rate $Q_ж$ – 20 m^3/day, 5 % water, dynamic level $H_д$ – 1600 m, developed after drilling without hydraulic fracturing, rapid decrease of inflow was observed.

A good example of the influence of the zone of tectonic deformations fixed on temporary seismic sections is the location of well No. 7240 (Figure 7). From the foundation rocks to SE-1 basement there is a discontinuous fault, some operating faults fade in the deposits of the middle Jurassic.

According to the results of the dynamic analysis, it was recommended to conduct sidetracking with horizontal well No. 7246G eastwards with a distance of 200 m from the mother hole and opening of productive rocks in the zone of excessive fissuring. The well was drilled and launched with a flow rate of 35 m^3/day of clean oil, a dynamic level at the well mouth. Waterless operation during 1.5 years of this well, 100 % water cut of adjacent production wells can indicate the presence of hydrocarbons in this area of faults.

In the eastern part of the section, after watering of well No. 7244 up to 100 % into the zone of detected excessive fissuring, the second horizontal borehole was drilled in the north-west direction between wells No. 7238, 7243. The flow rate was 60 m^3/day, 34 % water, oil flow rate – 35 t/day, $H_д$ – 580 m, in 5 months the water cut began to grow. The location of the zone of tectonic deformations was also confirmed here. In this area it is proposed to drill two lateral horizontal holes: No. 7243G with a distance of 200 m from the mother hole in the western direction, No. 7238G with a distance of 200 m in the azimuth of 290°, as well as to drill two new wells in the zones of fault tectonics development (Figure 3).

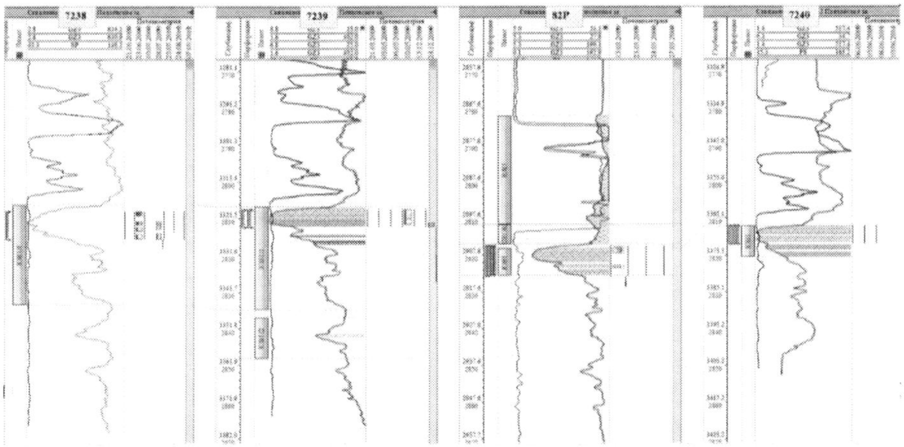

Figure 5. Geological section of formation SE-1 along line 1 of the West Povkhovsky section

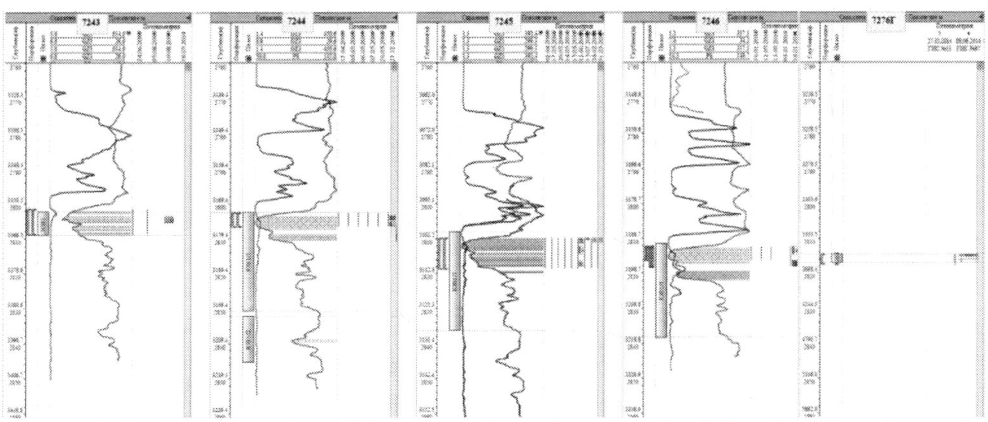

Figure 6. Geological section of formation SE-1 along line 2 of West Povkhovsky section

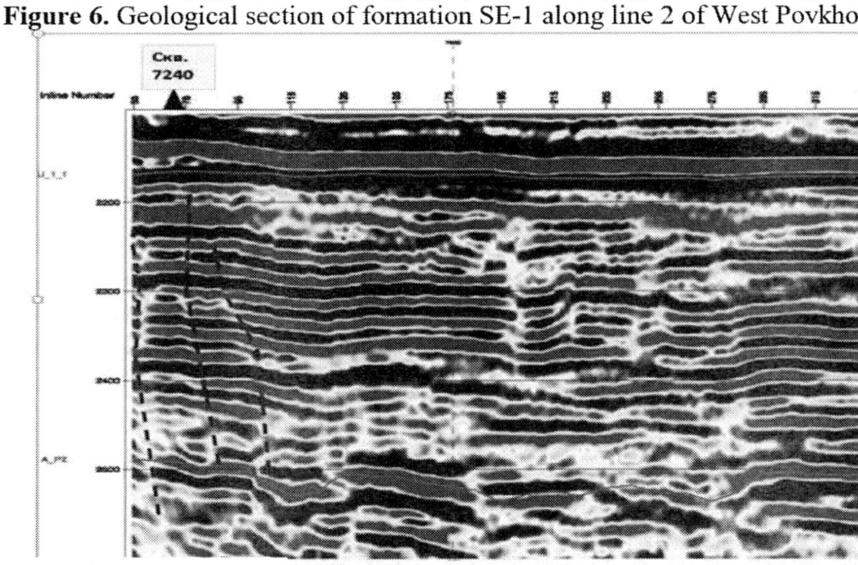

Legends:

Discontinuous faults identified in seismic section

Drilled well

Figure 7. Tectonic deformations in the area of well No. 7240 according to 3D seismic survey

It is planned to drill 12 new operational wells, 7 of which are dependent, 2 directional second wellbores (well No. 7214H, 7355H), 1 second horizontal wellbore (well No. 7354G) (Figure 8) near well No. 83P.

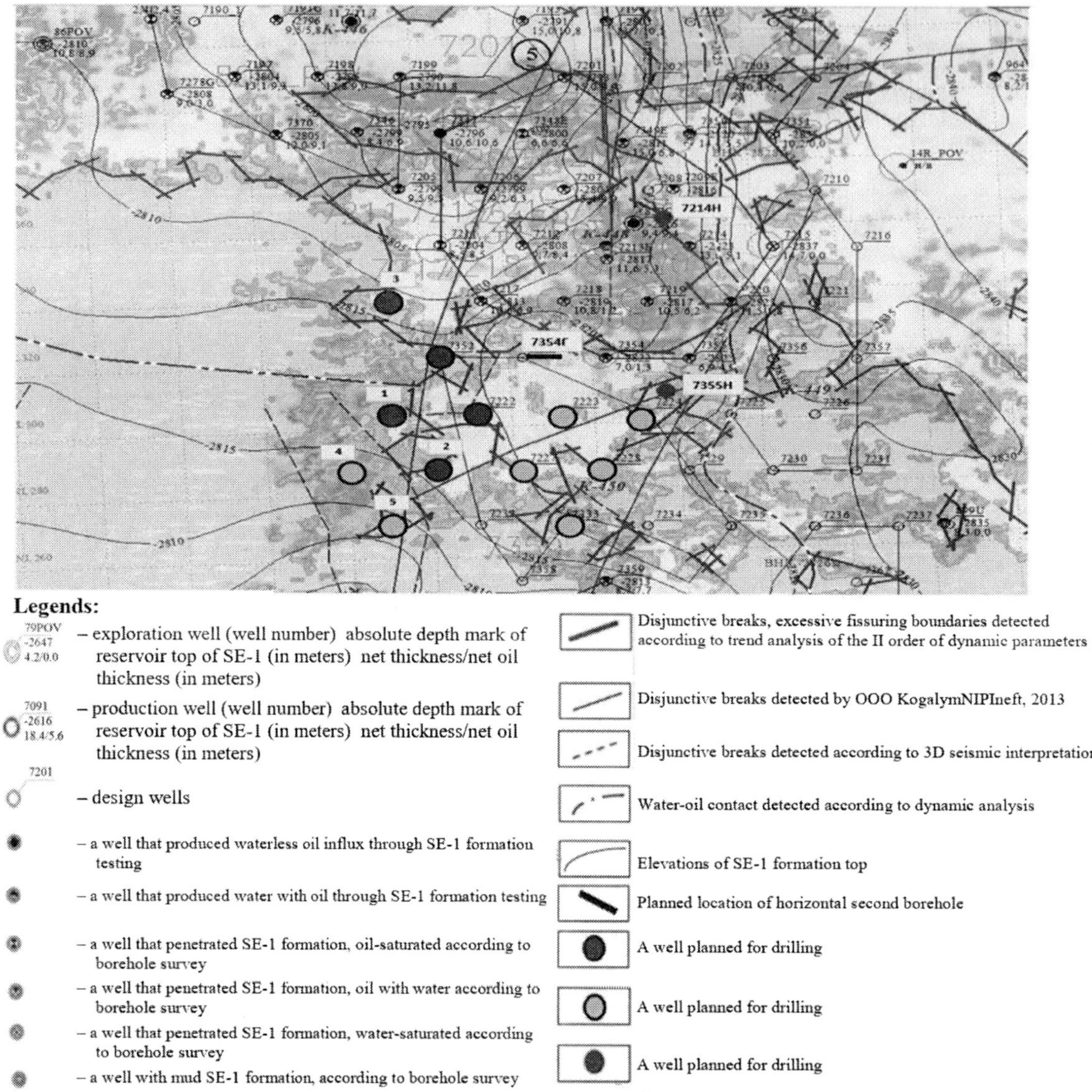

Figure 8. Mapping of planned measures on the distribution of the maximum absolute value of instantaneous acceleration of the seismic wave signal in the interval of SE-1 formation deposits in the area of well No. 83P of West Povkhovsky section

The dependence of well flow rates on the distance relative to tectonic deformations is observed in almost all fields of Western Siberia up to deposits of Aptian-Albian-Cenomanian complex of gas deposits of northern regions (Yamburgskow, West Tarkosalinskoe, Kharampurovskoe fields). Summarizing the available data on the development of oil deposits of the Upper Jurassic, the maximum distance of fault impact on well production was determined, which made 1.5 km. At West Povkhovsky section at a distance of 500 m from the detected fracture zone the well flow rate was on average 50 % lower, at a distance of 1.5 km the decrease reached up to 4–6 times.

The detection of elevated trough systems, faults and overlap folds, different disjunctive breaks on the surface of the foundation and inherited structures in the Jurassic sedimentary mantle allows approaching the geological exploration and planning operational drilling more reasonably. This approach reduces the share of wasteful expenditures, which is important at the stage of field development to obtain large volumes of hydrocarbons, to develop subsoil more rationally.

This approach is particularly relevant in areas with developed infrastructure, where drilling and development of small areas is accompanied by minimal capital costs, the access to which is not so problematic.

4. Conclusion

The detailed analysis of structural and tectonic structure of the foundation surface and Jurassic deposits of Western Siberia allows identifying complex, small-sized geological objects promising for the detection of small oil deposits, clarifying the location of exploration wells, searching for deposits in subvertical destruction zones in deposits of the middle and lower Jurassic, foundation rocks in territories with already discovered overlying oil deposits. The zones of excessive fissuring, connected, among other things, with the zones of horizontal shear, which made it possible to change the expected contours of existing deposits and to reveal perspective excessive fissuring zones.

References

[1] Rogachev M K and Mukhametshin V V 2018 Control and regulation of the hydrochloric acid treatment of the bottomhole zone based on field-geological data *J. of Mining Inst.* **231** 275–80. DOI: 10.25515/PMI.2018.3.275

[2] Andreev V E, Chizhov A P, Kotenev Yu A. et al 2019 Prediction of Gas Exposure in the Conditions of the Oil Fields of the Volga-Urals *Atlantis Highlights in Material Sciences and Technology (Int. Symp., dedicated to the 85th anniversary of H.I. Ibragimov, Engineering and Earth Sciences: Applied and Fundamental Research (ISEES 2019))* vol 1 pp 666–9. DOI: 10.2991/isees-19.2019.132

[3] Hartman R L, Lecerf B, Frenier W W, Ziauddin M E and Fogler H S 2006 Acid-Sensitive Aluminosilicates: Dissolution Kinetics and Fluid Selection for Matrix-Stimulation Treatments *SPE Product. & Operat.* **21(2)** 11. DOI: 10.2118/82267-PA

[4] Akhmetov R T and Mukhametshin V V 2018 Range of application of the Brooks-Corey model for approximation of capillary curves in reservoirs of Western Siberia *Advan. in Engineer. Res. (AER) Int. Conf. Actual Issues of Mechan. Engineer. (AIME 2018)* vol 157 pp 5–8. DOI: 10.2991/aime-18.2018.2

[5] Zakaria A S and Nasr-El-Din H A 2016 A Novel Polymer-Assisted Emulsified-Acid System Improves the Efficiency of Carbonate Matrix Acidizing *SPE J.* **21(3)** 14. DOI: 10.2118/173711-PA

[6] Mukhametshin V V and Andreev V E 2018 Increasing the efficiency of assessing the performance of techniques aimed at expanding the use of resource potential of oilfields with hard-to-recover reserves *Bull. of the Tomsk Polytech. Univer. Geo Assets Engineer.* **329(8)** 30–6

[7] Lestz R S, Clarke J N, Plattner D and Byrd A C 2002 Perforating for Stimulation: An Engineered Solution *SPE Drill. & Complet.* **17(1)** 8. DOI: 10.2118/76812-PA

[8] Mukhametshin V V 2018 Rationale for trends in increasing oil reserves depletion in Western Siberia cretaceous deposits based on targets identification *Bull. of the Tomsk Polytech. Univer. Geo Assets Engineer.* **329(5)** 117–24

[9] Khayredinov N Sh, Popov A M and Mukhametshin V Sh 1992Increasing the flooding efficiency of poor-producing oil deposits in carbonate collectors *Oil industry* **9** 18–20

[10] Kuleshova L S, Kadyrov R R, Mukhametshin V V and Akhmetov R T 2019 Auxiliary equipment for downhole fittings of injection wells and water supply lines used to improve their performance in winter *IOP Conf. Ser. Mater. Sci. and Engineer.* **560(1)** 1–6. DOI: 10.1088/1757-899X/560/1/012071

IOP Conf. Series: Materials Science and Engineering **905** (2020) 012087 doi:10.1088/1757-899X/905/1/012087

Assessment of professional risks in the operation of mud pumps

R R Shangareev

Department of Mechanics and Machine Building Technology, Ufa State Petroleum Technological University, Branch of the University in the City of Oktyabrsky, Oktyabrsky, Republic of Bashkortostan, Russia

E-mail: shagu64@mail.ru

Abstract. Assessment of professional risks in the performance of basic technological operations in drilling oil and gas wells is today an important scientific and practical task in the field of labor protection. The article discusses the main negative production factors that lead to damage to the health of workers during the operation of the mud pump. To exclude injuries, employees should have information about the main occupational risks in carrying out technological operations and have the competence in labor protection and production safety. The conclusion is drawn on the need for further scientific research in the field of analysis and assessment of professional risks in the drilling of oil and gas wells.

1. Introduction

The decline in occupational injuries because of accidents and accidents has long been one of the pressing problems throughout the world. Statistics show that in Russia in recent years there has been a steady trend towards a decrease in the number of cases of industrial injuries at enterprises (Figure 1).

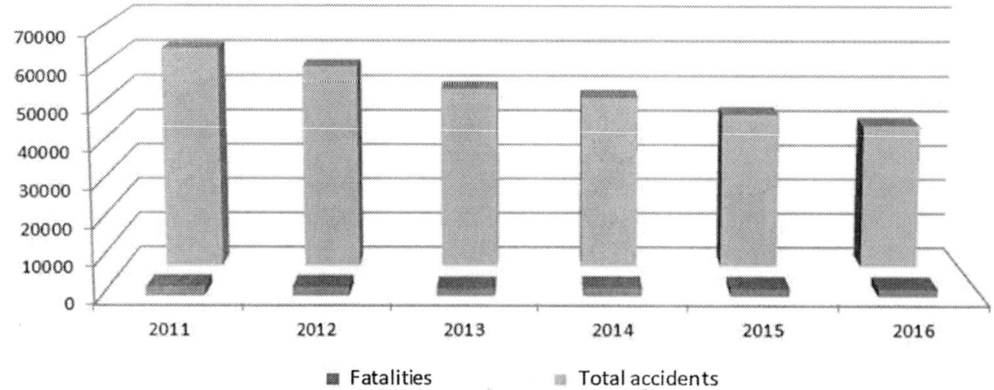

Figure 1. The dynamics of changes in injuries in Russia (according to the Ministry of Labor of the Russian Federation)

The main reasons for this are the mechanization of production, the development of new safe technologies, innovative solutions in the field of PPE. However, the proportion of accidents with severe and fatal outcomes remains high; the Russian economy loses 1.5 trillion rubles annually due to losses caused by occupational diseases, injuries and compensation [1].

To reduce the number of accidents and occupational diseases and, as a result, ensure production stability and financial efficiency of enterprises, it is necessary to search for new approaches and effective solutions in the field of labor protection. In the human-machine system, it is the individual who is the weak link, therefore, occupational risk management as a system for identifying, assessing, eliminating or minimizing production hazards is an important component in the formation of new approaches to improving the level of labor protection and safety culture in the enterprise, as it determines the management approach and personnel to preserve life and health and other values of the organization [2].

The results of investigations of most accidents tell us that their main reason is the inability of employees to anticipate, eliminate or minimize occupational risks in advance, although it is always easier to prevent an accident than to deal with the consequences. A common problem with standard professional risk management systems is the complexity of the methodology and isolation from reality, the lack of clear and effective methods that involve each employee in the process. Local staff are more aware than many managers of the possible dangers in the workplace, and what could lead to a failure in the performance of a particular technological operation.

The practice of developed countries shows that technological modernization of the economy alone is not able to effectively ensure industrial safety if, in the "man-machine" system, the human factor from the point of view of forecasting interaction remains a weak link. To solve problems related to the human factor, it is necessary to teach and, most importantly, motivate the employee to anticipate dangerous situations or working conditions analyze professional risks and prevent them in a timely manner. A special role in the formation of a conscious safety culture for a motivated transition from an attitude to labor protection "we were ordered – we do to complete the assignment on time" to the attitude "we do this because we need it to maintain life, health and receive decent wages" assigned to everyone in their workplace.

The most typical reason for intentional violations of labor protection requirements is the desire to achieve any benefits (make work easier, faster and easier). Despite the possible punishment, under such conditions, the employee will continue to violate until they cease to be a source of benefits. In such cases, for the effective application of the incentive mechanism for the labor protection worker, it should be done so that the costs of violating safety rules exceed the benefits received from this. Then it will be unprofitable to break them. At the same time, it is most expedient and effective to use not negative (punishment), but positive and individual (based on the motivation mechanism of the professional risk management system) incentives for safe work, which, according to international experience, is the most effective way to increase the level of labor protection. The safe behavior of the employee must be conscious, while his qualifications (knowledge), experience (skills) and skills that determine competence are the foundation of the labor protection system in the organization [3].

2. Assessment of professional risks during operation of drilling pumps

Technological safety is a way of managing facilities throughout their entire life cycle in order to prevent the inadvertent exit of hazardous materials or energy from the process. Process Safety Management (PSM) is part of the organization's occupational risk management system, giving the employee in-depth understanding and structured tools to prevent and manage hazards. Failures of equipment, deviations in technological processes can lead to tragic consequences associated with human lives and significant financial losses of the enterprise. A proactive approach to identifying and accounting for occupational risks associated with technological safety, the use of an advanced set of methods, approaches and tools for managing all types of processes that affect the safety of the enterprise, which will prevent the implementation of negative events (including design stages, the use of appropriate technical and operational standards and "barriers").

Assessment of professional risks when performing basic technological operations in drilling oil and gas wells is today an important scientific and practical task in the field of labor protection. According to the results of studies, working conditions for 96 % of the jobs of personnel servicing drilling rigs are harmful or dangerous [4]. The ability to identify hazards at any stage of the production process, determine their characteristics, determine the intensity and duration of exposure, the cause of the occurrence and the possible outcome of the negative impact, which will minimize the risk of accidents, is of great importance today to ensure the safety of personnel working on the rig .

Mud pumps are designed to pump flushing fluid into the well through the drill string and to remove the rock destroyed by the bit from the well [5]. In the vast majority of cases, a clay solution is used as a flushing fluid, the injection of which is carried out using piston mud pumps. The main hydraulic quantities – the necessary supply and pressure – determine the choice of type (brand) of mud pumps and their number. In turbine drilling, in addition to cleaning the bottom, the energy of the flushing fluid is used to rotate the turbodrill and the bit fixed on its shaft. When operating mud pumps, hazardous and harmful production factors can affect workers, which are divided into physical, chemical, psychophysiological and biological factors that create occupational hazards for working personnel. The largest number of accidents occurs in the following main groups of negative factors.

The operation of piston pumps, due to their kinematic features, is associated with uneven supply of fluid and the resulting pressures, which leads to shocks in the discharge line, significant structural vibrations, weakening and rupture of the joints and discharge line. An increase in pressure above the permissible level leads to rupture of the connections or the hydraulic part of the pump – cylinders, air cap, discharge line, valves, expansion joints, and mud hose. When workers are in the danger zone, such accidents can be accompanied by severe or fatal accidents [6].

In connection with the pulsation of liquid in the pipeline, instrumentation (in particular, pressure gauges) prematurely wears out, which affects the accuracy of its readings. Faultless operation of pressure gauges (there should be pressure gauges in the piping of the pumps: on the riser

- to monitor the pressure of the pumped fluid during drilling and in the pumping room
- to monitor the pressure change during the start-up of each pump and the operation of paired pumps in the system) is one of the decisive conditions safety when operating mud pumps.

The causes of pulsations and, as a consequence, increased pressure can be:

- wear of the working surfaces of the valve couples of the pump, the working surfaces of the pistons or pump sleeves or their unreliable reinforcement, the presence of a gap between them due to the ingress of a foreign object, a broken valve spring;
- a large number of gate valves on the discharge line, creating local resistance to the moving fluid flow, with the presence of sharp (90° or more) turns and contractions;
- failure or contamination of the downhole motor and bit, as well as the presence of sharp narrowing in the wellbore;
- overlapping channels through which flushing fluid circulates with ice plugs, cuttings, and foreign objects;
- fishing work;
- forcing the solution in the initial period of circulation;
- start-up of the pump with closed starting or passage valves, or untimely closing of the starting valve or start of the second pump until the circulation in the pump-well system and others is restored (human factor).

In order to suppress pulsations and equalize the flow of the drilling fluid in the injection line, pneumatic compensators are used in modern designs of mud pumps. When the pressure in the discharge line of the pump is greater than the pressure inside the gas, the pneumatic compensator works like an air cap. When carrying out the procedure for filling pneumatic compensators with air or inert gas, the possibility of oil and other combustible substances entering the cavity of the compensators should be excluded [7].

The discharge line from the riser to the pump outflow should be as rectilinear as possible or without sharp turns, should have slopes to ensure that the mud flows from the entire line when the pumps are stopped. This is especially important in areas with a cold climate, since freezing of the solution remaining in the line can lead to its rupture when starting the pumps. For safety reasons, to prevent breaks in the discharge pipe or pump when the pressure is higher than the permissible near the pump outlet, a plate-type safety device is mounted in the factory. The safety plate should only be a standard sample with a working pressure digit stamped on the peripheral section. Its groove should be directed towards the discharge of the solution. The drainpipe from the safety device is sent to the receiving tank. Such an installation eliminates the contamination of workplaces in the pump room and injuries to workers from a direct impact by a stream of clay solution or flying fragments of the plate during its destruction [10].

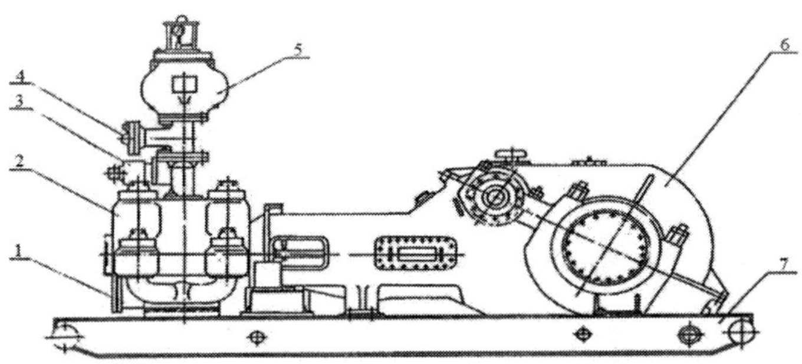

Figure 2. Scheme of the UNB-600 mud pump with safety devices: 1 – receiving box; 2 – hydraulic box; 3 – safety valve; 4 – discharge tee; 5 – compensator; 6 – drive part; 7 – bed

High-quality cleaning of washing liquid with the help of modern cleaning devices, closed external circulation system, installation of receiving tanks and maintaining the liquid level in them so that it does not fall below the axis of the piston bushings of the pump. The correct installation and operation of the intake and discharge lines of the pumps, the filter in at the end of the receiving line – the necessary measures to minimize dangerous situations from possible pulsations with increasing pressure and accidents leading to accidents.

During drilling during the operation of mud pumps, when the bottomhole pressure exceeds the static pressure of the liquid column in the well, oil and gas manifestations are possible. Especially dangerous is gas saturation during a long break in the operation of boreholes. When the circulation is restored, the output of packs of drilling fluid saturated with gas, an increase in the liquid level in the receiving tanks of the drilling pumps, is likely. Excessive pressure is achieved by weighting the drilling fluid, but sometimes this is not enough. The air of the working area in the pump room is polluted by oil and gas components (methane, ethane, propane, hydrogen sulfide, carbon dioxide, etc.). In the summer, the evaporation of harmful substances increases. Since pumps and mortar systems are located in sheltered blocks in the North, the content of hydrocarbons and other harmful gases in these conditions can be several times higher than the maximum permissible concentration, and therefore the possibility of acute poisoning, mainly hydrocarbons and hydrogen sulfide, is not ruled out. At elevated concentrations, chronic intoxication may also develop. It is known that in the process of working on the clothes of workers, static electricity accumulates. A discharge of static electricity, including in the presence of oil and gas manifestations, can lead to ignition of the oil or gas explosion and irreparable consequences. When a fire occurs during combustion, harmful substances are released, a large level of thermal radiation occurs, collapse of structures is possible, etc.

If horizons with a possible oil and gas manifestation are revealed during the operation of two pumps, it is necessary to provide for the possibility of their simultaneous operation from one tank. In the harness between the tanks of the circulation, system must be locking devices. Pressure pipelines, their parts and fittings after assembly at the plant, as well as after repair using welding, are subject to pressure testing, in other cases, the pressure of pressure testing should be equal to the working one, multiplied by the safety factor. Each shift must be controlled by an oil lubricator, regularly inspecting the installation equipment, wiping and cleaning the external surfaces from dust and dirt. Exhaust gases from diesel engines rotating the drill string, which pollute the air environment in the breathing zone of workers (they mainly contain hydrocarbons, sulfur dioxide and carbon monoxide), have a negative impact on the health of workers.

To protect labor and manage occupational risks, drilling workers should know and correctly apply personal respiratory protection equipment, tools and methods for monitoring the air of the working area in rooms and outdoor installations, an emergency response plan and conditional alarms for workers to communicate, techniques and methods providing first aid to victims.

During the operation of the mud pump, due to the abrasive action of sand in the mud, its cylinder bushings, pistons, rods, valves and seats wear out most quickly, which leads to disruption of the normal pump operation and a decrease in the flow of drilling mud to the bottom of the well. Intense wear occurs when working with heavier flushing fluid. It is necessary to check the condition of the surface of the stem, as well as the gaskets of the stuffing box packing. Worn or disconnected rods should be removed from the pump, since they exclude the possibility of creating a reliable seal [8].

The operation and repair of mud pumps is associated with the implementation of a large complex of labor-intensive and hazardous operations for their maintenance. Moving parts of the pump (pulley, V-belt drive, gears, crank mechanisms) create a danger to workers. The inspection hatches of the oil bath and chambers must be tightly closed with metal shields. Moving and rotating parts of the pump (shaft ends, V-belt drive, etc.) should be securely enclosed (the protruding end of the transmission shaft with a continuous casting, the V-belt drive with a mesh fence). When fencing V-belts should also provide fender frontal metal shields of sufficient rigidity and strength, able to withstand the impact of a broken texture belt.

Removing the pneumatic compensator, replacing a failed wickel is associated with significant physical effort and requires caution when releasing the residual amount of the working agent. Wickel (rubber element), as if vulcanized to the neck of the compensator, does not allow to release the residual amount of the working agent. In winter, the release of the agent from the pneumatic compensator block may be prevented by an ice plug formed at the bottom of the block from the washing liquid accumulated during the wickel rupture, which blocks the passage opening. It is necessary to replace the pneumatic compensator under pressure, which does not exclude the occurrence of an accident. To facilitate working conditions and reduce the risk when removing the cylinder of the pneumatic compensator should use special devices.

The labor intensity of workers is due to neuro-emotional overloads associated with high responsibility for the result of their own activities and the significance of the error, the likelihood of risk for their own lives and increased responsibility for the final result. Significant physical and neuro-emotional stresses lead to diseases of periarthritis of the shoulder and shoulder blade, shoulder epicondylitis, arthrosis of the cervical spine and polyneuritis of the upper extremities [9].

During the operation of the mud pumps, the worker has to tighten the stem seal of the stem with the sleeve by pressing the cover, tightening the nuts through the window of the bed. This operation is often repeated and, as a rule, is carried out during the operation of the pump, which is unacceptable, as this may result in an accident due to pressing the worker's hand with the cutter against the gland. When loosening the gaskets, the cylinder bushing may be pushed out of the pump. Periodically check the pressure of the cylinder liners. Valves and seats with significant wear should be replaced. Otherwise, erosion of the valve box and failure of the pump may occur. To safely change wear parts of a mud pump and increase the level of mechanization of operations to change parts of a hydraulic unit (valve, valve seat, stem with piston, piston, cylinder bushing), unscrew and tighten the retaining screw of the valve box, remove and install the stem seal housing, and remove crosshead finger it is necessary to use complex devices developed by VNIITB [10]. Before repairing any pump, the pressure in the piping should be reduced to atmospheric.

Sites and equipment of the drilling rig pumping unit are slippery, causing a danger of falling, including from a height during repair operations without special devices or personal protective equipment [1, 7]. Sharp edges, burrs and roughness on the surfaces of the mud pump elements can lead to cuts and infection of workers.

When personnel use electric motors to drive drilling pumps, non-conductive parts of electrical equipment in the event of an emergency can be energized, touching which becomes deadly if there is no or malfunctioning protective ground connecting the metal and current-carrying parts of the equipment to the ground. The cause of leaks should be fixed immediately. For electrical safety, employees must use tested dielectric gloves without damage, insulating supports before pressing the Start button or turning on the oil switch drive, and when the motor is turned off, the Stop button. There should be no people in the danger zone before giving a start signal. If at start-up of the electric motor any extraneous noises, crackles or obvious malfunctions of the electric motor are found, as a rule, the second assistant to the driller should immediately turn off the electric motor, then hang up a poster on the starting device: "Do not turn it on – people work!" and call electrical personnel.

The insufficient illumination of the working area of the drilling pump unit is one of the dangerous factors due to the technology of the drilling process. Industry standards for illuminating workplaces in drilling oil and gas wells (10–100 lux) are underestimated by 3-5 times in comparison with the requirements of SNiP 23-05-95 "Natural and artificial lighting". This is because the drilling rig is not considered as a production building, but as a construction site (table 1). In the territory of the drilling rig, in open areas for equipment during work, floodlighting with dustproof luminaires in explosion-proof design, equipped with a protective mesh from mechanical damage, is widely used.

Table 1. Pump unit workplace lighting rits at the drilling rig

Workplaces to be covered	Discharge and subdischarge- visual work	Lighting locations	Illumination rate, lx
Pump room	3	At a height of at least 3 m	75
Clay mixers	3	At a height of at least 3 m	75
Gutter system	5	At a height of at least 3 m along the entire length of the gutters	10

Lack of light and irrationally arranged production lighting can distort the information received by the employee through vision, impair spatial orientation, coordination of movements, speed of reactions, cause fatigue, which will lead to reduced attention, errors in the application of safe working methods and increase the risk of accidents and industrial injuries. According to statistics, on average, for various types of production activities, the number of accidents associated with poor lighting is 30–50 % of the total. In addition to injuries, adverse lighting conditions can cause fatigue of the visual analyzer (with systematic exposure, the development of visual defects) and lead to occupational diseases. On the contrary, rational, satisfying hygienic requirements, lighting has a positive psychophysiological effect on the staff, contributes to increased productivity and safety, is an important stimulator not only of the visual analyzer, but also of the body as a whole, makes the employee's movements more confident, which reduces his level of professional risk.

The leading place in the complex of harmful production factors that form professional risks during the operation of mud pumps belongs to industrial noise and vibration. Noise sources are rotating and moving parts of pump mechanisms. The excess of noise levels over the standards in accordance with GOST 12.1.003-83 in the pump compartment reaches up to 10 dBA, for vibrating screens – up to 18 dBA, which has a harmful effect on the human body and leads to irreversible hearing loss, disrupts the normal activity of the nervous, cardiovascular and digestive system, causes fatigue, decreased performance and slowed mental reactions.

The causes of vibration in the discharge system can be poor fastening of the pumps and their drive to the foundations, misalignment of the pump and drive, wear or weakening of individual parts of the pump (elements of the crank mechanism, rods, pistons, bushings, etc.), pulsation of the flushing fluid.

The most common two-piston pumps in drilling send the pumped fluid into the injection system with jerks. Therefore, with sixty double strokes in one minute, four powerful pulses occur in the discharge system within one second. The existing damping system for these pulses (pneumatic compensators) cannot completely smooth out the pressure in the pumped liquid. For example, the total vibration during maintenance of the vibrating screen reaches 112 dB at the workplace at the maximum permissible level of 92 dB. The harmful effects of vibration are expressed in the occurrence of a vibratory disease and a breakdown of the nervous system.

Table 2. Noise levels during personnel stay at work places of the pump drilling block

Workplace location	Noise Levels, dBA	Excess noise level, dBA
Pump compartment	88–90	8–10
Capacitive compartment	88–89	8–9
Circulation system for vibrating screens	91–98	11–18

It is customary to subdivide methods and means of noise control into methods of reducing noise along the path of its propagation from the source, in the source of its formation and personal protective equipment against noise. In order to reduce the harmful effects of noise and vibration on the drilling rig, it is necessary to carry out timely preventive inspection and repair, tighten the weakened joints, timely lubricate rotating parts. If it is impossible to suppress noise in the source of origin, then PP-80, PA/O, PA/S soundproofing and soundproofing screens should be used. To combat vibration, the following methods are used:

- suppression at the source of occurrence (centering, adjustment);
- change in design;
- the use of spring shock absorbers, vibration dampers.

Existing technologies for the operation and repair of equipment of mud pumps determine the performance of work under adverse effects (depending on the season) of the air temperature of the working area. At high temperatures, the employee's attention and reaction speed decrease, persistent changes in the cardiovascular system occur, at low – mobility of the extremities decreases, diseases of the respiratory tract and joints occur. Working in the field exposes the drilling personnel to the negative effects of biological factors (blood-sucking poisonous insects – ticks, midges, mosquitoes), which can lead to occupational diseases (tick-borne encephalitis).

3. Conclusion

An employee's professional risk can be estimated as the probability of harm to health as a result of exposure to hazardous and (or) harmful production factors in the performance of labor duties or other legitimate activities carried out in the interests of the organization or individual entrepreneur. Assessment of professional risks is the most effective preventive measure, which takes into account not only adverse events and accidents that occurred earlier, but also dangers that have not yet caused adverse effects. The basis of the risk assessment procedure is the identification of hazardous and harmful production factors in the performance of the labor functions of the organization's employees. If these dangers cannot be eliminated, their risk to the health and safety of personnel should be assessed in order to make informed decisions to increase the level of labor protection. The proposed measures should be specific and feasible. The constancy of the risk assessment also implies an assessment of the effectiveness of the implemented measures, constant monitoring of changes in risk levels and the effectiveness of communications.

Based on this brief scientific review of the main negative production factors that arise during the operation of mud pumps, we can conclude how relevant today are studies to identify and assess occupational risks, establish the nature of their impact on personnel, and develop measures to ensure the safety of work in well construction. The development of an effective mechanism for identifying and assessing professional risks, increasing the competence and motivation of employees will reduce the number of injuries caused by hazardous activities of personnel engaged in drilling oil and gas wells.

References

[1] Shangareev R R 2015 Relevance of assessment science research concerning occupational hazards of drilling operations *Ecol., Environ. and Conservat. Paper* **21** 175–80

[2] Khabibullin M Y, Suleimanov R I, Sidorkin D I and Arslanov I G 2017 Parameters of damping of vibrations of tubing string in the operation of bottomhole pulse devices *Chem. and Petrol. Engineer.* **53(5-6)** 378–84

[3] Prashanth I, Fernandez G J, Sunder R G and Boardman B 2017 Factors influencing safety barrier performance for onshore gas drilling operations *J. of Loss Prevent. in the Process Industr.* **49** 291–8

[4] Lootz E, Ovesen M, Tinmannsvik R K, Okstad E H and Carlsen I M 2013 Risk of major accidents: Causal factors and improvement measures related to well control in the petroleum industry *Society of Petrol. Engineers – SPE Amer. E and P Health, Safety, Security, and Environ. Conf.* pp 340–50

[5] Suleimanov R I, Gabdrakhimov M S, Khabibullin M Y et al 2018 The study of hydraulic hammer device in drilling tool assembly in hydraulical rotary drilling *Int. J. of Engineer. and Technol.* **7(2)** 28–30

[6] Zhao Q, Zhang H and Fan J 2011 Human factor risk quantification for oil and gas drilling operation *Proc. Engineer.* **18** 312–7

[7] Shangareev R R 2018 Role of employee motivation in an industrial occupational risk management system *IOP Conf. Ser. Earth and Environ. Sci.* **194(2)** 1–7

[8] Watson V 2010 A quantitative risk assessment approach *Society of Petrol. Engineers – SPE Int. Conf. on Health, Safety and Environ. in Oil and Gas Explorat. and Product.* **4** 2828–38

[9] Rana S 2008 Facts and data on environmental risks-oil and gas drilling operations *SPE Asia Pacific Oil and Gas Conf. and Exhibit. 2008 – Gas Now: Deliver. on Expectat.* pp. 1448–68

[10] Bradish R, Bui J and Nunez Y 2008 Global challenges facing the health, safety and environmental (HS&E) function within the oil and gas industry *Proc. – SPE Annual Techn.l Conf. and Exhibit.* **3** 2158–61

IOP Conf. Series: Materials Science and Engineering **905** (2020) 012088 doi:10.1088/1757-899X/905/1/012088

Increased production of oil reserves in low production deposits of West Siberian oil and gas province

A V Stenkin[1], Yu A Kotenev[2], Shamil Khanifovich Sultanov[2], V V Mukhametshin[3], L S Kuleshova[4] and A R Safiullina[4]

[1] TPE Urayneftegaz, LLC LUKOIL–Western Siberia, Uray, Tyumen region, Khanty-Mansi Autonomous Area – Yugra, Russia
[2] Department of Geology and Oil and Gas Field Exploration, Ufa State Petroleum Technological University, SASI Institute of Strategic Research of the Republic of Bashkortostan, Ufa, Republic of Bashkortostan, Russia
[3] Department of Oil and Gas Field Exploration and Development, Ufa State Petroleum Technological University, Ufa, Republic of Bashkortostan, Russia
[4] Department of Oil and Gas Field Exploration and Development, Ufa State Petroleum Technological University, Branch of the University in the City of Oktyabrsky, Oktyabrsky, Republic of Bashkortostan, Russia

E-mail: StenkinAV@yandex.ru, geokot@inbox.ru, ssultanov@mail.ru, vv@of.ugntu.ru, markl212@mail.ru, safiullina.a.r@yandex.ru

Abstract. The paper considers the issues of assessing the influence of various factors on the production of oil reserves of Jurassic deposits of Shaimsky oil and gas region. The study states the influence of the network of disjunctive tectonics dividing the deposit onto the network of blocks on the degree of oil recovery. Trace research was conducted. Dependencies showing that hydrodynamic connectivity between wells takes place along fault lines or close to them, which is caused by the formation of decompression zones with increased permeability, are obtained. It was found that in the central blocks of the deposit there are large effective oil-saturated thicknesses with higher reservoir properties, which does not allow faults with low displacement amplitude to disrupt the connectivity.

1. Introduction

The problem of increasing the efficiency of oil production of Jurassic deposits of the West Siberian oil and gas province has been and is being given the highest attention due to significant amount of oil reserves and low degree of their production [1–5]. One of such objects is the oil fields of Shaimsky region, brought into production in the 2000s and characterized by rapid decrease in fluid flow rates, increase in water cut and decrease in the energy potential of the deposits [6–10].

2. Methods and materials

One of the productive formations are deposits of the middle and lower Jurassic, namely the Tyumen (J_2 – J_6) and Sherkalinskaya (J_{10}^1 – J_{10}^2) suites. Productive formations are combined into two operational objects: the first J_2 – J_6, the second J_{10}^1 – J_{10}^2. The operational objects differ from each other in terms of capacity, reservoir properties, recoverable oil reserves, and there is also sharp lithologic-and-facies inhomogeneity of reservoir rocks both in along the horizontal and through the section.

IOP Conf. Series: Materials Science and Engineering **905** (2020) 012088 doi:10.1088/1757-899X/905/1/012088

The West Tugrovskoe oil field was chosen as the object of study, which is complicated by a dense network of tectonic disorders and high variation of geological and field indicators.

In order to understand the distribution of oil reserves, the geological-hydrodynamic model (GHDM) was developed. Geological-hydrodynamic modeling allowed estimating the distribution of initial and remaining mobile reserves (Figure 1). The largest density of oil reserves within the object of study is concentrated in the area of increased effective oil-saturated thicknesses, in the central part of the deposit.

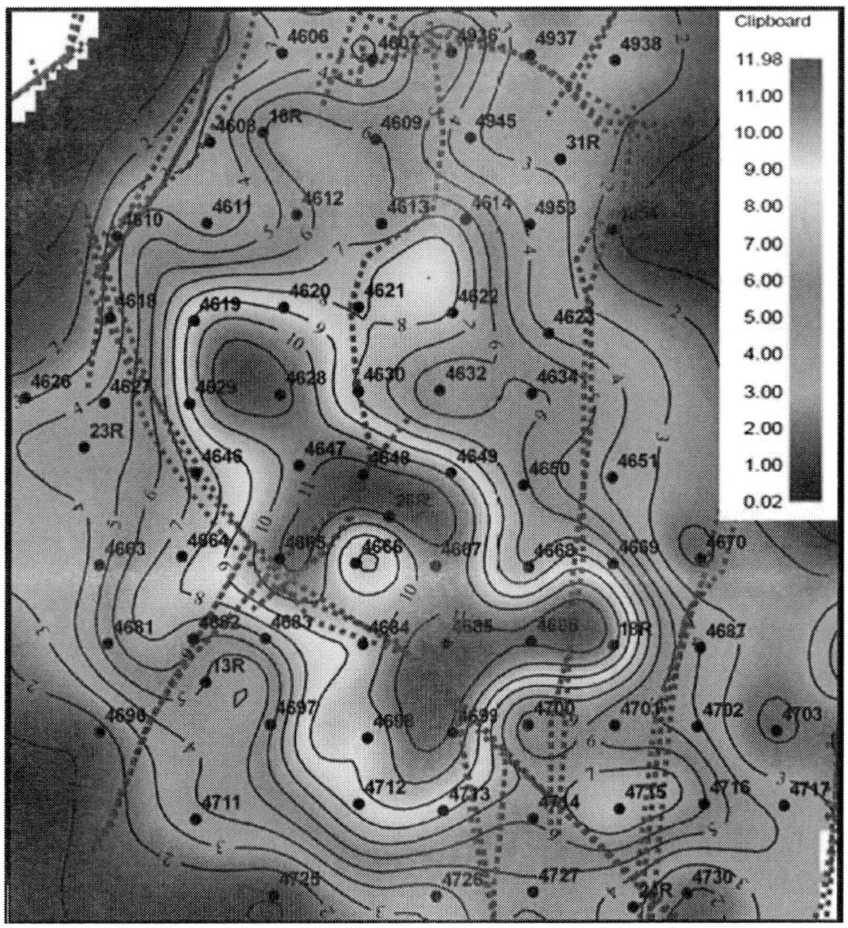

Figure 1. Extract from the density distribution map of remaining recoverable oil reserves of the objects under study

The distribution of effective thicknesses of Tyumen and Sherkalinskaya suites reveal $J_{2\text{-}3}$, J_4 and J_{10} formations, which are characterized by higher net oil thickness ($J_{2\text{-}3}$ – 6.8; J_4 – 4.3; J_{10}^1 – 3.6; J_{10}^2 – 5.4 m) in comparison with J_5 and J_6, which net oil thickness amounts to $\approx$ 2.0 m.

It should be noted that the maximum remaining oil reserves are related to high effective thickness of the Tyumen suite formations. With the total thickness of developed objects making more than 100 m, there are up to 38 interlayers in the section of wells, which makes it extremely difficult to solve the problem of their most complete involvement in the development process.

One of the most important reasons for weak production of deposit reserves is considerable variability of reservoir properties. According to A.A. Khanin, the reservoir rocks, which form the formations of Tyumen and Sherkalinskaya suites, mainly belong to reduced and low classes of permeability. The reservoirs of $J_{2\text{-}3}$ formation have (V–VI) class of permeability, J_4 – (III–VI) class of permeability, J_5 –

IOP Conf. Series: Materials Science and Engineering **905** (2020) 012088 doi:10.1088/1757-899X/905/1/012088

(IV–VI) class of permeability, J_6 – (V–VI) class of permeability, J_{10}^1 – (III–VI) class of permeability, J_{10}^2 – (I–VI) class of permeability.

According to permeability distribution on histograms (Figure 2), it can be noted that the permeability across the formations varies between 0.5 and $12 \cdot 10^{-3}$ μm², with a frequency greater than 85 %, the permeability of the remaining part is higher than $12 \cdot 10^{-3}$ μm². The formations have high degree of permeability anisotropy.

High porosity is observed in the central and northern parts of the deposit in the formations of Tyumen and Sherkalinskaya suites. The same areas have higher reserves depletion. The average porosity is 14 %.

The field has a 9-point well spacing system. Non-uniform distribution of water injected into the formation, including due to the lack of hydrodynamic connectivity between production and injection wells, which in turn is caused by non-uniformity of the productive part of the section and disjunctive tectonics [7, 8]. The main faults have north-south direction; smaller faults pass into the cross of the main ones (Figure 3).

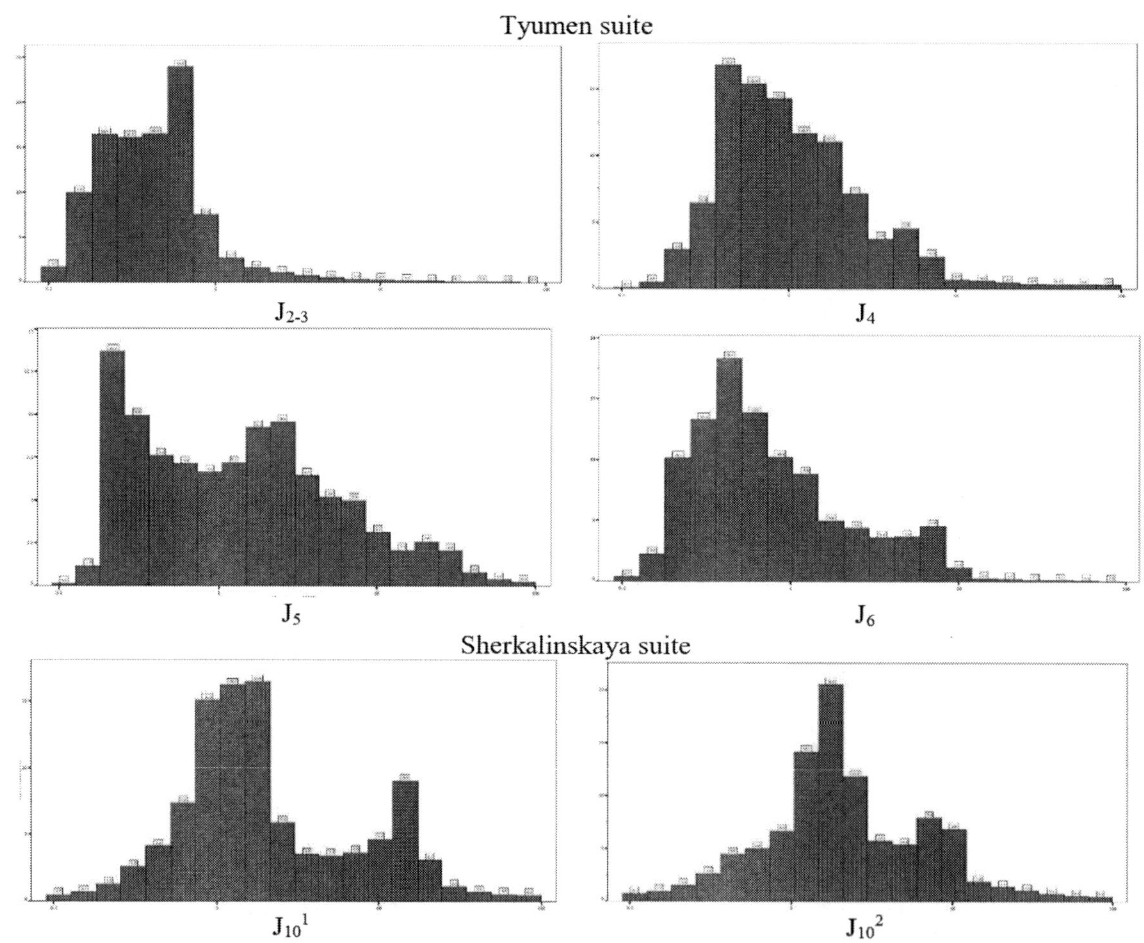

Figure 2. Histograms of permeability distribution across operational objects

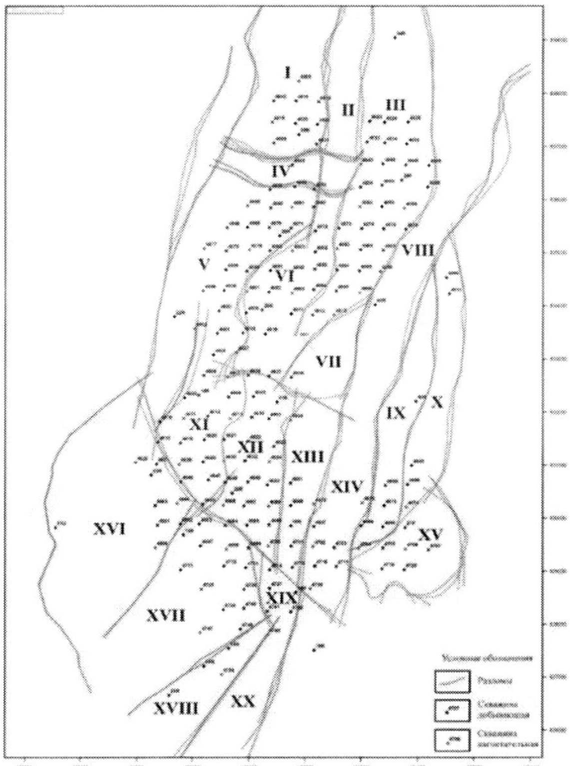

Figure 3. Block structure of West Tugrovskoe field

3. Results

In order to determine the degree of interaction of production and injection wells in the field, the tracer analysis was carried out for 4 wells.

The tracer analysis showed that in two wells, the best hydrodynamic connectivity is ensured along or close to the fault line. This is caused by the fact that as a result of tectonic movements, rocks are exposed to elevated pressures and temperatures, which in turn results in decompression zones that increase permeability. It should also be noted that the faults themselves do not fade at their termination point, as shown in Figure 3, but most likely continue further, diverging as branches and having a maximum field displacement amplitude.

Through two injection wells maintaining the formation pressure in different areas of the field, the marker is found in the samples of 10 producing wells.

An extensive hydrodynamically connected site is identified in the field. Thus, for example, tracer studies carried out on one injection well allowed recording the distribution of tracing fluid in 12 production wells. Figure 4 shows that the tracing fluid is detected in 7 observation wells located far from each other, through one or more transcurrent faults from the injection well. This made it possible to conclude that this site has the best reservoir properties. Besides, there is large effective thickness in these blocks, which in turn prevents faults with low displacement amplitude from disrupting the hydrodynamic connectivity.

The object of the study is characterized by uneven distribution of remaining oil reserves across the formations, as well as weak production. The share of recovered oil reserves from initial recoverable reserves does not exceed 10%. The reason for such indicators is high degree of heterogeneity both in the section and in the area of Tyumen and Sherkalinskaya suites, as well as uneven distribution of effective thicknesses.

High production of reserves is noted in the central section of the field near the folded part of the structure. This zone is characterized by high effective thickness and the best reservoir properties.

IOP Conf. Series: Materials Science and Engineering **905** (2020) 012088 doi:10.1088/1757-899X/905/1/012088

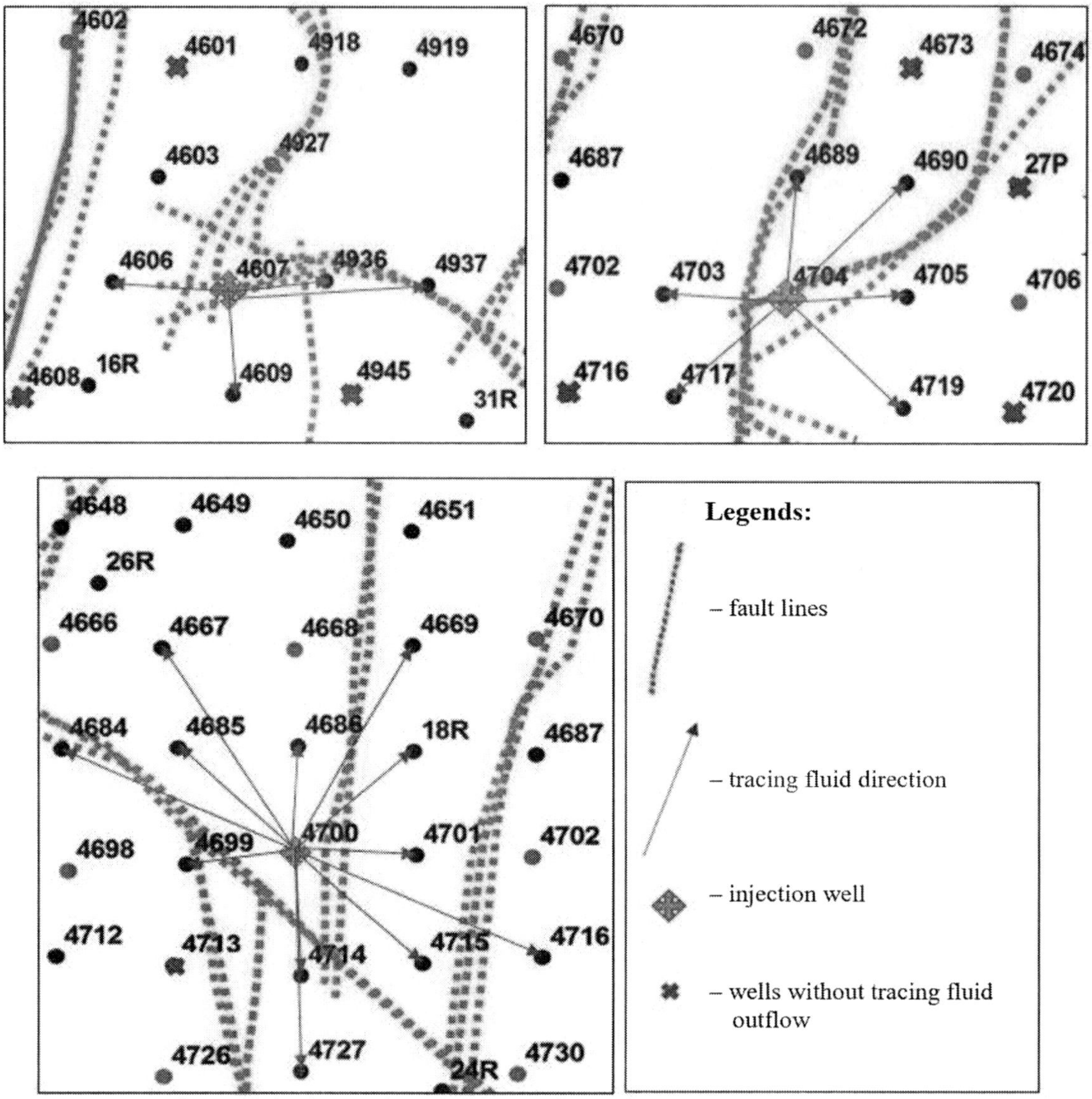

Figure 4. Interaction of injection and production wells

The comprehensive analysis of the state of remaining reserves allowed identifying the main technologies to develop the remaining oil reserves.

On the basis of the analysis of maps of remaining mobile oil reserves low production of J_5 formation is observed, which in a number of wells is traced as sandy and heterogeneous body in terms of capacity (Figure 5). The perforation works with subsequent hydraulic fracturing are recommended in order to involve this interval in development.

Due to low reservoir properties, high degree of heterogeneity of formations, as well as the reduction of productivity, it is recommended to carry out hydraulic fracturing in a number of wells for additional recovery of remaining reserves in the upper part of Tyumen suite from formations J_{2-3} and J_4.

In order to increase the efficiency of field development with poorly drained mobile oil reserves in certain sections of the production object characterized by high anisotropy, the cyclic flooding is envisaged.

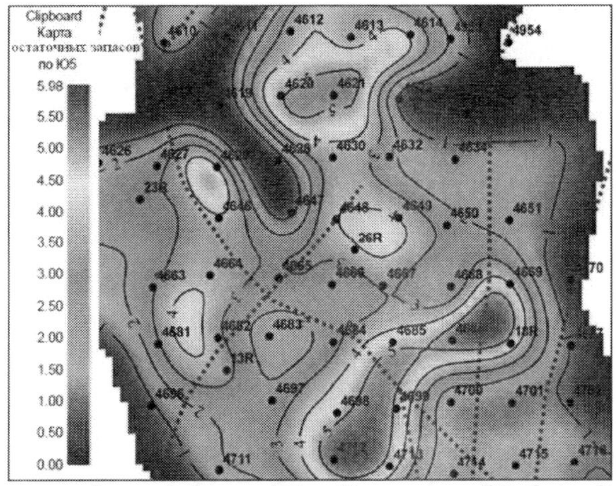

Figure 5. Extract from the map of remaining mobile oil reserves of J_5 formaiton

The area chosen for this treatment method represents a complex geological-hydrodynamic system with considerable non-uniformity of porosity and permeability, with developed system of formation pressure maintenance (FPM), limited from both sides by fault lines. The area is characterized by increased remaining oil reserves, which are concentrated mainly in J_5, J_{10}^1 and J_{10}^2 formations. The average oil flowrate for the area was 3.95 t/day with an average water cut of 78%. The injection capacity of wells varies within wide limits: from 49 to 133 m^3/day for Tyumen suite and from 0.7 to 91 m^3/day for Sherkalinskaya suite.

The calculations were made for the test section comprising 27 wells, 9 of which are injection wells, 18 producing wells, optimal periods and half-periods of injection were determined. According to these parameters, the wells operating at these intervals of operational objects were ranked. Thus, by the ratio of injection and production wells, the greatest coverage is obtained in the J_{10}^1 formation. The duration of cyclic flooding is 5–6 months.

4. Conclusion

Thus, the degree of hydrodynamic connectivity of wells, as well as the compensation of fluid extraction by water injection, was determined in the test site, which in turn allowed estimating the efficiency of cyclic flooding. It is worth noting the different nature of the influence of injection wells on reactive producing wells. The best hydrodynamic well connectivity for the site is identified in the south-west direction.

References

[1] Miller G A 2007 Some Perspectives on Various Methods of Oil Shale Extraction, Piceance Basin, Colorado *27th Oil Shale Symp. Colorado School of Mines Golden* (Colorado, 15–19 October 2007) p 14

[2] Polyakov V N, Zeigman Yu V, Kotenev Yu A et al 2018 System solution for technological problems of well construction completion *Nanotechnol. in Construct.* **10(1)** 72–87. DOI: 10.15828/2075-8545-2018-10-1-72-87

[3] Andreev V E, Chizhov A P, Kotenev Yu A et al 2019 Prediction of Gas Exposure in the Conditions of the Oil Fields of the Volga-Urals *Atlantis Highlights in Material Sciences and Technology (Int. Symp., dedicated to the 85th anniversary of H.I. Ibragimov, Engineering and Earth Sciences: Applied and Fundamental Research (ISEES 2019))* vol 1 pp 666–9. DOI: 10.2991/isees-19.2019.132

[4] Gomari K A R, Karoussi O and Hamouda A A 2006 Mechanistic Study of Interaction between Water and Carbonate Rocks for Enhancing Oil Recovery *SPE Europec/EAGE Annual Conf. and Exhibit.* (Vienna, Austria, 12–15 June) p 8. DOI: 10.2118/99628-MS

[5] Kuleshova L S and Mukhametshin V V 2019 Estimation of the wells hydrodynamic drag level based on wells geophysical survey data *Atlantis Highlights in Material Sciences and Technology (Int. Symp., dedicated to the 85th anniversary of H.I. Ibragimov, Engineering and Earth Sciences: Applied and Fundamental Research (ISEES 2019))* vol 1 pp 727–9. DOI: 10.2991/isees-19.2019.145

[6] Zeigman Yu V, Mukhametshin V Sh, Sergeev V V and Kinzyabaev F S 2017 Experimental study of viscosity properties of emulsion system with SiO_2 nanoparticles *Nanotehnol. in Construct.* **9(2)** 16–38. DOI: 10.15828/2075-8545-2017-9-2-16-38

[7] Rabie A I and Nasr-El-Din H A 2015 Effect of Acid Additives on the Reaction of Stimulating Fluids During Acidizing Treatments *SPE North Africa Techn. Conf. and Exhibit.* (Cairo, Egypt, 14–16 September 2015) p 27. DOI: 10.2118/175827-MS

[8] Sergeev V V, Belenkova N G, Zeigman Yu V and Mukhametshin V Sh 2017 Physical properties of emulsion systems with SiO_2 nanoparticles *Nanotechnol. in Construct.* **9(6)** 37–64. DOI: 10.15828/2075-8545-2017-9-6-37-64

[9] Mukhametshin V Sh, Zeigman Yu V and Andreev A V 2017 Rapid assessment of deposit production capacity for determination of nanotechnologies application efficiency and necessity to stimulate their development *Nanotechnol. in Construct.* **9(3)** 20–34. DOI: 10.15828/2075-8545-2017-9-3-20-34

[10] Imqam A, Bai B, Wei M, Elue H and Muhammed F A 2016 Use of Hydrochloric Acid To Remove Filter-Cake Damage From Preformed Particle Gel During Conformance-Control Treatments *SPE Product. & Operat.* **31(3)** 11. DOI: 10.2118/172352-PA

IOP Conf. Series: Materials Science and Engineering **905** (2020) 012089 doi:10.1088/1757-899X/905/1/012089

The use of variable frequency drive for the operation of the NBT mud pump

R I Suleimanov and M Ya Khabibullin

Department of Oil and Gas Field Machines and Equipment, Ufa State Petroleum Technological University, Branch of the University in the City of Oktyabrsky, Oktyabrsky, Russia

E-mail: rustamsul@rambler.ru, m-hab@mail.ru

Abstract. The article discusses the process of optimizing the operation of the NBT-600 mud pump by adjusting the frequency of the current supplied to the pump drive. As a result of a technical measure, efficient operation of the pump is ensured, the energy consumption of the pump unit is reduced, and the main parameters of the pump are increased.

1. Introduction

When constructing wells, the main factors of the technological regime that determine the productivity of drilling are the axial load on the bottom, the rotational speed of the projectile and the flushing mode, which substantially depends on the perfection of the mud pumps.

With the accumulation of individual scientific results, the need to develop the theory as a set of scientific ideas about the laws of hydraulic action and wear, simultaneously flowing in a piston mud pump during its operation and mainly determining the pump performance, is becoming increasingly obvious [1–3].

One of the common mud pumps is NBT-600, which is equipped with drilling rigs of grades 3-6. The NBT-600 drilling pump is a three-piston single action. A cylinder sleeve-piston with diameters of 120, 130, 140, 150, 160, 170, 180 mm is equipped with a number of pairs. The frequency of double strokes is 140 per minute, the ideal flow rate varies from $19.8 \cdot 10^{-3}$ to $44.5 \cdot 10^{-3}$ m^3/s, the pressure limit varies from 11.1 to 25 MPa, the stroke length of the rod is 250 mm.

The feed of the NBT-600 mud pump is determined by the formula:

$$Q = \frac{z \cdot \left(\frac{\pi \cdot D^2}{4} \right) \cdot S \cdot n}{60}, \tag{1}$$

where z is the number of cylinders, z = 3; n is the number of double strokes of the rod, n = 140 min^{-1}, D is the diameter of the piston; S is the stroke length of the rod, S = 0.25 m; The efficiency of the NBT-600 mud pump is on average 0.82.

The discharge pressure of the mud pump is determined by the formula:

$$P = \frac{N_{ПР} \cdot \eta}{Q}, \tag{2}$$

where $N_{ПР}$ – hydraulic power of the pump, W; η is the efficiency of the pump; Q – pump capacity, m^3/s.

IOP Conf. Series: Materials Science and Engineering **905** (2020) 012089 doi:10.1088/1757-899X/905/1/012089

Table 1. Drill pump characteristic NBT-600

№	Piston diameter mm	Pump performance, m^3/s	Pump discharge pressure, MPa
1	120	0.0198	25.0
2	130	0.0232	21.2
3	140	0.0269	18.3
4	150	0.0309	15.9
5	160	0.0352	13.9
6	170	0.0397	12.4
7	180	0.0445	11.1

2. Methods and materials

By installing a frequency converter on an induction motor, you can replace the DC drive. Adjusting the speed of the DC motor can be carried out by different methods, but the electric motor in this drive is the weakest point, due to its high cost and unreliability. In the process, the brushes spark and the collector wears out due to electroerosion. In addition, the use of such an electric motor is impossible in dusty and explosive areas [4, 5].

DC motors are inferior to induction motors in the simplicity and reliability of the design, due to the fact that, unlike DC motors, they have movable contacts. In addition, induction motors have a smaller size and weight, as well as lower cost at the same power. An additional advantage is the simplicity of manufacture and operation [6, 7].

The main disadvantage of induction motors is the difficulty of adjusting their speed. Standard methods are changing the supply voltage or introducing additional resistances into the winding circuit [8].

The theory of frequency regulation of an electric motor dates back to the thirties of the last century, however, controlling an asynchronous electric motor in a frequency mode was problematic until recently. The high cost of frequency converters seriously hampered the development of a frequency-controlled electric drive. Currently, due to the use of circuits with IGBT transistors, as well as the use of high-performance microprocessor control systems, frequency converters of European, Japanese and American manufacturers have a more affordable price [9, 10].

Traditional methods of adjusting the speed of rotation of mechanisms involve regulation using various devices: hydraulic couplings, resistors additionally introduced into the rotor or stator, mechanical variators, electromechanical frequency converters, static frequency converters.

However, the use of most of the above devices cannot provide high quality speed control, and most importantly, it is very expensive and requires high costs during installation and operation.

The most advanced asynchronous drive control devices are static frequency converters.

The principle of the frequency method for controlling the speed of an induction motor is that by changing the frequency f_1 of the supply voltage, it is possible in accordance with the expression:

$$\omega_0 = \frac{2\pi \cdot f_1}{p}, \tag{3}$$

for an unchanged number of pole pairs p change the angular velocity of the stator magnetic field.

A similar method allows for smooth speed control over a very wide range, while the mechanical characteristics have high rigidity.

Power losses with this regulation are minimal, since such adjustment is not accompanied by an increase in the slip of the induction motor.

It should be noted that in order to obtain high energy performance of an induction motor, namely power factor, efficiency, and overload capacity, the input voltage should be changed along with the frequency

The law of voltage change depends on the nature of the load moment M_c. With a constant load moment M_c=const the voltage on the stator should be regulated in proportion to the frequency:

$$\frac{U_1}{f_1} = const \tag{4}$$

For the fan nature of the load moment, this state has the form:

$$\frac{U_1}{f_1^2} = const \quad (5)$$

At a load moment inversely proportional to speed:

$$\frac{U_1}{\sqrt{f_1}} = const \quad (6)$$

Thus, for smooth stepless regulation of the shaft speed of an induction motor, the frequency converter must provide simultaneous regulation of the frequency and voltage at the stator of the induction motor.

The use of an adjustable electric drive provides energy saving and allows to obtain new qualities of systems and objects. Significant energy savings are achieved by regulating a technological parameter. If it is a conveyor or conveyor, then you can adjust the speed of its movement. If it is a pump or fan, you can maintain pressure or adjust the capacity. If it is a machine, then it is possible to smoothly adjust the feed rate or the main movement [11].

The prospect of frequency regulation is clearly visible from Figure 1.

The high economic effect of adjusting with the help of a frequency converter gives precisely when they are used at facilities that provide fluid supply (in this case, it is a mud pump). Today, the traditional way of regulating the performance of pumps is the so-called "throttling", which is the use of valves or control valves. However, at the moment, this method can be called obsolete, due to the fact that the frequency regulation of an asynchronous motor becomes available [12].

When controlling the valve, the flow of material that this valve holds back does not do any useful work. In the case of using frequency regulation, it is possible to set the required pressure or flow rate, which will ensure not only a reduction in energy consumption, but also reduce the loss of solution or water.

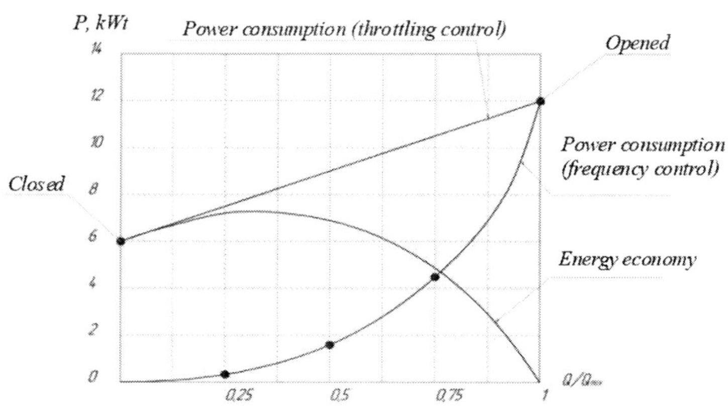

Figure 1. Separator slug catcher Comparative analysis between throttling control and frequency control

The structure of the frequency converter

Most of the existing designs of frequency converters are built according to the double conversion scheme. They are based on three main parts: a DC link, a pulsed power inverter, as well as a control system.

The alternating voltage of the supply network in the DC link passes through an uncontrolled rectifier and filter, as a result of which it is converted to DC voltage.

The power three-phase pulse inverter includes six transistor switches. Through the appropriate key, each motor winding is connected to the positive and negative terminals of the rectifier. The rectified voltage is converted into a three-phase variable in the inverter, and the voltage of the desired frequency and amplitude is obtained, which, in turn, is applied to the stator windings of the electric motor.

As a key on the output stages of the inverter, not thyristors are used, but more modern power IGBT transistors. Their advantage is that they have a higher switching frequency, due to which they produce a sinusoidal output signal and have minimal distortion.

The principle of operation of the frequency converter is important. The frequency converter consists of an uncontrolled diode power rectifier B, an autonomous inverter, a PWM control system, an automatic control system, an inductor $L_в$ and a filter capacitor $C_в$ (Figure 2). Regulation of the output frequency fout. and voltage U_{exp} is carried out in the inverter due to high-frequency pulse-width control.

Pulse-width control is characterized by a modulation period, inside which the stator winding of the electric motor is connected alternately to the positive and negative poles of the rectifier.

The duration of these states within the PWM period is modulated according to a sinusoidal law. At high (usually 2–15 kHz) PWM clock frequencies, sinusoidal currents flow in the motor windings, due to their filtering properties.

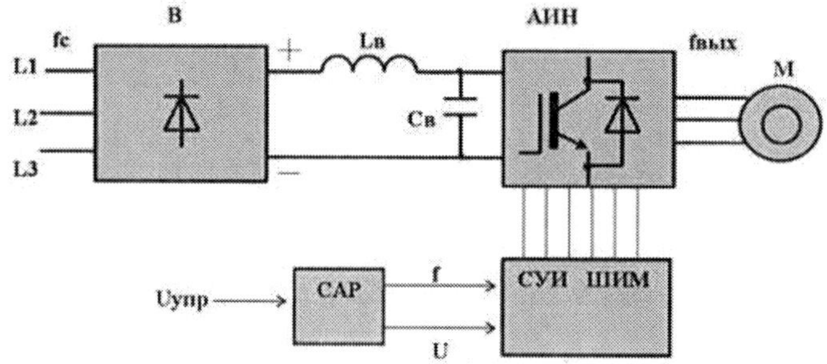

Figure 2. The principle of operation of the frequency converter

Thus, the shape of the output voltage curve is a high-frequency bipolar sequence of rectangular pulses (Figure 3).

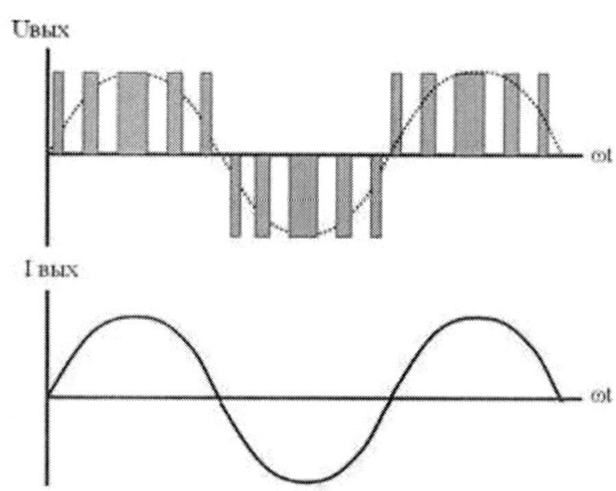

Figure 3. Graph of output voltage and current

The pulse frequency is determined by the PWM frequency, the duration (width) of the pulses during the period of the output frequency of the AIN is modulated according to a sinusoidal law. The shape of the output current curve (current in the windings of an induction motor) is almost sinusoidal.

IOP Conf. Series: Materials Science and Engineering **905** (2020) 012089 doi:10.1088/1757-899X/905/1/012089

The inverter output voltage can be regulated in two ways: amplitude (AR) by changing the input voltage $U_в$ and pulse-width (PWM) by changing the valve switching program V1-V6 with $U_в$ = const.

3. Results

The frequency converter, when used with the NBT-600 pump, can significantly save energy. This is achieved due to the cubic dependence of power consumption on the engine speed. The dependence is shown in Figure 4.

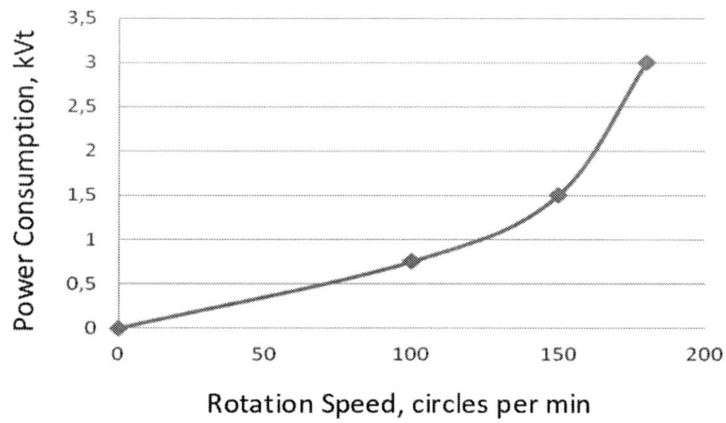

Figure 4. Graph of power consumption versus speed (H-n) using IF

When controlling throttling, the graph moves to the left according to the characteristics of the working mechanism, thereby obtaining the flow rate we need, the power consumed by the mechanism remains the same, and the flow rate decreases (Figure 5).

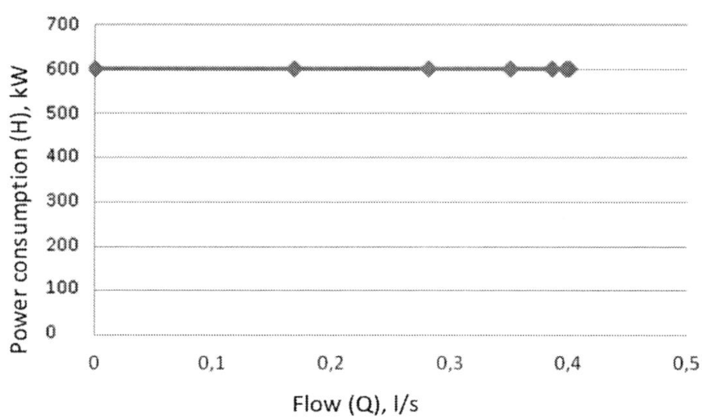

Figure 5. Graph of pump flow versus power consumption (Q-H)

When used to control the inverter, the following characteristics are obtained (Figure 6 and 7).

By changing the speed of the pumping mechanism using the inverter, the graph shifts to the left by the operating characteristic, while obtaining the flow rate we need, while the power consumption will be much less than during throttling. From the graphs it is obvious that when regulating the feed using a frequency converter, the power consumption will be much less than with throttling. The power consumed by the pumping mechanisms has a cubic dependence on the speed of rotation.

We have a pump load that consumes electric energy at a maximum rotation speed of 3 kW, reducing the rotation speed using a frequency converter by 20 %, we get tangible energy savings of 1.5 kW. With an unregulated drive, constant power mode is ensured by the use of a pair of "cylinder bushing - piston" of various diameters, which is characterized by the dependence of the productivity of the mud pump Q, discharge pressure P and piston diameter D.

Positions 7, 6, 5, 4, 3, 2, 1 (Figure 8) determine the ultimate pressure permissible from the condition of the strength of parts by pumps, we give different diameters of the pistons, or provide the maximum realization of drive power. The drilling process starts from position 0 at an injection pressure p < p7 with a piston of diameter D7. It is impossible to drill a well with a piston with a diameter of D7 above position 7, while the pump drive will be overloaded and the injection pressure will be higher than the permissible one, in connection with which, in position 7, a piston with a diameter of d7 is replaced with a piston with a diameter of D6, as a result, the pump flow will drop abruptly to pressure corresponding to position 8, and the pump pressure will decrease in proportion to the square of the supply. Subsequent drilling with a D6 piston as a result of a well being drilled with pressure will grow to p6 (the maximum permissible value), at which a piston with a diameter of D6 needs to be replaced with a piston with a diameter of D5, etc. With unregulated drives and changing pistons, the drilling process will be carried out under vertical segments 0- 7, 8-6, 9-5, 10-4, 11-3, 12-2, 13-1 (Figure 1), as a result, we can approach the equality P × Q = const, but the drive power will not be realized completely.

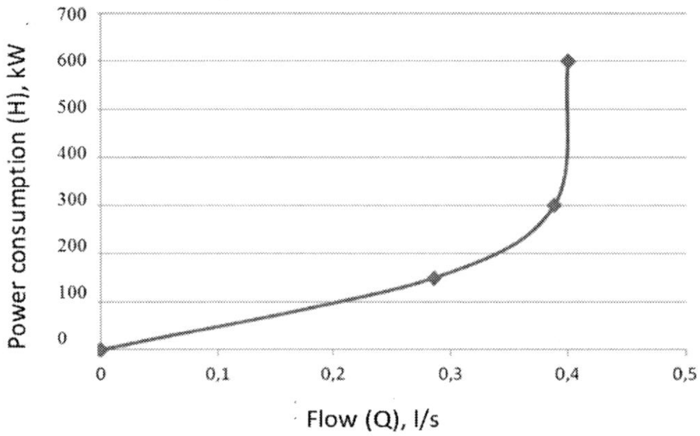

Figure 6. Graph of pump flow versus power consumption (Q-H) using inverter

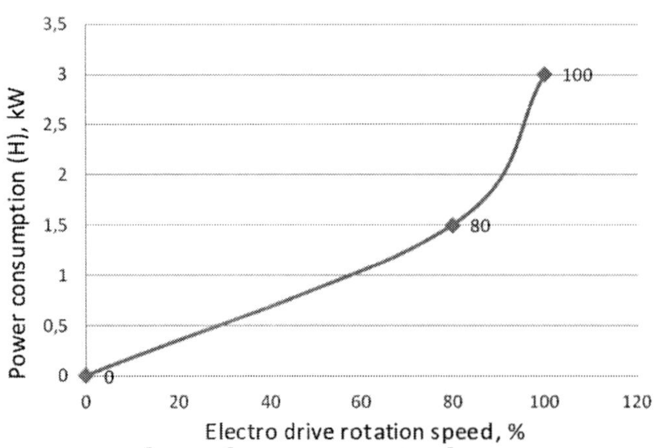

Figure 7. A graph of the dependence of the speed of rotation of the motor shaft on power consumption using IF

With an adjustable drive, it is possible to maintain the equality P × Q = const under equal conditions by increasing the discharge pressure by reducing the speed of the drive, but at the same time the volumetric efficiency of the pump will decrease, and as a result the flow.

The operating mode of the adjustable pump drive is as follows. Well drilling began at position 0 (Figure 1). As the penetration increases, the pressure rises and reaches the value p7 (item 7). With an adjustable drive, a piston change at a pressure of p7 is not required, because we reduce the speed of the drive at the rated torque on the shaft using frequency regulation. As a result, the pump flow will be reduced, while the discharge pressure will remain constant, corresponding to the allowable for the installed piston. This process will be carried out to position 8 '. Subsequent reduction in speed is inefficient, pump flow is reduced, piston replacement is required from diameter D7 to diameter D6. In the new mode, the drive frequency is brought to the nominal, and with an increase in penetration, the discharge pressure will increase to P6 (position 6), which is the limit for the piston diameter D6. A further reduction in the speed of the drive will ensure that the mode changes from position 6 to position 9', after which the piston is changed to diameter D5, etc. As a result of such an event, the discharge pressure and pump flow will change in the following sequence: 0-7-8'-6-9'-5-10'-4-11'-3-12'-2-13'-1.

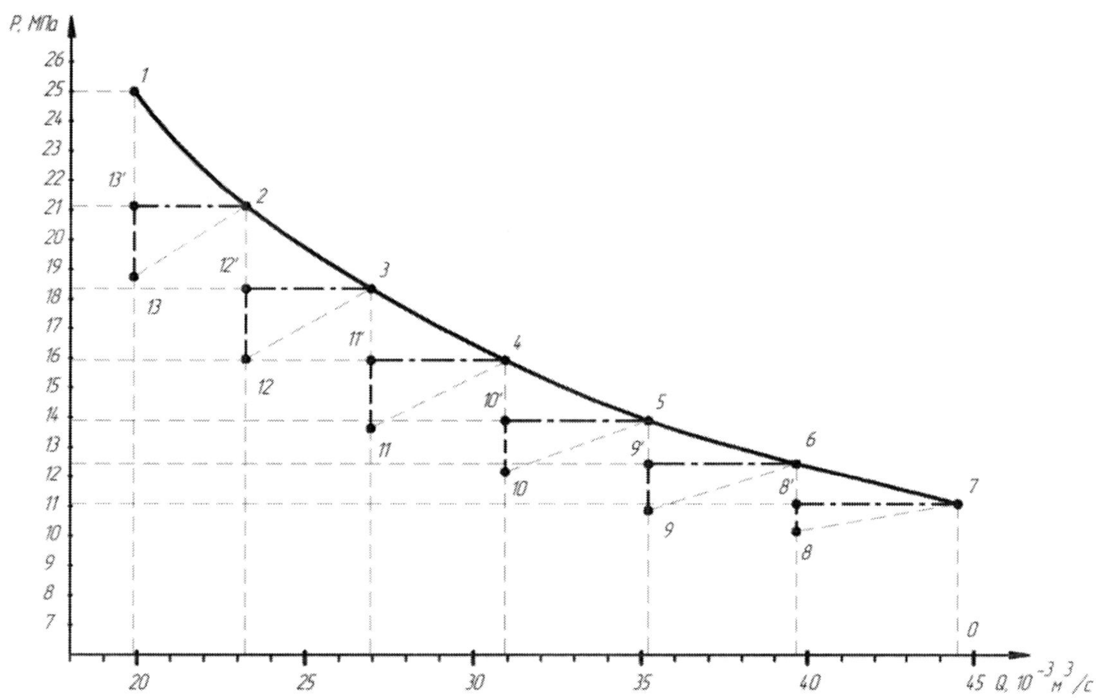

— - — - – adjustable drive (0-7-8 '; 6-9'; 5-10 '; 4-11'; 3-12 '; 2-13'; 1);
— — — – unregulated drive (0-7; 8-6; 9-5; 10-4; 11-5; 12-2; 13-1); Q – pump capacity, P – pump discharge pressure

Figure 8. Characterization of the NBT-600 mud pump

The total loss of discharge pressure (Figure 1) when comparing unregulated and variable drives is:

$$\Delta P = \Delta P8\text{-}8' + \Delta P9\text{-}9' + \Delta P10\text{-}10' + \Delta P11\text{-}11' + \Delta P12\text{-}12' + \Delta P13\text{-}13' =$$
$$= 0.95 + 1.61 + 1.79 + 2.29 + 2.44 + 2.48 = 11.56 \text{ MPa}.$$

4. Conclusion

Full implementation of the power of the mud pump with an adjustable drive ensures that the nominal pump flow is maintained during the entire construction cycle of the well.

IOP Conf. Series: Materials Science and Engineering **905** (2020) 012089 doi:10.1088/1757-899X/905/1/012089

References

[1] Zahraoui Y, Akherraz M, Fahassa C and Elbadaoui S 2019 Robust control of sensorless sliding mode controlled induction motor drive facing a large scale rotor resistance variation *ACM Int. Conf. Proc. Ser. a56*

[2] Nassif A B 2019 Assessing the Impact of Harmonics and Interharmonics of Top and Mudpump Variable Frequency Drives in Drilling Rigs *IEEE Transact. on Indust. Applicat.* **55(6(**8765746)) 5574–83

[3] Khabibullin M Ya, Suleimanov R I and Vasilyeva E R 2018 Express method of injection well treatment *Advan. in Engineer. Res. (AER) Int. conf. Actual issues of mechan. Engineer. (AIME 2018))* vol 157 pp. 280–3

[4] Wan Hong Y and Van Kuilenburg R 2018 Active mud line pumping, design, testing and future possibilities in rig design *IADC/SPE Drill. Conf. and Exhibit. (DC 2018)* (Society of Petroleum Engineers)

[5] Benoît D, Erik D, Eric C et al 2019 Automatic control of mud pumps, draw-works and top-drive on a floater *SPE/IADC Drill. Conf. Proc.* March

[6] Khabibullin M Ya and Suleimanov R I 2019 Automatic packer reliability prediction under pulsed transient flooding of hydrocarbon reservoirs *IOP Conf. Ser. Mater. Sci. and Engineer.* **560(1(012024)).** DOI: 10.1088/1757-899X/560/1/012024

[7] Zainagalina L Z, Petrova L V and Petrov V A 2019 Analysis of the efficiency of telemetric systems for drilling wells *IOP Conf. Ser. Mater. Sci. and Engineer.* **560(1(012096)).** DOI: 10.1088/1757-899X/560/1/012096

[8] Davydov A Yu 2018 Utilization of expert systems for screw pump sets with surface drive management *Advan. in Engineer. Res. (AER) Int. conf. Actual issues of mechan. Engineer.(AIME 2018)* vol 157 pp 129–33

[9] Abdyukova R Ya 2019 Studies on operation and types of drilling pump valves *IOP Conf. Ser. Mater. Sci. and Engineer.* **560(1(012050)).** DOI: 10.1088/1757-899X/560/1/012050

[10] Akhmetov R T, Mukhametshin V V and Kuleshova L S 2019 Grouping of objects using a limited number of parameters characterizing geological and physical properties of layers *Atlantis Highlights in Material Sciences and Technology (Int. Symp., dedicated to the 85th anniversary of H.I. Ibragimov, Engineering and Earth Sciences: Applied and Fundamental Research (ISEES 2019))* vol 1 pp 9–13. DOI: 10.2991/isees-19.2019.3

[11] Andreev V E, Chizhov A P, Kotenev Yu A et al 2019 Prediction of Gas Exposure in the Conditions of the Oil Fields of the Volga-Urals," *Atlantis Highlights in Material Sciences and Technology (Int. Symp., dedicated to the 85th anniversary of H.I. Ibragimov, Engineering and Earth Sciences: Applied and Fundamental Research (ISEES 2019))* vol 1 pp 666–9. DOI: 10.2991/isees-19.2019.132

[12] Liu W, Shi L, Zhou Y, Wang Y and Jiang H 2013 Development and application of pressure control drilling system (PCDS) for drilling complex problem // 2013 Society of Petroleum Engineers *Int. Petrol. Technolo. Conf. (IPTC 2013) Challeng. Technol. and Econ. Limits to Meet the Global Energy Demand* vol 7 pp 5391–402

IOP Conf. Series: Materials Science and Engineering **905** (2020) 012090 doi:10.1088/1757-899X/905/1/012090

Optimization of the design of the scrubber separator slug catcher

R I Suleimanov and M Ya Khabibullin

Department of Oil and Gas Field Machines and Equipment, Ufa State Petroleum Technological University, Branch of the University in the City of Oktyabrsky, Oktyabrsky, Russia

E-mail: rustamsul@rambler.ru, m-hab@mail.ru

Abstract. The article discusses the equipment of a system for collecting and preparing well products. The equipment is made in the form of a slug catcher separator, the design of which is the product inlet capacity. Its main element is a cyclone-type scrubber with a swirl, in which the product is separated into gas, liquid and impurities. As a result of design optimization, the entrainment coefficient and the degree of product purification are improved.

1. Introduction

In the process of lifting liquid from wells and transporting it to the central point of collection and preparation of oil, gas and water, pressure gradually decreases, and gas is released from the oil. The volume of gas released as the pressure in the system decreases, and usually several tens of times greater than the volume of liquid. Therefore, at low pressures, their joint storage, and sometimes collection, becomes inappropriate. Have to carry out their separate collection and storage.

The technological process of collecting and preparing oil and gas consists in the subsequent change in the state of production of an oil well and consists of several stages:

1) the collection of oil and gas;

2) bringing oil and gas to normalized properties.

The collection of oil and gas refers to their movement from the metering units to the points of their preparation. The separation of gas from oil is the process of separation from the liquid phase (oil) of the gaseous phase. Separation occurs when pressure decreases and temperature rises, as well as due to molecular diffusion of hydrocarbon and other components contained in oil into the space with their lower concentration located above the oil [1-3].

The separation of oil from gas and water in various separators is carried out with the aim of:

1) receiving petroleum gas, which is used as a chemical raw material or as fuel;

2) reduce the mixing of the oil and gas flow and reduce due to this hydraulic resistance;

3) reduce foaming (it is enhanced by the released gas bubbles);

4) reduction of pressure pulsations in pipelines during further oil transportation from separators of the first stage to the oil treatment unit (OTU) [4, 5].

2. Methods and materials

A two-phase hydrocarbon mixture is transported through the oil and gas pipeline, where the role of the liquid phase is played by oil and water, the role of the gas phase is associated petroleum gas. The basic structure of a gas-liquid flow is cork, which is formed due to the difference between the average velocities of the liquid and gas phases, that is, the velocity of the phase at which a homogeneous medium occupies the entire pipe section [6–8].

When operating a commercial oil and gas pipeline in a plug mode of product flow, it is necessary to determine the required volume of slug catcher at the point of reception of the hydrocarbon mixture. The maximum plug sizes will be observed when the scraper passes through the pipeline ("projectile" plugs), while the dimensions of regularly passing plugs have a significant effect on the vibrations and vibrations of above-ground pipelines. Thus, when designing the objects of oil and gas gathering systems, it is necessary to calculate the values of both "shell" and regular cork structures [9–12].

The separation of gas from oil is carried out under the influence of gravitational, inertial forces and due to the selective wettability of oil. Inertial separation occurs during sharp turns of the gas-oil flow. As a result, the liquid, as the most inertial, continues to move in a straight line, and the gas changes its direction. As a result, they are separated. Based on this principle, the operation of the slug catcher separator product inlet unit is constructed by feeding the gas-oil mixture to the scrubber, in which the liquid is discharged to the inner surface and then flows down into the slug catcher separator tank, and the gas moves in the center of the scrubber.

The slug catcher separator is designed for preliminary gas purification from dropping liquid and mechanical impurities, as well as for receiving volley flows of liquid carried by gas from lowered sections of gas pipelines.

The main elements of the slug catcher separator (Figure 1) are: product inlet capacity; separator; light output system; heavy fraction withdrawal system.

For maintenance, the unit is equipped with service platforms and stairs with fences.

The slug catcher is a vertical cyclone scrubber combined with a horizontal tank for receiving the liquid phase and with a system for removing solids

The formation mixture from the installation of the input loop blocks enters the cyclone scrubbers, in which the blades of the vortex tube rotate the gas-liquid flow. Inside each scrubber are placed: a vortex tube with blades, a conical chipper that prevents the entrainment of a droplet liquid with a stream of purified gas and anti-rotation plates that stop the rotation of the liquid.

Under the action of the emerging centrifugal force, droplets of liquid and solid particles are directed to the walls of the cyclone, where they form a liquid film moving to the lower part of the cyclone scrubber.

The most effective and technically advanced separator is one from which droplet liquid and gas bubbles cannot be carried out, while the delay time of the oil in the separator and the metal consumption for its manufacture should be minimal. In addition, a phase equilibrium between gas and oil should be established in such a separator.

When separating oil and gas in a separator, one should strive to create a large contact surface between the phases and not to allow excessive entrainment of the droplet liquid and acclaimed gas bubbles from the separator. The increase in the contact surface between oil and gas significantly reduces the time to achieve the equilibrium state of the system at a given temperature and pressure.

Therefore, the effective gas evolution from oil in the separator can only be in the finely dispersed state of the oil and gas mixture, which is provided, as a rule, with nozzle nozzles or with special dispersants. In this case, a finely dispersed oil and gas mixture is formed in the separator volume, consisting mainly of oil droplets ranging in size from 1 to 2 mm. The size of oil droplets in the separators is a function of the ratio $\sigma/\Delta\rho$, where σ is the interfacial tension, and $\Delta\rho$ is the difference in phase densities.

Droplets of this size are successfully captured in the drop trap section, which is set as a gas-dynamic module.

The highest degree of purification in gas pre-separators, with a louvred nozzle and inclined shelves at a relatively wide speed range. At high speeds of the medium (gas) being cleaned, a gas-dynamic module works very efficiently from liquid droplets, which can beat off a suspension of 10 microns in size. In third place in terms of efficiency is a hydrocyclone separator and in fourth is a separator with shutters, but without tilted shelves, and finally, the most ineffective separator is a gravitational type without shelves and louvred nozzles.

In a centrifugal separator, centrifugal force is widely used to separate oil from gas. Consider the centrifugal force and the separation factor of the mixture.

When the body rotates, a centrifugal force C in kg · m / s^2 arises, directed along the radius from the axis of rotation and equal to the product of the body mass m in kg by the square of the peripheral speed W in m / s divided by the radius of rotation r in meters.

3. Results

The analysis of the separator-slug catcher (Figure 1) revealed the following disadvantages:

1) a gas plug at the entrance to the product inlet tank has an impact on its entire structure. Cyclic impacts contribute to the accumulation of fatigue stresses, which leads to premature damage and subsequent destruction of the product input capacity

2) entrainment of the separated liquid into the gas outlet line from the product outlet tank by the ascending vortex flow of the separated gas, which reduces the separation efficiency.

3) insufficiently effective separation of the gas-liquid flow into gas and liquid.

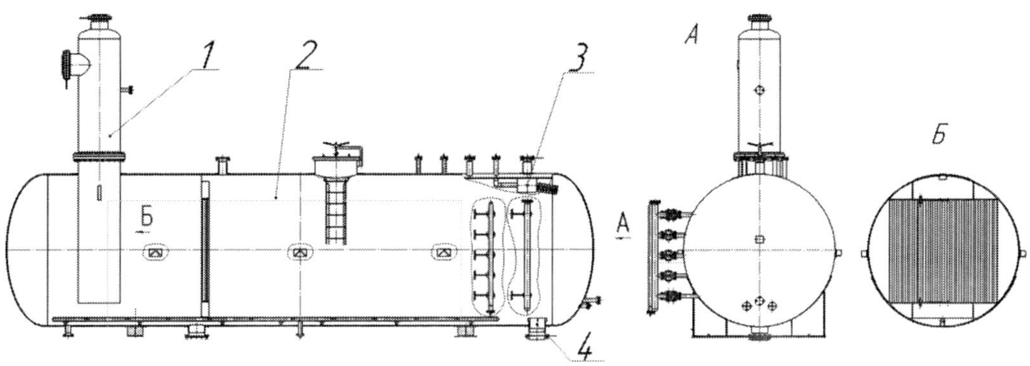

Figure 1. Separator slug catcher: 1 – product inlet capacity; 2 – separator; 3 – light output system; 4 – heavy fraction withdrawal system

To eliminate this problem, it is proposed to change the design of the product inlet container with a scrubber installation (Figure 2). The scrubber is used to separate the liquid from the gas stream and to protect equipment downstream, and also removes mechanical impurities from the stream.

As a result of the implementation of this technical solution, the pressure drop across the inlet capacity of the separator product was halved while maintaining the same degree of separation. The calculations showed that, despite the change in the geometric characteristics of the separator, the degree of separation efficiency has not changed.

The product inlet capacity is designed to capture gas plugs formed during oil transportation, gas-liquid mixture separation on CPF.

The capacity includes the following elements:

- nozzle type washer for sand;
- slowing down solid partitions;
- defoamer of low shear pulses at the input;
- defoaming perforated partition;
- a device for separating gas and water from oil.

The apparatus is a horizontal cylindrical vessel with hemispherical bottoms mounted on two saddle supports.

The optimized scrubber (Figure 3), by the principle of operation, is a cyclone type scrubber.

Scopes of a cyclone-type scrubber:

1) separation of liquid (water, hydrocarbons) from gas (associated gas, etc.),

2) downstream equipment protection (separator elements, etc.).

3) removal of mechanical impurities from the stream (dust, sand, etc.).

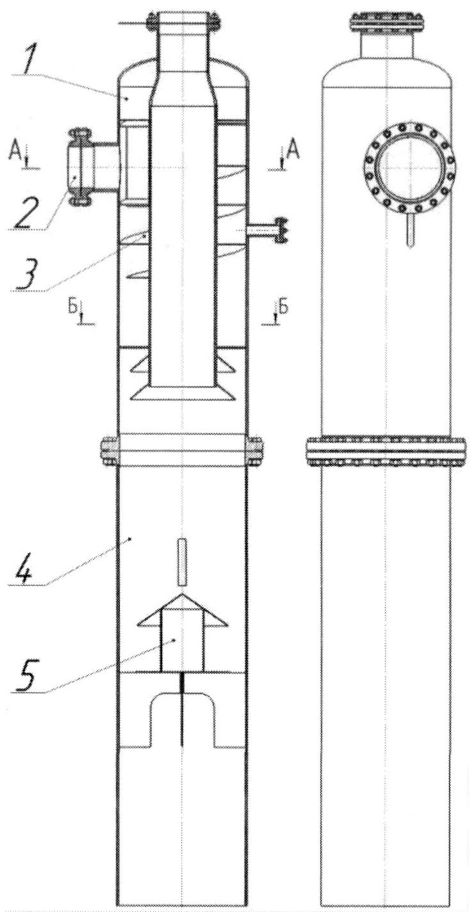

Figure 2. Product input capacity:
1 – capacity body; 2 – product entry;
3 – scrubber; 4 – shank; 5 – anti-swirl

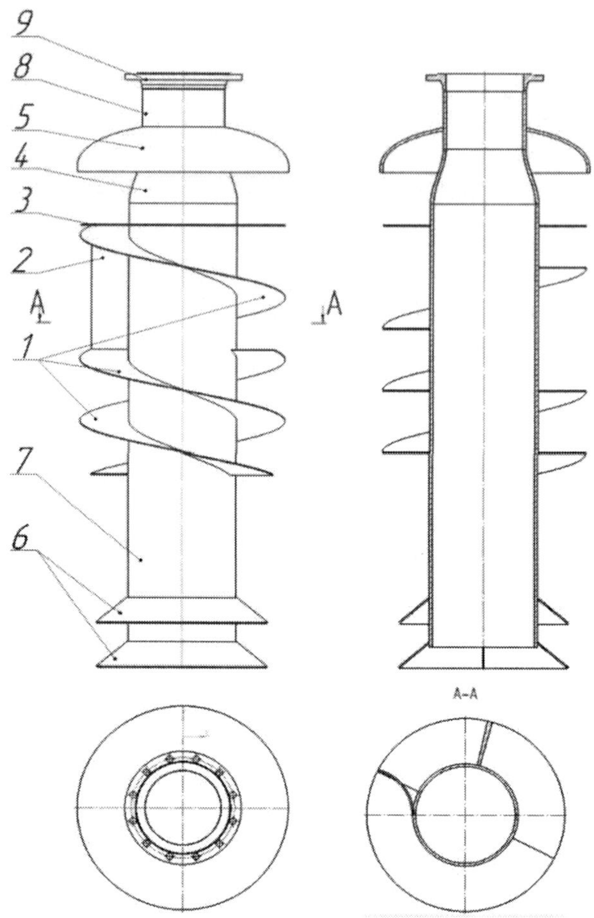

Figure 3. Scrubber: 1 – swirl; 2 – guide; 3 – disk; 4 – transition; 5 – bottom; 6 – chipper; 7 – a pipe; 8 – pipe; 9 – flange

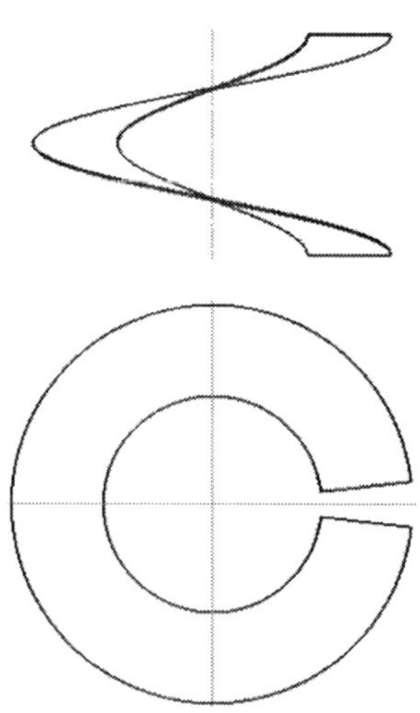

Figure 4. Swirl

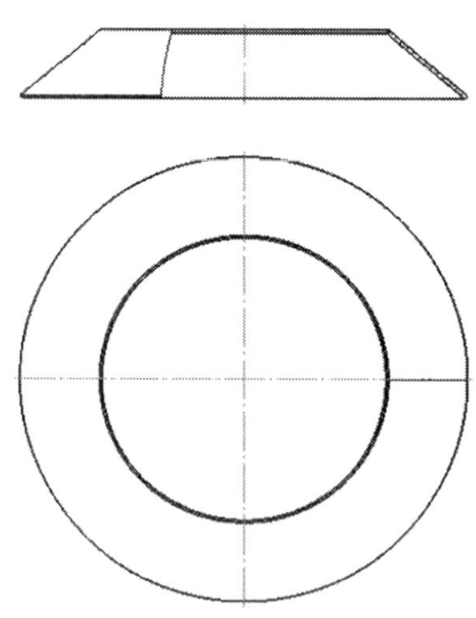

Figure 5. Chipper

The blades of the optimized geometric shape of the scrubber drive the combined phase flow into rotation.

The resulting centrifugal force directs liquid droplets and mechanical impurities to the walls of the apparatus, where they form a liquid film moving down the scrubber. Gas leaves the scrubber through the central gas outlet. The baffles at the bottom of the cyclone stop the rotation of the fluid and the secondary entrainment of the fluid by the gas stream. Thus, efficient phase separation is provided.

When using the optimized scrubber of the separator, a gas-liquid stream is supplied to the scrubber with its subsequent separation into gas, liquid and impurities. As a result, the gas-liquid flow feeds into the inlet pipe located on the scrubber body, after which they give the stream a rotational movement by feeding it to a spiral surface located between the inner surface of the body and the outlet pipe. Then, a rotating gas-liquid flow is directed to the walls of the scrubber body, thereby separating the liquid from the gas due to the density difference and the action of centrifugal and gravity forces, while ensuring that the separated liquid moves down the body walls, and the released gas moves upward through the outlet pipe. At the same time, to prevent the ingress of fluid flowing along the walls of the scrubber body and located in the center of the rotating flow, conical belts are used inside the outlet pipe, which are installed on the outer surface of the body of the outlet pipe in its inlet part and with the help of which liquid is drained from the inlet part.

The rotating stream is directed downward to the liquid collecting tank, providing during its movement from the inlet pipe to the lower part of the scrubber body the liquid is separated from the gas.

Then, to exclude the entrainment of liquid from the stream of purified gas, the gas-liquid stream is slowed down by expanding along the boundaries of the base of the cone, which is installed in the lower part of the housing, while liquid and splashes are not allowed to enter the inlet of the outlet pipe using this cone and conical belts installed on the outer surface of the body of the outlet pipe, then in the lower part of the scrubber body the flow is directed to the ribs of the device to prevent rotation of the flow, which are located in the lower parts of the casing, preferably below the cone, mainly on the inner wall of the casing parallel to the vertical axis of the casing, then the fluid flow having an axial velocity component is sent to the fluid collecting container, where it is accumulated, sand and solids are removed from the liquid by sedimentation under the action of gravity and the subsequent periodic removal of the collected fluid and these foreign particles.

During the analysis of scrubber optimization, it was revealed:

1) a change in the design of the scrubber allowed to reduce the pressure drop in the separator by 50 %;

2) the calculations showed that, despite the change in the geometric characteristics of the separator, the degree of separation efficiency has not changed.

Benefits of an optimized scrubber:

1) allows you to remove the restrictions imposed by the small dimensions of gas and liquid devices;

2) are not subject to clogging;

3) excellent work with slug flow mode;

4) small size and weight, which is especially important when used on offshore installations;

5) does not require regular maintenance;

6) quick payback by increasing system throughput

4. Conclusion

The optimized scrubber design has the following advantages:

1) the small size and weight of the apparatus, due to the high value of the permissible gas load;

2) ease of maintenance: the absence of channels and transfer pipes of small diameter, capable of clogging;

3) excellent cork flow;

4) large working range.

5) does not require regular maintenance;

6) quick payback by increasing system throughput

With the introduction of this technical solution, the average annual economic effect will amount to 175044 rubles, and the profit remaining at the disposal of the enterprise 140035 rubles.

References

[1] Jaworski A J and Meng G 2009 On-line measurement of separation dynamics in primary gas/oil/water separators: Challenges and technical solutions-A review *J. of petrol. Sci. and engineer.* **68(1–2)** 47–59

[2] Zhao Ch, Guo J C, Yang T J,Yang Z 2016 A Study of the Mechanism of Separation in a Multi-Cup Downhole Gas-Oil Separator *Chem. and technol. of fuels and oils* **52(3)** 261–71

[3] Zainagalina L Z, Suleimanov R I, Gabdrakhimov M S and Khabibullin M Ya 2018 Determining oscillating system dynamic parameters of a near-bit junk pulper *Advan. in Engineer. Res. (AER) Int. Conf. Actual issues of mechan. Engineer. (AIME 2018)* vol. 157 pp 642–5

[4] Frank M, Kamenicky R, Drikakis D et al 2019 Multiphase Flow Effects in a Horizontal Oil and Gas Separator *Energies* **12(11)**

[5] Suleimanov R I, Zainagalina L Z, Khabibullin M Ya et al 2018 Studying heat-affected zone deformations of electric arc welding *IOP Conf. Ser. Mater. Sci. and Engineer.* **327(3(032053))**. DOI: 10.1088/1757-899X/327/3/032053

[6] Skjefstad H S and Stanko M 2019 Experimental performance evaluation and design optimization of a horizontal multi-pipe separator for subsea oil-water bulk separation *J. of petrol. Sci. and engineer.* **176** 203–19

[7] Zainagalina L Z, Petrova L V and Petrov V A 2019 Analysis of the efficiency of telemetric systems for drilling wells *IOP Conf. Ser. Mater. Sci. and Engineer.* **560(1(012096))**. DOI: 10.1088/1757-899X/560/1/012096

[8] Davydov A Yu 2018 Utilization of expert systems for screw pump sets with surface drive management *Advan. in Engineer. Res. (AER) Int. conf. Actual issues of mechan. Engineer. (AIME 2018)* vol 157 pp 129–33

[9] Andreev V E, Chizhov A P, Kotenev Yu A et al 2019 Prediction of Gas Exposure in the Conditions of the Oil Fields of the Volga-Urals *Atlantis Highlights in Material Sciences and Technology (Int. Symp., dedicated to the 85th anniversary of H.I. Ibragimov, Engineering and Earth Sciences: Applied and Fundamental Research (ISEES 2019))* vol 1 pp 666–9. DOI: 10.2991/isees-19.2019.132

[10] Akhmetov R T, Kuleshova L S and Mukhametshin V V 2019 Absolute Permeability and Distribution of Pore Throats of the Productive Strata of Western Siberia," *Atlantis Highlights in Material Sciences and Technology (Int. Symp., dedicated to the 85th anniversary of H.I. Ibragimov, Engineering and Earth Sciences: Applied and Fundamental Research (ISEES 2019))* vol 1 pp 535–9. DOI: 10.2991/isees-19.2019.105

[11] Abdyukova R Ya 2019 Studies on operation and types of drilling pump valves *IOP Conference Series: Materials Science and Engineering* **560(1(012050))**. DOI: 10.1088/1757-899X/560/1/012050

[12] Almukhametova E M, Karimova A A, Almukhametov F F et al 2019 Study of water-oil mixture separation in bottomhole area via drift flow model *IOP Conf. Ser. Mater. Sci. and Engineer.* **560(1(012005))**. DOI: 10.1088/1757-899X/560/1/012005

Influence of complicating factors on the operation of an electric centrifugal pump installation

R I Suleimanov, M Ya Khabibullin and R I Suleimanov

Department of Oil and Gas Field Machines and Equipment, Ufa State Petroleum Technological University, Branch of the University in the City of Oktyabrsky, Oktyabrsky, Russia

E-mail: rustamsul@rambler.ru, m-hab@mail.ru

Abstract. The article analyzes the complicated stock of producing wells equipped with electro-centrifugal pump installations in an oil field. The study identified factors causing complications in oil production. In the course of the study, a complicated well stock was examined, and possible ways to prevent the formation of complications during the operation of oil producing wells were studied. The article presents well performance indicators that reflect the factors that complicate the operation of electric centrifugal pump installations. It developed recommendations to improve the condition of the complicated deposit fund.

Keywords: installation of an electric centrifugal pump, complicating factors, failure, operational efficiency, scaling, mechanical impurities, corrosion, gas content.

1. Introduction

The main method of operating oil wells in the Russian Federation is a mechanized pumping method.

A feature of this method of oil production in recent years is the manifestation of many complicated situations during the operation of wells associated with the commissioning of hard-to-recover oil reserves, as well as the intensification of oil production in existing fields. The number of oil wells operating low-productivity and heterogeneous reservoirs by a mechanized pumping method has increased, as well as the use of low-productivity pumps with relatively low efficiency indicators. The number of production wells introduced after the hydraulic fracturing technology has increased significantly. Because of the intensification of oil production, the content of free gas and the volume of mechanical impurities at the reception of electric centrifugal pumps increased. The developed head of electric centrifugal pumps increased, as well as the descent depths, because of which the operating temperature of submersible pumping equipment units' increased, favorable conditions for scaling appeared on working elements of electric centrifugal pumps. As a result of the above changes, the well stock, with complicated characteristics, increased and, consequently, the frequently repaired fund increased, the overhaul period and mean time between failures decreased [1, 2].

High quality and reliable operation of the mechanized pumping stock of wells determines the performance indicators of an oil-producing enterprise. High rates of oil recovery, capital and operating costs, costs of the enterprise for power supply depend on the efficiency and reliability of the mechanized pumping stock of wells [3, 4].

To increase the mean time between failures and the overhaul period of the mechanized pumping stock of wells, we created software products that facilitate the implementation of technological measures, that will ensure the technological effect during their implementation [5].

The analysis of field data on the occurrence of failures of submersible pumping equipment is carried out and probable problems and causes that reduce the reliability of wells equipped for the mechanized method are determined [6].

The main efforts when working with a mechanized well stock were focused on the following areas:

- improving the reliability of submersible equipment units;
- strengthening control over the implementation of all requirements of technological regulations and standards;
- constant work with staff to improve the technical level and involvement in the process of "work on the result" [7].

To create conditions for the long and uninterrupted operation of electric centrifugal pump installations in order to ensure a basic level of oil production at oil producing enterprises, we took measures to monitor the current technical condition in order to reduce the number of failures. The most common causes of failure of the centrifugal pump installations are: failure of the electric centrifugal pump supply, malfunction of the cable electric line and lack of insulation resistance, violation of the tightness of the tubing when lifting well products, jamming of rotating elements and the likelihood of deformation and subsequent breaking of the shaft of the electric centrifugal pump [8, 9].

In the structure of complications of the mechanized well stock, five dominant directions can be distinguished (in importance):

- removal from the reservoir of solid particles (rock and proppant after hydraulic fracturing) that clog and abrade the pump;
- salt deposition on the working parts of electric-centrifugal pumps;
- temperature effect;
- structural reliability of individual nodes and the entire ESP system;
- staff qualifications.

2. Methods and materials

In the field practice of the Vankorskoye field, technological and technical solutions are used to fight the listed complications, for example, the use of filters, inhibitor protection, the use of formation fastening measures that significantly reduce the impact of complications on the operation of the pump and the installation of the electric centrifugal pump as a whole. Complications themselves are not manifested in hierarchy, but in the simultaneous combination of different species among themselves, which requires the use of integrated protection technologies [10].

As a result, the company carries out a certain amount of routine maintenance, uses various technologies that require costs to eliminate the complications that arose. The company has certain problems associated with a complicated fund that need to be addressed, the most common of which are as follows:

- determination of the estimated period of routine maintenance;
- scheduling routine maintenance;
- maintaining documentation of work performed and document management;
- determination of the effectiveness of work performed on implemented technologies;
- determination of the effectiveness of the work of personnel conducting the work;
- accounting for complications arising from individual wells and the field as a whole;
- development of a plan of work performed, ordering technical equipment, materials for the future;
- determination of the cost of work performed and the technical and economic effect of the measures.

IOP Conf. Series: Materials Science and Engineering **905** (2020) 012091 doi:10.1088/1757-899X/905/1/012091

A number of measures are being taken by oil enterprises to create conditions for the quality work of electric centrifugal pump installations;

- periodic measurements of the operating parameters of the wells using modern methods of hydrodynamic research of wells;
- conducting continuous monitoring and evaluation of the obtained data on the operation of wells in order to maintain an optimal operating mode, the subsequent selection of geological and technical measures and ensuring performance indicators of wells;
- ensuring adjustment of the operating mode of wells equipped with electric centrifugal pump units.

Diagnostics of the condition of pumping equipment includes:

1) comparison of the planned indicator of the technical condition with its actual current state;
2) determining the coherence in the work of the underground and ground parts of electric centrifugal pump installations with an oil reservoir;
3) detection of negative factors that disrupt the operation of the underground and ground parts of the electric centrifugal pump units and the bottomhole zone of the well.

According to the data obtained during diagnostics, the necessary recommendations are developed for the required geological and technical measures, both in the current mode and subsequently in time. They recalculate well productivity, oil and fluid production rates for specific facilities and the field as a whole for any period of well operation. Timely diagnostics of the state of electric centrifugal pump installations allows us to effectively change the current situation and introduce optimal solutions that increase the efficiency of well operation [11].

Various kinds of complications during the operation of the mechanized pumping stock of wells affect the final result due to the failure volumes of electric centrifugal pump installations. The performance of the entire pumping system is simultaneously affected by many factors, while determining the dominant root cause of failure is the most important issue for the further selection of equipment protection methods. Wrong decisions in the diagnostic process when identifying the real cause of failure lead to their re-manifestation and additional overhead costs to eliminate the consequences.

In managing complications, the determining role is played by the identification of criteria for the application of various technological measures and the choice of the most effective one from a number of alternative solutions.

There are a number of unsolved problems that reduce the efficiency of the production cycle:

- The most acute problem is the removal of proppant from the formation after hydraulic fracturing. In such wells, the concentration of solid abrasive particles is tens of times higher than the permissible values for the normal operation of the ESP pump (150–450 mg/l), which reduces their production time by 2.5–4 times.
- Inflows with low rates (up to 30 m^3/day) that do not provide stable operation of electric centrifugal pumps in a constant mode. For such a well stock, it is necessary to purchase more low cost low-cost equipment (ESP 20–40) with a lower efficiency factor, the operating time of which is 45–50 % lower than the average operating time in these wells, equipped with electric centrifugal pump installations in periodic mode (ESP 45–90) The transfer to operation of a sucker rod pump will not allow preserving the existing level of production caused by restrictions on the depth of descent of pumping equipment.
- Constant work with personnel servicing the objects in question is required. Changing the design of pumping equipment, new, relevant modern requirements for technological processes, other various factors require constant attention to training personnel, any incorrect decisions of which negatively affect the result of work.

The process of oil and gas production is associated with a large number of complicating factors. For this reason, one of the main tasks facing the oil company is to reduce the degree of influence of these factors, which in turn will reduce costs and increase production efficiency.

3. Results

Dynamics of changes in the complicated well stock from 2011–2015 presented in figures 1 and 2.

The main complications by type:

- the harmful effects of gas;
- corrosion downhole pumping equipment;
- mechanical impurities;
- scaling.

As you can see the problem of ESP failure at the Vankor field due to complicating factors such as clogging, high gas factor, corrosion, and fur. impurities, the deposition of salts rises quite sharply (Figure 3).

The main reasons for the growth of complications:

- decrease in bottomhole pressure;
- increased water cut;
- decrease in reservoir pressure.

The increase in water cut in wells is affected by:
1) corrosion;
2) salts;
3) mechanical impurities.

The decrease in reservoir pressure is affected by:
1) gas;
2) mechanical impurities.

The decrease in bottomhole pressure is affected by:
1) gas;
2) mechanical impurities.

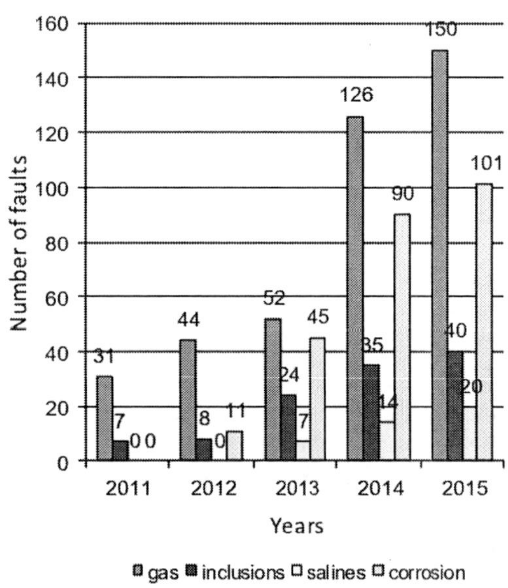

Figure 1. Distribution of ESP failures for reasons of occurrence: number of failures, years, gas, mechanical impurities, salts, corrosion

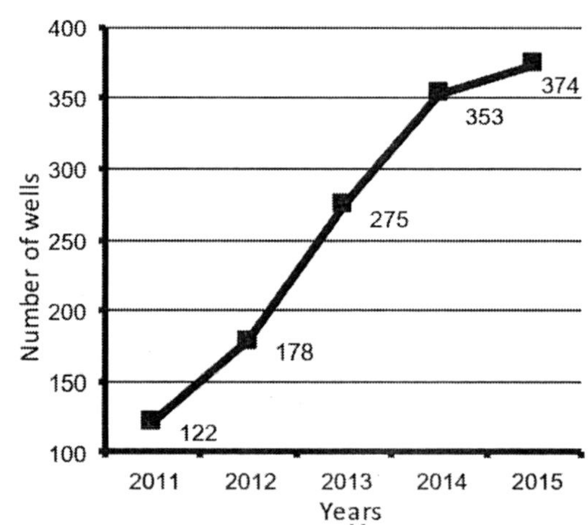

Figure 2. The dynamics of the ESP fund by years: number of wells, years

In the conditions of the Vankorskoye field, the saturation pressure is equal to the initial reservoir pressure.

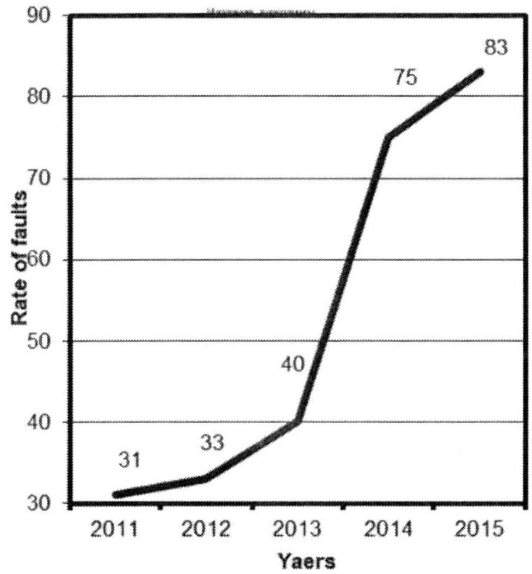

Figure 3. Dynamics of ESP failures by years: percent of failures

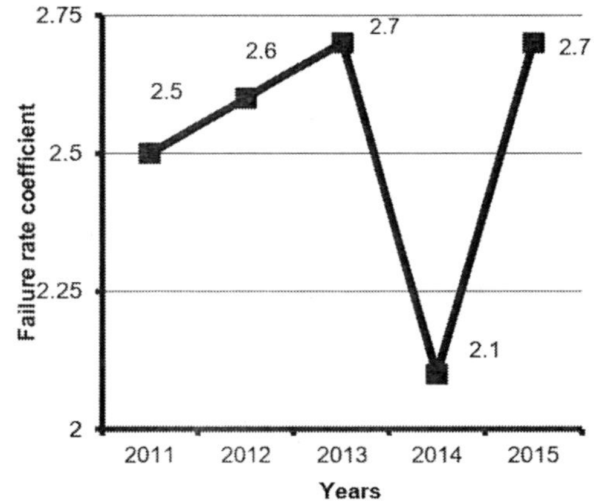

Figure 4. Dynamics of the ESP failure rate by years: failure rate, years

In this case, the saturation pressure is a multiple of the current bottomhole pressure, which leads to complications in the operation of the ESP associated with a large amount of free gas in the ESP.

In 2015, the decrease in MTBF was due to an increase in the number of failures.

Given the decrease in the number of new wells, the increase in the share of the complicated stock, as well as the late effect of the ongoing measures to combat complications, in 2015 the overhaul period and failure rate remained at the level of 2011-2014 (Figures 4 and 5).

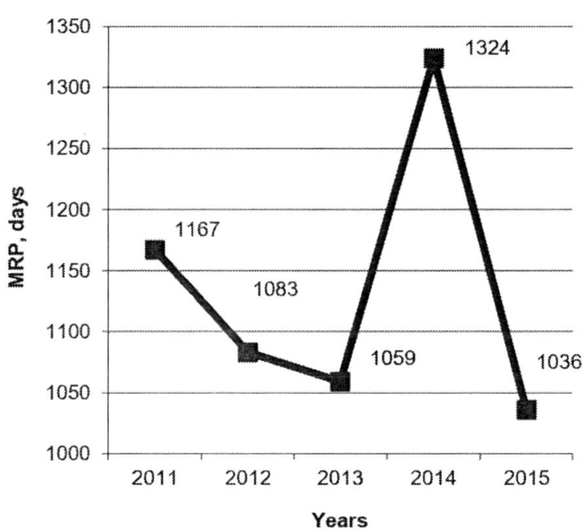

Figure 5. Dynamics of the overhaul period of wells by years: MRP per 24 hours, years

One of the main reasons for the growth of failures is a significant increase in the share of the complicated fund, and as a result, an increase in the negative impact of complicating factors on the operation of electric centrifugal pump installations, as well as an increase in the failure rate of electric centrifugal pump plants with high operating time, which is caused by an increase in the number and proportion of wells with high operating time (figure 6).

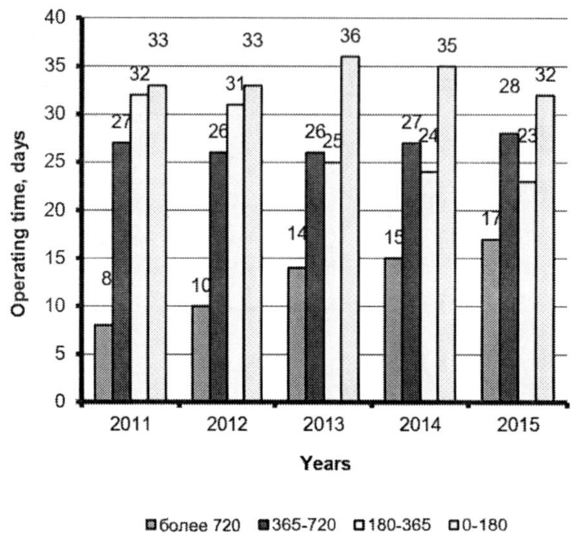

Figure 6. Dynamics of ESP operating time by years: profit per 24 hours, years

Measures to protect electric centrifugal pump installations from the negative impact of complicating factors used at the Vankor field are as follows:

1) by the complicating factor "Clogging with mechanical impurities";

- the use of separators of mechanical impurities and sand filters;
- operation of electric centrifugal pump installations with increased pressure at a reduced frequency;
- normalization of the face during the current and capital repair of wells;

2) by the complicating factor "Corrosion":

- introduction of suspensions of tubing with production material - steel with a chromium content of 13 %;
- the use of tubing corrosion-resistant performance with a protective silicate-enamel coating;
- the introduction of protectors to protect the submersible motor from the negative effects of corrosion;
- transfer of oil wells to constant dosing with a chemical reagent through a BDR reagent dosing unit, refusal from periodic treatments;
- applying a corrosion-resistant coating "Monel" and "Stainless steel" by the method of electric arc spraying on the casing of the installation of electric centrifugal pumps;

3) according to the complicating factor "Salt deposits":

- transfer of wells to constant dosing with a chemical reagent through a BDR reagent dosing unit, refusal of periodic treatments;
- the deposition inhibitor in the reservoir under pressure;

4) by the complicating factor "Effect of gas":

- the use of installations of electric centrifugal pumps of a smaller size - 400 series instead of 538;
- the use of cone assemblies of electric centrifugal pump installations with a retaining section of greater productivity at the pump intake;
- the use of electric centrifugal pump units with enlarged sections of the MVP multiphase pump;
- descent of gas separators;

- testing of devices for monitoring the performance of pumps during operation on the boundary of the supply stall due to the influence of gas.

4. Conclusion

In the period 2011–2015 the average failure rate remains at approximately the same level, with the exception of 2014 (a decrease of 24 %). At the same time, there is a dynamic of a smooth decrease in the overhaul period (also with the exception of 2014), which is typical for the initial stage of field operation. In 2014, there was a significant increase in the overhaul period, due to a significant increase in the fund and a decrease in the number of failures compared to the plan.

In general, the average failure rate in 2015 remains at the level of 2011–2013.

References

[1] Ji Hong, Sun Lei, Xu Dandan et al 2009 The efficiency analysis of an electric, motor-pump with port-plate centrifugal pump *7th Int. Conf. on Fluid Power Transmiss. and Control (ICFP 2009)* pp 183–8

[2] Sengpanich K, Bohez Erik L J, Thongkruer P and Sakulphan K 2019 New mode to operate centrifugal pump as impulse turbine *Renewable Energy* **140** 983–93

[3] Zainagalina L Z, Suleimanov R I, Gabdrakhimov M S and Khabibullin M Ya 2018 Determining oscillating system dynamic parameters of a near-bit junk pulper *Advan. in Engineer. Res. (AER) Int. Conf. Actual issues of mechan. Engineer. (AIME 2018)* vol 157 pp 642–5

[4] Kuleshova L S, Kadyrov R R, Mukhametshin V V and Safiullina A R 2019 Design changes of injection and supply wellhead fittings operating in winter conditions *IOP Conf. Ser. Mater. Sci. and Engineer.* **560(1(012072))** 1–5. DOI: 10.1088/1757-899X/560/1/012072

[5] Cortes B, Araujo L R and Penido D R R 2018 Passive Filters Design Applied to an Electrical Submersible Pump System *IEEE Latin Amer. Transact.* **16(7)** 1992–9

[6] Stan M, Pana I, Minescu M, Ichim A and Teodoriu C 2018 Centrifugal Pump Monitoring and Determination of Pump Characteristic Curves Using Experimental and Analytical Solutions *Processes* **6/2**

[7] Davydov A Yu 2018 Utilization of expert systems for screw pump sets with surface drive management *Advan. in Engineer. Res. (AER) Int. Conf. Actual issues of mechan. Engineer. (AIME 2018)* vol 157 pp 129–33

[8] Zainagalina L Z, Petrova L V and Petrov V A 2019 Analysis of the efficiency of telemetric systems for drilling wells *IOP Conf. Ser. Mater. Sci. and Engineer.* **560(1(012096))**. DOI: 10.1088/1757-899X/560/1/012096

[9] Akhmetov R T, Kuleshova L S and Mukhametshin V V 2019 Application of the Brooks-Corey model in the conditions of lower cretaceous deposits in terrigenous reservoirs of Western Siberia *IOP Conf. Ser. Mater. Sci. and Engineer.* **560(1(012004))** 1–4. DOI: 10.1088/1757-899X/560/1/012004

[10] Sun H, Yuan Sh, Luo Y et al 2017 Unsteady characteristics analysis of centrifugal pump operation based on motor stator current *Proc. of the Instit. of Mechan. Engineers Part A – J. of Power and Energy* **231(8)** 689–705

[11] Khabibullin M Ya and Suleimanov R I 2019 Automatic packer reliability prediction under pulsed transient flooding of hydrocarbon reservoirs *IOP Conf. Ser. Mater. Sci. and Engineer.* **560(1)** Art. no 012024. DOI: 10.1088/1757-899X/560/1/012024

IOP Conf. Series: Materials Science and Engineering **905** (2020) 012092 doi:10.1088/1757-899X/905/1/012092

Dynamics of the oscillatory system of the near-bit slurry grinder

L Z Zaynagalina

Department of Oil and Gas Field Machines and Equipment, Ufa State Petroleum Technological University, Branch of the University in the City of Oktyabrsky, Oktyabrsky, Russia

E-mail: zlz11@mail.ru

Abstract. The article proposes a near-bit device for vibrodynamic grinding of sludge in a horizontal section of a well – a near-bit sludge shredder, its advantages are given. Theoretical studies of the dynamics of the near-bit sludge grinder oscillatory system depending on the main geometric design parameters are presented. Based on the solution of the mathematical model of the operation of the oscillatory system of the slurry grinder, the law of the relative rotation of the drum-mesh and its angular velocity of rotation relative to the shaft of the turbodrill (bottom of the drilling tool) is established. According to the results of solving the equation of motion of the working elements near-bit slurry grinder, the dependence of the amplitude of the angular velocity of the relative vibrational rotation of the drum on the stiffness of the springs and on the moment of inertia of the drum is established.

1. Introduction

The bottoms of oil and gas wells during drilling, as a rule, become clogged with large drill cuttings, which leads to a deterioration in the process of rock destruction and a decrease in the performance of bits [1]. With an increase in drilling depth, more and more sediment remains in the face, consisting mainly of large-sized sludge [2]. The removal of sludge is complicated when drilling deep vertical, deviated and horizontal wells, as well as when drilling shallow wells, when the circulation of the flushing fluid does not create a sufficient upward flow for the removal of sludge. The most urgent is the problem of cleaning the horizontal wellbore from cuttings. Flushing a horizontal wellbore is characterized in that the flushing fluid flow velocity vector is perpendicular to the slurry particle deposition rate vector. The absence of a vertical component of the flow velocity vector causes sludge to settle on the bottom wall of the well [3–6]. If sludge particles are not removed efficiently enough, they will settle around the string, usually in the area of the bottom assembly of the drill string, blocking pipes and clamping the string until sticking occurs.

In horizontal drilling, all drill cuttings tend to sink to the bottom wall of the well. Two sludge transport mechanisms are possible:

1) At a sufficiently high flow velocity, turbulent pulsations (vortices) pick up sludge particles and transfer them to the region of high flow velocities. When the vortex decays, the particles begin to settle until they are picked up by the next vortex. Thus, the transport of particles in suspension.

2) If the average value of the transverse velocity pulsations is less than the sedimentation rate of solid particles, the solid phase will begin to settle on the bottom wall of the well, forming sediments. Settled particles could move over the surface of the sediment if the force exerted on the particle by the liquid is greater than the forces of resistance to the movement of the particle. Such a mechanism for moving a particle is called moving in an entrained state [7].

To prevent sedimentation of sediment and the formation of sediment, it is necessary that the particles of the cuttings are in a suspended rather than entrained state. But, an increase in the supply and an increase in the density of the drilling fluid can cause undesirable consequences: the mud of the well bottom; increase in excess pressure on permeable formations, up to their hydraulic fracturing; erosion of the walls of the well during drilling of unstable rocks [3, 8].

The condition for the transport of sludge particles along the bottom of the channel by a turbulent flow [2]:

$$Q = U * S_{кп} \geq \sqrt{\frac{5.3\Delta_{\rho} * d * f * (1 + 0.46e^{2.2})}{\lambda * \rho}} * S_{кп},$$

where Q is the flow rate of flushing fluid, U is the upward velocity of flushing fluid in the annular space, $S_{КП}$ is the largest area of the annular space, $\Delta\rho$ is the pressure drop in the eccentric annular space, d is the equivalent diameter of the solid particle, f is the coefficient of resistance to movement, e is the relative the eccentricity of the drill string, λ is the coefficient of hydraulic resistance, ρ is the density of the fluid.

Calculations show that it is possible to transport particles of sludge in size of not more than 1-2 mm in suspension. Larger particles, even with a turbulent flow regime, will settle on the bottom wall of the well and can only move in the entrained state [9, 10].

A new technical solution for cleaning vertical, deviated and horizontal wells from cuttings using a slurry grinder is proposed. A near-bit design was developed for crushing, sieving and slurry removal from the bottom when drilling vertical, deviated and horizontal wells (Figure 1).

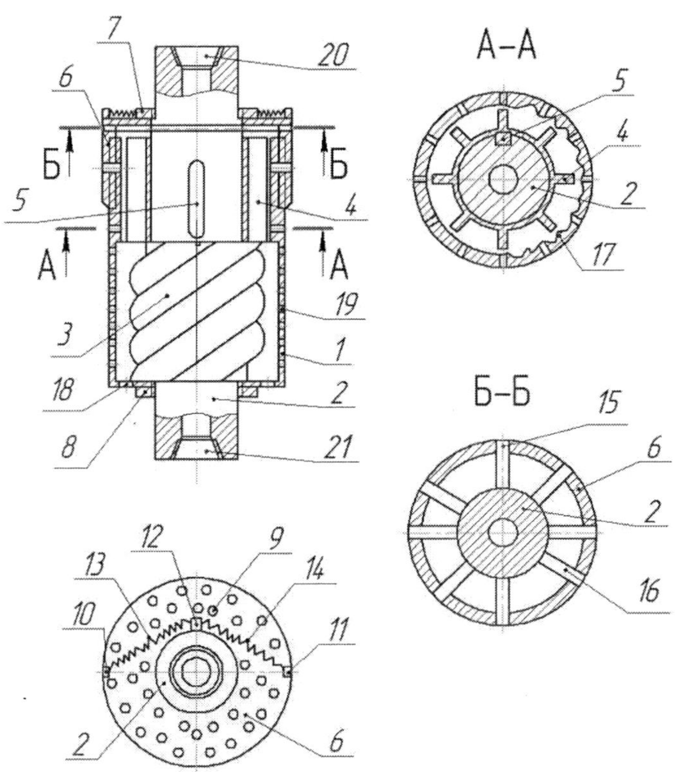

Figure 1. Structural diagram of a near-bit slurry grinder

The device is installed during rotary drilling (rotary method, downhole motors) above the bit. When drilling a well, the sub 2 rotates, drives the screw 3 and the rotor 4. The rotating sub 2 and the drum connected by springs 13 and 14 form an oscillating system. The auger directs the flushing liquid with slurry into the internal cavity of the device. The auger directs the liquid with slurry in the axial direction and in the radial direction, while the slurry is partially crushed due to impacts, small particles are sifted through the grid (mesh holes 2–3 mm), large particles fall into the rotor and are thrown radially to the reflector, further grinding of the slurry occurs. The final grinding occurs when large particles enter the channels 16 and crush them with the end face of the rotor. Purification of the solution is facilitated by torsional vibrations of the drum (mesh), which occur under the influence of the inertia of the drum, the fluid ejected by the rotor onto the eccentric reflector, and spring forces [9].

The device is designed for drilling soft and medium hard formations. Grinding sludge using a near-bit device will help prevent accumulation of sludge dune on the bottom wall of a horizontal well.

2. Methods and materials

To determine the rational geometrical dimensions of the near-bit sludge grinder elements that ensure the stable operation of the device, theoretical and experimental studies are necessary.

To determine the law of motion of the drum relative to the rotating shaft, analytical studies of the processes under study were carried out using the classical principles of modern theoretical mechanics.

3. Results

The rotating shaft of the turbo-drill and the drum, connected by springs, form an oscillatory system. When the turbo-drill shaft rotates due to the presence of springs, torsional vibrations of the drum are created, which ensures grinding and sieving of the sludge.

The rotating drum receives rotation from the shaft through an elastic spring of stiffness “c”. The rotation of the drum is coaxial with the shaft, on which the torque $M_{вр}$ acts. We believe that the axial moments of inertia of the drum $J_б$ and the shaft $J_в$ relative to the axis of rotation O. We neglect the spring mass.

The mechanical system (shaft, drum, spring) begins to work from a state of rest under the action of a torque $M_{вр}$. It is assumed that when the shaft rotates under the action of spring elasticity, the drum will oscillate rotationally with respect to the shaft. The nature of these oscillations will depend on the torque acting on the shaft, $M_{вр}=M_{вр}(t)$. The task is to determine the law of torsional vibrations of the drum relative to the rotating shaft.

We believe that as an assumption, and based on the principle of superposition of vibrations, in this case we will consider torsional vibrations without taking into account their influence on longitudinal vibrations, i.e. longitudinal and torsional vibrations do not affect each other (manifest independently of each other).

Consider the dynamic transmission of rotation from the leading link of the mechanism (shaft) to the driven link (drum) through an elastic spring (Figure 2).

When the shaft begins to rotate under the influence of the torque $M_{вр}$, the spring first twists (moving point A of the spring with the shaft) relative to the stationary drum (point B of the spring with the drum); in this case, an internal elastic moment $M_у$ (counteracting the moment $M_{вр}$), arises in the spring, which increases in proportion to the spring angle $\varphi_{пр}$. After overcoming the inertia of the rest of the drum, the last one begins to turn in the direction of $M_{вр}$ by the value of the angle $\varphi_б$ (moving – turning point B). Therefore, at a certain point in time t , the angle of rotation of the shaft (relative to the fixed space XOY) $\varphi_в$ will be determined by the sum of the angle of rotation of the drum $\varphi_б$ and the angle of rotation of the spring $\varphi_{пр}$, i.e. $\varphi_в=\varphi_б+\varphi_{пр}$. It follows that initially the rotation of the drum will lag behind the rotation of the shaft by the value of the angle of rotation of the spring $\varphi_{пр}$, and this is essentially the circular displacement of the drum relative to the shaft. The function $\varphi_{пр}(t)$ describes the relative rotation of the drum; and the function $\varphi_б(t)$ is the portable rotation of the drum (together with the shaft). Thus, the task is to determine the relative rotation of the drum, that is, the function of changing the angle of rotation of the spring $\varphi_{пр}(t)$ depending on the actual torque $M_{вр}$ on the shaft.

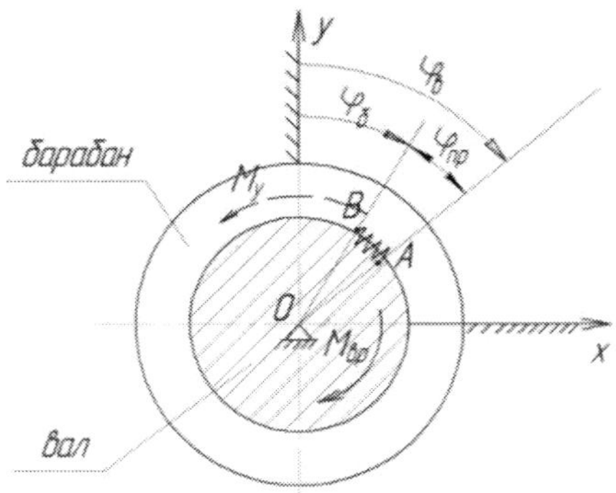

Figure 2. The dynamic transmission of rotation from the leading link of the mechanism (shaft) to the driven link (drum) through an elastic spring

The problem is solved on the basis of Lagrange equations of the second kind. The system has two degrees of freedom. The generalized coordinates q_1 and q_2 of the mechanical system are selected (independent parameters that uniquely determine the position of all elements of the mechanical system at any time.)

Accept:

- $q_1=\varphi_{\text{в}}(t)$ is the angle of rotation of the shaft;
- $q_2=\varphi_{\text{пр}}(t)$ is the angle of twisting of the spring (relative angle of rotation of the drum).

In the future, for the convenience of judgment, we take $\varphi_{\text{пр}}(t)= \varphi_{\text{бr}}(t)$ – the angle of rotation of the drum relative to the shaft (relative rotation of the drum).

Thus, we accept:

- $q_1=\varphi_{\text{в}}(t)$;
- $q_2=\varphi_{\text{бr}}(t)$.

The initial Lagrange equations in accordance with the degrees of freedom:

$$\frac{\mathrm{d}}{\mathrm{d}t}\left(\frac{\partial T}{\partial \dot{\varphi}_{\text{в}}}\right)-\frac{\partial T}{\partial \varphi_{\text{в}}}=Q_1;\ \frac{\mathrm{d}}{\mathrm{d}t}\left(\frac{\partial T}{\partial \dot{\varphi}_{\text{бr}}}\right)-\frac{\partial T}{\partial \varphi_{\text{бr}}}=Q_2, \tag{1}$$

where T is the kinetic energy of the entire mechanical system; Q_1, Q_2 are the generalized forces of a mechanical system corresponding to each generalized coordinate.

Next is the kinetic energy of the mechanical system created by the rotating shaft and drum (the spring is neglected due to the relatively small mass), taking into account that $\varphi_{\text{б}}(t)= \varphi_{\text{в}}(t)-\varphi_{\text{бr}}(t)$.

$$\begin{aligned} T &= T_{\text{в}}+T_{\text{б}}=\frac{1}{2}J_{\text{в}}\omega_{\text{в}}^2+\frac{1}{2}J_{\text{б}}\omega_{\text{б}}^2= \\ &=\frac{1}{2}\left[(J_{\text{в}}+J_{\text{б}})\cdot\dot{\varphi}_{\text{в}}^2+J_{\text{б}}\cdot\dot{\varphi}_{\text{б}2}^2-2J_{\text{б}}\cdot\dot{\varphi}_{\text{в}}\cdot\dot{\varphi}_{\text{б}2}\right] \end{aligned}, \tag{2}$$

where $T_{\text{в}}$ and $T_{\text{б}}$ are the kinetic energy of the shaft and drum, respectively, and $J_{\text{в}}$, $J_{\text{б}}$ are the axial moments of inertia of the shaft and drum.

The expression (2) determines the values of the derivatives for the Lagrange equation (1). Next, we find the expressions of the generalized forces Q_1 and Q_2 for differential equations (1). To determine the expressions Q_1 and Q_2 we consider a mechanical system as a system with ideal bonds (without friction in the supports). Therefore, the possible work will be performed by the torque $M_{\text{вр}}$ on the rotor shaft and the elastic moment M_y of the spring.

Based on this

$$Q_1=\frac{\delta A_1}{\delta q_1}=\frac{\delta A_1}{\delta \varphi_{\text{в}}}=M_{\text{вр}};\ Q_2=\frac{\delta A_2}{\delta q_2}=\frac{\delta A_2}{\delta \varphi_{\text{бr}}}=-c\varphi_{\text{бr}}, \tag{3}$$

where A is work; c – spring stiffness.

Taking into account expressions (2) and (3), the initial differential equations (1) take the form

$$\begin{cases} (J_в + J_б)\ddot{\phi}_в - J_б \cdot \ddot{\phi}_{бr} = M_{вр}; \\ J_б(\ddot{\phi}_{бr} - \ddot{\phi}_в) = M_{вр} \ . \end{cases} \tag{4}$$

From the system of Lagrange equations (4) we obtain the differential equation of relative rotation of the drum

$$\ddot{\varphi}_{бr} + k^2 \cdot \varphi_{бr} = aM_{вр}, \tag{5}$$

where $k = \sqrt{\left(\frac{1}{J_в} + \frac{1}{J_б}\right)c}$, $a = \frac{1}{J_в}$.

For torsional vibrations, the expression of torque for the general case has the form

$$M_{вр} = M_O + A\sin\omega t,$$

where M_o is the average value of the torque, A is the amplitude of the momentum oscillations $M_{вр}$, A, ω are constants.

In this case, differential equation (5) takes the form

$$\ddot{\varphi}_{бr} + k^2\varphi_{бr} = M + N\sin\omega t,$$

where $M = aM_o = \frac{M_o}{J_в} = const$, $N = aA = \frac{A}{J_в} = const$.

The solution to the obtained differential equation of forced torsional vibrations of the drum (5) is found at zero initial conditions: at $t=0$, $\varphi_{бr}=0$.

We obtain expressions for determining the relative rotation of the drum $\varphi_{бr}$ and its relative angular velocity of rotation $\omega_{бr}(t)$:

$$\varphi_{бr} = -\frac{N\omega}{k(k^2-\omega^2)}\sin kt - \frac{M}{k^2}\cos kt + \frac{M}{k^2} + \frac{N}{k^2-\omega^2}\sin\omega t, \tag{6}$$

where t is time.

The first two terms of expression (6) determine vibrations with a natural frequency k (natural vibrations), and the third term defines forced vibrations of a drum with a frequency ω of torque.

At the same time, we find the angular velocity of the relative vibrational rotation of the drum

$$\omega_{бr} = -\frac{N\omega}{k^2-\omega^2}\cos kt + \frac{M}{k}\sin kt + \frac{N\omega}{k^2-\omega^2}\cos\omega t \ , \tag{7}$$

where t is time.

When calculating the angular velocity of the drum relative to the shaft $\omega_{бr}$, the optimality criterion is the maximum swing of the angular velocity of the drum relative to the shaft $\Delta\omega_{бr}$:

$$\omega_{бrmax} - \omega_{бrmin} = \Delta\omega_{бr}. \tag{8}$$

From expression (6), one can obtain the equation (law) of the relative torsional vibrations of the drum for simpler expressions of the torque Mвр (special cases).

In the case of $M_{кр}=A\sin\omega t$.

$$\varphi^*_{бr} = -\frac{N\omega}{k(k^2-\omega^2)}\sin kt + \frac{N}{k^2-\omega^2}\sin\omega t$$

$$\omega_{бr} = \dot{\varphi}^*_{бr} = -\frac{N\omega}{k^2-\omega^2}\cos kt + \frac{N\omega}{k^2-\omega^2}\cos kt = \frac{N\omega}{k^2-\omega^2}(\cos\omega t - \cos kt).$$

In this case, the drum performs its own torsional vibrations with a frequency k (first terms) and forced oscillations with a frequency ω (second term).

In the case of $M_{кр}=M_o$=const.

$$\varphi^{**}_{бr} = \frac{M}{k^2}(1-\cos kt),)$$

those. the drum will only perform its own torsional vibrations (with frequency k) relative to the rotating shaft.

$$\omega_{бr} = \dot{\varphi}^{**}_{бr} = \frac{M}{k}\sin kt$$

It is of interest to determine the law (equation) of rotation of the shaft $\varphi_В(t)$ for a given expression of the moment of rotation $M_{вр}$. Having obtained the law of the relative rotational motion of the drum $\varphi б_r(t)$, we can find the law of rotation of the shaft $\varphi_В(t)$.

In the case of Mвр = Mo = const, the angle of rotation of the shaft is the sum of the angle of rotation of the drum together with the shaft (first term) and the relative rotation of the drum φбr(t), i.e. relative vibration of the drum

$$\varphi_в(t) = \frac{J_б}{(J_в + J_б)}\left[\frac{1}{2}Mt^2 + \varphi_{бr}(t)\right].$$

Find the angular velocity of the shaft

$$\omega_в = \dot{\varphi}_в = \frac{J_б}{J_б + J_в}\left[Mt + \dot{\varphi}_{бr}\right] = \frac{J_б}{J_б + J_в}\left[M \cdot t + \frac{M}{k}\sin kt\right]$$

In the case of Mв = Asinωt, the angle of rotation of the shaft

$$\varphi_в = \frac{A}{(J_в + J_в)\omega}\left(t - \frac{1}{\omega}\sin\omega t\right) + \frac{J_б}{(J_б + J_б)} * \varphi_{бr}.$$

Angular shaft speed

$$\omega_в = \varphi_в = -\frac{d\varphi_в}{dt} = \frac{A}{(J_в + J_в)\omega}(1 - \cos\omega t) + \frac{J_б}{(J_б + J_б)} * \varphi_{бr}.$$

By solving expression (8), we find the values Δ $\omega_{бr}$ for springs with different stiffness. The problem was solved under the following conditions: a bit of type 215,9T3-ГHR15, a turbodrill 3ТСШ1-195, an average flow rate of flushing fluid of 0.023–0.028 m^3/s, an average axial load on a bit of 180 kN, ω=6.28 rad/s – the frequency of change in torque, $J_в$=4.5 kg · m^2 – the moment of inertia of the bit with the turbo-drill rotor and the near-bit shaft of the device, $J_б$=0.346 kg·m^2, M_o=1131 N · m – the average value of the torque, A = 246 N · m – the vibration amplitude $M_{вр}$.

Based on the obtained values, a graph of the change in the magnitude of the angular velocity of the drum relative to the rotating shaft is constructed (Figure 3). With an increase in spring stiffness from 400 to 1000 N/m, the oscillation range decreases from 14 to 9.5 rad/s.

Let us determine the dependence of $\Delta\omega_{бr}$ on $J_б$. By solution (8) we find the values $\Delta\omega_{бr}$ for various values of $J_б$. The problem was solved under the following conditions: a bit of type 215,9T3-ГHR15, a turbodrill 3ТСШ1-195, an average flow rate of flushing fluid of 0.023-0.028 m^3/s, an average axial load on a bit of 180 kN, ω=6,28 rad/s – the frequency of rotation moment, $J_в$=4,5 kg·mN/m – spring stiffness – moment of inertia of a bit with a turbodrill rotor and near-bit device shaft, c=400 N/m – spring stiffness, M_0=1131 N · m – average torque value, A=246 N · m – the amplitude of the oscillations $M_{вр}$.

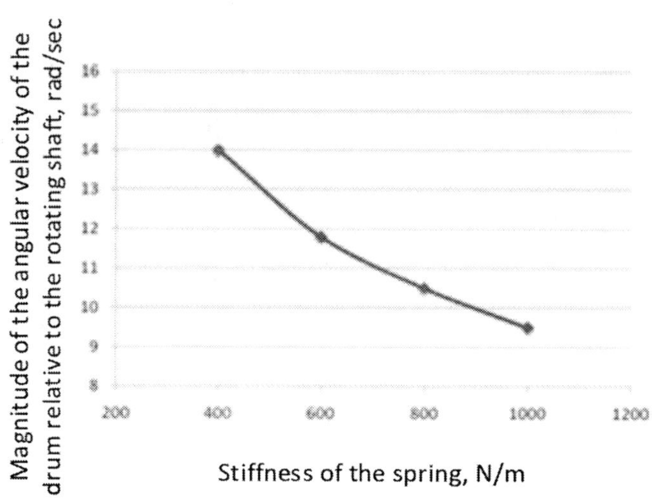

Figure 3. The dependence of the magnitude of the angular velocity of the drum relative to the rotating shaft from the stiffness of the springs

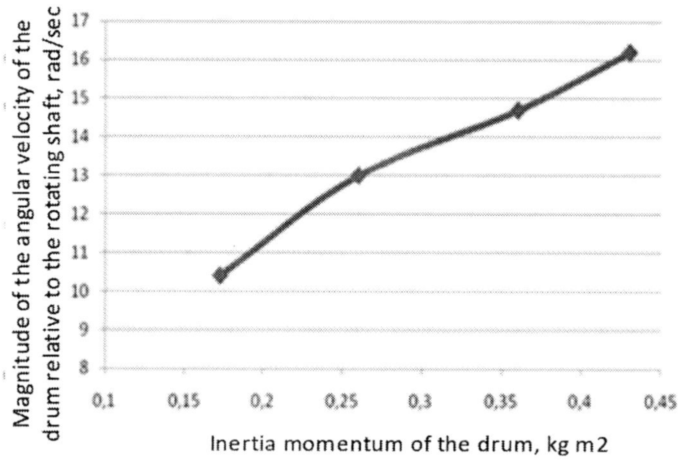

Figure 4. The dependence of the magnitude of the angular velocity of the drum relative to the rotating shaft from the moment of inertia of the drum

The dependence of the magnitude of the angular velocity of the drum relative to the shaft on the moment of inertia of the drum is shown in Figure 4. With an increase in the moment of inertia of the drum from 0.173 kg · m^2 to 0.43 kg · m^2, the amplitude of the relative vibrational rotation of the drum increases from 10.4 to 16.2 rad/s.

4. Conclusion

Based on the study of the developed mathematical model of the oscillatory motion of the drum near-bit slurry grinder, the dependencies of the relative angular velocity of the drum on the parameters of the oscillatory system – spring stiffness and drum inertia – are obtained. It was found that for a bit with a diameter of 215.9 mm, an increase in the moment of inertia of the drum from 0.173 to 0.43 kg·m^2 increases the magnitude of the angular velocity of the drum from 10.4 to 16.2 rad/s, while increasing the stiffness of the springs from 400 to 1000 N/m, the magnitude of the angular velocity decreases from 14 to 9.5 rad/s.

References

[1] Akhmetov R T, Kuleshova L S and Mukhametshin V V 2019 Absolute Permeability and Distribution of Pore Throats of the Productive Strata of Western Siberia *Atlantis Highlights in Material Sciences and Technology (Int. Symp., dedicated to the 85th anniversary of H.I. Ibragimov, Engineering and Earth Sciences: Applied and Fundamental Research (ISEES 2019))* vol 1 pp 535–9. DOI: 10.2991/isees-19.2019.105

[2] Naganawa S, Sato R and Ishikawa M Cuttings-transport simulation combined with large-scale-flow-loop experimental results and logging-while-drilling data for hole-cleaning evaluation in directional drilling *SPE Drill. and Complet.* **32(3)** 194–207

[3] Akhmetov R T., Mukhametshin V V and Kuleshova L S 2019 Grouping of objects using a limited number of parameters characterizing geological and physical properties of layers *Atlantis Highlights in Material Sciences and Technology (Int. Symp., dedicated to the 85th anniversary of H.I. Ibragimov, Engineering and Earth Sciences: Applied and Fundamental Research (ISEES 2019))* vol 1 pp 9–13. DOI: 10.2991/isees-19.2019.3

[4] Kuleshova L S and Mukhametshin V V 2019 Estimation of the wells hydrodynamic drag level based on wells geophysical survey data *Atlantis Highlights in Material Sciences and Technology (Int. Symp., dedicated to the 85th anniversary of H.I. Ibragimov, Engineering and Earth Sciences: Applied and Fundamental Research (ISEES 2019))* vol 1 pp 727–9. DOI: 10.2991/isees-19.2019.145

[5] Zainagalina L Z, Petrova L V and Petrov V A 2019 Analysis of the efficiency of telemetric systems for drilling wells *IOP Conf. Ser. Mater. Sci. and Engineer.* **560(1)**. DOI: 10.1088/1757-899X/560/1/012096

[6] Zainagalina L Z, Suleimanov R I, Gabdrakhimov M S and Khabibullin M Ya 2018 Determining oscillating sys-tem dynamic parameters of a near-bit junk pulper *Advan. in Engineering Res. (AER) Int. Conf. Actual issues of mechan. Engineer. (AIME 2018)* vol 157 pp 642–5

[7] Yagafarova Kh N, Arslanov I G and Teregulov R A 2018 Performance analysis of surface reducing gear of rod driven screw pump with involute gearing and Novikov gearing *Advan. in Engineering Res. (AER) Int. Conf. Actual issues of mechan. Engineer. (AIME 2018)* vol 157 pp 627–30

[8] Rogachev M K, Mukhametshin V V and Kuleshova L S 2019 Improving the efficiency of using resource base of liquid hydrocarbons in Jurassic deposits of Western Siberia *J. of Mining Inst.* **240** 711–5. DOI: 10.31897/PMI.2019.6.711

[9] Andreev V E, Chizhov A P, Kotenev Yu A et al 2019 Prediction of Gas Exposure in the Conditions of the Oil Fields of the Volga-Urals *Atlantis Highlights in Material Sciences and Technology (Int. Symp., dedicated to the 85th anniversary of H.I. Ibragimov, Engineering and Earth Sciences: Applied and Fundamental Research (ISEES 2019))* vol 1 pp 666–9. DOI: 10.2991/isees-19.2019.132

[10] Suleimanov R I, Zainagalina L Z, Khabibullin M Ya et al 2018 Studying heat-affected zone deformations of electric arc welding *IOP Conf. Ser. Mater. Sci. and Engineer.* **327(3)**. DOI: 10.1088/1757-899X/327/3/032053

IOP Conf. Series: Materials Science and Engineering **905** (2020) 012093 doi:10.1088/1757-899X/905/1/012093

On the development of an experimental design and field test of an upper bit tool

L Z Zaynagalina

Department of Oil and Gas Field Machines and Equipment, Oktyabrsky Branch of Ufa State Petroleum Technological University, Oktyabrsky, Russia

E-mail: zlz11@mail.ru

Abstract. The design of upper bit tools has been developed. Such tools provide crushing, sifting and removal of slurry from the bottom when drilling vertical, deviated and horizontal wells with a roller cone bit. The analysis of slurry samples from directional and horizontal wells was carried out to determine rational geometric dimensions of the working channels of the upper bit tool. The design requirements of upper bit tool, ensuring the stability of its operation, are presented.

1. Introduction

Bottoms of oil and gas wells during drilling, as a rule, become clogged with large drill cuttings, which leads to a deterioration in the process of rock destruction and a decrease in the performance of bits [3]. With increasing a drilling depth, an increasing amount of sediment, consisting mainly of large-sized sludge, remains in a bottom of well [1]. The removal of sludge is complicated when drilling deep vertical, deviated and horizontal wells, as well as when drilling shallow wells, i.e. when the circulation of the flushing fluid does not create sufficient upward flow for the removal of sludge. The most urgent problem is the cleaning of a horizontal wellbore from cuttings.

In the modern oil industry, horizontal drilling is the most effective way to develop oil and gas reserves. The task associated with the development of highly efficient equipment for cleaning the bottom of the well from sludge is relevant.

The eccentric location of the drill string in the borehole, the radial orientation of gravitational forces acting on the particles of the sludge determine the significant difference between the conditions for the removal of sludge by drilling fluid during drilling of horizontal wells from vertical ones [2, 4, 8].

Poor cleaning of wells from sludge leads to such complications as a decrease in the load on the bit, an unregulated change in the trajectory of the bore, difficulty in orienting the drilling tool, a change in the nature of the work of downhole assemblies, absorption of flushing fluid, and hydrodynamic sticking [3, 6].

In the narrowest part of annular space, the accumulation of sludge occurs, which causes complications when drilling horizontal wells [1, 5]. If the average value of transverse pulsations of the flow rate of washing liquid is less than the sedimentation rate of solid particles, the solid phase will begin to settle on the lower wall of the horizontal section of well, forming sediments. Calculations show that it is possible to transport particles of sludge of not more than 1-2 mm in suspended state [7].

To increase the efficiency of cleaning the bottom and the borehole from cuttings, various devices are used. These devices are built into the bit: ball vibrators, balancer, acoustic, jet pumps. Various upper bit devices are used to improve the cleaning of bottoms: a sub equipped with a screw, vibrators, jet pumps of various designs [11, 13]. These devices improve the cleaning of bottoms of well from a sludge to a certain extent, but do not provide the removal of large, solid sludge particles, allow re-crushing of sludge, the formation of sludge dunes when drilling horizontal wells [12].

2. Materials and methods

To increase the efficiency of cleaning wells from sludge, an upper bit device was developed, passing through which, large sludge particles are crushed to a size of 2 mm or less [7]. A schematic diagram of a upper bit sludge grinder is shown in Figure 1. The mesh screen 4 is mounted on the shaft 2 of the turbo-drill 1 and connected to the rotating shaft 2 by means of springs 5. The rotating shaft of the turbo-drill and drum connected by springs form an oscillating system. When the turbo-drill shaft rotates due to the presence of springs, torsional vibrations of the drum are created, which ensures grinding and sieving of the sludge.

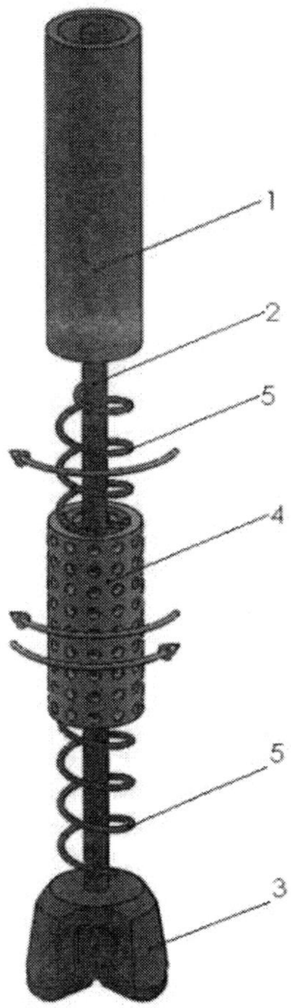

Figure 1. Schematic diagram of the drum drives near-bit slurry grinder: 1 – turbodrill; 2 – a shaft of a turbodrill; 3 – bit; 4 – drum; 5 – springs

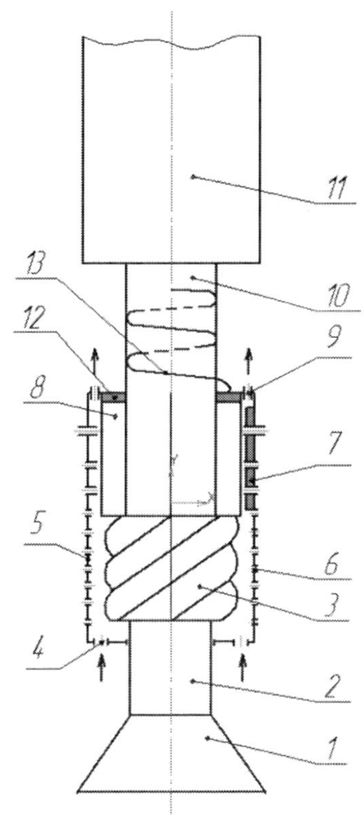

Figure 2. Schematic diagram of an upper bit sludge grinder during turbine drilling: 1 – bit; 2 – case; 3 – auger; 4 – inlet openings; 5 – drum; 6 – drum holes; 7 – chipper; 8 – rotor; 9 – output channel; 10 – a shaft of a turbodrill; 11 – turbodrill; 12 – ribbed surface of the drum; 13 – torsion spring

Figure 2 shows a schematic diagram of the operation of an upper bit slurry grinder during turbine drilling. The main device parts are the housing 2, which is equipped with a screw 3 and rotor 8, and the drum 5. The screw and rotor are installed inside the drum, which has input and output channels 4, 9, openings 6, bump 7. In the upper part, the drum is equipped with a ribbed surface 12, which mates with the end face of the rotor. The device is equipped with a screw, which sucks the sludge and directs it to the rotor, which in turn discards the sludge to the drum chipper 7. In this case, sludge is crushed, small particles of sludge are sifted through the drum, large particles are re-crushed by the end face of the rotor.

To determine the dimensions of working channels and holes of the upper bit sludge grinder drum, we analyzed the sludge samples from the following directional and horizontal wells [9, 10].

3. Results

Production horizontal wells No. 138Г, No. 140Г No. 141Г, *Shakshinskaya* area. The layout of the drilling tool: the *123.8 SL62P* bit, the screw downhole motor *DR-106*, telesystem *CTT*, *SBT-73*. The drilling was carried out by a clayless polymer-salt solution with a flow rate of 0.010 m^3 / s. The axial load on the bit is 6 tons. Sludge samples were taken during drilling of the horizontal part of the trunk (Figure 3), depth 2100–2250 m, Tournaisian layer, hard rocks.

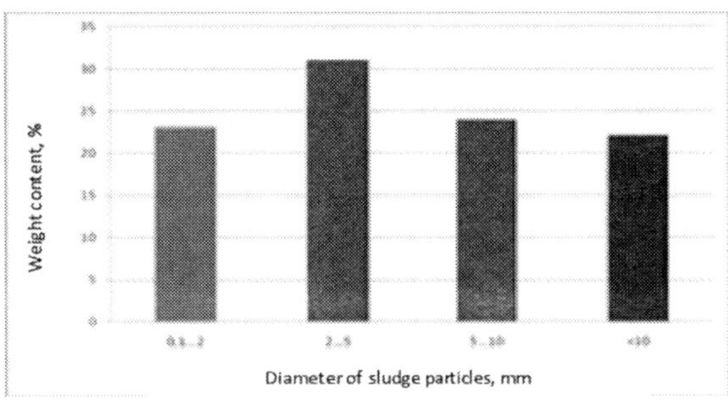

Figure 3. Size distribution of sludge particles (well No. 138Г, No. 140Г, No. 141Г)

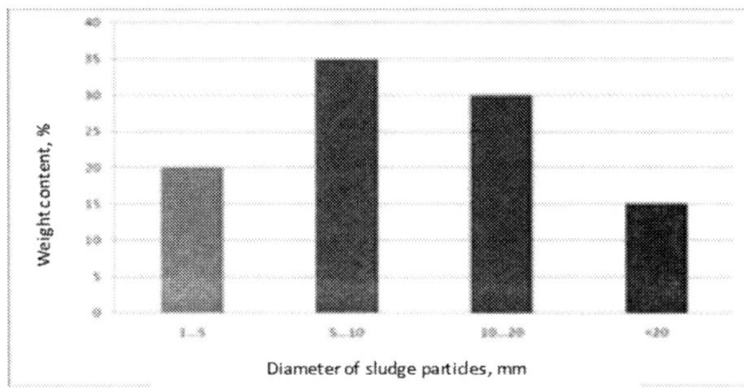

Figure 4. Size distribution of sludge particles (well No. 106, No. 107)

Production directional wells No. 106 and No. 107, area – Yubileinaya. The layout of the drilling tool: the *215.9 TD61AXLK* bit, the screw downhole motor *D5-195*, the weighted drill pipe *UBTS-178* 100 m long, SBT-127. The drilling was carried out by a clayless polysaccharide hydrophobizing low-carbonate solution with a flow rate of 0.028 m^3 / s. The axial load on the bit is 15 tons. Sludge samples (Figure 4) were taken from a depth of 1208–1210 m, *Kizel-Cherepetskiy* horizon, hard rocks.

Production well No. 486, area – Belebeyevskaya. Drilling cutting 347 m to a depth of 2015 m. The geological section is composed of hard rock. The layout of the drilling tool: the *215.9 HP62ALK* bit, *UBTS-178* 100 m long. The drilling was carried out by a polyglycol inhibited drilling fluid with the flow rate of 0.027 m^3 / s, using a downhole screw motor *D5-195* with a rotation speed of 90 rpm. Sludge samples (Figure 5) were taken from a depth of 2050–2052 m (*Mullinsky* horizon).

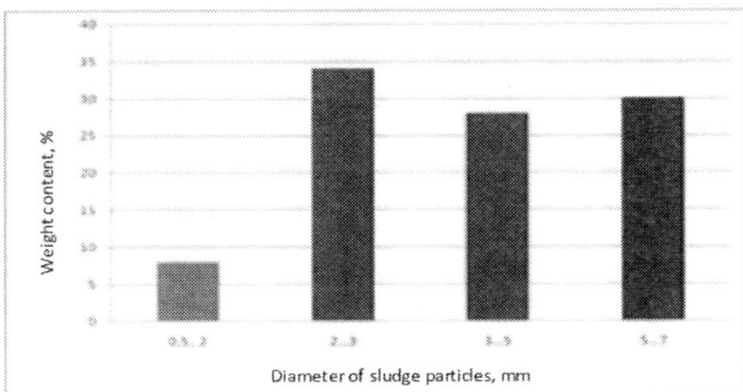

Figure 5. Size distribution of sludge particles (well No. 486)

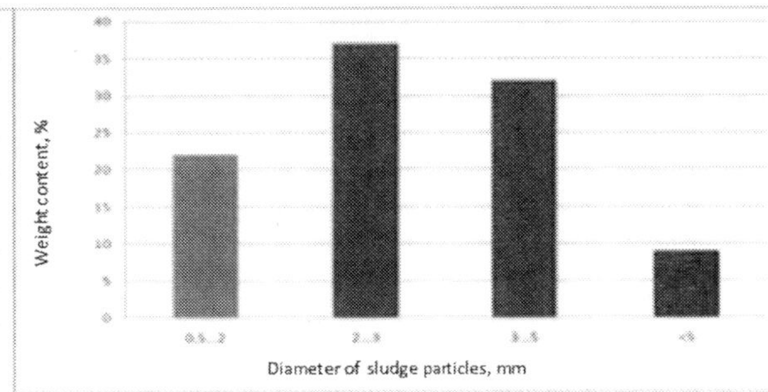

Figure 6. Size distribution of sludge particles (well No. 3114)

Production well No. 3114, area – Leonidovskaya. Drilling cutting 455 m to a depth of 1555 m. The geological section is composed of hard rock. The layout of the drilling tool: the *215.9 TZ-GAU R590* bit, UBTS-178 100 m long. The drilling was carried out by clayless polysaccharide hydrophobizing low-carbonate solution with the flow rate of 0.03 m^3 / s, using the downhole screw motor *D5-195* with the rotation speed of 90 rpm. Samples (Figure 6) of sludge were taken from a depth of 1546–1548 m (Nizhne-Famensky horizon).

Also, the sludge removal was investigated when drilling horizontal sections of wells No. 452 and No. 331 on *Tuimazinskaya* Square. The layout of the drilling tool: the *ENP 142.9 KS* bit, UBT-108 8 m long, SBT-73. The load on the bit was 30–50 kN, the pressure on the riser was 7 MPa and 10 MPa for wells No. 452 and No. 331, respectively. Drilling was carried out with a mud with a flow rate of 0.010–0.013 m^3 / s, using a rotor with a rotation speed of 72 rpm. The geological section is composed of hard and hard rocks. Sludge samples (Figure 7) were taken from a depth of 1730 m from well No. 331 and 1178 m from well No. 452.

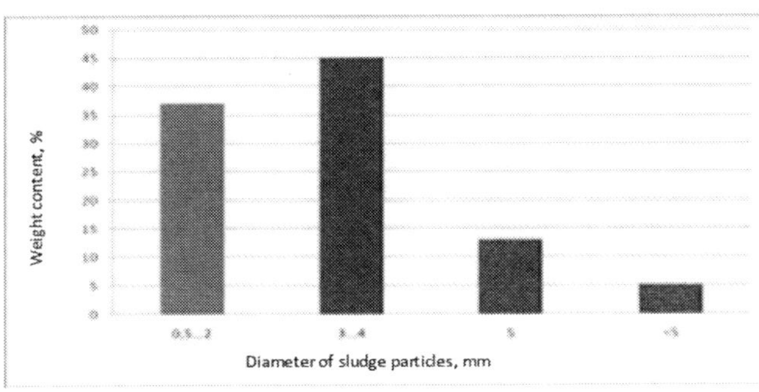

Figure 7. Size distribution of sludge particles (well No. 331 and No. 452)

4. Conclusion

According to the study results, the largest particle size of sludge was 25 mm. With a margin, we select the diameter of input channels of the upper bit device equal to 30 mm. The diameter of output channels (holes in the drum) is 2.5 mm. The geometric dimensions of elements of slurry grinder were determined using the results of bench tests of the model and the results of theoretical studies of the upper bit system of slurry grinder.

Applications upper-bit sludge grinder: for drilling deep vertical, deviated and horizontal wells. When drilling shallow wells – when the circulation of flushing fluid does not create a sufficient upward flow for the sludge removal.

Requirements for the construction of upper bit slurry grinder: 1) ensuring the reliability of work when performing all technological work on drilling wells; 2) providing a gap between the wall of well and drum; 3) length not more than 600 mm; 4) the outer diameter is less than the diameter of turbodrill (layout of the bottom of drilling tool); 5) connecting threads must allow installation above the bit; 6) the drum is mounted on a shaft with an axial and radial clearance of 2 mm; 7) minimum manufacturing cost.

The technical characteristics of upper bit device developed are given in the table.

The study of the effect of upper bit sludge grinder on the mechanical speed and drill bit penetration during drilling was carried out at well No. 27B/180 of the *Mellyaneft'* area drilled for technical water supply. In the process of drilling wells, the following parameters were monitored: flow rate of flushing fluid, standpipe pressure, mechanical drilling speed, penetration to the bit, dimensions of the slurry carried out by flushing fluid.

The well drilling was carried out by the drilling rig *URB 3A3*, the pump *NB-40*. The drilling was carried out with a clay solution (specific gravity 1.21 g/cm, viscosity 30 s) with a flow rate of 4 l/s, using the *P-25* rotor with a rotation speed of 120 rpm. The chisel load is the weight of the tool. The depth of well is 102.5 meters. Drill tool layout (Figure 8):

- bit 215.9 MFG;
- near-bit slurry shredder;
- 147 mm weighted drill pipes 4.65 m long.

Table 1. Technical characteristic of near-bit Sludge Grinder

Parameters of upper bit sludge grinder	For assembly with a chisel diameter 215,9 мм	For assembly with a chisel diameter 146 mm
Outer diameter mm	190	126
Mesh outer diameter, mm	185	121
Drum inner diameter, mm	168	109
Outer diameter of the drum, mm	123	70
Internal shaft diameter, mm	85	48
Overall length, mm	510	500
Mesh length, mm	264	264
Connecting thread, mm	3-117	Z-86-upper, Z-88-lower
The diameter of the grid cell, mm	2,5	2,5
Screw diameter, mm	157	103

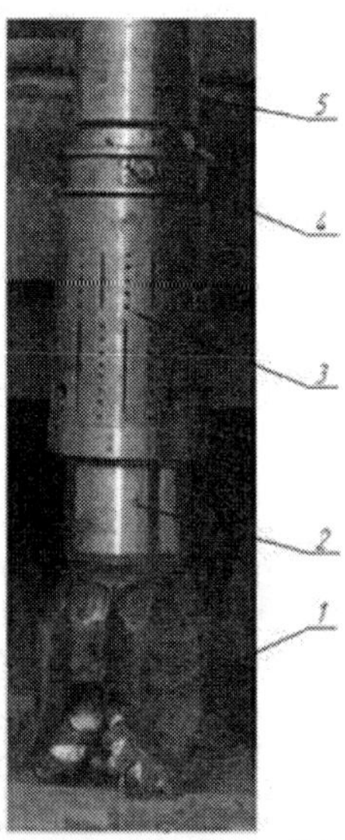

Figure 8. The bottom of the boring tool, equipped with a slurry grinder: 1 – chisel; 2 – case; 3– drum; 4 – a nut; 5 – sub

IOP Conf. Series: Materials Science and Engineering **905** (2020) 012093 doi:10.1088/1757-899X/905/1/012093

The geological section of drilled interval: grayish-brown sandstone, sandstone interbedding with marl and clay, brownish-gray clay with marl and sandstone interlayers, strong brown-red sandstone with clay interlayers, red-brown sandstone strong clay, thick reddish-brown clay, interbedding sandstone with marl and clay, clay is brownish-brown, sandstone is brown of medium strength.

Using a slurry grinder, 93.5 m were drilled in the interval 9.0–102.5 m in 26 hours. The mechanical speed was 3.59 m / h. Sludge, crushed to 2.5 mm, was stably carried to the surface by a stream of washing liquid. Bit wear is negligible.

As a result of experimental drilling, it was established:

- upper bit sludge shredder is operational: during the drilling of the well there were no failures (breakdowns of parts);
- large particles of crushed sludge are 2.5 mm;
- the mechanical drilling speed due to the stable removal of crushed sludge by the flushing fluid increased by 18 %;
- upper bit slurry shredder is recommended for widespread adoption to increase the efficiency of drilling vertical and horizontal wells.

References

[1] Akhmetov R T, Kuleshova L S and Mukhametshin V V 2019 Absolute Permeability and Distribution of Pore Throats of the Productive Strata of Western Siberia *Atlantis Highlights in Material Sciences and Technology (Int. Symp., dedicated to the 85th anniversary of H.I. Ibragimov, Engineering and Earth Sciences: Applied and Fundamental Research (ISEES 2019))* vol 1 pp 535–9. DOI: 10.2991/isees-19.2019.105

[2] Pang B, Wang S, Jiang X and Lu H 2019 Effect of orbital motion of drill pipe on the transport of non-Newtonian fluid-cuttings mixture in horizontal drilling annulus *J. of Petrol. Sci. and Engineer.* 201–215

[3] Akhmetov R T, Mukhametshin V V and Kuleshova L S 2019 Grouping of objects using a limited number of parameters characterizing geological and physical properties of layers *Atlantis Highlights in Material Sciences and Technology (Int. Symp., dedicated to the 85th anniversary of H.I. Ibragimov, Engineering and Earth Sciences: Applied and Fundamental Research (ISEES 2019))* vol 1 pp 9–13. DOI: 10.2991/isees-19.2019.3

[4] Kuleshova L S and Mukhametshin V V 2019 Estimation of the wells hydrodynamic drag level based on wells geophysical survey data *Atlantis Highlights in Material Sciences and Technology (Int. Symp., dedicated to the 85th anniversary of H.I. Ibragimov, Engineering and Earth Sciences: Applied and Fundamental Research (ISEES 2019))* vol 1 pp 727–9. DOI: 10.2991/isees-19.2019.145

[5] Zainagalina L Z, Petrova L V and Petrov V A 2019 Analysis of the efficiency of telemetric systems for drilling wells *IOP Conf. Ser. Mater. Sci. and Engineer.* **560(1)**. DOI: 10.1088/1757-899X/560/1/012096.

[6] Naganawa S, Sato R and Ishikawa M (2017) Cuttings-transport simulation combined with large-scale-flow-loop experimental results and logging-while-drilling data for hole-cleaning evaluation in directional drilling *SPE Drill. and Complet.* **32(3)** 194–207

[7] Zainagalina L Z., Suleimanov R I., Gabdrakhimov M S and Khabibullin M Ya 2018 Determining oscillating sys-tem dynamic parameters of a near-bit junk pulper *Advan. in Engineering Res. (AER) Int. Conf. Actual issues of mechan. Engineer. (AIME 2018)* vol 157 pp 642–5

[8] Ignatenko Y, Bocharov O, Gavrilov A and May R 2018 Steady-state cuttings transport simulation in horizontal borehole annulus *Proc. of the Int. Conf. on Offshore Mechan. and Arctic Engineer. – OMAE* vol 8

[9] Rogachev M K, Mukhametshin V V and Kuleshova L S 2019 Improving the efficiency of using resource base of liquid hydrocarbons in Jurassic deposits of Western Siberia *J. of Mining Inst.* **240** 711–5. DOI: 10.31897/PMI.2019.6.711

[10] Andreev V E, Chizhov A P, Kotenev Yu A et al 2019 Prediction of Gas Exposure in the Conditions of the Oil Fields of the Volga-Urals. *Atlantis Highlights in Material Sciences and Technology (Int. Symp., dedicated to the 85th anniversary of H.I. Ibragimov, Engineering and Earth Sciences: Applied and Fundamental Research (ISEES 2019))* vol 1 pp 666–9. DOI: 10.2991/isees-19.2019.132

[11] Davydov A Yu 2018 Utilization of expert systems for screw pump sets with surface drive management *Advan. in Engineering Res. (AER) Int. Conf. Actual issues of mechan. Engineer. (AIME 2018)* vol 157 pp 129–33

[12] Zeng C, Yan X, Zeng Z and Yang S 2018 The formation and broken of cuttings bed during reaming process in horizontal directional drilling *Tunnell. and Underground Space Technol.* **76** 21–9

[13] Khabibullin M Ya, Samigullina L Z and Zainagalina L Z 2018 Laboratory studies for the momentum of liquid optimal impact on hydrocarbon formation *Advan. in Engineering Res. (AER) Int. Conf. Actual issues of mechan. Engineer. (AIME 2018)* vol 157 pp 276–9

IOP Conf. Series: Materials Science and Engineering **905** (2020) 012094 doi:10.1088/1757-899X/905/1/012094

On the determination of design and geometric parameters of a upper bit slurry grinder

L Z Zaynagalina

Department of Oil and Gas Field Machines and Equipment, Oktyabrsky Branch of Ufa State Petroleum Technological University, Oktyabrsky, Russia

E-mail: zlz11@mail.ru

Abstract. Experimental and field studies have identified the rational geometric dimensions of elements of upper bit sludge grinder, ensuring its operation's stability. Laboratory studies of upper bit device model showed that the amplitude of drum oscillations depends on the shaft rotation frequency, spring stiffness, and drum inertia moment. To determine the dimensions of working channels and drum openings of a upper bit device, the analysis of slurry samples from directional and horizontal wells was carried out. Designs of a slurry grinder for drilling wells with cone bits with diameters of 146 and 215.9 mm have been developed. Field tests showed that the use of the developed design of near-bit sludge grinder allows to increase the mechanical speed by 18% compared to the conventional method of drilling.

1. Introduction

The increase in oil and gas productions is inextricably linked with the increase in technical and economic indicators of the well drilling process. Well construction time can be reduced by increasing drilling speeds and using new types of rock cutting tools, new types of flushing fluids, a circulation system equipment (mud pumps, equipment for preparing and cleaning flushing fluids) as well as by reasonably improving equipment operation. A significant reserve in increasing the efficiency of drilling operations is the improvement of drilling techniques and technology based on the introduction of effective equipment for cleaning bottom holes from sludge [6]. The degree of perfection of flushing the well and equipment for its implementation significantly affects the successful, trouble-free well drilling. The quality of cleaning well bottom from sludge affects the mechanical drilling speed, the quality of wellbore, sinking and durability of bit, turbo-drill, and drill tool layout elements.

In horizontal drilling, all drill cuttings tend to sink to the bottom wall of well. To prevent the sedimentation and formation of sediment, it is necessary that particles of cuttings be in a suspended rather than entrained state [5]. But, the increase in supply and density of drilling fluid leads to the mudding of well walls [4, 6]. According to the literature, it is known that suspended particles can be transported in drill particles of a size of not more than 1–2 mm [3, 10–14].

To increase the efficiency of cleaning wells from sludge, a upper bit device has been developed. The device (Figure 1) is installed during rotary drilling (rotary method, downhole motors) over a bit. When drilling a well, the sub 2 rotates, drives the screw 3 and the rotor 4. The rotating sub 2 and the drum connected by springs 13 and 14 form an oscillating system. The auger directs the flushing liquid with slurry into the internal device cavity. The auger directs the liquid with the slurry in the axial direction and the radial direction, while the slurry is partially crushed due to impacts, small particles are sifted through the grid (2–3 mm mesh openings), large particles fall into the rotor and are radially rejected onto the reflector, further grinding of the slurry occurs. The final grinding occurs when large particles enter the channels 16 and crush them with the end face of the rotor. The purification of solution is facilitated by torsional vibrations of drum (mesh), which occur under the action of drum inertia, the fluid ejected by the rotor onto the eccentric reflector and spring forces [1].

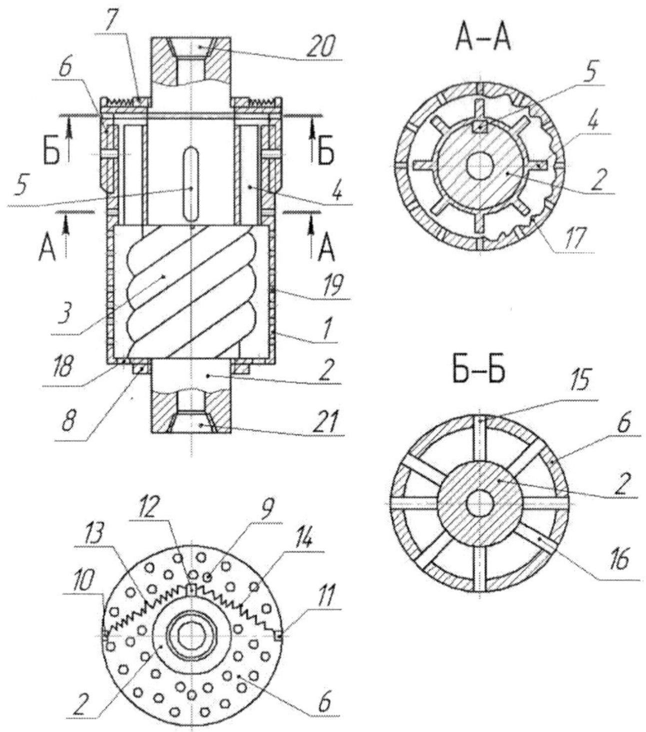

Figure 1. Structural diagram of a upper bit slurry grinder

The stand is equipped with instrumentation: the *ТЭ-М1* tachometer, the *Diana-2M* vibration analyzer. To process the measurement results, the *Atlant* and *Avrora-2000* software are used [9].

2. Methods and materials

The main dimensions of upper bit device model were determined by physical modeling, taking into account necessary conditions of geometric similarity and physical similarity of the model and nature [2, 8]. Some of the secondary processes that do not affect the operation of oscillatory system of sludge grinder were modeled approximately.

Figure 2 shows a diagram of a bench for researching a near-bit device model. The stand is designed to determine the optimal parameters of the near-bit oscillation system of the device: spring stiffness c, rotor speed n, moment of inertia of the drum Jb and the frequency and amplitude of drum vibrations depending on them. The stand consists of a frame 1, on which the supports are fixed 2. An electric motor 3, a speed controller 4 and a drum 5 are mounted on the supports 2. The engine transmits power to the speed controller 4 using a coupling 6. The drum is mounted on a shaft 7, which in turn is mounted on radial bearings 8. The rotation from the speed controller 4 to the drum 5 is transmitted by means of the coupling 9. The drum is mounted on the shaft with a clearance by means of centralizers 11. The drum 5 is connected to the centralizers by means of springs 12. For measuring the speed, the sensor 10, d To measure vibration – sensor 13.

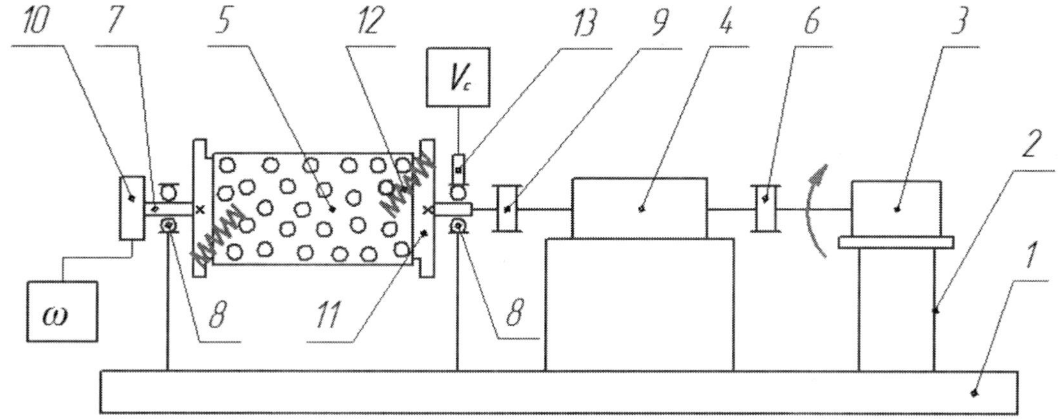

Figure 2. Scheme of the stand for researching the upper bit model of the device 1 – frame, 2 – support, 3 – electric motor, 4 – speed controller, 5 – drum, 6, 9 – coupling, 7 – shaft, 8 – radial bearing, 10 – tachometer sensor, 11 – centralizers, 12 – springs, 13 – vibration analyzer sensor

3. Results

The stand tests were carried out under various conditions, changing the shaft rotation frequency *n*, spring stiffness *c*, and drum inertia moment $J_б$.

When planning the experiment, the value of the root mean square value (RMS) of drum vibration velocity was chosen as the optimization parameter [7]. The RMS depends on the following factors: the number of shaft revolutions, spring stiffness, and the moment of drum inertia. Figures 3 and 4 show the results of measuring the vibration drum velocity.

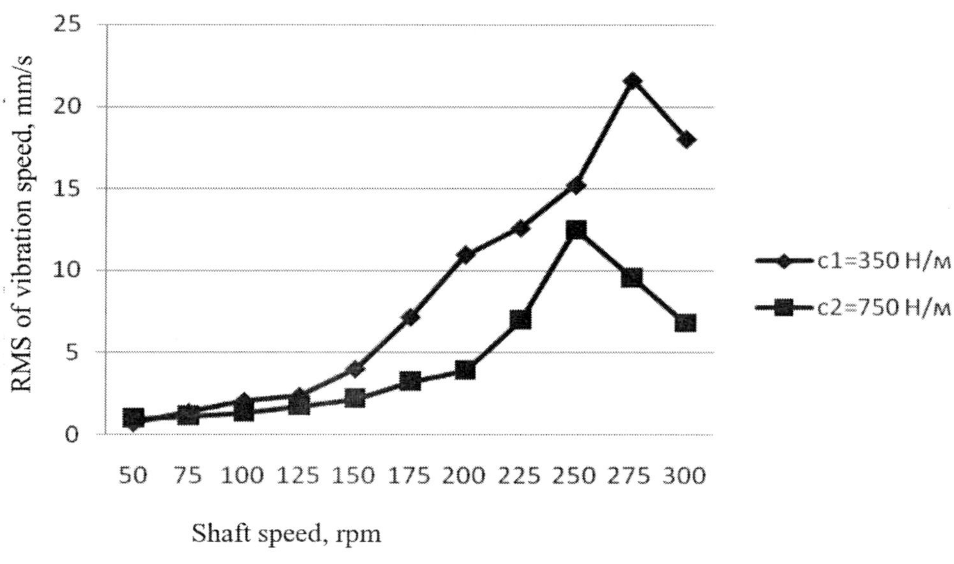

Figure 3. The graph of the dependence of the RMS vibrational velocity of the drum on the shaft rotation frequency at $J_б$=0,0036 kg·m^2 and springs of different stiffness

It can be seen from the graphs that when the shaft rotational speed increases to 200–275 rpm, the RMS vibration speed increases. At a shaft rotation frequency of 250–275 rpm, a resonance phenomenon is observed. Recommended values of the shaft speed of the model are 50–200 rpm. With increasing spring stiffness, the RMS vibration speed decreases. With increasing moment of inertia of the drum, the RMS vibration velocity increases. The most optimal drum is with a moment of inertia J_6= 0,0036 kg·m^2

Stand studies of the oscillatory system of slurry grinder confirm the operability of upper bit system of the grinder grinder. The results of theoretical and experimental studies show sufficient convergence. To verify the accuracy of determining design-geometric parameters and verify the operability of upper bit device, field trials are necessary.

Designs of a slurry grinder for drilling wells with cone bits with diameters of 146 and 215.9 mm have been developed (table).

Requirements for the construction of upper bit slurry grinder: 1) ensuring the work reliability when performing all technological work on drilling wells; 2) providing a gap between the well wall and the drum; 3) length not more than 600 mm; 4) the outer diameter is less than the diameter of turbodrill (layout of drilling tool bottom); 5) connecting threads should allow installation over a bit; 6) the drum is mounted on a shaft with an axial and radial clearance of 2 mm; 7) minimum manufacturing cost.

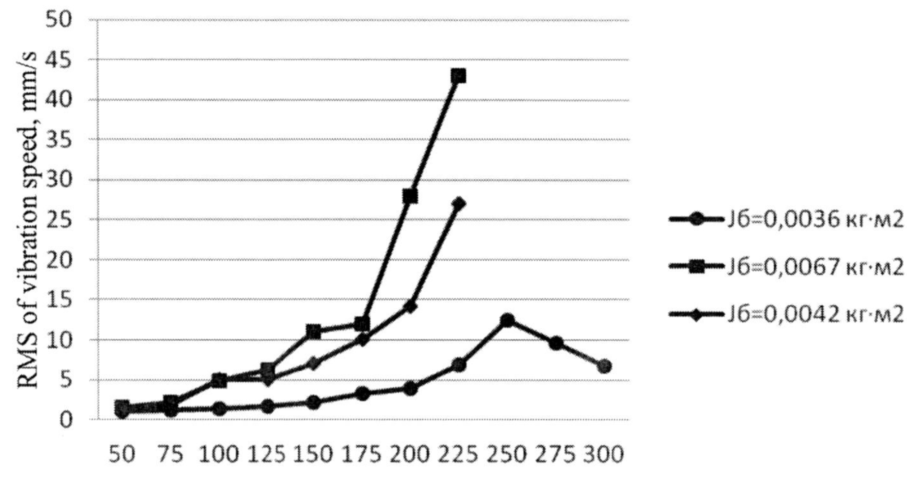

Figure 4. The graph of the dependence of the RMS vibrational velocity of the drum on the shaft rotation frequency at c = 750 N / m and various moments of inertia of the drum

Table 1. Technical characteristic of near-bit Sludge Grinder

Parameters of near-bit sludge grinder	For bit diameters 215.9 mm	For bit diameters 146 mm
Outer diameter, mm	190	126
The outer diameter of the drum , mm	185	121
Drum inner diameter, mm	168	109
The shaft's outer diameter, mm	123	70
Internal shaft diameter, mm	85	48
Overall length, mm	510	500
Drum length, mm	264	264
Connecting thread, mm	Z-117	Z-86-upper, Z-88-lower
Diameter of the drum cells, mm	2.5	2.5
Screw diameter, mm	157	103

The geometric dimensions of slurry grinder elements were determined using the results of stand tests of the model and the results of theoretical studies of the upper bit system of slurry grinder.

The diameter of drum was determined:

$$D_б = D_м \cdot K_м,$$

where $D_м$ is the outer diameter of drum of model, $K_м$ is the simulation coefficient.

The inner and outer diameters of shaft, screw and the inner diameter of the drum were determined in a similar way.

The linear size of slurry grinder drum was determined:

$$L_б = L_м \cdot K_l,$$

where $L_м$ is the drum length of the model, K_l is the simulation coefficient.

The stiffness of the spring connecting the shaft to the drum is determined by:

$$C_{пр} = C_м \cdot K_c,$$

where $C_м$ is the stiffness of the model spring, K_c is the coefficient of spring simulation.

To determine the size of working channels and holes of the upper bit sludge grinder drum, we analyzed the sludge samples from the following directional and horizontal wells.

Horizontal production wells No. 138Г, No. 140Г No. 141Г, Shakshinskaya area, the *Bashneft'* oil joint stock company. The layout of drilling tool: the *123.8 SL62P* bit, the *DR-106* screw downhole motor, the *CTT* telesystem, SBT-73. The drilling was carried out by a clayless polymer-salt solution with a flow rate of 0.010 m3 / s. The axial load on the bit is 6 tons. Sludge samples were taken during drilling of the horizontal part of the trunk, depth 2100–2250 m, Tournaisian layer, hard rocks.

Production directional wells No. 106 and No. 107, area – Yubileynaya, the *Bashneft'* oil joint stock company. The layout of drilling tool: the *215.9 TD61AXLK* bit, the *D5-195* screw downhole motor, the *UBTS-178* weighted drill pipes 100 m long, SBT-127. The drilling was carried out by a clayless polysaccharide hydrophobizing low-carbonate solution with a flow rate of 0.028 m^3 / s. The axial load on bit is 15 tons. Sludge samples were taken from a depth of 1208-1210 m, Kizel-Cherepetsk horizon, hard rocks.

Production well No. 486, area – Belebeyevskaya, the *Bashneft'* oil joint stock company. Drilling cutting 347 m to a depth of 2015 m. The geological section is composed of hard rock. The layout of the drilling tool: the *215.9 HP62ALK* bit, UBTS-178 100 m long. Drilling was carried out by a polyglycol inhibited drilling fluid with a flow rate of 0.027 m^3/s, using the *D5-195* downhole screw motor with a rotation speed of 90 rpm. The sludge samples were taken from a depth of 2050–2052 m (Mullinsky horizon).

Production well No. 3114, area – Leonidovskaya, the *Bashneft'* oil joint stock company. Drilling cutting 455 m to a depth of 1555 m. The geological section is composed of hard rock. The layout of the drilling tool: the *215.9 TZ-GAU R590* bit, UBTS-178 100 m long. Drilling was carried out by a clayless polysaccharide hydrophobizing low-carbonate solution with a flow rate of 0.03 m3 / s, using the *D5-195* screw downhole motor with a rotation speed of 90 rpm min Sludge samples were taken from a depth of 1546–1548 m (Nizhne-Famensky horizon).

Also, sludge removal was investigated when drilling horizontal sections of wells No. 452 and No. 331 on Tuimazinskaya Square of the *Bashneft'* oil joint stock company. The layout of drilling tool: the *ENP 142.9 KS* bit, UBT-108 8 m long, SBT-73. The bit load was 30–50 kN, the riser pressure was 7 and 10 MPa for wells No. 452 and No. 331, respectively. The drilling was carried out with a clay solution with a flow rate of 0.010–0.013 m^3/s, using a rotor with a rotation speed of 72 rpm. The geological section is composed of hard and hard rocks. Sludge samples were taken from a depth of 1730 m from well No. 331 and 1178 m from well No. 452.

According to the study results, the largest particle size of sludge was 25 mm. With a margin, we select the diameter of input channels of the upper bit device equal to 30 mm. The diameter of output channels (holes in the drum) is 2.5 mm.

The study of the effect of upper bit sludge grinder on the mechanical speed and drill bit penetration during drilling was carried out at well No. 27B / 180 of the *Mellyaneft'* area of the *Tatneft'* oil joint stock company drilled for technical water supply. The well drilling was carried out by the *УРБ 3А3* drilling rig, the *NB-40* pump. Drilling was carried out with a clay solution with a flow rate of 4 l/s, using the *R-25* rotor with a rotation speed of 120 rpm. The bit load is the weight of the tool. The well depth is 102.5 meters. The layout of drilling tool: the *215.9 MFG* bit; the upper bit slurry shredder with a diameter of 190 mm; weighted drill pipes with a diameter of 147 mm and a length of 4.65 m.

Using a slurry grinder, 93.5 m were drilled in the interval 9.0–102.5 m in 26 hours. The mechanical speed was 3.59 m / h. Sludge, crushed to 2 mm, was stably carried to the surface by a stream of washing liquid. The wear of bit and the working elements of grinder is negligible (Figure 5).

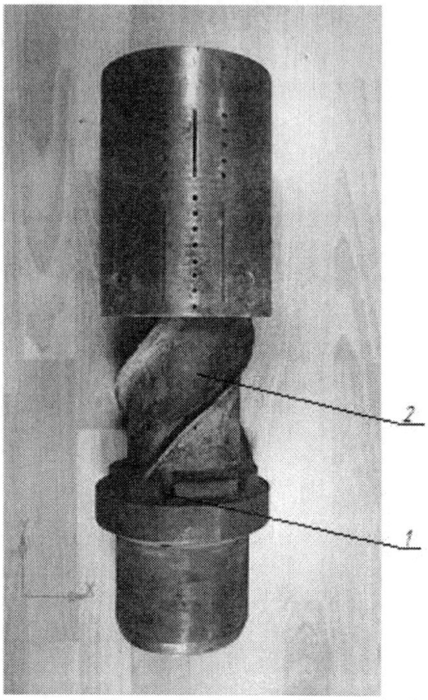

Figure 5. The main details of near-bit sludge grinder after field tests: 1 – hole for sludge suction; 2 – auger

4. Conclusion

Experimental and field studies have identified the rational geometric dimensions of upper bit sludge grinder elements, ensuring its operation's stability. An upper bit sludge shredder is efficient: during the well drilling process there were no failures (parts breakdowns). Large particles of crushed sludge are 1.7–2.5 mm. The mechanical drilling speed due to the stable removal of crushed sludge by the flushing fluid increased by 18%. upper bit slurry shredder is recommended for widespread adoption to increase the efficiency of drilling vertical, deviated and horizontal wells.

References

[1] Zainagalina L Z 2019 Analysis of the efficiency of telemetric systems for drilling wells *IOP Conf. Ser. Mater. Sci. and Engineer.* **560(1)**. DOI: 10.1088/1757-899X/560/1/012096

[2] Andreev V E, Chizhov A P, Kotenev Yu A et al 2019 Prediction of Gas Exposure in the Conditions of the Oil Fields of the Volga-Urals *Atlantis Highlights in Material Sciences and Technology (Int. Symp., dedicated to the 85th anniversary of H.I. Ibragimov, Engineering and Earth Sciences: Applied and Fundamental Research (ISEES 2019))* vol 1 pp 666–9. DOI: 10.2991/isees-19.2019.132

[3] Zainagalina L Z, Suleimanov R I, Gabdrakhimov M S and Khabibullin M Ya 2018 Determining oscillating system dynamic parameters of a near-bit junk pulper *Advan. in Engineering Res. (AER) Int. Conf. Actual issues of mechan. Engineer. (AIME 2018)* vol 157 pp 642–5

[4] Konnov Yu D 2018 Mechanization of technological process of round-trip operations in well servicing and workover *SOCAR Proc.* **2** 15–24

[5] Khabibullin M Y, Suleimanov R I, Sidorkin D I and Arslanov I G 2017 Parameters of damping of vibrations of tubing string in the operation of bottomhole pulse devices *Chem. and Petrol. Engineer.* **53(5-6)** 378–84

[6] Khabibullin M Ya 2018 Laboratory studies for the momentum of liquid optimal impact on hydro-carbon formation *Advan. in Engineering Res. (AER) Int. Conf. Actual issues of mechan. Engineer. (AIME 2018)* vol 157 pp 276–9

[7] Golovina N Y, Akhmetov L G, Vikharev A N and Arslanov I G 2017 Analysis on compressor blading conditions of helicopter's gas-turbine engine working in polluted environment *Int. J. of Appl. Engineer. Res.* **12(3)** 293–6

[8] Chin W C 1990 *Exact cuttings transport correlations developed for high angle wells* (Offshore) 67 p

[9] Larsen T I, Pilehvari A A and Azar J J 1997 *Development of a new cutting transport model for high-angle wellbores including horizontal wells* (SPEDC) 129 p

[10] Nguyen D and Rahman S S 1998) *A three-layer hydraulic program for effective cuttings transport and hole cleaning in highly deviated and horizontal wells* (SPEDE) 182 p

[11] Akhmetov R T, Kuleshova L S and Mukhametshin V V 2019 Absolute Permeability and Distribution of Pore Throats of the Productive Strata of Western Siberia *Atlantis Highlights in Material Sciences and Technology (Int. Symp., dedicated to the 85th anniversary of H.I. Ibragimov, Engineering and Earth Sciences: Applied and Fundamental Research (ISEES 2019))* vol 1 pp 535–9. DOI: 10.2991/isees-19.2019.105

[12] Akhmetov R T., Mukhametshin V V and Kuleshova L S 2019 Grouping of objects using a limited number of parameters characterizing geological and physical properties of layers *Atlantis Highlights in Material Sciences and Technology (Int. Symp., dedicated to the 85th anniversary of H.I. Ibragimov, Engineering and Earth Sciences: Applied and Fundamental Research (ISEES 2019))* vol 1 pp 9–13. DOI: 10.2991/isees-19.2019.3

[13] Kuleshova, L.S., Mukhametshin, V.V. (2019) Estimation of the wells hydrodynamic drag level based on wells geophysical survey data. *Atlantis Highlights in Material Sciences and Technology (Int. Symp., dedicated to the 85th anniversary of H.I. Ibragimov, Engineering and Earth Sciences: Applied and Fundamental Research (ISEES 2019))* vol 1 pp 727–9. DOI: 10.2991/isees-19.2019.145

[14] Rogachev M K, Mukhametshin V V and Kuleshova L S 2019 Improving the efficiency of using resource base of liquid hydrocarbons in Jurassic deposits of Western Siberia *J. of Mining Inst.* **240** 711–5. DOI: 10.31897/PMI.2019.6.711

IOP Conf. Series: Materials Science and Engineering **905** (2020) 012095 doi:10.1088/1757-899X/905/1/012095

The choice of the correlating function of capillary pressure curves under conditions of reservoirs in Western Siberia

R T Akhmetov[1], V V Mukhametshin[2], L S Kuleshova[1], Sh G Mingulov[1] and P M Malyshev[1]

[1] Department of Oil and Gas Field Exploration and Development, Ufa State Petroleum Technological University, Branch of the University in the City of Oktyabrsky, Oktyabrsky, Republic of Bashkortostan, Russia

[2] Department of Oil and Gas &Oil Field Development and Operation, Ufa State Petroleum Technological University, Ufa, Republic of Bashkortostan, Russia

E-mail: vv@of.ugntu.ru, markl212@mail.ru

Abstract. The article considers the approximation of capillary pressure curves of productive formations in Western Siberia. The relationship between the correlating functions of capillary pressure curves is shown: the Leverett J-function and the Brooks-Corey function. In this case, the Brooks-Corey function is the same Leverett function but used for water saturation of the effective part of the void space (reduced water saturation). The authors recommend choosing a correlation function for a given reservoir, depending on the tightness of the correlation between capillary pressure and volumetric or reduced water saturation. In this case, the Leverett function should be preferred, since it depends only on two parameters (permeability and porosity).

1. Introduction

Any operational facility is characterized by the following basic parameters: porosity, permeability, and residual water saturation [1–4].

The analysis shows that capillary pressure curves of core samples relate all these basic parameters of reservoir formations. Therefore, the study of capillary characteristics, as a rule, is a common laboratory analysis, as well as the determination of reservoir properties of productive formations [5].

According to capillarimetry data of a particular field, the following main tasks are solved:

- quantitative assessment of distribution of pore channels in size;
- construction of curves of relative permeabilities for oil and water;
- distribution of water saturation along the height of oil reservoir.

2. Methods and materials

In this article, the results of laboratory studies of core samples were used. The relationship between the correlating functions of capillary pressure curves is investigated: the Leverett J-function and the Brooks-Corey function.

Currently, there are a number of mathematical models that allow approximate the entire set of capillary curves and transform them into the formula for the capillary pressure of water saturation.

The most popular among them are the Leverett J-function and the Brooks-Corey model [6].

In the first case, the correlation function J is used, taking into account physical properties of rock and liquids [6]:

$$J(К_в) = \frac{P_к}{\sigma cos\theta}\sqrt{\frac{К_{пр}}{К_п}}, \tag{1}$$

where $P_к$ is capillary pressure; σ is surface tension; $К_{пр}$ – permeability; $К_п$ – porosity; $К_в$ – water saturation.

The dependence of the J-function on the current water saturation, as a rule, has the following form:

$$J = A \cdot К_в^{-n}, \tag{2}$$

where A and n are fixed coefficients for a given reservoir.

Many researchers believe that the Leverett J-function does not always allow satisfactory approximation of capillary pressure curves of reservoir layers with pronounced variability in the structure of void space.

In this regard, when modeling capillary pressure curves, it is considered more preferable to use the Brooks-Corey function [7–11]:

$$К_в = К_{во} + (1 - К_{во})\left(\frac{P_н}{P_к}\right)^{\frac{1}{n}}, \tag{3}$$

where $К_{во}$ is the residual water saturation; $P_н$ – initial (inlet capillary pressure); n is the coefficient of curvature of the capillary curves.

Compare the Leverett J-function with the model proposed by researchers Brooks and Corey.

To do this, we equate the right parts of formulas (1) and (2) and obtain the expression for capillary pressure:

$$\frac{P_к}{\sigma cos\theta}\sqrt{\frac{К_{пр}}{К_п}} = A \cdot К_в^{-n}.$$

From here we have:

$$P_к = \frac{A\sigma cos\theta}{\sqrt{К_{пр}/К_п}} К_в^{-n}.$$

Further, let the water saturation coefficient $К_в$ tends to unity, then

$$P_к \to \frac{A\sigma cos\theta}{\sqrt{К_{пр}/К_п}} = P_н,$$

where $P_н$ – initial (input) capillary pressure.

Then we finally get a simple formula for capillary pressure:

$$P_к = P_н К_в^{-n}.$$

Hence

$$К_в = \left(\frac{P_н}{P_к}\right)^{\frac{1}{n}}.$$

Now I get into the Brooks-Corey formula. First, we transform the formula (3) in the following form:

$$\frac{К_в - К_{во}}{1 - К_{во}} = \left(\frac{P_н}{P_к}\right)^{\frac{1}{n}}. \tag{4}$$

Further we denote

$$К_в^* = \frac{К_в - К_{во}}{1 - К_{во}}.$$

Then we finally get

$$К_в^* = \left(\frac{P_н}{P_к}\right)^{\frac{1}{n}}, \tag{5}$$

IOP Conf. Series: Materials Science and Engineering **905** (2020) 012095 doi:10.1088/1757-899X/905/1/012095

where $K_{\text{в}}^{*}$ is the reduced water saturation of reservoir void space, i.e. water saturation of the effective volume of pore channels.

3. Results

Thus, if you use the reduced amount of void space, then the Brooks-Corey model automatically goes into the Leverett model.

In other words, the Brooks-Corey model takes into account the fact that the residual water does not participate in fluid movement and therefore is more justified [12, 13].

Figure 1 shows the results of comparing the J-function with the volumetric water saturation (a) and reduced water saturation (b) for the terrigenous reservoir АВ_1^3 of *Las-Eganskoe* field in Western Siberia.

The review of the graphs shows that this parameter is closely related to the reduced water saturation, while its relationship with the volumetric water saturation is rather weak.

Figure 2 shows similar dependencies for the productive reservoir ЮВ_1^1 of *Povkhovskoe* field.

The analysis of the graphs shows that the J-function is quite closely related both to the volumetric reduced water saturation (Figure 2, a) and to the volumetric water saturation (Figure 2, b)

Thus, in some layers, a capillary pressure curve is well described by the Leverett J-function, and in others – by the Brooks-Corey model.

In the first case, the rock is probably relatively homogeneous; and residual water, mainly, is capillary-retained and film adsorbed on the surface of the pore channels, while the clay content is small and scattered in the void space volume.

In the second case, the rock contains a large amount of bound water held by clay particles (layered clay), while even at high pressures, clay remains in the clay layers.

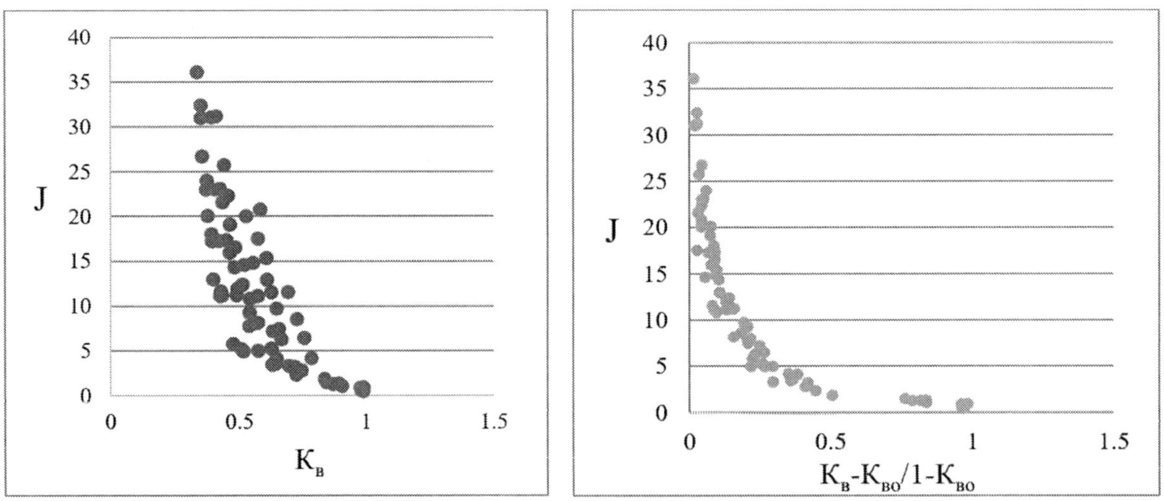

a *b*

Figure 1. The dependence of the J-function: a – from the volumetric water saturation; b – from the reduced water saturation

It should be noted that the Leverett J-function is determined by two parameters: absolute permeability and open porosity.

The Brooks-Corey model is described by three parameters: in addition to the above two parameters, one more parameter is added: residual water saturation, which is a drawback of this model.

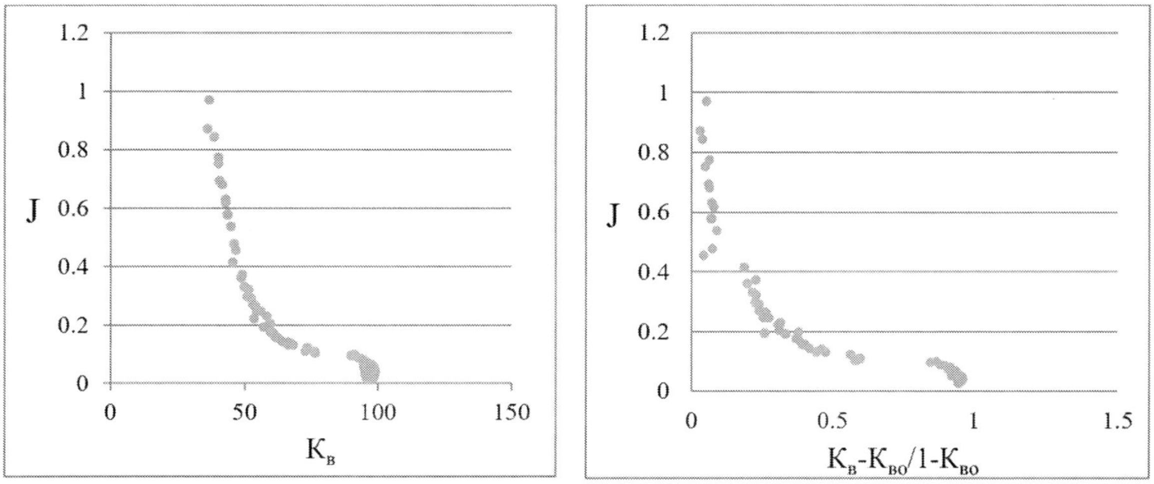

a *b*

Figure 2. The dependence of the J-function: a – from the volumetric water saturation; b – from the reduced water saturation

4. Conclusion

1. The J-function and the Brooks-Corey function have the same meaning. However, the first describes the entire void space of formation, and the second – only the effective part (reduced water saturation).

2. When choosing a correlating function for a given formation, it is necessary:

a) according to laboratory studies, build two dependencies: capillary pressure and current water saturation, and capillary pressure and reduced water saturation

b) choose a schedule that is characterized by a greater tightness of communication (less scatter of points). With the same scatter of points, the Leverett function should be preferred, since it depends on only two parameters (permeability and porosity).

References

[1] Andreev V E, Chizhov A P, Kotenev Yu A et al 2019 Prediction of Gas Exposure in the Conditions of the Oil Fields of the Volga-Urals *Atlantis Highlights in Mater. Sci. and Technol.* **1** 666–9. DOI: 10.2991/isees-19.2019.132

[2] Sergeev V V, Belenkova N G, Zeigman Yu V and Mukhametshin V Sh 2017 Physical properties of emulsion systems with SiO_2 nanoparticles *Nanotechnol. in Construct.* **9(6)** 37–64. DOI: 10.15828/2075-8545-2017-9-6-37-64

[3] Mukhametshin V Sh, Zeigman Yu V and Andreev A V 2017 Rapid assessment of deposit production capacity for determination of nanotechnologies application efficiency and necessity to stimulate their development *Nanotechnol. in Construct.* **93** 20–34. DOI: 10.15828/2075-8545-2017-9-3-20-34

[4] Zeigman Y V, Mukhametshin V Sh, Sergeev V V and Kinzyabaev F S 2017 Experimental study of viscosity properties of emulsion system with SiO_2 nanoparticles *Nanotehnol. in Construct.* **9(2)** 16–38. DOI: 10.15828/2075-8545-2017-9-2-16-38

[5] Gudok N S 1970 *Studying the physical properties of porous media* (Moscow: Nedra) 204 p

[6] Khabarov A V and Volokitin Ya E 2009 Procedure for combined analysis of core and log data for lithologic classification of terrigenous reservoirs *Well Logger.* **12(189)** 83–128

[7] Brooks R H and Corey A T 1964 Hydraulic Properties of Porous Media *Colorado State Univer. Hydrology* **3**

[8] Adams S J and Van den Oord R J 1993 Capillary Pressure and Saturation-Height Functions *Report EP 93-0001 SIPM BV*

[9] Tiab D and Donaldson E C 1999 *Petrophysics: Theory and Practice of Measuring Reservoir Rock and Fluid Transport Properties* (Houston, TX: Gulf Publ.) 608 p
[10] Inc. Baker Hughes 1995 *Introduction to Wireline Log Analysis* (Baker Atlas)
[11] Akhmetov R T, Mukhametshin, V V and Kuleshova L S 2019 Grouping of objects using a limited number of parameters characterizing geological and physical properties of layers *Atlantis Highlights in Mater. Sci. and Technol.* **1** 9–13. DOI: 10.2991/isees-19.2019.3
[12] Akhmetov R T, Kuleshova L S and Mukhametshin V V 2019 Absolute Permeability and Distribution of Pore Throats of the Productive Strata of Western Siberia *Atlantis Highlights in Mater. Sci. and Technol.* **1** 535–9. DOI: 10.2991/isees-19.2019.105
[13] Akhmetov R T and Mukhametshin V V 2018 Range of application of the Brooks-Corey model for approximation of capillary curves in reservoirs of Western Siberia *Advan. in Engineer. Res. (AER)* **157** 5–8. DOI: 10.2991/aime-18.2018.2

AUTHOR INDEX

Institute of Physics
Dirac House, Temple Back
Bristol BS1 6BE UK

ISSN: 1757-8981
ISBN 978-1-7138-4753-3

9 781713 847533

CURRAN ASSOCIATES INC.
proceedings.com